AF327692

PURINERGIC APPROACHES IN EXPERIMENTAL THERAPEUTICS

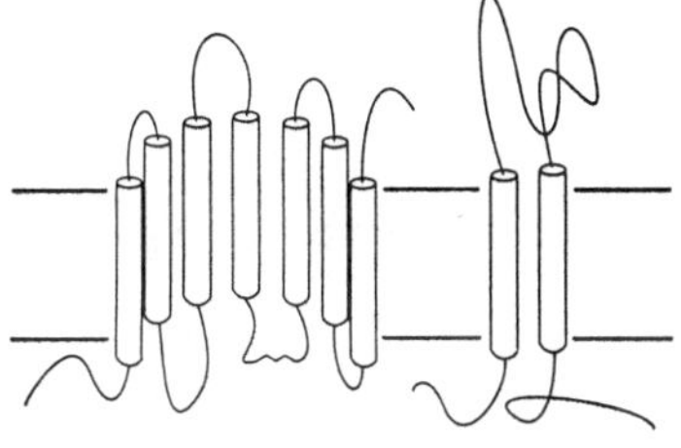

PURINERGIC APPROACHES IN EXPERIMENTAL THERAPEUTICS

Edited by

Kenneth A. Jacobson
National Institute of Diabetes, Digestive and Kidney Diseases

Michael F. Jarvis
Abbott Laboratories

WILEY-LISS
A JOHN WILEY & SONS, INC., PUBLICATION
New York • Chichester • Weinheim • Brisbane • Singapore • Toronto

Address All Inquiries to the Publisher
Wiley-Liss, Inc., 605 Third Avenue, New York, NY 10158-0012

Copyright © 1997 Wiley-Liss, Inc.

Printed in the United States of America.

Under the conditions stated below the owner of copyright for this book hereby grants permission to users to make photocopy reproductions of any part or all of its contents for personal or internal organizational use, or for personal or internal use of specific clients. This consent is given on the condition that the copier pay the stated per-copy fee through the Copyright Clearance Center, Incorporated, 222 Rosewood Drive, Danvers, MA 01923, as listed in the most current issue of "Permissions to Photocopy" (Publisher's Fee List, distributed by CCC, Inc.), for copying beyond that permitted by sections 107 or 108 of the US Copyright Law. This consent does not extend to other kinds of copying, such as copying for general distribution, for advertising or promotional purposes, for creating new collective works, or for resale.

While the authors, editors, and publisher believe that drug selection and dosage and the specifications and usage of equipment and devices, as set forth in this book, are in accord with current recommendations and practice at the time of publication, they accept no legal responsibility for any errors or omissions, and make no warranty, express or implied, with respect to material contained herein. In view of ongoing research, equipment modifications, changes in governmental regulations and the constant flow of information relating to drug therapy, drug reactions and the use of equipment and devices, the reader is urged to review and evaluate the information provided in the package insert or instructions for each drug, piece of equipment or device for, among other things, any changes in the instructions or indications of dosage or usage and for added warnings and precautions.

Library of Congress Cataloging-in-Publication Data

Purinergic approaches in experimental therapeutics / edited by Kenneth A. Jacobson and Michael F. Jarvis.
p. cm.
ISBN 0-471-14071-6 (alk. paper)
1. Purines—Receptors—Agonists. 2. Purines—Receptors—Antagonists. 3. Adenosine—Receptors—Agonists. 4. Adenosine—Receptors—Antagonists. I. Jacobson, Kenneth Alan, 1953– II. Jarvis, Michael F.
[DNLM: 1. Receptors, Purinergic—physiology. 2. Adenosine—therapeutic use. 3. Adenine Nucleotides—therapeutic use. QU 57 P985 1997]
QP801.P8P86 1997
615′.7—dc21
DNLM/DLC
for Library of Congress 96-38941
CIP

The text of this book is printed on acid-free paper.

10 9 8 7 6 5 4 3 2 1

CONTENTS

CONTRIBUTORS

Maria P. Abbracchio, Institute of Pharmacological Sciences, University of Milan, 20133 Milan, Italy [**383**]

Tony R. Bai, University of British Columbia Pulmonary Research Laboratory, St. Paul's Hospital, Vancouver, BC V6Z 1Y6 [**315**]

Luiz Belardinelli, Departments of Medicine and Pharmacology, University of Florida, Gainesville, FL 32610 [**185**]

Barry Bertolet, Department of Medicine, University of Florida, Gainesville, FL 32610 [**185**]

Gyslaine Bertrand, Laboratoire de Pharmacologie, Faculté de Medécine, Montpellier, France [**253**]

Geoffrey Burnstock, Department of Anatomy and Embryology, University College London, London WC1 6BT, United Kingdom [**3**]

David Carley, University of Illinois College of Medicine at Chicago, Chicago, IL 60612 [**515**]

Jeannie Chapal, Laboratoire de Pharmacologie, Faculté de Medécine, Montpellier, France [**253**]

Fedias L. Christofi, Department of Anesthesiology, College of Medicine, Ohio State University, Columbus, OH 43210 [**261**]

Michael V. Cohen, Department of Medicine, University of South Alabama, Mobile, AL 36688 [**153**]

Michael A. Cook, Department of Pharmacology and Toxicology, Faculty of Medicine, University of Western Ontario, London, Ontario N6A 5C1, Canada [**261**]

Bruce N. Cronstein, New York University Medical Center, New York, NY 10016 [**285**]

Anne B. Curtis, Department of Medicine, University of Florida, Gainesville, FL 32610 [**185**]

James M. Downey, Department of Physiology, University of South Alabama, Mobile, AL 36688 [**153**]

Thomas V. Dunwiddie, Department of Pharmacology/Program in Neuroscience, University of Colorado Health Sciences Center, Denver, and Veterans Administration Medical Research Service, Denver, CO 80262 [**359**]

The numbers in brackets are the opening page numbers of the contributors' articles.

Jeffrey S. Fedan, Pathology and Physiology Research Branch, Health Effects Laboratory Division, National Institute for Occupational Safety and Health, Morgantown, WV 26505; Department of Pharmacology and Toxicology, Robert C. Byrd Health Sciences Center, West Virginia University, Morgantown, WV 26505 [**333**]

Theresa M. Filtz, Department of Pharmacology, School of Medicine, University of North Carolina at Chapel Hill, Chapel Hill, NC 27599-7365 [**39**]

Gary S. Firestein, UCSD School of Medicine, La Jolla, CA 92093 [**301**]

Bertil B. Fredholm, Department of Physiology and Pharmacology, Section of Molecular Neuropharmacology, Karolinska Institutet, S-171 77 Stockholm, Sweden [**359**]

Atsuo F. Fukunaga, Department of Anesthesiology, Harbor-UCLA Medical Center, Torrance, CA 90509 [**471**]

Jonathan D. Geiger, Department of Pharmacology and Therapeutics, University of Manitoba Faculty of Medicine, Winnipeg, Manitoba R3E 0W3, Canada [**55**]

T. Kendall Harden, Department of Pharmacology, School of Medicine, University of North Carolina at Chapel Hill, Chapel Hill, NC 27599-7365 [**39**]

Dominique Hillaire-Buys, Laboratoire de Pharmacologie, Faculté de Medécine, Montpellier, France [**253**]

Robert G. Humphries, Department of Pharmacology, Astra Charnwood, Loughborough, Leicestershire LE11 5RH, United Kingdom [**203**]

Ad P. IJzerman, Leiden/Amsterdam Center for Drug Research, Division of Medicinal Chemistry, 2300 RA Leiden, The Netherlands [**129**]

Edwin K. Jackson, Departments of Pharmacology and Medicine, Center for Clinical Pharmacology, University of Pittsburgh Medical Center, Pittsburgh, PA 15213-2582 [**217**]

Kenneth A. Jacobson, Molecular Recognition Section, Laboratory of Bioorganic Chemistry, National Institute of Diabetes, Digestive and Kidney Diseases, National Institutes of Health, Bethesda, MD 20892 [**101**]

Marlene A. Jacobson, Department of Pharmacology, Merck Research Laboratories, West Point, PA 19486 [**315**]

Michael F. Jarvis, Neurological and Urological Diseases Research, Pharmaceutical Products Division, Abbott Laboratories, Abbott Park, IL 60064-3500 [**405**]

Charles Kennedy, Department of Physiology and Pharmacology, University of Strathclyde, Royal College, Glasgow G1 1XW, United Kingdom [**173**]

Lars J. S. Knutsen, MedChem Research I, Health Care Discovery, Novo Nordisk A/S, Novo Nordisk Park, DK 2760, Malov, Denmark [**423**]

Elizabeth A. Kowaluk, Neurological and Urological Diseases Research, Pharmaceutical Products Division, Abbott Laboratories, Abbott Park, IL 60064-3500 [**55**]

Paul Leff, Department of Pharmacology, Astra Charnwood, Loughborough, Leicestershire LE11 5RH, United Kingdom [**203**]

Joel Linden, University of Virginia Health Sciences Center, Charlottesville, VA 22908 [**85**]

Guang-Shung Liu, Department of Physiology, University of South Alabama, Mobile, AL 36688 [**153**]

Marie-Madeleine Loubatières-Mariani, Laboratoire de Pharmacologie, Faculté de Medécine, Montpellier, France [**253**]

Gerald J. McLaren, Department of Physiology and Pharmacology, University of Strathclyde, Royal College, Glasgow G1 1XW, United Kingdom [**173**]

Thomas F. Murray, College of Pharmacy, Oregon State University, Corvallis, OR 97331 [**423**]

Robert A. Nicholas, Department of Pharmacology, School of Medicine, University of North Carolina at Chapel Hill, Chapel Hill, NC 27599-7365 [**39**]

Fiona E. Parkinson, Department of Pharmacology and Therapeutics, University of Manitoba Faculty of Medicine, Winnipeg, Manitoba R3E 0W3, Canada [**55**]

Pierre Petit, Laboratoire de Pharmacologie, Faculté de Medécine, Montpellier, France [**253**]

Eliezer Rapaport, Worcester Foundation for Biomedical Research, Shrewsbury, MA 01545 [**545**]

Miodrag Radulovacki, University of Illinois College of Medicine at Chicago, Chicago, IL 60612 [**515**]

Scott A. Rivkees, Department of Pediatrics, Yale Medical School, New Haven, CT 06520 [**527**]

Mark J. Robertson, Department of Pharmacology, Astra Charnwood, Loughborough, Leicestershire LE11 5RH, United Kingdom [**203**]

Sanna Rosengren, Gensia, Inc., San Diego, CA 92121 [**301**]

Jana Sawynok, Department of Pharmacology, Dalhousie University, Halifax, Nova Scotia B3H 4H7, Canada [**495**]

Peter Sneddon, Department of Physiology and Pharmacology, University of Strathclyde, Royal College, Glasgow G1 1XW, United Kingdom [**173**]

Gary L. Stiles, Duke University Medical Center, Durham, NC 27710 [**29**]

Nora M. van der Wenden, Leiden/Amsterdam Center for Drug Research, Division of Medicinal Chemistry, 2300 RA Leiden, The Netherlands [**129**]

A. Michiel van Rhee, Molecular Recognition Section, Laboratory of Bioorganic Chemistry, National Institute of Diabetes, Digestive and Kidney Diseases, National Institutes of Health, Bethesda, MD 20892 [**101**]

Dag K. J. E. von Lubitz, Laboratory of Bioorganic Chemistry, Molecular Recognition Section, NIH/NIDDK/LBC, Bethesda, MD 20892 [**449**]

Michael Williams, Pharmaceutical Products Division, Abbott Laboratories, Abbott Park, IL 60064 [**3**]

PREFACE

An extensive body of scientific data defining the role(s) of extracellular purines and pyrimidines in cellular homeostasis and disease etiology has evolved over the past 40 years, principally due to the vision and determination of Professor Geoffrey Burnstock, the "father" of purinergic neurotransmission. In the past five years, with the cloning and expression of these receptors, there has been an exponential increase in research interest in the area such that we are now at an important crossroads in terms of understanding the complexities of the receptors' involvement in mammalian tissue physiology and pathophysiology and the potential for using this information to develop novel therapeutic modalities. The diversity and multitude of the cellular targets that recognize adenosine, ATP, and related nucleotides—four distinct P1 receptors sensitive to adenosine and at least 15 different P2 receptors sensitive to ATP and UTP—is reminscent of the early days of serotonin receptor research in the 1970s. In the latter area, the coupling of molecular diversity to innovative medicinal chemistry has led to major new drugs for the treatment of depression, chemotherapy-evoked emesis, and migraine. And there is every reason to anticipate that still-to-be-identified purinergic and pyrimidinergic receptor ligands as well as modulators of purine metabolism will be important research tools in defining system function as well as having the potential to be leads for new drugs in areas where there is considerable unmet medical need.

In the present volume, authoritative updates from many of the world's leading researchers active in the study of the role(s) of adenosine, ATP, and UTP in cellular function provide an important focus and resource for both the newcomer to this rapidly expanding area as well as the expert. Thus the first third of the book provides a concise and timely historical perspective of the role(s) of adenosine and ATP in cellular communication processes, as well as reviews of the ligands for, and molecular pharmacology of, purine and pyrimidine receptors. The remaining two thirds of the book are devoted to a critical analysis of the role(s) of adenosine and ATP, and to a lesser extent, given the current state of knowledge, UTP, in the physiology of the central nervous, cardiovascular, respiratory, renal, and immune systems, as well as what is known regarding the pathophysiology of these systems. A unique feature of this volume is that the editors have invited scientists from different laboratories, many of whom had not previously collaborated, to co-author chapters. This approach provides a more balanced perspective, with active dialogue during the writing process providing a more synergistic and balanced approach to each of the chapter topics.

Over the last twenty years, there has been a significant focus, given advances in medicinal chemistry in the late 1970s, for the therapeutic targeting of P1 purinergic ligand interventions for cardiovascular and central nervous system targets. With the exception of adenosine itself, used as a treatment for supraventricular tachycardia, these drug discovery efforts have been singularly disappointing, in part because of the therapeutic targets chosen and in part because of class-related side-effect liabilities.

However, it is important to appreciate that the majority of these early endeavors were undertaken in the absence of the current understanding of the molecular diversity of adenosine and ATP receptor subtypes, with poor pharmacological probes for the various receptors (especially in the P2 area) that had limited (and continue to limit) *in vivo* usefulness as well as the ubiquitous involvement of adenosine and ATP in central and peripheral tissue function. For instance, the last five years have seen an explosion of information in the context of the role of adenosine as a modulator of inflammatory responses with diverse and synergistic effects on free radical and cytokine formation, neutrophil and macrophage function, and cell adhesion processes. In a similar vein, ATP and related nucleotides have been shown to have important roles as trophic factors and modulators of apoptosis.

In the present volume, the respective roles of adenosine and ATP and their respective recognition sites are considered over a broad range of therapeutic areas, including cardioprotection; thrombosis; pulmonary, renal and gastintestinal function; diabetes; inflammation; cancer; epilepsy; neurodegeneration; ischemia; anesthesia; and pain. As the reader proceeds through the book, the editors hope that a sense of the excitement and opportunity regarding the therapeutic implications of our continued study of the constellation of adenosine- and ATP-mediated responses in cellular regulation is communicated.

The editors would like to extend their sincere thanks to the authors for their excellent contributions and their patience during the inevitable time considerations of the editorial process. The authors would also like to express their indebtedness to Fiona Stevens, their editor at Wiley, for her guidance, assistance, and persistence during the evolution of this project into reality, and to Mike Williams, who initially suggested the idea for this much-needed addition to the purinergic literature.

Kenneth A. Jacobson
Michael F. Jarvis

PART I

HISTORICAL PERSPECTIVE

CHAPTER 1

Purinergic Neurotransmission and Neuromodulation: A Historical Perspective

MICHAEL WILLIAMS and GEOFFREY BURNSTOCK

Pharmaceutical Products Division, Abbott Laboratories, Abbott Park, IL 60064-3500 (M.W.); Department of Anatomy and Embryology, University College London, London WC1 6BT, United Kingdom (G.B.)

INTRODUCTION

The seminal study of Drury and Szent-Gyorgyi (1992) showed that adenosine (ADO), adenosine monophosphate (AMP) and adenosine triphosphate (ATP) were coronary vasodilators and bradycardic agents and that these actions led to the hypotensive actions of the nucleoside. This study has been the most frequently cited of the early work in purine-related research, probably because of its accessibility in the English language. However, a number of other papers in German preceded the work of Drury and Szent-Gyorgi. Following the observation of Bass (1914) that adenine was present in blood (probably as AMP), a similar substance was shown to have depressor activity (Freund, 1920). Thannhauser and Bommes (1914) had also shown that ADO, unlike adenine, was nontoxic when administered subcutaneously to humans. ATP was not identified until the late 1920s (Embden and Zimmerman, 1927; Lohmann, 1929; Fiske and Subbarow, 1929).

The therapeutic potential for ADO was first evaluated in the 1930s (Honey et al., 1930; Jezer et al., 1933). At that time, its short plasma half-life (3–6 seconds; Paterson et al., 1987) limited any meaningful efficacy measures and potential usefulness. Interestingly, the short half-life of ADO has made it an ideal compound for the treatment for supraventricular tachycardia, for which use it has been approved in man (Belardinelli et al., 1995).

The 1950s brought renewed scientific interest in the potential role of purines as modulators of cellular and tissue function and saw the publication of reports on the effects of purines on CNS function (Green and Stoner; 1950), together with the study

Purinergic Approaches in Experimental Therapeutics, Edited by Kenneth A. Jacobson and Michael F. Jarvis
ISBN 0-471-14071-6 © 1997 Wiley-Liss, Inc.

of Holton (1959) demonstrating ATP release following antidromic sensory nerve stimulation of rabbit ear artery. Furthermore, the suggestion that ADO could function as a physiological regulator of coronary flow (Lindner and Rigler, 1931) was supported by studies from the groups of Berne (1963) and Gerlach et al. (1963), which indicated that ADO could regulate coronary blood flow under ischemic or hypoxic conditions, e.g., during reactive hyperemia. This recognition of the ADO-mediated linkage of oxygen availability to the metabolic demands of the heart led to ADO being termed a "retaliatory metabolite" (Newby, 1984) and to the evaluation of a generalized framework in which the homeostatic effects of ADO were viewed as the "signal of life" (Engler, 1991). However, Burnstock and Ralevic (1994) showed that hypoxia-induced vasodilation resulted primarily from endothial ATP release that in turn stimulated NO (nitrous oxide) production via P2Y receptor activation, the ADO produced from ATP breakdown contributing to a later stage P1 receptor–mediated vasodilatory response.

Born et al. (1965) had shown that when ADO and other ADO analogues, especially 2-chloroADO (2-CADO; Figure 1) were administered into the brachial artery of normal human subjects, forearm vasodilation ensued and was accompanied by pain, the latter being comparable to the pain associated with angina pectoris or ischemia (Sylven et al., 1986, 1988). In the 1970s, the stable ADO analogue, R-N^6-phenylisopropylADO (R-PIA; Figure 1) was demonstrated to have antilipolytic actions in man (May and Leinweber, 1971). More recently, ADO administered by infusion (Sollevi et al., 1995) and R-PIA administered intrathecally (Karlsten and Gordh, 1996) have been reported to have unexpectedly long acting analgesic actions in patients with chronic neuropathic pain, analgesic effects that contrast with the ADO effects seen in normal subjects (Born et al., 1965; Sylven et al., 1986, 1988). This may be due to the route of administration and the use of acute bolus dosing regimens.

Concommitant with the studies of the effects of ADO on coronary function, Burnstock et al. (1970) evaluated ATP as a candidate for the role of effector in autonomic nonadrenergic, noncholinergic (NANC) transmission processes, while Sattin and Rall (1970) identified ADO-sensitive adenylate cyclase activity in mammalian nervous tissue. These observations provided evidence for the concept of purinergic neurotransmission (Burnstock, 1972). Subsequent biochemical, pharmacological and physiological data provided a framework for the description of two distinct purinoceptor subtypes (Burnstock, 1978; Table 1): the P1 purine receptors sensitive to ADO; and the P2 receptor, sensitive to ATP and related nucleotides. Additional pharmacological data accumulated over the next two decades (Gordon, 1986; Burnstock and Kennedy, 1985; Dubyak and El-Moatassim, 1993; Abbracchio and Burnstock, 1994) provided, together with more definitive evidence in the form of identified molecular clones (Burnstock and King, 1996), an important impetus to delineating receptor subtypes. It also initiated, in earnest, a search for receptor-selective compounds to define purinergic receptor subtype function and to address potential therapeutic utility (Humphries et al., 1996; Williams, 1996).

The identification of a separate class of P2 receptor, similar in molecular structure to members of the P2Y superfamily but preferentially sensitive to the pyrimidine nucleotide, UTP (Lazorowski and Harden, 1994; Harden et al., 1995; Communi et al., 1995, 1996; Nguyen et al., 1995; Nicholas et al., 1996), led to the concept of a separate family of pyrimidinoceptors termed P_{2U}. However, the close structural similarities of these receptors to members of the P2Y family (Boeynaems et al., 1996) has resulted in the renaming of P2 purinoceptors (Fredholm et al., 1994) as P2 receptors to encompass both purine- and pyrimidine-sensitive entities (Fredholm et al., 1997).

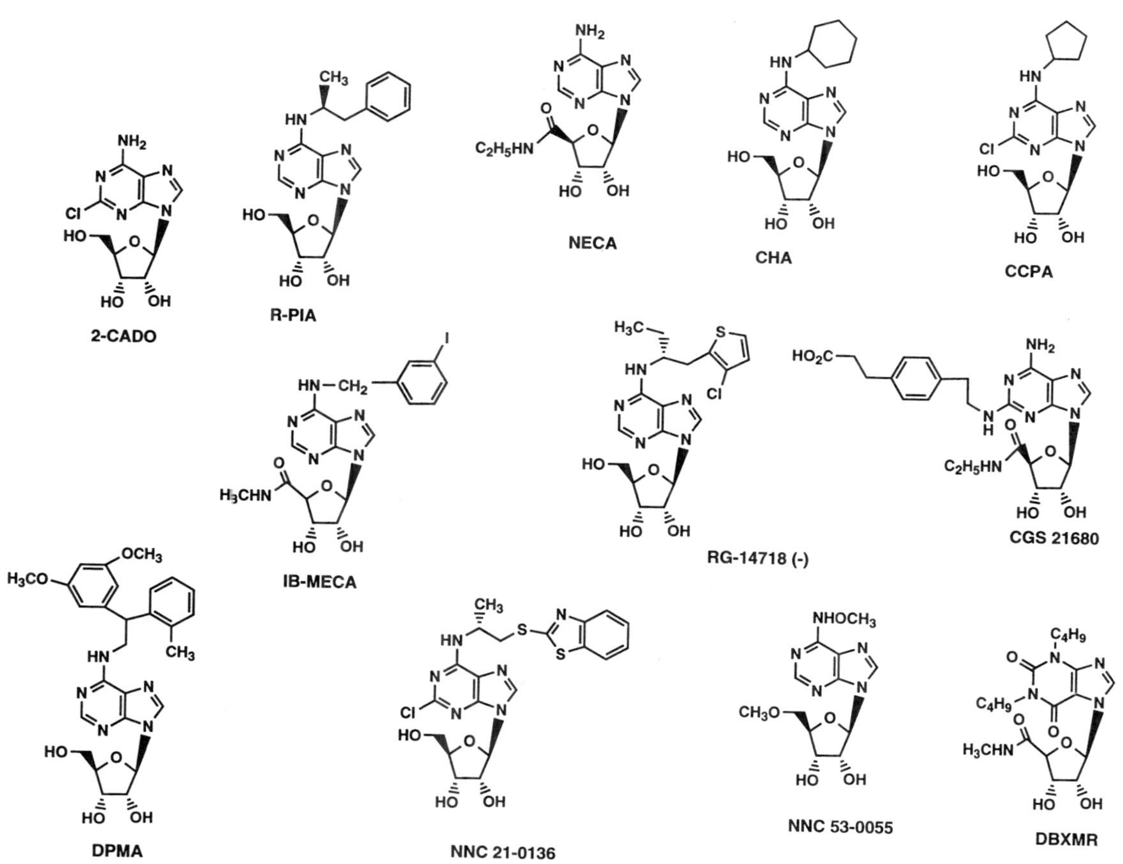

2-CADO

R-PIA

NECA

CHA

CCPA

IB-MECA

RG-14718 (-)

CGS 21680

DPMA

NNC 21-0136

NNC 53-0055

DBXMR

FIGURE 1. P1 receptor agonists.

TABLE 1 Purinergic Receptor Classification Based on Functional Cloning

Receptor	Tissue of Origin *GenBank Accession #*	Pharmacological Profile	Transduction System(s)
A_1	Human brain-*L22214*	Agonist: CCPA, CPA Antagonist: DPCPX, CPX, BIIP 20	Adenylate cyclase inhibition
A_{2A}	Human brain-*S46950*	Agonist: CGS 21680 Antagonist: KF 17837, ZM 241385, SCH 58251	Adenylate cyclase activation
A_{2B}	Human brain-*X68487*	Agonist: NECA Antagonist: Enprofylline (?)	Adenylate cyclase activation
A_3	Human brain-*L20463*	Agonist: IB-MECA, DBXMR Antagonist: MRS 1222, L-249,313, L-268,605	IP3
$P2X_1$	Rat vas deferens-*X80477*	2-MeSATP ≥ ATP > α,β meATP	$I_{Na/K/Ca}$
	Urinary bladder-human *X83688*	ATP > α,β meATP	
	Mouse-*X84896*		$I_{Na/K/Ca}$
$P2X_2$	Rat PC12 cell-*U14414*	2-MeSATP > ATP, α,β meATP inactive	$I_{Na/K/Ca}$
$P2X_{2-1}$	Rat cochlea-short form-*L43511*		
$P2X_3$	Rat dorsal root ganglion-*X90651*	2-MeSATP > ATP > α,β meATP	$I_{Na/K}$
	Rat dorsal root ganglion-*X91167*		
$P2X_4$	Rat hippocampus-*X91200*	ATP > 2-MESATP > α,β meATP	$I_{Na/K}$
	Rat DRG-*X87763*	ATP > 2-MeSATP > α,β meATP	
	Rat/human brain-*U32497/ X93565*		
	Rat endocrine tissue-*U47031*		
$P2X_5$	Rat celiac ganglia-*X92069*	ATP > 2-MeSATP > ADP	
$P2X_{5a}$	Human-*U49395*		
$P2X_{5b}$	Human-*U49396*		
$P2X_6$	Rat brain, rat cervical ganglion-*X92070*	ATP > 2-MeSATP > ADP	
$P2X_7$	Mouse macrophage-*X95882*	BzATP > ATP > UTP	$I_{Na/K}$ or I_{ca}
$P2Y_1$	Chick brain-*X73268*	2-MeSATP > ATP > ADP, UTP inactive	PLCβ/IP_3/Ca^{2+}
	Turkey brain-*U09842*	2-MeSATP > ATP > ADP, UTP inactive	
	Mouse insulinoma cell-*U22829*	2-MeSATP ≥ 2Cl-ATP ≥ ATP, α,β meATP inactive	PLCβ/IP_3Ca^{2+}
	Rat insulinoma cell-*U22830*	2-MeSATP > ATP ≫ UTP	
	Human placenta-*Z49205*		PLCβ/IP_3Ca^{2+}
	Bovine endothelium-*X87628*		
	Human HEL cell-*U42029*		PLCβ/IP_3/Ca^{2+}
	Human brain, prostate and ovary		

TABLE 1 ***(Continued)***

Receptor	Tissue of Origin *GenBank Accession #*	Pharmacological Profile	Transduction System(s)
$P2Y_2$	Mouse NG-108-1-*L14751*	ATP = UTP ≫ 2-MeSATP	$PLC\beta/IP_3/Ca^{2+}$
	Human CT/43 cells-*U07225*	ATP = UTP ≫ 2-MeSATP	
	Rat lung-*U09402*	ATP = UTP	$PLC\beta/IP_3/Ca^{2+}$
	Rat pituitary-*L46865*		$PLC\beta/IP_3/Ca^{2+}$
	Wistar Kyoto rat-*U56839*		
$P2Y_3$	Chick brain-*X98283*	ADP > UTP > ATP = UDP	$PLCb/IP_3/Ca^{2+}$
$P2Y_4$	Human placenta-*X91852*	UTP = UDP > ATP = ADP	$PLC\beta/IP_3/Ca^{2+}$
	Rat brain-*X91852*		
$P2Y_5$	Human activated T cells-*P32250/L06109*	ATP > ADP > 2-MeSATP ≫ UTP, α,β,-meATP	
$P2Y_6$	Aortic smooth muscle-*D63665*	UDP > 5-Br-UTP > UTP > ADP > 2-MeSATP > ATP	$PLC\beta/IP_3/Ca^{2+}$
$P2Y_7$	HEL cells-*U41070*	ATP > ADP = UTP	
XL-P2Y ($P2Y_8$)	Xenopus neutral plate-*X99953*	ATP = UTP = ITP = CTP = GTP	$PLC\beta/IP_3/Ca^{2+}$

Purine Receptor Nomenclature

P1 Receptors ADO receptors are divided into three major classes, A_1–A_3 (Olah and Stiles, 1995), with two subclasses of the A_2 receptor being designated "A" and "B" (Fredholm et al., 1994, 1997). A novel ADO receptor termed the A_4, designated on the basis of binding data (Cornfeld et al., 1992), was subsequently demonstrated to be a thermodynamically constrained form of the A_{2A} receptor (Luthin and Linden, 1995).

P2 Receptors P2 receptor nomenclature has evolved from a physiological/pharmacological basis (Abbracchio and Burnstock, 1994) to one relating receptor structure to function (Burnstock and King, 1996). The nomenclature is thus confusing to the newcomer. For instance, the $P2Y_2$ receptor that is responsive to both ATP and UTP is also known as the P_{2U} receptor (Williams and Bhagwat; 1996). The International Union of Pharmacology (IUPHAR) guidelines have recommended (Vanhoutte et al., 1996) that receptors which are defined from a pharmacological basis be shown in italics, e.g., for purine receptors P_{2Y1}; that cloned receptors where *only* the amino acid sequence is known be shown in lower case, e.g., p2y(x); and that receptors for which *both* sequence *and* pharmacology are known be shown in upper case, e.g., $P2Y_1$. However, the IUPHAR Purine Nomenclature Subcommittee (Fredholm et al., 1997) has favored the use of only the latter nomenclature. Thus members of the two P2 receptor superfamilies are being numbered in the sequence in which they are identified based on structural relationships via a newsletter developed by the Burnstock group (Burnstock and King, 1996). To avoid confusion, P1 and P2 are used generically to describe the receptor families. By convention, members of the P1 purine receptor family are shown as A_1, A_{2A}, A_{2B}, and A_3.

The P2X receptor family is a ligand-gated ion channel (LGIC) receptor family (Surprenant et al., 1995; Humphrey et al., 1995), seven of which ($P2X_{1-7}$) have been cloned to date (Table 1). The P2Y is a G protein–coupled receptor (GPCR) family (Boarder et al., 1995; Simon et al., 1995) with eight reported members, designated $P2Y_{1-8}$. Two receptors selective for UTP have been cloned as the $P2Y_4$ and $P2Y_6$ pyrimidine receptors, respectively (Communi et al., 1995, 1996). The P_{2Z} receptor, an ATP-gated ion pore sensitive to 2′- and 3′-O-(4-benzoylbenzoyl)ATP (Di Vigillio, 1995) has been cloned as the $P2X_7$ receptor (Surprenant et al., 1996), while the P_{2T} receptor has not yet been cloned.

$P2Y_1$ and $P2Y_3$ receptors have been cloned from chick brain (Barnard et al., 1994; Simon et al., 1995; Janssens et al., 1996; Webb et al., 1996a); the $P2Y_2$ receptor from the NG 108-15 neuroblastoma/glioma cell line (Lustig et al., 1993); the $P2Y_5$ receptor from activated chick T cells (Communi et al., 1995; Webb et al., 1996b); the $P2Y_6$ from human placenta; and the $P2Y_8$ from the frog neural crest (Burnstock and King, 1996). The $P2X_1$ receptor has been cloned from rat vas deferens (Valera et al., 1994), the $P2X_2$ from PC12 neuroblastoma cells (Brake et al., 1994); the $P2X_3$ from rat dorsal root ganglion (DRG; Chen et al., 1995; Lewis et al., 1995); the $P2X_4$ from rat hippocampus (Bo et al., 1995); and the $P2X_5$ and $P2XY_6$ from rat celiac and cervical ganglion, respectively (Collo et al., 1996).

The structure of P2X receptors is atypical, being unlike other LGICs in that it consists of two transmembrane-spanning regions related to the epithelial amiloride-sensitive sodium channel (North, 1996). The $P2X_1$ receptor is sensitive to α,β-MeATP (Figure 1) and undergoes rapid desensitization. In contrast, $P2X_2$ receptors are insensitive to α,β-MeATP and do not readily desensitize (Surprenant et al., 1995). The $P2X_1$ receptor is identical to RP-2, a cDNA clone, found in apoptotic thymocytes, that may play a role in programmed cell death (Brake et al., 1994). However, functional evidence regarding this is controversial (Zheng et al., 1991; Jiang et al., 1996). P2X receptors exist as multimers in transfected oocytes (Evans et al., 1995; Lewis et al., 1995), with their pharmacological diversity and function being defined on the basis of whether and which homo- and hetero-oligomeric forms are present. Sensory neurons, for instance, express LGICs composed of $P2X_2$ and $P2X_3$ subunits (Lewis et al., 1995). At the present time, the relationship between hetero-oligomeric structure and receptor function has not been defined but, as with other LGICs, it will no doubt be extremely complex. The $P2X_1$ receptor reportedly requires four or five subunits for functional activity (Nicke et al., 1996).

The $P_{2D}/P2Y_7$ receptor is a GPCR activated by diadenosine polyphosphates such as Ap4A (Figure 2) and is found in both the cardiovascular (Hilderman et al., 1991) and nervous (Pintor and Miras Portugal, 1995) systems. Ap4A is equipotent to ATP at the P_{2U} receptor (Lazorowski et al., 1995). Its ability to block platelet aggregation (P_{2T} receptor; Zamecnik et al., 1992) indicates that this type of ligand is not selective for a single class of P2 receptor.

Purines as Modulators of Cellular Homeostasis

ADO has been described as a prototypic homeostatic modulator (Williams, 1989) that acts endogenously to protect tissues under adverse conditions. ADO can be formed from ATP in the extracellular space via the action of extracellular ectonucleotidases.

Both P1 and P2 receptors are widely distributed in mammalian tissues, although several of the recently identified P2 receptor subtypes, like the $P2X_3$ show a discrete tissue localization that may presage a unique functional role (Chen et al., 1995).

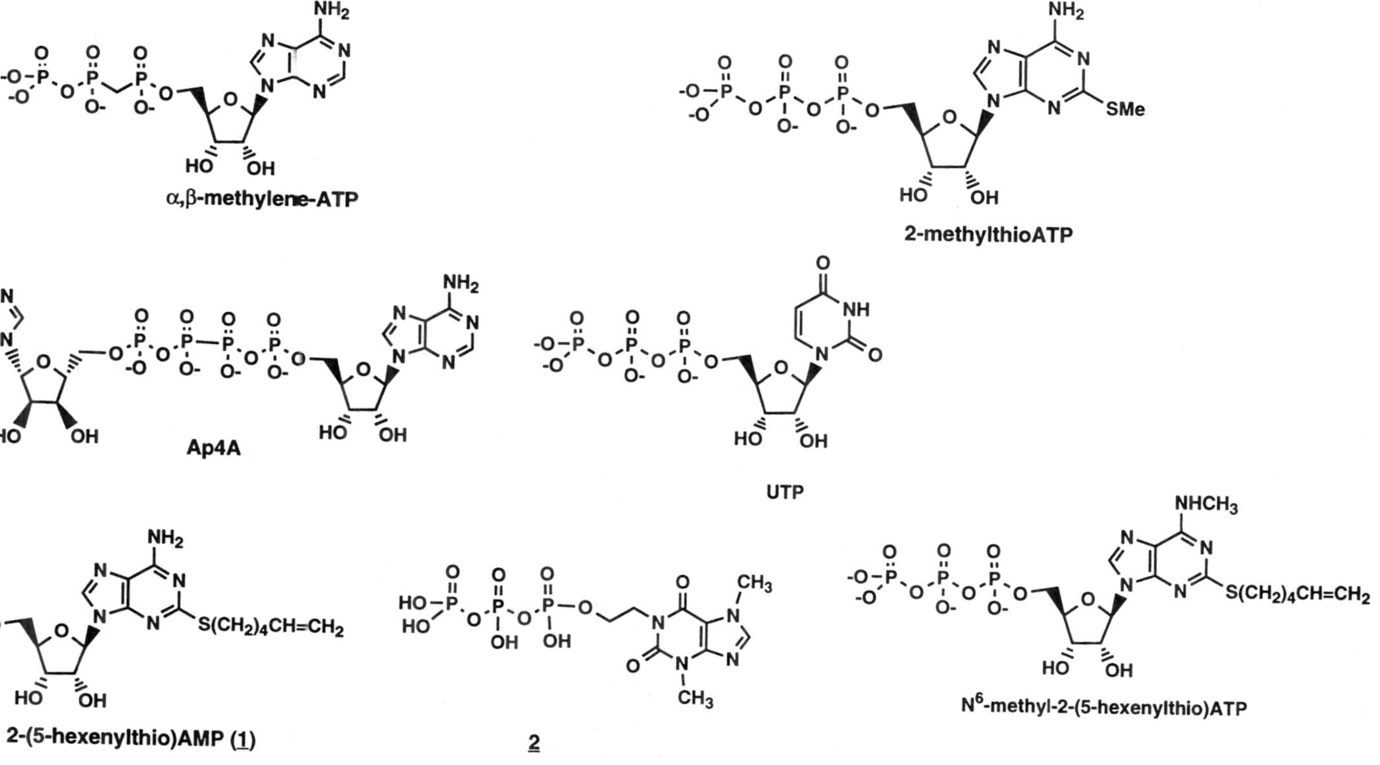

FIGURE 2. P2 receptor agonists.

In terms of its intracellular origin and complex function, ATP is considered an unusual neurotransmitter candidate. Its well-defined role in cellular energy processes (Arch and Newsholme, 1978) has required the cloning of the P2 receptor superfamily (Burnstock and King, 1996; Humphrey et al., 1995) to substantiate the large body of pharmacological and physiological data amassed to support its neurotransmitter role (Abbracchio and Burnstock, 1994). Phylogenetically, however, ATP has been described as an effector agent in amoebae, sea anemone, leech, snail, marine molluscs, starfish, elasmobranchs, and teleosts (Burnstock, 1996a,b).

Each ATP molecule used extracellularly results in the loss of three high-energy phosphate bonds at an energy "cost" of approximately −30 kcal/mol. The thermodynamic properties of ATP have been discussed in the context of its role as a neuromodulator (Burnstock and Barnard, 1996). Barnard has argued that the energy dissipated when the chemical bonds of complex neurotransmitter molecules (e.g., peptides, acetylcholine) are hydrolyzed differs markedly from the energy contained in the phosphate bonds of ATP. This difference relates to the group transfer potential, the ability of the energy in any single high-energy ATP phosphate bond to be transferred to other anabolic reactions, which is a unique property of nucleotide phosphate bonds, and one first recognized by Lippmann (1941). Energy usage is not the sole criteria used to assess the role of a molecule as a chemical messenger, a role that ultimately must be viewed in the overall context of information transfer in maintaining cellular and tissue function and viability.

The neurotransmitter role of ATP has an additional layer of complexity in that, in addition to interacting with the various members of the P2 receptor superfamily, ATP undergoes hydrolysis by synaptic ectonucleotidases to form ADP, AMP, and ADO compounds. Each of these has its own receptor activity, ADO being a neuromodulator at P1 receptors (Williams, 1995). There is also the potential for crosstalk between the two purinergic receptor classes. ATP can block ADP effects on platelet aggregation (Humphries et al., 1995) and also modulates ADO interactions at the A_1 receptor (Williams and Braunwalder, 1986; Zhang et al., 1996). ADO can also potentiate the activation of $P2X_{2/3}$ receptors in pain pathways (Burnstock and King, 1996). Yet another facet of the neurotransmitter role of ATP is that its release, as well as degradation and resynthesis (the latter potentially in cells different from that from which ATP is released), can contribute to changes in the dynamics of the cellular "energy charge" (Atkinson, 1977). ATP thus has a significant, but as yet poorly defined, physiological potential to act as a multifunctional effector agent in a cascade-like manner, producing effects that are complex and potentially mutually antagonistic. ATP is a short-lived mediator of fast excitatory neurotransmission via P2X receptor activation (Edwards et al., 1992; Evans et al., 1992), while the hydrolysis product, ADO, is a potent and selective inhibitor of excitatory transmitter release via the activation of presynaptic ADO A_1 receptors (Williams, 1995). The degree to which such mechanisms are physiologically important depends on intrinsic tissue ectonucleotidase activity (Kennedy and Leff, 1995), as well as the array of purinergic receptors and cell types present in a given tissue environment. Guanosine triphosphate (GTP; Neary et al., 1996) and diadenosine tetraphosphate (Ap4A; Fig. 2, Pintor and Miras-Portugal, 1995) also modulate cellular function via receptor-mediated mechanisms.

Finally, P2 receptors can undergo cell cycle–dependent isotype switching (Dubyak et al., 1996). For instance in leukocytes, the $P2Y_2$ receptor density is decreased as TNF-α (tissue necrosis factor α) is increased in response to inflammatory activation. At the same time, the $P2Y_1$ receptor density is increased.

Purine Pharmacophores

Early work on the medicinal chemistry of both P1 and P2 receptors focused on the chemical modification of the agonists, ADO and ATP, to produce more stable analogs that had the potential to be more selective and efficacious at the various receptor subtypes. Attempts were also made to modify these agonist pharmacophores to develop antagonists (Black, 1989). For ADO, substituents in the 2-, 5′-, and N^6-positions resulted in lead compounds like NECA, CHA, and 2-CADO (Figure 1) that either permitted the development of selective radioligands (Williams and Jacobson, 1990) or provided important research tools that aided in the characterization of P1 receptor systems.

P1 Agonists CHA and related N^6-substituted adenosine analogs are selective for the A_1 receptor, with the 2-chloro analog, CCPA (Figure 1) being among the most selective of these. (−)-RG 14718 is the most potent A_1 receptor agonist yet identified (6 pM) and is over 2000-fold selective for A_1 versus A_{2A} receptors (Fink et al., 1992). The 5′-substituted uronamide, NECA, was the first potent agonist for the A_2 receptor family, although the compound is nonselective in its activity at A_1 and A_2 receptors. CGS 21680C, an N^6-substituted NECA analog, was the first selective A_{2A} agonist with DPMA (N^6-[2(3,5 dimethoxyphenyl)-2-(2-methylphenyl)ethyl]adenosine) (Daly and Jacobson, 1995) maintaining A_{2A} selectivity without the need for the 5′-substitution. IB-MECA and DBXRM (1,3 dibutylxanthine-7-riboside-5′-N-methylcarboxamide) were the first agonists described with reasonable selectivity for the A_3 receptor (Jacobson et al., 1995). NNC 53-0055 (Figure 1) is yet another A_3 receptor–selective agonist (Bowler et al., 1996). While NECA is active at A_{2B} receptors in addition to its activity at A_1 and A_{2A} receptors, efforts in medicinal chemistry have yet to yield A_{2B} receptor–selective agonists. Receptor mechanisms that are more sensitive to NECA than to known A_1, A_{2A}, or A_3 receptor agonists are usually defined as A_{2B} by a process of elimination.

P2 Agonists All known agonist ligands for the P2 receptor family are variations on the purine nucleotide pharmacophore. Efforts to establish structive–activity relationships over the past 20 years have focused on increasing the stability of the polyphosphate side chain using bioisotere replacement strategies, e.g., substitution of the phosphoester bridging oxygen with methylene or dihalomethylene groups, or the introduction of a terminal thio group to increase compound stability (Cusack, 1993; Jacobson et al., 1995; Williams and Bhagwat, 1996). However, in assessing the agonist activity of newer ATP analogs, their potential to inhibit ectonucleotidase activity and thus indirectly alter the potency of endogenous ATP, as well as the rank order agonist potency of diverse ATP analogs, needs to be taken into consideration in the context of the pharmacological characterization of a given tissue (Kennedy and Leff, 1995). Bioisoteric replacement of the labile phosphate has resulted in compounds like $\alpha\beta$-MeATP (Figure 2) that are important research tools. There has been only limited emphasis on substitutions in the purine ring, e.g., 2-methylthioATP (Fig. 2). More recently, however, Jacobson et al. (1995) have identified 2-thioether AMP analogs such as **1** (Figure 2) that are more potent than ATP at erythrocyte P2Y receptors (Williams and Jacobson, 1995). The ability of a monophosphate to be as effective as the parent triphosphate may be due to the long chain at a site distal to the phosphate group that can reduce nucleotidase activity. This leaves the 2-thioether to act as a secondary anchor for receptor binding.

Modification of ATP analogs at N^6 enhances P2Y selectivity. N^6-methyl ATP (Figure 2) is equipotent with ATP at taenia coli P2Y receptors but is inactive or vascular P2Y receptors and smooth muscle P2X receptors. N^6-methyl-2-(5-hexenylthio)-ATP (Figure 2) is active at erythrocyte and taenia coli P2Y receptors and inactive at P2X receptors. Bioisosteric replacements in UTP have apparently been less successful (Pendergast et al., 1996). A series of phosphorylated theobromine derivatives that includes **2** (Figure 2) function as P2Y receptor agonists, providing important structural information regarding the role of the ribose group in determining agonist activity (Fischer et al., 1996).

P1 Antagonists The identification of theophylline as the prototypic, albeit weak, ADO receptor antagonist (Sattin and Rall, 1970) led to the identification of a number of 8-substituted alkyl- and arylxanthines as selective and potent (nanomolar) ADO receptor antagonists (Daly, 1982; Jacobson et al., 1992; Jacobson and Daly, 1995; Baraldi et al., 1996). This provided a major advance in the study of P1 receptor function, although these agents were limited by solubility and some instances potency, precluding or limiting their use in *in vivo* situations (Bruns et al., 1988). The compounds CPX, KFM 19 (BIIP 20, apatoxfylline), MDL 102,503, and KW 3902 (Figure 3) are representatives of large number of A_1 selective xanthine antagonists; KF 17837 and CSC are xanthine antagonists selective for the A_{2A} receptor, and enprofylline has been provisionally characterized as a selective antagonist for the A_{2B} receptor (Feoktistov et al., 1996). Among the first nonxanthine ADO antagonists were CGS 15943 and CP 68,247 followed by ZM 241385 (Keddie et al., 1996) and SCH 58251 (Baraldi et al., 1994) as potent and selective nonxanthine A_{2A} receptor antagonists. The flavone MRS 1067 and the dihydropyridines, MRS 1097 and MRS 1222, are the first of a series of new A_3 receptors antagonists (Karton et al., 1996; van Rhee et al., 1996) that also includes the triazonapthyridine, L-249,313 and the thiazolopyrimidine L-268,605 (Figure 3). These latter compounds are potent and selective A_3 antagonists that have been derived from chemical library sources (M. Jacobson et al., 1996).

P2 Antagonists P2 receptor antagonists include compounds like the polysulfonated aromatic antitrypanocidal drug suramin, PPADS (pyridoxalphosphate-6-azaphenyl-2′,4′-disufonic acid), and various dyes, including reactive blue 2 (Burnstock, 1996a; Figure 4). However these agents lack selectivity and efficacy at P2 receptors and have other activities that limit their *in vivo* use. For instance, suramin also interacts with bFGF and NMDA receptors and can inhibit the activity of ectonucleotidases and various protein kinases. The newer suramin analogs XAMR 0721 and NF023 (Figure 4) lack ectonucleotidase inhibitory activity (van Rhee et al., 1994; Ziyal et al., 1996). However, suramin and NF023 can also affect receptor–G protein coupling processes, their IC_{50} values in uncoupling dopamine D_2 receptor–G protein interactions being in the 500-nM range (Beindl et al., 1996). The purity of the various dyes used to block P2 receptors has also been a problem in defining their utility. Thus a major challenge for the future, both in terms of providing research tools and in developing leads for drug discovery efforts will be the identification of selective antagonists of the P2 receptor subtype.

Site- and Event-Specific Modulation of Purine Receptor Function The concept of "site and event specific" modulation of purinergic events was first proposed by Engler (1987) in the context of inhibitors of ADO transport and/or metabolism as potential therapeutic agents. This "site and event specific" approach focused on the

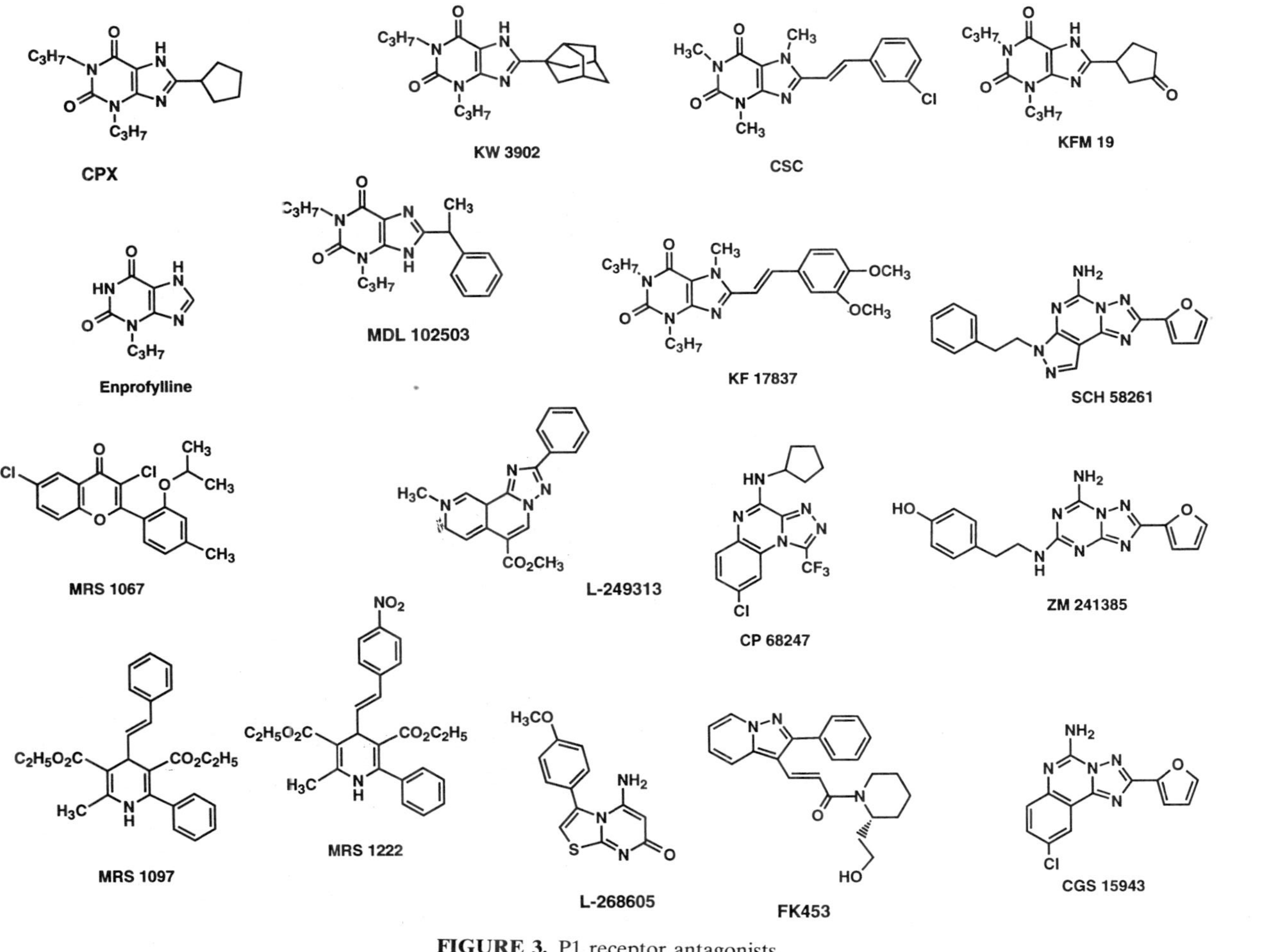

FIGURE 3. P1 receptor antagonists.

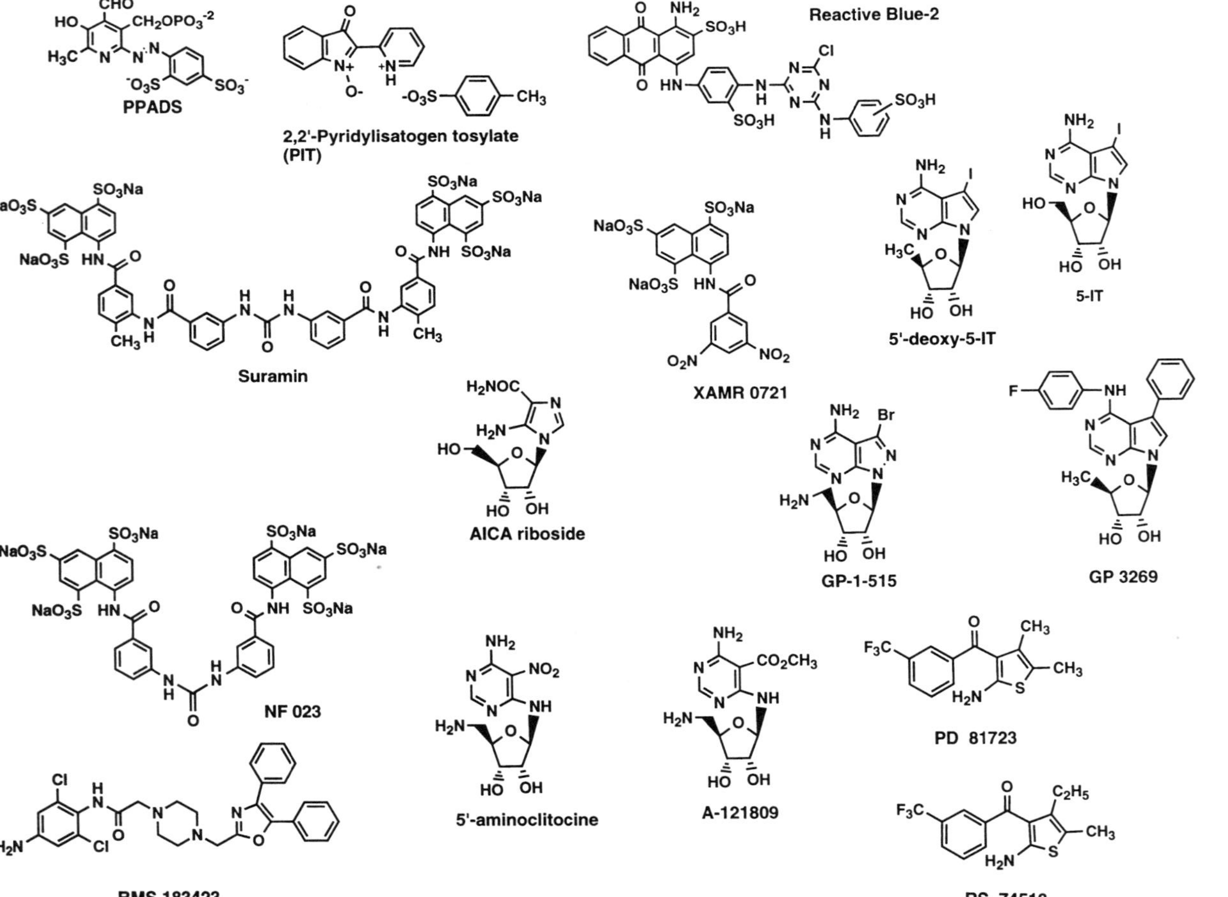

FIGURE 4. P2 receptor antagonists and adenosine potentiating agents.

ability of certain classes of compounds (ADO potentiating agents) to potentiate the physiological actions of localized increases in extracellular ADO caused by tissue trauma. Mechanistically, this occurs by blockade of ADO uptake, phosphorylation (ADO kinase; AK), or deamination (ADO deaminase; ADA).

The utility of this approach has been recently demonstrated in animal models with unilateral kainic acid–induced (KA-induced) lesions of the striatum (Britton et al., 1996). The prototypic AK inhibitor, 5′-d-IT (5′-deoxyiodotubercidin; Figure 4) has been shown to potentiate localized increases in extracellular ADO produced on the KA-lesioned side, as measured by microdialysis, but has no effect on endogenous ADO levels in the nonlesioned striatum.

The first "site and event specific" compound to enter clinical trials was AICA riboside (Acadesine™; Mullane and Young, 1993; Figure 4), a relatively weak inhibitor of ADA, AK, and ADO uptake. Acadesine showed some potential for the treatment of reperfusion injury following coronary angioplasty bypass graft surgery (CABG) but failed to show an overall positive effect on outcomes in pivotal Phase III trials. As a result, attention has been more recently focused on novel modulators of ADO uptake such as BMS 183423 (Figure 4; Moon et al., 1995), and on specific inhibitors of AK. The latter include GP 5-515 (Firestein et al., 1995) and GP 3269 (Mullane et al., 1996), which have been shown to produce beneficial effects in animal models of pain, stroke, and septic shock by virtue of their ability to enhance the actions of the ADO produced extracellularly in these models. 5-Aminoclitocine and A-121809 (Figure 4) are novel AK inhibitors (Lee et al., 1996).

Allosteric Modulators The concept of allosteric modulation of receptor function is based on the now classical enzyme models described by Monod-Wyman-Changeux and Koshland et al. (Williams et al., 1995). Compounds that produce their effects by indirectly modulating the activity of a receptor, in essence modulating the gain of an ongoing signal transduction system, are generally considered to be potentially safer, e.g., less prone to side-effect liability than direct agonists or antagonists. An example of this approach is the $GABA_A$/benzodiazepine (BZ) receptor complex. Directly acting $GABA_A$ agonists have not to date proven to be good drug candidates but the BZs are safe and effective drugs as a result of their ability to modulate $GABA_A$ receptor function.

There is considerable evidence from NMDA, $GABA_A$/BZ, and nicotinic LGICs for physiologically relevant allosteric modulation of receptor function. One of the first P2 receptor antagonists to be described, PIT(2,2′-pyridylisatogen; Figure 4), can enhance $P2Y_1$ receptor–mediated functional responses in transfected oocytes (King et al., 1996). The nicotinic agonist, *d*-tubocurarine enhances binding to the transfected $P2Y_4$ receptor (Michel et al., 1996).

The potential for the allosteric modulation of GPCRs is an emerging concept. The peptide adenoregulin (Daly et al., 1992) and the benzylthiophene PD 81723 (Bruns and Fergus, 1990; Figure 4) represent the first modulators of A_1 purine receptor binding. While adenoregulin appears to be a G protein modulator rather than binding specifically to the ADO A_1 receptor, PD 81723 has been shown, in a variety of functional systems, to be effective in modulating A_1 receptor function. RS 74513 (Figure 4) is another thiophene that can allosterically modulate ADO A_1 receptor function in ileum (Leung et al., 1995).

A major issue in targeting agonists for potential disease treatment is that tolerance tends to develop to agonist effects as a function of prolonged treatment. When the

A_{2A} agonist CGS 21680C is given via osmotic minipumps, its hypotensive effects disappear over a 2-week period and there is a corresponding decrease in brain A_{2A} receptor density (Webb et al., 1993). In a similar study involving systemic administration of ADO agonists, Costrati et al. (1994) found that only A_1 receptor–mediated responses were attenuated by prolonged agonist treatment, a possible reflection of differential receptor selectivity and/or agonist pharmacokinetics.

Interestingly, Bhattacharya and Linden (1996) have shown that the allosteric modulator, PD 81724, shows less functional desensitization *in vivo* than a directly acting ADO agonist, in this instance N^6-cyclopentylADO (CPA). Similarly, in animal models of anticonvulsant efficacy, repeated administration of the AK inhibitor, GP 3269 (Mullane et al., 1996), shows only a slight reduction in functional efficacy.

The acute and chronic effects of ADO agonists and antagonists administered *in vivo* have been found to be very distinct from one another, a phenomenon termed "effect inversion" (K. A. Jacobson et al., 1996). Thus chronic treatment with ADO receptor antagonists produces responses that are similar to those seen with acute treatment with ADO agonists while the converse is true for chronic ADO agonist treatment. Acute administration of ADO analogs has neuroprotective activity in animal models, while chronic (15-day) treatment can exacerbate neuronal injury. The molecular mechanisms for the reported "effect inversion" are unclear and do not appear to consistently involve changes in agonist- or antagonist-induced receptor density. They may be related to differences in the dosing regimens used. Long-term caffeine treatment can cause phenotypic changes in neurons related to immediate early gene expression that involve the transcription factor AP-1 (Svenningsson et al., 1995). AP-1 is also involved in the ADO-related modulation of TNF-α production in inflammatory disease models (Firestein et al., 1995).

THERAPEUTIC TARGETS FOR PURINE RECEPTOR LIGANDS

The major effort in drug targeting for purine receptor ligands has involved the P1 receptor family (Jacobson et al., 1992), with the central nervous and cardiovascular systems being historically the most attractive targets. However, despite the considerable effort expended in developing potent and selective ligands for the various P1 receptor subtypes (A_1, A_{2A}, and A_3), there has been very little progress in terms of *bona fide* clinical candidates that are agonists for these receptors (Williams, 1993). Thus both Cl 936, an A_1 receptor agonist that entered clinical trials as a novel antipsychotic, and the A_{2A} agonist CGS 21680, which was initially targeted as a novel antihypertensive agent, were compounds entering clinical arenas where the molecular mechanisms of the disease were poorly understood (schizophrenia) or where there was little in the way of perceived unmet medical need (hypertension). In addition, both compounds were limited by a marked propensity for side effect liability unrelated to their primary therapeutic targets. NNC 21-0136 (Figure 1) is a novel A_1 receptor agonist that appears to lack class-related side effects (Knutsen, 1996). The compound, like other ADO agonists (Rudolphi et al., 1992) is active in animal models of stroke; it prevents the neuronal cell loss associated with ischemia and, due to some unusual properties that do not appear to be class-related, is reportedly devoid of limiting cardiovascular side effects. Thus the majority of directly acting ADO agonists produce effects on the central nervous, cardiovascular, and renal systems. Compounds targeted for CNS conditions therefore have cardiovascular side effects, while novel antihypertensives have central effects such as sedation.

On the antagonist side, a number of xanthine-based compounds are currently in advanced-stage clinical trials for the ameloration of cognitive defects (apatoxfylline, MDL 102,234), renal dysfunction (FK453, KW 3905; Figure 3), and Parkinson's disease (KF 17837, KW 6002, and related compounds; see Williams, 1993, Spedding and Williams, 1996 for review).

The failure to develop therapeutic agents acting via P1 receptor modulation, despite a considerable wealth of basic knowledge related to the effect of ADO on tissue function, obviously raises questions regarding the continued focus on P1-based therapeutics and the developing opportunities for P2 receptor–based approaches to drug discovery.

In the past decade, however, as the concept of P2 receptor function and diversity has advanced, there has been a paradigm shift in approaches to drug discovery and the enabling technologies used to facilitate the process. In the late 1970s and early 1980s, when industrial ADO-related research was at its peak, the tools to identify and characterize ADO receptors were limited. Furthermore, the health care environment was distinctly different, with little competition from the biotechnology industry (Williams and Gordon, 1996; Williams and Reed, 1997).

The therapeutic targets in the purine area were also ill-defined, being more idiosyncratic extrapolations of basic research findings than commercially viable targets. For instance, purines have been implicated in the mechanism of action of anxiolytics and antidepressants on the basis of very limited and circumstantial animal data (Williams, 1995), in areas where the basic disease mechanisms are poorly understood. More recently, however, substantial newer data has been generated to support the potential for purinergic ligands in the treatment of chronic pain, various inflammatory disorders, diabetes, drug abuse, and sleep disorders.

There is a considerable body of data supporting a role of ADO and ATP in pain mechanisms (Sawynok et al., 1989; Burnstock, 1996c; Burnstock and Wood, 1996), although the molecular tools available to define these roles in man have been limited. While ADO causes ischemia-like pain in man (Born et al., 1965; Sylven et al., 1986, 1988), ATP has been described as a mediator for primary pain afferents from mammalian skin (Fyffe and Perl, 1984), and ADO and related analogs have demonstrated remarkable effects in the treatment of chronic pain in the clinic (Sollevi et al., 1995; Karlsten and Gordh, 1996), albeit it in open trials with limited patient numbers. These effects appear to be mediated via A_1 receptors in the spinal cord (Lee and Yaksh, 1996). Furthermore, the localization of the $P2X_3$ receptor to the dorsal root ganglion and its destruction by capsacin (Chen et al., 1995) reinforces a role for purines in the control of the chronic pain associated with nervous system trauma, and the neuropathies associated with HIV infection, diabetes, and cisplatin and taxol chemotherapy.

The role of ADO as an anti-inflammatory agent (Cronstein and Weismann, 1995; Cronstein et al., 1996) has been slowly evolving over the past 5 years. Thus ADO has been shown to inhibit neutrophil-free radical production via A_{2A} receptor mechanisms (Cronstein et al., 1990), decrease cytokine expression via an A_3 receptor mechanism (Sajjadi et al., 1996), inhibit neutrophil adhesion, an effect mimicked by AK inhibition (Firestein et al., 1995), and regulate $\beta 2$ integrin expression and L-selectin shedding from human lymphocytes via an $A2_A$ mechanism (Thiel et al., 1996).

The role of purines in inflammation is underlined by recent work (Denlinger et al., 1996) showing that 2 MeS-ATP, acting via a P2Y receptor, can inhibit lipopolysaccharide-induced (LPS-induced) inducible nitric oxide synthase (NOS) production in addition to its effects in reducing TNF-α and IL-1α levels in response to LPS (Proctor et al., 1994). Activation of A_{2B} receptors on synoviocytes inhibits matrix metalloproteinase (MMPase) gene expression (Boyle et al., 1996), activation of the latter being a major cause of the bone and cartilage damage associated with arthritic disease process. While studies on the role of purines in inflammatory mechanisms are relatively recent compared to the extensive studies on purines and CNS function, both ADO and ATP appear to have considerable potential as novel, and much needed, approaches to the treatment of a broad spectrum of inflammatory disorders.

The slow-wave sleep (SWS) induced by the infusion of prostaglandin D_2 (PGD_2) into the subarachnoid space of rats (the PGD_2-sensitive sleep-promoting zone in the rostral basal forebrain) can be blocked by the selective A_{2A} receptor antagonist KF 17837 (Satoh et al., 1996). ADO is a potent CNS depressant (Williams, 1995), and it is consequently of interest that the sleep-inducing effects of PGD_2 can be mimicked by the A_{2A} agonist CGS 21680, a finding that indicates a pivotal role for A_{2A} receptor mechanisms in SWS. The molecular relationships between the effects of PGD_2 and CGS 21680 remain unclear.

In a number of studies, ADO receptor mechanisms have been shown to be involved in aspects of the acute and chronic effects of opiates, although conflicting data exists on the opposing effects of ADO agonists and antagonists. In morphine-dependent mice, R-PIA (A_1 agonist) reduced the wet dog shakes and diarrhea associated with withdrawal, while CGS 21680 (A_{2A} agonist) inhibited teeth chattering and forepaw treading (Kaplan and Sears, 1996). In contrast, ADO antagonists exacerbate these withdrawal symptoms, suggesting that ADO agonists may be effective in the treatment of opiate withdrawal in humans, a major social and health problem.

ADO, ADP and ATP have been shown to modulate insulin secretion from pancreatic B cells (Loubatieres-Mariani et al., 1995; Nichols et al., 1996). ATP stimulates insulin secretion via a P2Y receptor mechanism and can potentiate the effects of tolbutamide. While ADO also increases insulin secretion, R-PIA was found to decrease insulin secretion. ADO could also increase glucagon secretion, although NECA was up to 800 times more potent. Given the increasing incidence of diabetes as a major health problem of the aging population, purinergic approaches to the modulation of insulin and glucagon secretion may represent important therapeutic opportunities.

FUTURE PROSPECTS

With the current molecular focus on purine and pyrimidine receptors and on increased efforts in medicinal chemistry to design selective ligands for the various receptor subtypes, the potential for developing novel therapeutic agents acting via P1 and P2 receptor families has never been more promising, especially in chronic disease states where normal tissue function is compromised by changes in the sensitivity of receptor and signal transduction systems. The more recent data showing trophic, necrotic, and apoptotic effects of ADO, ATP, and GTP also suggests an important role for purines in tissue development, remodeling, and senesence, and extends the scope of P1 and P2 receptor function to basic mechanisms of tissue function and repair.

An interesting consideration related to the role of UTP as a transmitter is why there have been few reports of uridine functioning as a neuromodulator in a manner similar to that of ADO as a product of ATP hydrolysis. Uridine has been reported to affect dopaminergic responses in animal models (Agnati et al., 1989; Myers et al.,

1994). However, attempts to identify uridine receptors using radioligand binding have, to date, been unsuccessful (M. F. Jarvis, unpublished data).

REFERENCES

Abbracchio M, Burnstock G (1994): Purinoceptors: Are there families of P2X and P2Y purinoceptors? Pharmacol Ther 64:445–475.

Arch JRS, Newsholme EA (1978): The control of the metabolism and the normal role of adenosine. Essays Biochem 14:82–123.

Atkinson DE (1977): "Cellular Energy Metabolism and Its Regulation," San Diego, CA, Academic Press.

Agnati LR, Fuxe K, Ruggeri M, Merlo-Pich E, Benefenati F, Volterra V, Ungerstedt U, Zini I (1989): Effects of chronic uridine on striatal dopamine release and dopamine related behaviors in the absence or presence of chronic treatment with haloperidol. Neurochem Int 15:107–113.

Baraldi PG, Cacciari B, Spalluto G, Borioni A, Viziano M, Dionisotti S, Ongini E (1996): Current development of A_{2a} adenosine receptor antagonists. Curr Med Chem 2:707–722.

Baraldi PG, Manfredini S, Simoni D, Zappaterra L, Zocchi C, Dionisotti S, Ongini E (1994): Synthesis of new pyrazolol (4,3-e) 1,2,4-olo-(1,5-e) pyrimidine and 1,2,3-triazolo (4,5-e) 1,2,4-triazolo(1,5-c) pyrimidine displaying potent and selective activity as A_{2a} adenosine receptor antagonists. Bioorganic Med Chem Letts 21:2539–2544.

Barnard EA, Burnstock G, Webb TE (1994): G-protein-coupled receptors for ATP and other nucleotides: A new receptor family. Trends Pharmacol Sci 15:772–778.

Bass R (1914): Uber die purinkorper des menschlichen blutes und den wirkungsmodus der 2-phenyl-4- cholincarbonsaure (Atophan). Arch Exp Pathol Pharmakol 76:40.

Beindl W, Mitterauer T, Hohenegger M, Ijzerman AP, Nanoff C, Freissmuth M (1996): Inhibition of receptor/G protein coupling by surmain analogues. Mol Pharmacol 50:415–523.

Bellardinelli L, Shyrock JC, Song Y, Wang D, Shinivas M (1995): Ionic basis of the electrophysiological actions of adenosine on cardiomyocytes. FASEB J 9:359–365.

Berne RM (1963): Cardiac nucleotides in hypoxia: Possible role in the regulation of coronary blood flow. Am J Physiol 204:317–322.

Bhattacharya S, Linden J (1996): Effects of long-term treatment with the allosteric enhancer, PD81,723 on Chinese Hamster Ovary cells expressing recombinant human A_1 adenosine receptors. Mol Pharmacol 50:104–111.

Black JW (1989): Drugs from emasculated hormones: The principle of synoptic antagonism. Science 245:486–492.

Bo X, Zhang Y, Nassar M, Burnstock G, Shoepfer R (1995): A P2X purinoceptor cDNA conferring a novel pharmacological profile. FEBS Lett 375:129–133.

Boarder MR, Weisman GA, Turner JT, Wilkinson GF (1995): G-protein coupled P_2 purinoceptors: From molecular biology to functional responses. Trends Pharmacol Sci 16:133–139.

Boeynaems J-M, Communi D, Pirotton S, Motte S, Parmentier M (1996): Involvement of distinct receptors in the actions of extracellular uridine nucleotides. CIBA Found Symp 198:266–277.

Born GVR, Haslam RJ, Goldman N, Lowe RD (1965): Comparative effects of adenosine analogues on the inhibition of blood platelet aggregation and as vasodilators in man. Nature 205:678–680.

Bowler AN, Olsen UB, Thomsen C, Knutsen LJS (1996): New adenosine A_3 ligands controlling cytokines. Drug Dev Res 37:173.

Boyle DL, Sajjadi FG, Firestein GS (1966): Adenosine receptor stimulation inhibits synoviocyte collagenase gene expression. Arthritis Rheum 39:923–930.

Brake AJ, Wagenbach MJ, Julius D (1994): New structural motif for ligand-gated ion channels defined by an ionotropic ATP receptor. Nature 371:519–523.

Britton DR, Mikusa J, Lee L, Williams M, Kowaluk E (1996): Adenosine kinase inhibition enhances kainic acid-induced increases in strital adenosine. Soc Neurosci Abstr (in press).

Bruns RF, Fergus JH (1990): Allosteric enhancement of adenosine A_1 receptor binding and function by 2-amino-3-benzoyl-thiophenes. Mol Pharmacol 38:939–949.

Bruns RF, Davis RE, Ninteman FW, Poschel BP, Wiley JN Heffner TG (1988): Adenosine antagonists as pharmacological tools. In Paton DM (ed): "Adenosine and Adenine Nucleotides: Physiology and Pharmacology." London, Taylor & Francis, pp 39–49.

Burnstock G (1972): Purinergic nerves. Pharmacol Rev 24:509–581.

Burnstock G (1978): A basis for distinguishing two types of purinergic receptor. In Straub RW, Bolis L (eds): "Cell Membrane Receptors for Drugs and Hormones: A Multidisciplinary Approach." New York: Raven Press, pp 107–118.

Burnstock G (1996a): P2 purinoceptors: Historical perspective and classification. CIBA Found Symp 198:1–34.

Burnstock G (1996b): Purinoceptors: Ontogeny and phylogeny. Drug Dev Res 39:204–242.

Burnstock G (1996c): A unifying purinergic hypothesis for the initiation of pain. Lancet 347:1604–1605.

Burnstock G, Barnard EA (1966): ATP as a neurotransmitter. CIBA Found Symp 198:261–265.

Burnstock G, Kennedy C (1985): Is there a basis for distinguishing two types of P_2-purinoceptor? Gen Pharmacol 16:433–440.

Burnstock G, King BF (1996): The numbering of cloned P_2 purinoceptors. Drug Dev Res 38:67–71.

Burnstock G, Ralevic V (1994): New insights into the local regulation of blood flow by perivascular nerves and endothelium. Br J Plas Surg 47:527–543.

Burnstock G, Wood JN (1996): Purinergic receptors: Their role in nociception and primary afferent neurotransmission. Curr Opin Neurobiol 6:526–532.

Burnstock G, Campbell G, Satchell D, Smythe A (1970): Evidence that adenosine triphosphate or a related nucleotide is the transmitter substance released by non-adrenergic inhibitory nerves in the gut. Br J Pharmacol 40:668–688.

Chen CC, Akopian AN, Sivilotti L, Colquhoun D, Burnstock G, Wood JN (1995): A P_{2X} purinoceptor expressed by a subset of sensory neurons. Nature 377:428–431.

Collo G, North RA, Kawashima E, Merlo-Pich E, Neidhart S, Surprenant A, Buell G (1996): Cloning of $P2X_5$ and $P2X_6$ receptors, and the distribution and properties of an extended family of ATP-gated ion channels. J Neurosci 16:2495–2500.

Communi D, Pirotton S, Parmentier M, Boeynaems JM (1995): Cloning and functional expression of a human uridine nucleotide receptor. J Biol Chem 270:30849–30852.

Communi D, Parmentier M, Boeynaems JM (1996): Cloning, functional expression and tissue distribution of the human $P2Y_6$ receptor. Biochem Biophys Res Comm 222:303–308.

Cornfield LJ, Hu S, Hurt SD, Sills MA (1992): (^{3}H)-2-phenylaminoadenosine ((^{3}H)-CV 1808) labels a novel adenosine receptor in rat brain. J Pharmacol Exp Ther 363:552–561.

Costrati C, Monopoli A, Dionisotti S, Zocchi C, Bonizzoni E, Ongini E (1994): Repeated administration of selective adenosine A_1 and A_2 receptor agonists in the spontaneously hypertensive rat: Tolerance develops to A_1-mediated hemodynamic effects. J Pharmacol Exp Ther 268:1506–1511.

Cronstein BN, Bouma MG, Becker BF (1996): Purinergic mechanisms in inflammation. Drug Dev Res 39:426–435.

Cronstein BN, Weissmann G (1995): Targets for antiinflammatory drugs. Ann Rev Pharmacol Toxicol 35:449–462.

Cronstein BN, Daguma L, Nicholls D, Hutchison AJ, Williams M (1990): The adenosine/neutrophil paradox resolved: Human neutrophils possess both A_1 and A_2 receptors which promote chemotaxis and inhibit O^{2-} generation respectively. J Clin Invest 85:1150–1157.

Cusack NJ (1993): P2 receptor: Subclassification and structure-activity relationships. Drug Dev Res 28:244–252.

Daly JW (1982): Adenosine receptors: Targets for future drugs. *J Med Chem* 25:197–207.

Daly JW, Jacobson KA (1995): Adenosine receptors: Selective agonists and antagonists. In Bellardinelli L, Pelleg A, (eds.): "Adenosine and Adenine Nucelotides: From Molecular Biology to Integrative Physiology, Boston: Kluwer, pp 157–166.

Daly JW, Caceres J, Moni RW, Gusovsky F, Moos Jr M, Seamon KB, Milton K, Myers CW (1992): Frog sections and hunting magic in the upper Amazon: identification of a peptide interacting with an adenosine receptor. Proc Natl Acad Sci USA 89:10960–10963.

Denlinger LC, Fisette PL, Garis KA, Kwon G, Vazquez-Torres A, Simon AD, Nguyen B, Proctor RA, Bertics PJ, Corbett JA (1996): Regulation of inducible nitric oxide synthase expression by macrophage purinoceptors and calcium. J Biol Chem 271:337–342.

Di Virgillio F (1995): The purinergic P2Z receptor: An intriguing role in immunity, inflammation and cell death. Immunol Today 16:524–528.

Drury AN, Szent-Gyorgyi A (1929): The physiological activity of adenine compounds with special reference to their effects upon the mammalian heart. J Physiol 68:213–237.

Dubyak GR, Martin KA, Humphreys BD, Clifford EE (1996): Expression of multiple ATP receptor–based signaling cascade during development and inflammatory activation of human mononuclear phagocytes. Drug Dev Res 37:110.

Dubyak GR, El-Moatassim C (1993): Signal transduction via P2-purinergic receptors for extracellular ATP and other nucelotides. Am J Physiol 265:577C–505C.

Edwards FA, Gibb AJ, Colquhoun D (1992): ATP receptor-mediated synaptic currents in the central nervous system. Nature 359:144–147.

Embden C, Zimmerman G (1927): Uber die bedeutung der adenylsaure. Hoppe-Seyler's Z Physiol Chem 167:137–140.

Engler R (1987): Consequences of activation and adenosine-mediated inhibition of granulocytes during myocardial ischemia. Fed Proc 46:2407–2412.

Engler R (1991): Adenosine: The signal of life? Circulation 84:951–954.

Evans RJ, Derkach V, Surprenant A (1992): ATP mediates fast synaptic transmission in mammalian neurons. Nature 357:503–505.

Evans RJ, Lewis C, Buell G, Valera S, North RA, Surprenant A (1995): Pharmacological characterization of heterologously expressed ATP-gated cation channels ($P2_X$) purinoceptors. Mol Pharmacol 48:178–183.

Feoktistov I, Vallejo V, Biagionni I (1996): Adenosine A2b receptors: G-protein coupling and calcium signaling. Drug Dev Res 37:121.

Fink CA, Spada AP, Colussi D, Rivera L, Merkel L (1992): Synthesis of a potent A_1 selective adenosine agonist: N^6-(1-R(3-chloro-2-thienyl)methyl)propyl) adenosine RG 14719 (−). Nucleosides Nucleotides 11:1077–1088.

Firestein GS, Bullough DA, Erion MD, Jiminiez R, Ramirez-Weinhouse M, Barankiewicz J, Smith CW, Gruber HE, Mullane KM (1995): Inhibition of neutrophil adhesion by adenosine and an adenosine kinase inhibitor. J Immunol 154:326–344.

Fischer B, Yefidol R, Boyer JL, Harden TK, Jacobson KA (1996): Novel nonnucleotide P_2-purinoceptor agonists. Drug Dev Res 37:132.

Fiske CH, Subbarow Y (1929). Phosphorous compounds of muscle and liver. Science 70:381–382.

Fredholm BB, Burnstock G, Abbracchio MP, Daly JW, Hardin TK, Jacobson KA, Williams M (1994): Nomenclature and classification of purinoceptors: A report from the IUPHAR Subcommittee. Pharmacol Rev 46:143–156.

Fredholm BB, Burnstock G, Dubyak G, Harden K, Schwabe U, Spedding MF, Williams M (1997): Interim report of the IUPHAR Committee on purine and pyrimidine receptor nomenclature. Trends Pharmacol Sci (in press).

Freund H (1920): Uber die pharmakologischen wirkungen des defibriniertin blutes. Arch Exp Pathol Pharmakol 86:267.

Fukunaga AF, Miyamoto TA, Kikuta Y, Kaneko Y, Ichinohe T (1995): Role of adenosine and adenosine triphosphate as anesthetic adjuvants. In Bellardinelli L, Pelleg A (eds): "Adenosine and Adenine Nucleotides: From Molecular Biology to Integrative Physiology." Boston: Kluwer, pp 511–523.

Fyffe REW, Perl ER (1984): Is ATP a central synaptic mediator for certain primary afferent fibers from mammalian skin? Proc Natl Acad Sci USA 81:6890–6893.

Gerlach E, Deuticke B, Driesbach RH (1963): Der nucleotid-abbau in herzmuskel bei sauorstoff-mangel und seine mogliche bedeutung fur die coronar durch blutung. Naturwissenschaft 50:228–229.

Gordon JL (1986): Extracellular ATP: Effects, sources and fate. Biochem J 233:309–319.

Green RM, Stoner HB (1950): "Biological Actions of Adenosine Nucleotides." Lewis, London.

Harden TK, Boyer JL, Nicholas RA (1995): P_2-purinergic receptors: Subtype–associated signaling responses and structure. Ann Rev Pharmacol Toxicol 35:541–579.

Hilderman RH, Martin M, Zimmerman JK, Pivorun EB (1991): Identification of a unique membrane receptor for adenosine 5′,5′′′-P_1 P_4 tetraphosphate. J Biol Chem 266:6915–6918.

Holton P (1959): The liberation of adenosine triphosphate on antidromic stimulation of sensory nerves. J Physiol 145:494–504.

Honey RM, Ritchie WT, Thompson WAR (1930): The action of adenosine upon the human heart. Quart J Med 23:485–490.

Humphries RG, Robertson MJ, Leff P (1995): A novel series of P2T purinoceptor antagonists: Definition of the role of ADP in arterial thrombosis. Trends Pharmacol Sci 16:179–181.

Humphrey PPA, Buell G, Kennedy I, Khakh BS, Michel AD, Surprenant A, Triezise DJ (1995): New insights on P_{2X} purinoceptors. Naunyn-Schmideberg's Arch Pharmacol 352:585–596.

Jacobson KA, van Galen PJM, Williams M (1992): Adenosine receptors: Pharmacology, structure–activity relationships, and therapeutic potential. J Med Chem 35:407–415.

Jacobson KA, Fisher B, Malliard M, Boyer, JL, Hoyle CHV, Hardin TK, Burnstock G (1995): Novel ATP agonists reveal receptor heterogeneity within P_{2x} and P_{2Y} subtypes. In Bellardinelli L, Pelleg A (eds): "Adenosine and Adenine Nucleotides: From Molecular Biology to Integrative Physiology." Boston: Kluwer pp 149–156.

Jacobson KA, von Lubitz DKJE, Daly JW, Fredholm BB (1996): Adenosine receptor ligands: Differences with acute versus chronic treatment. Trends Pharmacol Sci 17:108–113.

Jacobson MA, Chakravarty PK, Johnson RG, Norton R (1996): Novel selective non-xanthine A3 adenosine receptor agonists. Drug Dev Res 37:131.

Janssens R, Communi D, Pirroton S, Samson M, Parmentier M, Boeynaems J-M (1996): Cloning and tissue distribution of human P2Y1 receptor. Biochem Biophys Res Comm 221:588–593.

Jezer A, Oppenheimer BS, Schwartz SP (1933): The effect of adenosine of cardiac irregularities in man. Am Heart J 9:252–258.

Jiang SN, Kull B, Fredholm BB, Orrenius S (1996): P2X purinoceptor is not important in thymocyte apoptosis. Immunol Lett 49:197–201.

Kaplan GB, Sears MT (1996): Adenosine receptor agonists attenuate and adenosine receptor antagonists exacerbate opiate withdrawal signs. Psychopharmacology 123:64–70.

Karlsten R, Gordh T Jr. (1996): An A_1-selective adenosine agonist abolishes allodynia elicited by vibration and touch after intrathecal injection. Anesth Analg 80:844–847.

Karton Y, Jiang J-L, Ji X-D, Melman N, Olah ME, Stiles GL, Jacobson KA (1996): Synthesis and biological activities of flavenoid derivatives as A3 adenosine receptor antagonists. J Med Chem 39:2293–2301.

Keddie JR, Poucher SM, Shaw GR, Brooks R, Collis MG (1996): In vivo characterization of ZM 241385, a selective adenosine $A2_A$ receptor antagonist. Eur J Pharmacol 301:107–113.

Kennedy C, Leff P (1995): How should P2X purinoceptors be classified pharmacologically? Trends Pharmacol Sci 16:168–174.

Kidd EJ, Grahames BA, Simon J, Michel AD, Barnard EA, Humphrey PPA (1995): Localization of P2X purinoceptor transcripts in the rat nervous system. Mol Pharmacol 48:569–573.

King BF, Dacquet C, Ziganshin AU, Weetman, DF, Burnstock G, Vanhoutte PM, Spedding M, Burnstock G (1996): Potentiation by 2,2′-pyridylisatogen tosylate of ATP-responses at a recombinant $P2Y_1$ purinoceptor. Br J Pharmacol 117:1111–1118.

Knutsen LJS (1996): Selective agents for modulation of adenosine A_1 adenosine receptors. Drug Dev Res 37:111.

Lazorowski ER, Harden TK (1994); Identification of a uridine-nucleotide-selective G-protein-linked receptor that activates phospholipase C. J Biol Chem 269:11830–11836.

Lazarowksi ER, Watt WC, Stuts MJ, Boucher RC, Harden TK (1995): Pharmacological selectivity of the cloned human P_{2U}-purinoceptor: Potent activation by diadenosine tetraphosphate. Br J Pharmacol 116:1619–1627.

Lee C-H, Jiang M, Daanen J, Kohlass KL, Alexander KM, Yu H, Kowaluk EA, Bhagwat SS (1996): Synthesis and biological evaluation of clitocine analogs as adenosine kinase inhibitors. Abstracts 212th American Chemical Society Meeting, Orlando, FL, August, 1996, MEDI 156.

Lee Y-H, Yaksh TL (1996): Pharmacology of the spinal adenosine receptor which mediates the allodynic action of intrathecal adenosine agonists. J Pharmacol Exp Ther 277:1642–1648.

Leung E, Walsh LKM, Flippin LA, Kim EJ, Lazar DA, Seran CS, Wong EHF, Eglen RM (1995): Enhancement of adenosine A_1 receptor functions by benzothiophenes in guinea pig tissue in vitro. Naunyn-Schimedberg's Arch Pharmacol 353:206–212.

Lewis C, Neldhart S, Holy C, North RA, Buell G, Surprenant A (1995): Coexpression of $P2X_2$ and $P2X_3$ receptor subunits can account for ATP-gated currents in sensory neurons. Nature 377:432–435.

Lippmann F (1941): Metabolic generation and utilization of phosphate bond energy. Enzymology 1:99.

Lindner F, Rigler R (1931); Uber die beeinflussung der weita der herzteranzgefasse durch produckte des zelkern stoffwechsels. Pflugers Arch 226:697–707.

Lohmann K (1929): Uber die pyrophosphatfraktion in muskel. Naturwissenschaften 17:624–625.

Loubatieres-Mariani MM, Petit P, Chapal J, Hillaire-Buys D, Bertrand G, Ribes G (1995): Effects of purinoceptor agonists on insulin secretion. In Bellardinelli L, Pelleg A (eds): "Adenosine and Adenine Nucleotides: From Molecular Biology to Integrative Physiology. Boston: Kluwer, pp 337–345.

Lustig KF, Shaiu AK, Brake AJ, Julius D (1993): Expression cloning of an ATP receptor from mouse neuroblastoma cells. Proc Natl Acad Sci USA 90:5113–5117.

Luthin DR, Linden J (1995): Comparison of A_4 and A_{2a} binding sites in the striatum and COS cells transfected with adenosine A_{2a} receptors. J Pharmacol Exp Ther 272:511–518.

May B, Leinweber W (1971): Antilipolytic action of phenylisopropyl adenosine in man. Nauyn-Schimedberg's Arch Pharmacol 269:467–468.

Michel AD, Miller KJ, Buell G, Lundsrom K, Humphrey PPA (1996): Direct labeling of P2X purinoceptor subtypes. Drug Dev Res 37:113.

Moon SL, Signor L, Myers RA, Matos F, Lomabrado L, Hibbard J, Ortiz A, Reid S, Balasubramanian N (1995): Adenosine re-uptake inhibitors are neuroprotective. Abstr Soc Neurosci 21:999.

Mullane KM, Young M (1993): Acadesine: Prototype adenosine regulating agent for treating myocardial ischemia-reperfusion injury. Drug Dev Res 28:336–343.

Mullane KM, Foster AC, Weisner JB (1966): GP3269: A selective adenosine kinase inhibitor as a novel approach to anticonvulsant therapy. Drug Dev Res 37:175.

Myers CS, Fisher H, Wagner GC (1994): Uridine potentiates haloperidol's disruption of conditioned avoidance responding. Brain Res 651:194–198.

Neary JT, Rathbone MP, Cattabeni F, Abbracchio MP, Burnstock G (1996): Trophic actions of extracellular nucleotides and nucleosides on glial and neuronal cells. Trends Neurosci 19:13–18.

Newby A (1984): Adenosine and the concept of a retaliatory metabolite. Trends Biochemical Sci 9:42–44.

Nicholas RA, Watt WC, Lazorowski ER, Li Q, Harden TK (1996): Uridine nucleotide selectivity of three phospholipase C-activating P_2 receptors: Identification of a UDP-selective, a UTP-selective and an ATP- and UTP-specific receptor. Mol Pharmacol 50:224–229.

Nichols CG, Shyng SL, Nestorowicz A, Glaser B, Clemont JP IV, Gonzalez G, Aguilar-Bryan L, Permitt MA, Bryan J (1996): Adenosine diphosphate as an intracellular regulator of insulin secretion. Science 272:1785–1787.

Nicke A, Rettinger J, Eichele A, Lambrecht G, Mutschler E, Schmalzing G (1996): Possible quaternary structure of the P2X1 purinoceptor. Drug Dev Res 37:115.

North RA (1996): P2X purinoceptor plethora. Seminars Neurosci 8:187–194.

Nguyen T, Erb L, Weisman GA, Marchese A, Heng HHQ, Garrad RC, George SR, Turner JT, O'Dowd BF (1995): Cloning, expression, and chromosomal localization of the human uridine nucleotide receptor gene. J Biol Chem 270:30845–30848.

Olah ME, Stiles GL (1995): Adenosine receptor subtypes: Characterization and therapeutic regulation. Ann Rev Pharmacol Toxicol 35:581–606.

Paterson ARP, Jakobs ES, Ng CYC, Odegard RD, Adjei AA (1987): Nucleoside transport inhibition in vitro and in vivo. In Gerlach E, Bekcer B (eds): "Topics and Perspectives in Adenosine Research." Berlin: Springer-Verlag, pp 89–101.

Pendergast W, Siffiqi SM, Rideout JL, James MK, Dougherty RW (1996): Stabilized uridine triphosphate analogs as agonists of the $P2Y_2$ purinergic receptor. Drug Dev Res 37:133.

Pintor J, Miras-Portugal MT (1995): P_2 purinergic receptors for diadenosine polyphosphates in the nervous system. Gen Pharmacol 26:229–235.

Proctor RA, Denlinger L, Leventhal PS, Daugherty SK, Van Der Loo J-W, Tanake T, Firestein GS, Bertics PJ (1994): Protection of mice from endotoxic death by 2-methylthio-ATP. Proc Natl Acad Sci USA 91:6017–6020.

Rudolphi K, Schubert P, Parkinson FE, Fredholm BB (1992): Neuroprotective role of adenosine in cerebral ischemia. Trends Pharmacol Sci 13:439–445.

Sajjadi FG, Takabayashi K, Foster A, Domingo R, Firestein GS (1996): Inhibition of TNF-alpha gene expression by adenosine: The role of A3 adenosine receptors. J Immunol 156:3435–3442.

Satoh S, Matsumura H, Suzuki F, Hayaishi O (1996): Promotion of sleep mediated by the A_{2a}-adenosine receptor and possible involvement of this receptor in the sleep induced vt prostaglandin D_2 in rats. Proc Natl Acad Sci 93:5980–5984.

Sattin A, Rall TW (1970): The effects of adenosine and adenine nucleotides on the cyclic adenosine 3′-5′-phosphate content of guinea pig cerebral cortex. Mol Pharmacol 6:13–23.

Sawynok J, Sweeney MI, White TD (1989): Adenosine release may mediate spinal analgesia by morphine. Trends Pharmacol Sci 10:186–189.

Simon J, Webb TE, Barnard EA (1995): Characterization of a P2Y purinoceptor in the brain. Pharmacol Toxicol 76:302–307.

Sollevi A, Belfrage M, Lundeberg T, Segerdahl M, Hansson P (1995): Systemic adenosine infusion: a new treatment modality to alleviate neuropathic pain. Pain 61:155–158.

Spedding M, Williams M. (1996):Development in purine and pyrimidine receptor based therapeutics. Drug Dev Res 39:436–441.

Surprenant A, Buell G, North RA (1995): P2X receptors bring new structure to ligand-gated ion channels. Neurosci 18:224–229.

Surprenant A, Kawashima E, North RA, Buell G (1996): The cytolytic P_{2Z} receptor for extracellular ATP identified as a P2X receptor ($P2X_7$). Science 272:735–738.

Svenningsson P, Nomikos GG, Fredholm BB (1995): Biphasic changes in locomotor behavior and in expression of mRNA for NGFI-A and NGFI-B in rat striatum following acute caffeine administration. J Neurosci 15:7612–7624.

Sylven C, Beermann B, Jonzon B, Brant R (1986): Angina pectoris-like pain provoked by intravenous adenosine in healthy volunteers. Br Med J 293:227–230.

Sylven C, Jonzon B, Fredholm BB, Kaijser L (1988): Adenosine injection into the brachial artery produces ischaemia like pain or discomfort in the forearm. Cardiovasc Res 22:674–678.

Thannhauser SJ, Bommes A (1914): Experimentelle studies ober den nucleinstoffwechsel, II. Mittelung. Stoffwechselversuche mit adenosin and guanosin. Hoppe-Seyler's Z Physiol Chem 91:336–434.

Thiel M, Chambers JD, Chouker A, Fischer S, Zouredildis C, Bardenheuer HJ, Arfors KE, Peter K (1996): Effect of adenosine on the expression of $\beta 2$ integrins and L-selectin of human polymorphonuclear leukocytes in vitro. J Leukocyte Biol 59:671–682.

Valera S, Hussy N, Evans RJ, Adami N, North RA, Surprenant A, Buell G (1994): A new class of ligand-gated channel defined by P2x receptor for extracellular ATP. Nature 371:516–519.

van Rhee AM, Jiang J-L, Melman N, Olah ME, Stiles GL, Jacobson KA (1996): Interaction of 1,4-dihydropyridine and pyridine dervatives with adenosine receptors: Selectivity for A_3 receptors. J Med Chem 39:2980–2989.

van Rhee AM, Van der Heijed MPA, Beukers MW, Ijzerman AP, Soudijin W, Nickel P (1994): A novel competitive antagonists for P2 purinoceptors. Eur J Pharmacol 268:1–7.

Vanhoutte PM, Humphrey PPA, Spedding M (1996): International Union of Pharmacology recommendations for nomenclature of new receptor subtypes. Pharmacol Rev 48:1–2.

Webb RL, Sills MA, Chovan JP, Peppard JV, Francis JE (1993): Development of tolerance to the antihypertensive effects of highly selective adenosine A_2 agonists upon chronic administration. J Pharmacol Exp Ther 267:287–295.

Webb TE, Henderson D, King BF, Wang S, Simon J, Bateson AN, Burnstock G, Barnard EA (1996a): A novel G protein–coupled P_2 purinoceptor (P_{2Y3}) activated preferentially by nucleoside diphosphates. Mol Pharmacol 50:258–265.

Webb TE, Kaplan MG, Barnard EA (1996b): Identification of 6H1 as a P2Y purinoceptor: $P2Y_5$. Biochem Biophys Res Comm 219:105–110.

Webster JM, Heseltine L, Taylor R (1996): In vitro effects of adenosine agonist GR 79236 on the insulin sensitivity of glucose utilisation in rat soleus and human rectus abdominus muscle. Biochim Biophys Acta Mol Basis Dis 1316:109–113.

Weisman GA, Turner JT, Fedan JS (1996): Structure and function of P_2 purinoceptors. J Pharmacol Exp Ther 277:1–9.

Williams M (1989): Adenosine: The prototypic neuromodulator. Neurochem Int 14:249–264.

Williams M (1993): Purinergic drugs: Opportunities in the 1990s. Drug Dev Res 28:438–444.

Williams M (1995): Purinergic Receptors. In Bloom FE, Kupfer DJ (eds): "Psychopharmacology, The Fourth Generation of Progress." New York: Raven, pp 643–655.

Williams M (1996): Challenges in developing P2-purinoceptor-based therapeutics. CIBA Found Symp 198:309–321.

Williams M, Bhagwat SS (1996): P_2 purinoceptors: A family of novel therapeutic targets. Ann Rep Med Chem 31:21–30.

Williams M, Braunwalder A (1986): Purine nucleotide effects on the binding of (^{3}H)-cyclopentyladenosine to adenosine A-1 receptors in rat brain membrane. J Neurochem 47:88–97.

Williams M, Gordon EM (1996): Drug discovery: An overview. In Krogsgaard-Larsen P, Liljefors T, Madsen U "A Textbook of Drug Design and Development," 2nd ed. Chur/London: Harwood Academic, pp 1–34.

Williams M, Jacobson KA (1990): Radioligand binding assays for adenosine receptors. In Williams M (ed): "Adenosine and Adenosine Receptors." Clifton, NJ: Humana, pp 17–55.

Williams M, Jacobson KA (1995): P_2 purinoceptors: Advances and therapeutic opportunities. Exp Opin Invest Drugs 4:925–934.

Williams M, Reed R (1997): The Pharmaceutical Industry in the 21st Century. In Williams M, McAfee DA (eds): "Pharmaceutical Research and Development: Applied Biomedical Research in the Drug Discovery Process." New York: Wiley-Liss, in press.

Williams M, Deecher DC, Sullivan JP (1995): Drug receptors. In Wolf ME (ed): "Burger's Medicinal Chemistry and Drug Discovery, Fifth Ed., Vol. 1: Principles & Practices." New York: John Wiley & Sons, pp 349–397.

Zamecnik PC, Byung K, Gao MJ, Taylor G, Blackburn GM (1992): Analogues of diadenosine 5′,5′′′-P1, P4-tetraphosphate (Ap4A) as potential anti-platelet-aggregation agents. Proc Natl Acad Sci USA 89:2370–2373.

Zhang Y, Palmblad J, Fredholm BB (1996): Biphasic effect of ATP on neutrophil functions mediated by P_{2U} and adenosine A_{2A} receptors. Biochem Pharmacol 51:957–965.

Zheng LM, Zychlinsky A, Liu C-C, Ojcius DJ, Young JD-E (1991): Extracellular ATP as a trigger for apoptosis or programmed cell death. J Cell Biol 112:279–283.

Ziyal R, Ralevic V, Ziganshin AU, Nickel P, Ardanuy U, Mutschler E, Lambrecht G, Burnstock G (1996): NF023, a selective P2X-purinoceptor antagonist in rat, hamster and rabbit isolated blood vessels. Drug Dev Res 37:113.

PART II

MOLECULAR PHARMACOLOGY

CHAPTER 2

Adenosine Receptor Subtypes: New Insights From Cloning and Functional Studies

GARY L. STILES

Duke University Medical Center, Durham, NC 27710

INTRODUCTION

The physiological importance of adenosine has been recognized for almost 70 years. Drury and Szent-Gyorgyi (Palmer and Stiles, 1995; Olah and Stiles, 1995) demonstrated that adenosine played a major role in the regulation of cardiac rate and vascular tone. Since that time it has become abundantly clear that effects of adenosine are not limited to the cardiovascular system and that adenosine appears to modulate physiological functions in almost every organ and tissue in the body. The myriad effects have been reviewed recently (Palmer and Stiles, 1995; Olah and Stiles, 1995). For more than 20 years it has been recognized that the physiological effects of adenosine were mediated by specific membrane proteins, which have subsequently been identified as members of the family of G protein–coupled receptors. Burnstock originally classified the receptors as P1 and P2 purinergic receptors depending upon their preference for adenosine or adenine nucleotides (Burnstock, 1978). The relative potency series for the P1 site was adenosine > AMP ≥ ADP ≥ ATP, while the potency series for P2 sites were ATP ≥ ADP ≥ AMP ≥ adenosine. This review will discuss only the P1 purinergic receptors, and the reader is referred to other chapters in this book for a description of P2 receptors.

The adenosine-sensitive P1 sites have been further subdivided into adenosine receptor subtypes on the basis of the differential selectivity for a variety of adenosine analogs (Palmer and Stiles, 1995; Olah and Stiles, 1995). The potency series for the A_1 receptor is (−)-N^6-[(R)-phenylisopropyl]adenosine (R-PIA) > 5′-N-ethylcarboxamidoadenosine (NECA) > (+)-N^6-[(S)-phenylisopropyl]adenosine (S-PIA). The A_2 receptor potency series is NECA > R-PIA > S-PIA. For many years it was thought there were only

Purinergic Approaches in Experimental Therapeutics, Edited by Kenneth A. Jacobson and Michael F. Jarvis
ISBN 0-471-14071-6 © 1997 Wiley-Liss, Inc.

two subtypes of adenosine receptors (ARs), namely the A_1AR and the A_2AR. In most tissues A_1 adenosine receptors are coupled to inhibition of adenylyl cyclase, while A_2 adenosine receptors mediate activation of the enzyme. Adenosine receptors are known to modulate adenylyl cyclase activity via G proteins, specifically G_s and G_i for the A_2- and A_1-adenosine receptors, respectively. As will be described below the A_1 receptor is now known to be able to couple to a variety of effector systems other than adenylyl cyclase.

In 1983, Daly et al. proposed the concept that the A_2 receptors could be subdivided into A_{2a} and A_{2b} receptors based on differences observed for agonist potencies in a variety of tissues. The A_{2a} receptor was considered to have relatively high affinity for a variety of agonists, while the A_{2b} receptor had relatively lower affinity for these same compounds. As will be described below, subsequent cloning studies have indeed found two types of A_2 receptors, confirming Daly's original hypothesis. To complete the current set of adenosine receptor subtypes, Meyerhof and coworkers in 1991 published a sequence for a G protein–coupled receptor clone isolated from a rat testis cDNA library, which encoded a protein that displayed 40% to 50% overall amino acid identity to the canine A_1 and A_2 adenosine receptors (Meyerhof et al., 1991). The authors, however, were not able to identify it as any specific receptor, nor did they carry out any pharmacological or biochemical analysis of the clone. In 1992, Zhou and coworkers independently identified a similar clone and were able to characterize it using a variety of adenosine receptor ligands in CHO cells stably transfected with this receptor. It was determined that this receptor had high affinity for adenosine receptor ligands and its pharmacological properties were unique and distinct from any known adenosine receptor subtype. It was found that activation of this receptor resulted in a pertussis toxin–sensitive inhibition of forskolin-stimulated adenylyl cyclase activity. A very interesting property of this receptor was its relative insensitivity to xanthine derivatives known to be adenosine receptor antagonists for other adenosine receptor subtypes. Based on all its properties, the clone and its protein product were termed the A_3 adenosine receptor (A_3AR). Since 1992 no new, clearly defined adenosine receptor subtypes have been identified. A variety of species homologs have been described and these will be discussed below.

RECEPTOR–EFFECTOR COUPLING

As mentioned above, all the adenosine receptors are members of the superfamily of G protein–coupled receptors. Much work has been undertaken in recent years to understand which G proteins the adenosine receptors couple to and which effector systems are modulated. It has become abundantly clear that many of these receptor systems are promiscuous in that they can interact with a variety of G protein and effector systems. These interactions are summarized in Table 1. As shown, the A_1 receptor has been demonstrated to interact with pertussis toxin–sensitive G proteins, particularly $G_{i1,2,3}$ and G_o in reconstituted systems (Freissmuth et al., 1991). However, whether these interactions all occur *in vivo* remains largely unknown. Studies on the regulation of A_1ARs have documented that activation of the A_1ARs can indeed modulate the levels of $G_{i\alpha1}$ and $G_{i\alpha2}$. Whether these modulations are the direct result of the receptor interacting with these G proteins, or whether other molecules must act on the G proteins, remain to be determined. It is also unclear at this time whether the A_1 receptor can act through the G_q-like proteins in a pertussis toxin–insensitive

TABLE 1 Receptor–Effector Coupling

Receptor	G Protein	Effector
A_1AR	$G_{i\alpha 1,2,3}$	↓ Adenylyl cyclase
	$G_{q\alpha}$?	↑ Phospholipase C
	$G_{o\alpha}$?	↑ K^+ channels
$A_{2a}AR$	$G_{s\alpha}$	↑ Adenylyl cyclase
$A_{2b}AR$	$G_{s\alpha}$	↑ Adenylyl cyclase
		↑ Ca^{++} channels
A_3AR	$G_{i\alpha 2,3}$	↓ Adenylyl cyclase
		↑ Phospholipase C

manner. The A_1ARs have clearly been documented to interact with at least three effector systems, namely adenylyl cyclase, phospholipase C, and potassium channels (Palmer and Stiles, 1995; Olah and Stiles, 1995).

The A_{2a} receptor is known to interact with the G_s protein, which mediates activation of adenylyl cyclase. The A_{2b} receptor has been documented to interact with the G_s protein, in which case it can either increase adenylyl cyclase activity or open calcium channels. Finally, the A_3AR has been documented to interact with and activate both $G_{i\alpha 2}$ and $G_{i\alpha 3}$, but whether it is able to interact with other G proteins remains to be determined (Palmer et al., 1995). This receptor predominantly interacts in a inhibitory manner with adenylyl cyclase and in a stimulatory manner with phospholipase C. This interaction appears to be pertussis toxin–sensitive, suggesting that it is interacting through the G_i-like proteins rather than the G_q-type proteins.

THE CHARACTERIZATION OF ADENOSINE RECEPTOR SUBTYPES

A_1 Adenosine Receptors

For information on the pharmacological and biochemical characterization of A_1ARs the reader is referred to a number of recent reviews (Palmer and Stiles, 1995; Olah and Stiles, 1995). To date, the A_1AR has been cloned from a variety of species, including the rat, human, bovine, and rabbit (Palmer and Stiles, 1995; Olah and Stiles, 1995). The overall structures of these clones and the expressed protein are remarkably similar in that each encodes a protein of ~326 amino acids, which represents a protein with a molecular weight of 36,000–37,000 daltons. Pharmacologically these receptors are similar to each other, except for bovine A_1AR, which has a slightly different agonist potency series, such that R-PIA is more potent than S-PIA, which is more potent than NECA. In addition, agonists and antagonists have a relatively higher affinity for this receptor than the other A_1 receptors. A recent study by Tucker et al. (1994) suggests that specific amino acids within the transmembrane domain 7, namely, Ile-270 and Thr-277, are responsible for the differences in the agonist potency series.

Insight into the structure–function relationships within the A_1ARs is being provided by a number of mutation studies and the creation of chimeric receptors. Olah et al. (1992) have directly mutated His-251 in transmembrane domain 6 and His-274 in transmembrane domain 7 of the bovine A_1 receptor and replaced these with leucine residues. Subsequent expression and binding studies reveal that mutation of His-274 resulted in a complete loss of both agonist and antagonist binding. Work by Townsend-

Nicholson (personal communication) has demonstrated that a similar mutation in the human A_1AR also results in a loss of agonist and antagonist binding, while immunocytochemical analysis with receptor-specific antibodies demonstrated that the receptor was, indeed, expressed in membranes. This suggests that His-274 may be very important in the docking of agonists and antagonists into the receptor. The replacement of His-251 in transmembrane domain 6 results in a fourfold decrease in antagonist affinity, while agonist affinities were unchanged (Olah et al., 1992). Townsend-Nicholson and Schofield (1994) have demonstrated that mutation of Thr-277 to alanine in transmembrane domain 7, which lies very close to His-274, produced a 400-fold decrease in NECA affinity, while the affinity of an antagonist radioligand (DPCPX) and R-PIA and S-PIA was relatively unchanged.

Additional insights have been provided by the creation of chimeric receptors in which various regions of A_1ARs and A_3ARs are interchanged to create unique receptors that contain components of each of the receptor subtypes. An interesting chimeric receptor was created in which the sequence of 11 amino acids of the distal portion of the second extracellular loop of the A_1AR is placed into the A_3AR. This results in a chimeric receptor that displays ligand affinities much closer to the A_1AR receptor than the A_3AR. Specifically, affinities for agonists are increased 30-fold, and antagonist affinity is increased $>$ 500 fold compared to the wild-type A_3 receptor. These results strongly suggest that the second extracellular loop, particularly the distal one third, is important in binding of ligands into the receptor (Olah et al., 1994c). Whether this represents a site for direct interaction with ligands or whether these changes represent conformational changes in the overall architecture of the receptor remains to be determined.

Using the chimeric receptor approach, Olah et al. (1994b) have also been able to demonstrate that a 6-amino-acid segment of transmembrane domain 5 at the extracellular membrane interface is important in the binding of 5′ substituted agonists. For example, the replacement of the transmembrane domain 5 of the A_1AR receptor with the analogous region of the A_3 receptor results in binding parameters identical to the wild-type A_1 receptor with regard to antagonist binding and the interaction of those agonists that have a substitution only at the N^6 position. However, the binding of 5′ substituted agonists is significantly enhanced at the chimeric receptor compared to the wild-type A_1 receptor.

Thus multiple regions within the receptor appear to be important for both agonist and antagonist binding, and these regions include transmembrane domains 5, 6, and 7 as well as the second extracellular loop. Further studies are clearly going to define various other regions within the transmembrane domain important for agonist and antagonist binding.

A_{2a} Adenosine Receptors

The A_{2a} adenosine receptor ($A_{2a}AR$) has been cloned from a variety of species including canine, rat, and human (Olah and Stiles, 1995). $A_{2a}ARs$ are typically larger than the A_1ARs and range in size from 410–412 amino acids corresponding to a protein of ~45,000 daltons. The most striking difference between A_1 and A_{2a} is in the size of the carboxy terminal tail; the A_{2a} tail is much larger than the carboxy terminal tail found in A_1ARs. The structural and functional importance of this tail remains largely unknown. The presence in the tail of multiple serines and threonines suggests that this region may be important in the regulation of the receptor by phosphorylation following agonist stimulation. This, however, seems unlikely, since mutated canine A_{2a} in which stop codons were inserted at amino acids 309 or 316 lead to markedly truncated carboxy terminal tails as defined by photoaffinity labelling. These mutations failed to alter either the binding of ligands to this receptor or the receptors ability to stimulate adenylyl cyclase. Further studies will be needed to understand the importance of this unique structural feature of A_{2a} receptors.

Site-directed mutagenesis studies have now been undertaken with the human A_{2a}AR. Kim et al. (1995) have studied a wide range of individual point mutations and their effect on agonist and antagonist binding. In order to document whether there were differences between actual loss in binding and expression of the receptor on the surface, they have incorporated a specific hemagglutinin epitope tag into their receptor so that they can identify the expressed protein immunologically in the cell membranes by the use of monoclonal antibodies. Following expression of the tagged receptors they studied the functional properties of the mutant receptors by measuring stimulation of adenylyl cyclase and also determined specific binding using ^{3}H CGS21680 (agonist) and ^{3}H XAC (antagonist) radioligands. They explored a number of regions of the receptor and found that mutation of the histidines to alanine in transmembrane domains 6 and 7 led to a loss of both agonist and antagonist binding. In addition mutation to alanine of Phe-182 in transmembrane domain 5, Asp-253 in transmembrane domain 6, Ile-274 and and Ser-281 in transmembrane domain 7, in each case led to loss of both agonist and antagonist binding. Each of the above noted mutants, however, were well expressed in the membrane as determined by the epitope tag. Many of these same regions were found to be important for binding of agonists in the A_1AR, thus highlighting the similarity of A_1 and A_2 receptors in terms of agonist and antagonist interactions. In addition, it was found that the hydroxyl group of Ser-277 was required for high-affinity binding of agonists but was not important for antagonist binding. It is also of interest that although mutants I274A, S277A, and H278A failed to show any binding as determined by ^{3}H-CGS21680 or ^{3}H-XAC, when high concentrations of cold CGS21680 were used, full stimulation of adenylyl cyclase activity occurred. This is not surprising, since shifts in agonist affinity of $>$ 50-fold would result in loss of radioligand binding, but the receptor would still be able to be activated with high concentrations of an agonist. This was the first detailed study of the A_{2a} receptor, and the authors have developed a model in which agonists and antagonists might dock into the receptor. The reader is referred to the original publication for detailed description of the model (Kim et al., 1995).

A_{2b} Adenosine Receptors

The A_{2b}AR has been cloned from both rat and human libraries and expressed in a variety of cells (Olah and Stiles, 1995). The A_{2b} receptor is much smaller than the A_{2a} and is much more similar in size to the A_1ARs and A_3ARs. Studies on the A_{2b} receptor have been few in number, by and large due to the fact that there are no useful radioligands, either agonist or antagonist, available to study this interesting receptor. The only radioligand that has had some use is [^{3}H]NECA and its relatively low affinity precludes detailed studies.

A_3 Adenosine Receptors

This fascinating receptor, whose identity was made known only through the use of cloning techniques, remains somewhat of an enigma in terms of where it fits into the overall adenosine receptor picture. This receptor has now been cloned from a number of species including human, rat, and sheep (Olah and Stiles, 1995). These clones code for a protein of ~320 amino acids and display between 40% and 50% overall amino acid identity to the A_1- and A_2ARs. The first characterized A_3 receptor was the rat A_3AR (Zhou et al., 1992). It had striking characteristics, as described in the introduction, in that adenosine receptor agonists had relatively low affinity compared to other ARs and, as might be anticipated, a very different potency series than that observed for A_1 or A_2ARs. In addition, the rat A_3 receptor had an extremely low affinity for any of the alkylxanthine antagonists. The sheep and human A_3ARs were subsequently cloned and displayed ~72% overall amino acid identity to the rat A_3AR. The pharmacological properties of the sheep and human were very similar and quite distinct from the rat A_3AR, in that the alkylxanthines had much higher affinity for the sheep and human A_3AR than they did for the corresponding rat receptor, and the affinity of agonists was also significantly increased. Using knowledge of the agonist affinities for these receptors, a new radioligand ^{125}I-AB-MECA was developed as a high-affinity radioligand for the A_3AR. This ligand's only drawback is that it is not highly subtype-selective. This radioligand will also bind to A_1ARs (Olah et al., 1994a). The marked species differences between the A_3AR's pharmacology as well as the low sequence homology between species homologs begs the question of whether the rat A_3 and human and sheep A_3ARs are not, in fact, species homologs but distinct subtypes of the A_3AR. No definitive studies are available thus far to differentiate these two possibilities. Another aspect that leads credence to the idea that these may be more than just species homologs is a significant difference in the tissue distribution of the mRNAs for the A_3ARs in these species. The most abundant mRNA for the rat A_3AR is found in the testis, with lesser amounts in the lung, kidney, and heart; and much lower levels in various brain regions (Zhou et al., 1992). In contrast, as reported by Linden et al. (1993), the distribution of mRNA differs in the sheep, greatest abundance in the lung and spleen, and with much lower levels in various brain regions and the kidney and testis. Much additional information regarding the expression and pharmacological characteristics of these receptors as well as their genomic structure should make it possible to discern whether the subtypes are species homologs or different subtypes. The use of chimeric receptors to investigate the structure–function relationship of A_3 versus A_1 has previously been described under the A_1AR section.

GENOMIC STRUCTURE

Thus far the only detailed information on the gene structure for any of the adenosine receptor subtypes has been for the human A_1AR. Our laboratory has been able to document that the human gene of the A_1AR contains multiple introns and exons. There are at least two distinct promoters, which act as the transcriptional unit to create two distinct mRNAs having alternative exons in the 5′ region (Ren and Stiles, 1994a, 1995) (see Figure 1). These two messages have different translatability and appear to act to control the level of receptor abundance in a given tissue (Ren and Stiles, 1994b). This represents a new mechanism to explain why various tissues express such different levels of receptors. It was found that the message that contains exons 1A, 2, and 3 is found in tissues that have a high level of receptor expression, while the alternative message containing exons 1B, 2, and 3 is found in all tissues that contain any level of adenosine receptor thus far studied (Ren and Stiles, 1995). *In vitro* analysis of these

HUMAN A1AR GENE STRUCTURE

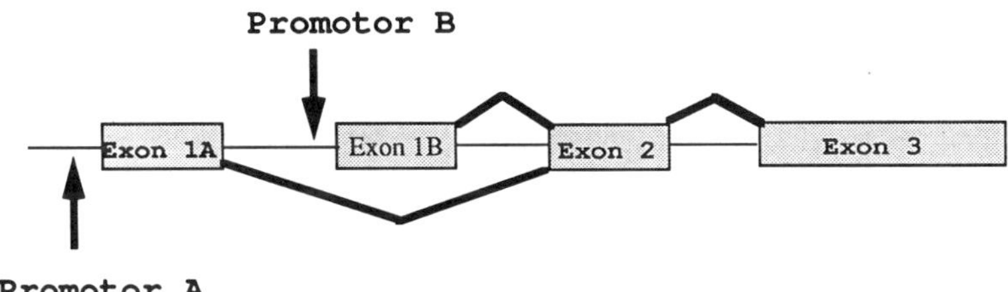

Promotor A -----> Transcript A (Ex 1A.2.3)

Promotor B -----> Transcript B(Ex 1B.2.3)

FIGURE 1. Structure of the human A_1AR gene, located on chromosome 1.

messages, and mutants derived therefrom, has documented that there are two 5′ upstream ATG start codons in exon 1B but none within exon 1A (Ren and Stiles, 1994b). Both of these Kozak consensus sequences are excellent consensus sites for translation initiation. It has been found that, when constructs containing exons 1A, 2, and 3 (or 1B, 2, and 3) were expressed in COS 7 cells, the construct Ex 1A.2.3 expressed receptors at a level 3–8 times higher than that found with the construct containing Ex 1B.2.3. The appropriate controls were carried out to insure that there was equal expression of mRNA within the cells. This finding, in association with the knowledge that exon 1B contained these two ATG start sites and that these sequences have been shown in protooncogenes to suppress translation, led us to study mutations of these ATG sites. We have mutated singly and in combination the two ATGs to GGGs, leaving the remainder of the construct unaltered, and then expressed them in COS 7 cells. It was found that individually mutating the ATGs led to a significant increase in the receptor abundance but this did not reach the level of expression found with the construct containing exon 1A.2.3; however, when both ATGs were mutated simultaneously the levels of expressions seen with Ex 1A.2.3 and Ex 1B.2.3 were exactly the same. Thus, this suggests that the two ATG (or AUG) codons are inhibitory to translation and may indeed be a mechanism whereby the gene regulates receptor abundance.

In addition, as noted above, it was found that there are two promoters, one of which, termed promoter A, acts to produce the transcript containing Ex 1A.2.3 message. Promoter B drives the message containing Ex 1B.2.3, and the appropriate splicing takes place to eliminate the introns and appropriate exons. It should be noted that we have never found a message that contains both exons 1A and 1B in the same transcript. Much additional work will need to be carried out to understand how these promoters are controlled and what specific transcriptional binding factors may be binding to give tissue-specific expression.

In summary, there are now four known adenosine receptor subtypes, all of which display unique properties in terms of structure, function, and regulation. The last few years have seen a dramatic increase in our knowledge of the binding pocket of each of these adenosine receptors except (the A_{2b} receptor), thanks to the use of mutant and chimeric receptors. Much new information needs to be gathered to fully understand how agonists and antagonists dock into the receptor, and how the receptors may be manipulated genetically to alter their function and binding properties.

It has now been several years since the last new adenosine receptor subtype has been described and it will be interesting to see whether the A_3 adenosine receptors are indeed species homologs or receptor subtypes and whether any new adenosine receptor subtypes will in fact be discovered. The development of new and highly subtype-selective agonists and antagonists should bring about the potential for new therapeutic interventions. Adenosine receptor research has come a long way in the past 10 years and I believe significant new data will be forthcoming in the near future.

REFERENCES

Burnstock G (1978): A basis for distinguishing two types of purinergic receptors. In Bolis L, Straub RW (eds): "Cell Membrane Receptors for Drugs and Hormones: A Multidisciplinary Approach". New York: Raven Press, pp 107–118.

Daly JW, Butts-Lamb P, Padgett W (1983): Subclones of adenosine receptors in the central nervous system: Interaction with caffeine and related methylxanthines. Cell Mol Neurobiol 1:69–80.

Freissmuth M, Schutz W, Linder ME (1991): Interactions of the bovine brain A_1-adenosine receptor with recombinant G-protein alpha-subunits. J Biol Chem 266:17778–17783.

Kim J, Wess J, van Rhee AM, Schöneberg T, Jacobson KA (1995): Site-directed mutagenesis identifies residues involved in ligand recognition in the human A_{2a} adenosine receptor. J Biol Chem 270:13987–13997.

Linden J, Taylor HE, Robeva AS, Tucker AL, Stehle JH, Rivkees SA, Fink JS, Reppert SM (1993): Molecular cloning and functional expression of a sheep A_3 adenosine receptor with widespread tissue distribution. Mol Pharmacol 44:524–532.

Meyerhof W, Muller-Brechlin R, Richter D (1991): Molecular cloning of a novel putative G-protein coupled receptor expressed during rat spermiogenesis. FEBS Lett 284:155–160.

Olah ME, Stiles GL (1995): Adenosine receptor subtypes: Characterization and therapeutic regulation. Annu Rev Pharmacol Toxicol 35:581–606.

Olah ME, Gallo-Ridriguez C, Jacobson KA, Stiles GL (1994a): ^{125}I-4-Aminobenzyl-5′-N-methylcarboxamidoadenosine, a high affinity radioligand for the rat A_3 adenosine receptor. Mol Pharmacol 45:978–982.

Olah ME, Jacobson KA, Stiles GL (1994b): Identification of an adenosine receptor domain specifically involved in binding of 5′-substituted adenosine agonists. J Biol Chem 269:18016–18020.

Olah ME, Jacobson KA, Stiles GL (1994c): Role of the second extracellular loop of adenosine receptors in agonist and antagonist binding. J Biol Chem 269:24692–24698.

Olah ME, Ren H, Ostrowski J, Jacobson KA, Stiles GL (1992): Cloning, expression, and characterization of the unique bovine A_1 adenosine receptor studies on the ligand binding site by site-directed mutagenesis. J Biol Chem 267:10764–10770.

Palmer TM, Stiles GL (1995): Review: Neurotransmitter receptors, VII. Adenosine receptors. Neuropharmacology 34:683–694.

Palmer TM, Gettys TW, Stiles GL (1995): Differential interaction with and regulation of multiple G-proteins by the rat A_3 adenosine receptor. J Biol Chem 270:16895–16902.

Ren H, Stiles GL (1995): Separate promoters in the human A_1 adenosine receptor gene direct the synthesis of distinct mRNAs which regulate receptor abundance. Mol Pharmacol 48:975–980.

Ren H, Stiles GL (1994a): Characterization of the human A_1 adenosine receptor gene: Evidence for alternative splicing. J Biol Chem 269:3104–3110.

Ren H, Stiles GL (1994b): Posttranscriptional mRNA processing as a mechanism for regulation of human A_1 adenosine receptor expression. Proc Natl Acad Sci USA 91:4864–4866.

Townsend-Nicholson A, Schofield PR (1994): A threonine residue in the seventh transmembrane domain of the human A_1 adenosine receptor mediates specific agonist binding. J Biol Chem 269:2373–2376.

Tucker AL, Robeva AS, Taylor HE, Holeton D, Bockner M, Lynch KR, Linden J (1994): A_1 adenosine receptors: Two amino acids are responsible for species differences in ligand recognition. J Biol Chem 269:27900–27906.

Zhou Q-Y, Li C, Olah ME, Johnson RA, Stiles GL, Civelli O (1992): Molecular cloning and characterization of an adenosine receptor: The A_3 adenosine receptor. Proc Natl Acad Sci USA 89:7432–7436.

CHAPTER 3

Structure, Pharmacological Selectivity, and Second Messenger Signaling Properties of G Protein–Coupled P2-Purinergic Receptors

THERESA M. FILTZ, T. KENDALL HARDEN, and ROBERT A. NICHOLAS

Department of Pharmacology, School of Medicine, University of North Carolina at Chapel Hill, Chapel Hill, NC 27599-7365

INTRODUCTION

Receptors for adenosine and adenine nucleotides exist as two distinct classes of signaling proteins, the P1- and P2-purinergic receptors. P1-purinergic receptors display a high affinity for adenosine and methylxanthine analogues, and are well-characterized both at the molecular and pharmacological level. P2-purinergic receptors originally were subclassified into P2X and P2Y receptors based on agonist potencies of several structural analogues of ATP in various contractile tissues. P2X receptors were proposed to exhibit a pharmacological selectivity of α,βMeATP (α,β-methyleneadenosine 5′-triphosphate) > ATP > 2MeSATP (2-methylthioadenosine 5′-triphosphate), whereas P2Y receptors were proposed to exhibit a pharmacological selectivity of 2MeSATP > ATP ≫ α,βMeATP (Burnstock and Kennedy, 1985). P2X and P2Y receptors subsequently were more clearly defined on the basis of their functional properties. P2X purinergic receptors are cation-selective ion channels (Dubyak and El-Moatassim, 1993; Evans et al., 1992; Edwards et al., 1992), whereas P2Y receptors couple to G proteins and either activate phospholipase C (PLC) (Charest et al., 1985; Boyer et al., 1989; Berrie et al., 1989; Forsberg et al., 1987; Pirotton et al., 1987) or inhibit adenylyl cyclase (AC) (Okajima et al., 1987; Pianet et al., 1989; Valeins et al., 1992; Yamada et al., 1992; Boyer et al., 1993).

The field of G protein–coupled P2-purinergic receptors has expanded to include subtypes based on differences in pharmacological selectivity in various tissues. The so-called P2U receptors can be distinguished from P2Y receptors by virtue of the low

Purinergic Approaches in Experimental Therapeutics, Edited by Kenneth A. Jacobson and Michael F. Jarvis
ISBN 0-471-14071-6 © 1997 Wiley-Liss, Inc.

potency and efficacy of 2MeSATP. Also, whereas UTP is not an agonist at the originally studied P2Y receptors, it is equipotent with ATP for activation of P2U receptors (reviewed by O'Connor et al., 1991). A uridine nucleotide–specific receptor that is not activated by adenine nucleotides has been identified on a subclone of C6 glioma cells (Lazarowski and Harden, 1994). Additionally, platelets possess a G protein–coupled P2-purinergic receptor subtype, the P2T receptor, at which ADP is an agonist and ATP is a competitive antagonist (reviewed by Hourani and Hall, 1994). Other P2-purinergic receptor subtypes have been suggested on the basis of differences in potencies of nucleotide analogues (e.g., diadenosine tetraphosphate) among different tissues.

While it is likely that many more subtypes of P2-purinergic receptors exist, the occurrence of multiple receptors or cell subtypes within tissues complicates the study of receptor subtype–specific responses. Delineation of multiple P2-purinergic receptors has been particularly difficult because (i) extensive interconversion and breakdown of nucleotides occurs due to the presence of extracellular nucleotide-metabolizing enzymes, (ii) few receptor subtype–selective agonists have been available, and (iii) no P2-purinergic receptor–selective antagonists have been identified. Molecular cloning of P2-purinergic receptor sequences largely circumvents these problems. Cloned P2-purinergic receptors can be expressed in a cell line lacking endogenous P2-purinergic receptors and studied under conditions in which the breakdown and release of nucleotides can be both quantitated and minimized. This allows the pharmacological selectivity and second messenger signaling specificity to be accurately and unambiguously established for receptors of defined structure.

At least four P2Y (denoted $P2Y_n$ where n = 1,2,3, etc.) and five P2X (denoted $P2X_n$) receptor subtypes have been cloned and their sequences reported. The P2Y subtypes are G protein–coupled receptors and are discussed in detail below. The amino acid sequences of the P2X receptors connote ion channels that show homology in their putative pore-forming domains to potassium channels (Valera et al., 1994; Brake et al., 1994; Lewis et al., 1995; Chen et al., 1995; Bo et al., 1995). It is notable that expression of cloned P2X receptors in *Xenopus laevis* oocytes revealed a pharmacological selectivity that was inconsistent with that historically associated with P2X receptors. That is, oocyte–expressed P2X receptors exhibited a selectivity of 2MeSATP > ATP > α,βMeATP, which is much more akin to that classically associated with a P2Y receptor (Valera et al., 1994). Although the reasons for this discrepancy are unclear, hydrolysis or interconversion of nucleotides is a contributing factor. As discussed below, accurate determination of agonist selectivities requires attention to the initial purity of nucleotides and their potential breakdown or interconversion during incubation with cells or tissues. These problems can be expected to be magnified in studies using tissue preparations, such as those utilized originally to delineate the pharmacological selectivity of P2X and P2Y receptors.

Nucleotide Purity, Degradation, Bioconversion, and Release Complicate the Characterization of Nucleotide-Responsive Receptors

Pharmacological studies of P2Y receptors often have resulted in highly misleading conclusions. Commercial preparations of nucleotides are predictably impure; diphosphates contaminate commercial sources of nucleotide triphosphates and vice versa. Release of adenine nucleotides from cultured cells contributes a second (and usually unaccounted for) source of nucleotides in studies of P2Y receptors (Lazarowski et al., 1995). For example, 1321N1 human astrocytoma cells, which have been widely utilized for expression of cloned P2Y receptors, release ATP during physical movement of the culture dish and release ATP in large amounts during changes in medium. Most cells also express very active ectonucleotidases in the extracellular environment, and the recent discovery of significant nucleoside diphosphokinase activity on the extracellular face of many cell types adds a heretofore unappreciated concern (E. R. Lazarowski, and W. C. Watt, unpublished observation). For example, a significant amount of UDP is converted to UTP by 1321N1 cells in the presence of exogenously added ATP or following a change of medium (which releases ATP). Taken together, the combined contributions of impure preparations of nucleotides, release of nucleotides from cells, and the bioconversions of nucleotides by extracellular enzymes complicate the determination of the selectivities of P2Y receptors for di- versus triphosphate, and adenine versus uridine nucleotides.

To minimize these problems, the absolute purity of nucleotides should be established prior to assay, nucleotide release should be limited during the assay, and the stability of nucleotides during incubation should be directly quantitated. To minimize the contribution of nucleoside diphosphates to the agonist activity of nucleoside triphosphates, all commercially available preparations of nucleoside triphosphates need to be purified by high-performance liquid chromatography (HPLC) prior to use. To minimize the contribution of nucleoside triphosphates to the agonist activity of nucleoside diphosphates, the contaminating nucleoside triphosphate must be converted to its corresponding diphosphate. To catalyze this conversion, stock solutions of the nucleoside diphosphates are incubated with hexokinase and glucose prior to assay, and hexokinase also is included during incubation of nucleoside diphosphates with cells.

PHARMACOLOGICAL SELECTIVITY, SECOND MESSENGER SIGNALING, AND STRUCTURAL ASPECTS

Cloned P2Y Purinergic Receptors

$P2Y_1$ Receptors The first cDNA (complementary DNA) of a prototypic P2Y receptor was isolated from a chick brain library screened at low stringency with a probe derived from the guinea pig vasoactive intestinal peptide receptor. Characterization of positive clones was carried out in cRNA-injected *Xenopus* oocytes, and one clone, named the chick $P2Y_1$ receptor, elicited a slowly developing inward current in response to application of ATP or 2MeSATP (Webb et al., 1993). This receptor displayed highest homology to receptors for interleukin-8, thrombin, vasoactive intestinal peptide, platelet activating factor, angiotensin II, and neuropeptide Y; homology of the $P2Y_1$ receptor to P1 receptors for adenosine was relatively low. Since the relative nucleotide selectivity and second messenger signaling pathway of the chick $P2Y_1$ receptor were not clearly established, our laboratory cloned the turkey homolog of this receptor and stably expressed it in 1321N1 human astrocytoma cells (Filtz et al., 1994). ATP and 2MeSATP increased inositol phosphate accumulation, but no effect on adenylyl cyclase activity was observed. ADP and 2MeSADP also were highly potent agonists at the turkey $P2Y_1$ receptor, whereas UTP had little effect. The pharmacological selectivity of the cloned turkey $P2Y_1$ receptor was essentially identical to that of the extensively characterized PLC-linked P2Y receptor endogenously expressed in turkey erythrocytes (Boyer et al., 1989), and ribonuclease protection analysis of various turkey tissue RNAs indicated that $P2Y_1$ receptor mRNA was present at the highest levels in blood, lung, and brain. These results suggested that the cloned turkey $P2Y_1$ receptor is the receptor natively expressed in turkey erythrocytes.

The human homolog of this receptor has been cloned and is 93% homologous to the turkey $P2Y_1$ receptor (Schachter et al., 1996). The human (as well as rat, murine, and bovine) receptor contains an additional 11 amino acids in the putative extracellular N-terminus not found in the avian sequence, and several amino acid differences exist between the mammalian and avian sequences within the putative transmembrane domains of the receptor (Schachter et al., 1996; Tokuyama et al., 1995; Henderson et al., 1995). However, the relative pharmacological selectivity and second messenger coupling specificity of both the avian and human receptors were identical when stably expressed in 1321N1 cells. Complete pharmacological characterization of the mouse and rat $P2Y_1$ receptors remains to be accomplished, but based on their high homology with the human homolog remarkable differences in drug selectivity would be surprising.

Using intracellular Ca^{2+} levels as a measure of receptor activation, the bovine $P2Y_1$ receptor exhibited differences in pharmacological selectivity compared to the turkey $P2Y_1$ receptor (Henderson et al., 1995). Notably, ADP was relatively more potent and ATP relatively less potent at the bovine $P2Y_1$ receptor. Since the turkey and human $P2Y_1$ receptor homolog display identical agonist selectivities when assayed in the same cell line, it is not likely that the differences in agonist potencies observed with bovine $P2Y_1$ receptors are due to differences in amino acid sequence. The observed differences are more likely a consequence of the different cell types (Jurkat vs 1321N1 cells) and second messenger responses (Ca^{2+} vs inositol phosphates) assayed in these two studies.

$P2Y_2$ Receptors An expression cloning approach was used to isolate cDNA encoding the $P2Y_2$ (formerly P2U) receptor. Injection of poly(A^+) mRNA from mouse neuroblastoma cells into *Xenopus* oocytes conferred an ATP- and UTP-stimulated Ca^{2+}-activated Cl^- current. Progressive division of pools of mouse neuroblastoma cDNA, which were transcribed, injected into oocytes, and analyzed for a nucleotide-induced current, eventually resulted in the cloning of the coding sequence for the $P2Y_2$ receptor (Lustig et al., 1993). Application of ATP, UTP, and ATPγS to oocytes expressing the receptor produced an inward rectifying current under voltage clamp; neither the classical P2Y receptor agonist 2MeSATP nor ADP were agonists. This pharmacological selectivity was typical of that described for $P2Y_2$ receptors and was replicated in assays of intracellular Ca^{2+} levels following receptor expression in K-562 human leukemia cells, a cell line lacking an endogenous UTP response (Erb et al., 1993). The human homolog of this receptor, which is 89% identical to the mouse receptor, was cloned from CF/T43 human airway epithelial cells (Parr et al., 1994). Messenger RNA encoding $P2Y_2$ receptors is expressed in many tissue types, including skeletal muscle, placenta, liver, heart, and lung (Parr et al., 1994). In 1321N1 cells expressing the human $P2Y_2$ receptor, both inositol phosphate accumulation and intracellular Ca^{2+} levels were increased following incubation with ATP or UTP. Although both ADP and UDP were initially reported to be lower-potency full agonists at the $P2Y_2$ receptor (Parr et al., 1994), these nucleotide diphosphates have no effect if assays are carried out in the presence of hexokinase and glucose to eliminate any contributions of the triphosphate forms of the nucleotides. Thus, the $P2Y_2$ receptor is activated by UTP and ATP, but not by UDP or ADP (Table 1).

$P2Y_6$ Receptors Chang et al. (1995) recently reported the cloning of an additional uridine nucleotide–responsive receptor. This receptor has been designated the $P2Y_6$

TABLE 1 Properties of P2Y receptors

Receptor Subtype	Nucleotide Selectivity	Signaling Response	Clonal Status
$P2Y_1$	2MeSATP > ADP > ATP	↑ PLC	Cloned
$P2Y_2$	ATP = UTP	↑ PLC	Cloned
$P2Y_6$	UDP	↑ PLC	Cloned
$P2Y_4$	UTP > ATP	↑ PLC	Cloned
P2Y (C6)	2MeSATP > ADP > ADP	↓ AC	Not Cloned

receptor even though it was the third functional P2Y receptor whose sequence was published.* The $P2Y_6$ receptor was cloned from a rat aortic smooth muscle cDNA library by screening with an oligonucleotide based on the homology observed in the third transmembrane domains of the $P2Y_1$ and $P2Y_2$ receptors (Figure 1). The rat $P2Y_6$ receptor is 38% homologous to the human $P2Y_1$ receptor and 44% homologous to the human $P2Y_2$ receptor. Northern analyses of various tissues identified significant levels of $P2Y_6$ receptor mRNA in aorta, mesentery, lung, spleen, stomach, and intestine. Stable expression of the $P2Y_6$ receptor cDNA in C6 cells by Chang et al. (1995) conferred a Ca^{2+} response to nucleotides with an agonist potency of UTP > 2MeSATP = ADP. ATP, ADPβS, and ATPγS were partial agonists, and α,βMeATP was ineffective. The $P2Y_6$ receptor was shown to couple to stimulation of phospholipase C but exhibited no effect on adenylyl cyclase activity.

We have cloned the rat $P2Y_6$ receptor by PCR amplification of genomic DNA and have stably expressed this signaling protein in 1321N1 cells (Nicholas et al., 1996). As was previously reported by Chang et al. (1995), UTP apparently was a full agonist at the $P2Y_6$ receptor. However, UDP was over 100-fold more potent than UTP. Treatment of UDP with hexokinase and glucose to eliminate any contribution of UTP had no effect on its potency. Conversely, the effect of UTP at the $P2Y_6$ receptor was apparently due to contamination by small amounts of UDP, since UTP was essentially inactive following HPLC purification. Likewise, 5BrUDP was a potent full agonist at the $P2Y_6$ receptor, but 5BrUTP was inactive if it was purified by HPLC prior to assay. In contrast to the results reported by Chang et al. (1995), ATP and 2MeSATP were essentially inactive at the $P2Y_6$ receptor. These data indicate that the $P2Y_6$ receptor is a UDP-selective receptor that is not activated by adenine nucleotides.

Communi et al. (1996) have reported the cloning of a human homolog of the $P2Y_6$ receptor that displays 88% amino acid identity to the rat $P2Y_6$ receptor. When the human $P2Y_6$ receptor was expressed in 1321N1 cells, UDP was the most potent stimulator of inositol lipid hydrolysis, whereas ATP was inactive.

Lazarowski and Harden (1994) reported the existence of a phospholipase C-linked receptor on rat C6-2B cells that is activated by uridine nucleotides. The nucleotide selectivity of this receptor was UDP > 5BrUTP > UTP; ATP and other adenine nucleotides had no effect. The pharmacological selectivity of the expressed $P2Y_6$ receptor was essentially identical to that of the receptor on C6-2B cells when the nucleotides were used without prior purification and in the absence of hexokinase and glucose, as was done in the original study of the receptor on C6-2B cells (Lazarowski

* The general recommendations of the IUPHAR Committee on Receptor Nomenclature were to number sequentially functional P2Y receptor subtypes based on the chronology of publication. However, this receptor has been referred to as $P2Y_6$ by others (see website http://mgddk1.niddk.nih.gov:8000/nomenclature.html) and this nomenclature is used here.

```
                                                                       TM 1
P2Y2                   MAADLGPWNDTINGT--WDGDELGYRCRF-NEDFKYVLLPVSYGVVCVLGLCLNAVGLYIFLCRLKTWNASTTYM
P2Y4                   MASTESSLLRSLGLSPGPGSSEVELDCWF-DEDFKFILLPVTYAVVFVLGLGLNAPTLWLFIFRLRPWDATATYM
P2Y6                        MERDNGTIQAPGLPPTT--CVY-REDFKRLLLPPVYSVVLVVGLPLNVCVIAQICASRRTLTRSAVYT
P2Y1  MTEVLWPAVPNGTDAAFLAGPGSS--WGNSTVASTAAVSSSFKCALTKTGFQFYYLPAVYILVFIIGFLGNSVAIWMFVFHMKPWSGISVYM

        TM 2                              TM 3                                         TM 4
P2Y2  FHLAVSDALYAASLPLLVYYYARGDHWPFSTVLCKLVRFLFYTNLYCSILFLTCISVHRCLGVLRPLRSL-RWGRARYARRVAGAVWVLVLA
P2Y4  FHLALSDTLYVLSLPTLIYYYAAHNHWPFGTEICKFVRFLFYWNLYCSVLFLTCISVHRYLGICHPLRAL-RWGRPRLAGLLCLAVWLVVAG
P2Y6  LNLALADLLYACSLPLLIYNYARGDHWPFGDLACRLVRFLFYANLHGSILFLTCISFQRYLGICHPLAPWHKRGGRRAAWVVCGVVWLVVTA
P2Y1  FNLALADFLYVLTLPALIFYYFNKTDWIFGDAMCKLQRFIFHVNLYGSILFLTCISAHRYSGVVYPLKSLGRL-KKKNAICISVLVWLIVVV

                                          TM 5
P2Y2  CQAPVLYFVTTSARGP--L-TCHDTSAPELFSRFVAYSSVMLGLLFAVPFAVILVCYVLMARRLLKPAYGTSGGLPRAKRKSVRTIAVVLAV
P2Y4  CLVPNLFFVTTSNKGTT-V-LCHDTTRPEEFDHYVHFSSAVMGLLFGVPCLVTLVCYGLMARRLYQPLPGSAQSSSRLRSLR--TIAVVLTV
P2Y6  QCLPTAVFAATGIQRNR-T-VCYDLSPPILSTRYLPYGMALTVIGFLLPFTALLACYCRMARRLCRQDGPAGPVAQERRSKAARMAVVVAAV
P2Y1  AISPILFYSGTGVRKNKTI-TCYDTTSDEYLRSYFIYSMCTTVAMFCVPLVLILGCYGLIVRALIYKDLDNSP----LRRKSIYLVIIVLTV

         TM 6                                TM 7
P2Y2  FALCFLPFHVTRTLYYSFRSL-DLS--CHTLNAINMAYKVTRPLASANSCLDPVLYFLAGQRL-VRFARDAKPPT-GPSPATPARRTLGLRR
P2Y4  FAVCFVPFHITRTIYYLARLL-EAD--CRVLNIVNVVYKVTRPLASANSCLDPVLYLLTGDKY-RRQLRQLCGGG-KPQPRTAASSLALVSL
P2Y6  FVISFLPFHITKTAYLAVRSTPGVS--CPVLETFAAAYKGTRPFASANSVLDPILFYFTQQKF-RRQPHDL------LQKLTAKWQRQRV@
P2Y1  FAVSYIPFHVMKTMNLRARLDFQTPAMCAFNDRVYATYQVTRGLASLNSCVDPILYFLAGDTFRRRLSRATRKAS-RRSEANLQSKSEDMTL


P2Y2  SDRTDMQRIGDVLGSSEDSRRTESTPAGSENTKDIRL@
P2Y4  PEDSSCRWAATPQDSSCSTPRADRL@
P2Y6
P2Y1  NILPEFKQNGDTSL@
```

FIGURE 1. Sequence alignment of four subtypes of P2Y receptors. Amino acid identities between at least two of the four subtypes are highlighted in grey. Dark grey highlights indicate amino acid identity pairs where the four subtypes divide into two groups.

and Harden, 1994; Nicholas et al., 1996). Moreover, multiple independent $P2Y_6$ receptor clones have been identified in a C6-2B cell cDNA library probed with the rat $P2Y_6$ receptor sequence. These data indicate that the $P2Y_6$ receptor is natively expressed in rat C6-2B cells and apparently accounts for the uridine nucleotide-selective receptor originally defined in these cells.

$P2Y_4$ Receptors The $P2Y_4$ receptor was cloned independently by two groups. Communi et al. (1995) isolated a partial sequence of the $P2Y_4$ receptor by PCR amplification of genomic DNA using degenerate primers based on conserved sequences in the $P2Y_1$ and $P2Y_2$ receptors. This partial sequence was used to clone the full-length sequence. Nguyen et al. (1995) also isolated a partial $P2Y_4$ receptor clone by PCR amplification of genomic DNA with primers based on conserved sequences in opioid and somatostatin receptors, and this clone was used to isolate a full-length genomic clone encoding the $P2Y_4$ receptor. The nucleotide sequences reported from the two laboratories differ at five bases, which translates into three amino acid changes. The deduced amino acid sequence is 55% homologous to the human $P2Y_2$ receptor, 38% homologous to the human $P2Y_1$ receptor, and 40% homologous to the rat $P2Y_6$ receptor (Table 2). The receptor couples to stimulation of phospholipase C when exogenously expressed in 1321N1 astrocytoma cells. Communi et al. (1995) reported that UTP and UDP were equipotent full agonists for eliciting inositol phosphate accumulation, whereas ATP and ADP were much less potent partial agonists. In contrast, Nguyen et al. (1995), quantitating Ca^{2+} levels as a measure of receptor activity, reported that UTP was the most potent agonist, UDP was a less potent full agonist, and ATP was not an agonist. Although data were not shown, Nguyen et al. (1995) reported that ATP was an antagonist at the $P2Y_4$ receptor.

To more clearly define the nucleotide selectivity of the $P2Y_4$ receptor and to compare its selectivity to that of the other uridine nucleotide–responsive receptors, we amplified the coding sequence of the $P2Y_4$ receptor from human genomic DNA and expressed

Table 2. Percent sequence identities among cloned P2Y receptors[a]

	Human $P2Y_1$[b]	Human $P2Y_2$[c]	Human $P2Y_6$[d]	Human $P2Y_4$[e]
Human $P2Y_1$	100	37	32	38
Human $P2Y_2$		100	40	55
Human $P2Y_6$			100	40
Human $P2Y_4$				100

[a] Based on deduced amino acid sequences and calculated using a Pam250 matrix with MacVector software (Eastman Kodak).
[b] Schachter et al., 1996.
[c] Parr et al., 1994.
[d] Communi et al., 1996.
[e] Communi et al., 1995.

the receptor in 1321N1 cells (Nicholas et al., 1996). The nucleotide sequence of the $P2Y_4$ receptor was identical to that reported by Communi et al. (1995). As was previously reported, UTP was the most potent agonist for promotion of inositol phosphate production (Table 1). However, in contrast to the results of Communi et al. (1995) and Nguyen et al. (1995), UDP was not an agonist when assays were carried out in the presence of hexokinase and glucose to eliminate any contribution of UTP. ATP was a full agonist at the $P2Y_4$ receptor, although it was 50-fold less potent than UTP. Furthermore, ATP neither added to nor blocked the effects of UTP, which indicates that ATP is a full agonist with no antagonist activity. The differences in efficacy of ATP observed in our study versus the study of Communi et al. (1995) may be due to differences in levels of receptor expression. We also observed no effect of ADP or 2MeSATP. Thus, in contrast to the selectivities reported previously, the $P2Y_4$ receptor is activated most potently by UTP, less potently by ATP, and not at all by UDP or ADP; i.e., the $P2Y_4$ receptor is nucleoside triphosphate–specific.

In Vivo Agonists of P2Y Receptors

The significance of ATP and its breakdown products as extracellular signaling molecules was brought into focus by the work of Burnstock in the 1960s and 1970s. ATP is released in a regulated fashion from neurons, platelets, macrophages, and other cells, and also appears extracellularly in most tissues as a consequence of stress- or hypoxia-induced release or cell death. In contrast to the physiological importance of adenine nucleotides, regulated release of uridine nucleotides is less well established (Goetz et al., 1971; Richards et al., 1993; Enomoto et al., 1994). Perhaps the strongest evidence comes from the identification of three G protein–coupled receptors that are activated by uridine nucleotides. The fact that two of these receptors, the $P2Y_6$ and $P2Y_4$ receptors, are selectively or preferentially activated by uridine nucleotides is strong support for a physiological function of extracellular uridine nucleotides. Unambiguous evidence for regulated release of UTP is now needed to establish a role for uridine nucleotides as extracellular signaling molecules. The fact that the $P2Y_6$ receptor is activated specifically by UDP (and is not activated by UTP or adenine nucleotides) suggests that release of UDP also may have an important physiological function.

A role for diadenosine polyphosphates (A_2P_4, A_2P_5, and A_2P_6) as extracellular signaling molecules has been suggested (reviewed by Pintor and Miras-Portugal, 1993). Castro et al. (1992) observed that A_2P_4 evoked increases in intracellular Ca^{2+} levels in adrenal medullary endothelial cells through a receptor that displayed a pharmacological selectivity similar to the $P2Y_2$ receptor. This result was followed by the observation that A_2P_4 is a full agonist at the cloned $P2Y_2$ receptor expressed in 1321N1 cells and exhibits a potency that is similar to that of UTP and ATP (Lazarowski et al., 1995). A_2P_2 is a low-potency full agonist at the $P2Y_1$ receptor expressed in 1321N1 cells, A_2P_4 is a more potent partial agonist, and A_2P_5 and A_2P_6 have no effect (Schachter et al., 1996). A_2P_4 also is a low-potency agonist at both the $P2Y_6$ and $P2Y_4$ receptors (E. R. Lazarowski, and W. C. Watt, unpublished observation). These data suggest, but do not prove, that diadenosine phosphates may be endogenous activators of the $P2Y_2$ receptor and also may regulate other P2 receptors.

Structural Aspects of the Cloned P2Y Receptors

The deduced amino acid sequences of all of the cloned P2Y receptors contain seven hydrophobic domains that are typical of the G protein–coupled receptor superfamily and that are thought to represent the seven transmembrane domains observed in the known structure of bacteriorhodopsin (Henderson et al., 1990). Based upon studies of β-adrenergic receptors, the amino terminus of G protein–coupled receptors is thought to be extracellular and the carboxyl terminus intracellular (Dohlman et al., 1987; Wang et al., 1989). An alignment of the amino acid sequences of the four cloned P2Y receptors is shown in Figure 1. The greatest homology among these receptors occurs within their transmembrane domains, a feature that is also observed for other G protein–coupled receptor families. The P2Y receptors display highest homology to each other (Table 2), followed by G protein–coupled receptors for peptide hormones such as somatostatin, platelet activating factor, angiotensin II, and neuropeptide Y. Homology of all four P2Y receptors to the P1-purinergic receptors is surprisingly low. The four cloned P2Y receptors exhibit several structural features typical of G protein–coupled receptors, such as conserved cysteine residues in the second and third extracellular loops that are thought to form a disulfide bridge, and N-linked glycosylation sites in the amino terminus. An exception to the latter is the $P2Y_4$ receptor, which lacks an N-linked glycosylation site in its amino terminus.

The location and identity of residues important for nucleotide binding in P2Y receptors is unknown. The binding site of most G protein–coupled receptors for small molecules is located within the plane of the bilayer. The seven helices of the receptors are postulated to form a binding pocket, and several residues within transmembrane domains (TM) 2, 5, and 7 have been implicated in the binding of catecholamines and other biogenic amines. Since nucleotides are small molecules, one might expect that the residues important for binding would reside within the TM domains of P2Y receptors. However, peptide hormone receptors apparently utilize extracellular loops and the amino terminus as well as TM domains to form the binding site for their cognate peptide ligands. The similarity of P2 receptors to peptide hormone receptors suggests the possibility that the binding site for nucleotides may be located within the bilayer as well as in extracellular regions.

On the basis of charge interaction, it is reasonable to assume that several positively charged amino acids might be involved in tethering the negatively charged phosphate groups of nucleotides. Using this rationale, Erb et al. (1995) designed a series of point mutations in the $P2Y_2$ receptor at residues that were predicted to reside within the membrane bilayer and that potentially might interact with the phosphate moieties of nucleotides. Mutation of Lys-107 and Arg-110 in TM3 of the $P2Y_2$ receptor to either Ile or Leu had no effect on the potency of ATP or UTP to stimulate Ca^{2+} mobilization in 1321N1 cells stably expressing the mutant receptors. This result suggests that these residues do not participate in nucleotide binding. In contrast, mutation to Leu residues of His-262 and Arg-265 in TM6 and of Arg-292 in TM7 greatly decreased the potency of both ATP and UTP for stimulation of Ca^{2+} mobilization in transfected cells. The authors concluded that these residues likely were involved in stabilization of agonist binding and/or receptor activation. Finally, whereas mutation of Lys-289 to Ile had no effect on the potency of ATP or UTP, the conservative substitution of Lys-289 to Arg changed the order of agonist potency such that ADP and UDP became more potent than ATP and UTP. Lys-289 is conserved in $P2Y_6$ and $P2Y_4$ receptors, but is Gln in the $P2Y_1$ receptor. Although these results are intriguing, they must be viewed with some caution since they were based primarily on "loss of function," there was no quantitation of receptor expression, and detailed analyses of responses to nucleotides were not presented.

A signature sequence, LFLTCIS, is conserved in TM3 of all P2Y subtypes. This sequence is followed by HRY in $P2Y_1$ and $P2Y_4$ receptors, QRY in the $P2Y_6$ receptor, and HRC in $P2Y_2$ receptors. This triad motif in TM3 is DRY in most other G protein–coupled receptors. The effect of substitution for Asp in the DRY motif in P2Y receptors is unknown, but this Asp residue has been shown to be necessary for G protein coupling in several other receptors (Strader et al., 1988; Fraser et al., 1988; Wang et al., 1989). An Asp residue in TM2 (Asp-97 in the human $P2Y_1$ receptor) is conserved in virtually all G protein–coupled receptors (including all P2Y receptor subtypes) and is necessary for agonist binding and activation in β_2-adrenergic, m1-muscarinic, and lutropin receptors (Strader et al., 1988; Fraser et al., 1989; Ji and Ji, 1991).

Van Rhee and coworkers (1995) have modeled the avian $P2Y_1$ receptor using the electron density map of rhodopsin as a template. Their model predicts that four basic residues near the extracellular surface are involved in docking the phosphate chain of ATP: His-121 in TM3, His-266 and Lys-269 in TM6, and Arg-299 in TM7. The last three residues correspond to residues that were implicated in nucleotide binding to the $P2Y_2$ receptor by mutational analysis (Erb et al., 1995). His-121 and Lys-269 are not conserved in the other P2Y receptors, whereas His-266 and Arg-299 are conserved. Other residues in the $P2Y_1$ receptor were proposed by van Rhee et al. (1995) to play a role in binding of the adenine moiety (Gln-296 in TM7) and ribose moiety (Ser-303 and Ser-306 in TM7). Tyr-125 in TM3 also was proposed to be important in coordinating the triphosphate chain. Several of these residues coincide with conserved positions that fulfill key roles in binding of ligands to other G protein–coupled receptors (reviewed by Strader et al., 1995). The accuracy of this model for the $P2Y_1$ receptor awaits testing by site-directed mutagenesis and in-depth pharmacological analyses.

Additional P2Y Receptors

All four of the cloned P2Y receptors activate phospholipase C, resulting in increases in inositol lipid hydrolysis and elevation of intracellular Ca^{2+} levels. We and others have established that a P2Y receptor exists on C6 glioma cells and that it is coupled to inhibition of adenylyl cyclase (Boyer et al., 1993). No evidence for P2Y receptor–mediated regulation of inositol lipid hydrolysis or Ca^{2+} mobilization was observed in C6 cells under conditions where muscarinic cholinergic receptor activation markedly increased inositol phosphate accumulation. Pretreatment of C6 cells with pertussis toxin blocked the inhibitory effects of P2Y receptor agonists, suggesting the involvement of G_i (Boyer et al., 1993). The pharmacological selectivity of a series of analogs of ATP and ADP was similar but not identical to that of the $P2Y_1$ receptor. PPADS (pyridoxyl phosphate 6-azophenyl-2′,4′-disulfonic acid) was a competitive antagonist at the turkey $P2Y_1$ receptor but had no effect at the P2Y receptor on C6 glioma cells (Boyer et al., 1994). These results suggested that a novel P2Y receptor subtype with distinct signaling properties exists on C6 cells.

To confirm that these differences in signaling specificity are due to two distinct P2Y receptors and are not a result of the different cells in which the P2Y receptors are expressed, the human $P2Y_1$ receptor also was stably expressed in C6 glioma cells and was shown to confer 2MeSATP-stimulated inositol lipid hydrolysis to these cells (Schachter et al., 1996). The phospholipase C–activating $P2Y_1$ receptor could be distinguished from the endogenous adenylyl cyclase–coupled P2Y receptor of C6 glioma cells using the P2 receptor antagonist PPADS. PPADS competitively antagonized

2MeSATP-promoted inositol phosphate accumulation, but had no effect on 2MeSATP-promoted inhibition of cyclic AMP accumulation (Boyer et al., 1994; Schachter et al., 1996). P2Y receptor agonists that exhibited different selectivities for the two receptors expressed in different cell lines also exhibited differential selectivities for activation of second messenger responses in the C6 glioma cells engineered to express both the endogenous and heterologous P2Y receptor (Schachter et al., 1996). These data confirm that the adenylyl cyclase–inhibiting P2Y receptor natively expressed in C6 glioma cells is a different receptor than the cloned phospholipase C–activating $P2Y_1$ receptor. The P2Y receptor natively expressed in C6 glioma cells remains to be cloned.

Matsuoka et al. (1995) have described an ATP-induced increase in cyclic AMP accumulation in NG108-15 cells that is not secondary to an elevation of Ca^{2+} levels. This putative P2Y receptor on NG108-15 cells has a unique pharmacological profile with several xanthine analogues acting as antagonists. Elevations of cyclic AMP levels in response to extracellular adenine nucleotides have been observed in other preparations (Griese et al., 1991; Tokumitsu et al., 1991; Gailly et al., 1993; Henning et al., 1993); however, adenosine, as a breakdown product of ATP, may contribute to the stimulatory effect.

Additional receptors for adenine nucleotides remain to be cloned. The properties of the P2T receptor of platelets have been widely reviewed and will not be reiterated here. However, this is a physiologically important receptor that also has potentially large importance as a therapeutic target. It is unclear from second messenger signaling responses whether single or multiple P2T receptors exist. The P2Z receptor forms a nonselective pore passing molecules of <1000 daltons in response to ATP^{4-}. Surprenant et al. (1996) have cloned a P2X receptor ($P2X_7$) that not only functions as an ATP-gated cation channel but also is capable of forming a pore permeable to dyes of large molecular size. Thus, it appears that the P2Z receptor of macrophages is similar in sequence and ion conductance to P2X receptors, but with an additional cytolytic function.

CONCLUSIONS AND FUTURE DIRECTIONS

By analogy with other families of G protein–coupled receptors, there are likely several additional P2Y receptor subtypes that have not been cloned. At a minimum, the ADP receptor(s) of platelets and a G_i/adenylyl cyclase–linked P2Y receptor remain to be isolated. The development of receptor subtype–selective agonists and antagonists may uncover yet more subtypes of P2Y receptors. Association of P2 receptor subtypes of known structure with specific physiological responses to nucleotides has not yet been accomplished. As discussed above, careful consideration of nucleotide metabolism in all receptor assay systems needs to be applied to establish receptor identities as well as to reveal differences between receptor subtypes. A large portion of the work to be accomplished with P2Y receptors would be aided by the availability of high-affinity antagonists for these receptors. Lead competitive antagonist molecules have been identified in our laboratory and work proceeds on further structural refinement (Boyer et al., 1996). Analysis of the relationship between agonist specificity and receptor structure is at a nascent stage, and novel agonist synthesis in concert with carefully directed mutagenesis studies should eventually define the binding pocket of this unique family of G protein–coupled receptors.

ACKNOWLEDGMENTS

This work was supported by USPHS grants GM38213, GM16918, and HL32322 and American Heart Association Grant-in-Aid 93010670. R. A. N. is an established investigator of the American Heart Association. The authors thank Brenda Lindsey Asam for typing corrections to the manuscript. The authors also would like to acknowledge José Boyer, Eduardo Lazarowski, Qing Li, Joel Schachter, Gary Waldo, and William Watt, who carried out much of the work outlined in this chapter.

REFERENCES

Berrie CP, Hawkins PT, Stephens LR, Harden TK, Downes CP (1989): Phosphatidylinositol 4,5-bisphosphate hydrolysis in turkey erythrocytes is regulated by P_{2Y}-purinoceptors. Mol Pharmacol 35:526–532.

Bo X, Zhang Y, Nassar M, Burnstock G, Schoepfer R (1995): A P2X purinoceptor cDNA conferring a novel pharmacological profile. FEBS Lett 375:129–133.

Boyer JL, Downes CP, Harden TK (1989): Kinetics of activation of phospholipase C by P_2 purinergic receptor agonists and guanine nucleotides. J Biol Chem 264:884–890.

Boyer JL, Lazarowski ER, Chen X-H, Harden TK (1993): Identification of a P_{2Y}-purinergic receptor that inhibits adenylyl cyclase but does not activate phospholipase C. J Pharmacol Exp Ther 267:1140–1146.

Boyer JL, Schachter JL, Romero T, Harden TK (1996): Identification of partial agonist/competitive antagonist ligands for P_{2Y}-purinergic receptors. Mol Pharmacol 50:1323–1328.

Boyer JL, Zohn I, Jacobson KA, Harden TK (1994): Differential effects of putative P_2-purinergic receptor antagonists on adenylyl cyclase- and phospholipase C-coupled P_{2Y}-purinergic receptors. Br J Pharmacol 113:614–620.

Brake AJ, Wagenbach MJ, Julius D (1994): A new structural motif for ligand-gated ion channels defined by an ionotropic ATP receptor. Nature 371:519–523.

Burnstock G, Kennedy C (1985): Is there a basis for distinguishing two types of P_2-purinoceptors? Gen Pharmacol 16:433–440.

Castro E, Pintor J, Miras-Portugal MT (1992): Ca^{2+}-stores mobilization by diadenosine tetraphosphate, Ap4A, through a putative P_{2Y} purinoceptor in adrenal chromaffin cells. Br J Pharmacol 106:833–837.

Chang K, Hanaoka K, Kumada M, Takuwa Y (1995): Molecular cloning and functional analysis of a novel P_2 nucleotide receptor. J Biol Chem 270:26152–26158.

Charest R, Blackmore PF, Exton JH (1985): Characterization of responses of isolated rat hepatocytes to ATP and ADP. J Biol Chem 260:15789–15794.

Chen C-C, Akoplan AN, Sivilotti L, Colquhoun D, Burnstock G, Wood JN (1995): A P2X purinoceptor expressed by a subset of sensory neurons. Nature 377:428–431.

Communi D, Parmentier M, Boeynaems JM (1996): Cloning, functional expression and tissue distribution of the human $P2Y_6$ receptor. Biochem Biophys Res Commun 222:303–308.

Communi D, Pirotton S, Parmentier M, Boeynaems JM (1995): Cloning and functional expression of a human uridine nucleotide receptor. J Biol Chem 270:30849–30852.

Dohlman HG, Bouvier M, Benovic J, Caron MG, Lefkowitz RJ (1987): The multiple membrane spanning topography of the β2-adrenergic receptor. J Biol Chem 262:14282–14288.

Dubyak GR, El-Moatassim C (1993): Signal transduction via P_2-purinergic receptors for extracellular ATP and other nucleotides. Am J Physiol 265:C577–C606.

Edwards FA, Gibb AJ, Colquhoun D (1992): ATP receptor-mediated synaptic currents in the central nervous system. Nature 359:144–147.

Enomoto K, Furuya K, Yamagishi S, Oka T, Maeno T (1994): The increase in the intracellular Ca^{2+} concentration induced by mechanical stimulation is propagated via release of pyrophosphorylated nucleotides in mammary epithelial cells. Pflugers Arch 427:533–542.

Erb L, Garrad R, Wang Y, Quinn T, Turner JT, Weisman GA (1995): Site-directed mutagenesis of P_{2U}-purinoceptors: Positively charged amino acids in transmembrane helices 6 and 7 affect agonist potency and efficacy. J Biol Chem 270:4185–4188.

Erb L, Lustig KD, Sullivan DM, Turner JT, Weisman GA (1993): Functional expression and photoaffinity labeling of a cloned P_{2U} purinergic receptor. Proc Natl Acad Sci USA 90:10449–10453.

Evans RJ, Derkach V, Suprenant A (1992): ATP mediates fast synaptic transmission in mammalian neurons. Nature (London) 357:503–505.

Filtz TM, Li Q, Boyer JL, Nicholas RA, Harden TK (1994): Expression of a cloned P_{2Y}-purinergic receptor that couples to phospholipase C. Mol Pharmacol 46:8–14.

Forsberg EJ, Feverstein G, Shohami E, Pollard HB (1987): Adenosine triphosphate stimulates inositol phospholipid metabolism and prostacyclin formation in adrenal medullary endothelial cells by means of P_2-purinergic receptors. Proc Natl Acad Sci USA 84:5630–5634.

Fraser CM, Chung FZ, Wang CD, Venter JC (1988): Site-directed mutagenesis of human beta-adrenergic receptors: Substitution of aspartic acid-130 by asparagine produces a receptor with high-affinity agonist binding that is uncoupled from adenylate cyclase. Proc Natl Acad Sci USA 85:5478–5482.

Fraser CM, Wang CD, Robinson DA, Gocayne JD, Venter JC (1989): Site-directed mutagenesis of m1 muscarinic receptors: Conserved aspartic acids play important roles in receptor function. Mol Pharmacol 36:840–847.

Fredholm BB, Abbracchio MP, Burnstock G, Daly JW, Harden TK, Jacobson KA, Leff P, Williams M (1994): Nomenclature and classification of purinoceptors. Pharmacol Rev 46:143–156.

Gailly P, Boland B, Paques C, Himpens B, Casteels R, Gillis JM (1993): Postreceptor pathway of the ATP-induced relaxation in smooth muscle of the mouse vas deferens. Br J Pharmacol 110:326–330.

Goetz U, DaPrada M, Pletscher A (1971): Adenine-, guanine- and uridine-5′-phosphonucleotides in blood platelets and storage organelles of various species. J Pharmacol Exp Ther 178:210–215.

Griese M, Gobran LI, Rooney SA (1991): A_2 and P_2 purine receptor interactions and surfactant secretion in primary cultures of type II cells. Am J Physiol 261:L140–L147.

Henderson DJ, Elliot DG, Smith GM, Webb TE, Dainty IA (1995): Cloning and characterization of a bovine P_{2Y} receptor. Biochem Biophys Res Commun 212:648–656.

Henderson R, Baldwin JM, Cesca TA, Zemlin F, Beckman E, Downing KH (1990): Model for the structure of bacteriorhodopsin based on high-resolution electron cryo-microscopy. J Mol Biol 213:899–929.

Henning RH, Duin M, den Hertog A, Nelemans A (1993): Characterization of P_2-purinoceptor mediated cyclic AMP formation in mouse C2C12 myotubes. Br J Pharmacol 110:133–138.

Hourani SMO, Hall DA (1994): Receptors for ADP on human blood platelets. Trends Pharmacol Sci 15:103–108.

Ji I, Ji TH (1991): Asp383 in the second transmembrane domain of the lutropin receptor is important for high affinity hormone binding and cAMP production. J Biol Chem 260:14953–14957.

Lazarowski ER, Harden TK (1994): Identification of a uridine nucleotide-selective G-protein-linked receptor that activates phospholipase C. J Biol Chem 269:11830–11836.

Lazarowski ER, Watt WC, Stutts MJ, Boucher RC, Harden TK (1995): Pharmacological selectivity of the cloned human phospholipase C-linked P_{2U}-purinergic receptor. Potent activation by diadenosine tetraphosphate. Br J Pharmacol 116:1619–1627.

Lewis C, Neidhart S, Holy C, North RA, Buell G, Suprenant A (1995): Coexpression of P2X2 and P2X3 receptor subunits can account for ATP-gated currents in sensory neurons. Nature 377:432–435.

Lustig KD, Shiau AK, Brake AJ, Julius D (1993): Expression cloning of an ATP receptor from mouse neuroblastoma cells. Proc Natl Acad Sci USA 90:5113–5117.

Matsuoka I, Zhou Q, Ishimoto H, Nakanishi H (1995): Extracellular ATP stimulates adenylyl cyclase and phospholipase C through distinct purinoceptors in NG108-15 cells. Mol Pharmacol 47:855–862.

Nguyen T, Erb L, Weismann GA, Marchese A, Heng HHQ, Garrad RC, George SR, Turner JT, O'Dowd BF (1995): Cloning, expression, and chromosomal localization of the human uridine nucleotide receptor gene. J Biol Chem 270:30845–30848.

Nicholas RA, Watt WC, Lazarowski ER, Li Q, Harden TK (1996): The uridine nucleotide selectivity of three phospholipase C-activating P2 receptors: Identification of a UDP-selective, a UTP-selective, and an adenine and uridine triphosphate-specific receptor. Mol Pharmacol 50:224–229.

O'Connor SE, Dainty IA, Leff P (1991): Further subclassification of ATP receptors based on agonist studies. Trends Pharmacol Sci 12:137–141.

Okajima F, Tokumitsu Y, Kondo Y, Ui M (1987): P_2-purinergic receptors are coupled to two signal transduction systems leading to inhibition of cAMP generation and to production of inositol triphosphate in rat hepatocytes. J Biol Chem 262:13483–13490.

Parr CE, Sullivan DM, Paradiso AM, Lazarowski ER, Burch LH, Olsen JC, Erb L, Weisman GA, Boucher RC, Turner JT (1994): Cloning and expression of a human P_{2U} nucleotide receptor, a target for cystic fibrosis pharmacology. Proc Natl Acad Sci USA 91:3275–3279.

Pianet I, Merle M, Labouesse J (1989): ADP and, indirectly, ATP are potent inhibitors of cAMP production in intact isoproterenol-stimulated C6 glioma cells. Biochem Biophys Res Commun 163:1150–1157.

Pintor J, Miras-Portugal MT (1993). Diadenosine polyphosphates (Ap_xA) as new neurotransmitters. Drug Dev Res 28:259–262.

Pirotton S, Raspe E, Demeolle D, Erneux C, Boeynaems JM (1987): Involvement of inositol 1,4,5-trisphosphate and calcium in the action of adenine nucleotides on aortic endothelial cells. J Biol Chem 262:17461–17466.

Richards SM, Dora KA, Rattigan S, Colquhoun EQ, Clark MG (1993): Role of extracellular UTP in the release of uracil from vasoconstricted hindlimb. Am J Physiol 264:H233–H237.

Schachter JL, Li Q, Boyer JL, Nicholas RA, Harden TK (1966): Second messenger cascade specificity and pharmacological selectivity of the human $P2Y_1$ receptor. Br J Pharmacol 118:167–173.

Strader CD, Sigal IS, Candelore MR, Rands E, Hill WS, Dixon RAF (1988): Conserved aspartic acid residues 79 and 113 of the β-adrenergic receptor have different roles in receptor function. J Biol Chem 263:10267–10271.

Strader CD, Fong TM, Graziano MP, Tota MR (1995): The family of G-protein-coupled receptors. FASEB J 9:745–754.

Surprenant A, Rassendren F, Kawashima E, North RA, Buell G (1996): The cytolytic P2Z receptor for extracellular ATP identified as a P2X receptor ($P2X_7$). Science 272:735–738.

Tokumitsu Y, Yanagawa Y, Nomura Y (1991): Stimulation of DNA synthesis in Jurkat cells by synergistic action between adenine and guanine nucleotides. FEBS Lett 288:81–85.

Tokuyama Y, Hara M, Jones EMC, Fan Z, Bell GI (1995): Cloning of rat and mouse P_{2Y} purinoceptors. Biochem Biophys Res Commun 211:211–218.

Valeins H, Merle M, Labouesse J (1992): Pre-steady state study of the β-adrenergic and purinergic receptor interaction in C6 cell membranes: Undelayed balance between positive and negative coupling to adenylyl cyclase. Mol Pharmacol 42:1033–1041.

Valera S, Hussy N, Evans RJ, Adami N, North RA, Suprenant A, Buell G (1994): A new class of ligand-gated ion channel defined by P_{2X} receptor for extracellular ATP. Nature 371:516–519.

van Rhee AM, Fischer B, van Galen PJM, Jacobson KA (1995): Modelling the P_{2Y} purinoceptor using rhodopsin as template. Drug Design Discov 13:133–154.

Wang H, Lipfert L, Malbon CC, Bahouth S (1989): Site-directed anti-peptide antibodies define the topography of the β-adrenergic receptor. J Biol Chem 264:14424–14431.

Webb TE, Simon J, Krishek BJ, Bateson AN, Smart TG, King BF, Burnstock G, Barnard EA (1993): Cloning and functional expression of a brain G-protein-coupled ATP receptor. FEBS Lett 324:219–225.

Yamada M, Hamamori Y, Akita H, Yokoyama M (1992): P_2-purinoceptor activation stimulates phosphoinositide hydrolysis and inhibits accumulation of cAMP in cultured ventricular myocytes. Circ Res 70:477–485.

CHAPTER 4

Regulators of Endogenous Adenosine Levels as Therapeutic Agents

JONATHAN D. GEIGER, FIONA E. PARKINSON and ELIZABETH A. KOWALUK

Department of Pharmacology and Therapeutics, University of Manitoba Faculty of Medicine, Winnipeg, Manitoba R3E 0W3, Canada (J.D.G., F.E.P.); Neurological and Urological Diseases Research, Pharmaceutical Products Division, Abbott Laboratories, Abbott Park, IL 60064-3500 (E.A.K.)

INTRODUCTION

Physiological actions of adenosine almost certainly result from the binding of adenosine to specific adenosine receptors and activation of signal transduction systems. The levels of adenosine available for stimulation of adenosine receptors are controlled by processes related to its production, release, re-uptake (transport), and metabolism—all of which are closely inter-related. Therefore, regardless of whether adenosine originates endogenously or exogenously, its effects appear to be inexorably linked to its levels at adenosine receptor sites. As a result, over the past two decades, an approach has emerged to the development of adenosine pharmaceuticals that is an alternative to the development of adenosine receptor agonists that mimic and antagonists that interfere with the widespread regulatory actions of adenosine. In so doing, agents continue to be identified that influence adenosine transport, metabolism, and/or release and thereby alter intracellular or extracellular levels of endogenous adenosine and, subsequently, adenosine receptor–mediated events.

Two desirable therapeutic objectives might be achieved by developing agents that are **r**egulators of **e**ndogenous **a**denosine **l**evels (REAL agents). First, levels of endogenous adenosine might be selectively elevated by such agents only where and when adenosine is being produced/released. Adenosine produced/released locally should act locally because of its extremely short half-life. Secondly, side effects for REAL agents may be expected to be less than those of metabolism-resistant, longer-acting, adenosine receptor agonists.

LEVELS AND ACTIONS OF ADENOSINE

Adenosine affects a variety of physiological processes (see Phillis, 1991). The levels and actions of endogenous adenosine are amenable to pharmacological manipulation,

Purinergic Approaches in Experimental Therapeutics, Edited by Kenneth A. Jacobson and Michael F. Jarvis
ISBN 0-471-14071-6 © 1997 Wiley-Liss, Inc.

and therapeutic advantage could be taken of its role—for example, in the heart as an endogenous antiarrhythmic (Conti et al., 1995) and cardioprotectant (Yang and Mehta, 1994); in the brain as a neuroprotectant to reduce seizure incidence and severity, as well as to protect against ischemic neuronal injury (Heurteaux et al., 1995); and throughout the body as an anti-inflammatory autocoid (Cronstein et al., 1995). Development of REAL agents can proceed in at least two ways: (i) By identifying compounds that enhance actions attributable to endogenous adenosine and then determining mechanisms that enable these actions; and (ii) By identifying compounds that selectively affect mechanisms and then determining the extent to which the compounds increase, in more integrated systems, levels and actions of endogenous adenosine. Both approaches continue to be used today.

For development of adenosine pharmaceuticals, measurements of endogenous adenosine levels directly adjacent to its site(s) of action (adenosine receptors) would be most informative. This remains a desirable, but as yet unobtained, objective. Nevertheless, purine levels in tissues and in extracellular fluids have been measured and provide useful information relevant to the identification, development, and characterization of REAL agents.

An important caveat to consider with REAL agents is that although the vast majority of studies have shown that adenosine produces protective responses, a few studies have now reported that adenosine, at high levels, exhibits either no appreciable protective effects or deleterious effects in central and peripheral tissues (Fujiwara et al., 1994; Leone and Merrill, 1995). Adenosine was also recently shown to induce apoptosis in chick embryonic sympathetic neurons (Wakade et al., 1995).

Tissue Levels

For precise and accurate measurements of endogenous adenosine levels, adenosine release, production/metabolism, and re-uptake must be inhibited completely and virtually instantaneously. Experimentally, it has been well documented that sampling methods can have profound effects on measured levels of purines. For example, tissue levels of adenosine change rapidly and markedly due to postmortem breakdown of adenine nucleotides. Therefore, accurate determinations of basal levels of purines can only be obtained when postmortem-induced increases in adenosine and decreases in cellular energy charge ($\frac{1}{2}$ ADP + ATP) ÷ (AMP + ADP + ATP) or ATP/ADP ratios are prevented by rapid inactivation of enzymes responsible for adenine nucleotide breakdown and adenosine production. Importantly, tissue levels of adenosine do correlate well with adenosine actions.

Brain For brain, postmortem-induced changes in tissue levels of purines were minimized when animals were killed by blow-freezing (Winn et al., 1979) and focused microwave irradiation (a method for killing animals that has been approved by the Canada Council on Animal Care) (Delaney and Geiger, 1995, 1996; Phillis et al., 1995). For example, adenosine levels in whole brain of rats killed by decapitation were 1130 pmol/mg protein and by blow-freezing or microwave irradiation were 10 to 20 pmol/mg protein (Winn et al., 1979; Delaney and Geiger, 1995, 1996; Phillis et al., 1995). Even with microwave animal sacrifice systems (the largest commercially available unit is 10 kW), postmortem breakdown of adenine nucleotides may occur as levels of adenosine and AMP increased, and ATP decreased with irradiation power levels less than 10 kW (Delaney and Geiger, 1996).

Microwave irradiation is used rather than blow-freezing when measurements of adenosine are required in discrete brain regions. In rats killed by microwave irradiation at 10 kW, adenosine levels varied 17-fold among six brain regions tested. In comparison, levels in brain regions of decapitated rats, or brains of rats killed with less powerful irradiation sacrifice systems, varied up to three-fold (see Delaney and Geiger, 1996). Thus, it appears that large differences are observed in adenosine levels among brain regions only in the absence of extensive postmortem metabolism.

Blood In other tissues, as with brain, great care must be taken to ensure that adenosine is neither being formed nor lost during the sampling period. In plasma, for example, basal levels of adenosine ranged from 0.3 to 1.7 μM when experimenters used a "stop" solution of dilazep to inhibit adenosine uptake into and release from red blood cells, erythro-9-(2-hydroxy-3-nonyl)adenine (EHNA) to block ADA activity, and indomethacin to inhibit nucleotide release from platelets (Zhang et al., 1991a). However, when the "stop" solution contained dipyridamole (to inhibit adenosine uptake) and EHNA along with ethylene glycol-bis (β-amino ethyl ether) N,N,N′,N′-tetraacetic acid (EGTA), ethylene diaminetetraacetic acid (EHNA), and d,l-alpha-glycerophosphate (to inhibit platelet activation, alkaline phosphatase, 5′-ectonucleotidase and nonspecific phosphatase), measured levels were 79 nM in rat (Phillis et al., 1992) and 35 nM in human (Linden et al., 1992). These lower (nanomolar levels) may be more physiologically relevant, as vasodilatation started to occur at 10 nM and was maximal at about 5 μM (Collis, 1991). In dog, where the half-life is over 10 times longer than in human, less elaborate precautions were necessary to measure levels in the range of 13–20 nM (Miura et al., 1991). Thus, it appears clear that aggressive use of pharmacological agents is necessary to measure nanomolar levels of adenosine in blood.

Other Tissues Where measures were taken to prevent metabolism-induced changes in endogenous adenosine, levels in the nanomolar range have been consistently noted. For example, levels of endogenous adenosine in heart were 80 nM (Deussen et al., 1988), and in jejunum/ileum were 100–200 nM (Proctor and Langkamp-Henken, 1991). Even though "stop" solutions were used, levels of adenosine in the CSF varied from 40 to 600 nM between studies (Walter et al., 1988; Laudignon et al., 1990).

Some variability between studies may arise because of circadian variations. During a 24-hour cycle, levels ranged in liver from 0.2 to 1.2 μM, in blood from 0.25 to 0.75 μM, and in brain from 1 to 12 μM (Chagoya de Sanchez, 1995). However, the accuracy of these values may be questioned because these measurements were made from animals that were stunned and decapitated with no apparent precautions taken to reduce postmortem changes to adenosine.

Extracellular Fluid Levels

In attempting to determine levels of adenosine near cell surface adenosine receptors, extracellular levels of adenosine in tissue transudates have been measured. With microdialysis in brain, nanomolar concentrations of adenosine were measured only when the large amounts of adenosine produced by tissues that were damaged with insertion of the microdialysis probe were allowed to dissipate and stabilize (Ballarin et al., 1991). With closed cranial window and cortical cup methods in brain, care must be taken because of the high volume:surface area ratio that might lead to artifically low levels (Van Wylen et al., 1991). Nevertheless, levels of interstitial adenosine in brain were about 160 nM using the closed cranial window technique (Meno et al., 1991) and 70 nM using the cortical cup technique (Phillis et al., 1991). These appear to be similar to those of, for example, about 40–70 nM in rat striatum (Ballarin et al., 1991; Pazzagli et al., 1995) obtained using microdialysis. Levels of adenosine determined in other tissues with microdialysis, were, for example, 1600 nM in swine heart (Schulz et al., 1995), 490 nM in dog heart (Wang et al., 1994), and 200 nM in rat kidney (Baranowski and Westenfelder, 1994).

Summary

To accurately measure endogenous adenosine, it is necessary to decide first whether to measure total tissue levels or extracellular concentrations, then to sample near the relevant site(s) of action, and to take proper steps to limit pre- and postsampling production/metabolism of adenosine.

MECHANISMS FOR REGULATING ENDOGENOUS ADENOSINE LEVELS

Production

Adenosine receptors are predominantly located on cell surfaces. Therefore, adenosine, whether formed intracellularly or extracellulary through tightly regulated and complex enzymatic pathways (Figure 1) (Meghji, 1991), must have access to these sites. Adenosine formed extracellularly activates adenosine receptors. Adenosine formed intracellularly may activate cell surface receptors; however, the extent to which and the mechanisms by which intracellular adenosine accesses cell surface adenosine receptors are not always clear, and much has been inferred from studies with other tissue sources and systems. These are important issues to resolve in understanding the physiological actions of endogenous adenosine as well as the actions of REAL agents.

As illustrated in Figure 1, adenosine can be produced by the breakdown of nucleotides (including ATP, ADP, and cAMP) to AMP followed by the actions of 5′-nucleotidase on AMP. The latter route for adenosine production appears to be especially active during periods when cellular energy demands are high and oxygen supplies are low. Adenosine levels rise when the demand for oxygen is greater than its supply, as in situations such as ischemia and hypoxia (Schrader, 1991). Increased levels of adenosine will result in a readjustment of the energy supply to demand ratio, and accordingly adenosine has been referred to as a "retaliatory metabolite" (Newby, 1991).

Intracellular Formation Cytosolic 5′-nucleotidase (5′-N) metabolizes AMP to adenosine. As ATP is the major source of AMP, it follows therefore that levels of intracellular adenosine are tightly linked to the ratio of energy supply and demand. Two 5′-N enzymes have been characterized. One, the so called AMP-specific 5′-N, accepts with almost equal affinity the substrates AMP and inosine 5′-monophosphate (IMP); the other 5′-N has been termed IMP-specific. The AMP-specific 5′-N, which has been found in rat, pigeon, and rabbit hearts, is thought to be the main contributor to cytosolic adenosine formation during periods of nucleotide breakdown. Although IMP-specific 5′-N has a greater than 10-fold preference for IMP versus AMP, it has been shown that this enzyme contributes significantly, at least in some tissues, to adenosine production (Worku and Newby, 1983). Together, IMP- and AMP-specific 5′-N appear to be the

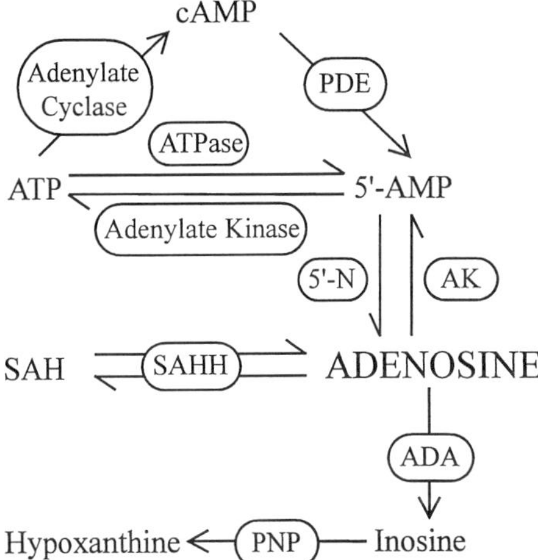

FIGURE 1. Pathways and enzymes involved in adenosine production and metabolism. Abbreviations are as follows: ADA, adenosine deaminase; AK, adenosine kinase; 5′-nucleotidase, 5′-N; SAH, S-adenosyl homocysteine; SAHH, SAH hydrolase; PDE, cAMP phosphodiesterase; PNP, purine nucleoside phosphorylase.

main sources of intracellular adenosine (Fredholm and Sollevi, 1986). Neither the IMP- nor the AMP-specific 5′-N enzymes are inhibited by α,β-methylene ADP (AOPCP), and much more work is needed to determine the tissue distribution and physiological significance of these enzymes in tissues where adenosine-mediated actions are studied. Alternatively, adenosine can be formed in an oxygen-independent manner by metabolism of S-adenosylhomocysteine (SAH) by SAH hydrolase (SAHH) (Schrader, 1991). Significant contributions of this pathway to levels of endogenous adenosine have been observed in, for example, heart, liver, and endothelial cells under normoxic conditions (Schrader, 1991; Kochan et al., 1994). Of course, once it is formed intracellularly, adenosine must gain access to cell surface adenosine receptors (see section on adenosine release).

Extracellular Formation The best-characterized enzymatic source for adenosine is ecto-5′-N. This enzyme has been cloned and sequenced, and its distribution, biochemistry, and pharmacology have been widely studied. Because AOPCP and inhibitory antibodies to ecto-5′-N are available as pharmacological tools to study this enzyme, it is possible to distinguish between intracellular and extracellular 5′-N contributions to adenosine production. A delay in adenosine formation from released ATP may be observed because ADP inhibits 5′-N activity (Plesner, 1995). To a lesser extent, cAMP released from platelets, muscle cells, and neural cells may represent a physiologically important source of AMP upon which 5′-N acts (see Mi and Jackson, 1995).

Metabolism

Adenosine can be catabolized by deamination via ADA, phosphorylation via AK, or incorporation into S-adenosylhomocysteine via SAHH (Figure 1).

Adenosine Deaminase (ADA) As detailed elsewhere (Geiger et al., 1991), ADA (EC 3.5.4.4) is a member of a family of enzymes capable of deaminating purine nucleosides. ADA is a 36-kDa protein that, along with what has been termed ADA-binding proteins (ADA-BP), forms a 280-kDa polypeptide. Although, ADA activity has been thought to be mainly localized in the cytoplasm, recent evidence suggests that ecto-ADA activity is present in a number of tissues including erythrocytes, lung, nerves, brain, lymphocytes, cell lines, and endothelial cells (see Geiger et al., 1991; Martin et al., 1995), and it has been suggested that ADA participates in adenosine transport [Centelles and Franco, 1990]. Recently, ADA-BP was found to be analogous to CD26 and binds ADA at the cell surface [Kameoka et al., 1993), and this has opened up the possibility that ADA may be involved in diverse cellular functions.

ADA is heterogeneously distributed among tissues and ADA activity is highest during early stages of development [Geiger and Nagy, 1990). Accordingly, it has been argued that ADA may not be a constitutively expressed housekeeping enzyme (Geiger et al., 1991). While ADA can decrease (and inhibitors of ADA can increase) levels of endogenous adenosine, it is not yet clear whether ADA's primary role is to decrease adenosine, form the physiologically active nucleoside inosine, or participate in other, as yet undefined, cellular functions (Geiger et al., 1991). Further, it is still not clear whether, or the degree to which, cells that express especially high levels of ADA activity preferentially accumulate and/or release purines and thereby have special intra-or extracellular requirements for purines (Nagy et al., 1990).

Adenosine Kinase (AK) Adenosine kinase (AK), (ATP: adenosine 5′-phosphotransferase, EC 2.7.1.20) is a cytosolic enzyme that catalyzes the phosphorylation of adenosine to AMP. AK has broad tissue and species distribution, and has been isolated from yeast (Leibach et al., 1971), a variety of mammalian sources (e.g., Miller et al., 1979a, Palella et al., 1980; Yamada et al., 1980; Rottlan and Miras-Portugal, 1985), and certain microorganisms (e.g., Lobelle-Rich and Reeves, 1983; Datta et al., 1987). It is present in virtually every human tissue assayed, including kidney, liver, brain, spleen, placenta, and pancreas (Andres and Fox, 1979). Phosphorylation of adenosine to AMP is known to be mediated by AK in the presence of ATP, as well as by a phosphotransfer exchange mechanism in the absence of ATP. Both reactions are inhibited by 5-iodotubercidin (ITU) (Sayos et al., 1994). AK appears to be monomeric, with a molecular weight in the range of 38–56 kDa. The AK gene has been reported to reside on chromosome 10 in humans (Chan et al., 1978) and on chromosome 14 in mouse (Samuelson and Farber, 1985). The cloning and expression of AK cDNA was reported recently (Spychala et al., 1996). The enzyme lacks a "classic" ATP binding motif and although it appears to be structurally distinct from other well-characterized nucleoside kinases, the enzyme contains two regions with sequence similarity to several plant and microbial sugar kinases.

ATP is generally considered to be the preferred phosphate source for the reaction catalyzed by AK. However, the enzyme appears to have a fairly broad phosphate donor specificity, which may vary depending on its source. In general, purine nucleoside triphosphates appear to be more efficient than their pyrimidine counterparts, although GTP may represent a widely accepted alternative phosphate source (Miller et al., 1979a; Yamada et al., 1980). Magnesium is required for AK activity, and the true AK substrate is probably complexed $MgATP^{2-}$ (Palella et al., 1980). Free Mg^{2+} and free ATP^{4-} appear to inhibit AK (Miller et al., 1970b); inhibition of rat brain AK by free ATP^{4-} was competitive with respect to adenosine [Kowaluk and Cowart, 1994].

Adenosine can inhibit mammalian AK under conditions where the intracellular concentrations of adenosine are only slightly higher than the apparent K_m value of AK for adenosine. The mechanism(s) for this are not well understood, but it has been proposed that two adenosine binding sites exist on AK—a catalytic site with high affinity for adenosine and a low-affinity regulatory (inhibitory) site (Hawkins and Bagnara, 1987; Lin et al., 1988). Recent data using bovine liver AK suggest that the low-affinity (inhibitory) adenosine binding site might be the ATP (catalytic) binding site (Elalaoui et al., 1994). The onset of substrate inhibition by adenosine appears to be promoted by high concentrations of free Mg^{2+}. The adenosine concentration at which optimal AK activity is observed therefore varies among studies; values ranged from around K_m (at low free Mg^{2+} concentration) to about 10 times K_m (at high free Mg^{2+} concentrations) (Miller et al., 1979b; Kowaluk and Cowart, 1994). Adenosine inhibition of AK may also be pH-dependent, with higher adenosine concentrations required to induce inhibition at more acid pH (Arch and Newsholme, 1978; Yamada et al., 1980). Thus, in studies aimed at determining the relative importance of AK and ADA to the maintenance of endogenous adenosine levels, it is important to consider the levels of adenosine, free Mg^{2+}, and pH in the environment of the enzymes and to recognize that the answers obtained may be quite different, depending on the levels of these substances and the physiological state of the cells and tissues. Furthermore, because AK is subject to substrate inhibition, the effectiveness of AK inhibitors may be overestimated somewhat, because inhibitor-induced increases in adenosine levels might lead to enzyme inhibition if adenosine levels reach sufficiently high concentrations.

While the kinetic mechanism of AK has been studied by a number of workers it has not been clearly defined. The majority of studies indicate a sequential mechanism in which both adenosine and $MgATP^{2-}$ bind before product release. Studies with partially purified AK from Erlich ascites cells (Henderson et al., 1972) and with purified rat liver AK (Mimouni et al., 1994) suggest that $MgATP^{2-}$ may bind first to the enzyme. In contrast, studies with AK from rat brain (Kowaluk and Cowart, 1994), human placenta (Palella et al., 1980), human erythrocytes (Hawkins and Bagnara, 1987), bovine adrenal medulla (Rottlan and Miras Portugal, 1985), and *Leishmania donovani* (Datta et al., 1987) appear to exclude a mechanism in which $MgATP^{2-}$ binds first. It has also been suggested that AK may act via a ping-pong mechanism involving the formation of a phosphoryl enzyme as an obligatory intermediate (Chang et al., 1983). However, a phosphorylated intermediate has not been isolated (Richard et al., 1980; Chang et al., 1983; Mimouni et al., 1994). It is unclear why differing reaction mechanisms have been described for AK. The possibility of isozymes could be considered, however AK from rat liver and brain appear to be identical (Yamada *et al.,* 1980, 1981), yet are suggested to have differing kinetic mechanisms (Kowaluk and Cowart, 1994; Mimouni et al., 1994).

In comparison to ADA, less is known about the distribution of AK among tissues and cells, especially in the CNS (Geiger and Nagy, 1990; Nagy et al., 1990; Geiger et al., 1991). In Table 1, the distribution of AK activity among 13 CNS areas is listed; AK activity varied by about six-fold between olfactory bulb, the area with the highest activity, and posterior hypothalamus, the area with the lowest activity (J.D. Geiger and S.M. Delaney, unpublished observation). Figure 2 compares the activity levels of ADA and AK in these 13 CNS regions. While a significant positive correlation (r^2 = 0.533, p = 0.007) was observed among the 13 areas, cerebellum and ventral spinal cord, which had high levels of AK activity, were among the areas with the lowest

TABLE 1 Distribution of Adenosine Kinase Activity in Rat CNS

CNS Region	AK Activity[a] (nmoles/5 min/mg protein)
Olfactory bulb	35.6 ± 6.5
Cerebellum	22.7 ± 5.8
Ventral spinal cord	18.9 ± 1.8
Habenula	17.1 ± 3.8
Anterior hypothalamus	16.7 ± 5.1
Dorsal spinal cord	15.4 ± 4.5
Striatum	13.4 ± 1.8
Cerebral cortex	12.7 ± 0.9
Hippocampus	12.0 ± 2.2
Superior colliculus	12.0 ± 1.4
Spinal cord white matter	10.0 ± 1.8
Septum	6.7 ± 0.7
Posterior hypothalamus	5.6 ± 0.7

[a] AK activity was measured as previously described (Gu et al., 1991) using 1 μM [^{3}H]adenosine and 500 nM EHNA to inhibit endogenous ADA activity. Incubations were for 5 minutes at 37°C. Values represent mean ± S.E.M. from five experiments, each conducted in duplicate.

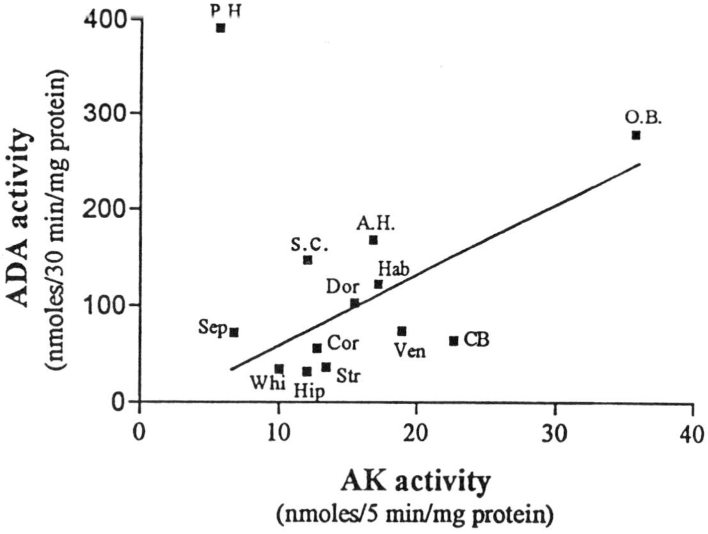

FIGURE 2. Correlation between activities of ADA and AK in 13 CNS regions. Values for ADA were previously published (Geiger and Nagy, 1986) and values for AK were determined using a method previously described (Gu et al., 1991). Abbreviations are as follows; P.H., posterior hypothalamus; Sep, septum; Whi, white matter; Hip, hippocampus; Str, striatum; Cor, cerebral cortex; S.C., superior colliculus; Dor, dorsal spinal cord; Ven, ventral spinal cord; A.H., anterior hypothalamus; Hab, habenula; CB, cerebellum; and O.B., olfactory bulbs.

levels of ADA activity; and superior colliculus and posterior hypothalamus, which had very high levels of ADA activity, had low levels of AK activity. This distribution pattern, the most complete one reported to date, may help investigators interpret results of experiments where relative contributions of ADA and AK to adenosine metabolism are being determined in brain.

The activity of mammalian AK appears to be controlled by a variety of interrelated factors (Figure 3) including: (i) substrate availability and inhibition by adenosine, (ii) inhibition by free ATP and free magnesium, (iii) pH, which affects both the enzyme and the formation of enzyme/ATP-Mg^{2+} complexes, (iv) product inhibition by ADP and AMP, and (v) the relative activities of ADA and AK.

S-Adenosylhomocysteine hydrolase (SAHH) S-Adenosylhomocysteine (SAH) is potentially an important source of or "sink" for adenosine. It has been suggested that metabolism of SAH by SAHH (EC 3.3.1.1) may represent an important source of adenosine, but tissue levels of SAH are low to minimize inhibition of cellular transmethylation reactions (Snyder, 1985). Indeed, Wagner et al. (1995) showed that SAH was not a source of adenosine in isolated rabbit cardiomyocytes under either basal or ATP-depleted conditions. Furthermore, SAH does not appear to be an important "sink" for adenosine, even though SAH formation is the favored direction for the reversible enzyme SAHH, because homocysteine levels are low and tightly controlled (see Geiger and Nagy, 1990). Thus, even though the K_m of SAHH for adenosine is similar to that of AK, adenosine disposition in many tissues seems to be most tightly linked to ADA and AK, and not to SAHH.

Summary Determining the relative importance of ADA, AK, and SAHH in the metabolism of endogenous adenosine has relied on measures of their Michaelis-Menten kinetics, inhibition by potent and selective enzyme inhibitors, and estimates of levels of endogenous adenosine. The affinities (K_m values) of AK, ADA, and SAHH for adenosine range from 0.2–2.0 μM, 20–100 μM, and 1–5 μM, respectively (Geiger and

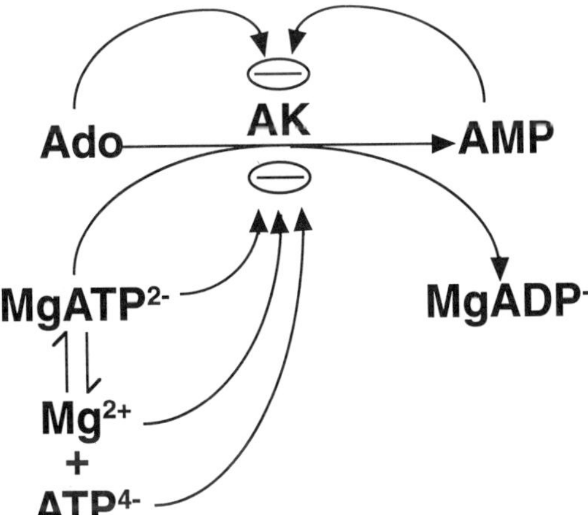

FIGURE 3. Substrate, cosubstrate, and product inhibition may regulate the activity of mammalian AK.

Nagy, 1990). Thus, considering levels of endogenous adenosine are in the nanomolar to low micromolar range, it might be expected that phosphorylation of adenosine by AK would be the predominant mode of adenosine disposition. However, AK, has a much lower capacity than does ADA; the ADA : AK ratio for V_{max} values, for example, was about 18:1 in rat and 6:1 in human heart (Tavenier et al., 1995). Furthermore, AK is nearly saturated at physiological concentrations of adenosine, and its activity is inhibited by even slightly elevated (>0.5 μM) levels of endogenous adenosine. Accordingly, in such diverse tissues as isolated rabbit cardiomyocytes, human cultured umbilical vein endothelial cells, rat hepatocytes, intact animal brains, brain slice preparations, brain vascular tissue, and CSF, adenosine was found to be mainly rephosphorylated by AK under basal conditions (Geiger and Nagy, 1990), and deaminated by ADA under ATP-depleted conditions, where adenosine levels were increased (Wu and Phillis, 1984; Meghji, 1991; Bontemps et al., 1993; Sciotti and Van Wylen, 1993; Wagner et al., 1994; Smolenski et al., 1994; Lloyd and Fredholm, 1995). Similarly when high concentrations of adenosine (10 μM) were added to whole blood *in vitro,* its metabolism was significantly slowed by inhibition of ADA with 2′-deoxycoformycin (DCF), but not by inhibition of AK with 5-iodotubercidin (ITU) (Bullough et al., 1994). It might seem, therefore, that ADA inhibition may provide greater therapeutic benefit than AK inhibition. However, the situation appears to be more complex. In *in vivo* studies where AK inhibition was found to be more efficacious than ADA inhibition (Keil and DeLander, 1992; Zhang et al., 1993), adenosine levels may not have risen sufficiently to evoke significant ADA activity. Furthermore, complex regulatory mechanisms for AK may make it difficult to predict its behavior *in vivo* based on *in vitro* measurements.

Transport (Uptake)

Nucleoside transporters (i) regulate transmembrane movements of purine nucleosides and may be uniquely involved in the expression of the physiological actions of adenosine, (ii) control the entry of nucleosides needed for *de novo* synthesis of nucleotides, and (iii) control the entry of a variety of nucleoside analogs that are used as therapeutically important chemotherapeutic agents. As such, nucleoside transporters may represent important targets for pharmaceuticals.

Prior to 1970, nucleosides were thought to enter cells by an unsaturable passive diffusion process. The early 1970s marked the discovery of saturable carrier-mediated nucleoside transporters (Plagemann and Richey, 1974; Berlin and Oliver, 1975). In the mid-1980s, functionally distinct nucleoside transporters were characterized, equilibrative transporters of erythrocytes were purified to homogeneity, and more recently, with the benefit of modern molecular biological techniques, structurally distinct transporters have been cloned and functionally expressed. It is now known that nucleoside transporters are functionally diverse proteins with differing Michaelis-Menten kinetics, substrate affinities, tissue distribution patterns, species specificity, structures, and sensitivities to blockade by pharmacological agents.

Equilibrative Nucleoside Transporters Equilibrative nucleoside transporters are commonly referred to as being equilibrative-sensitive (*es*) or equilibrative-insensitive (*ei*) based on their relative sensitivities to inhibition by nitrobenzylthioinosine (NBI). The *es* transporters are inhibited by NBI at low nanomolar concentrations, and *ei* transporters are inhibited by NBI at micromolar concentrations. Accordingly, it is more appropriate to refer to *ei* transporters as equilibrative inhibitor-resistant. However, inhibitor resistance appears largely confined to NBI because nucleoside transport inhibitors like dipyridamole and dilazep, which are structurally unrelated to nucleosides, inhibit *es* and *ei* transporters at least in some types of cells (Belt and Noel, 1985; Plagemann and Kraupp, 1986). Both types of equilibrative transporters accept as permeants a broad spectrum of purine and pyrimidine nucleosides, albeit with different kinetics and substrate specificity. As with other equilibrative carrier–mediated (so called facilitated diffusion) transport systems, *es* and *ei* transporters can potentially translocate permeants bidirectionally across plasma membranes (Cass, 1995).

Equilibrative nucleoside transporters are 45- to 65-kDa proteins with structural similarity to facilitated glucose transporters. The purification and subsequent structural determination of *es* transporters remains a difficult task, given that the equilibrative nucleoside and glucose (GLUT1) transporters co-elute chromatographically and that *es* transporters are about 25 times less abundant (Cass, 1995). Nevertheless, *es* transporters are known to be glycosylated and to be immunologically distinct from GLUT1. Functional NBI-sensitive *es* transporters have been expressed in *Xenopus* oocytes, antibodies to human *es* transporters have identified certain clones from a cDNA library, and the primary sequence of *es* transporter proteins was described recently (Boumah et al., 1994; Griffiths et al., 1997).

In various tissues and species, substantial differences in kinetic properties, function, and apparent sizes of *es* transporter proteins have been noted, and the possibility exists that multiple *es* transporter isoforms exist within a single species and tissue (Barros et al., 1995). The *es* transporter polypeptides of pig erythrocytes may be larger than *es* proteins from human erythrocytes, and monoclonal antibodies raised against pig erythrocyte *es* transporters do not cross-react with *es* transporter polypeptides from a number of other tissues and species (Good et al., 1987). Nevertheless, similarities among *es* transporters have been observed (Kwong et al., 1993).

The study of *es* nucleoside transporters has been aided by the use of NBI, a nucleoside analog that potently inhibits nucleoside transport, apparently by binding with high affinity (Kd = $\leq$1 nM) to a site on the extracellular face of the plasma membrane at or near the permeant-binding site (Jarvis et al., 1992). Because NBI is photoreactive, and binds semi-irreversibly and with high specificity to *es* transporters, it has proven to be especially useful in identifying, purifying, and quantifying transporter proteins. It has been generally assumed that each transporter has associated with it one NBI binding site; accordingly, the number of NBI binding sites approximates the density of *es* transporters (Cass et al., 1974). Measured transporter densities vary from 10^2 sites per cell in dog erythrocytes to $>10^7$ sites per cell in certain human cultured cells (Boumah et al., 1994).

In contrast, although little is known about structural features of *ei* transporters, information is available on kinetic properties for permeants and the relative resistance to inhibition by NBI and other nucleoside transport inhibitors. Indeed, it is not yet clear whether *es* and *ei* are different protein entities or the same protein that has been conformationally shifted (Hammond and Clanachan, 1985).

Concentrative Nucleoside Transporters Five subclasses of concentrative nucleoside transporters have been identified based on their dependence on transmembrane Na^+ gradients, permeant selectivity, and blockade by NBI: N1/*cif* transporters are concentrative NBI-insensitive and prefer purines (formycin B); N2/*cit* transporters are concentrative NBI-insensitive and prefer pyrimidines (thymidine); N2/*cit*-like (N4) transporters recognize guanosine as a permeant; N3/*cib* transporters are concentrative NBI-insensitive and exhibit a broad selectivity to purines and pyrimidines; and *cs* transporters are concentrative and sensitive to inhibition by low nanomolar concentrations of NBI (Cass, 1995). Except for the *cs* system, selective inhibitors for Na^+/nucleoside transporters have not been identified. These transport systems appear to function as symporters; one nucleoside is transported for each Na^+ ion in the case of N1/*cif*, N2/*cit*, and the N2/*cit*-like (N4) transporter, whereas the ratio is 1:2 for N3/*cib*.

There appears to be a heterogeneous distribution of Na^+/nucleoside transporters among cells and tissues. Activity of N1/*cif* was found in splenocytes, macrophages, and hepatocytes, as well as in several types of cultured cells (Cass, 1995). A cDNA for an N2/*cit* transporter has been cloned from rat jejunum (Huang et al., 1994) and was termed rCNT1. Activity of N2/*cit* was found in mouse enterocytes and in kidney brush-border vesicles from bovine, rat, and rabbit epithelial cells (Cass, 1995). Cloned cDNAs for N1/*cit* transporters have been obtained from rat liver (Che et al., 1995) and rat jejunum (Yao et al., 1996) and were termed SPNT and rCNT2, respectively. Activity of N3/*cib* was found in choroid plexus (Wu et al., 1994), and a cDNA, termed SNST1, was cloned from rabbit kidney (Pajor and Wright, 1992). While rCNT1 and rCNT2 (SPNT) show significant sequence homology and thus appear to be members of a single gene family, SNST1 has no significant homology to either rCNT1 or rCNT2 (SPNT). Of possible importance to adenosine pharmacologists was the finding that the adenosine receptor agonist 2-chloroadenosine inhibited nucleoside transport by N3/*cib* with an inhibition constant of approximately 21 μM (Gutierrez and Giacomini, 1993; Wu et al., 1994).

The sequence for SNST1 shows significant sequence homology to the Na^+/glucose cotransporter termed SGLT1 (Pajor and Wright, 1992). Expression of SNST1 in *Xenopus* oocytes results in Na^+/nucleoside cotransport activity characteristic of N3/*cib*, and the predicted sequence suggests a protein of 672 amino acids, 3 potential N-linked glycosylation sites, 12 transmembrane domains, and 2 binding sites for Na^+. In examining tissue distribution, SNST1 mRNA was found in kidney and heart, but not in liver or intestine.

The sequence for rCNT1 shows some homology (27%) with a bacterial H^+/nucleoside cotransporter (Huang et al., 1994). Expression of rCNT1 in *Xenopus* oocytes resulted in characteristic features of N2/*cit*, including Na^+/nucleoside co-transport and transport of the anti-HIV drugs azidothymidine and dideoxycytidine (Huang et al., 1994). The predicted sequence of rCNT1 suggests a protein of 648 amino acids, 3 N-linked and 4 O-linked glycosylation sites, 4 protein kinase C–dependent phosphorylation sites, and 14 transmembrane domains. rCNT1 mRNA was found in intestine and kidney, but not in heart, spleen, lung, liver, or skeletal muscle (Huang et al., 1994).

The sequence for rCNT2 (SPNT) shows significant sequence homology with rCNT1. Expression of rCNT2 (SPNT) in *Xenopus* oocytes resulted in Na^+/nucleoside cotransport activity characteristic of N2/*cit*, and the predicted sequence suggests a protein of 659 amino acids, 5 potential N-linked glycosylation sites, 14 transmembrane domains, one ATP/GTP binding site, and binding sites for protein kinases A and C. Regarding transporter distribution, rCNT2 (SPNT) mRNA was found in liver, jejunum, heart, brain, and skeletal muscle (Che et al., 1995).

A small component of Na^+-dependent nucleoside transport was observed in dissociated brain cells from rat and guinea pig, but not from mouse (Johnston and Geiger, 1989; Geiger and Fyda, 1991). Primary cultures of neurons and astrocytes from rat demonstrated Na^+-dependent transport (Hösli and Hösli, 1988), and those from mouse demonstrated concentrative transport, although the Na^+-dependence was not examined (Hertz, 1991). Additional evidence has been presented that Na^+-dependent adenosine transporters may be present on brain astrocytes (Bender et al., 1994) and dorsal brain stem synaptosomes (Lawrence et al., 1994). Recently, mRNA for rCNT1 and rCNT2 (SPNT) was detected using reverse transcriptase polymerase chain reaction in total RNA isolated from various regions of rat brain including choroid plexus, posterior hypothalamus, superior colliculus, brain stem, striatum, hippocampus, cerebellum, and cortex (Anderson et al., 1996). While mRNA for rCNT2 (SPNT) appeared to be of uniform abundance, mRNA levels for rCNT1 varied widely among the brain regions tested. Thus, if mRNA levels mirror the distribution and abundance of transporter proteins, Na^+-coupled nucleoside transport may form a significant component of adenosine uptake in some neural cells.

Functional Considerations Functional aspects of nucleoside transport have been extensively reviewed elsewhere (Plagemann and Wohlhueter, 1980; Paterson et al., 1981; Young and Jarivs, 1983; Wu and Phillis, 1984; Geiger and Fyda, 1991). The lack of inhibitors for *ei* equilibrative transporters and Na^+/nucleoside transporters has delayed progress in determining the degree to which these systems participate in the regulation of levels and actions of endogenous adenosine. Consequently, most studies focus on *es* transporters for which potent and selective inhibitors have been identified. At least in some cells, adenosine influx through transport processes, appears to be more rapid than its subsequent intracellular metabolism (Paterson et al., 1985; Geiger and Fyda, 1991) and if initial rate conditions are not met, phosphorylation of adenosine by AK, with its high affinity for adenosine, will lead to trapping of radiolabeled adenosine as phosphorylated membrane-impermeable nucleotides. Therefore, under prolonged incubation conditions, transporter affinity (K_T) values for adenosine accumulation may reflect the activity of AK. Alternatively, deamination of adenosine by ADA may affect transport measurements whereby intracellularly accumulated adenosine and inosine may act as competitive substrates for transport processes, thereby affecting both influx and efflux of adenosine. Inhibitors of AK and ADA can be used to stop the metabolism and aid in measurements of transport. However, because some of these inhibitors (e.g., DCF and ITU) may serve as inhibitors of or permeants for nucleoside transporters (Parkinson and Geiger, 1996), it is necessary to differentiate between a metabolic effect and competitive inhibition for the transport process. Alternate, more metabolically stable permeants, including formycin B, the inosine analog, and L-adenosine, the enantiomer of physiological D-adenosine, have been used. Although L-adenosine may be a substrate for passive and, to a far lesser extent, facilitated diffusion systems in mouse erythrocytes and L-1210 cells (Gati et al., 1989), more recent studies have shown that L-adenosine uptake by rat brain synaptosomes or syrian hamster smooth muscle DDT_1-MF_2 cells is inhibited by NBI and DPR (Gu et al., 1991; 1995; Gu and Geiger, 1992; Foga et al., 1996).

Nucleobase Transport Purine and pyrimidine nucleobases (hypoxanthine and adenine) are transported into cells and play an important role in purine and pyrimidine salvage. However, in contrast to nucleosides, far less is known about nucleobase transport systems. Nevertheless, it is clear that, as with nucleoside transporters, a family of nucleobase transporters are expressed in cells and tissues. Nucleobase transporters, including equilibrative and Na^+-dependent systems, can exhibit various degrees of permeant selectivity.

A number of Na^+-independent equilibrative systems have been characterized in astrocytes, endothelial cells, erythrocytes, cell lines, and brush-border membrane vesicles (Barros, 1994; Ceballos and Rubio, 1995). The K_m value for Na^+-independent hypoxanthine transport was 124 μM (Griffith and Jarvis, 1993). Hypoxanthine transport was inhibited by the nucleoside transport inhibitors papaverine, dilazep, and DPR with IC_{50} values in the low μM range (Plagemann and Kraupp, 1986; Kraupp et al., 1994).

The Na^+-dependent system, at least in cultured renal epithelial cells, appears to have a higher affinity for hypoxanthine; K_m value for Na^+-dependent transport was 0.79 μM (Griffith and Jarvis, 1993). Na^+-dependent transport had a hypoxanthine : Na^+ coupling stoichiometry of 1 : 1 in cultured renal epithelial cells and 2 : 1 in renal brush-border vesicles (Griffith and Jarvis, 1993, 1994). Na^+-dependent hypoxanthine transport was blocked by low micromolar concentrations of DPR and dilazep, and high micromolar concentrations of the inhibitor of Na^+-dependent glucose transport, phloridzin (Griffith and Jarvis, 1993). Sodium-dependent hypoxanthine transport in guinea pig placenta perfused *in situ* was blocked by high concentrations of papaverine [Barros, 1994]. Thus, multiple nucleobase transporters appear to be expressed in animal cells, and nucleoside transport inhibitors can block nucleobase transport at pharmacologically relevant concentrations.

Release

Adenosine release remains an often studied, yet ill-defined, process. With respect to adenosine release as a target for REAL agents, it is important to recognize that:

1. Adenosine is released under basal conditions and by a variety of stimuli.
2. Adenosine can be released from virtually all cells and tissues.
3. Adenosine release may be either calcium-dependent or calcium-independent.
4. Adenosine may be released as such, or extracellular adenosine may be derived from released/metabolized adenine nucleotides.
5. The degree to which extracellular adenosine originates as released adenosine or adenine nucleotides varies both qualitatively and quantitatively.

Adenine Nucleotide Release The nucleotide ATP, and to a lesser degree cAMP, represent sources for extracellular adenosine. Nucleotides may be dephosphorylated intracellularly and adenosine *per se* can be released, or alternatively, adenine nucleotides may be released by a variety of mechanisms and then dephosphorylated extracellularly. ATP may be released from secretory vesicles in nerves, where it is stored with other transmitter-like substances; or it can be released from muscle under physiological and ischemic conditions through the multidrug resistance (mdr1) gene product P-glycoprotein, as well as other members of the ABC family of transporter proteins (Plesner, 1995). Further, ATP can be released as a result of what have been referred to as nonspecific permeability changes, as well as from cells undergoing lysis. The release of cAMP occurs, at least in brain astrocytes, by a probenecid-sensitive mechanism that can be inhibited by DPR (Rosenberg et al., 1994).

Adenosine Release Adenosine can be released from cells and tissues in response to such diverse signals as high K^+, veratridine, glutamate receptor agonists, electrical stimulation, glucose and oxygen deprivation, ischemia, and hypertonic NaCl (White and Hoehn, 1991; Manzoni et al., 1994; Baudourin-Legros et al., 1995). Although at least a portion of adenosine release from nerve preparations has been shown to be calcium-dependent, adenosine has not been shown to be contained in nor released from synaptic vesicles (Cahill et al., 1993). The calcium required for calcium-dependent adenosine release originates from extracellular as well as intracellular stores (Ballerini et al., 1993, 1995). The mechanism most commonly implicated in adenosine release is efflux through bidirectional nucleoside transporters, although the bidirectional nature of nucleoside flux has been characterized in relatively few cells and tissues, and most of this work predates the discovery of many of the nucleoside transporters now identified.

Bidirectional flux studies of nucleosides have shown that the maximum velocities for efflux and influx may be equal and that permeants of a transporter can trans-accelerate the release of other permeants. However, rates of efflux have been observed that are greater than rates of influx (Cabantchik and Ginsburg, 1977). Such findings raise the possibility that directional symmetry or lack thereof may vary depending on physiological and pathological conditions. Furthermore, where symmetry of transport is present, extracellular adenosine originating from dephosphorylated adenine nucleotides may trans-accelerate the release of intracellular adenosine and thus act as an adenine nucleotide–induced adenosine release mechanism.

Adenosine efflux can occur via *es* transporters. For example, nucleoside influx and efflux occurred via *es* transporters on human, sheep, and guinea pig erythrocytes, and NBI and DPR inhibited nucleoside influx and efflux with equal potency (Jarvis et al., 1982). It is not known whether nucleoside efflux occurs through *ei*-mediated transporters, or the extent to which it occurs. The Na^+-dependent transporters identified to date appear to function as symporters, mediating Na^+-coupled inward movements of nucleosides. However, by reducing the transmembrane Na^+ gradient, nucleoside release can be observed (Borgland and Parkinson, in press). In addition, it remains possible that as-yet-unidentified Na^+-nucleoside transporters may function as antiporters.

Differentiating Between Release of Adenine Nucleotide and Adenosine Release The sources for basal as well as stimulation-induced increases in levels of extracellular adenosine have been determined using nucleoside transport and ecto-5′-nucleotidase inhibitors. Decreased adenosine release following treatment with the prototypic nucleoside transport inhibitors DPR, NBI, mioflazine, and dilazep has been interpreted as evidence for efflux of adenosine per se through what have been presumed to be bidirectional symmetrical nucleoside transporters. Decreased release of adenosine following application of AOPCP has been interpreted as evidence of extracellular metabolism of released adenine nucleotides to adenosine. However, some care must be taken with such pharmacological approaches. Consider, for example, that DPR has limited solubility and has multiple mechanisms of action, especially at higher concentrations; its potency in blocking transport (influx) in rat is much less than in human, guinea pig, and rabbit; and rarely have the relative potencies for inhibition of uptake and release been determined in the same preparation. Similarly, AOPCP may not completely inhibit ecto-5′-N [Meghji et al., 1988; MacDonald and White, 1985).

Summary Although nucleoside transporters may be mediating adenosine release, the following should be considered:

1. Nucleoside transport processes may not be symmetrical and inhibitors may selectively inhibit adenosine influx or efflux.

2. Multiple transport processes, not equally sensitive to the inhibitors, may coexist on a single cell.
3. Influx (Na^+-dependent) and bidirectional (Na^+-independent) processes may coexist.
4. Poorly characterized nonvesicular release processes for adenine nucleotides may be inhibited by nucleoside transport inhibitors (as seen with the probenecid- and DPR-sensitive cAMP release from glia).
5. Extracellular adenosine may be derived from different sources depending on tissues and physiological conditions.

REGULATORS OF ENDOGENOUS ADENOSINE LEVELS

A few different strategies have been employed in attempting to increase levels of endogenous adenosine and thereby its actions. In terms of a therapeutic strategy, increased levels of adenosine would best occur where and when cells and tissues need it most. Thus, the beneficial actions of adenosine might be amplified. Here, we review in brief some of the literature on the use of inhibitors of ADA, AK, or adenosine transport as REAL agents.

Therapeutic Utility of AK and ADA Inhibitors

A number of compounds have been reported to inhibit AK. The most potent of these are active at nanomolar concentrations, and include the purine nucleoside, 5′-amino-5′-deoxyadenosine (5′-NH_2dAdo) (Miller et al., 1979b), and the pyrrolopyrimidine nucleosides ITU (Wotring et al., 1979) and 5′-deoxy-5-iodotubercidin (d-ITU) (Davies et al., 1984). ITU is also a substrate for AK, and the phosphorylated form may inhibit other enzymes of nucleotide metabolism. d-ITU, which lacks the 5′-hydroxyl group, is unlikely to undergo phosphorylation by AK, and may therefore be a more selective AK inhibitor. More recently, several series of pyrrolo- and pyrazolo pyrimidine nucleosides related to ITU and d-ITU have been described (Cottam et al., 1993). Although ITU is a potent (5 nM) and specific inhibitor of AK, recent findings that ITU inhibited Ser/Thr specific protein kinases (IC_{50} value 0.4 μM) and insulin receptor tyrosine kinase (IC_{50} value 28 μM) suggest that ITU may cause a general decrease in cellular phosphorylation reactions (Massillon et al., 1994). Further, ITU was recently found to inhibit nucleoside transport in DDT_1 MF_2 smooth muscle cells with IC_{50} values in the low micromolar range (Parkinson and Geiger, 1996).

The most widely used inhibitors of ADA are DCF (Pentostatin™) and erythro-9-(2-hydroxy-3-nonyl) adenine (EHNA) (Agarwal et al., 1979; Schaeffer and Schwender, 1974). DCF inhibits ADA at picomolar concentrations and EHNA at nanomolar concentrations; both are quite selective for ADA (Skolnick et al., 1978; Padua et al., 1990; Geiger et al., 1990). DCF is a noncompetitive, tight-binding, and essentially irreversible inhibitor (Skolnick et al., 1978; Agarwal, 1979) that has been used for the treatment of patients with hairy cell leukemia (Kane et al., 1992). EHNA is a semi-tight-binding competitive inhibitor of ADA (Skolnick et al., 1978). Analogs of EHNA with ADA inhibitory activity have been described (Harriman et al., 1992; Vargeese et al., 1994).

The *in vivo* activity of AK inhibitors has been described using d-ITU (Davies et al., 1984, 1986). This compound potently inhibited [^{3}H]adenosine accumulation in guinea pig brain slices *in vitro* by virtue of its ability to inhibit AK. It did not inhibit ADA activity. Both d-ITU and/or ITU have been shown to decrease spontaneous locomotor activity and body temperature in mice, and to reduce heart rate and blood pressure in rats (Davies et al., 1986). These effects were blocked by adenosine receptor antagonists. In addition, low doses of d-ITU, which by themselves had no effect, potentiated the effects of adenosine on heart rate and blood pressure (Davies et al., 1986). These findings are consistent with the notion that AK inhibition potentiates physiological actions in the body that are governed, at least in part, by endogenous adenosine.

In terms of neuroprotection, administration of 5′-NH_2dAdo and ITU protected against bicuculline methiodide–induced seizures in rat, and AK inhibition was found to be more efficacious than was ADA inhibition with DCF (Zhang et al., 1993). However, the maximal efficacy of DCF was increased about three times when it was co-injected with 5′-NH_2dAdo at a dose that by itself had no effect on seizure activity. Thus, even minimal blockade of AK activity resulted in shunting of adenosine metabolism towards ADA. Systemically administered d-ITU, but not ITU or 5′-NH_2dAdo, protected mice and rats against pentylenetetrazol (PTZ)-induced seizures; this effect was blocked by the centrally acting A_1 adenosine receptor antagonist, cyclopentyltheophylline (Kowaluk et al., 1996). The lack of effect of ITU and 5′-NH_2dAdo may result from their inability to penetrate the blood-brain barrier. Electroshock seizures, and to a lesser extent PTZ-induced seizures, were inhibited by the AK inhibitor GP683 (Weisner et al., 1995); however, both GP683 and d-ITU induce hypothermia (mouse > rats) and impair spontaneous locomotor activity at anticonvulsant doses, and the separation of these effects remains a challenge. ITU did not protect against cerebral ischemic damage in gerbil (Phillis and Smith-Barbour, 1993); however, ITU may not penetrate the blood-brain barrier as noted above. Some evidence has been presented that DCF effectively protects against focal ischemic damage in rat when administered prior to, but not after onset of, ischemia (Lin and Phillis, 1992). These data suggest that inhibitors of AK and ADA may find use in treatment of seizure and ischemia-related disorders.

Adenosine kinase inhibitors may be more effective than ADA inhibitors as antinociceptive agents. Intrathecal administration of 5′-NH_2dAdo induced antinociception and enhanced opioid-induced antinociception in mice; these effects were blocked by theophylline (Keil and DeLander, 1992). Although DCF and EHNA did not produce significant antinociception nor did they enhance opioid-induced antinociception (Keil and DeLander, 1992, 1994), both 5′-NH_2dAdo and DCF enhanced adenosine-induced antinociception (Keil and DeLander, 1994). 5′-NH_2dAdo, but not DCF, significantly reduced formalin-induced nociceptive responses in rat (Poon and Sawynok, 1995). Overall, these observations appear to suggest that, at least in terms of pain sensation, ADA assumes greater importance when adenosine levels are artificially elevated by the administration of adenosine. It remains to be determined whether systemically administered enzyme inhibitors show therapeutic promise in the absence of undesirable side-effects.

Adenosine kinase inhibitors may also act as potent anti-inflammatory agents. GP-1-515 significantly improved rodent survival following endotoxin-induced and bacterial peritonitis–induced septic shock via multifaceted mechanisms that included adenosine-mediated inhibition of neutrophil adhesion, TNF-α production, and generation of free radicals (Firestein et al., 1994). GP-1-515 also effectively decreased carrageenan-induced paw edema and skin lesions induced by a variety of inflammatory agents (Rosengren et al., 1995). GP-1-515-induced inhibition of neutrophil accumulation in inflamed skin was blocked by the adenosine A_2 receptor antagonist 3,7-dimethyl-1-propargylxanthine (DMPX), but not by the adenosine A_1 receptor antagonist 8-cyclopentyl-1,3-dipropylxanthine (DPCPX), observations that suggest that the effects of GP-1-515 were mediated by endogenous adenosine acting at A_2 receptors. GP-1-515 apparently exerted these anti-inflammatory effects at doses that did not affect cardiovascular functions, thus lending support to the concept of site- and event-specific elevation of adenosine via AK inhibition (Rosengren et al., 1995). A novel series of pyrazolo[3,4-*d*]pyrimidine AK inhibitors with anti-inflammatory activity was also recently described (Cottam et al., 1993). In the immune system, ADA deficiency is associated with severe combined immunodeficiency disease (SCID) apparently resulting from the accumulation of lymphotoxic concentrations of 2′-deoxyadenosine and deoxy-ATP (Giblett et al., 1972; Hirschorn et al., 1980). This forms the basis for the use of DCF in lymphoproliferative disorders. Methotrexate, sulfasalazine, sodium salicylate, and azathioprine may also exert their anti-inflammatory activity, at least in part, by enhancing extracellular adenosine levels (Cronstein et al., 1995). Inhibition of AICAR transformylase might represent the site of action.

The effects of ADA inhibition have been widely evaluated for a variety of conditions with mixed results (see Geiger and Nagy, 1990; Geiger et al., 1990, 1991). Where protective actions have been noted it is not always clear that the effects were mediated through increased actions of endogenous adenosine working through cell surface adenosine receptors. EHNA enhanced the preservation of function after hypothermic storage of isolated rat hearts (Zhu et al., 1994). DCF augmented levels of endogenous adenosine in myocardial interstitial fluid, decreased neutrophil accumulation, and reduced postischemic dysfunction of the heart resulting from cardiopulmonary bypass and 30-minute global ischemia in dogs (Hudspeth et al., 1994). Administration of DCF pre- or postinsult protected against CNS damage induced by a hypoxic/ischemic insult in perinatal rats (Gidday et al., 1995). DCF also significantly reduced infarct size in a swine coronary artery occlusion/reperfusion model (Martin et al., 1993), and reduced infarct size in Langendorff rabbit hearts (Freeman et al., 1994). At least in the latter study, the effects of DCF were not blocked by the adenosine receptor antagonist 8-sulfophenyltheophylline, and this suggests that the effects of DCF were not mediated through cell surface adenosine receptors. Although DCF has been shown to increase levels of adenosine in dog cardiac interstitial fluid during regional myocardial ischemia/reperfusion, DCF failed to reduce myocardial infarct size (Silva et al., 1995).

Overall, AK and ADA inhibitors have been shown to increase levels of endogenous adenosine and under certain circumstances augment the actions of adenosine. It appears clear that the relative contribution of AK and ADA to the maintenance of the levels of endogenous adenosine vary with cell type and condition. Pharmaceutical development of inhibitors of ADA and AK will require some resolution of these issues.

Therapeutic Utility of Adenosine Transport Inhibitors

Inhibitors of nucleoside transport processes have been proposed as therapeutic agents to increase the levels and actions of endogenous adenosine. However, even though adenosine transport inhibitors have been shown to enhance the apparent effects of adenosine and, at least under certain conditions, increase the levels of endogenous adenosine, the extent to which the actions of these inhibitors are adenosine receptor–mediated is not always clear. The proliferating "family" of nucleoside transporters both complicates attempts to manipulate specific transporters pharmacologically and opens up a new frontier of pharmaceutical development aimed at selectively manipulating these transporters to achieve specific therapeutic benefit.

Nucleoside transport inhibitors represent a structurally diverse group of compounds sharing at least one common mechanism of action. Structurally, there are at least five distinct types of nucleoside transport inhibitors: purine ribosides (e.g., NBI), pyrimidopyrimidine derivatives (e.g., DPR), substituted piperazines (e.g., mioflazine), tertiary amine diazepines (e.g., dilazep), and xanthines (e.g., propentofylline).

The purine derivative NBI (S^6-(4-nitrobenzyl)-6-thioinosine) continues to be an important tool for the study and characterization of nucleoside transporters. NBI is not a clinically used drug, and chronically administered NBI may lead to the accumulation of its immunosuppressive and toxic metabolite 6-mercaptopurine (see Geiger and Fyda, 1991). It is a very potent and selective inhibitor of the *es*-type nucleoside transporter.

The pyrimidinopyrimidine derivative DPR (2,6-*bis*(diethanolamino)-4,8-diperidino-pyrimido-(5,4-*d*)-pyrimidine) continues to be used clinically for dilatation of coronary and cerebral arteries, platelet antiaggregation, and in adjunct therapy to enhance the pharmacological effects of modified nucleosides, including 3′-azido-3′-deoxythymidine (AZT) and 2′,3′-dideoxycytidine (ddC). DPR, although best known as an inhibitor of carrier-mediated nucleoside transport, has been shown to have other actions, including inhibition of cGMP phosphodiesterase activity (see Bult et al., 1991) and reactive oxygen species formation (Iuliano et al., 1995).

Dilazep (1,4-*bis*[3-(3,4,5-trimethoxy benzoyloxy)propyl]perhydro-1,4-diazepine) vasodilates, decreases coronary and total vascular resistance, and increases coronary blood flow (Marzilli et al., 1984). Hydrolase inhibitors must be used in conjunction with dilazep to slow metabolism by ester hydrolases (Ijzerman and Voorschuur, 1990). Although quite a selective inhibitor of nucleoside transport, dilazep has also been shown to be a calcium antagonist (Tonini et al., 1982).

Mioflazine (3-(aminocarbonyl)-4-[4,4-*bis*(4-fluorophenyl)butyl]-N-(2,6-dichlorophenyl)-1-piperazineacetamide • 2HCl) inhibits adenosine transport, vasodilates, and (possibly through its ability to cross the blood-brain barrier) causes sedation and increases sleep (Wauquier et al., 1987).

Propentofylline (3-methyl-1-(5′-oxohexyl)-7-propylxanthine) inhibits most potently *es* transporters, but at much higher concentrations blocks *ei* and N1/*cif* transport (Parkinson et al., 1993). In addition to inhibition of adenosine transport, propentofylline is an antagonist at A_1, A_{2a}, and A_{2b} adenosine receptors, inhibits rolipram-sensitive cAMP phosphodiesterase IV, inhibits free radical formation, and stimulates the production of nerve growth factor. Propentofylline protects against ischemic neuronal injury (Parkinson et al., 1994).

A great deal of work has been conducted on the *in vivo* and *in vitro* pharmacological effects of nucleoside transport inhibitors, which are reviewed in part elsewhere (Geiger and Fyda, 1991). Adenosine transport inhibitors have been shown to potentiate a variety of physiological processes controlled, at least in part, by actions of adenosine and adenosine receptors. These include blood flow, cardiac function, animal locomotion, seizure activity, and neuronal firing. Of course, much depends on the specificity of the inhibitor's actions and the degree to which adenosine is principally involved in mediating the physiological response.

The levels of endogenous adenosine have been found to be increased, decreased, or unaffected by adenosine transport inhibitors. Transport inhibitors, by blocking adenosine influx, might be expected to enhance the extracellular levels of adenosine if the extracellular adenosine originated as released adenine nucleotide, or if the inhibitor was selective in decreasing the uptake but not the release of adenosine through bidirectional transporters. Alternatively, transport inhibitors might decrease extracellular levels of adenosine if adenosine is formed intracellularly and released through bidirectional transporters, and if the inhibitor blocked release as well as uptake. For example, adenosine transport inhibitors including DPR, lidoflazine, propentofylline, and NBI increased basal levels of endogenous adenosine in brain (Ballarin et al., 1991; Pazzagli et al., 1993; Park and Gidday, 1990). They also enhanced peak interstitial adenosine concentrations following a brief ischemic episode and slowed the postischemia drop in adenosine levels (Andine et al., 1990; Park and Gidday, 1990), inhibited the uptake to a greater extent than the release (Fredholm et al., 1994), and increased levels in coronary sinus plasma (Conti et al., 1995). Moreover, adenosine transport inhibitors increased interstitial levels of adenosine in cardiac dialysate (Wang et al., 1992), guinea pig heart exudates (Decking et al., 1988; Mohrman and Hiller, 1990), cultured cardiac myocytes (Mustafa et al., 1979; Bukoski and Sparks, 1986), and in isolated perfused guinea pig hearts by up to 29-fold (Wangler et al., 1989). However, studies using cultured myocytes (Meghji et al., 1988) and brain preparations (see Geiger and Fyda, 1991) have shown that adenosine transport inhibitors can decrease adenosine efflux.

In brain, adenosine transport inhibitors, at pharmacologically relevant concentrations, modulate a number of processes measured *in vitro* and *in vivo.* These include seizures, animal locomotion, neuronal firing, blood flow, and antinociception (see Geiger and Fyda, 1991). More recently, it was found that propentofylline decreased infarct size and behavioral manifestations following focal ischemia (Park and Rudolphi, 1994), and protected against hypoxic/ischemic injuries to CNS tissues by a theophylline-sensitive mechanism (Fern et al., 1994; Gidday et al., 1995). Further support for the notion that the effects of transport inhibitors were mediated through adenosine receptors comes from findings that theophylline blocked DPR-induced cerebral vasodilatation and hypotension in normotensive and hypertensive animals (Hegedus et al., 1993).

In peripheral tissues, nucleoside transport inhibitors exhibit a number of effects that are therapeutically relevant: for example, dilazep decreased ischemia- and reperfusion-induced myocardial damage as well as lysophosphatidylcholine-induced myocardial damage in heart (Hoque et al., 1995a,b); NBI protected against reperfusion injury (Abd-Elfattah et al., 1993); R75231 induced a preconditioning-like protection against ischemia-induced cardiac dysfunction (Kirkeboen et al., 1994); DPR, dilazep, and R75231 mimicked the infarct-reducing effects of cardiac preconditioning (Itoya et al., 1994); propentofylline and DPR inhibition of adenosine transport enhanced the inhibitory actions of adenosine on neutrophil activation as measured by H_2O_2 production (Zhang and Fredholm, 1994). That these effects were mediated through adenosine receptors is suggested by work that showed, for example, that NBI-induced functional recovery of hypothermically stored rat hearts and the effects of dilazep on blood flow were mediated by adenosine A_1 receptors (Zhang et al., 1991b,c; Yang et al., 1994; Fremes et al., 1995). In humans, draflazine potentiated adenosine-induced vasodilation and increased respiratory ventilation, and caffeine blocked draflazine-induced increases in heart rate, blood pressure, and plasma levels of norepinephrine and epinephrine (Rongen et al., 1995a,b). However, some evidence has been presented that the cardioprotective effects of R75231 and lidoflazine were not mediated through adenosine A_1 receptors (Grover and Sleph, 1994).

Summary

The identification of REAL agents and elucidation of their mechanisms of action is made difficult because of the complexities of purine biochemistry, the ill-defined nature of adenosine release processes, and the ever-increasing number of identified adenosine transporters. Nevertheless, progress continues to be made in the development of REAL agents, and this must be viewed as encouraging for the pharmaceutical development of purinergic agents and for the research community whose efforts are focused on determining mechanisms of action.

ACKNOWLEDGMENTS

J.D.G. would like to thank Ms. Suzanne Delaney and Ms. Irene Foga for their assistance in various aspects related to the preparation of this manuscript. J.D.G. and F.E.P. gratefully acknowledge operating grant support from the Medical Research Council of Canada (MRC), the University of Manitoba Faculty Fund, the Health Science Centre Research Foundation, and the Manitoba Health Research Council. J.D.G. is a Scientist and F.E.P. is a Scholar of the MRC.

REFERENCES

Abd-Elfattah AS, Ding M, Dyke CM, Wechsler AS (1993): Protection of the stunned myocardium. Selective nucleoside transport blocker administered after 20 minutes of ischemia augments recovery of ventricular function. Circulation 88:336–343.

Agarwal RP (1979): Recovery of 2′-deoxycoformycin inhibited adenosine deaminase of mouse erythrocytes and leukemia L1210 in vivo. Cancer Res 39:1425–1426.

Agarwal RP, Spector T, Parks RE (1977): Tight binding inhibitors, IV. Inhibition of adenosine deaminases by various inhibitors. Biochem Pharmacol 26:359–367.

Anderson CM, Xiong W, Young JD, Cass CE, Parkinson FE (1996): Demonstration of the existence of mRNAs encoding N1/*cif* and N2/*cit* sodium/nucleoside cotransporters in rat brain. Mol Brain Res 42:358–361.

Andine P, Rudolphi KA, Fredholm BB, Hagberg H (1990): Effect of propentofylline (HWA 285) on extracellular purines and excitatory amino acids in CA1 of rat hippocampus during transient ischemia. Br J Pharmacol 100:814–818.

Andres CM, Fox IH (1979): Purification and properties of human placental adenosine kinase. J Biol Chem 254:11388–11393.

Arch JRS, Newsholme EA (1978): Activities and some properties of 5′-nucleotidase, adenosine kinase, and adenosine deaminase in tissue from vertebrates and invertebrates in relation to the control of the concentration and the physiological role of adenosine. Biochem J 174:965–977.

Ballarin M, Fredholm BB, Ambrosio S, Mahy N (1991): Extracellular levels of adenosine and its metabolites in the striatum of awake rats: Inhibition of uptake and metabolism. Acta Physiol Scand 142:97–103.

Ballerini P, Ciccarelli R, Di Iorio P, Giuliani P, Caciagli F (1995): Influence of Ca^{2+} channel modulators on [^{3}H]purine release from rat cultured glial cells. Neurochem Res 20:697–704.

Ballerini P, Ciccarelli R, Di Iorio P, Giuliani P, Francano D, Fano G, Caciagli F (1993): TMB-8 and thapsigargin modulate purine release from dissociated primary cultures of rat brain astrocytes. Res Commun Chem Pathol Pharmacol 82:167–174.

Baranowski RL, Westenfelder C (1994): Estimation of renal interstitial adenosine and purine metabolites by microdialysis. Am J Physiol 267:F174–F182.

Barros LF (1994): Hypoxanthine transport in the guinea pig and human placenta is a carrier-mediated process that does not involve nucleoside transporters. Am J Obstet Gynecol 171:111–117.

Barros LF, Yudilevich DL, Jarvis SM, Beaumont N, Young JD, Baldwin SA (1995): Immunolocalization of nucleoside transporters in human placental trophoblast and endothelial cells: Evidence for multiple transporter isoforms. Pflugers Arch Eur J Physiol 429:394–399.

Baudourin-Legros M, Badou A, Paulais M, Hammet M, Teulon J (1995): Hypertonic NaCl enhances adenosine release and hormonal cAMP production in mouse thick ascending limb. Am J Physiol 268:F103–F109.

Belt JA, Noel LD (1985): Nucleoside transport in Walker 256 rat carcinosarcoma and S49 mouse lymphoma cells. Biochem J 232:681–688.

Bender AS, Woodbury DM, White HS (1994): Ionic dependence of adenosine uptake into cultured astrocytes. Brain Res 661:1–8.

Berlin RD, Oliver JM (1975): Membrane transport of purine and pyrimidine bases and nucleosides in animal cells. Int Rev Cytol 42:287–336.

Bontemps F, Vincent MF, van den Berghe G (1993): Mechanisms of elevation of adenosine levels in anoxic hepatocytes. Biochem J 290:671–677.

Borgland SL, Parkinson FE: Uptake and release of [^{3}H]formycin B via sodium-dependent nucleoside transporters in mouse leukemic L1210/MA27.1 cells. J Pharmacol Exp Ther (in press).

Boumah CE, Harvey CM, Paterson ARP, Baldwin SA, Young JD, Cass CE (1994): Functional expression of the nitrobenzylthioinosine-sensitive nucleoside transporter of human choriocarcinoma (BeWo) cells in isolated oocytes of Xenopus laevis. Biochem J 299:769–773.

Bukoski RD, Sparks HV Jr (1986): Adenosine production and release by adult rat cardiocytes. J Mol Cell Cardiol 18:595–605.

Bult H, Fret HRL, Jordaens FH, Herman AG (1991): Dipyridamole potentiates the antiaggregating and vasodilator activity of nitric oxide. Eur J Pharmacol 199:1–8.

Bullough DA, Zhang C, Montag A, Mullane KM, Young MA (1994): Adenosine-mediated inhibition of platelet aggregation by acadesine. A novel antithrombic mechanism in vitro and in vivo. J Clin Invest 94:1524–1532.

Cabantchik ZI, Ginsburg H (1977): Transport of uridine in human red blood cells. Demonstration of a simple carrier-mediated process. J Gen Physiol 69:75–96.

Cahill CM, White TD, Sawynok J (1993): Influence of calcium on the release of endogenous adenosine from spinal cord synaptosomes. Life Sci 53:487–496.

Cass CE (1995): Nucleoside transport. In Georgopapadakou NH (ed): "Drug Transport in Antimicrobial and Anticancer Chemotherapy." New York: Marcel Dekker, pp 403–451.

Cass CE, Gaudette LA, Paterson ARP (1974): Mediated transport of nucleosides in human erythrocytes. Specific binding of the inhibitor nitrobenzylthioinosine to nucleoside transport sites in the erythrocyte membrane. Biochim Biophys Acta 345:1–10.

Ceballos G, Rubio R (1995): Coculture of astroglial and vascular endothelial cells as apposing layers enhances the transcellular transport of hypoxanthine. J Neurochem 64:991–999.

Centelles JJ, Franco R (1990): Is adenosine deaminase involved in adenosine transport? Med Hypotheses 33:245–250.

Chagoya de Sanchez V (1995): Circadian variation of adenosine and of its metabolism. Could adenosine be a molecular oscillator for circadian rhythms? Can J Physiol Pharmacol 73:339–355.

Chan TS, Creagan RP, Reardon MP (1978): Adenosine kinase as a new selective marker in somatic cell genetics: Isolation of adenosine kinase-deficient mouse cell lines and human-mouse hybrid cell lines containing adenosine kinase. Somatic Cell Genet 4:1–12.

Chang CH, Cha S, Brockman RW, Bennett LL Jr (1983): Kinetic studies of adenosine kinase from L1210 cells: A model enzyme with a two-site ping-pong mechanism. Biochemistry 22:600–611.

Che M, Ortiz DF, Arias IM (1995): Primary structure and functional expression of a cDNA encoding the bile canalicular, purine-specific Na^+ nucleoside cotransporter. J Biol Chem 270:13596–13599.

Collis MG (1991): Effect of adenosine on the coronary circulation. In Phillis JW (ed): "Adenosine and Adenine Nucleotides as Regulators of Cellular Function." Boca Raton, FL: CRC Press, pp 171–180.

Conti JB, Belardinelli L, Utterback DB, Curtis AB (1995): Endogenous adenosine is an antiarrhythmic agent. Circulation 91:1761–1767.

Cottam HB, Wasson DB, Shih HC, Raychaudhuri A, Di Pasquale G, Carson DA (1993): New adenosine kinase inhibitors with oral antiinflammatory activity: Synthesis and biological evaluation. J Med Chem 36:3424–3430.

Cronstein BN (1995): A novel approach to the development of anti-inflammatory agents: Adenosine release at inflamed sites. J Invest Med 43:50–57.

Datta AK, Bhaumik D, Chatterjee R (1987): Isolation and characterization of adenosine kinase from Leishmania donovani. J Biol Chem 262:5515–5521.

Davies LP, Baird-Lambert J, Marwood JF (1986): Studies on several [2,3-d]pyrimidine analogues of adenosine which lack significant agonist activity at A1 and A2 receptors but have potent pharmacological activity in vivo. Biochem Pharmacol 35:3021–3029.

Davies LP, Jamieson DD, Baird-Lambert JA, Kazlauskas R (1984): Halogenated pyrrolopyrimidine analogues of adenosine from marine organisms: Pharmacological activities and potent inhibition of adenosine kinase. Biochem Pharmacol 33:347–355.

Decking UKM, Jeungling E, Kammermeier H (1988): Interstital transudate concentration of adenosine and inosine in rat and guinea pig hearts. Am J Physiol 254:H1125–H1132.

Delaney SM, Geiger JD (1995): NMDA-induced increases in levels of endogenous adenosine are enhanced by inhibitors of adenosine deaminase and adenosine transport in striatum of rats. Brain Research 702:72–76.

Delaney SM, Geiger JD (1996): Brain regional levels of adenosine and adenine nucleotides in rats killed by high energy focused microwave irradiation. J Neurosci Meth 64:151–156.

Deussen A, Borst M, Schrader J (1988): Formation of S-adenosylhomocysteine in the heart, I. An index of free intracellular adenosine. Circ Res 63:240–249.

Elalaoui A, Divita G, Maury G, Imbach JL, Goody RS (1994): Intrinsic tryptophan fluoresence of bovine liver adenosine kinase, characterization of ligand binding sites and conformational changes. Eur J Biochem 221:839–846.

Fern R, Waxman SG, Ransom BR (1994): Modulation of anoxic injury in CNS white matter by adenosine and interaction between adenosine and GABA. J Neurophysiol 72:2609–2616.

Firestein GS, Boyle D, Bullough DA, Gruber HE, Sajjadi FG, Montag A, Sambol B, Mullane KM (1994): Protective effect of an adenosine kinase inhibitor in septic shock. J Immunol 152:5853–5859.

Foga IO, Geiger JD, Parkinson FE (1996): Nucleoside transporter mediated uptake and release of $[^3H]$*L*-adenosine in DDT_1MF_2 smooth muscle cells. Eur J Pharmacol 318:455–460.

Fredholm BB, Sollevi A (1986): Cardiovascular effects of adenosine. Clin Physiol 6:1–21.

Fredholm BB, Lindstrom K, Wallman-Johansson A (1994): Propentofylline and other adenosine transport inhibitors increase the efflux of adenosine following electrical or metabolic stimulation of rat hippocampal slices. J Neurochem 62:563–573.

Freeman S, Mullane K, Appleman J, Young M (1994): Limitation of infarct size in rabbit hearts by adenosine deaminase inhibition is not mediated via adenosine receptors. J Amer Coll Cardiol 232A.

Fremes SE, Furukawa RD, Zhang J, Li RK, Tumiati LC, Weisel RD, Mickle DAG (1995): Cardiac storage with University of Wisconsin solution and a nucleoside-transport blocker. Ann Thorac Surg 59:1127–1133.

Fujiwara F, Hirai H, Tamaki N, Okada Y (1994): Adenosine enhances neuronal damage during deprivation of oxygen and glucose in guinea pig superior collicular slices. Neurosci Lett 182:283–286.

Gati WP, Dagnino L, Paterson ARP (1989): Enantiomeric selectivity of adenosine transport systems in mouse erythrocytes and L1210 cells. Biochem J 263:957–960.

Geiger JD, Fyda D (1991): Adenosine transport systems in the CNS. In Stone T (ed): "Adenosine in the Nervous System." London: Academic Press, pp 1–23.

Geiger JD, Nagy JI (1986): Distribution of adenosine deaminase activity in rat brain and spinal cord. J Neurosci 6:2707–2714.

Geiger JD, Nagy JI (1990): Adenosine deaminase and [^{3}H]nitrobenzylthioinosine as markers of adenosine metabolism and transport in central purinergic systems. In Williams M (ed): "Adenosine and Adenosine Receptors," Clifton, NJ: Humana Press, pp 225–288.

Geiger JD, Padua R, Nagy JI (1990): Adenosine deaminase and transport systems in rat CNS. In Jacobson KA, Daly JW, Manganiello V (eds): "Purines in Cellular Signalling: Targets for New Drugs." New York: Springer-Verlag, pp 20–25.

Geiger JD, Padua R, Nagy JI (1991): Adenosine deaminase regulation of purine actions. In Phillis JW (ed): "Purinergic Regulation of Cell Function." Boca Raton, FL, CRC Press, pp 71–84.

Giblett ER, Anderson JE, Cohen F, Pollara B, Meunissen HJ (1972): Adenosine-deaminase deficiency in two patients with severely impaired cellular immunity. Lancet 2:1067–1069.

Gidday JM, Fitzgibbons JC, Shah AR, Kraujailis MJ, Park TS (1995): Reduction in cerebral ischemic injury in the newborn rat by potentiation of endogenous adenosine. Ped Res 38:306–311.

Good AH, Craik JD, Jarvis SM, Kwong FY, Young JD, Paterson ARP, Cass CE (1987): Characterization of monoclonal antibodies that recognize band 4.5 polypeptides associated with nucleoside transport in pig erythrocytes. Biochem J 244:749–755.

Griffith DA, Jarvis SM (1993): High affinity sodium-dependent nucleobase transport in cultured renal epithelial cells (LLC-PK1). J Biol Chem 268:22085–22090.

Griffith DA, Jarvis SM (1994): Characterization of a sodium-dependent concentrative nucleobase-transport system in guinea-pig kidney cortex brush-border membrane vesicles. Biochem J 303:901–905.

Griffiths M, Beaumont N, Yao SYM, Sundaram M, Boumah CE, Davies A, Kwong FYP, Coe I, Cass CE, Young JD, Baldwin SA (1997): Cloning of a human nucleoside transporter implicated in the cellular uptake of adenosine and chemotherapeutic drugs. Nature Med 3:89–93.

Grover GJ, Sleph PG (1994): The cardioprotective effects of R75231 and lidoflazine are not caused by adenosine A1 receptor activation. J Pharmacol Exp Ther 268:90–97.

Gu JG, Geiger JD (1992): Transport kinetics and metabolism of accumulated [^{3}H]D-adenosine and [^{3}H]L-adenosine in rat cerebral cortical synaptoneurosomes. J Neurochem 58:1699–1705.

Gu JG, Delaney S, Sawka AN, Geiger JD (1991): [^{3}H]L-Adenosine, a new metabolically stable enantiomeric probe for adenosine transport systems in rat brain synaptoneurosomes. J Neurochem 56:548–552.

Gu JG, Foga I, Parkinson F, Geiger JD (1995): Involvement of bidirectional adenosine transporters in the release of L-[^{3}H]adenosine from rat brain synaptosomal preparations. J Neurochem 64:2105–2110.

Gutierrez MM, Giacomini KM (1993): Substrate selectivity, potential sensitivity and stoichiometry of Na^+-nucleoside transport in brush border membrane vesicles from human kidney. Biochim Biophys Acta 1149:202–208.

Gutierrez MM, Giacomini KM (1994): Expression of a human renal sodium nucleoside cotransporter in Xenopus laevis oocytes. Biochem Pharmacol 48:2251–2253.

Hammond JR, Clanachan AS (1985): Species differences in the binding of [^{3}H]nitrobenzylthioinosine in the nucleoside transport system in mammalian central nervous system membranes: Evidence for interconvertible conformations of the binding site/tranporter complex. J Neurochem 45:527–535.

Harriman GCB, Poirot AF, Abushanab E, Midgett RM, Stoeckler JD (1992): Adenosine deaminase inhibitors. Synthesis and biological evaluations of Cl′ and nor-Cl′ derivatives of (+)-erythro-9-(2(S)-hydroxy-3(R)-nonyl)adenine. J Med Chem 35:4180–4184.

Hawkins CF, Bagnara AS (1987): Adenosine kinase from human erythrocytes: Kinetic studies and characterization of adenosine binding sites. Biochemistry 26:1982–1987.

Hegedus K, Fekete I, Molnar L (1993): Effects of dipyridamole on hypertensive rabbits with diffuse chronic cerebral ischemia. Eur J Pharmacol 237:293–298.

Henderson JF, Mikoshiba A, Chu SY, Caldwall IC (1972): Kinetic studies of adenosine kinase from Ehrlich ascites tumor cells. J Biol Chem 247:11972–11975.

Hertz L (1991): Active transport of adenosine into primary cultures of brain cells and its methodological consequences. In Harkness RA (ed): "Purine and Pyrimidine Metabolism in Man VII." New York: Plenum Press, pp 399–402.

Heurteaux C, Lauritzen I, Widmann C, Lazdunski M (1995): Essential role of adenosine, adenosine A_1 receptors, and ATP-sensitive K^+ channels in cerebral ischemic preconditioning. Proc Natl Acad Sci USA 92:4666–4670.

Hirschorn R, Roegner V, Rubinstein A, Papageorgious PS (1980): Plasma deoxyadenosine, adenosine and erythrocyte deoxyATP are elevated at birth in an adenosine deaminase-deficient child. J Clin Invest 65:768–771.

Hösli E, Hösli L (1988): Autoradiographic studies on the uptake of adenosine and on binding of adenosine analogues in neurons and astrocytes of cultured rat cerebellum and spinal cord. Neuroscience 24:621–628.

Hoque ANE, Hoque N, Hashizume H, Abiko Y (1995a): A study on dilazep, I. Mechanism of anti-ischemic action of dilazep is not coronary vasodilation but decreased cardiac mechanical function in the isolated, working rat heart. Jpn J Pharmacol 67:225–232.

Hoque ANE, Hoque N, Hashizume H, Abiko Y (1995b): A study on dilazep, II. Dilazep attenuates lysophosphatidylcholine-induced mechanical and metabolic derangements in the isolated, working rat heart. Jpn J Pharmacol 67:233–241.

Huang QQ, Yao SYM, Ritzel MWL, Paterson ARP, Cass CE, Young JD (1994): Cloning and functional expression of a complementary DNA encoding a mammalian nucleoside transport protein. J Biol Chem 269:17757–17760.

Hudspeth DA, Williams MW, Zhao ZQ, Sato H, Nakanishi K, McGee DS, Hammon JW, Vinten-Johansen J, Van Wylen DGL (1994): Pentostatin-augmented interstitial adenosine prevents post cardioplegia injury in damaged hearts. Ann Thorac Surg 58:719–727.

IJzerman AP, Voorschuur AH (1990): The relationship between ionization and affinity of nucleoside transport inhibitors. Naunyn Schmiedebergs Arch Pharmacol 342:336–341.

Itoya M, Miura T, Sakamoto J, Urabe K, Iimura O (1994): Nucleoside transport inhibitors enhance the infarct size-limiting effect of ischemic preconditioning. J Cardiovasc Pharmacol 24:846–852.

Iuliano L, Pedersen J, Rotilio G, Ferro D, Violi F (1995): A potent chain-breaking antioxidant activity of the cardiovascular drug dipyridamole. Free Radical Biology and Medicine 18:239–247.

Jarvis SM, McBride D, Young JD (1982): Erythrocyte nucleoside transport: Asymmetrical binding of nitrobenzylthioinosine to nucleoside permeation sites. J Physiol 324:31–46.

Johnston ME, Geiger JD (1989): Sodium-dependent uptake of nucleosides by dissociated brain cells from rat. J Neurochem 52:75–81.

Kameoka J, Tanaka T, Nojima Y, Schlossman SF, Morimoto C (1993): Direct association of adenosine deaminase with a T cell activation antigen CD26. Science 261:466–469.

Kane BJ, Kuhn JG, Roush MK (1992): Pentostatin: An adenosine deaminase inhibitor for the treatment of hairy cell leukemia. Ann Pharmacother 26:393–947.

Keil GJ, DeLander GE (1992): Spinally-mediated antinociception is induced in mice by an adenosine kinase-, but not by an adenosine deaminase-inhibitor. Life Sci 51:171–176.

Keil GJ, DeLander GE (1994): Adenosine kinase and adenosine deaminase inhibition modulate spinal adenosine- and opioid agonist-induced antinociception in mice. Eur J Pharmacol 271:37–46.

Kirkeboen KA, Ilebekk A, Tonnessen T, Leistad E, Naess PA, Christensen G, Aknes G (1994): Cardiac contractile function following repetitive brief ischemia: Effect of nucleoside transport inhibition. Am J Physiol 267:H57–H65.

Kochan Z, Smolenski RT, Yacoub MH, Seymour AML (1994): Nucleotide and adenosine metabolism in different cell types of human and rat heart. J Mol Cell Cardiol 26:1497–1503.

Kowaluk EA, Cowart M (1994): The purification and characterization of adenosine kinase from rat brain. Drug Dev Res 31:287.

Kowaluk EA, Kohlhaas KL, Daanen J, Lee L (1996): Anticonvulsant effects of adenosine kinase inhibitors. Drug Dev Res 37:190.

Kraupp M, Paskutti B, Schon C, Marz R (1994): Inhibition of purine nucleobase transport in human erythrocytes and cell lines by papaverine. Biochem Pharmacol 48:41–47.

Kwong FYP, Wu JSR, Shi MM, Fincham HE, Davies A, Henderson PJF, Baldwin SA, Young JD (1993): Enzymic cleavage as a probe of the molecular structures of mammalian equilibrative nucleoside transporters. J Biol Chem 268:22127–22134.

Laudignon N, Beharry K, Rex J, Aranda JV (1990): Effect of adenosine on total and regional cerebral blood flow of the newborn piglet. J Cerebral Blood Flow Metab 10:392–398.

Lawrence AJ, Castillo-Melendez M, Jarrott B (1994): [^{3}H]Adenosine transport in rat dorsal brain stem using a crude synaptosomal preparation. Neurochem Int 25:221–226.

Leibach TK, Spiess GI, Neudecker TJ, Peschke GJ, Puchwein G, Hartmann GR (1971): Purification and properties of adenosine kinase from dried brewer's yeast. Hoppe-Seylers Z Physiol Chem 352:328–344.

Leone RJ, Merrill GF (1995): Inhibition of adenosine deaminase and administration of adenosine increase hypoxia induced ventricular ectopy. Basic Res Cardiol 90:234–239.

Lin BB, Hurley MC, Fox IH (1988): Regulation of adenosine kinase by adenosine analogs. Mol Pharmacol 34:501–505.

Lin Y, Phillis JW (1992): Deoxycoformycin and oxypurinol: Protection against focal ischemia brain injury in the rat. Brain Res 571:272–280.

Linden J, Taylor HE, Feldman MD, Woodward EB, Ayers CR, Ripley ML, Iflah S, Patel A (1992): The precise radioimmunoassay of adenosine: Minimization of sample collection artifacts and immuno cross reactivity. Anal Biochem 201:246–254.

Lloyd HGE, Fredholm BB (1995): Involvement of adenosine deaminase and adenosine kinase in regulating extracellular adenosine concentration in rat hippocampal slices. Neurochem Int 26:387–395.

Lobelle-Rich PA, Reeves RE (1983): The partial purification and characterization of adenosine kinase from entamoeba histolytica. Am J Trop Med Hyg 32:976–979.

Manzoni OJ, Manabe T, Nicoll RA (1994): Release of adenosine by activation of NMDA receptors in the hippocampus. Science 265:2098–2101.

Martin M, Aran JM, Colomer D, Huguet J, Centelles JJ, Vives-Corrons JL, Franco R (1995): Surface adenosine deaminase. A novel B-cell marker in chronic lymphocytic leukemia. Human Immunol 42:265–273.

Martin BJ, McClanahan TB, Steinkampf RW, Klohs WD, Gallagher KP (1993): Inhibition of adenosine deaminase with pentostatin reduces infarct size in swine. FASEB J 7:418A.

Marzilli M, Simonetti I, Levantesi D, Trivella MG, De Nes M, Perissonotto A, Puntoni R, Buzzigoli G, Boni C, Michelassi C (1984): Effects of dilazep on coronary and systemic hemodynamics in humans. Am Heart J 108:276–285.

Massillon D, Stalmans W, van de Werve G, Bollen M (1994): Identification of the glycogenic compound 5-iodotubercidin as a general protein kinase inhibitor. Biochem J 299:123–128.

Meghji P (1991): Adenosine production and metabolism. In Stone T (ed): "Adenosine in the Nervous System." London: Academic Press, pp 25–42.

Meghji P, Rubio R, Berne RM (1988): Intracellular adenosine formation and its carrier-mediated release in cultured embryonic chick heart cells. Life Sci 43:1851–1859.

Meno JR, Ngai AC, Ibayashi S, Winn HR (1991): Adenosine release and changes in pial arteriolar diameter during transient cerebral ischemia and reperfusion. J Cerebral Blood Flow Metab 11:986–993.

Mi Z, Jackson EK (1995): Metabolism of exogenous cyclic AMP to adenosine in the rat kidney. J Pharmacol Exp Therap 273:728–733.

Miller RL, Adamczyk DL, Miller WH (1979a): Adenosine kinase from rabbit liver. J Biol Chem 254:2339–2345.

Miller RL, Adamczyk DL, Miller WH, Koszalka GW, Rideout JL, Beacham LM, Chao EY, Haggerty JJ, Krenitsky TA, Elion GB (1979b): Adenosine kinase from rabbit liver, II. Substrate and inhibitor specificity. J Biol Chem 254:2346–2352.

Mimouni M, Bontemps F, van den Berghe G (1994): Kinetic studies of rat liver adenosine kinase. Explanation of exchange reaction between adenosine and AMP. J Biol Chem 269:17820–17825.

Miura K, Okumura M, Yukimura T, Yamamoto K (1991): Fluorometric determination of plasma adenosine concentrations using high-performance liquid chromatography. Analyt Biochem 196:84–88.

Mohrman DE, Hiller LJ (1990): Transcapillary adenosine transport in isolated guinea pig and rat hearts. Am J Physiol 259:H772–H783.

Mustafa SJ (1979): Effects of coronary dilator drugs on the uptake and release of adenosine in cardiac cells. Biochem Pharmacol 280:2617–2634.

Nagy JI, Geiger JD, Staines WA (1990): Adenosine deaminase and purinergic neuromodulation. Neurochem Int 16:211–221.

Newby AC (1991): Adenosine: Origin and clinical roles. In R.A. Harkness (ed): "Purine and Pyrimidine Metabolism in Man, Part VII, Part A." New York: Plenum Press, pp 265–270.

Padua R, Geiger JD, Dambock S, Nagy JI (1990): 2′-Deoxycoformycin inhibition of adenosine deaminase in rat brain: In vivo and in vitro analysis of specificity, potency and enzyme recovery. J Neurochem 54:1169–1178.

Pajor AM, Wright EM (1992) Cloning and functional expression of a mammalian Na^+/nucleoside cotransporter. A member of the SGLT family. J Biol Chem 267:3557–3560.

Palella TD, Andres CM, Fox IH (1980): Human placental adenosine kinase, kinetic mechanism and inhibition. J Biol Chem 255:5264–5269.

Park CK, Rudolphi KA (1994): Antiischemic effects of propentofylline (HWA285) against focal cerebral infarction in rats. Neurosci Lett 178:235–238.

Park TS, Gidday JM (1990): Effect of dipyridamole on cerebral extracellular adenosine level in vivo. J Cereb Blood Flow Metab 10:424–427.

Parkinson FE, Geiger JD (1996): Effects of iodotubercidin on adenosine kinase activity and nucleoside transport in DDT_1MF_2 smooth muscle cells. J Pharmacol Exp Ther 277:1397–1401.

Parkinson FE, Paterson ARP, Young JD, Class CE (1993): Inhibitory effects of propentofylline on [^{3}H]adenosine influx: a study of three nucleoside transport systems. Biochem Pharmacol 46:891–896.

Parkinson FE, Rudolphi KA, Fredholm BB (1994): Propentofylline: A nucleoside transport inhibitor with neuroprotective effects in cerebral ischemia. Gen Pharmacol 25:1053–1058.

Paterson ARP, Harley ER, Cass CE (1985): Measurement and inhibition of membrane transport of adenosine. In Paton EM (ed): "Methods in Pharmacology, Volume 6." New York: Plenum Press, pp 165–180.

Paterson ARP, Kolassa N, Cass CE (1981): Transport of nucleoside drugs in animal cells. Pharmacol Ther 12:515–536.

Pazzagli M, Corsi C, Fratti S, Pedata F, Pepeu G (1995): Regulation of extracellular adenosine levels in the striatum of aging rats. Brain Res 684:103–106.

Pazzagli M, Pedata F, Pepeu G (1993): Effects of K^+ depolarization, tetrodotoxin, and NMDA receptor inhibition on extracellular adenosine levels in rat striatum. Eur J Pharmacol 234:61–65.

Phillis JW (ed) (1991): "Adenosine and Adenine Nucleotides as Regulators of Cellular Function." Boca Raton, FL, CRC Press, pp 3–436.

Phillis JW, Smith-Barbour M (1993): The adenosine kinase inhibitor, 5-iodotubercidin, is not protective against cerebral ischemic injury in the gerbil. Life Sci 53:497–502.

Phillis JW, Perkins LM, Smith-Barbour M, O'Regan MH (1995): Oxypurinol-enhanced postischemic recovery of the rat brain involves preservation of adenine nucleotides. J Neurochem 64:2177–2184.

Phillis JW, O'Regan MH, Perkins LM (1992): Measurement of rat plasma adenosine levels during normoxia and hypoxia. Life Sci 51:149–152.

Plagemann PGW, Kraupp M (1986): Inhibition of nucleoside and nucleobase transport and nitrobenzylthioinosine binding by dilazep and hexobendine. Biochem Pharmacol 35:2559–2567.

Plagemann GW, Richey DP (1974): Transport of nucleosides, nucleic acid bases, choline and glucose by animal cells in culture. Biochim Biophy Acta 344:263–305.

Plagemann PGW, Wohlhueter RM (1980): Permeation of nucleosides, nucleic acid bases and nucleotides in animal cells. Curr Top Membr Transp 14:225–330.

Plesner L (1995): Ecto-ATPases: Identities and functions. Int Rev Cytol 158:141–214.

Poon A, Sawynok J (1995): Antinociception by adenosine analogs and an adenosine kinase inhibitor: Dependence on formalin concentration. Eur J Pharmacol 286:177–184.

Proctor KG, Langkamp-Henken B (1991): Adenosine and blood flow through the gastrointestinal tract. In Phillis JW (ed.) "Adenosine and Adenine Nucleotides as Regulators of Cellular Function." Boca Raton, FL, CRC Press, pp 203–212.

Richard JP, Carr MC, Ives DH, Frey PA (1980): The stereochemical course of thiophosphoryl group transfer catalyzed by adenosine kinase. Biochem Biophys Res Commun 30:1052–1056.

Rongen GA, Smiths P, Bootsma G, Ver Donck K, de Vries A, Thien T (1995a): High-grade nucleoside transport inhibition stimulates ventilation in humans. J Clin Pharmacol 35:357–361.

Rongen GA, Smits P, Ver Donck K, Willemsen JJ, De Abreu RA, Van Belle H, Thien T (1995b): Hemodynamic and neurohumoral effects of various grades of selective adenosine transport inhibition in humans. J Clin Invest 95:658–668.

Rosenberg PA, Knowles R, Knowles KP, Li Y (1994): Beta receptor-mediated regulation of extracellular adenosine in cerebral cortex in culture. J Neurosci 14:2953–2965.

Rosengren S, Bong GW, Firestein GS (1995): Anti-inflammatory effects of an adenosine kinase inhibitor. Decreased neutrophil accumulation and vascular leakage. J Immunol 154:5444–5451.

Rottlan P, Miras-Portugal MT (1985): Adenosine kinase from bovine adrenal medulla. Eur J Biochem 151:365–371.

Samuelson LC, Farber RA (1985): Cytological localization of adenosine kinase, nucleoside phosphorylase-1, and esterase-10 genes on mouse chromosome 14. Somat Cell Mol Genet 11:157–165.

Sayos J, Solsona C, Mallol J, Lluis C, Franco R (1994): Phosphorylation of adenosine in renal brush-border membrane vesicles by an exchange reaction catalyzed by adenosine kinase. Biochem J 297:491–496.

Schaeffer HJ, Schwender CF (1974): Enzyme inhibitor, XXVI. Bridging hydrophobic and hydrophilic regions on adenosine deaminase with some 9-(2-hydroxy-3-alkyl)adenines. J Med Chem 17:6–8.

Schrader J (1991): Formation and metabolism of adenosine and adenine nucleotides in cardiac tissue. In Phillis JW (ed): "Adenosine and Adenine Nucleotides as Regulators of Cellular Function." Boca Raton, FL, CRC Press, pp 55–69.

Sciotti VM, Van Wylen DGL (1993): Increases in interstitial adenosine and cerebral blood flow with inhibition of adenosine kinase and adenosine deaminase. J Cereb Blood Flow Metab 13:201–207.

Schulz R, Rose J, Post H, Heusch G (1995): Involvement of endogenous adenosine in ischemic preconditioning in swine. Pflugers Arch Eur J Physiol 430:273–282.

Silva PH, Dillon D, Van Wylen DG (1995): Adenosine deaminase inhibition augments interstitial adenosine but does not attenuate myocardial infarction. Cardiovasc Res 29:616–623.

Skolnick P, Nimitkitpaisan Y, Stalveya L, Daly JW (1978): Inhibition of brain adenosine deminase by 2′-deoxycoformycin and erythro-9-(2-hydroxy-3-nonyl)adenine. J Neurochem 30:1579–1582.

Smolenski RT, Kochan Z, McDouall R, Page C, Seymour AL, Yacoub MH (1994): Endothelial nucleotide catabolism and adenosine production. Cardiovasc Res 28:100–104.

Snyder SH (1985): Adenosine as a neuromodulator. Annu Rev Neurobiol 8:103–124.

Spychala J, Datta NS, Takabayashi K, Datta M, Fox IH, Gribbin T, Mitchell BS (1996): Cloning of human adenosine kinase cDNA: Sequence similarity to microbial ribokinases and fructokinases. Proc Natl Acad Sci USA 93:1232–1237.

Tavenier M, Skladanowski AC, DeAbreu RA, deJong JW (1995): Kinetics of adenylate metabolism in human and rat myocardium. Biochim Biophys Acta 1244:351–356.

Tonini M, Perucca E, Manzo L, Marcolli M, D'Angelo L, Saltarelli P, Onori L. (1982). Dilazep: An inhibitor of adenosine uptake with intrinsic calcium antagonistic properties. J Pharm Pharmacol 35:434–439.

Van Wylen DGL, Sciotti VM, Winn HR (1991): Adenosine and the regulation of cerebral blood flow. In Phillis JW (ed): "Adenosine and Adenine Nucleotides as Regulators of Cellular Function." Boca Raton, FL, CRC Press, pp 191–202.

Vargeese C, Sarma MSP, Pragnacharyulu PVP, Adushanab E, Li SY, Stoeckler JD (1994): Adenosine deaminase inhibitors. Synthesis and biological evaluation of putative metabolites of (+)-erythro-9-(2(S)-hydroxy-3(R)-nonyl)adenine. J Med Chem 37:3844–3849.

Wagner DR, Bontemps F, van den Berghe G (1994): Existence and role of substrate cycling between AMP and adenosine in isolated rabbit cardiomyocytes under control conditions and in ATP depletion. Circulation 90:1343–1349.

Wakade TD, Palmer KC, McCauley R, Przywara DA, Wakade AR (1995): Adenosine-induced apoptosis in chick embryonic sympathetic neurons: A new physiological role for adenosine. J Physiol 488:123–138.

Walter GA, Phillis JW, O'Regan MH (1988): Determination of rat cerebrospinal fluid concentrations of adenosine, inosine, hypoxanthine, xanthine and uric acid by high performance liquid chromatography. J Pharm Pharmacol 40:140–142.

Wang T, Mentzer RM Jr, Van Wylen DGL (1992): Interstitial adenosine with dipyridamole: Effect of adenosine receptor blockade and adenosine deaminase. Am J Physiol 263:H552–H558.

Wang T, Sodhi J, Mentzer RM Jr, Van Wylen DGL (1994): Changes in interstitial adenosine during hypoxia: Relationship to oxygen supply : demand imbalance, and effects of adenosine deaminase. Cardiovasc Res 28:1320–1325.

Wangler RD, Gorman MW, Wang CY, Dewitt DF, Chan IS, Bassingthwaighte JB, Sparks HV (1989): Transcapillary adenosine transport and interstitial adenosine concentration in guinea pig hearts. Am J Physiol 257:H89–H106.

Wauquier A, Van Belle H, Van den Broeck WA, Janssen PA (1987): Sleep improvement in dogs after oral administration of mioflazine, a nucleoside transport inhibitor. Psychopharmacology 91:434–439.

Weisner JB, Zimring ST, Castellino AJ, Ugarkar BG, Shea M, Erion MD, Foster AC (1995): Anticonvulsant profile of a novel adenosine kinase inhibitor, GP683, in rodent. Epilepsia 36:86.

White TD, Hoehn K (1991): Adenosine and adenine nucleotides in tissues and perfusates. In Phillis JW (ed): "Adenosine and Adenine Nucleotides as Regulators of Cellular Function." Boca Raton, FL, CRC Press, pp 109–120.

Winn HR, Rubio R, Berne RM (1979): Brain adenosine production in the rat during 60 seconds of ischemia. Circ Res 45:486–492.

Worku Y, Newby AC (1983): The mechanism of adenosine production in rat polymorphonuclear leucocytes. Biochem J 214:325–330.

Wotring LL, Townsend LB (1979): Study of the cytotoxicity and metabolism of 4-amino-3-carboxamido-1-(β-D-ribofuranosyl)pyrazolo[3,4-*d*]pyrimidine using inhibitors of adenosine kinase and adenosine deaminase. Cancer Res 39:3018–3023.

Wu PH, Phillis JW (1984): Uptake by central nervous tissues as a mechanism for the regulation of extracellular adenosine concentrations. Neurochem Int 6:613–632.

Wu X, Gutierrez MM, Giacomini KM (1994): Further characterization of the sodium-dependent nucleoside transporter (N3) in choroid plexus from rabbit. Biochim Biophys Acta 1191:190–196.

Yang BC, Mehta JL (1994): Platelet-derived adenosine contributes to the cardioprotective effects of platelets against ischemia-reperfusion injury in isolated rat heart. J Cardiovasc Pharmacol 24:779–785.

Yang X, Zhu Q, Claydon MA, Hicks GL Jr, Wang T (1994): Enhanced functional preservation of cold-stored rat heart by a nucleoside transport inhibitor. Transplantation 58:28–34.

Yamada Y, Goto H, Ogasawara N (1980): Purification and properties of adenosine kinase from rat brain. Biochem Biophys Acta 616:199–207.

Yamada Y, Goto H, Ogasawara N (1981): Adenosine kinase from human liver. Biochim Biophys Acta 660:36–43.

Yao SYM, Ng AML, Ritzel MWL, Cass CE, Young JD (1996): Transport of adenosine by recombinant purine and pyrimidine selective sodium/nucleoside cotransporters from rat jejunum expressed in *Xenopus* oocytes. Mol Pharmacol 50:1529–1535.

Young JD, Jarvis SM (1983): Nucleoside transport in animal cells. Biosci Rep 3:309–322.

Zhang Y, Fredholm BB (1994): Propentofylline enhancement of the actions of adenosine on neutrophil leukocytes. Biochem Pharmacol 48:2025–2032.

Zhang G, Franklin PH, Murray TF (1993): Manipulation of endogenous adenosine in the rat prepiriform cortex modulates seizure susceptibility. J Pharmacol Exp Ther 264:415–424.

Zhang Y, Geiger JD, Lautt WW (1991a): Improved high-pressure liquid chromatographic-fluorometric assay for measurement of adenosine in plasma. Am J Physiol 260:G658–G664.

Zhang Y, Legare DJ, Geiger JD, Lautt WW (1991b): Dilazep potentiation of adenosine-mediated superior mesenteric arterial vasodilation. J Pharmacol Exp Therap 258:767–771.

Zhang Y, Geiger JD, Legare DJ, Lautt WW (1991c): Dilazep-induced vasodilation is mediated through adenosine receptors. Life Sci 49:129–133.

Zhu Q, Yang X, Claydon MA, Hicks GL, Wang T (1994): Adenosine deaminase inhibitor in cardioplegia enhanced function preservation of the hypothermically stored rate heart. Transplantation 57:35–40.

CHAPTER 5

Allosteric Enhancement of Adenosine Receptors

JOEL LINDEN
University of Virginia Health Sciences Center, Charlottesville, VA 22908

ALLOSTERIC ENHANCERS

Allosteric enhancers of receptors are defined as compounds that potentiate responses to agonists and bind to an *allosteric* site distinct from the *orthosteric* binding site of the naturally occurring ligand. An example of a well-characterized class of compounds that is thought to act allosterically is the benzodiazepines, including diazepam. These compounds are enhancers of $GABA_A$ receptors. They potentiate the response to GABA (γ-aminobutyric acid) by stabilizing a high-affinity state of the receptor and shifting the dose–response curve of GABA to the left (Barker et al., 1986). The clinical advantages of allosteric enhancers over agonists is illustrated by the fact that benzodiazepines have relatively few side effects and are used clinically. In contrast, direct-acting $GABA_A$ agonists produce numerous side effects and are not used clinically. Muscarinic cholinergic receptors also are known to bind gallamine to an allosteric site, resulting in noncompetitive receptor antagonism (Ehlert et al., 1994). In this chapter we discuss compounds that have been shown to be allosteric enhancers of A_1 adenosine receptors. Currently available adenosine enhancers are suboptimal in that they have low binding affinities, and at high concentrations they possess competitive antagonist properties. Hence, the currently available compounds are not suitable as therapeutic agents. However, given the widespread protective effects of adenosine that are mediated by A_1 adenosine receptors and the problems with desensitization that occur with the chronic use of agonists, it is possible that there will be therapeutic applications for a future generation of allosteric enhancers of adenosine receptors.

ADENOREGULIN

It is notable that a 33-amino-acid peptide isolated from frog skin and named adenoregulin was reported to be an adenosine receptor enhancer (Daly et al., 1992). However,

Purinergic Approaches in Experimental Therapeutics, Edited by Kenneth A. Jacobson and Michael F. Jarvis
ISBN 0-471-14071-6 © 1997 Wiley-Liss, Inc.

molecular cloning of the adenoregulin gene revealed it to be a member of the dermaseptin antimicrobial peptide family (Amiche et al., 1993; Mor et al., 1994). Additional experimentation has revealed that the effects of this peptide are not selective for adenosine receptors, and adenoregulin is now thought to produce nonspecific effects as a consequence of its ability to interact with and disrupt membrane lipid bilayers.

BENZOYLTHIOPHENE ALLOSTERIC ENHANCERS OF A_1 ADENOSINE RECEPTORS

Binding Studies To Brain Membranes

During a broad drug screening program undertaken at Parke-Davis, 2-amino-3-benzoylthiophenes and some structurally similar compounds were unexpectedly found to increase the binding of the agonist, $[^3H]N^6$-cyclohexyladenosine ($[^3H]$CHA), to A_1 receptors of rat brain membranes by 10%–30% when added to a concentration of 10 μM. The first report of the existence of these allosteric enhancers of A_1 adenosine receptors was published in 1989 (Bruns and Fergus). A more in-depth characterization followed in 1990 (Bruns and Fergus). It was noted that the effect of these compounds on the binding of $[^3H]$CHA to rat brain membranes was biphasic, suggesting that at high concentrations these compounds are competitive antagonists, in addition to acting as allosteric enhancers, Bruns et al. measured slowing of the dissociation of $[^3H]$CHA from rat brain membranes as a selective measure of the allosteric effects of these compounds (Bruns and Fergus, 1989; Bruns et al., 1990). The rationale for this type of analysis is based on binding theory, which holds that competitive antagonists have no influence on the dissociation rate of a prebound ligand. In this kinetic assay system, some of the enhancers (100 μM) were shown to decrease the dissociation rate of $[^3H]$CHA from brain membranes as much as five-fold.

Even in these early studies the enhancers were shown to be highly selective for A_1 adenosine receptors. The adenosine A_1 enhancers were found to have little or no effect on the binding of agonists to A_{2A} adenosine, M_2 muscarinic, α_2 adrenergic, or δ-opiate receptors. These findings suggest that the enhancers bind directly to the A_1 adenosine receptor and not to a GTP-binding protein or to some other ancillary site that indirectly influences receptor binding or function.

Structure–Activity Profile

A structure–activity analysis of the enhancer activity of 88 benzodiazepine-like compounds that had been synthesized in the 1970s was published in 1990 (Bruns). The conclusions were limited because second-generation enhancer compounds were not synthesized at that time. Nevertheless, it was possible to conclude that the keto carbonyl and the 2-amino group are required for enhancer activity, and at least one hydrogen on the amino group (Figure 1). Alkyl substitution at the 4-position of the thiophene ring increases activity. Activity also is increased by various substitutions on the phenyl ring, with 3-(trifluoromethyl) showing optimal activity as in PD81,723 (Figure 1). Since the binding of antagonists is not increased by the 2-amino-3-benzoylthiophenes, competition by 2-amino-3-benzoylthiophenes for the binding of $[^3H]$8-cyclopentyl-1,3-dipropylxanthine ($[^3H]$CPX) was developed as a specific measure of the antagonist properties of these compounds. Structure–activity relationships for allosteric enhance-

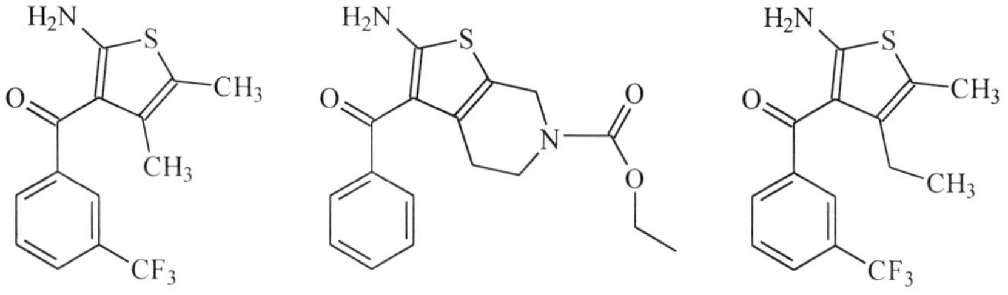

PD 81,723 PD 78,416 RS 74,513

FIGURE 1. The structures of PD81,723, 2-amino-4,5-dimethyl-3-thienyl-[3-(trifluromethyl)-phenyl]-methanone, also designated LY202472; PD78,416, thieno[2,3-c]pyridine-6(5H)-carboxylic acid, 2-amino-3-benzoyl-4,7-dihydro-ethyl ester; and RS74513, 2-amino-4-ethyl-5-methyl-3-(3-trifluoro-methyl-benzoly) thiophene.

ment and competitive antagonism are distinct. These data suggest that the two activities are produced by different portions of the 2-amino-3-benzoylthiophene molecules and that it might be possible to develop new molecules that are more selective for producing enhancer actions.

Most studies of 2-amino-3-benzoylthiophene function have utilized PD81,723, identified as the optimal enhancer in this first-generation series of compounds. An identical compound has been synthesized by Eli Lilly and Company, and is referred to as LY202472. In some studies, similar compounds such as PD78,416 have been characterized (Figure 1). More recently, a novel benzoylthiophene designated RS74513 (see Figure 1) was synthesized by Syntex Discovery Research (Leung et al., 1995).

Functional Assays

FRTL-5 Thyroid Cells As an initial functional assay for allosteric enhancers of A_1 adenosine receptors, Bruns and Fergus demonstrated that these compounds cause a leftward shift in the dose–response curve for N^6-cyclopentyladenosine to inhibit forskolin-stimulated cyclic AMP accumulation in cultured FRTL-5 rat thyroid cells that possess an endogenous A_1 adenosine receptor (Bruns and Fergus, 1989; Bruns and Fergus, 1990). Besides shifting the N^6-cyclopentyladenosine concentration–response curve to the left, the addition of enhancers alone partially lowered cyclic AMP in the absence of N^6-cyclopentyladenosine, suggesting that the enhancers are partial agonists, or that cultured cells release some endogenous adenosine that elicits a response that is magnified by the addition of enhancers. In accordance with one of the models of allosterism (Monod et al., 1963), Bruns and Fergus proposed that the 2-amino-3-benzoylthiophenes are agonists that bind to the A_1 adenosine receptor at an allosteric site different from the orthosteric adenosine binding site (Bruns and Fergus, 1990). According to this model, at low concentrations allosteric enhancers will potentiate the response to agonists that bind to the orthosteric binding pocket.

Cardiac Effects In isolated rat atria, PD81,723 was found to enhance the negative inotropic and chronotropic effects of N^6-cyclopentyladenosine (Mudumbi et al., 1993). PD81,723 also was found to potentiate the A_1 adenosine receptor–mediated negative dromotropic effects of adenosine added to isolated perfused *in situ* guinea pig hearts.

Adenosine production is increased in hypoxic myocardium, suggesting that during hypoxia, the effects of endogenous adenosine released by the tissue might be enhanced. This was shown to be true in the *in situ* guinea pig heart, where PD81,723 increased the effects of hypoxia to slow atrioventricular conduction velocity (Kollias-Baker et al., 1994b). To establish the A_1 versus A_{2A} adenosine receptor selectivity of PD81,723 in the guinea pig heart, Amoah-Apraku et al. (1993) demonstrated that there was enhancement of the negative dromotropic effect of adenosine that is mediated by A_1 receptors, but no enhancement of the dilation of coronary arteries, a response mediated by A_{2A} adenosine receptors.

CNS Effects Brain ischemia triggers the release of excitatory amino acids including glutamate, aspartate, and GABA, from rat cerebral cortex. In a model of global cerebral ischemia in the rat, PD81,723 was found to depress glutamate efflux (Phillis et al., 1994). Janusz and Berman (1993) investigated the effects of PD81,723 on epileptiform bursting in the hippocampal brain slice. Extracellular recordings were obtained from the CA3 pyramidal cell layer while electrically stimulating the stratum radiatum during low magnesium perfusion. Adenosine or PD81,723 reduced the duration of triggered epileptiform bursting, consistent with activity by these compounds as anticonvulsant agents. In another study, PD81,723 added to hippocampal brain slices in the presence of adenosine caused a reduction in the amplitude of the population spike recorded from the CA1 cell layer, and enhanced paired-pulse facilitation (Janusz et al., 1991). These results demonstrate that in addition to enhancing binding of agonists to A_1 adenosine receptors, PD81,723 also enhances the functional activity of adenosine in the hippocampal slice.

Guinea Pig Ileum Whereas PD81,723 and PD78,416 enhance the negative inotropiceffect of N^6-cyclopentyladenosine on electrically paced guinea pig heart, RS-74513, which is structurally similar to PD81,723 (Figure 2), has no enhancing effect on the guinea pig heart (Leung et al., 1995). In contrast, PD78,416 failed to enhance the ability of N^6-cyclopentyladenosine to inhibit electrically induced contraction of the

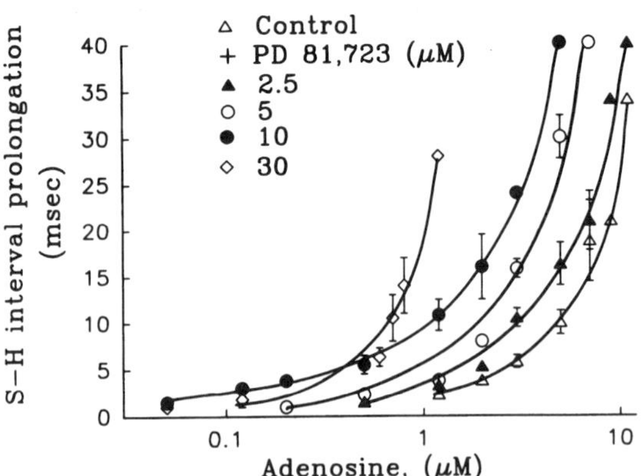

FIGURE 2. Potentiation by PD81,723 of the negative dromotropic effect of adenosine. The stimulus to His bundle (S-H) interval was measured in isolated guinea pig hearts. Values are mean ± SEM of 10 hearts (From Kollias-Baker et al., 1994a.)

guinea pig ileum, although 3 μM RS-74513 caused a three-fold leftward shift in the N^6-cyclopentyladenosine dose–response curve in the ileum (Leung et al., 1995). Thus, PD78,416 appears to act like PD81,723 as an A_1 adenosine receptor enhancer in guinea pig heart, but not ileum, and RS-74513 is an enhancer in ileum, but not heart. Since there is no evidence to suggest that A_1 adenosine receptors of heart and ileum are different, it is possible that tissue differences in the effects of these enhancer compounds may be due to differences in the G proteins that are coupled to A_1 adenosine receptors in various tissues.

Recombinant Human A_1 Adenosine Receptors The effects of PD81,723 on recombinant human A_1 receptors stably expressed in CHO-K1 cells has been investigated (Bhattacharya and Linden, 1995). PD81,723 has effects on recombinant A_1 receptors that are very similar to its effects on rat FRTL-5 A_1 adenosine receptors (Bruns and Fergus, 1990). It possesses both enhancer and competitive antagonist activities. The enhancer component of activity was assessed by measuring the dissociation rate of a prebound agonist, ^{125}I-N^6-aminobenzyladenosine (^{125}I-ABA); and the antagonist component of activity was selectively assessed by measuring the effect of PD81,723 in competition assays with the antagonist, [^{3}H]CPX. The studies with recombinant human receptors have provided some novel insights about the mechanism of action of the enhancer compounds. PD81,723 produces only a modest increase in the affinity of ^{125}I-ABA for receptors uncoupled from G proteins; the predominant effect on equilibrium binding is to increase the number of receptors that bind agonists with high affinity, i.e., to increase receptor coupling to heterotrimeric G proteins. Figure 3 shows the effect of PD81,723 (20 μM) on the equilibrium binding of ^{125}I-ABA to membranes of CHO-K1 cells expressing recombinant human A_1 adenosine receptors. The effect of 20 μM PD81,723 on the concentration dependence of ^{125}I-ABA binding measured at equilibrium is shown. ^{125}I-ABA binds to two affinity states of recombinant A_1 adenosine receptors, PD81,723 was found to significantly increase the B_{max} of the high-affinity binding site, while having little detectible effect on binding affinity.

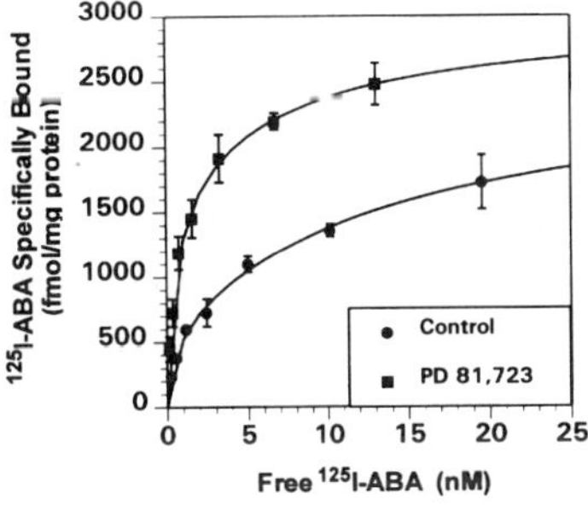

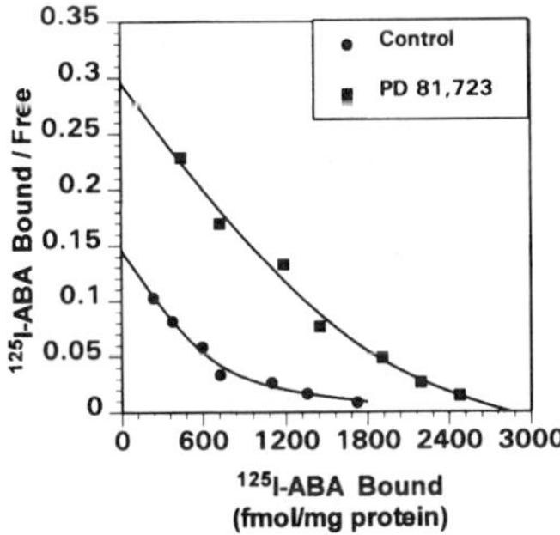

FIGURE 3. Effects of PD81,723 on equilibrium binding of ^{125}I-ABA to membranes prepared from CHO-K1 cells stably transfected with recombinant human A_1 adenosine receptors. ^{125}I-ABA equilibrium binding was measured in the absence (●) and presence (■) of 20 μM PD81,723. Specific binding is shown: untransformed (*left*) and Scatchard (Scatchard, 1949) transformed (*right*). (From Bhattacharya and Linden, 1995.)

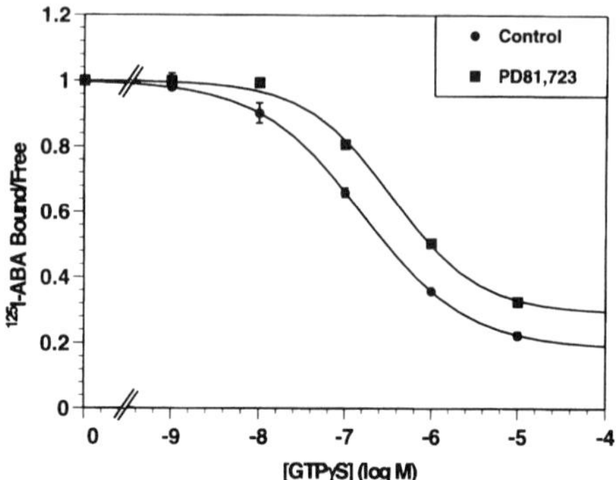

FIGURE 4. Effects of GTPγS on ^{125}1-ABA binding to human A_1 adenosine receptors. Membranes from stably transfected CHO-K1 cells were incubated with 0.5 nM ^{125}I-ABA $\pm$ 20 μM PD,81,723 and various concentrations of GTPγS. Each point is the mean $\pm$ SEM of data pooled from three experiments, each assayed in triplicate. (From Bhattacharya and Linden, 1995.)

Effects of Enhancers on Receptor–G Protein Coupling The effects of PD81,723 on equilibrium ^{125}I-ABA binding suggested that the enhancers stabilize agonist -receptor–G protein ternary complexes. Since guanine nucleotides such as GTPγS [guanosine 5′-*O*-(3-thiophosphate] destabilize ternary complexes upon binding to G protein α subunits, we investigated the possibility that PD81,723 might influence guanine nucleotide exchange from receptor–G protein ternary complexes. Figure 4 demonstrates that the enhancer increases the concentration of the stable GTP (guanosine triphosphate) analog, GTPγS, necessary to uncouple receptor–G protein complexes. The half-maximal concentration of GTPγS required to decrease ^{125}I-ABA binding is increased in the presence of PD81,723 (20 μM) by two- to threefold. PD81,723 also has been found to decrease the potency of another GTP analog, 5′-guanylylimidodiphosphate, in reducing the binding of [^{3}H]CHA to membranes prepared from guinea pig heart and brain (Kollias-Baker et al., 1994a). These findings may help to account for the observation that there is some variability in the magnitude of the increase in agonist binding to membranes prepared from various tissues in response to enhancer compounds. We speculate that the amount of GTP present in membranes may have a strong influence on enhancer action as measured in broken cell systems. Since the enhancer compounds reduce the ability of guanine nucleotides to inhibit the binding of agonists to A_1 adenosine receptors, the effect of the enhancer compounds on agonist binding is likely magnified in crude membrane preparations that contain relatively high levels of GTP as compared to more purified or dilute membrane preparations that contain relatively low levels of GTP.

THERAPEUTIC TARGETS OF ALLOSTERIC ENHANCERS

Acute Protection

The allosteric enhancers have great potential as therapeutic agents. Adenosine has been clearly established as an agent that protects against damage due to inadequate tissue oxygenation, as in ischemia. Numerous studies have demonstrated that adenosine is cardioprotective during periods of ischemia and hypoxia (see review by Ely and Berne, 1992). A similar protective action of adenosine has been demonstrated in brain (Mitchell et al., 1995; Schubert et al., 1994). Adenosine-mediated protection has been noted in other tissues, including liver (Hernandez-Munoz et al., 1984) and intestine (Kaminski and Proctor, 1989; Kaminski and Proctor, 1992). Adenosine, or agents known to slow adenosine metabolism, have been used to preserve tissues during organ transplantation. This use has been most extensively studied in cardiac muscle (Hudspeth et al., 1994; Harden et al., 1995).

Although adenosine itself is used clinically to treat supraventricular arrhythmias (DiMarco et al., 1990; Roelke and Yurchak, 1992), adenosine agonists are not yet used therapeutically, in part due to side effects, such as heart block. Moreover, adenosine agonists are likely to cause desensitization *in vivo,* as they do *in vitro* (Shryock et al., 1989; Lee et al., 1993). In contrast, side effects and tachyphylaxis are likely to be limited with allosteric enhancers.

Chronic Protection

The appeal of A_1 adenosine receptor enhancers has increased in recent years because of evidence that adenosine acts not only by producing acute protection of ischemic tissues, but also by contributing to preconditioning, a process by which transient ischemia protects tissues from damage during subsequent prolonged ischemic episodes (Parratt, 1994). In cardiac muscle, adenosine-mediated preconditioning appears to be mediated by A_1 receptors, or possibly a combination of A_1 and A_3 receptors (Auchampach and Gross, 1993; Liu et al., 1994; Parratt, 1994). Recently, A_1 adenosine receptors have been shown to mediate a substantial preconditioning response in a rat brain model of ischemia (Heurteaux et al., 1995). The involvement of A_1 receptors in these protective responses suggests that PD81,723 might enhance preconditioning responses as it does other A_1 receptor–mediated responses. Very recent data indicate that PD81,723 increases agonist binding to recombinant canine A_1, but not to recombinant canine A_3 adenosine receptors; furthermore PD81,723 enhances preconditioning in canine cardiac muscle (Mizumura et al., 1996). These data suggest that allosteric enhancers of A_1 adenosine receptors are potentially useful to enhance myocardial preconditioning responses, and that at least part of the preconditioning response in the dog heart is mediated by A_1 adenosine receptors. Thus, it is possible to envisage the use of enhancers of A_1 adenosine receptors to preserve tissues; accelerate the termination of acute disturbances in excitable tissues, e.g., in heart (arrhythmias) and brain (seizures); and to promote the protective consequences of recurrent transient ischemic events, as in angina and stroke. The enhancers also may have some utility for distinguishing responses mediated by A_1 adenosine receptors from responses mediated by other adenosine receptor subtypes.

METABOLIC ENHANCERS OF ADENOSINE ACTION

Some compounds that "enhance" adenosine act by interfering with adenosine metabolism. Intracellular adenosine can be phosphorylated by adenosine kinase to form AMP, or deaminated by adenosine deaminase to form inosine. A third relatively minor source of adenosine is via S-adenosylhomocystein-hydrolase. Specific inhibitors of these enzymes (see Table 1) have in some instances been found to enhance adenosine

TABLE 1. Compounds Reported to Enhance Adenosine Action by Interfering with Nucleoside Metabolism

Category	Compounds[1]	References
Adenosine transport Inhibitors	Dipyridamole	Phillis et al., 1989
	NBTI (NBMPR)	Conant and Jarvis, 1994
	Soluflazine	Phillis et al., 1989
Adenosine deaminase Inhibitors	EHNA	Lloyd and Fredholm, 1995; Keil and DeLander, 1994; Zhu et al., 1994
	2′-deoxycoformycin	Zhang et al., 1993; Keil and DeLander, 1994
Adenosine kinase Inhibitors	5′-amino-5′-deoxyadenosine	Zhang et al., 1993
	5′-iodotubercidin	Zhang et al., 1993; Keil and DeLander, 1994
Purine synthesis Modulators	AICAR (Acadesine) methotrexate	Galinanes et al., 1992; Kingma, Jr. et al., 1994; Vincent et al., 1992; Cronstein et al., 1993

Abbreviations: NBTI, S(p-nitrobenzyl)-6-thioinosine; EHNA, erythro-9-(2-hydroxy-3-nonyl)adenine; AICAR, 5-Aminoimidazole-4-Carboxamide Riboside.

production, and secondarily to increase various adenosine receptor–mediated actions. Among the effects of adenosine produced when the nucleoside accumulates in the extracellular space are inhibition of neuronal activity and an antinociceptive action. Since compounds that interfere with adenosine metabolism cause levels of endogenous adenosine to increase, such compounds are nonselective among the four known adenosine receptor subtypes, A_1, A_{2A}, A_{2B}, and A_3 (Tucker and Linden, 1993). In this regard "metabolic" enhancers differ from allosteric enhancers of receptors that have been found to be selective for the A_1 adenosine receptor subtype. Also, to the extent that metabolic enhancers interfere with normal purine metabolism, they may be prone to produce toxic side effects unrelated to receptor-mediated actions of adenosine.

Adenosine Transport Inhibitors

Adenosine is transported bidirectionally across cell membranes on facilitated nucleoside transport proteins. Since most adenosine is metabolized intracellularly, inhibition of transport usually is found to increase interstitial levels of adenosine and enhance its actions. The adenosine transport inhibitor dipyrimadole is used clinically as a coronary vasodilator.

Adenosine Kinase, Adenosine Deaminase, and S-Adenosylhomocystein Hydrolase Inhibitors

The addition of iodotubercidin, an inhibitor of adenosine kinase, to rat hippocampal slices, increases release of adenosine from the slices and inhibits population spike discharges and hyperpolarization of pyramidal cells, mimicking the effects of exogenously applied adenosine (Pak et al., 1994). The importance of adenosine kinase in adenosine metabolism is underscored by the finding that neither adenosine dialdehyde, an inhibitor of S-adenosylhomocystein hydrolase, or erythro-9(2-hydroxy-3-nonyl)adenine (EHNA), an inhibitor of adenosine deaminase mimic the effects of iodotubercidin in hippocampal slices (Pak et al., 1994). Intrathecal administration of another adenosine kinase inhibitor, 5′-amino-5′-deoxyadenosine, enhances adenosine- and opioid agonist–induced antinociception in mice (Keil and DeLander, 1994). Administration of adenosine into the rat prepiriform cortex protects against bicuculline methiodide–induced seizures; these anticonvulsant effects are significantly potentiated by treatment with 5′-amino-5′-deoxyadenosine, as well as by adenosine transport inhibitors (dilazep or S(*p*-nitrobenzyl)-6-thioinosine). These compounds are more effective at inhibiting seizure susceptibility than the adenosine deaminase inhibitor 2′-deoxycoformycin (Poucher et al., 1994).

The nucleoside transport inhibitor S(*p*-nitrobenzyl)-6-thioinosine was found to reduce infarct size in anesthetized ferrets when given prior to occlusion of the left anterior descending coronary artery (Poucher et al., 1994). The addition of the adenosine deaminase inhibitor, erythro-9-(2-hydroxy-3-nonyl)adenine (EHNA) to cardioplegia solution improves functional preservation by an action not involving ATP conservation (Zhu et al., 1994).

5-Aminoimidazole-4-Carboxamide Riboside (AICAR)

The compound 5-aminoimidazole-4-carboxamide ribonucleotide (ZMP) is an intermediate in the *de novo* synthesis of purine nucleotides. Incubation of tissues or cells with the nucleoside precursor of this compound, AICAR (also called acadesine), has been reported to increase cellular ZMP (Corton et al., 1995), to increase tissue adenosine during ischemia, and to reduce myocardial reperfusion injury (Kingma et al., 1994). AICAR also improves the ability of the heart to recover from ischemia and reperfusion when administered before ischemia or with cardioplegia (Galinanes et al., 1992), and prolongs myocardial preconditioning (Tsuchida et al., 1994). Although these protective effects of AICAR have been attributed to alterations in adenosine metabolism, recent evidence suggests that the increase in ZMP following AICAR administrations may produce a protective effect by allosterically stimulating an AMP-activated protein kinase (Corton et al., 1995). This kinase is believed to protect cells against environmental stress, e.g., heat shock, by switching off biosynthetic pathways when AMP becomes elevated.

In the metabolic pathway of *de novo* purine nucleotide synthesis, ZMP is formylated by N^{10}-formyltetrahydrofolate (THF). Methotrexate has an anti-inflammatory action, blocks THF-mediated methyl-transferase reactions, and has been shown to cause AICAR accumulation and adenosine release from inflamed tissues. The anti-inflammatory actions of methotrexate are blocked by adenosine receptor antagonists. On the basis of these observations it has been proposed that the anti-inflammatory effects of methotrexate are mediated by adenosine, which inhibits the adherence of neutrophils to connective tissue cells (Cronstein et al., 1991; Cronstein et al., 1993).

CONCLUSIONS

The 2-amino-3-benzoylthiophenes in general, and PD81,723 in particular, modestly increase the binding of adenosine agonists to adenosine receptors. Interest in these compounds in growing due to recent studies confirming that they enhance functional effects of adenosine in isolated cells and tissues, including human cells. The enhancer compounds have therapeutic appeal inasmuch as their actions are likely to be selectively targeted to hypoxic or ischemic tissues where adenosine concentrations are locally increased. They are less likely to produce side effects or desensitization than are orthosteric agonists, or compounds that interfere with adenosine metabolism.

The mechanism of action of the 2-amino-3-benzoylthiophenes is of interest. For reasons that are not yet understood, these compounds appear to be highly selective in their actions for A_1 adenosine receptors, with little or no effects on other adenosine receptor subtypes. They seem to act to stabilize the interaction between A_1 adenosine receptors and GTP-binding proteins. Understanding how this occurs may provide insights into factors involved in the control of signal transduction. Since high concentrations of 2-amino-3-benzoylthiophenes behave as competitive antagonists, they apparently can occlude the orthosteric binding pocket of unliganded receptors. Nevertheless enhancers clearly also can bind to agonist-occupied receptors. These considerations suggest that the enhancer and antagonists activities of these compounds are separable, and that it may be possible to design enhancer compounds that are devoid of antagonist activity.

REFERENCES

Amiche M, Ducancel F, Lajeunesse E, Boulain J-C, Menez A, Nicolas P (1993): Molecular cloning of a cDNA encoding the precursor of adenoregulin from frog skin: Relationships with the vertebrate defensive peptides, dermaseptins. Biochem Biophys Res Commun 191:983–900.

Amoah-Apraku B, Xu J, Lu JY, Pelleg A, Belardinelli L (1993): Selective potentiation by an A_1 adenosine receptor enhancer of the negative dromotropic action of adenosine in the guinea pig heart. J Pharmacol Exp Ther 266:611–617.

Auchampach JA, Gross GJ (1993): Adenosine A_1 receptors K_{ATP} channels, and ischemic preconditioning in dogs. Am J Physiol Heart Circ Physiol 264:H1327–H1336.

Barker JL, Harrison NL, Mariani AP (1986): Benzodiazepine pharmacology of cultured mammalian CNS neurons. Life Sci 39:1959–1968.

Bhattacharya S, Linden J (1995): The allosteric enhancer, PD81,723, stabilizes human A_1 adenosine receptor coupling to G proteins. Biochim Biophys Acta 1265:15–21.

Bruns RF (1990): Structure activity relationship for adenosine antagonists. In Jacobson KA, Daly JW, Manganiello V (eds): "Purines in Cellular Signaling: Targets for New Drugs." Berlin: Springer-Verlag, pp 126–135.

Bruns RF, Fergus JH (1989): Allosteric enhancers of adenosine A_1 receptor binding and function. In Ribeiro JA (ed): "Adenosine Receptors in the Nervous System." London: Taylor & Francis, pp 53–60.

Bruns RF, Fergus JH (1990): Allosteric enhancement of adenosine A1 receptor binding and function by 2-amino-3-benzoylthiophenes. Mol Pharmacol 38:939–949.

Bruns RF, Fergus JH, Coughenour L, Courtland GG, Puggsley TA, Dodd JH, Tinney FJ (1990): Structure-activity relationships for enhancement of adenosine A_1 receptor binding by 2-amino-3-benzoylthiophenes. Mol Pharmacol 38:950–958.

Conant AR, Jarvis SM (1994): Nucleoside influx and efflux in guinea-pig ventricular myocytes. Inhibition by analogues of lidoflazine. Biochem Pharmacol 48:873–880.

Corton JM, Gillespie JG, Hawley SA, Hardie DG (1995): 5-aminoimidazole-4-carboxamide ribonucleoside. A specific method for activating AMP-activated protein kinase in intact cells? Euro J Biochem 229:558–565.

Cronstein BN, Eberle MA, Gruber HE, Levin RI (1991): Methotrexate inhibits neutrophil function by stimulating adenosine release from connective tissue cells. Proc Natl Acad Sci USA 88:2441–2445.

Cronstein BN, Naime D, Ostad E (1993): The antiinflammatory mechanism of methotrexate. Increased adenosine release at inflamed sites diminishes leukocyte accumulation in an in vivo model of inflammation. J Clin Invest 92:2675–2682.

Daly JW, Caceres J, Moni RW, Gusovsky F, Moos M Jr, Seamon KB, Milton K, Myers CW (1992): Frog secretions and hunting magic in the upper Amazon: Identification of a peptide that interacts with an adenosine receptor. Proc Nat Acad Sci USA 89:10960–10963.

DiMarco JP, Miles W, Akhtar M, Milstein S, Sharma AD, Platia E, McGovern B, Scheinman MM, Govier WC, The Adenosine for PSVT Study Group (1990): Adenosine for paroxysmal supraventricular tachycardia: Dose ranging and comparison with verapamil. Assessment in placebo-controlled, multicenter trials. Ann Intern Med 113:104–110.

Ehlert FJ, Roesky WR, Yamamura HI (1994): Muscarinic receptors and novel strategies for the treatment of age-related brain disorders. [Review]. Life Sci 55:2135–2145.

Ely SW, Berne RM (1992): Protective effects of adenosine in myocardial ischemia. [Review]. Circulation 85:893–904.

Galinanes M, Mullane KM, Bullough D, Hearse DJ (1992): Acadesine and myocardial protection. Studies of time of administration and dose-response relations in the rat. Circulation 86:598–608.

Harden TK, Boyer JL, Nicholas RA (1995): P_2-purinergic receptors: Subtype-associated signaling responses and structure. Annu Rev Pharmacol Toxicol 35:541–579.

Hernandez-Munoz R, Glender W, Diaz Munoz M, Adolfo J, Garcia-Sainz JA, Chagoya de Sanchez V (1984): Effects of adenosine on liver cell damage induced by carbon tetrachloride. Biochem Pharmacol 33:2599–2604.

Heurteaux C, Lauritzen I, Widmann C, Lazdunski M (1995): Essential role of adenosine, adenosine A1 receptors, and ATP-sensitive K^+ channels in cerebral ischemic preconditioning. Proc Natl Acad Sci USA 92:4666–4670.

Hudspeth DA, Williams MW, Zhao ZQ, Sato H, Nakanishi K, McGee DS, Hammon JW Jr, Vinten-Johansen J, Van Wylen DG (1994): Pentostatin-augmented interstitial adenosine prevents postcardioplegia injury in damaged hearts. Ann Thor Surg 58:719–727.

Janusz CA, Berman RF (1993): The adenosine binding enhancer, PD81,723, inhibits epileptiform bursting in the hippocampal brain slice. Brain Res 619:131–136.

Janusz CA, Bruns RF, Berman RF (1991). Functional activity of the adenosine binding enhancer, PD81,723, in the in vitro hippocampal slice. Brain Res 567:181–187.

Kaminski PM, Proctor KG (1989): Attenuation of no-reflow phenomenon, neutrophil activation, and reperfusion injury in intestinal microcirculation by topical adenosine. Circ Res 65:426–435.

Kaminski PM, Proctor KG (1992): Extracellular and intracellular actions of adenosine and related compounds in the reperfused rat intestine. Circ Res 71:720–731.

Keil GJ, DeLander GE (1994): Adenosine kinase and adenosine deaminase inhibition modulate spinal adenosine- and opioid agonist-induced antinociception in mice. Eur J Pharm 271:37–46.

Kingma JG Jr, Simard D, Rouleau JR (1994): Timely administration of AICA riboside reduces reperfusion injury in rabbits. Cardiovasc Res 28:1003–1007.

Kollias-Baker C, Ruble J, Dennis D, Bruns RF, Linden J, Belardinelli (1994a): Allosteric enhancer PD81,723 acts by novel mechanism to potentiate cardiac actions of adenosine. Circulation Res 75:961–971.

Kollias-Baker C, Xu J, Pelleg A, Belardinelli L (1994b): Novel approach for enhancing atrioventricular nodal conduction delay mediated by endogenous adenosine. Circulation Res 75:972–980.

Lee HT, Thompson CI, Hernandez A, Lewy JL, Belloni FL (1993): Cardiac desensitization to adenosine analogues after prolonged R-PIA infusion in vivo. Am J Physiol 265:H1916–H1927.

Leung E, Walsh LKM, Flippin LA, Kim EJ, Lazar DA, Seran CS, Wong EHF, Elgen RM (1995): Enhancement of adenosine A1 receptors functions by benzoylthiophenes in guinea pig tissues in vitro. Naunyn Schmiedbergs Arch Pharmacol 352:206–212.

Liu Gs, Richards SC, Olsson RA, Mullane K, Walsh RS, Downey JM (1994): Evidence that the adenosine A_3 receptor can mediate preconditioning's protection in isolated rabbit hearts. Cardiovas Res 28:1057–1061.

Lloyd HG, Fredholm BB (1995): Involvement of adenosine deaminase and adenosine kinase in regulating extracellular adenosine concentration in rat hippocampal slices. Neurochemistry International 26:387–395.

Mitchell HL, Frisella WA, Brooker RW, Yoon K-W (1995): Attenuation of traumatic cell death by an adenosine A_1 agonist in rat hippocampal cells. Neurosurgery 36:1003–1008.

Mizumura T, Auchamphach JA, Linden J, Bruns RF, Gross GJ (1996): PD81,723, an allosteric enhancer of the A_1 adenasine receptor, lowers the threshold for ischemic preconditioning in dogs. Circ Res 79:415–423.

Monod J, Changeux J-P, Jacob F (1963): Allosteric proteins and cellular control systems. J Mol Biol 6:306–329.

Mor A, Amiche M, Nicolas P (1994): Structure, synthesis, and activity of dermaseptin b, a novel vertebrate defensive peptide from frog skin: Relationship with adenoregulin. Biochemistry 33:6642–6650.

Mudumbi RV, Montamat SC, Bruns RF, Vestal RE (1993): Cardiac functional responses to adenosine by PD81,723, an allosteric enhancer of the adenosine A1 receptor. Am J Physiol 264:H1017–H1022.

Pak MA, Haas HL, Decking UK, Schrader J (1994): Inhibition of adenosine kinase increases endogenous adenosine and depresses neuronal activity in hippocampal slices. Neuropharmacology 33:1049–1053.

Parratt JR (1994): Protection of the heart by ischaemic preconditioning: Mechanisms and possibilities for pharmacological exploitation. Trends Pharmacol Sci 15:19–25.

Phillis JW, O'Regan MH, Walter GA (1989): Effects of two nucleoside transport inhibitors, dipyridamole and soluflazine, on purine release from the rat cerebral cortex. Brain Res 481:309–316.

Phillis JW, Smith-Barbour M, Perkins LM, O'Regan MH (1994): Characterization of glutamate, aspartate, and GABA release from ischemic rat cerebral cortex. Brain Res Bull 34:457–466.

Poucher SM, Brooks R, Pleeth RM, Conant AR, Collis MG (1994): Myocardial infarction and purine transport inhibition in anaesthetised ferrets. Eur J Pharmacol 252;19–27.

Roelke M, Yurchak PM (1992): Adenosine and supraventricular tachycardia. N Engl J Med 326:1221.

Scatchard G (1949): The attractions of proteins for small molecules and ions. Ann NY Acad Sci 51:660–672.

Schubert P, Rudolphi KA, Fredholm BB, Nakamura Y (1994): Modulation of nerve and glial function by adenosine—role in the development of ischemic damage. [Review]. Int J Biochem 26:1227–1236.

Shryock J, Patel A, Belardinelli L, Linden J (1989): Downregulation and desensitization of A_1-adenosine receptors in embryonic chicken heart. Am J Physiol 25:H321–H327.

Tsuchida A, Yang XM, Burckhartt B, Mullane KM, Cohen MV, Downey JM (1994): Acadesine extends the window of protection afforded by ischaemic preconditioning. Cardiovasc Res 28:379–383.

Tucker AL, Linden J (1993): Cloned receptors and cardiovascular responses to adenosine. Cardiovasc Res 27:62–67.

Vincent MF, Bontemps F, Van den Berghe G (1992): Inhibition of glycolysis by 5-amino-4-imidazolecarboxamide riboside in isolated rat hepatocytes. Biochem J 281:267–272.

Zhang G, Franklin PH, Murray TF (1993): Manipulation of endogenous adenosine in the rat prepiriform cortex modulates seizure susceptibility. J Pharmacol Exp Ther 264:1415–1424.

Zhu Q, Yang Z, Claydon MA, Hicks GL, Jr, Wang T (1994): Adenosine deaminase inhibitor in cardioplegia enhanced function preservation of the hypothermically stored rat heart. Transplantation 57:35–40.

PART III

MEDICINAL CHEMISTRY

CHAPTER 6

Development of Selective Purinoceptor Agonists and Antagonists

KENNETH A. JACOBSON and A. MICHIEL VAN RHEE

Molecular Recognition Section, Laboratory of Bioorganic Chemistry, NIDDK, National Institutes of Health, Bethesda, MD 20892

INTRODUCTION

The purine adenosine is an important regulator of the homeostasis of the brain, heart, kidney, and other organs. Four distinct subtypes of the adenosine (P_1) receptors (termed A_1, A_{2A}, A_{2B}, and A_3) have been defined, based on pharmacological and genetic differences (Stiles, 1997). Theophylline, caffeine, and other alkylxanthines are nonselective antagonists at adenosine receptors, and many selective derivatives thereof have now been synthesized. Furthermore, purine and pyrimidine nucleotide (P_2) receptors, which have not been well explored from a medicinal chemistry perspective, are now coming into focus concurrently with the cloning of numerous new subtypes (Filtz et al., 1997). It is clear that the P_2 receptors are structurally distinct from P_1 receptors; thus they have completely different ligand binding requirements. The recently reorganized nomenclature of ATP and UTP receptors divides them into ligand-gated ion channels (P2X, of which 7 subtypes have already been cloned) and G protein–coupled subtypes (P2Y, of which at least 4 subtypes have already been cloned). Further information on purinoceptor nomenclature and structure is available on the Internet, e.g., at the following URL:http://mgddk1.niddk.nih.gov:8000/.

The cloning of the A_3 adenosine receptor (Zhou et al., 1992; Salvatore et al., 1993) has opened new therapeutic vistas in the adenosine field. The A_3 receptor has a unique SAR (structure activity relationship) profile (Jacobson et al., 1995b), tissue distribution (Linden, 1994), and effector coupling (Palmer et al., 1995a). Activation of A_3 receptors requires relatively high, ie., pathological, concentrations of adenosine; the K_i value of adenosine at the rat A_3 receptor has been estimated to be ~1 μM, versus 10 and

Purinergic Approaches in Experimental Therapeutics, Edited by Kenneth A. Jacobson and Michael F. Jarvis
ISBN 0-471-14071-6 © 1997 Wiley-Liss, Inc.

30 nM at rat A_1 and A_{2A} receptors, respectively (Jacobson et al., 1995b). Thus, the physiological role of A_3 receptors may be very different from that of A_1 and A_{2A} subtypes.

STRUCTURE–ACTIVITY RELATIONSHIPS AT ADENOSINE RECEPTORS

Adenosine Receptor Agonists

There is a tremendous impetus for the development of therapeutic agents based on selective interactions with one of the four subtypes of adenosine receptors. Adenosine agonists, which are almost exclusively derivatives of adenosine (Table 1), have been sought as potential antiarrhythmic (Belardinelli et al., 1997), cerebroprotective (von Lubitz, 1997), and cardioprotective agents (Liu et al., 1997) (via A_1), and as hypotensive and antipsychotic agents (Bridges et al., 1988) (via A_{2A}). A_3 agonists have been found to have potential as prophylactic cerebroprotective agents (Jacobson et al., 1995b) and possibly to have potential in modulating immune function and in treating inflammation (Sajjadi et al., 1996).

Many highly selective ligands for adenosine A_1 and A_{2A} receptors (Jacobson et al., 1992a) have been designed, while the first selective agonists at A_3 receptors were only recently introduced (Jacobson et al., 1995b). In general, for adenosine agonists (Table 1), numerous modifications of the N^6-position with hydrophobic moieties (such as CPA, N^6-cyclopentyladenosine, **8;** CHA, N^6-cyclohexyladenosine, **14;** and *R*-PIA, (*R*)-N^6-phenylisopropyladenosine, **31**) provide selectivity for A_1 receptors. The affinities of these N^6-substituted adenosine derivatives at A_3 receptors are often intermediate between their respective A_1 and A_{2A} affinities. SPA (N^6-*p*-sulfophenyladenosine, **17**) is a selective agonist for peripheral A_1 receptors, due to its diminished ability to cross biological membranes (Jacobson et al., 1992b). NNC 21-0136 (2-chloro-N^6-[(*R*)-(benzothiazolyl)thio-2-propyl]adenosine, **40**) is somewhat selective for central A_1 receptors and is under development as a neuroprotective agent (Knutsen and Murray, 1997). Although most N^6-substituted adenosine derivatives are A_1-selective, the series of N^6-substituted adenosine analogs includes the potent A_{2A}-agonist DPMA (N^6-[2-(3,5-dimethoxyphenyl)-2-(2-methylphenyl)ethyl]adenosine, **27**), the racemate of which is 30-fold selective for the A_{2A} vs the A_1-receptor (Bridges et al., 1988). Few purine ring modifications are tolerated, but 1-deazaadenosines retain high affinity for adenosine receptors. Both classical medicinal chemistry approaches and a functionalized congener approach (Jacobson et al., 1992a) have been utilized in purinoceptor ligand development efforts. By the latter approach, a chemically functionalized chain is incorporated at a specific site on a pharmacophore (e.g., ADAC, **18**). The site of attachment must correspond to a region of relaxed steric requirements at or near the receptor binding site, and targeting accessory sites of favorable interaction on the receptor may enhance the affinity and/or selectivity of the ligands.

The SAR at the 2-position of adenosine has also been extensively probed. The 2-arylamino adenosine analog, CV 1808 (**41**) is only moderately potent ($K_i \sim$ 100 nM) at A_{2A} receptors and has a modest fivefold selectivity for A_{2A}-versus the A_1-receptor. Further structural modifications at the 2-position led to development of the 2-alkoxyadenosines such as **44** (Ueeda et al., 1991) and the 2-alkynyladenosines such as **46.** The 2-alkynyl-5′-uronamide derivative, HE-NECA (Cristalli et al., 1995) is a potent A_{2A} agonist (K_i = 2.2 nM).

Among ribose modifications, amide substitution at the 5′-position as in NECA (adenosine-5′-N-ethyluronamide, **57**) provides increased potency at A_{2A} receptors. A_{2A}-selective adenosine agonists have been developed more recently. Evaluation of 2-position modifications of NECA (**57**) led to the identification of CGS 21680 (**58**; 2-[4-[(2-carboxyethyl)phenyl]ethylamino]-5′-N-ethylcarboxamidoadenosine), which is 140-fold selective for the A_{2A} versus the A_1 receptor, with a K_i value of 21 nM at A_{2A} receptors (Hutchison et al., 1989). Its ethylenediamine conjugate APEC (**58**), a functionalized congener, can be radioiodinated to obtain the conjugate iodo-PAPA-APEC (**60**) (Jacobson et al., 1992a). Unlike CGS 21680 (**58**) which apparently does not cross the blood-brain barrier, APEC is highly potent as a locomotor depressant when administered peripherally. Both CGS 21680 (**58**) and APEC (**59**) are inactive at A_{2B} receptors (Hide et al., 1992). A thio-substitution of the 4′-oxygen of 2-chloroadenosine was found to enhance affinity selectively at A_{2A} receptors (Siddiqi et al., 1995a). Other ribose modifications at 2′- and 3′-positions are generally not well tolerated by the binding site (Siddiqi et al., 1995a), although the 2′-methyl ether of CHA, SDZ WAG 994 (Wagner et al., 1995), was orally active and surprisingly potent *in vivo*.

Very recently A_3 selective agonists have appeared (Jacobson et al., 1995b). APNEA (**33**) had been used previously as an A_3 agonist but it is actually selective for A_1 receptors by a factor of eight (Kim et al., 1994b). One principle of achieving true A_3 selectivity among adenosine derivatives is the combination of optimal substitutions at the N^6-and 5′-positions of adenosine (van Galen et al., 1994). Specifically, among N^6-substituents, a benzyl group is favored, due to its diminished potency at A_1 and A_{2A} receptors, and these A_3-selectivity enhancing effects are additive with the A_3-affinity enhancing effects of the 5′-uronamido group. The first such hybrid molecule to show A_3 selectivity was N^6-benzyl-NECA (**64**). In a comparison of various 5′-uronamido groups in monosubstituted adenosine derivatives, the 5′-N-methyluronamide (**47**) had particularly favorable A_3 receptor versus A_1/A_{2A} affinity. Empirical and QSAR studies (Siddiqi et al., 1995b) of substituent effects on the N^6-benzyl group have shown that substitution at the 3-position of the phenyl ring with sterically bulky groups, such as the iodo group, is optimal, leading to the development of the highly potent A_3 agonists N^6-(3-iodobenzyl)-adenosine-5′-N-methyluronamide (IB-MECA; **50**), which is selective (Gallo-Rodriguez et al., 1994) for A_3 versus either A_1 or A_{2A} receptors by a factor of 50 *in vitro* and appears to be highly selective for A_3 *in vivo.* Although not as selective for A_3 versus A_1 receptors as IB-MECA, $[^{125}I]$I-AB-MECA (**51**) is widely used as a high-affinity radioligand for A_3 receptors (Olah et al., 1994).

Substitution at the 2-position in combination with modifications at N^6 and 5′-positions was found to further enhance A_3 selectivity. Cl-IB-MECA (2-chloro-N^6-(3-iodobenzyl)-adenosine-5′-N-methyluronamide, **54**), which displayed a K_i value of 0.33 nM at A_3 receptors, showed 2500- and 1400-fold selectivity for A_3 versus A_1 and A_{2A} receptors, respectively (Kim et al., 1994b). Certain N^6-benzyl derivatives of adenosine also inhibit the Na^+-independent adenosine transporter, yet IB-MECA and Cl-IB-MECA were shown to have very low affinity at this site (Kim et al., 1994b). 2-Methylthio-N^6-(3-iodobenzyl)adenosine-5′-N-methyluronamide (**55**) and 2-methylamino-N^6-(3-iodobenzyl)-adenosine-5′-N-methyluronamide (**56**) were slightly less potent at A_3 receptors, but nearly as selective as Cl-IB-MECA.

Nearly all of the adenosine agonists yet reported are primary or secondary amine derivatives of the 6-position. Unexpectedly, acylation of the exocyclic amine, as in the

TABLE 1 Affinities (nM) of Adenosine Derivatives at Rat Brain A_1, A_{2A}, and A_3 Receptors[a]

R_4 = H, R_5 = H, Z = H, X = O, and Y = N, unless noted.

Compound	R_1	R_2	R_3	[K_i (nM)] A_1	A_{2A}	A_3	Ref.
1. 5′-dADO	CH_3	H	H	269	596	2830	1
2. ADO	$HOCH_2$	H	H	10	30	1000	2
3. R_4 = Br	$HOCH_2$	H	H	~100,000	22,700	>100,000	1
4. R_4 = NH_2	$HOCH_2$	H	H	~100,000	24,100	>100,000	3
5. R_4 = NHBu	$HOCH_2$	H	H	11,400	11,500	>100,000	3
6.	$HOCH_2$	H	$HO(CH_2)_3$	7.0	4920		4
7.	$HOCH_2$	H	$H_2N(CH_2)_9$	0.32 (bovine)			5
8. CPA	$HOCH_2$	H	Cyclopentyl	0.59	462	240[b]	4
9. CCPA	$HOCH_2$	Cl	Cyclopentyl	0.6	950	237[b]	4,6
10. Y = CH	$HOCH_2$	Cl	Cyclopentyl	1.6	13,200		7
11. R_4 = 8-Cyclopentyl NH	$HOCH_2$	H	Cyclopentyl	1090			8
12. ENBA	$HOCH_2$	H	S-endo-Norbornyl	0.59	462	240[b]	4
13. GR79236	$HOCH_2$	H	HO (2-hydroxycyclopentyl structure)	3.1	1300	[b]	13
14. CHA	$HOCH_2$	H	Cyclohexyl	1.3	514	167[b]	4
15.	$HOCH_2$	H	Phenyl	4.62	663	802[b]	1

16. Z = $CH_2CH = CH_2$ BM 11. 189	$HOCH_2$	H	Phenyl				10
17. SPA	$HOCH_2$	H	4-Sulfophenyl	1.3	514	167[b]	1
18. ADAC	$HOCH_2$	H	$-CH_2$-p-Ph-CONH-p-Ph-CONH-$(CH_2)_2$NH(-H)	0.85	210	281[b]	1
19.	$HOCH_2$	H	$CH_3(CH_2)_{16}$CO-ADAC	0.22	8400		4
20. p-DITC-ADAC	$HOCH_2$	H	ADAC-NHCSNH-p-Ph-NCS	0.47	191		4
21. Biotinyl-ADAC	$HOCH_2$	H	See ref.	11.4			11
22. FITC-ADAC	$HOCH_2$	H	See ref.	7.1			4
23.	$HOCH_2$	H	Benzyl	120	285	120[b]	1
24. Metrifudil	$HOCH_2$	H	2-Methylbenzyl	59.6	24.1	360	3
25. I-ABA	$HOCH_2$	H	4-Amino-3-iodobenyzl	1.3	514	167[b]	6
26. Cl-936	$HOCH_2$	H	$CH_2CH(Ph)_2$	6.8	25		4
27. DPMA	$HOCH_2$	H	$CH_2CH(o\text{-}CH_3\text{-}Ph)(m\text{-}(OCH_3)_2Ph)$	142	4.4	3570	1
28.	$HOCH_2$	H	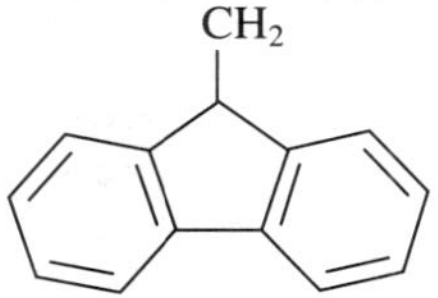 CH_2 (9-fluorenylmethyl structure)	5.2	4.9		4
29. AMG-1	$HOCH_2$	H	5-Hydroxy-2-pyridylmethyl				12
30.	$HOCH_2$	H	Phenylethyl	12.7	161	240[b]	1
31. R-PIA	$HOCH_2$	H	R-Phenylisopropyl	1.2	124	158[b]	1
32. S-PIA	$HOCH_2$	H	S-Phenylisopropyl	49.3	1820	920[b]	1
33. APNEA	$HOCH_2$	H	4-Aminophenylethyl	14	172	116	6
34. I-APNEA	$HOCH_2$	H	3-Iodo-4-aminophenylethyl	2.1		15.5	
				5.9	28	10,400	
				9.3	63	1890	3
37. X = S	$HOCH_2$	Cl	H	300	19.8	1090	3
38. NNC 90-1515	$HOCH_2$	Cl	N-Piperidinyl	4.2	1490		13
39. NNC 21-0041	$HOCH_2$	Cl	$CH(CH_3)CH_2OPh$	43	1200		13

TABLE 1 *(Continued)*

Compound	R_1	R_2	R_3	[K_i (nM)] A_1	A_{2A}	A_3	Ref.
40. NNC 21-0136	$HOCH_2$	Cl	CH_3 S N S	11	660		13
41. CV1808	$HOCH_2$	Phenylamino	H	560	119		4
42.	$HOCH_2$	Hexynyl	H	150	4.1		14
43.	$HOCH_2$	Phenylethoxy	H	130	17		14
44.	$HOCH_2$	Cyclohexyl-Eto	H	1600	22		14
45.	$HOCH_2$	Cyclohexyl-EtNH	H	6000	11		14
46.	$HOCH_2$	Cyclohexyl-CH_2-C ≡ C	H	210	65		14
47. MECA	CH_3NHCO	H	H	83.6	66.8	72[b]	18
48.	CH_3NHCO	H	Benzyl	898	597	16	3
49. R_5 = Me	CH_3NHCO	H	Benzyl	62.4	53.6	604	3
50. IB-MECA	CH_3NHCO	H	3-Iodobenzyl	54	56	1.1	2
51. I-AB-MECA	CH_3NHCO	H	3-I-4-NH_2-Benzyl	18	197	1.3	2
52. MRS1163	CH_3NHCO	H	3-NCS-Benzyl	145	272	10.0	15
53.	CH_3NHCO	H	3-(L-Boc-Ala-NH)-benzyl	4500	1960	46.7	16
54. Cl-IB-MECA	CH_3NHCO	Cl	3-Iodobenzyl	820	470	0.33	2
55. MRS544	CH_3NHCO	SCH_3	3-Iodobenzyl	2140	3210	2.3	2
56. MRS537	CH_3NHCO	$NHCH_3$	3-Iodobenzyl	4890	4120	3.12	2
57. NECA	C_2H_5NHCO	H	H	6.3	10.3	113[b]	1
58. CGS21680	C_2H_5NHCO	$-NH(CH_2)_2$-*p* $Ph(CH_2)_2$-COOH	H	2600	15	584[b]	1
59. APEC	C_2H_5NHCO	$-NH(CH_2)_2$-p-$Ph(CH_2)_2CO$–$NH(CH_2)_2NH_2$	H	400	5.7	50	6
60. I-PAPA-APEC	C_2H_5NHCO	$-NH(CH_2)_2$-p-$Ph(CH_2)_2CO$–$NH(CH_2)_2NHCO$(3-I-4-NH_2-Ph)	H		1.0		4

61. p-DITC-APEC	C_2H_5NHCO	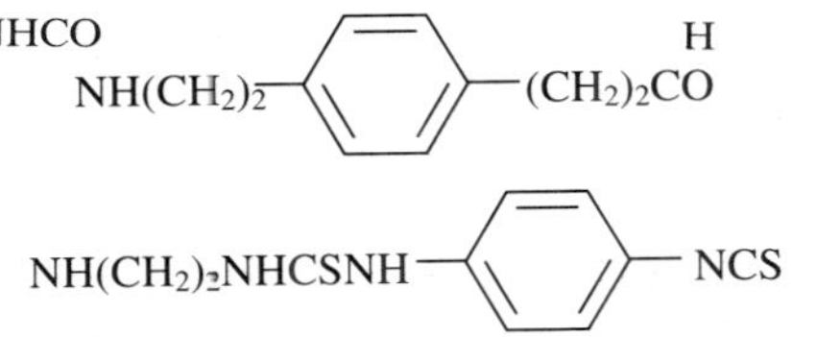 $NH(CH_2)_2$–C6H4–$(CH_2)_2CO$ / $NH(CH_2)_2NHCSNH$–C6H4–NCS	H	280	7.1		4
62. HE-NECA	C_2H_5NHCO	Hex-1-ynyl	H	130	2.2	25.6	
				0.43	170	16.0	
				87.3	95.3	6.8[b]	1
65.	C_2H_5NHCO	H	3-Cl-Ph-CO-	45	420	4.4	
				112	55	103[b]	18
67. UP 202-32	cyclopropyl-NHCO	H	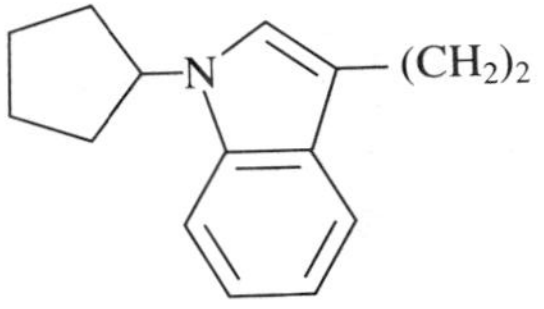 $(CH_2)_2$	110	350		19
68. NNC21-0113	CH_2Cl	Cl	$-OCH_3$	170	13,000	26	9

[a] $K_i \pm$ S.E.M. determined in radioligand binding assays expressed in μM, using the following radioligands: A_1, [^{3}H]N^6-R-phenylisopropyladenosine, **31,** in rat cortical membranes; A_{2A}, [^{3}H]CGS 21680, **58,** binding in rat striatal membranes; A_3, [^{125}I]AB-MECA, **51.** Binding, unless noted, measured in membranes of CHO cells stably transfected with the rat A_3-cDNA.

[b] A_3 affinity determined versus [^{125}I]APNEA (N^6-(4-amino-(2-phenylethyl))adenosine) binding.

References: 1, van Galen et al., 1994; 2, Jacobson et al., 1995b; 3, Siddiqi et al., 1995a; 4, Jacobson et al., 1992a; 5, van Galen et al., 1987; 6, Kim et al., 1994b; Cristalli et al., 1988; 8, Roelen et al., 1996; 9, Bowler et al., 1996; 10, Wieser and Pendleton, 1979; 11, Jacobson et al., 1985; 12, Watanabe et al., 1990; 13, Knutsen et al., 1997; 14, Daly and Jacobson, 1995; 15, Ji et al., 1994; 16, Siddiqi et al., 1995b; 17, Baraldi et al., 1996; 18, Gallo-Rodriquez et al., 1994; 19, Gungor et al., 1994.

urea derivative **65,** was shown to lead to potent agonists, some of which displayed A_3 selectivity (Baraldi et al., 1996).

The second messenger effects of A_3 agonists have also been explored. At cloned rat A_3 receptors expressed in CHO cells, Cl-IB-MECA inhibited adenylyl cyclase with an IC_{50} of 67 nM (Kim et al., 1994b). Coupling of adenosine A_3 receptors to the activation of phosphatidylinositol-4,5-bisphosphate-specific phospholipase C (PLC) has also been demonstrated. In both rat striatal and hippocampal slices selective A_3 adenosine agonists, such as IB-MECA, stimulated PLC in a concentration-dependent manner (Abbracchio et al., 1995). In rat striatum, the potency order for adenosine agonists was identical to their potency order in binding at cloned rat A_3 receptors.

Adenosine Receptor Antagonists

Xanthine Antagonists Adenosine receptor antagonists, of which xanthines (Table 2) and numerous classes of fused heterocyclic compounds (Table 3 and Figure 1) are representative, have been under development as antiasthmatics (Francis et al., 1988), antidepressants (Sarges et al., 1990), antiarrhythmics (Belardinelli et al., 1995), and renoprotective (Suzuki et al., 1992), anti-Parkinson (Kanda et al., 1994), and cognition-enhancing (Schingnitz et al., 1991) drugs.

Theophylline (1,3-dimethylxanthine, **70**) and caffeine (1,3,7-trimethylxanthine, **72**) have only moderate affinity ($K_i \geq 10\ \mu M$) and are essentially nonselective for A_1/A_{2A} receptors. Increasing chain length at the N1- and N3-positions, for example, 1,3-dibenzyl substitution (**76**) results in an increase in A_1 receptor selectivity. Selective antagonists for A_1 receptors principally include many 8-aryl and 8-cycloalkyl xanthine derivatives, such as the xanthine amine congener XAC (8-[4-[[[[(2-aminoethyl)amino]-carbonyl]methyl]oxy]phenyl]-1,3-dipropylxanthine, **108**) and CPX (1,3-dipropyl-8-cyclopentylxanthine, **92**). CPX (also known as DPCPX) shows ~500-fold selectivity for A_1 versus A_{2A} receptors in the rat (Jacobson et al., 1992a). XAC is A_1 selective in the rat (Jacobson et al., 1992a), but nonselective in human and rabbit tissue. Recently, ENX (1,3-dipropyl-8-[2-(5,6-epoxy)norbornyl)xanthine, **99**) was found to be extremely selective for A_1 receptors in binding and functional assays (Belardinelli et al., 1995). Other 8-substituted xanthines include KW-3902 (**98**), which has potent diuretic properties (Suzuki et al., 1992), 8-(dicyclopropylmethyl)-1,3-dipropylxanthine (KF15372, **97**), which is even more potent and A_1-selective than CPX in guinea pig forebrain (Suzuki et al., 1992), and KFM-19 (**95,** (±)8-(3-oxocyclopentyl)-1,3-dipropylxanthine), a potent A_1-selective compound with sufficient aqueous solubility to display good bioavailability (Schingnitz et al., 1991). The *S*-isomer of KFM-19, known as BIIP20 (**96**), is the more potent isomer. KF15372 and KFM-19 are currently under development as cognition enhancers. 8-*p*-Sulfophenyltheophylline (8-PST, **80**) is useful as a peripherally acting antagonist, but is nonselective.

The A_1 receptor has considerably more bulk tolerance than the A_{2A} receptor for substitution at the N^3-position of xanthines. For example, BWA-522 (1-propyl-3-(4-amino-3-iodophenethyl)-8-[(4-carboxymethyloxy)phenyl]xanthine, **107**) shows 19-fold and 32-fold A_1-selectivity versus A_{2A} and A_3 receptors, respectively (all in the rat) (Linden, 1994; Kim et al., 1994a). Substitutions at the N7-position of xanthines are usually not favorable, while N9-substitutions are detrimental to affinity. High receptor affinity is maintained in xanthines with a thio substitution in the carbonyl group at the 2-position, but not at the 6-position. Selectivity for A_{2A} receptors in xanthines has been more difficult to achieve. However, 8-styrylxanthines, such as KF17837 (**85;** Kanda et al., 1994), and CSC (8-(3-chlorostyryl)caffeine, **86**), display a degree of selectivity for A_{2A} versus A_1 receptors.

The parallel in A_1 affinity between CPX and CPA, which have cyclopentyl substituents in common, and between XAC and ADAC, which both have long functionalized chains attached, suggests a common mode of binding of xanthines and adenosines in the receptor binding site. A computer-generated model of the antagonist binding site of the adenosine receptor assumes that N^6-substituents of agonists and C8-substituents of xanthine antagonists bind to the same region of the receptor (van der Wenden et al., 1995).

The A_3 adenosine receptor cloned from rat (Zhou et al., 1992) was shown to be unique among the subtypes in that agonist action is not antagonized by xanthines, such as theophylline. A_3-selective antagonists for general use in pharmacological studies across species are currently under development. Typical K_i values at rat A_3 receptors of roughly 100 μM have been determined for many xanthines that have nearly nanomolar potency at the A_1 or A_{2A} subtypes (Kim et al., 1994a), and these xanthines do not effectively antagonize A_3 agonist–elicited inhibition of adenylyl cyclase. An anionic group attached to the xanthine tended to diminish the affinity at A_1 and A_{2A} receptors. One such compound, **83** had a K_i value at rat A_3 receptors of 9.4 μM and was 7-fold selective for rat A_3 vs A_{2A} receptors. The A_3 receptor affinity of most xanthines is highly species dependent. BWA-522 (**107**), for instance, is more potent in binding to the human (K_i = 18 nM) and sheep (K_i 3 nM) homologs (Salvatore et al., 1993) of the A_3 receptor than to the rat A_3 receptor (K_i = 1.17 μM). XAC (**108**) has been used in pharmacological experiments *in vitro* and *in vivo* in rodents for distinguishing A_3 receptors (K_i = 29 μM, rat) from A_1 and A_{2A} receptors, at which it is much more potent.

It has been proposed (Kim et al., 1994c) that the ribose moiety of adenosine, which is relatively more important for high-affinity binding to A_3 receptors than at other subtypes, is coordinated to a conserved histidine in the seventh transmembrane helical domain of adenosine receptors. Consequently, we tested the hypothesis that a means of anchoring xanthines in the A_3 binding site is adding a sugar moiety at the N^7-position to form xanthine-7-ribosides (Figure 2). At rat brain A_3 receptors, 1,3-dibutylxanthine-7-riboside (DBXR) (**117**), was found to bind with a K_i value of 6.03 μM, whereas the parent xanthine, 1,3-dibutylxanthine (**114**), displayed a K_i value of 143 μM. Thus, the presence of the ribose moiety enhances affinity of xanthines at rat A_3 receptors, while at A_1 receptors the xanthine-7-riboside derivatives are, as a rule, less potent than the parent xanthines. Functionally, DBXR (**117**), as a structural hybrid of classical A_1/A_{2A} agonist (i.e., nucleoside) and antagonist (i.e., xanthine) molecules, appeared to act as a partial agonist at rat A_3 receptors, providing hope that this was a means of designing antagonists. However, upon structural modification that increased the potency and selectivity of the xanthine ribosides at A_3 receptors, full agonism was achieved. Specifically, the structural parallel between adenosine derivatives and the xanthine-7-ribosides was maintained with respect to A_3 receptors affinity. This parallel lead to the design of 1,3-dibutylxanthine-7-riboside-5′-N-methylcarboxamide (DBXRM, **118** Figure 2), which had a K_i value of 229 nM at A_3 receptors, and had 160-fold and >400-fold selectivity for A_3 versus A_1 and A_{2A} receptors, respectively (Jacobson et al., 1995b). The selectivity of this compound is a result of incorporation of the 5′-methyluronamide group, found to enhance A_3 selectivity in IBM-MECA (**50**), and optimization of the alkyl chain length at N′- and N^3-positions. Unlike DBXR, DBXRM acted as a full agonist in the rat A_3 receptor–mediated inhibition of adenylyl

TABLE 2 Affinities (nM) of Xanthine Derivatives in Radioligand Binding Assays at Rat Brain A_1, A_{2A}, and A_3 Receptors[a–d]

X = O, unless noted.[e]

Compound	R_1	R_3	R_7	R_8	K_i (nM) (A_1)[a]	(A_{2A})[b]	(A_3)[c]	Refs.
69. Theobromine	H	CH_3	CH_3	H	83,400	187,000	>100,000 20%(10^{-4})	1
70. Theophylline[d]	CH_3	CH_3	H	H	8,500	25,000	>100,000	2,4
71. [d]	CH_3	CH_3	H	Cl	>100,000 30%(10^{-4})	>100,000 25%(10^{-4})	>100,000	1
72. Caffeine[d]	CH_3	CH_3	CH_3	H	29,000	48,000	~100,000	2
73. Paraxanthine	CH_3	H	CH_3	H	30,000	19,400	>100,000	1
74. IBMX	CH_3	$CH_2CH(CH_2)_3$	H	H	7,000	16,000	~100,000	1
75. DMPX	Propargyl	CH_3	CH_3	H	45,000	16,000		3
76. [d]	Benzyl	Benzyl	H	H	2000	>100,000		2
77.	$CH_3(CH_2)_3$	Benzyl	H	H	668	1200	74,200	2
78. CPT	CH_3	CH_3	H	Cyclopentyl	24	1400	~100,000	3,1
79. 8-PT	CH_3	CH_3	H	Phenyl	86	850	>100,000	1
80. SPT	CH_3	CH_3	H	Phenyl-4-SO_3H	4500	850		3
81. CHC	CH_3	CH_3	CH_3	Cyclohexyl	28,000	9300	~100,000	3,1
82.	CH_3	CH_3	H	$(CH_2)_3COOH$	>100,000 >100,000	>100,000	93,400	2
83. X = S	CH_3	CH_3	H	$(CH_2)_3COOH$	5700	67,800	9,360	2
84.	CH_3	CH_3	CH_3	CH = CHCOOH[e]	>100,000	42,000	130,000	2
85. KF17837	CH_3	CH_3	CH_3	3,4-Dimethoxystyryl	2600	24		3
86. CSC	CH_3	CH_3	CH_3	3-Chlorostyryl[e]	28,200[f]	9.08[g]	>100,000	2
87.	CH_3	CH_3	CH_3	Styryl-3-NH-CO$(CH_2)_2$COOH[e]	20,500[g]	12.9[g]	189,000[g]	2
88.	CH_3	CH_3	CH_3	Styryl-3-NH-CO$(CH_2)_4$COOH[e]	6510[g]	7.22[g]	234,000[g]	2

89. DPX	CH_3CH_2	$CH_3(CH_2)_2$	H	Phenyl	44.5	863		5
90.	allyl	CH_3	CH_3	Phenyl-4-SO_3H	>100,000	27,000		6
91. X = S	$CH_3(CH_2)_2$	$CH_3(CH_2)_2$	H	Cyclopentyl	0.655	314		
92. CPX or DPCPX	$CH_3(CH_2)_2$	$CH_3(CH_2)_2$	H	Cyclopentyl	0.46	340	5290	2
93. BWA 844U	$CH_3(CH_2)_2$	$(CH_2)_2$-(3-I-4-$NH_2\phi$)	H	Cyclopentyl	0.23	2000		7
94. FSCPX	$CH_3(CH_2)_2$	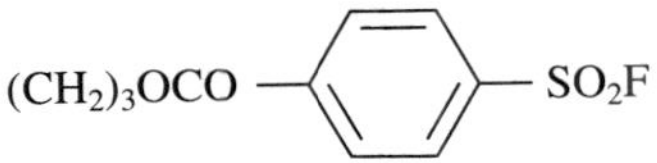 $(CH_2)_3OCO$— —SO_2F	H	Cyclopentyl	1.2	136		8
95. KFM19	$CH_3(CH_2)_2$	$CH_3(CH_2)_2$	H	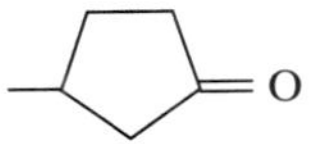	15	2700		3
96. BIIP20	$CH_3(CH_2)_2$	$CH_3(CH_2)_2$	H	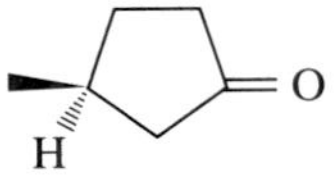				
97. KF15372	$CH_3(CH_2)_2$	$CH_3(CH_2)_2$	H	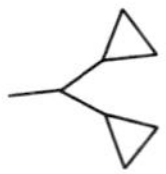	3	430		9
98. KW-3902	$CH_3(CH_2)_2$	$CH_3(CH_2)_2$	H	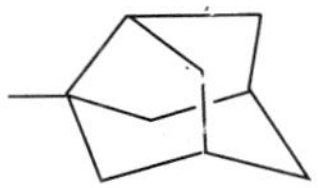	1.3	380		9
99. ENX	$CH_3(CH_2)_2$	$CH_3(CH_2)_2$	H		0.95	380		10
100. 8-PX	$CH_3(CH_2)_2$	$CH_3(CH_2)_2$	H	Phenyl	10	180		3
101. PACPX	$CH_3(CH_2)_2$	$CH_3(CH_2)_2$	H	4-Cl-2-NH_2-Phenyl	2.5	92		9
102. SPX	$CH_3(CH_2)_2$	$CH_3(CH_2)_2$	H	Phenyl-4-SO_3H	140	790	90,100	2
103.	$CH_3(CH_2)_2$	$CH_3(CH_2)_2$	H	Phenyl-4-COOH	200	637	45,100	2
104. BWA1433	$CH_3(CH_2)_2$	$CH_3(CH_2)_2$	H	Phenyl-4-CH=CHCOOH[e]	15	800	15,000	2

TABLE 2 *(Continued)*

Compound	R_1	R_3	R_7	R_8	K_i (nM) $(A_1)^a$	$(A_{2A})^b$	$(A_3)^c$	Refs.
105. XCC	$CH_3(CH_2)_2$	$CH_3(CH_2)_2$	H	Phenyl-4-OCH_2COOH	58	2200	75,700	2
106. X = S	$CH_3(CH_2)_2$	$CH_3(CH_2)_2$	H	Phenyl-4-OCH_2COOH	53.8	226	6770	2
107. I-ABOPX	$CH_3(CH_2)_2$	CH_2(3-I-4-$NH_2\phi$)	H	Phenyl-4-OCH_2COOH	37	700	1170	2
108. XAC	$CH_3(CH_2)_2$	$CH_3(CH_2)_2$	H	Phenyl-4-OCH_2CO-$NH(CH_2)_2NH_2$	1.2[j]	63	29,000	2
109. PD 115,199	$CH_3(CH_2)_2$	$CH_3(CH_2)_2$	H	Phenyl-4-$OCH_2CONH(CH_2)_3N(CH_3)_2$	140	26		3
110. I-PAPA-XAC	$CH_3(CH_2)_2$	$CH_3(CH_2)_2$	H	Phenyl-4-$OCH_2CONH(CH_2)_2NHCO$-phenyl-3-I-4-NH_2	0.1			9
111. m-DITC-XAC	$CH_3(CH_2)_2$	$CH_3(CH_2)_2$	H	Phenyl-4-OCH_2CO-$NH(CH_2)_2NHCSNH$-phenyl-3-NCS				9
112. MDL 102,234	$CH_3(CH_2)_2$	$CH_3(CH_2)_2$	H	[structure: C_2H_5]	23.2	3510		11

113. MDL 102,503	$CH_3(CH_2)_2$	$CH_3(CH_2)_2$	H	CH_3	6.9	1600		11
					6.9	1600		11
114. DBX[d]	$CH_3(CH_2)_2$	$CH_3(CH_2)_2$	H	H	500	29,300	143,000	2
115.	$CH_3(CH_2)_2$	$CH_3(CH_2)_2$	H	CH = CHCOOH[e]	3.37	16.5	73.7	2
116.	$CH_3(CH_2)_2$	$CH_3(CH_2)_2$	CH_3	CH = CHCOOH[e]	5.91	23.2	127	2
117. DBXR[d]	$CH_3(CH_2)_2$	$CH_3(CH_2)_2$	H	Ribose	4190	19,500	6030	12
118. DBXRM	$CH_3(CH_2)_2$	$CH_3(CH_2)_2$	H	Ribose-5′-N-methyluronamide	37,300	>100,000	229	12
119. [d]	$CH_3(CH_2)_2$	$CH_3(CH_2)_2$	H	H	1260	>10,000	>10,000	2

[a] Displacement of specific [^{3}H]R-PIA binding, unless noted, in rat brain membranes.

[b] Displacement of specific [^{3}H]CGS 21680 binding, unless noted, in rat striatal membranes.

[c] Displacement of specific [^{125}I]AB-MECA binding, unless noted, in membranes of CHO cells stably transfected with the rat A_3-cDNA.

[d] Values from van Galen et al., 1994: A_3 affinity vs. [^{125}I]APNEA.

[e] CH = CH group, when present in R_8 substituent, is always trans.

[f] The binding assay was carried out under fluorescent room light.

[g] The assay was carried out in the dark.

References: 1, van Galen et al., 1994; 2, Kim et al., 1994a; 3, Daly and Jacobson, 1995,; 4, Linden et al., 1994; 5, Bruns et al., 1986; 6, Jacobson et al., 1993; 7, Patel et al., 1988; 8, Scammels et al., 1994; 9, Jacobson et al., 1995b; 10, Belardinelli et al., 1995; 11, Peet et al., 1993; 12, Jacobson et al., 1995b.

TABLE 3 Classes of Adenosine Receptor Antagonists[m]

Chemical Class	Example
Adenines	N-0861 (**122**)[a]; N^6-butyl-8-phenyladenine
Adenosines, ribose-modified	5′-Deoxy-5′-methylthioadenosine
Barbiturates	DMBB
Benzimidazoles	1-Methyl-2-phenylimidazole
Benzo[1,2-c;5,4-c′]dipyrazoles	1,7-Dihydro-3,5,8-trimethylbenzo[1,2-c;5,4-c′]dipyrazole
Benzo[b]furans	5-(3-Hydroxypropyl)-7-methoxy-2-(3′-methoxy-4′-hydroxyphenyl)-3-benzo[b]furancarbaldehyde
Benzo[g]pteridine-2,4-diones	Alloxazine (**127**)[b]
β-Carbolines	DMCM[c]
Cytochalasins	Cytochalasin B[c]
7-Deazadenines	See pyrrolo[2,3-d]pyrimidines
2,4-Diaminopteridines	See Siddiqi et al., 1996
Dibenz[b,f]azepines	Carbamazepine
Dihydropyridines	Nicardipine, MRS 1191 (**129**)[i]
Flavanones	Sakuranetin[d]
Flavones	Hispidulin[d], MRS 1067 (**128**)
Flavonols	Galangin
Imidazo[1,2-a]pyrazines	SC-12
Imidazo[4,5-b]pyridines	Sulmazole
Imidazo[4,5-c]quinolines	CPPIQA
Imidazo[4,5-e][1,4]diazepine-5,8-diones	4,7-Dipropyl-1-benzyl-4,5-tetrahydro-6H-imidazo-[4,5-e][1,4]diazepine-5,8-dione
Imidazo[4,5-f]quinazoline-7,9-diones	*Prox*-benzotheophylline
Imidazo[4,5-g]quinazoline-6,8-diones	*Lin*-benzotheophylline[c]
Imidazolidines	DPI
Isoflavones	Genistein[d]
5-Oxoimidazo[1,5-c]pyrimidines	See Siddiqi et al., 1996
7-Phenyllumazine	See Siddiqi et al., 1996
Phenylpyrazolo[1,5-a]pyridines	FK453 (**123**)[j]
Pteridine-2,4-diones	Lumazine[c]
Pyrazolo[3,4-b]pyridines	Cartazolate; ethazolate; tracazolate
Pyrazolo[3,4-d]pyrimidines	DJB-KK; APPP
Pyrazolo[4,3-d]pyrimidines	5-(2-Amino-4-chlorophenyl)-1,3-dimethyl-pyrazolo[4,3-d]pyrimidin-7-one
Pyrazolo[1,5-a]quinazolines	See Siddiqi et al., 1996
Pyrazolo[4,3-c]quinolines	CGS 8216
Pyrazolo[4,3-e]triazolo[1,5-c]pyrimidines	SCH 52861 **(125)**[k]
Pyrimidines	Amiloride
Pyrimido[4,5-b](tetrahydro)indoles	4-Amino-9-phenyl-9H-pyrimido[4,5-b]indole
Pyrrolo[2,3-d]pyrimidines (7-deazaadenines)	ADPEP
Quinazolines	ADQZ
Tetrahydrobenzothiophenones	BTH_4 **(121)**[e]
Thiazolo[2,3-b]quinazolines	HTQZ
Thiazdopyrimidines	L-268605 **(131)**
Thiazolo[4,5-d]pyrimidine-5,7-diones	4,6-Dimethyl-2-phenyl-4,5,6,7-tetrahydrothiazolo[4,5-d]pyrimidine-5,7-dione
Thiazolo[5,4-d]pyrimidine-5,7-diones	DJB-W
Triazolonaphthyridines	L-249313

TABLE 3 *(Continued)*

Chemical Class	Example
[1,2,4]Triazolo[4,3-b]pyridazines	CL 218872
[1,2,4]Triazolo[1,5-c]quinazolines	CGS 15943 **(120)**[f]
[1,2,4]Triazolo[4,3-a]quinoxalines	CP 66713 **(126)**; CP 68247[g]
Triazolo[2,3-a][1,3,5]triazines	ZM 241385 **(124)**[l]
Xanthines	See Table 2
Xanthines, benzo-separated	See imidazo[4,5]quinazolinediones
Xanthines, mesoionic	Anhydro-6,8-di-n-propyl-5-hydroxy-7-oxothiazolo[3,2-a]pyrimidinium hydroxide[c]
Xanthine-7-ribosides	1,3-Dibutylxanthine-7-riboside[h]

[a] Belardinelli et al., 1995.
[b] Daly and Jacobson, 1995.
[c] Siddiqi et al., 1996.
[d] Ji et al., 1996.
[e] van Rhee et al., 1996a.
[f] Francis et al., 1988.
[g] Sarges et al., 1990.
[h] Kim et al., 1994c.
[i] van Rhee et al., 1996b.
[j] Akahare et al., 1996.
[k] Baraldi et al., 1995.
[l] Poucher et al., 1995.
[m] Jacobson et al., 1992b.

cyclase. Thus, there was a tendency towards an increase of efficacy as the affinity increased within the same series of compounds.

Nonxanthine Antagonists Numerous structurally diverse nonxanthine antagonists (Table 3, Figure 1) have also been identified, many of which are not well defined in terms of SAR. Among the first classes of heterocycles found to antagonize the effects of adenosine agonists were the "tricyclic" nonxanthine antagonists, including the triazoloquinazolines (Francis et al., 1988), the triazoloquinoxalines (Sarges et al., 1990), and the imidazoquinolines (van Galen et al., 1991). Other potent and A_1-selective antagonists have been derived from adenine and include the N^6-substituted 9-methyladenines (Thompson et al., 1991), such as N-0861 (**122,** (±)N^6-*endo*norbornyl-9-methyladenine; A_1: 10 nM; A_{2A}: 6100 nM in bovine brain) (Belardinelli et al., 1995, Figure 1). An attempt to design A_3-selective adenine derivatives by applying the structural principles of recognition derived from adenosine agonists resulted in loss of potency and selectivity (Jacobson et al., 1995).

FK453, **123,** is a novel non-xanthine antagonist that is 657-fold selective for binding to A_1 versus A_{2A} receptors and is under development for the treatment of hypertension and renal failure (Akahani et al., 1996). The triazoloquinazoline, CGS 15943 (9-chloro-2-(2-furyl)[1,2,4]triazolo[1,5-*c*]quinazoline-5-amine, **120**), is a potent adenosine receptor antagonist with seven-fold selectivity for A_{2A} receptor versus A_1 receptors, and an IC_{50} of 3 nM at the A_{2A} receptor (Francis et al., 1988). Another series of tricyclic compounds, the triazoloquinoxalines (Sarges et al., 1990), which for a time were in clinical trials as antidepressants, were later found to be adenosine receptor antagonists. The selectivity depended strongly on the ring substitution, resulting in both A_1- and A_{2A}-selective adenosine receptor antagonists, such as CP 66713 (**126**). Recent reports

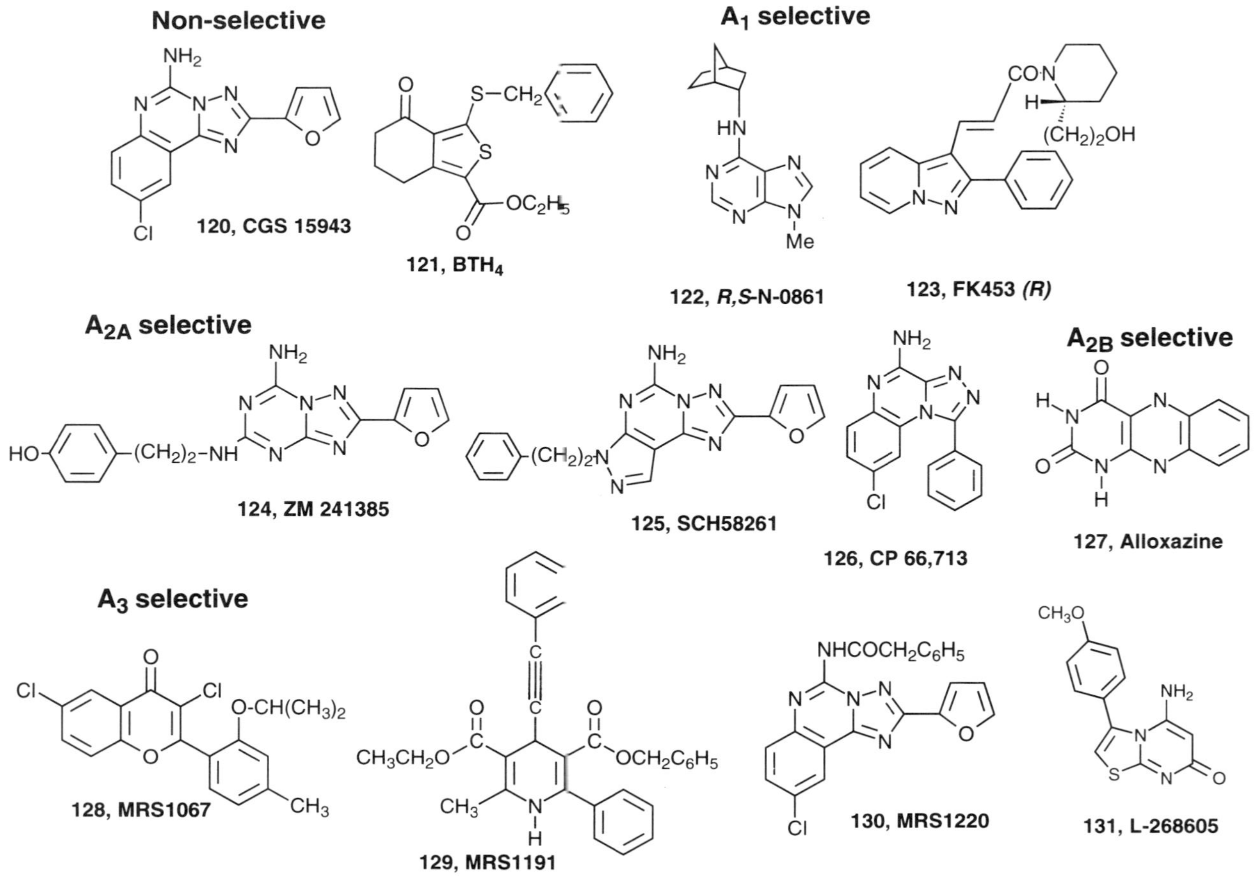

Non-selective

120, CGS 15943

121, BTH_4

A_1 selective

122, *R,S*-N-0861

123, FK453 *(R)*

A_{2A} selective

124, ZM 241385

125, SCH58261

126, CP 66,713

A_{2B} selective

127, Alloxazine

A_3 selective

128, MRS1067

129, MRS1191

130, MRS1220

131, L-268605

FIGURE 1. Non-xanthine adenosine antagonists.

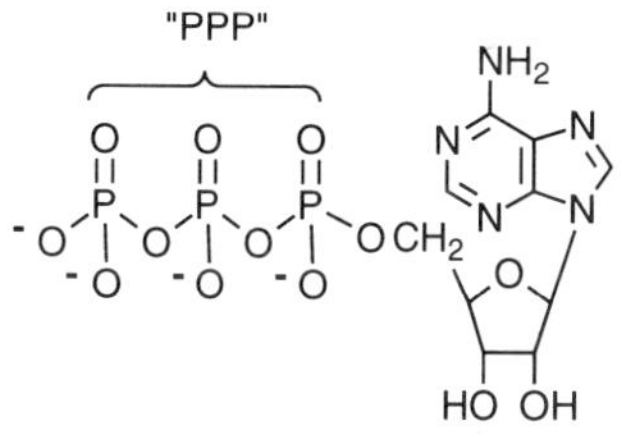

132, ATP 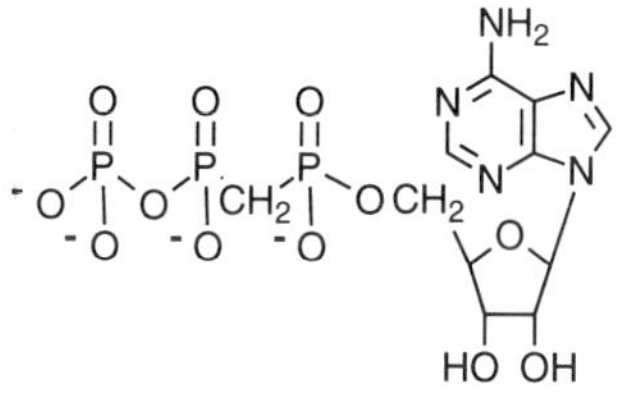 **133, α,βMeATP** 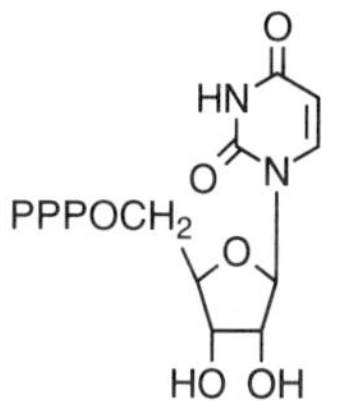 **134, UTP**

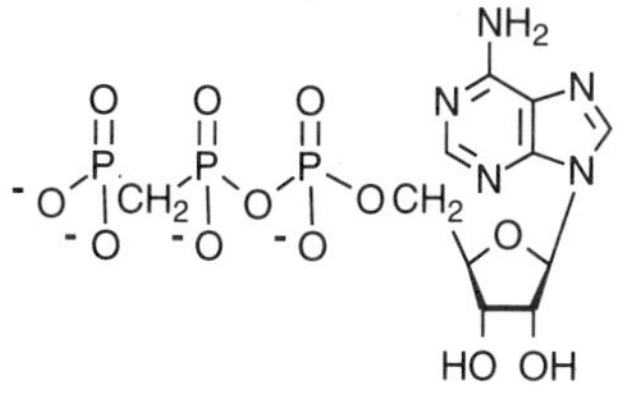

135, β,γMeATP 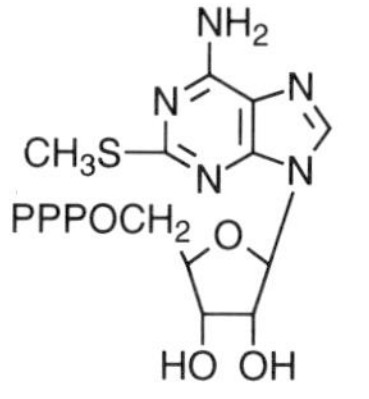 **136, 2MeSATP** 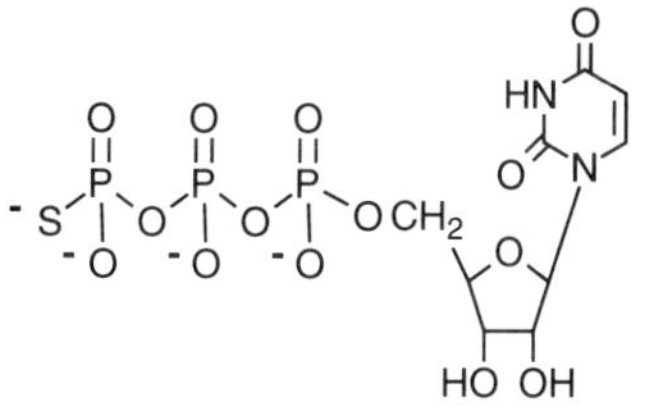 **137, UTP-γS**

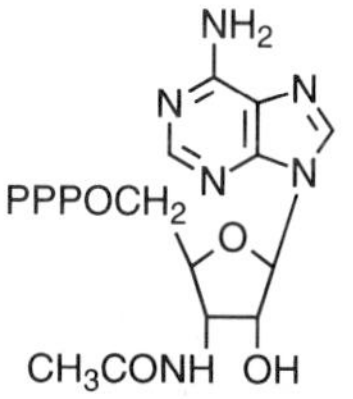

138, 3'-deoxy-3'-acetylamino-ATP 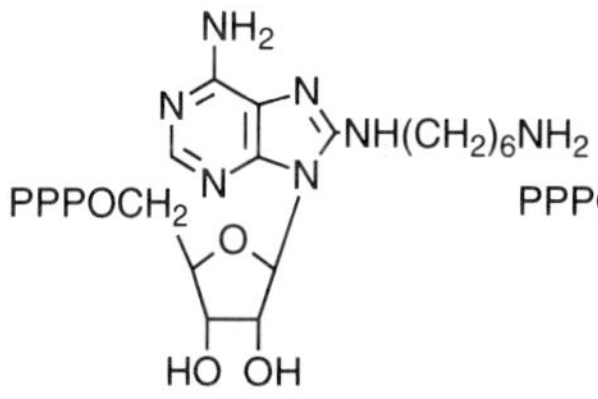 **139, 8-(6-aminohexyl)-amino-ATP** 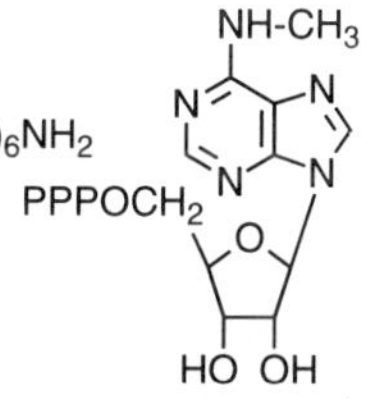 **140, N^6-Methyl-ATP** 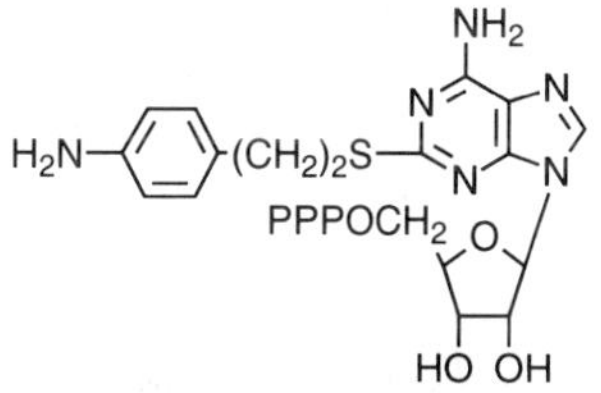 **141, 2-(4-aminophenylethylthio)ATP**

FIGURE 2. P2 receptor agonists.

(Baraldi et al., 1995; Poucher et al., 1995) of the highly potent and selective nonxanthine A_{2A} receptor antagonists SCH58261 (**125**) and ZM241385 (**124**), have removed a major obstacle in the characterization of the function of adenosine A_{2A} receptors. The nonxanthine ZM241385 (**124**) is the most selective A_{2A} antagonist (6800-fold) yet reported. Radioiodination of ZM241385 has provided a highly potent and selective A_{2A} antagonist radioligand (Palmer et al., 1995b).

There are no potent and selective antagonists for the A_{2B} receptor. However, alloxazine (**127**) is 10-fold selective for A_{2B} versus either A_1 or A_{2A} adenosine receptors (Daly and Jacobson, 1995).

Recently we identified novel adenosine receptor ligands as a result of screening of a natural products library (Ji et al., 1996; Siddiqi et al., 1996). A number of classes of nonxanthine adenosine antagonists have already been reported, including various nitrogen heterocycles and several classes of non-nitrogen heterocycles. For example we recently reported that tetrahydrobenzothiophenones, e.g., ethyl 3-benzylthio-4,5,6,7-tetrahydrobenzo[*c*]thiophen-4-one-1-carboxylate (**121;** Figure 1, BTH_4), bind to adenosine receptors in the micromolar range (van Rhee et al., 1996a). One member of this series, 4,5-dihydro-1-(methylthio)-benzo[*c*]thiophen-4-one-1-carboxylic acid hydrazide, bound with slight selectivity for A_3 receptors. Diverse structures that have displayed some degree of selectivity in binding to rat A_3 receptors (Siddiqi et al., 1996) include folic acid, a pyridopyrimidinone, cytochalasin H, 11-hydroxytetracarbazolenine, dipyridamole, and certain sulfonylpiperazines (e.g., HA-100).

Flavonoids, a group of phenolic compounds that are widespread in all vascular plants, have also provided leads for A_3 selectivity (Ji et al., 1996). The naturally occurring galangin binds nonselectivity to adenosine receptors, but chemical modification has resulted in analogs such as MRS 1067 (**128**), which is 200-fold selective for human A_3 versus A_1 receptors (Karton et al., 1996).

Exploration of the structure–activity relationship of dihydropyridines at adenosine receptors has resulted in MRS 1191 (**129**), which is 1300-fold selective for human A_3 versus rat A_1 receptors (van Rhee et al., 1996, Jiang et al., 1996). Although much less potent and selective at rat A_3 receptors, MRS 1191 nevertheless was used as selective receptor probe in the rat hippocampus (Dunwiddie et al., 1997). The A_3 antagonist L-268605, **131** was identified by M. Jacobson et al. at Merck.

STRUCTURE–ACTIVITY RELATIONSHIPS AT ATP RECEPTORS

P_2 Receptor Agonists

Nearly all of the P_2 receptor agonists are derivatives of adenosine- and uridine-5′-triphosphate. The pharmacological profiles have been reported for numerous ATP analogs synthesized with modifications at the triphosphate, ribose 2′ or 3′, purine C2 or C8, or at the purine N^6 positions (Figure 2) (Cusack et al., 1990; Jacobson et al., 1995a). The structure–activity relationships for a variety of adenine nucleotide analogs at P_{2X} and P_{2Y} receptors have been studied classically in smooth muscle preparations. For example, P_{2Y} receptors mediate the relaxation of smooth muscle in three different preparations (guinea pig taenia coli, rabbit aorta, and rabbit mesenteric artery). ATP-induced relaxation of the aorta occurs via an endothelium-dependent P_{2Y} receptor, and relaxation of the mesenteric artery occurs via an endothelium-independent P_{2Y} receptor. Activity at P_{2X} receptors has been established by measurement of the contraction of rabbit saphenous artery and of guinea pig vas deferens and urinary bladder. More recently, it has been possible to screen for activity using cloned P_2 receptors. The development of selective ligands for P_2 receptors has lagged behind that for most other receptors due to the difficulty in interpreting classical pharmacological studies, which are characterized by receptor heterogeneity and ligand instability, the lack of potent antagonists, and, until recently, the lack of cloned receptors.

Triphosphate Modifications Modification of the triphosphate group in the form of replacement of the bridging oxygen atoms with methylene units or of the charged oxygen atoms with sulfur has in some cases resulted in potent analogs having stability towards nucleotidases, α,β-MeATP (**133,** α,β-methylene adenosine 5′-triphosphate), in particular is noteworthy for its potency at P_{2X} receptors and its stability (Kennedy and Leff, 1995). The thio substitution at the terminal phosphate also provides stability, leading to such analogues as ATPγS (adenosine 5′-O-(3-thiotriphosphate)), ADPβS (adenosine 5′-O-(2-thiodiphoshate)), and UTPγS (**137,** uridine 5′-O-(3-thiotriphosphate)), a potent agonist at P_{2U} receptors.

Ribose Modifications The weak agonist 2′-deoxy-ATP was selective for taenia coli P_{2Y} receptors versus either vascular P_{2Y} receptors or P_{2X} receptors (Burnstock et al., 1994). 2′,3′-Isopropylidene-AMP acted selectively at endothelial rabbit aorta P_{2Y} receptors versus P_{2Y} receptors in the taenia coli, mesenteric artery, and erythrocytes and was inactive at P_{2X} receptors (Burnstock et al., 1994).

3′-Benzylamino-3′-deoxyl-ATP was very potent at P_{2X} receptors in the guinea pig vas deferens and slightly less potent in the urinary bladder. This nucleotide was inactive at rabbit saphenous artery P_{2X} receptors and at all P_{2Y} receptors (Burnstock et al., 1994). The P_{2X}-potency noted was approximately an order of magnitude greater than that of α,β-MeATP, which is used widely in studies of P_{2X} receptors. 3′-Benzylamino-3′-deoxy-ATP caused the same profile of contractile response as ATP or α,β-MeATP, i.e., a transient twitchlike contraction. This suggests that it, too, would cause rapid desensitization.

Adenine Modifications Functionalized congeners of 2-methylthioadenosine-5′-triphosphate (2-MeSATP), a potent P_{2Y} receptor agonist, were synthesized as receptor probes (Fischer et al., 1993). The strategy consisted of attaching substituted alkyl thio chains at the C2-position. Chain elongation at this site preserved high potency at P_{2Y} receptors, thus proving that this is a position on the receptor having great tolerance for structural variation of the ligand. Activity of 2-thioether derivatives of ATP at P_{2Y} receptors varied somewhat, depending on the distal structural features, and activity at P_{2X} receptors varied to an even greater degree. All 11 of the synthesized 2-thioethers of ATP (Fischer et al., 1993) stimulated phospholipase C in turkey erythrocyte membranes and led to the production of inositol phosphates with $K_{0.5}$ values between 1.5 and 770 nM. In smooth muscle assay systems for activity at P_{2Y} receptors, 2-thioethers of ATP displayed EC_{50} values in the range between 10 nM and 1 μM. There was a significant correlation for the 2-thioether compounds between the concentrations required for inositol phosphate production in turkey erythrocyte membranes and those for relaxation mediated via the P_{2Y} receptors in the guniea pig taenia coli, but not for vascular P_{2Y} receptors or for P_{2X} receptors. At P_{2X} receptors, no activity was observed in the rabbit saphenous artery, but variable degrees of activity were observed in the guniea pig vas deferens and bladder depending on distal substituents of the 2-thioether moiety.

2-(7-Cyanohexylthio)-ATP is more potent than 2-MeSATP at the taenia coli P_{2Y} receptor. A *p*-aminophenethylthio ether, intended as a receptor group for radioiodination and potential cross-linking to the receptor, displayed the highest affinity of all the analogs at turkey erythrocyte P_{2Y} receptors. A *p*-nitrophenethylthioether was relatively weak at P_{2Y} receptors, but provided selectivity for a subset (vas deferens) of the P_{2X} receptor class. 7-Aminoheptylthio- and 7-thioheptylthioethers synthesized for the ease of the further derivatization by acylation or alkylation and to probe potential accessory binding sites on the receptor, displayed $K_{0.5}$ values of 73 and 770 nM, respectively, at erythrocyte P_{2Y} receptors.

The addition of a functionalized chain at the 2-position allowed for truncation of the triphosphate group with retention of affinity, thus circumventing one of the major complications in interpreting ATP pharmacological results: i.e., the action of ectonucleotidases. AMP, itself, was inactive at P_{2Y} receptors. 2-Thioether analogs of adenosine monophosphate were full agonists at erythrocyte P_{2Y} receptors (Fischer et al., 1993; Boyer et al., 1996), although generally several orders of magnitude less potent than the corresponding 2-thioether triphosphate analog. For example, a 2-hexenylthioether of AMP was 8 times more potent than ATP itself in the stimulation of phospholipase C, but was 33 times less potent than the corresponding triphosphate.

Thus, the long chain may act as a distal anchor of the ligand at an accessory binding site on the receptor. Also, several adenosine diphosphate 2-thioether analogs proved equipotent to the corresponding ATP analogs at erythrocyte P_{2Y} receptors. The endogenous agonists ADP and ATP were also nearly equipotent in this assay.

A further benefit of the presence of a long chain thioether group at the 2-position was increased stability of the triphosphate group at the 5′-position (Fischer et al., 1993). It is likely that the long chains, although at a site on the molecule distal to the triphosphate group, interfere with the binding site of ectonucleotidases. The apparent stability of these compounds combined with a high potency suggests the synthesis of radiolabeled analogs in this series.

Modifications of ATP other than 2-thioethers also resulted in unexpected selectivity (Burnstock et al., 1994). Certain analogs displayed selectivity or specificity within the P_{2X} or P_{2Y} receptor superfamilies, suggesting possible subclasses. Selectivity was achieved for P_{2Y} receptors of the mesenteric artery, the aorta, or the P_{2Y} receptor of the taenia coli and erythrocytes. 8-(6-Aminohexylamino)ATP (**139**) acted selectively at endothelial rabbit aorta P_{2Y} receptors versus P_{2Y} receptors in the taenia coli, mesenteric artery, and erythrocytes, and was inactive at P_{2X} receptors. The potent agonist N^6-methyl ATP was selective for taenia coli P_{2Y} receptors versus either vascular P_{2Y} receptors or P_{2X} receptors (Fischer et al., 1993). N^6-Methyl ATP was approximately equipotent to ATP at taenia coli P_{2Y} receptors. The N^6-modification may prove to be a general means of increasing P_{2Y} selectivity, since it was compatible with other modifications. A hybrid N^6-methyl and 2-thioether ATP derivative, N^6-methyl-2-(5-hexenylthio)-ATP, was synthesized and found to be very potent at erythrocyte and taenia coli P_{2Y} receptors and inactive at P_{2X} receptors (Fischer et al., 1993).

In summary, long-chain 2-thioethers have been found to enhance the potency (particularly at P_{2Y} receptors) or selectivity (particularly within the P_{2X}-class) of ATP analogs. The longer chain members tended to be stable to ectonucleotidases. The corresponding monophosphatase are full agonists at P_{2Y} receptors. Groups for radiolabeling and attachment of larger reporter moieties, such as aliphatic and aryl amines, were introduced on the 2-thioether chain. A modification of the ribose 2′-position and certain purine modifications of ATP other than 2-thioethers have resulted in P_{2Y} selectivity. N^6-Methyl-ATP, its congeners, and 2′-deoxy-ATP are selective agonists at taenia coli P_{2Y} receptors. 8-(6-Aminohexylamino)ATP and ATP N^1-oxide are selective for endothelial P_{2Y} receptors. However modifications at the ribose 3′-position result in highly variable receptor selectivity. Thus, 3′-deoxy-ATP is a weak, but selective P_{2X} agonist, while 3′-acetamido-3′-deoxy-ATP (**138**) is active at both P_{2X} receptors and mesenteric artery P_{2Y} receptors. 3′-Benzylamino-3′-deoxy-ATP is highly potent and selective for P_{2X} receptors.

P_2 Receptor Antagonists

The first antagonist described for P_2 receptors is probably the alkaloid quinidine (Burnstock, 1972), although it was neither very potent nor selective for P_2 receptors versus adrenoceptors. Throughout the evolution of the P_2 purinergic field there have been reports of antagonism of the effects of ATP by a number of substances, usually highly negatively charged high-molecular-weight organic molecules (Figure 3). Among these substance were histochemical dyes, such as reactive blue 2 (**145,**) a mixture of cibachrome blue and basilen blue, itself a mixture of isomers; Burnstock and Warland, 1987), brilliant blue G (Soltoff et al., 1989); Evans blue (Bültmann et al., 1993), trypan

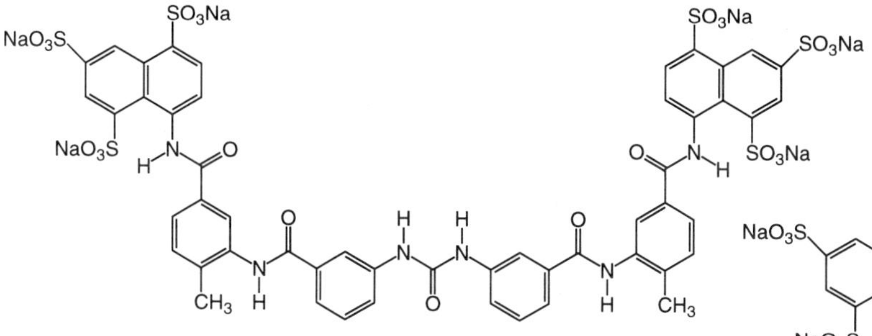

142, suramin

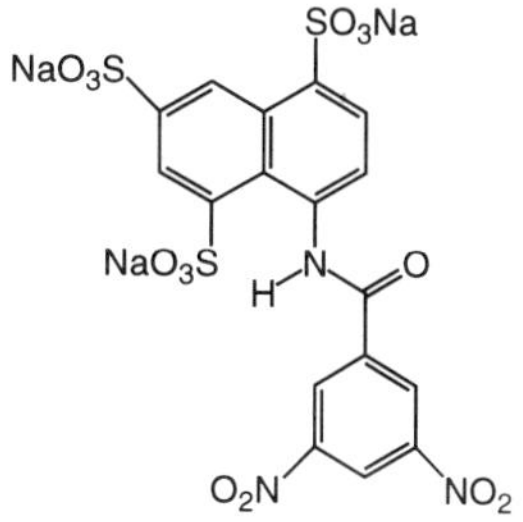

143, XAMR0721

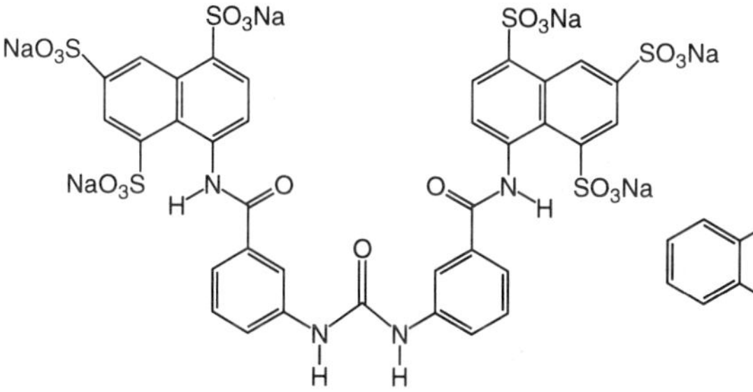

144, NF023

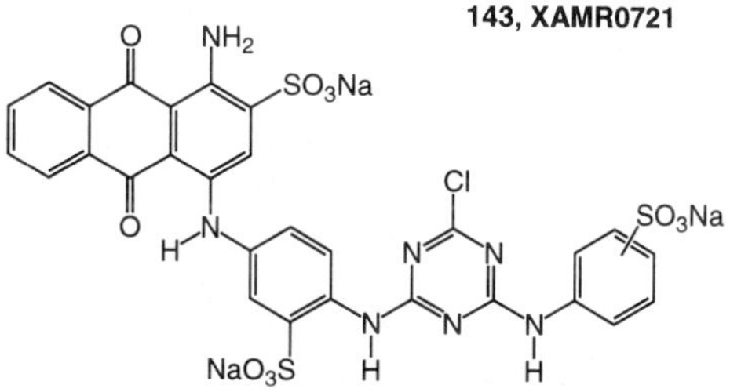

145, Reactive Blue 2

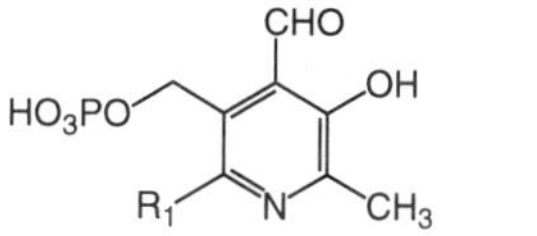

R_1 = H **146, pyridoxal phosphate**

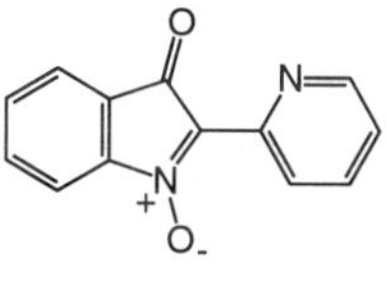

149, 2,2'-Pyridylisatogen

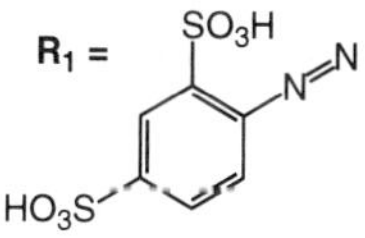

R_1 =

147, PPADS

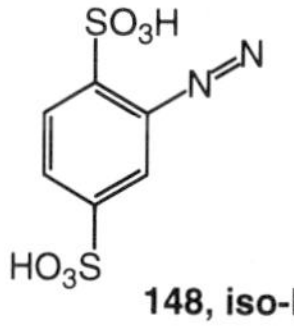

148, iso-PPADS

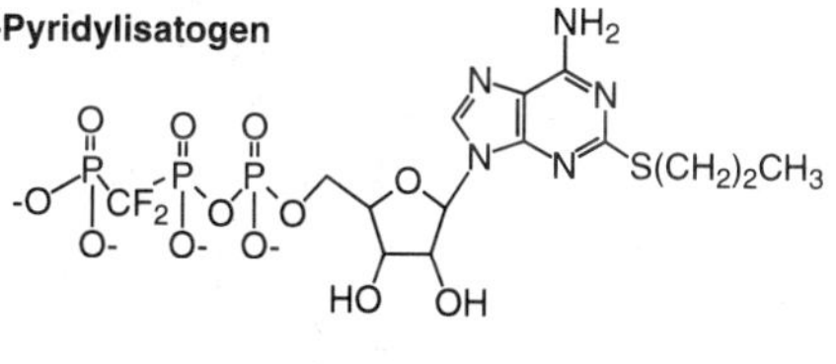

150, ARL 66,096

FIGURE 3. P2 receptor antagonists.

blue (Bültmann et al., 1994a), and reactive red 2 (Bültmann et al., 1995). Other structural classes included the pyridylisatogen tosylates (**149**), weak P_2 receptor antagonists that also displayed strong receptor-independent vasorelaxant effects (Chahl, 1979; Foster et al., 1983), and the bee venom peptide apamin. Initially, apamin was thought to be an ATP antagonist, but was later shown to be unsatisfactory since it turned out to be a channel blocker rather than a competitive receptor antagonist (Maas et al., 1980).

The naphthylsulfonate derivative suramin (**142**) has been developed as a trypanocidal drug (Gray, 1966), for treatment of *Onchocerciasis* (Duke, 1968), and as a HIV-RT inhibitor (Jentsch et al., 1987). Suramin was also found to be a competitive antagonist at several subtypes of P_2 receptors (e.g., Dunn and Blakely, 1988; Boyer et al., 1994). Since suramin is a large complex polysulfonated molecule, an effect was made to identify the minimal pharmacophore with respect to P_2 receptors. Truncated versions of suramin were tested as antagonists in binding and functional assays in hepatocytes and turkey erythrocytes (van Rhee et al., 1994). A number of the compounds proved to be selective for P_2 receptors versus ectonucleotidases (Beukers et al., 1995). One suramin analogue, NF023 (**144**), is selective for P_{2X} receptors by approximately an order of magnitude in either rabbit vas deferens or rat urinary bladder versus the P_{2Y} receptors in guinea-pig taenia coli and rat duodenum (Ziyal et al., 1994). Another related compound, GR200282 (4,4′-[carbonylbis(imino-3-benzoylimino)]-bis[5-hydroxy-naphthalene-2,7-disulfonic acid]), has been described as antagonizing P_{2X} receptor-mediated responses, but its selectivity is not known (Khakh et al., 1994).

A diazo derivative of the coenzyme pyridoxal phosphate (**146**), pyridoxalphosphate-6-azophenyl-2′,4′-disulfonic acid (**147,** PPADS), was recently shown to be an ATP antagonist at the vas deferens P_{2X} receptor. PPADS has now been examined in a variety of P_2 receptor systems and its selectivity for P_{2X} receptors was found to be 10 to 20-fold (Windscheif et al., 1995). Nevertheless, PPADS was demonstrated to antagonize ATP-agonist effects at several P_{2Y} receptors, including those natively present on rat C6 glioma cells and turkey erythrocytes (Boyer et al., 1994). Initially, no distinction was made between PPADS (pyridoxalphosphate-6-azophenyl-2′,4′-disulfonic acid) and isoPPADS (**148,** pyridoxalphosphate-6-azophenyl-2′,5′-disulfonic acid) due to a mistake of the suppliers. The compounds have distinct pharmacological profiles (e.g., Connolly, 1995), which makes the analysis of data with regard to these compounds burdensome.

The irreversible antagonist $ANAPP_3$ (3′-O-(2-[N-(4-azido-2-nitrophenyl)amino]-propionyl)ATP) has been used to analyze P_{2X} receptor–mediated contractions in guinea pig vas deferens (Hogaboom et al., 1980), but because it is a direct derivative of ATP its usefulness is limited by the same side effects as the ATP-derived agonists.

The ATP transport inhibitor DIDS (4,4′-diisothiocyanostilbene-2,2′-disulfonic acid) was initially used to antagonize P_{2Z} receptor–mediated responses in parotid acinar cells (IC_{50} = 35 μM; Soltoff et al., 1993), but is particularly useful as an antagonist for distinguishing bladder and PC12 P_{2X} purinoceptors, where it has IC_{50} values of ~1 and >100 μM, respectively (Bültmann et al., 1994b).

Thienopyridines have been developed as antagonists of the ADP-induced aggregation of platelets, but they are mechanistically unrelated to the effects observed by P_2 receptor stimulation (Savi et al., 1994).

The stable nucleotide derivative ARL67085, **150,** antagonizes the effects of ADP at the yet uncloned P_{2T} receptor in platelets, and it is 30,000-fold selective for this subtype. ARL67085 is under development as an antithrombotic agent (Humphries et al., 1995).

REFERENCES

Abbracchio MP, Brambilla R, Ceruti S, Kim HO, von Lubitz DKJE, Jacobson KA, Cattabeni F (1995): G-protein-dependent activation of phospholipase-C by adenosine A_3 receptors in rat-brain. Mol Pharmacol 48:1038–1045.

Baraldi PG, Cacciari B, Spalluto G, Ji XD, Olah ME, Stiles G, Dionisotti S, Zocchi C, Ongini E, Jacobson KA (1996): Novel N^6-(substituted-phenylcarbamoyl)adenosine-5′-uronamides as potent agonists for A_3 adenosine receptors. J Med Chem 39:802–806.

Baraldi PG, Cacciari B, Spalluto G, Borioni A, Viziano M, Dionisotti S, Ongini E (1995): Current developments of A_{2a} adenosine receptor antagonists. Curr Med Chem 2:707–722.

Belardinelli L, Curtis AB, Bertolet B (1997): Cardiac electrophysiology of adenosine: Antiarrhythmic and proarrhythmic actions, this volume.

Belardinelli L, Shryock JC, Zhang Y, Scammells PJ, Olsson R, Dennis D, Milner P, Pfister J, Baker SP (1995): 1,3-dipropyl-8-[2-(5,6-epoxy)norbornyl]xanthine, a potent, specific and selective A_1 adenosine receptor antagonist in the guinea-pig heart and brain and in DDT_1MF-2 cells. J Pharmacol Exp Ther 275:1167–1176.

Beukers MW, Kerkhof CJ, van RM, Ardanuy U, Gurgel C, Widjaja H, Nickel P, IJzerman AP, Soudijn W (1995): Suramin analogs, divalent cations and ATP-γ-S as inhibitors of ecto-ATPase. Naunyn Schmiedebergs Arch Pharmacol 351:523–528.

Bowler AN, Olsen UB, Thomsen C, Knutsen LJS (1996): New adenosine A_3 ligands controlling cytokines. Drug Devel Res 37:173.

Boyer JL, Siddiqi S, Fischer B, Romera-Avila T, Jacobson KA, Harden TK (1996): Identification of potent P_{2Y} purinoceptor agonists that are derivatives of adenosine 5′-monophosphate. Br J Pharmacol 118:1959–1964.

Boyer JL, Zohn IE, Jacobson KA, Harden TK (1994): Differential effects of P_2-purinoceptor antagonists on phospholipase C- and adenylyl cyclase-coupled P_{2Y}-purinoceptors. Br J Pharmacol 113:614–620.

Bridges AJ, Bruns RF, Ortwine DF, Priebe SR, Szotek DL, Trivedi BK (1988): N^6-[2-(3,5-dimethoxyphenyl)-2-(2-methylphenyl)ethyl]adenosine and its uronamide derivatives. Novel adenosine agonists with both high affinity and high selectivity for the adenosine A_2 receptor. J Med Chem 31:1282–1285.

Bruns RF, Lu GH, Pugsley TA (1986): Characterization of the A_2 adenosine receptor labeled by [^{3}H]NECA in rat striatal membranes. Mol Pharmacol 29:331–346.

Bültmann R, Starke K (1993): Evans blue blocks P_{2X}-purinoceptors in rat vas deferens. Naunyn Schmiedebergs Arch Pharmacol 348:684–687.

Bültmann R, Trendelenburg M, Starke K (1994a): Blockade of P_{2X}-purinoceptors by Trypan Blue in rat vas deferens. Br J Pharmacol 113:349–354.

Bültmann R, Starke K (1994b): Blockade by 4,4′-diisothiocyanatostilbene-2,2′-disulphonate (DIDS) of P_{2X}-purinoceptors in rat vas deferens. Br J Pharmacol 112:690–694.

Bültmann R, Starke K (1995): Reactive Red 2—a P_{2Y}-selective purinoceptor antagonist and an inhibitor of ecto-nucleotidase. Naunyn Schmiedebergs Arch Pharmacol 352:477–482.

Burnstock G (1972) Purinergic nerves. Pharmacol Rev 24:509–581.

Burnstock G, Warland JJ (1987): P_2-purinoceptors of two subtypes in the rabbit mesenteric artery: Reactive Blue 2 selectively inhibits responses mediated via the P_{2Y}- but not the P_{2X}-purinoceptor. Br J Pharmacol 90:383–391.

Burnstock G, Fischer B, Hoyle CHV, Maillard M, Ziganshin AU, Brizzolara AL, von Isakovics A, Boyer JL, Harden, TK, Jacobsen KA (1994): Structure-activity relationships for derivatives of adenosine-5′-triphosphate as agonists at P_2 purinoceptors—heterogeneity within P_{2X} and P_{2Y} subtypes. Drug Devel Res 31:206–219.

Chahl LA (1979): The effect of 2,2′-pyridylisatogen tosylate on the increase in capillary permeability produced by ATP. J Pharm Pharmacol 31:189–191.

Connolly GP (1995): Differentiation by pyridoxal 5-phosphate, PPADS and isoPPADS between responses mediated by UTP and those evoked by α,β-methylene-ATP on rat sympathetic-ganglia. Br J Pharmacol 114:727–731.

Cristalli G, Camaioni E, Vittori S, Volpini R, Borea PA, Conti A, Dionisotti S, Ongini E, Monopoli A (1995): 2-Aralkynyl and 2-heteroalkynyl derivatives of adenosine-5′-N-ethyluronamide as selective A_{2a} adenosine receptor agonists. J Med Chem 38:1462–1472.

Cristalli G, Franchetti P, Grifantini M, Vittori S, Klotz KN, Lohse MJ (1988): Adenosine receptor agonists: synthesis and biological evaluation of 1-deaza analogues of adenosine derivatives. J Med Chem 31:1179–1183.

Cusack NJ, Hourani SMO (1990): Structure activity relationships for adenine nucleotide receptors on mast cells, human platelets, and smooth muscle. In Jacobson KA, Daly JW, Manganiello V (eds): "Purines in Cellular Signalling: Targets for New Drugs," New York: Springer, pp 254–259.

Daly JW, Jacobson KA (1995): Adenosine receptors: Selective agonists and antagonists. In Bellardinelli L, Pelleg A, (eds): "Adenosine and Adenine Nucleotides: From Molecular Biology to Integrative Physiology." Boston: Kluwer, pp 157–166.

Duke BO (1968): The effects of drugs on *Onchocerca volvulus,* 3. Trials of suramin at different dosages and a comparison of the brands Antrypol, Moranyl and Naganol. Bull World Health Organ 39:157–167.

Dunn PM, Blakely AGH (1988): Suramin: A reversible P_2-purinoceptor antagonist in the mouse vas deferens. Br J Pharmacol 93:243–245.

Dunwiddie TV, Diao L, Kim HO, Jiang JI, Jacobson KA (1997): Activation of hippocampal adenosine A_3 receptors produces a heterologous desensitization of A_1 receptor mediated responses in rat hippocampus. J Neurosci (in press).

Fischer B, Boyer JL, Hoyle CHV, Ziganshin AU, Brizzolara AL, Knight GE, Zimmet J, Burnstock G, Harden TK, Jacobson KA (1993): Identification of potent, selective P_{2Y}-purinoceptor agonists—structure activity relationships for 2-thioether derivatives of adenosine 5′-triphosphate. J Med Chem 36:3937–3946.

Foster H, Hooper M, Imam SH, Lovett GS, Nicholson J, Swain CJ, Sweetman AJ, Weetman DF (1983): Increased inhibitory action against adenosine 5′-triphosphate in the isolated taenia of the guinea-pig caecum by substitution in the A-ring of 2-phenylisatogen. Br J Pharmacol 79:273–278.

Francis JE, Cash WD, Psychoyos S, Ghai G, Wenk P, Friedmann RC, Atkins C, Warren V, Furness P, Hyun JL, et al. (1988): Structure-activity profile of a series of novel triazoloquinazoline adenosine antagonists. J Med Chem 31:1014–1020.

Gallo-Rodriguez C, Ji XD, Melman N, Siegman BD, Sanders LH, Orlina J, Fischer B, Pu QL, Olah ME, van Galen PJM, Stiles GL, Jacobson KA (1994): Structure-activity-relationships of N^6-benzyladenosine-5′-uronamides as A_3-selective adenosine agonists. J Med Chem 37:636–646.

Gray AR (1966): Antigenic variation in clones of *Trypanosoma brucei,* II. The drug-sensitivities of variants of a clone and the antigenic relationships of trypanosomes before and after drug treatment. Ann Trop Med Parasitol 60:265–275.

Gungor T, Malabre P, Teulon JM, Camborde F, Meignen J, Hertz F, Virone OA, Caussade F, Cloarec A (1994): N^6-substituted adenosine receptor agonists. Synthesis and pharmacological activity as potent antinociceptive agents. J Med Chem 37:4307–4316.

Filtz TM, Harden TK, Nicholas RA (1997): Structure, pharmacological selectivity and second messenger properties of G protein–coupled P2 purinergic receptors, this volume.

Hide I, Padgett WL, Jacobson KA, Daly JW (1992): A_{2A} adenosine receptors from rat striatum and rat pheochromocytoma PC12 cells—characterization with radioligand binding and by activation of adenylate cyclase. Mol Pharmacol 41:352–359.

Hogaboom GK, O'Donnell JP, Fedan JS (1980): Purinergic receptors: Photoaffinity analog of adenosine triphosphate is a specific adenosine triphosphate antagonist. Science 208:1273–1276.

Humphries RG, Tomlinson W, Clegg JA, Ingall AH, Kindon ND, Leff P (1995) Pharmacological profile of the novel P_{2T}-purinoceptor antagonist, FPL-67085 in-vitro and in the anesthetized rat in-vivo. Br J Pharmacol 115:1110–1116.

Hutchison AJ, Webb RL, Oei HH, Ghai GR, Zimmerman MB, Williams M (1989): CGS 21680C, an A_2 selective adenosine receptor agonist with preferential hypotensive activity. J Pharmacol Exp Ther 251:47–55.

Jacobson KA, Fischer B, Maillard M, Boyer JL, Hoyle CHV, Harden TK, Burnstock G (1995a): Novel ATP agonists reveal receptor heterogeneity within P_{2X} and P_{2Y} subtypes. In Bellardinelli, L, Pelleg A. (eds), "Adenosine and Adenine Nucleotides: From Molecular Biology to Integrative Physiology," Boston: Kluwer, pp 149–156.

Jacobson KA, Kim HO, Siddiqi SM, Olah ME, Stiles G, von Lubitz DKJE (1995b): A_3 adenosine receptors: Design of selective ligands and therapeutic prospects. Drugs of the Future 20:689–699.

Jacobson KA, Kirk KL, Padgett W, Daly JW (1985): Probing the adenosine receptor with adenosine and xanthine biotin conjugates. FEBS Lett 184:30–35.

Jacobson KA, Nikodijevic O, Ji XD, Berkich DA, Eveleth D, Dean RL, Hiramatsu KI, Kassell NF, van Galen PJM, Lee KS, Bartus R, Daly JW, LaNoue KF, Maillard M (1992b): Synthesis and biological activity of N^6-p-sulfophenylalkyl and N^6-sulfoalkyl derivatives of adenosine: Water soluble and peripherally selective adenosine agonists. J Med Chem 35:4143–4149.

Jacobson KA, Shi D, Gallo RC, Manning MJ, Muller C, Daly JW, Neumeyer JL, Kiriasis L, Pfleiderer W (1993): Effect of trifluoromethyl and other substituents on activity of xanthines at adenosine receptors. J Med Chem 36:2639–2644.

Jacobson KA, van Galen PJM, Williams M (1992a): Perspective. Adenosine receptors: Pharmacology, structure-activity relationships and therapeutic potential. J Med Chem 35:407–422.

Jentsch KD, Hunsmann G, Hartmann H, Nickel P (1987): Inhibition of human immunodeficiency virus type I reverse transcriptase by suramin-related compounds. J Gen Virol 68:2183–2192.

Ji XD, Gallo-Rodriguez C, Jacobson KA (1994): A selective agonist affinity label for A_3 adenosine receptors. Biochem Biophys Res Commun 203:570–576.

Ji XD, Melman N, Jacobson KA (1996): Interactions of flavonoids and other phytochemicals with adenosine receptors. J Med Chem 39:781–788.

Jiang J.-L., van Rhee AM, Melman N, Ji XD, Jacobson KA (1996): 6-Phenyl-1,4-dihydropyridine derivatives as potent and selective A_3 adenosine receptor antagonists. J Med Chem 39:4667–4675.

Kanda T, Shiozaki S, Shimada J, Suzuki F, Nakamura J (1994): KF17837—a novel selective adenosine A_{2a} receptor antagonist with anticataleptic activity. Eur J Pharmacol 256:263–268.

Karton Y, Jiang JL, Ji XD, Melman N, Olah ME, Stiles GL, Jacobsen KA (1996): Synthesis and biological activities of flavonoid derivatives as A_3 adenosine receptor antagonists. J Med Chem 39:2293–2301.

Kennedy C, Leff P (1995): How should P_{2X} purinoceptors be classified pharmacologically. Trends Pharmacol Sci 16:168–174.

Khakh BS, Michel A, Humphrey PPA (1994): Estimates of antagonist affinities at P_{2X} purinoceptors in rat vas deferens. Eur J Pharmacol 263:301–309.

Kim YC, Ji XD, Jacobson KA (1996): Derivatives of the triazoloquinazoline adenosine antagonist (CGS15943) are selective for the human A_3 receptor subtype. J Med Chem 39:4142–4148.

Kim HO, Ji XD, Melman N, Olah ME, Stiles GL, Jacobsen KA (1994a): Structure activity relationships of 1,3-dialkylxanthine derivatives at rat A_3 adenosine receptors. J Med Chem 37:3373–3382.

Kim HO, Ji XD, Siddiqi SM, Olah ME, Stiles GL, Jacobson KA (1994b): 2-Substitution of N^6-benzyladenosine-5′-uronamides enhances selectivity for A_3 adenosine receptors. J Med Chem 37:3614–3621.

Kim HO, Ji XD, Melman N, Olah ME, Stiles GL, Jacobson KA (1994c): Selective ligands for rat A_3 adenosine receptors—structure activity relationships of 1,3-dialkylxanthine 7-riboside derivatives. J Med Chem 37:4020–4030.

Knutsen LJS, Lau J, Petersen H, Thomsen C, Weis JU, Shalmi M, Judge ME, Hansen AJ, Sheardown MJ (1997): Novel neuroprotective adenosine agonists with diminished hypotensive effects. J Med Chem (in press).

Knutsen LJS, Murray TF (1997): Adenosine and ATP in epilepsy, this volume.

Linden J (1994): Cloned adenosine A_3 receptors—pharmacological properties, species differences and receptor functions. Trends Pharmacol Sci 15:298–306.

Liu G-S, Downey JM, Cohen MV (1997): Adenosine, ischemia and preconditioning, this volume.

Maas AJ, Den Hartog A, Ras R, van den Akker J (1980): The action of apamin on guinea-pig taenia caeci. Eur J Pharmacol 67:265–274.

Olah ME, Gallo-Rodriguez C, Jacobsen KA, Stiles GL (1994): ^{125}I-4-aminobenzyl-5′-N-methylcarboxamido-adenosine, a high-affinity radioligand for the rat A_3 adenosine receptor. Mol Pharmacol 45:978–982.

Palmer TM, Gettys TW, Stiles GL (1995a): Differential interaction with and regulation of multiple G-proteins by the rat A_3 adenosine receptor. J Biol Chem 270:16895–16902.

Palmer TM, Poucher SM, Jacobsen KA, Stiles GL (1995b): ^{125}I-4-(-[7-amino-2-(2-furyl)(1,2,4)triazolo(2,3-a)(1,3,5)triazin-5-yl-amino]ethyl)phenol, a high-affinity antagonist radioligand selective for the A_{2a} adenosine receptor. Mol Pharmacol 48:970–974.

Patel A, Craig RH, Daluge SM, Linden J (1988): ^{125}I-BW-A844U, an antagonist radioligand with high affinity and selectivity for adenosine A_1 receptors and ^{125}I-azido-BW-A844U, a photoaffinity label. Mol Pharmacol 33:585–591.

Peet NP, Lentz NL, Dudley, MW, Ogden AM, McCarty DR, Racke MM (1993): Zanthines with C8 chiral substituents as potent and selective adenosine A_1 antagonists. J Med Chem 36:4015–4020.

Poucher SM, Keddie JR, Singh P, Stoggall SM, Caulkett PWR, Jones G, Collis MG (1995): The in-vitro pharmacology of ZM241385, a potent, nonxanthine, A_{2a} selective adenosine receptor antagonist. Br J Pharmacol 115:1096–1102.

Roelen H, Veldman N, Spek AL, von Frijtag Drabbe Künzel J, Mathot RAA, IJzerman, AP (1996): N-6,C8-disubstituted adenosine derivatives as partial agonists for adenosine A(1) receptors. J Med Chem 39:1463–1471.

Sajjadi FG, Takabayashi K, Foster AC, Domingo RC, Firestein GS (1996): Inhibition of TNF-alpha expression by adenosine. J Immunol 156:3435–3442.

Salvatore CA, Jacobson MA, Taylor HE, Linden J, Johnson RG (1993): Molecular cloning and characterization of the human A_3 adenosine receptor. Proc Natl Acad Sci USA 90:10365–10369.

Sarges R, Howard HR, Browne RG, Lebel LA, Seymour PA, Koe BK (1990): 4-Amino[1,2,4]triazolo[4,3-*a*]quinoxalines. A novel class of potent adenosine receptor antagonists and potential rapid-onset antidepressants. J Med Chem 33:2240–2254.

Savi P, Laplace MC, Maffrand JP, Herbert JM (1994): Binding of [^{3}H]-2-methylthio ADP to rat platelets—effect of clopidogrel and ticlopidine. J Pharmacol Exp Ther 269:772–777.

Scammells PJ, Baker SP, Belardinelli L, Olsson RA (1994): Substituted 1,3-dipropylxanthines as irreversible antagonists of A1 adenosine receptors. J Med Chem 37:2704–2712.

Schingnitz G, Küfner-Mühl U, Ensinger H, Lehr E, Kuhn FJ (1991): Selective A_1-antagonists for treatment of cognitive deficits. Nucleosides Nucleotides 10:1067–1076.

Siddiqi SM, Jacobson KA, Esker JL, Olah ME, Ji XD, Melman N, Tiwari KN, Secrist JA, Schneller SW, Cristalli G, Stiles GL, Johnson CR, IJzerman AP (1995a): Search for new purine-modifed and ribose-modified adenosine-analogs as selective agonists and antagonists at adenosine receptors. J Med Chem 38:1174–1188.

Siddiqi SM, Ji XD, Melman N, Olah ME, Jain R, Evans P, Glashofer M, Padgett WL, Cohen LA, Daly JW, Stiles GL, Jacobson KA (1996): A survey of nonxanthine derivatives as adenosine receptor ligands. Nucleosides Nucleotides 15:693–717.

Siddiqi SM, Pearlstein RA, Sanders LH, Jacobson KA (1995b): Comparative molecular-field analysis of selective A_3 adenosine receptor agonists. Bioorg Med Chem 3:1331–1343.

Soltoff SP, McMillan MK, Talamo BR, Cantley LC (1993): Blockade of ATP binding site of P_2 purinoceptors in rat parotide acinar cells by isothiocyanate compounds. Biochem Pharmacol 45:1936–1940.

Stiles GL (1997): Adenosine receptor subtypes: New insights from cloning and functional studies, this volume.

Suzuki F, Shimada J, Mizumoto H, Karasawa A, Kubo K, Nonaka H, Ishii A, Kawakita T (1992): Adenosine-A_1 antagonists, 2. Structure-activity relationships on diuretic activities and protective effects against acute renal failure. J Med Chem 35:3066–3075.

Akahane A, Katayama H, Mitsunaga T, Kita Y, Kusunoki T, Terai T, Yoshida K, Shiokawa Y (1996): Discovery of FK453, a novel nonxanthine adenosine A_1 receptor antagonist. Bioorg Med Chem Letters 6:2059–2062.

Thompson RD, Secunda S, Daly JW, Olsson RA (1991): N^6,9-Disubstituted adenines—potent, selective antagonists at the A_1-adenosine receptor. J Med Chem 34:2877–2882.

Ueeda M, Thompson RD, Arroyo LH, Olsson RA (1991): 2-Alkoxyadenosines: Potent and selective agonists at the coronary artery A_2 adenosine receptor. J Med Chem 34:1334–1339.

van der Wenden EM, Price SL, Apaya RP, IJzerman AP, Soudijn W (1995): Relative binding orientations of adenosine-A_1 receptor ligands—a test-case for distributed multipole analysis in medicinal chemistry. J Comp Aided Mol Design 9:44–54.

van Galen PJM, IJzerman AP, Soudijn W (1987): Adenosine derivatives with N^6-alkyl, -alkylamine or -alkyladenosine substituents as probes for the A_1-receptor. FEBS Lett 223:197–201.

van Galen PJM, Nissen P, van Wijngaarden I, IJzerman AP, Soudijn W (1991): 1H-imidazo[4,5-c]quinolin-4-amines: Novel non-xanthine adenosine antagonists. J Med Chem 34:1202–1206.

van Galen PJM, van Bergen AH, Gallo-Rodriguez C, Melman N, Olah ME, IJzerman AP, Stiles GL, Jacobson KA (1994): A binding site model and structure-activity relationships for the rat A_3 adenosine receptor. Mol Pharmacol 45:1101–1111.

van Rhee AM, Jiang J-I, Melman N, Olah ME, Stiles GL, Jacobson KA (1996b): Interaction of 1,4-dihydrophyridine and pyridine derivatives with adenosine receptors: Selectivity for A_3 receptors. J Med Chem 39:2980–2989.

van Rhee AM, Siddiqi SM, Melman N, Shi D, Padgett WL, Daly JW, Jacobson KA (1996a): Tetrahydrobenzothiophenone derivatives as a novel class of adenosine receptor antagonists. J Med Chem 39:398–406.

van Rhee AM, van der Heijden MPA, Beukers MW, IJzerman AP, Soudijn W, Nickel P (1994): Novel competitive antagonists for P_2 purinoceptors. Eur J Pharmacol 268:1–7.

von Lubitz DKJE (1997): Adenosine and acute treatment of cerebral ischemia and stroke—put out more flags, this volume.

Wagner H, Milaveckrizman M, Gadient F, Menninger K, Schoeffter P, Tapparelli C, Pfannkuche HJ, Fozard JR (1995): General pharmacology of SDZ WAG-994, a potent selective and orally active adenosine A_1 receptor agonist. Drug Devel Res 34:276–288.

Watanabe N, Obuchi T, Tamai M, Araki H, Omura S, Yang JS, Yu DQ, Liang XT, Huan JH (1990): A novel N^6-substituted adenosine isolated from mi huan jun (Armillaria mellea) as a cerebral-protecting compound. Planta Med 56:48–52.

Wieser PB, Pendleton TS (1979): Effects of an N6-disubstituted adenosine derivative (BM 11.189) on fat cell metabolism. Biochem Pharmacol 28:693–694.

Windscheif U, Pfaff O, Ziganshin AU, Hoyle CHV, Bäumert HG, Mutschler E, Burnstock G, Lambrecht G (1995): Inhibitory-action of PPADS or relaxant responses to adenine nucleotides or electrical-field stimulation in guinea-pig taenia coli and rat duodenum. Br J Pharmacol 115:1509–1517.

Zhou QY, Li CY, Olah ME, Johnson RA, Stiles GL, Civelli O (1992): Molecular cloning and characterization of an adenosine receptor—the A_3 adenosine receptor. Proc Natl Acad Sci USA 89:7432–7436.

Ziyal R, Pfaff O, Windscheif U, Bo X, Nickel P, Adanuy U, Burnstock G, Mutschler E, Lambrecht G (1994): A novel P_2-purinoceptor ligand which displays selectivity for the P_{2X}-subtype. Drug Devel Res 31:336.

CHAPTER 7

Modulators of Adenosine Uptake, Release, and Inactivation

AD P. IJZERMAN and NORA M. VAN DER WENDEN

Leiden/Amsterdam Center for Drug Research, Division of Medicinal Chemistry, 2300 RA Leiden, The Netherlands

INTRODUCTION

The formation, degradation, release, and uptake of adenosine are the four important parameters that determine the extra- and intracellular concentration of this nucleoside. Under physiological and pathophysiological conditions the parameters may vary considerably, giving rise to a complicated, well-tuned regulatory system. The formation and degradation of adenosine are controlled by various enzymes, whereas its release from and uptake in cells is governed by membrane-bound transport proteins. Figure 1 summarizes the enzymes that are involved in intracellular adenosine metabolism. All these enzymes have been assigned EC numbers by the nomenclature committee of the International Union of Biochemistry. Essential and updated information on these proteins can be found on the World Wide Web (e.g., http:\\www.genome.ad.jp).

The relative importance of every single enzyme is difficult to assess. One way to do this is to use computerized data bases. For example, a search through the Medline™ database (1994–1995) yielded no hits with respect to the enzymes 4, 5, 7, 11, and 12 in Figure 1. Consequently, we concluded that those enzymes that are immediately relevant for adenosine formation and breakdown as well as the subject of recent research are adenosine deaminase (1), adenosine kinase (2), 5′-nucleotidase (3), and adenosylhomocysteinase (6).

The aim of this chapter is to discuss modulators (inhibitors, activators) for these enzymes, as well as ligands that interfere with adenosine's transport over the cell membrane.

Purinergic Approaches in Experimental Therapeutics, Edited by Kenneth A. Jacobson and Michael F. Jarvis

ISBN 0-471-14071-6 © 1997 Wiley-Liss, Inc.

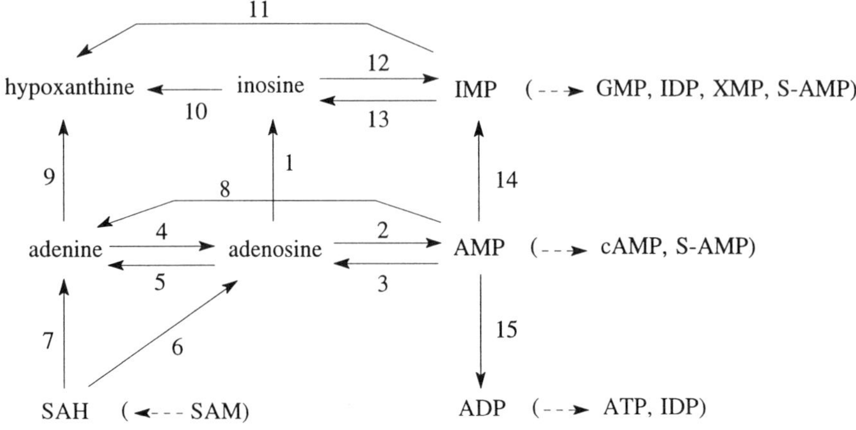

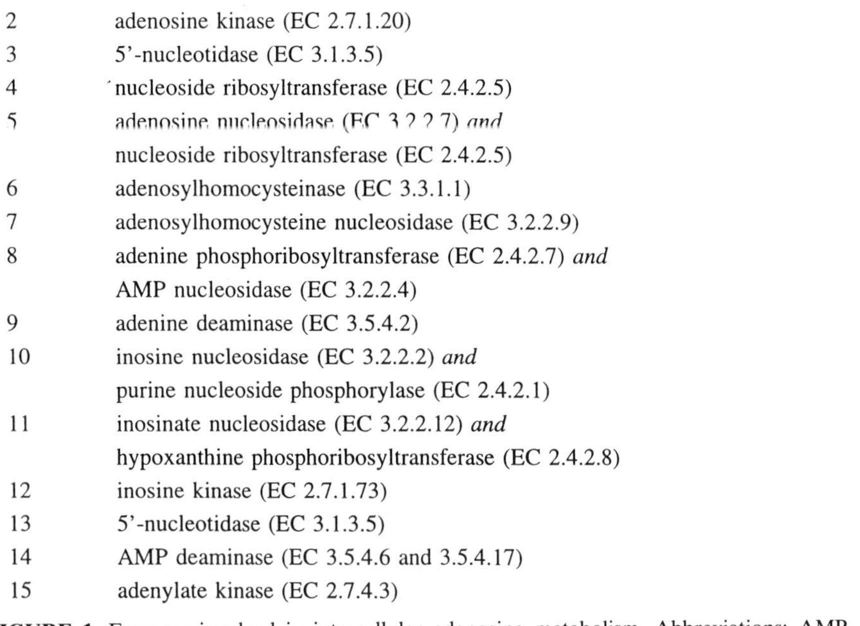

1	adenosine deaminase (EC 3.5.4.4)
2	adenosine kinase (EC 2.7.1.20)
3	5'-nucleotidase (EC 3.1.3.5)
4	nucleoside ribosyltransferase (EC 2.4.2.5)
5	adenosine nucleosidase (EC 3.2.2.7) *and*
	nucleoside ribosyltransferase (EC 2.4.2.5)
6	adenosylhomocysteinase (EC 3.3.1.1)
7	adenosylhomocysteine nucleosidase (EC 3.2.2.9)
8	adenine phosphoribosyltransferase (EC 2.4.2.7) *and*
	AMP nucleosidase (EC 3.2.2.4)
9	adenine deaminase (EC 3.5.4.2)
10	inosine nucleosidase (EC 3.2.2.2) *and*
	purine nucleoside phosphorylase (EC 2.4.2.1)
11	inosinate nucleosidase (EC 3.2.2.12) *and*
	hypoxanthine phosphoribosyltransferase (EC 2.4.2.8)
12	inosine kinase (EC 2.7.1.73)
13	5'-nucleotidase (EC 3.1.3.5)
14	AMP deaminase (EC 3.5.4.6 and 3.5.4.17)
15	adenylate kinase (EC 2.7.4.3)

FIGURE 1. Enzymes involved in intracellular adenosine metabolism. Abbreviations: AMP, adenosine-5′-monophosphate; ADP, adenosine-5′-diphosphate; ATP, adenosine-5′-triphosphate; cAMP, adenosine 3′:5′-cyclic monophosphate; GMP, guanosine-5′-monophosphate; IMP, inosine-5′-monophosphate; IDP, inosine-5′-diphosphate; SAH, S-adenosyl-L-homocysteine; SAM, S-adenosyl-L-methionine; S-AMP, adenylosuccinic acid; XMP, xanthosine-5′-monophosphate.

ADENOSINE DEAMINASE

Adenosine deaminase (ADA, EC 3.5.4.4.) catalyzes the breakdown of adenosine (and some related nucleosides) by hydrolytic deamination, resulting in the formation of inosine. The inhibition of ADA can be useful in chemotherapy, and can potentiate the effects of antileukemic or antiviral nucleosides by decreasing the inactivation of these compounds that are also substrates for this enzyme.

Adenosine deaminase inhibitors have been designed by modifying either the purine ring or the ribose moiety of the adenosine molecule. The best-known ADA inhibitors are 2′-deoxycoformycin (also called pentostatin or covidarabine), a transition-state analog or tight-binding inhibitor; EHNA (*erythro*-9-(2′-hydroxyl-3′-nonyl)adenine), a semi-tight-binding inhibitor; and DHMPR (1,6-dihydro-6-hydroxymethyl purine riboside), a reversible inhibitor of ADA (see also Figure 2).

More insight into the mechanism of deamination and enzyme inhibition was obtained from the crystal structures of this enzyme complexed with inhibitors. Two crystal structures of ADA have been described in the literature. The first crystal structure was of ADA bound to a transition state analog, the hydrated form of purine ribonucleoside, 6(R)-hydroxy-1,6-dihydropurine ribonucleoside (Wilson et al., 1991) This crystal structure demonstrated that ADA requires a zinc ion as a cofactor, which was inconsistent with the previous idea that no cofactor was required for ADA. When binding to ADA, purine ribonucleoside can accept a hydroxyl group from a water molecule, which is coordinated by the zinc ion in the binding site, but the molecule lacks a leaving group. The second crystal structure was of ADA incorporating 1-deazaadenosine. The latter is a pretransition state mimic because this molecule cannot accept a hydrogen atom at the 1-position, and therefore cannot form a hydrogen bond to Glu 217. As a consequence, the hydroxyl group coordinated by the zinc ion cannot bind to C6 (Wilson and Quiocho, 1993).

The probable transition state of adenosine bound to ADA is schematically represented in Figure 3.

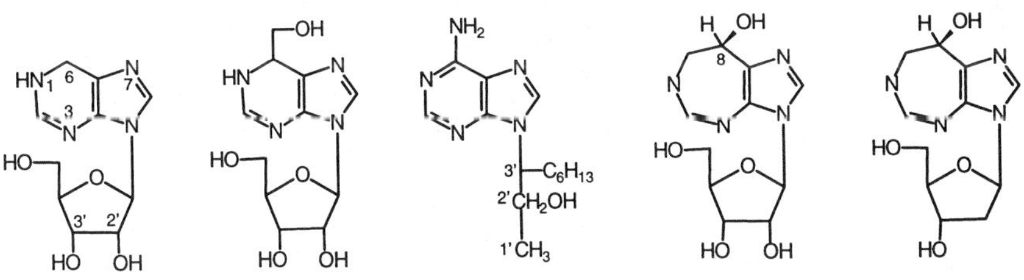

	purine riboside	DHMPR	EHNA	coformycin	2'-deoxy-coformycin
K_i:	7.5 μM	0.7 μM	7.0 nM	10 pM	4.5 pM

FIGURE 2. Chemical structures of some potent inhibitors of adenosine deaminase.

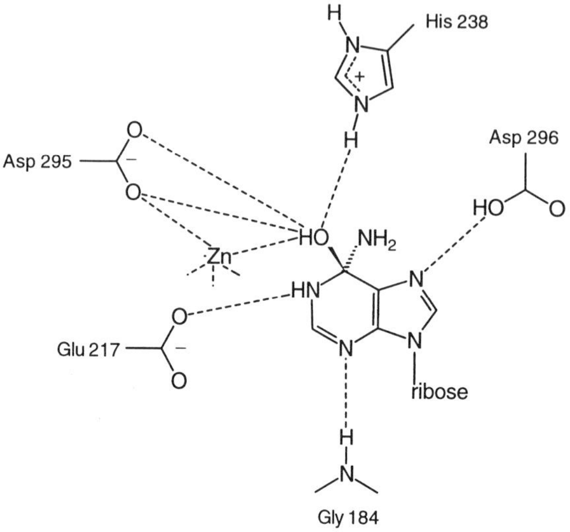

FIGURE 3. Schematic representation of the active site of adenosine deaminase with adenosine in its transition state.

Analogs of Adenosine

Analogs of adenosine can be substrates and/or inhibitors of ADA. These two features are not exclusive, but may depend on the concentration of a given compound. For example 2-aza-3-deazaadenosine and its 8-aza analog were described as both poor substrates and weak inhibitors of ADA (Bussolari et al., 1993).

The nitrogen atoms at the 1, 3, and 7 position of adenosine are necessary both for recognition by ADA and for subsequent hydrolysis (Table 1) (Ikehara and Fukui, 1974; Cristalli et al., 1991a). Compounds lacking these prerequisites, such as 1- and 3-deazaadenosine can bind to ADA, but are not hydrolyzed by the enzyme. They are therefore inhibitors, 3-deazaadenosine having a 500-fold lower potency than 1-deazaadenosine. A further removal of the nitrogen atom at the 1-position in 3-deazaadenosine hardly affects the K_i-value. 7-Deazaadenosine and 1,3,7-trideaza-adenosine have no inhibitory activity.

2-Substitution of 2′,3′-dideoxyadenosine leads to weak ADA inhibitors in the case of iodo, cyano, thiomethyl, or trifluoromethyl substituents; or to poor substrates with amino or ethyl substituents (Nair et al., 1991). N^6-hydroxyadenosine (HAPR) has antitumor activity but is rapidly deaminated by ADA. Derivatives of HAPR include 2-chloro,N^6-hydroxyadenosine, which was not active as an ADA inhibitor, and 1-deaza,N^6-hydroxyadenosine with a K_i value for calf intestine ADA of 1.2 μM (Cristalli et al., 1991b). Removal of the 2′-hydroxyl group of the latter compound led to a further four-fold increase in affinity.

Other adenosine analogs with inhibitory activity towards ADA include adenine, the 2′,3′-isopropylidene,5′-monoacetate analog of adenosine, α-adenosine, 5′-deoxy,5′-methylthioadenosine, 5′-deoxy,5′-iodoadenosine (Yoshida and Aikawa, 1993).

TABLE 1 K_i Values of Adenosine, EHNA, and Their Deaza Analogs[a]

X	Y	Z	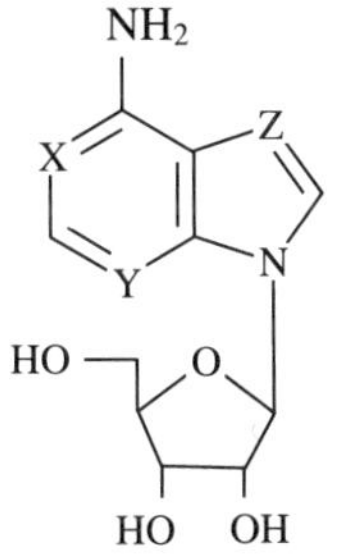 Adenosine analog (9-ribosyl) K_i (nM)	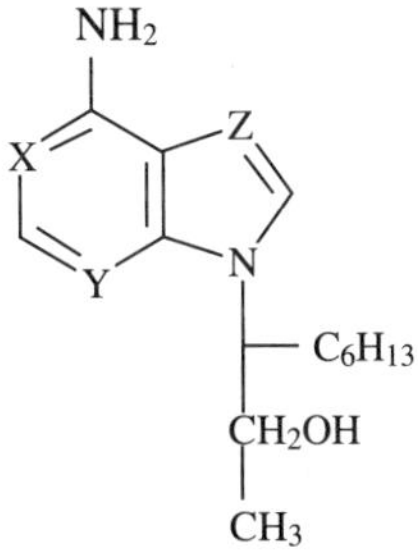 EHNA analog K_i (nM)
N	N	N	Substrate (K_m 33 μM)	7.0
CH	N	N	660	1,600
N	CH	N	360,000	10
N	N	CH	Not active	400,000
CH	CH	N	110,000	71,000

[a] Antonini et al., 1984; Cristalli et al., 1988; Cristalli et al., 1991a.

Analogs of EHNA

The inhibitor EHNA was first synthesized by Schaeffer and Schwender (1974) as one in a series of 9-(2′-hydroxy-3′-alkyl)adenines. These investigators showed that the inhibitory activity of these compounds was higher with large alkyl chains, and that the *erythro* isomers were far better inhibitors than the *threo* isomers. Several research groups have studied the structure–activity relationships for analogs of EHNA. However, most compounds have not been tested as pure stereoisomers, but as a combination of the *erythro* pair, 2′*R*,3′*S* and 2′*S*,3′*R*. Porter and Abushanab (1992) and Harriman et al. (1994) reported a 300-fold higher potency of (+)-EHNA, the 2′*S*,3′*R* form, than of the other isomer.

The nitrogen atoms at positions 1 and 3 in the purine ring are not necessary for inhibition of ADA (Table 1) (Antonini et al., 1984). Removal of the nitrogen atom at the 1-position reduces the inhibitory activity of EHNA 23-fold, and 3-deazaEHNA is equipotent to EHNA (Cristalli et al., 1991a). A simultaneous removal of the nitrogen atoms at both the 1- and 3-positions affects binding much more than would be expected from the effect of removing of either nitrogen atom singly, increasing the K_i-value 10,000-fold. This finding is indicative of a different binding mode of EHNA compared to the purine ribonucleosides (Table 1), because removal of the nitrogen atom at the 3-position in adenosine or purine ribonucleoside strongly affects the binding, whereas the 1-deaza analogs are not affected by an analogous change. The 6-amino group has an important contribution to the binding of EHNA, provided that the nitrogen atom at the 1-position is present (Antonini et al., 1984). Removal of the nitrogen atom at the 7-position strongly affects the binding, as it increases the K_i value more than 10,000-fold (Cristalli et al., 1988).

Changes in the 2′-hydroxyl-3′-nonyl chain of EHNA decrease the inhibitory activity. The inhibitory activity of the C1′-OH and the nor-C1′ analogs was decreased 10-fold in comparison with EHNA (Harriman et al., 1992). Substitution at the 1′-position leads to (sub)micromolar K_i values, but deletion of the 1′-methyl leads to a strong decrease in affinity. Ring closure within the alkyl chain or to N3 also led to micromolar affinities. Adding a 9′-hydroxyl substituent to EHNA leads to 5-fold increase, an 8′-hydroxyl to an 8-fold increase, and combining these substituents to a 20-fold increase in K_i value (Vargeese, et al., 1994).

Cristalli and coworkers synthesized and tested several ring-opened analogs, such as imidazole analogs and azole derivatives with a 2′-hydroxy-3′-nonyl tail. Substitution of 1-(2′-hydroxy-3′-nonyl)imidazole with 4-carboxamide yielded an inhibitor with a K_i value of 35 nM, slightly less than values found for EHNA and 3-deaza-EHNA. Azole derivatives were less potent than the imidazole analogs (Cristalli et al., 1991a, 1994).

Analogs of Coformycin and Purine Riboside

Coformycin is a nucleoside of microbial origin. The inhibitory activity of this compound was first reported in 1967 (Sawa et al.), and its structure was described in 1974 (Nakamura et al.). Its analog 2′-deoxycoformycin was isolated and its structure elucidated in the same year (Woo et al., 1974). Baker and Putt (1979) described the synthesis of this compound 5 years later.

Analogs of coformycin are few. The naturally occurring *R*-form of 2′-deoxycoformycin, a transition state analog, is 10^7 times as potent as the *S*-form, which is a competitive inhibitor of ADA (Schramm and Baker, 1985). The activity of 8-keto-2′-deoxycoformycin equals that of *S*-deoxycoformycin. Thus, *S*-deoxycoformycin and 8-keto-2′-deoxycoformycin lack the appropriate stereochemistry to induce a change in conformation from the ground state to the transition state of the enzyme. The addition of a nitrogen atom analogous to the 8-position of the purine system, resulting in 2-azacoformycin, reduces the activity 12-fold.

Purine riboside (nebularine) is also a transition state analog for ADA. It binds to the enzyme and can incorporate a water molecule, as does adenosine itself. However, lacking a good leaving group, the hydrated purine riboside is trapped in the enzyme in the transition state form. The K_i value for the transition state form of purine riboside was estimated to be 0.3 pM, an affinity 10-times that of purine riboside itself (Jones et al., 1989). Removal of the nitrogen atom at the 1-position of purine riboside increases the K_i value 7-fold (Kurz et al., 1992). The K_i value of 3-deazapurine riboside is much higher than that of the 1-deaza analog. 8-Azapurine riboside is approximately 400-times as active as purine riboside in inhibiting ADA (Shewach et al., 1992). Addition of a 2-amino substituent to purine riboside slightly decreases its activity. The 2′,3′-isopropylidene and the 2′,3′-isopropylidene-5′-monoacetate analogs of purine riboside are also inhibitors for ADA (Yoshida and Aikawa, 1993).

ADENOSINE KINASE

Adenosine kinase (AK), also called adenosine-5′-phosphotransferase (EC 2.7.1.20), catalyzes the phosphorylation of the 5′-hydroxyl group of adenosine by transfer of the

γ-phosphoryl group of either ATP or GTP. Inhibiting adenosine kinase prevents the rapid metabolism of adenosine.

Modulators of AK described in the literature are all analogs of adenosine. Lin et al. (1988) have described compounds that stimulate AK at high concentrations of adenosine, but inhibit AK at lower concentrations. The adenosine receptor ligands N^6-cyclohexyladenosine (CHA), N^6-R-phenylisopropyladenosine (R-PIA), and 2-Cl-adenosine are examples. A description of the AK modulatory potency of such compounds is complicated, because they may also be substrates for AK. The authors postulated that AK may have two binding sites for AK: a catalytic site with high affinity for adenosine and a regulatory site with lower affinity for adenosine to inhibit AK activity. Other N^6-substituted adenosines also display significant inhibitory potency. For instance, α,ω-di(adenosin-N^6-yl)alkanes, with an alkyl bridge at the N^6-position of 1 to 14 methylene units, inhibit AK, with the K_i value decreasing with increasing chain length (Prescott et al., 1989). 1,12-Di(adenosin-N^6-yl)dodecane, which had a K_i value of 75 nM, behaved as a competitive inhibitor with respect to adenosine, but as a noncompetitive inhibitor with respect to ATP. It should be mentioned that these "double" adenosines are also potent adenosine receptor agonists, and thus lack selectivity (Van Galen et al., 1987).

Removal of the nitrogen atom at the 7-position in adenosine, yielding 7-deazaadenosine or tubercidin, decreased binding. However, substitution of the 7-position of tubercidin with hydrophobic substituents led to a marked increase in AK affinity. An iodo substituent proved optimal (K_i 30 nM, Cottam et al., 1993; but concentration inhibiting 50% of enzyme activity 1.8 μM, Davies et al., 1995), which made 7-iodotubercidin (Figure 4) the reference inhibitor of AK. Removal of the 5′-hydroxyl group led to compounds even more active than the parent compounds. The 8-aza analog has a similar K_i value, but in this compound most 5′-substituents have been described to reduce affinity.

Removal of the 5′-hydroxyl group in adenosine or replacing this group by other substituents, such as carboxyl, chloro, azido, or amino, also leads to inhibitors for AK. In particular, 5′-deoxy-5′-aminoadenosine is a strong inhibitor for the enzyme (Cottam et al., 1993).

A combination of features that led to high-affinity inhibitors, the substituted 8-aza-7-deaza fragment and the 5′-amino substituent, gave GP-1-515 (Figure 4), which has a K_i value of 5 nM (Firestein et al., 1995).

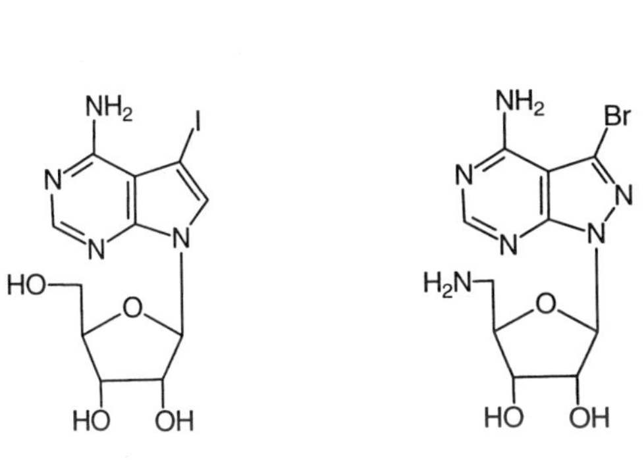

NH2 I N N N HO O HO OH

NH2 Br N N N N H2N O HO OH

7-iodotubercidine K_i 30 nM GP-1-515 K_i 5 nM

FIGURE 4. Chemical structures of potent adenosine kinase inhibitors.

Structural features for the binding of 128 adenosine analogs to AK of *Toxoplasma gondii* were described by Iltzsch et al. (1995) in an attempt to discriminate this organism's activity from the mammalian form of AK. However, 7-iodotubercidin was also most active here.

SAH HYDROLASE

The enzyme S-adenosyl-L-homocysteine hydrolase (adenosylhomocysteinase; EC 3.3.1.1) catalyzes the breakdown of S-adenosyl-L-homocysteine (SAH) into adenosine and L-homocysteine. The mechanism of catalysis was elucidated by Palmer and Abeles (1979). Three subsequent steps are discerned (Figure 5): (i) oxidation and elimination of the substrate to yield 4′,5′-didehydro-3′-keto-adenosine, which is converted into adenosine through (ii) addition of water at the 5′-position, and (iii) an NADH-dependent reduction of the keto function. Interestingly, the reverse synthetic route (adenosine → SAH) is also catalyzed by the enzyme; but under physiological conditions in which intracellular levels of adenosine and L-homocysteine are low, the reaction proceeds in the hydrolytic direction only (Ueland, 1982).

Inhibition of the enzyme would lead to diminished adenosine levels with a concomitant increase in SAH. This latter aspect has been the main impetus to search for potent and selective enzyme inhibitors as antiviral agents. The accumulation of

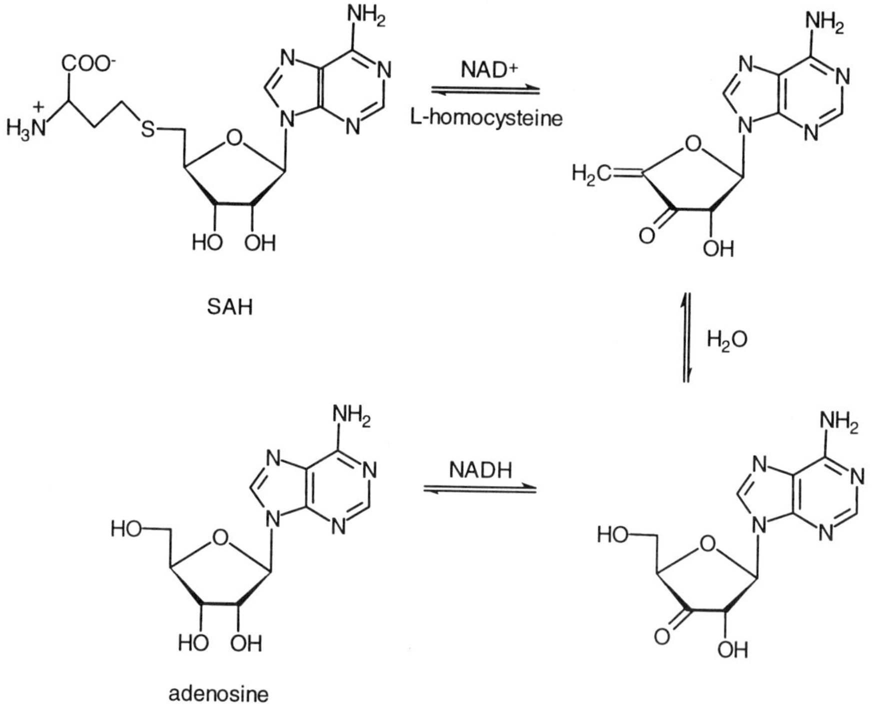

FIGURE 5. Substrate handling by S-adenosylhomocysteine hydrolase.

TABLE 2 Averaged Inhibitory Constants of "First Generation" Inhibitors of SAH Hydrolase from Different Sources[a]

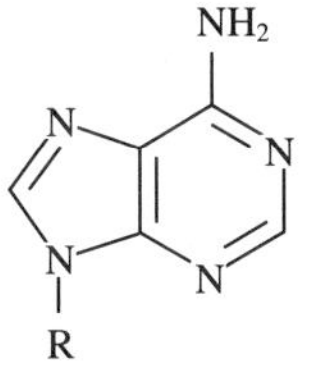

NH_2 N N N N R

R	Compound	K_i (nM)
$CH_2CHOHCH_2OH$	(S)-DHPA	1900
$CH_2CHOHCHOHCOOH$	*d*-Eritadenine	3
$CH_2CHOHCOOH$	(*RS*)-AHPA	78
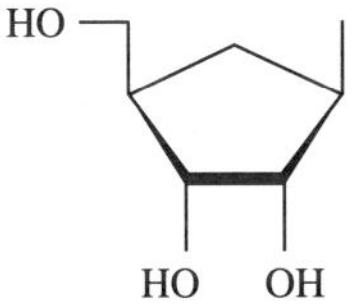 HO; HO OH	Aristeromycin (carbocyclic adenosine)	5
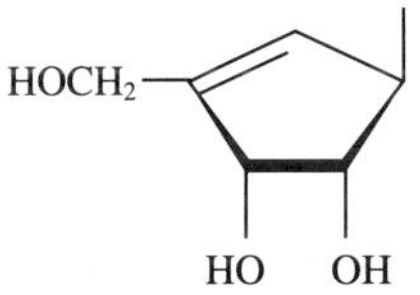 $HOCH_2$; HO OH	Neplanocin	5

[a] Compiled from De Clercq, 1987.

SAH results in feedback inhibition of S-adenosyl-L-methionine (SAM) dependent transmethylation reactions, such as the capping of viral mRNA, essential for its translation into proteins (Wolfe and Borchardt, 1991). Therefore, it is not surprising that substantial effort has been put into the design and synthesis of such inhibitors. Table 2 shows a selection of "first generation" compounds with significant inhibitory properties (De Clercq, 1987).

Despite the considerable potency of at least some of these compounds, their apparent cytotoxicity was a major drawback. From subsequent studies it appeared that most compounds are also substrates for adenosine kinase yielding the formation of the corresponding 5′-monophosphates, and through further metabolism the 5′-triphosphates. The latter nucleotides are considered as an important cause of cytotoxicity, although other metabolites may also be damaging to the cell. Thus, any chemical manipulation focused on poor substrate activity towards both adenosine kinase and adenosine deaminase would be useful, yielding a prolonged presence of the intact inhibitor in the cell. This notion led to the design of "second generation" SAH hydrolase inhibitors (Table 3).

Several strategies proved successful in this respect. First, the 3-deaza analogs of neplanocin and aristeromycin remained active as SAH hydrolase inhibitors, although they do not serve as substrates for adenosine kinase and adenosine deaminase (Tseng et al., 1989; Montgomery et al., 1982). In fact, 3-deazaneplanocin is the most potent SAH hydrolase inhibitor known to date. Second, the removal of the hydroxymethyl

TABLE 3 "Second Generation" Inhibitors of SAH Hydrolase with Improved Metabolic Stability[a]

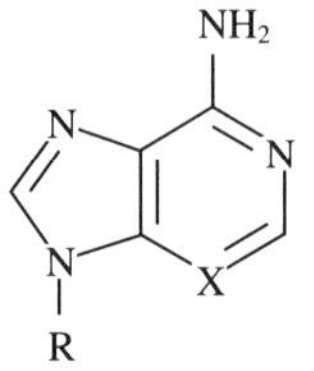

R	X	Compound	K_i (nM)
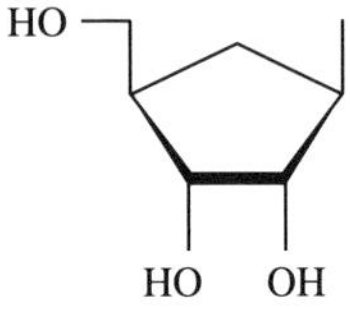	CH	3-Deazaaristeromycin	6
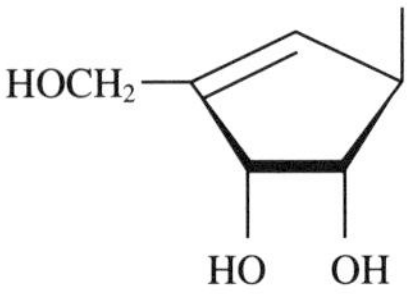	CH	3-Deazaneplanocin	0.05
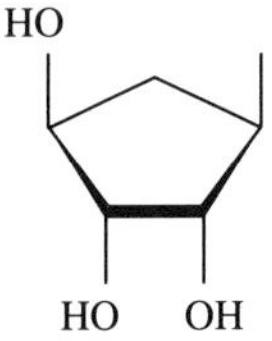	N	5′-Noraristeromycin	11
	N	Dihydroxycyclopentyladenine	12
	CH	Dihydroxycyclopentyl-3-deaza-adenine	10
	N	Dihydroxycyclopentenyladenine	41
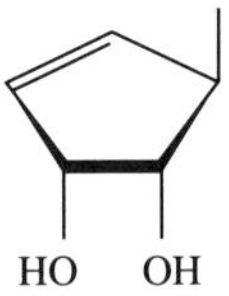	CH	Dihydroxycyclopentenyl-3-deaza-adenine	35

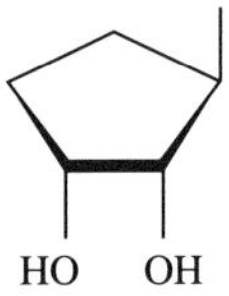

[a] Compiled from Wolfe and Borchardt, 1991, and references therein.

substituent in both neplanocin and aristeromycin did not affect SAH hydrolase inhibitory activity. Obviously, phosphorylation to the corresponding nucleotides is not possible in this case. Moreover, the OH group in the hydroxymethyl substituent is also required for substrate activity towards adenosine deaminase (Bloch et al., 1967). The resulting derivatives (dihydroxycyclopent(en)yladenines) and their 3-deaza analogs proved effective as SAH hydrolase inhibitors with a substantially increased "therapeutic window" (Wolfe and Borchardt, 1991). Replacement of the hydroxymethyl substitu-

TABLE 4 Some Properties of the Different 5′-Nucleotidases[a]

Type	Apparent K_M		Endogenous Action[b]			Metal Cofactor
	AMP	IMP	ATP	ADP	P_i	
Ectoenzyme	1–50 μM	10–50 μM	i	i	—	Zn^{2+}
Cytosolic (AMP preferring)	2–8 mM	—	—	a	i	Mg^{2+}
Cytosolic (IMP preferring)	1–15 mM	0.1–0.6 mM	a	a	i	Mg^{2+}

[a] Complied from Zimmermann, 1992.
[b] i, inhibition; a, activation.

ent rather than removal appeared less successful, although 5′-noraristeromycin preserved considerable activity (Patil and Schneller, 1992). A methyl or a vinyl substituent was tolerated in the neplanocin series, although overall potency was seriously affected (Wolfe, et al., 1992).

Current research is focused on the design of compounds selective for one of the three steps in the substrate handling. A recent example is (*E*)-5′,6′-didehydro-6′-deoxy-6′-fluorohomoadenosine (EDDFHA), which is a good substrate for the hydrolytic activity of the enzyme, but not for the oxidative capacity (Yuan et al., 1994). The aim now is to gain further insight in the enzyme's mechanism of action rather than developing another powerful inhibitor.

5′-NUCLEOTIDASE

The enzyme 5′-nucleotidase (EC 3.1.3.5) catalyzes the removal of the phosphate group in nucleoside-5′-monophosphates. The attack of a water molecule on the phosphate ester bond provokes this transition from nucleotide to nucleoside. In a thorough review Zimmerman (1992) makes a plea for a further subdivision of this enzymic activity. Indeed soluble and membrane-bound enzymes are distinct entities with large differences in substrate and inhibitor profiles. The cytosolic enzymes can be further subdivided into two forms. One enzyme prefers AMP as the substrate ($K_m \approx 5$ mM), the other IMP ($K_m \approx 0.5$ mM). In the presence of ADP both K_m values are shifted to lower values, suggesting that ADP activates the cytosolic enzymes.

Thus cytosolic enzymes behave very differently from the membrane-bound ectoenzyme that is inhibited by ADP. The ectoenzyme has higher affinity for nucleoside-5′-monophosphates ($K_{m,AMP} = 1–50$ μM) than the cytosolic enzymes. From a physiological point of view this means that extracellular AMP levels are kept low, since the catalytic site of the enzyme is facing the extracellular space. The enzyme is attached to the cell membrane via a GPI (glycosyl phosphatidyl inositol) anchor. The chain can be cleaved, however, to provide another soluble form of 5′-nucleotidase with characteristics similar to those of the ectoenzyme. This soluble form very much resembles the 5′-nucleotidase present in certain snake venoms. In Table 4 some properties of the various enzymes are summarized.

The differences apparent in Table 4 are also reflected in the availability of inhibitors. Few studies have addressed the inhibition of the cytosolic enzymes. Newby and coworkers (Skladanowski et al., 1989; Skladanowski and Newby, 1990) purified both the AMP- and IMP-selective forms from various tissues, and studied several purine nucleosides

TABLE 5 K_i Values of Inhibitors of Snake Venom and Rat Heart 5′-Nucleotidase

	Snake Venom		Rat Heart	
Compound	pH	K_i (μM)	pH	K_i (μM)
8-Br-AMP	8.2	10		
AMPαS	8.2	7	7.5	2.3
ADP	8.1	7	7.5	0.08
8-Br-ADP	8.1	10		
ADPβS	7.0	0.3	7.5	1.4
AOPCP	8.1	0.06	7.5	0.006
ATP	8.1	28	7.5	3
AOPCPOP	8.0	0.5	7.5	0.066
Pyrophosphate	8.1	1200	7.5	200

as potential inhibitors. They selected 5′-deoxy-5′-isobutylthioadenosine (IBTA) and 5′-deoxy-5′-isobutylthioinosine (IBTI), as these two compounds were the more effective. However, their selectivity and activity were still very modest. As an example, IBTI inhibited the AMP-selective cytosolic enzyme in a noncompetitive way with an apparent IC_{50} value of only 6–7 mM.

In contrast, potent inhibitors have been identified for the ectoenzyme. Burger and Lowenstein (1970) found that nucleoside-5′-diphosphates, particularly TDP and ADP, were very effective in inhibiting the 5′-nucleotidase prepared from pig intestine. Chemical modification of the phosphate chain afforded α,β-methylene ADP (AOPCP), which is even more potent. α,β-Methylene ATP was as effective as ADP, and more active than ATP and other 5′ triphosphates. The ionization characteristics of the nucleotides and the various pH optima of the enzyme complicate the determination of the "true" inhibitors' affinities to some extent. In Table 5 the K_i values of several inhibitors are reported for snake venom (Wierzchowski et al., 1984) and rat heart 5′-nucleotidase (Naito and Lowenstein, 1985).

Due to the relatively high ionization constant of the secondary hydroxyl group in the phosphate chain, AOPCP retains its high inhibitory potency over a wide pH range. These properties make this compound a very useful research tool.

NUCLEOSIDE TRANSPORT PROTEINS

Different proteins mediate the passage of nucleobases and nucleosides across the cell membrane (Plagemann et al., 1988). Two major classes are discerned based on their mechanisms of action. Active, sodium ion–dependent transport, predominantly in kidney and small intestine epithelium, is mediated by at least four different proteins (Pajor, 1995). No inhibitors have been found yet for this active transport system.

More widespread are proteins involved in passive transport by means of facilitated diffusion. They can also be divided into subclasses, based on their sensitivity towards the prototypic inhibitor nitrobenzylthioinosine (NBTI, NBI, or NBMPR, Table 6), being in the nanomolar or micromolar range, respectively (Plagemann et al., 1988). The relatively insensitive transporters are mainly found in transformed cells.

Active transport promotes the influx of adenosine into the (epithelial) cell due to the sodium dependency. The passive transport, however, can be bidirectional depending on intra- and extracellular nucleoside concentrations. Under physiological conditions the intracellular adenosine concentration is under strict enzymatic control and kept very low, evoking an influx of adenosine. By blockade of these proteins the extracellular concentration of adenosine will rise, leading to an enhanced interaction with, for example, membrane-bound adenosine receptors.

In this paragraph structure–activity relationships of various classes of nucleoside transport inhibitors will be discussed. It should be noted that the biological data all refer to the NBI-sensitive transport protein.

NBI and Analogues

Paterson and colleagues identified several purine ribosides with S- (i.e., thioinosines) or N-substituents (i.e., adenosines) at the C6-position as nucleoside transport inhibitors (Paul et al., 1975). Among the adenosine derivatives the introduction of a *p*-nitro group to N^6-benzyladenosine induced an almost 400-fold increase in activity. Since 6-benzylthioinosine proved more potent than N^6-benzyladenosine, the corresponding 6-(4-nitrobenzyl)thioinosine (NBI) was also prepared. It was active in the nanomolar range. Similarly active were the corresponding thioguanosine derivative and the 6-(2-hydroxy-5-nitrobenzyl)thioinosine and -guanosine analogs (Table 6). The ribose moiety does not appear essential for inhibitory activity; a simple N^9-butyl substituent will do (Paul et al., 1975). Also modification at the 5′-position of the sugar is allowed. This feature has been used to develop fluorescent and affinity column probes based on the structure of NBI (Agbanyo et al., 1990; Wiley et al., 1991). The introduction of tritiated NBI as a radioligand has greatly facilitated the characterization of the (NBI-sensitive) nucleoside transporter, particularly when it was found that NBI could also be used as a photoaffinity ligand (Jarvis and Young, 1987; IJzerman et al., 1989a).

TABLE 6 Nucleoside Transport Inhibition in Human Erythrocytes by NBI and Analogs[a]

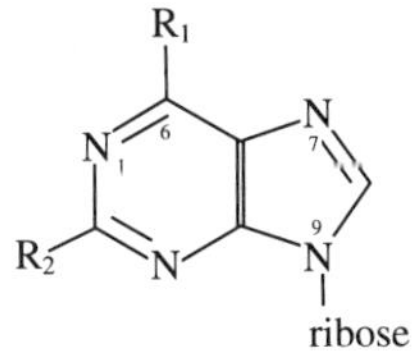

Compound	R_1	R_2	IC_{50} (nM)
N^6-Benzyladenosine	$NHCH_2C_6H_5$	H	22,000
6-Benzylthioinosine	$SCH_2C_6H_5$	H	3,500
6-Benzylthioguanosine	$SCH_2C_6H_5$	NH_2	11,000
N^6-(4-Nitro)benzyladenosine	$NHCH_2C_6H_4$-4-NO_2	H	58
6-(4-Nitro)benzylthioinosine (NBI)	$SCH_2C_6H_4$-4-NO_2	H	15
6-(4-Nitro)benzylthioguanosine	$SCH_2C_6H_4$-4-NO_2	NH_2	45
6-(2-Hydroxy-5-nitro)benzylthioinosine	$SCH_2C_6H_3$-2-OH,5-NO_2	H	69
6-(2-Hydroxy-5-nitro)benzylthioguanosine	$SCH_2C_6H_3$-2-OH,5-NO_2	NH_2	6

[a] Data from Paul et al. 1975.

Dipyridamole and Analogues

The coronary vasodilator dipyridamole has long been known to inhibit adenosine uptake as its principal mechanism of action. Several analogs have been synthesized, including related pteridin derivatives (Gerlach et al., 1965). Some of these compounds were tested in radioligand binding studies with [^{3}H]NBI as the radioligand (Table 7; IJzerman et al., 1989b). All compounds were active in the nanomolar range, RE 224 being the most active member (IC_{50} = 1.8 nM). Dipyridamole has also been tritiated to provide another radioligand for the nucleoside transporter (Marangos et al., 1985).

Dilazep, Hexobendine and Analogs

These piperazine-like esters from a third class of nucleoside transport blockers. The affinities of the compounds for the nucleoside transporter as present on calf lung tissue are gathered in Table 8. The symmetry in the molecules is important, since "half" molecules are not active (data not shown). The optimal chain length between basic nitrogen and the aromatic trimethoxyphenyl ring is three carbon atoms. The compounds appear structurally unrelated to the former two classes, due to the basic nitrogens present in the structures. However, studies in which the effects of pH changes on biological activity were examined, revealed that the uncharged, neutral compound was the bioactive species (Gati and Paterson, 1989; IJzerman and Voorschuur, 1990).

TABLE 7 K_i Values of Dipyridamole Analogs in [^{3}H]NBI Displacement Studies on Calf Lung Membranes

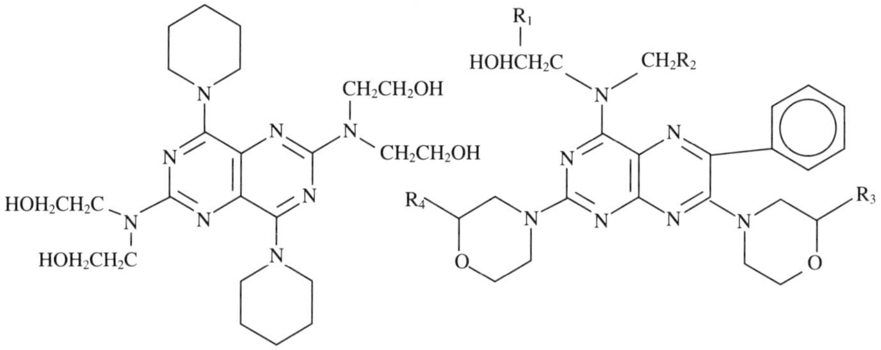

Dipyridamole — RE compounds

Compound	R_1	R_2	R_3	R_4	K_i (nM)
RE 57	H	H	H	H	8.1 ± 0.3
RE 61	H	CH_3	H	H	4.3 ± 1.1
RE 102	CH_3	$CH(OH)CH_3$	H	H	2.6 ± 0.9
RE 244	CH_3	CH_2OH	CH_3	CH_3	1.8 ± 0.6
Dipyridamole					6.2 ± 0.1

TABLE 8 K_i Values of Dilazep, Hexobendine, and Analogs in [^{3}H]NBI Displacement Studies on Calf Lung Membranes

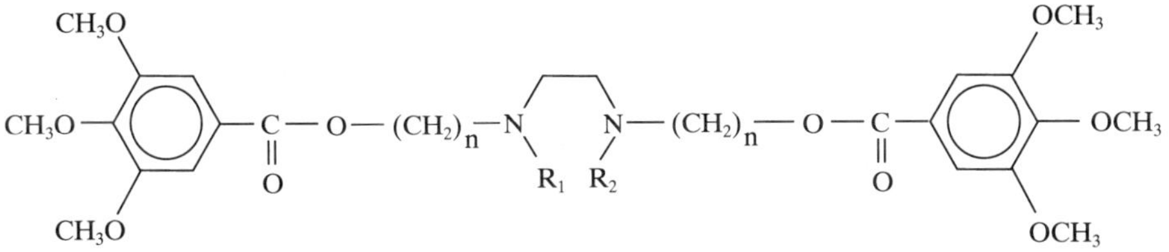

Compound	n	R_1	R_2	K_i (nM)
Dilazep	3	$-CH_2-CH_2-CH_2-$		1.8 ± 0.2
Hexobendine	3	$-CH_3$	$-CH_3$	6.0 ± 3.0
ST 7092	3	$-CH_2-CH_2-$		3.4 ± 0.6
ST 7094	4	$-CH_2-CH_2-$		11 ± 3
ST 7095	5	$-CH_2-CH_2-$		45 ± 15
ST 7135	3	$-CH_2-CH_3$	$-CH_2-CH_3$	14 ± 4

Mioflazine and Related Compounds

A relatively new series of compounds is based on the structure of lidoflazine. Modification of this classical calcium antagonist led to the synthesis of mioflazine. Detailed structure–activity relationships disclosed that introduction of a carboxamido substituent on the piperazine ring led to a further selectivity for the nucleoside transporter. Again, careful analysis of the ionization charactristics of these compounds revealed that at physiological pH almost all of the substance is in the uncharged, bioactive form (IJzerman and Voorschuur, 1990). A representative selection of the many compounds synthesized and tested is summarized in Table 9.

Due to the carboxamido group a chiral center is introduced in the compounds. The (−)-isomer of R75231, R88021 or draflazine, was chosen for further development as a cardioprotective agent and has entered clinical trials (Van Belle, 1995). As expected, draflazine proved twice as potent as R75231 in radioligand binding studies, with a stereoselectivity index of over 30 compared to the (+)-isomer, R88016 (Beukers et al., 1994). There is one peculiar aspect observed in the behavior of these compounds when they displace [^{3}H]NBI from its binding sites. The curves are steep, with pseudo–Hill coefficients significantly higher than unity. With tritiated R75231 as the radioligand, a "normal" behavior was observed, which suggests that the binding sites for NBI and the mioflazine analogs on the nucleoside transport protein are not identical (IJzerman et al., 1992).

CONCLUSION

The extra- and intracellular concentrations of adenosine are tightly controlled by a series of enzymes and transport proteins. Under physiological conditions adenosine concentrations are kept low. From a pharmacological point of view it is possible to interfere with the action of most enzymes and transport proteins. However, for some proteins, in particular cytosolic 5′-nucleotidases, no effective inhibitors are currently

TABLE 9 K_i Values of Mioflazine and Analogs in [^{3}H]NBI Displacement Studies on Calf Lung Membranes

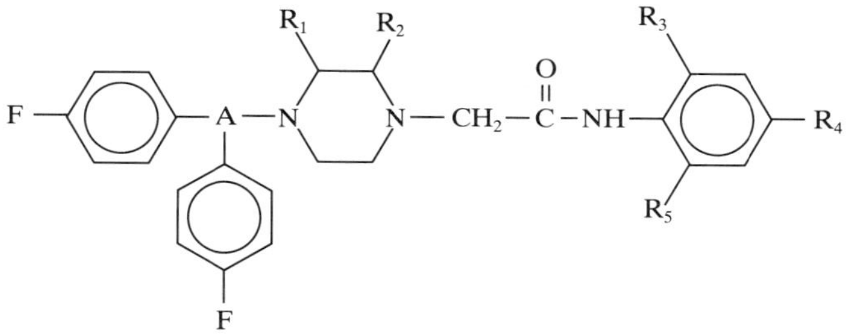

Compound	A	R_1	R_2	R_3	R_4	R_5	K_i (nM)
R7904 (lidoflazine)	$CH(CH_2)_3$	H	H	CH_3	H	CH_3	12 ± 2
R51469 (mioflazine)	$CH(CH_2)_3$	$CONH_2$	H	Cl	H	Cl	3.6 ± 0.3
R51975	$CH(CH_2)_3$	H	$CONH_2$	Cl	H	Cl	3.4 ± 0.4
R53531	$CH(CH_2)_4$	$CONH_2$	H	Cl	H	Cl	1.3 ± 0.1
R75231	$CH(CH_2)_4$	H	$CONH_2$	Cl	NH_2	Cl	1.3 ± 0.2
R73796	$CH(CH_2)_4$	H	$CONH_2$	Cl	$COCH_3$	Cl	1.4 ± 0.2

available. Another concern is that the selectivity of the many agents used has not been clearly established in all cases.

REFERENCES

Agbamyo FR, Cass CE, Paterson ARP (1988): External location of sites on pig erythrocyte membranes that bind nitrobenzylthioinosine. Mol Pharmacol 33:332–337.

Antonini I, Cristalli G, Franchetti P, Grifantini M, Martelli S, Lupardi G, Riva F (1984): Adenosine deaminase inhibitors. Synthesis of deaza analogues of erythro-9-(2-hydroxy-3-nonyl)adenine. J Med Chem 27:274–278.

Baker DC, Putt SR (1979): A total synthesis of pentostatin, the potent inhibitor of adenosine deaminase. J Am Chem Soc 101:6127–6128.

Beukers MW, Kerkhof CJM, IJzerman AP, Soudijn W (1994): Nucleoside transport inhibition and platelet aggregation in human blood: R75231 and its enantiomers, draflazine and R88016. Eur J Pharmacol - Mol Pharmacol Sect 266:57–62.

Bloch A, Robins MJ, McCarthy JR Jr (1967): The role of the 5′-hydroxyl group of adenosine in determining substrate specificity of adenosine deaminase. J Med Chem 10:908–912.

Burger RM, Lowenstein JM (1970): Preparation and properties of 5′-nucleotidase from smooth muscle of small intestine. J Biol Chem 245:6274–6280.

Bussolari JC, Ramesh K, Stoeckler JD, Chen S-F, Panzica RP (1993): Synthesis and biological evaluation of N[4]-substituted imidazo- and μ-triazolo[4,5-*d*]pyridazine nucleosides. J Med Chem 36:4113–4120.

Cottam HB, Wasson DB, Shih HC, Raychaudhuri A, Di Pasquale G, Carson DA (1993): New adenosine kinase inhibitors with oral antiinflammatory activity: Synthesis and biological evaluation. J Med Chem 36:3424–3430.

Cristalli G, Eleuteri A, Franchetti P, Grifantini M, Vittori S, Lupardi G (1991a): Adenosine deaminase inhibitors: Synthesis and structure-activity relationships of imidazole analogues of *erythro*-9-(2-hydroxy-3-nonyl)adenine. J Med Chem 31:1187–1192.

Cristalli G, Eleuteri A, Volpini R, Vittori S, Camaioni E, Lupidi G (1994): Adenosine deaminase inhibitors: Synthesis and structure-activity relationships of 2-hydroxy-3-nonyl derivatives of azoles. J Med Chem 37:201–205.

Cristalli G, Franchetti P, Grifantini M, Vittori S, Lupidi G, Riva F, Bordoni T, Geroni C, Verini MA (1988): Adenosine deaminase inhibitors. Synthesis and biological evaluation of deze analogues of *erythro*-9-(2-hydroxy-3-nonyl)adenine. J Med Chem 31:390–393.

Cristalli G, Vittori S, Eleuteri A, Grifantini M, Volpini R, Lupardi G, Capolongo L, Pesenti E (1991b): Purine and 1-deazapurine ribunucleosides and deoxyribonucleosides: synthesis and biological activity. J Med Chem 34:2226–2230.

Davies LP, Cook AF (1995): Inhibition of adenosine kinase and adenosine uptake in guinea-pig CNS tissue by halogenated tubercidin analogues. Life Sci 56:345–349.

De Clercq E (1987): S-Adenosylhomocysteine hydrolase inhibitors as broad-spectrum antiviral agents. Biochem Pharmacol 36:2567–2575.

Firestein GS, Bullough DA, Erion MD, Jimenez R, Ramirez-Weinhouse M, Barankiewicz J, Smith CW, Gruber HE, Mullane KM (1995): Inhibition of neutrophil adhesion by adenosine and an adenosine kinase inhibitor. J Immunol 154:326–334.

Gati WP, Paterson ARP (1989): Interaction of [^{3}H]dilazep at nucleoside transporter–associated binding sites on S49 lymphoma cells. Mol Pharmacol 36:134–141.

Gerlach E, Deuticke B, Koss FW (1965): Einfluss von Pyrimidopyrimidin- und Pteridin-Derivaten auf Phosphat- und Adenosin-Permeabilität menschlicher Erythrocyten. Arzneim-Forsch/Drug Res 15:558–563.

Harriman GCB, Abushanab E, Stoeckler JD (1994): Adenosine deaminase inhibitors. Synthesis and biological evaluation of 4-amino-1-(2-(*S*)-hydroxy-3(*R*)-nonyl)-1*H*-imidazo[4,5-*c*]pyridine (3-deaza-(+)-EHNA) and certain Cl′ derivatives. J Med Chem 37:305–309.

Harriman GCB, Poirot AF, Abushanab E, Midgett RM, Stoeckler JD (1992): Adenosine deaminase inhibitors. Synthesis and biological evaluation of Cl′ and nor-Cl′ derivatives of (+)-*erythro*-9-(2(*S*)-hydroxy-3(*R*)-nonyl)adenine. J Med Chem 35:4180–4184.

IJzerman AP, Voorschuur AH (1990): The relationship between ionization and affinity of nucleoside transport inhibitors. Naunyn Schmiedebergs Arch Pharmacol 342:336–341.

IJzerman AP, Kruidering M, Van Weert A, Van Belle H, Janssen C (1992): [^{3}H]R75231—a new radioligand for the nitrobenzylthioinosine sensitive nucleoside transport proteins. Characterization of (±)-[^{3}H]R75231 binding to calf lung membranes, stereospecificity of its two stereoisomers, and comparison with [^{3}H]nitrobenzylthioinosine binding. Naunyn Schmiedebergs Arch Pharmacol 345:558–563.

IJzerman AP, Menkveld GJ, Thedinga KH (1989a): A refined method for the photoaffinity labelling of the nitrobenzylthioinosine-sensitive nucleoside transport protein: Application to cell membranes of calf lung tissue. Biochim Biophys Acta 979:153–156.

IJzerman AP, Thedinga KH, Custers AFCM, Hoos B, Van Belle H (1989b): Inhibition of nucleoside transport by a new series of compounds related to lidoflazine and mioflazine. Eur J Pharmacol - Mol Pharmacol Sect 172:273–281.

Ikehara M, Fukui T (1974): Studies of nucleosides and nucleotides, LVIII. Deamination of adenosine analogs with calf intestine adenosine deaminase. Biochim Biophys Acta 338:512–519.

Iltzsch MH, Uber SS, Tankersley KO, Kouni MH (1995): Structure-activity relationship for the binding of nucleoside ligands to adenosine kinase from *Toxoplasma Gondii.* Biochem Pharmacol 49:1501–1512.

Jarvis SM, Young JD (1987): Photoaffinity labelling of nucleoside transporter polypeptides. Pharmacol Ther 32:339–359.

Jones W, Kurz LC, Wolfenden R (1989): Transition state stabilization by adenosine deaminase: 1,6-addition of water to purine ribonucleoside, the enzyme's affinity for 6-hydroxy-1,6-dihydropurine ribonucleoside, and the effective concentration of substrate water at the active site. Biochemistry 28:1242–1247.

Kurz LC, Moix L, Riley MC, Frieden C (1992): The rate of formation of transition-state analogues in the active site of adenosine deaminase is encounter-controlled: Implications for the mechanism. Biochemistry 31:39–48.

Lin BB, Hurley MC, Fox IH (1988): Regulation of adenosine kinase by adenosine analogs. Mol Pharmacol 34:501–505.

Marangos PJ, Houston M, Montgomery P (1985): [^{3}H]Dipyridamole: A new ligand probe for brain adenosine uptake sites. Eur J Pharmacol 117:393–394.

Montgomery JA, Clayton SJ, Thomas HJ, Shannon WM, Arnett G, Bodner AJ, Kion IK, Cantoni GL, Chiang PK (1982): Carbocyclic analogue of 3-deazaadenosine: A novel antiviral agent using S-adenosylhomocysteine hydrolase as a pharmacological target. J Med Chem 25:626–629.

Nair V, Buenger GS, Sells TB (1991): Inhibition of mammalian adenosine deaminase by novel functionalized 2′,3′-dideoxyadenosines. Biochim Biophys Acta 1078:121–123.

Naito Y, Lowenstein JM (1985): 5′-Nucleotidase from rat heart membranes. Biochem J 226:645–651.

Nakamura H, Koyama G, Iitaka Y, Ohno M, Yagasawa N, Kondo S, Maeda K, Umezawa H (1974): Structure of coformycin, an unusual nucleoside of microbial origin. J Am Chem Soc 96:4327–4328.

Pajor AM (1995): Molecular cloning and expression of SNST1, a renal sodium/nucleoside cotransporter. In Belardinelli L, Pelleg A (eds): "Adenosine and Adenine Nucleotides: From Molecular Biology to Integrative Physiology." Boston: Kluwer, pp 49–54.

Palmer JL, Abeles RH (1979): The mechanism of action of S-adenosylhomocysteinase. J Biol Chem 254:1217–1226.

Patil SD, Schneller SW (1992): Synthesis and antiviral properties of (±)-5′-noraristeromycin and related purine carbocyclic nucleosides. A new lead for anti-human cytomegalovirus agent design. J Med Chem 35:3372–3377.

Paul B, Chen MF, Paterson ARP (1975): Inhibitors of nucleoside transport. A structure–activity study using human erythrocytes. J Med Chem 18:968–973.

Plagemann PG, Wohlhueter RM, Woffendin C (1988): Nucleoside and nucleobase transport in animal cells. Biochim Biophys Acta 947:405–443.

Porter DJT, Abushanab E (1992): Kinetics of inhibition of calf intestinal adenosine deaminase by (+)- and (−)-*erythro*-9-(2-hydroxy-3-nonyl)adenine. Biochemistry 31:8216–8220.

Prescott M, McLennan AG, Agathocleous DC, Bulman Page PC, Cosstick R, Galpin IJ (1989): The inhibition of adenosine kinase by α,ω-di-(adenosin-N^6-yl)alkanes. Nucleosides Nucleotides 8:297–303.

Sawa T, Fukagawa Y, Homma H, Takeuchi T, Umezawa H (1967): Mode of inhibition of coformycin on adenosine deaminase. J Antibiot Ser A 20:227–231.

Schaeffer HJ, Schwender CF (1974): Enzyme inhibitors, 26. Bridging hydrophobic and hydrophilic regions on adenosine deaminase with some 9-(2-hydroxy-3-alkyl)adenines. J Med Chem 17:6–8.

Schramm VL, Baker DC (1985): Spontaneous epimerization of (*S*)-deoxycoformycin and interaction of (*R*)-deoxycoformycin (*S*)-deoxycoformycin, and 8-ketodeoxycoformycin with adenosine deaminase. Biochemistry 24:641–646.

Shewach DS, Krawazuk SH, Acevedo OL, Townsend LB (1992): Inhibition of adenosine deaminase by azapurine ribonucleoside. Biochem Pharmacol 44:1697–1700.

Skladanowski AC, Newby AC (1990): Partial purification and properties of an AMP-specific soluble 5′-nucleotidase from pigeon heart. Biochem J 268:117–122.

Skladanowski AC, Sala GB, Newby AC (1989): Inhibition of IMP-specific cytosolic 5′-nucleotidase and adenosine formation in rat polymorphonuclear leucocytes by 5′-deoxy-5′-isobutylthio derivatives of adenosine and inosine. Biochem J 262:203–208.

Tseng CKH, Marquez VE, Fuller RW, Goldstein BM, Haines DR, McPherson H, Parsons JL, Shannon WM, Arnett G, Hollingshead M, Driscoll JS (1989): Synthesis of 3-deazaneplanocin A, a powerful inhibitor of S-adenosylhomocysteine hydrolase with potent and selective in vitro and in vivo antiviral activities. J Med Chem 32:1442–1446.

Ueland PM (1982): Pharmacological and biochemical aspects of S-adenosylhomocysteine and S-adenosylhomocysteine hydrolase. Pharmacol Rev 34:223–253.

Van Belle H (1995): Adenosine uptake blockers for cardioprotection. In Belardinelli L, Pelleg A (eds): "Adenosine and Adenine Nucleotides: From Molecular Biology of Integrative Physiology Boston: Kluwer, pp 373–378.

Van Galen PJM, IJzerman AP, Soudijn W (1987): Adenosine derivatives with N^6-alkyl, alkylamine or -alkyladenosine substituents as probes for the adenosine A_1 receptor. FEBS Lett 223:197–201.

Vargeese C, Sarma MSP, Pragnacharyulu PVP, Abushanab E (1994): Adenosine deaminase inhibitors. Synthesis and biological evaluation of putative metabolites of (+)-*erythro*-9-(2S-hydroxy-3*R*-nonyl)adenine. J Med Chem 37:3844–3849.

Wierzchowski J, Lassotta P, Shugar D (1984): Continuous fluorimetric assay of 5′-nucleotidase with formycin 5′-phosphate as substrate, and its application to properties of substrates and inhibitors. Biochim Biophys Acta 786:170–178.

Wiley JS, Brocklebank AM, Snook MB, Jamieson GP, Sawyer WH, Craik JD, Cass Ce, Robins MY, McAdam DP, Paterson ARP (1991). A new fluorescent probe for the equilibrative inhibitor-sensitive nucleoside transporter. Biochm J 273:667–672.

Wilson DK, Rudolph FB, Quiocho FA (1991): Atomic structure of adenosine deaminase complexed with a transition-state analog: Immunodeficiency mutations. Science 252:1278–1284.

Wilson DK, Quiocho FA (1993): A pre-transition-state mimic of an enzyme: X-ray structure of adenosine deaminase with bound 1-deazaadenosine and zinc-activated water. Biochemistry 32:1689–1694.

Wolfe MS, Borchardt RT (1991): S-Adenosyl-L-homocysteine hydrolase as a target for antiviral chemotherapy. J Med Chem 34:1521–1530.

Wolfe MS, Lee Y, Bartlett WJ, Borcherding DR, Borchardt RT (1992): 4′-Modified analogues of aristeromycin and neplanocin A: Synthesis and inhibitory activity toward S-adenosyl-L-homocysteine hydrolase. J Med Chem 35:1782–1791.

Woo PKW, Dion HW, Lange SM, Dahl LF, Durham LJ (1974): A novel adenosine and ara-A deaminase inhibitor. (R)-3-(2-deoxy-β-D-*erythro*-pento-furanosyl)-3,6,7,8-tetrahydroimidazo[4,5-*d*][1,3]diazepin-8-ol. J Heterocycl Chem 11:641–643.

Yoshida S-Y, Aikawa T (1993): Effect of substituent ribose of adenosine analogs on binding to the adenosine deaminase in the scallop (*Patinopecten Yessoensis*) midgut gland. Comp Biochem Physiol 105B:63–67.

Yuan CS, Wnuk SF, Liu S, Robins MJ, Borchardt RT (1994): (E)-5′,6′-didehydro-6′-deoxy-6′-fluorohomoadenosine: A substrate that measures the hydrolytic activity of S-adenosylhomocysteine hydrolase. Biochemistry 33:12305–12311.

Zimmerman H (1992): 5′-Nucleotidase: Molecular structure and functional aspects. Biochem J 285:345–365.

PART IV

THERAPEUTIC IMPLICATIONS

1. Cardiology

CHAPTER 8

Adenosine, Ischemia, and Preconditioning

GUANG-SHUNG LIU, JAMES M. DOWNEY, and MICHAEL V. COHEN

Departments of Physiology (G.-S.L., J.M.D.) and Medicine (M.V.C.), University of South Alabama, Mobile, AL 36688

ADENOSINE AND ADENOSINE RECEPTORS

In recent years the cardioprotective effects of adenosine and adenosine agonists have been well established by various investigators. In myocardium, the pathways of endogenous adenosine formation include enzymatic dephosphorylation of 5′-adenosine monophosphate (5′-AMP) by 5′-nucleotidase (including both membrane-bound and cytosolic) and the hydrolysis of S-adenosylhomocysteine (SAH) by SAH-hydrolase.

Adenosine and other nucleosides traverse the cell membrane via a nucleoside carrier system. Adenosine produced locally in myocardium is either rapidly washed out into the circulating blood or deaminated to inosine. Both endogenous and exogenous adenosine can also be phosphorylated by adenosine kinase to AMP. Because the K_m for adenosine kinase is 100 times lower than that for adenosine deaminase, the preferential pathway for adenosine metabolism is salvage by phosphorylation to AMP.

The effects of adenosine are mainly mediated by extracellular membrane purinergic type 1 (P_1) receptors, which are further classified as either A_1 (A_1AR) or A_2 (A_2AR) based on radioligand binding studies and the specific pharmacological responses to adenosine and its analogues (Olsson and Pearson, 1990). In the heart, A_1ARs are found on cardiomyocytes and they decrease heart rate, reduce myocardial contractility and catecholamine release, and increase glucose uptake and glycolysis (Finegan et al., 1993, Wyatt et al., 1989). A_1ARs couple to pertussis toxin–sensitive inhibitory G proteins (G_i) to inhibit adenylate cyclase, modulate the activity of several ion channels (K^+, Ca^{2+}, and Na^+/Ca^{2+} exchanger) (Stiles, 1992), and probably activate phospholipase C or D, which in turn increases the production of diacylglycerol, the activator of protein kinase C. While mounting evidence indicates that a small population of A_2ARs may also exist on the cardiomycocytes (Stein et al., 1993), A_2ARs are primarily located

Purinergic Approaches in Experimental Therapeutics, Edited by Kenneth A. Jacobson and Michael F. Jarvis
ISBN 0-471-14071-6 © 1997 Wiley-Liss, Inc.

on the coronary vessels. A_2ARs dilate vascular smooth muscle. They also inhibit neutrophil function and depress generation of free radicals by leukocytes (Schrier et al., 1990; Cronstein et al., 1990). A_2ARs on vascular endothelium and smooth muscle activate adenylate cyclase and produce cAMP by coupling to stimulatory G proteins (G_s).

Recently a new adenosine A_3 receptor (A_3AR) has been described in the rat (Zhou et al., 1992), sheep (Linden et al., 1993), and human (Sajjadi and Firestein, 1993). The A_3AR binds to some A_1 AR selective agonists, but binds poorly to methylxanthine-derived A_1 receptor antagonists such as xanthine amine congener (XAC) or 8-cyclopentyl-1-3-dipropylxanthine (DPCPX). A_3ARs also couple to pertussis toxin–sensitive G_i proteins to inhibit adenylate cyclase (Zhou et al., 1992). However, the physiological effect of A_3AR is still poorly understood. It has been reported that A_3AR stimulation can cause hypotension in pithed rats (Fozard and Carruthers, 1993) as well as activation of mast cells (Ramkumar et al., 1993).

ADENOSINE PRODUCTION BY THE ISCHEMIC HEART

The release of adenosine from myocardium is related to myocardial metabolism. In the heart, high-energy phosphate stores are limited and exist predominantly in the form of ATP and creatine phosphate (PCr). During normoxia the hydrolysis of SAH is the primary source of adenosine. During myocardial ischemia or hypoxia a large quantity of the energy-rich ATP, which is present in the heart muscle in millimolar quantities, is degraded to AMP. Further breakdown of this AMP by 5′-nucleotidase is thought to be primarily responsible for the increased adenosine production during ischemia or hypoxia (Smolenski et al., 1992).

The extent of any myocardial ischemic injury is proportional to the duration and severity of the ischemic insult, the presence and degree of any collateral circulation, and the area of myocardium at risk (Reimer and Jennings, 1979). As the duration of ischemia increases, cardiac cells are first subject to arrhythmias, then contractile dysfunction (stunning), and finally infarction. The heart normally derives nearly 100% of its energy from oxidative metabolism of nutrients consisting primarily of lipids and lactate. During a period of limited blood flow, the delivery of oxygen becomes rate-limiting, and ATP production can no longer keep up with the demand. The metabolic response of the heart to ischemia or hypoxia involves an initial increase in glucose utilization, which helps to bolster myocardial ATP levels by enhanced glycolysis. As the duration and severity of ischemia increase, however, the glycolytic rate decreases as a result of the intracellular accumulation of NADH, hydrogen ion, and lactate (Jennings and Reimer, 1981).

Adenosine release from the myocardium is increased about 50-fold during ischemia and hypoxia and, in certain circumstances discussed below, can act as an endogenous protector of the ischemic heart. It has been reported that adenosine slows the rate of metabolism and delays the accumulation of H^+ during ischemia (Vander Heide et al., 1993b). The mechanisms most often cited for the cardioprotective effect of adenosine are increased coronary blood flow, reduced ventricular contractility, and enhanced ATP repletion by purine salvage. More than likely, however, the real underlying mechanisms of adenosine's cardioprotection are the result of factors that are yet to be discovered.

ANTIARRHYTHMIC EFFECT OF ADENOSINE

More than six decades ago, Drury and Szent-Gyorgyi showed for the first time that extracellular adenosine and related compounds exerted a negative chronotropic effect and negative dromotropic effect on myocardium. Adenosine's negative chronotropic effect on cardiac pacemakers, including the sinus node and junctional and ventricular pacemakers, is the result of A_1AR activation. The negative dromotropic effect has been used clinically to treat supraventricular tachycardias. Cardiac production of adenosine increases with increasing degrees of oxygen deprivation. The resulting increase in myocardial adenosine concentration can produce AV block and even cardiac arrest. This conduction block may act as a negative feedback mechanism to rectify the oxygen supply:demand imbalance by slowing the ventricular rate during supraventricular tachycardias.

In several different experimental models, it has been reported that adenosine prevents ischemia-induced ventricular arrhythmias. Parratt (1993) has reported that adenosine could reduce the severity of ischemia-induced ventricular arrhythmias in anesthetized rats. Intravenous infusion of adenosine (5-25 $\mu g\ kg^{-1}\ min^{-1}$) significantly reduced the number of ventricular premature beats and the incidence and duration of ventricular tachycardia, without having any hemodynamic effects. At the higher doses adenosine completely prevented ventricular fibrillation. Blockade of adenosine receptors during ischemia with 8-phenyltheophylline significantly hastened the onset of ventricular tachycardia.

Although much evidence supports the antiarrhythmic effect of adenosine, the exact mechanism is still unknown. Adenosine activates a distinct subtype of K^+ channel in atrial tissue, thereby inducing an outward current that may shorten the action potential duration in atrial myocytes and hyperpolarize sinus node cells. Adenosine also attenuates the presynaptic release of catecholamines, which are believed to be a major arrhythmogenic factor (Schömig and Richardt, 1990). Also, because adenosine A_1 receptors activate G_i, they oppose the effects of β-adrenergic stimulation. This antiadrenergic effect may play an important role in the ventricle. Only a mild attenuation of Ca^{2+} inward current (I_{Ca}) is caused by adenosine in unstimulated atrial myocytes (Visentin et al., 1990; Conti et al., 1995). Under conditions of catecholamine stimulation, however, adenosine can attenuate catecholamine-induced I_{Ca} markedly in both atrial and ventricular myocytes (Isenberg and Belardinelli, 1984).

ADENOSINE IMPROVES POSTISCHEMIC MECHANICAL FUNCTION

When the heart is subjected to a transient period of ischemia, the contractility of myocardium after reperfusion is severely impaired. This dysfunction is due to a combination of the death of myocytes and of stunning. Stunning is a reversible contractile defect in cells that have had a sublethal ischemic insult (Kusuoka and Marban, 1992). Although in nonischemic myocardium adenosine causes a negative inotropic effect by lowering cAMP levels, it actually improves functional recovery after resolution of myocardial ischemia. The studies of Lasley, Mentzer, and colleagues (Lasley et al., 1990) showed that the presence of adenosine or A_1AR-selective analogs in the perfusate prior to ischemia significantly improved recovery of postischemic left ventricular developed pressure in isolated rat hearts. They also found higher tissue ATP levels in adenosine-treated hearts after ischemia (Ely et al., 1985). Similar beneficial effects on functional recovery and myocardial ATP content were obtained when myocardial adenosine concentrations were increased during ischemia by inhibiting the metabolism of adenosine with the adenosine deaminase inhibitor erythro-9-(2-hydroxy-3-nonyl)adenosine (EHNA) (Brown et al., 1986). At the end of ischemia, EHNA-treated hearts had an elevated tissue adenosine content compared to control hearts.

More recently Bolli's group (Zughaib et al., 1992) found that combined administration of EHNA and a nucleoside transport blocker was remarkably effective in augmenting myocardial adenosine levels during regional ischemia and reperfusion in open-chest dogs. This treatment resulted in a significant and sustained attenuation of myocardial stunning. The attenuation of stunning was not related to ATP repletion nor to a nonspecific action on hemodynamic variables or coronary flow. EHNA plus transport blockade effectively prevent adenosine's degradation to urate. Adenosine metabolism to uric acid is catalyzed by xanthine oxidase and is a major source of free radicals in the reperfused dog heart (Downey et al., 1988). Indeed xanthine oxidase inhibition has also been shown to prevent stunning in the dog model (Charlat et al., 1987). Thus it is not clear whether observed protection by EHNA and the nucleoside transport blocker was due to increased intracellular adenosine and/or suppression of xanthine oxidase–derived free radicals.

Acadesine, the adenosine regulating agent, promotes adenosine's tissue effects, particularly during ischemia (Gruber et al., 1989). It also has been reported to improve postischemic recovery of function in both dogs (Hori et al., 1990) and rabbits (Mazur et al., 1991). 8-(*p*-sulfophenyl)theophylline (SPT), a potent nonselective adenosine receptor antagonist, blocked the beneficial effect of acadesine (Mullane, 1993).

EFFECT OF ADENOSINE ON DEVELOPMENT OF ISCHEMIC CONTRACTURE

Adenosine not only improves functional recovery but also delays the onset of ischemic contracture. Ischemic contracture is the increase in resting tension that occurs in the ventricle during ischemia. Ischemic contracture is thought to reflect the loss of ATP in the myocytes. Accordingly, the A_1AR-selective agonist N^6-l-(phenyl-2R-isopropyl) ADO (r-PIA) significantly delayed the onset of ischemic contracture, decreased the peak amplitude of ischemic contracture, and preserved myocardial ATP during ischemia in rats (Lasley et al., 1990). A similar result was obtained by Janier et al. (1993) in isolated rabbit hearts. The above studies suggest that adenosine, via the A_1AR, protects the heart by preserving myocardial ATP content during the ischemic period. However, it is still unresolved how A_1AR activation enhances postischemic function.

EFFECT OF ADENOSINE AT REPERFUSION

Some studies indicate that adenosine given at reperfusion may be beneficial. Hori et al. (1989) found that the coronary hyperemic effect was associated with improved function, a finding that indicated a role for the A_2AR. Forman's group (Babbitt et al., 1989) reported a dramatic increase in the recovery of regional function when adenosine was administered into the coronary artery just prior to reperfusion in dogs. However, Sekili et al. (1995) found that an intracoronary infusion of adenosine could dramatically improve postischemic segmental function in a canine stunning model *but not when it was begun at reperfusion.* As discussed below, A_1 receptor stimulation also failed to limit infarct size in rabbits when administered just prior to reperfusion (Thornton et al., 1992). Thus, the evidence that adenosine can protect the heart when given at reperfusion is controversial.

PROTECTION AGAINST INFARCTION BY ADENOSINE

Toombs et al. (1992) proposed that endogenous adenosine released by the ischemic myocardium limits infarction naturally in the rabbit heart, suggesting an inherent protection by the endogenous chemical. Zhao et al. (1995) found that blockade of adenosine receptors by SPT during ischemia and reperfusion significantly increased infarct size. They further proposed that adenosine had an impact on both ischemic injury (principally by an A_1AR effect) and reperfusion injury (principally by an A_2AR effect). They reported that acadesine, which increases the production of endogenous adenosine, also limited infarct size. However, work from our laboratory does not support the concept that endogenous adenosine protects against infarction. In a number of studies, blockade of adenosine receptors was never seen to increase infarct size in the non-preconditioned rabbit heart (Liu et al., 1991; Thornton et al., 1993a; Tsuchida et al., 1994a; Liu et al., 1995a). These data suggest that endogenous adenosine released during ischemia is unable to protect against infarction despite the fact that A_1AR are probably saturated within several minutes of the onset of ischemia. Furthermore, in our hands acadesine also has not limited infarct size in the non-preconditioned rabbit (Tsuchida et al., 1993, Burckhartt et al., 1995). Thus, we conclude that endogenous adenosine released by the ischemic non-preconditioned heart offers little, if any, detectable protection against infarction.

THE ROLE OF ADENOSINE IN ISCHEMIC PRECONDITIONING

For several decades there has been an active search for a pharmacological agent that could render the myocyte more tolerant to ischemia. Although some agents such as antioxidants, calcium antagonists, and β-blockers were thought to offer protection against infarction, none provided consistent infarct size limitation in preclinical trials. The first indication that such an intervention was even possible came with the description of ischemic preconditioning by Murry et al. (1986). If the heart were first preconditioned with a brief period of ischemia followed by reperfusion, it paradoxically became resistant to infarction from a subsequent ischemic insult. It has been shown that protection by preconditioning occurs in a variety of species including dogs (Murry et al., 1986; Li et al., 1990), pigs (Schott et al., 1990; Schulz et al., 1993), rabbits (Liu et al., 1991), rats (Liu and Downey, 1992; Li and Kloner, 1993) and, most importantly, humans (Deutsch et al., 1990, Yellon et al., 1993). Ischemic preconditioning is considered to be the most powerful intervention against myocardial infarction described to date (Lawson and Downey, 1993).

Although preconditioning's efficacy has been repeatedly confirmed by various investigators, the exact mechanism of this protection remains a mystery. In previous studies it had been shown that preconditioning does not involve an increase in coronary collateral blood flow (Murry et al., 1986) or synthesis of a new protective protein (Thornton et al., 1990). Cyclooxygenase inhibitors failed to block the anti-infarct effect of preconditioning, making prostaglandins unlikely candidates for the mediator (Liu et al., 1992). We also found that preconditioning's protection is not related to a change in the amount of antioxidants in the myocardium (Turrens et al., 1992). Based on their observation that the rate of ATP depletion during ischemia is slowed, Murry et al. (1990) proposed that there is a reduction in energy demand in preconditioned ischemic myocardium.

Brief ischemia depresses myocardial contractile function by stunning the heart, which could cause a reduced energy demand during ischemia. Several studies, however, have found that the efficacy of preconditioning correlates poorly with the degree of stunning (Miura et al., 1991). Finally, although stunning persists for many hours after ischemia, the protection of preconditioning wears off in about 1 hour (Van Winkle et al., 1991). Therefore, protection by preconditioning is unlikely to be mediated by a reduction of energy demand caused by myocardial stunning.

ENDOGENOUS ADENOSINE AS A TRIGGER OF CARDIAC PROTECTION

Evidence from our laboratory has indicated that adenosine released during the brief period of ischemia triggers the changes associated with ischemic preconditioning (Liu et al., 1991). We found that preconditioning's protection against infarction from 30 minutes of regional ischemia could be abolished by pretreatment with the nonselective adenosine receptor blockers, SPT or N-[2-(dimethylamino)ethyl]-N-methyl-4-(2,3,6,7-tetrahydro-2,6-dioxo-1,3-dipropyl-1H-purin-8-yl)benzenesulfonamide (PD 115,199) in open-chest rabbits. Those rabbits were preconditioned with 5 minutes of ischemia followed by 10 minutes of reperfusion prior to the subsequent 30-minute coronary occlusion. Figure 1 reveals that pretreatment with SPT or PD 115,199 blocked the protective effect of preconditioning, but had no significant effect on non-preconditioned hearts.

If adenosine released during the brief ischemia were the trigger of preconditioning, we would expect that exposure of the heart to exogenous adenosine should mimic the protection from preconditioning. To test this hypothesis we infused adenosine directly into coronary arteries of isolated blood-perfused hearts. When adenosine was infused for 5 minutes at a total dose of 1.4 mg followed by 10 minutes of drug-free washout prior to a subsequent period of regional ischemia, it protected the hearts to the same degree as ischemic preconditioning (Figure 2). To identify which adenosine receptor subtype was responsible for the protection, PIA, the A_1AR-selective agonist, was infused over 5 minutes into the coronary artery. We found that PIA also protected the myocardium from infarction, indicating that the A_1AR was responsible (Liu et al., 1991). Activation of A_2AR with CGS 21680 failed to protect the heart, a result that further implicated the A_1AR (Thornton et al., 1992). All of these data indicate that in the rabbit heart adenosine released during the preconditioning ischemia activates A_1AR, which triggers the biochemical changes associated with preconditioning.

Since these studies there has been general confirmation of adenosine's role in preconditioning in rabbit (Tsuchida et al., 1992), pig (Schwarz et al., 1991; Van Winkle et al., 1994), dog (Auchampach and Gross, 1993) and in human cardiomyocytes (Ikonomidis et al., 1995). Interestingly, adenosine receptor blockade does not block protection from preconditioning in rat heart (Liu and Downey, 1992; Li and Kloner, 1993).

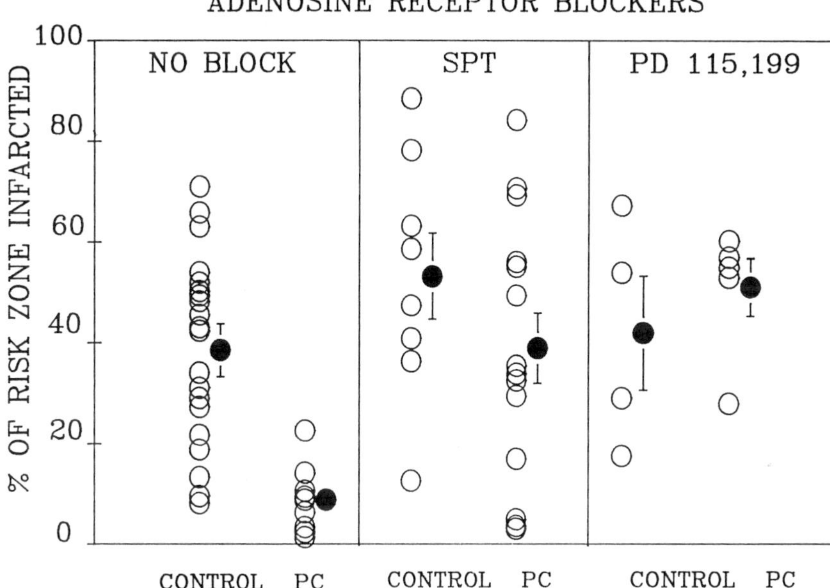

FIGURE 1. Adenosine's involvement in preconditioning is shown in this plot of infarct size, normalized as a percent of the risk zone infarcted, for rabbits *in situ*. All hearts experienced a 30-minute ischemic insult followed by 2 hours of reperfusion. Open circles indicate individual experiments, while filled circles show mean ± S.E.M. for each group. Neither of the nonselective adenosine blockers SPT nor PD 155,199 modulated infarct size in non-preconditioned hearts; however, both blockers abolished preconditioning's protection. PC: preconditioning; SPT: 8-*p*-sulfophenyltheophylline; PD: PD 115,199. Reproduced with permission from Liu et al., 1991.

INABILITY OF INTRAVENOUS ADENOSINE TO PROTECT THE HEART

A series of studies have been done by Forman and coworkers on the ability of adenosine to limit infarct size when given at reperfusion. They reported that intravenously infused adenosine started just prior to reperfusion significantly limited infarct size in dogs (Norton et al., 1992) and rabbits (Norton et al., 1991). A study by Goto et al. (1991), however, failed to show a reduction in infarct size in rabbits with intravenous adenosine infused just prior to reperfusion. Vander Heide et al. (1993a) failed to confirm the protective effect of intravenous adenosine administered at reperfusion in dogs. We tried to precondition rabbits by infusing adenosine intravenously at a dose of 2.5 mg/kg for 5 minutes, starting 15 minutes prior to the onset of ischemia, but this was not protective either (Liu et al., 1991). The animals would not tolerate a higher dose because of severe hypotension. All of these negative findings are not surprising. Interstitial adenosine quickly reaches the micromolar range with the onset of myocardial ischemia. Hypotension due to vasodilation (primarily an A_2 effect) prevents blood adenosine levels from rising to anywhere near this concentration following exogenous administration. Thus, the amount of adenosine delivered to the heart from a slightly elevated plasma adenosine titer would be negligible compared to what the heart is

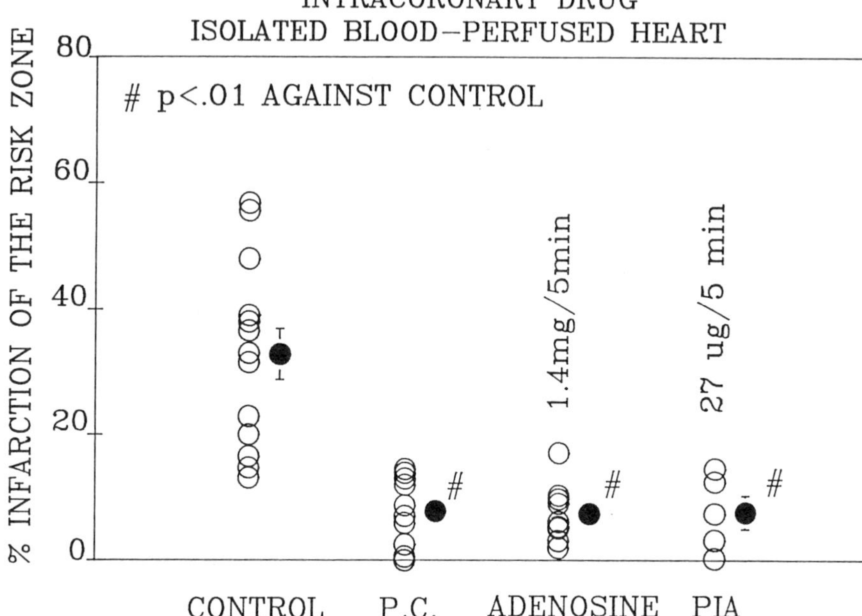

FIGURE 2. Intracoronary infusion of adenosine or PIA (5-minute infusion followed by 10-minute washout) prior to ischemia was as protective as ischemic preconditioning in isolated blood-perfused rabbit hearts. PC: ischemic preconditioning; AD: adenosine; PIA: N^6-1-(phenyl-2R-isopropyl)adenosine (r-PIA). Reproduced with permission from Liu et al., 1991.

already producing. A large intravenous bolus of adenosine can momentarily raise adenosine levels in the heart and actually cause cardiac standstill, but as the adenosine is metabolized the heart quickly recovers. We found that this protocol also failed to protect the heart against infarction (unpublished observation), suggesting that adenosine receptors must necessarily be occupied for a prolonged period of time before protection is realized. It is the opinion of the authors that there is no convincing evidence that an intravenous infusion of adenosine can achieve a plasma concentration high enough to trigger a limitation in infarct size.

HEART PROTECTION BY INTRAVENOUS ADMINISTRATION OF A_1AR-SELECTIVE AGENTS

Protection was seen in the rabbit model, however, when the A_1AR-selective agonists PIA and 2-chloro-N^6-cyclopentyladenosine (CCPA) were administered intravenously prior to ischemia (Thornton et al., 1992). These agents still caused some hypotension which was primarily related to the induced bradycardia. Reversing the bradycardia with pacing did not eliminate the protection. Our data strongly indicate that cardiac adenosine receptors must be stimulated before ischemia starts. Absolutely no protection was seen when a protective dose of PIA was given 5 minutes prior to reperfusion in the rabbit model (Thornton et al., 1992). Also, adenosine blockers did not prevent protection when given to preconditioned rabbits immediately prior to the end of the 30-minute ischemic period (Thornton et al., 1993a). It would appear that adenosine exerts its anti-infarct effect early during the prolonged ischemic period rather than at reperfusion.

PARTICIPATION OF A_3AR IN ADENOSINE-MEDIATED CARDIAC PROTECTION

Most of the experiments in rabbits, pigs, and dogs suggest that the A_1AR mediates the cardioprotection of preconditioning. One discrepancy, however, is that in the rabbit the highly selective A_1AR antagonist DPCPX does not prevent the anti-infarct effect of ischemic preconditioning. To solve this puzzle, we tested whether the newly described A_3AR might be involved in preconditioning's protection in rabbit hearts. An isolated rabbit heart model of regional ischemia and infarction was used. In this study we found that the potent A_3AR agonist N^6-[2-(4-aminophenyl)ethyl] adenosine (APNEA) was as potent as adenosine or ischemia in conferring protection. DPCPX (200 nM), which blocks A_1AR but not A_3AR, could not abort protection from either ischemic preconditioning, adenosine, or APNEA; but 100 μM SPT blocked it in all cases (Liu et al., 1994a). Furthermore, BW A1433, which was reported to be a potent blocker of sheep A_3AR (Linden et al., 1993), completely prevented protection from ischemic preconditioning. CCPA, which is thought to bind poorly to A_3AR, protected rabbit hearts, and its salutary affect could be completely blocked by 200 nM DPCPX (Liu et al., 1996). These data suggest that rabbit hearts have both A_1AR and A_3AR receptors, and that either receptor can precondition the heart. Similar results were also reported for a rabbit isolated myocyte model by Armstrong and Ganote (1994).

CARDIAC PROTECTION BOTH MEDIATED AND TRIGGERED BY ADENOSINE

Thornton demonstrated that PD 115,199, a nonselective adenosine receptor blocker, could block the effect of preconditioning not only when given before the preconditioning ischemia but also when given shortly before the prolonged ischemic insult (Thornton et al., 1993a). That study revealed that a continued or renewed occupancy of adenosine receptors during the prolonged ischemic period must occur before protection can be expressed in the preconditioned heart. Giving the PD 115,199 just prior to reperfusion had no effect on protection. Therefore, adenosine appears to be both a mediator and a trigger of ischemic preconditioning.

This observation is consistent with others concerning adenosine's protection. As discussed above, we have found no evidence that endogenous adenosine, which quickly saturates cardiac A_1AR during ischemia, is protective to the non-preconditioned heart. In a recent study, Silva and colleagues (1995) gave dogs an adenosine deaminase inhibitor prior to coronary occlusion. With the onset of ischemia interstitial adenosine levels rose almost 100-fold over those in untreated dogs. Yet infarct sizes in the two groups were not different. This again illustrates that the non-preconditioned heart does not appear to be affected by the adenosine it releases after the onset of ischemia. On the other hand, it appears that if the heart has been previously preconditioned, then it becomes very sensitive to its own adenosine. We would propose that the primary difference between a preconditioned heart and a non-preconditioned heart is that the former has established a coupling between the adenosine receptor and the end-effector of protection, which normally is not present.

ACTIVATION OF PKC BY A_1AR TO PROTECT AGAINST INFARCTION

A_1AR couples to pertussis toxin–sensitive G proteins. Indeed, ribosylation of G proteins with pertussis toxin blocked the protective effect of preconditioning in rabbits (Thornton et al., 1993b). G_i most notably acts to inhibit adenylyl cyclase, thus lowering cAMP levels in the cell. While cAMP is reportedly lower in preconditioned hearts, raising the cAMP level with forskolin does not abolish protection (Sandhu et al., 1994). Adenosine A_1 receptors in many tissues are known to also couple to protein kinase C (PKC), and several reports suggest that this may be the case in the heart as well (Strasser et al., 1994; Kohl et al., 1990). We tested whether activation of PKC might be part of the protective pathway for preconditioning. Figure 3 (middle panel) shows the effect of blocking PKC with staurosporine in an open-chest rabbit (Ytrehus et al., 1994). Staurosporine had no effect on infarct size in the non-preconditioned heart, but selectively blocked protection in the preconditioned animals. The right panel shows the converse experiment. Activation of PKC by 5 minutes of intracoronary phorbol myristate acetate (PMA) (0.15 nM) followed by 10 minutes of washout mimicked

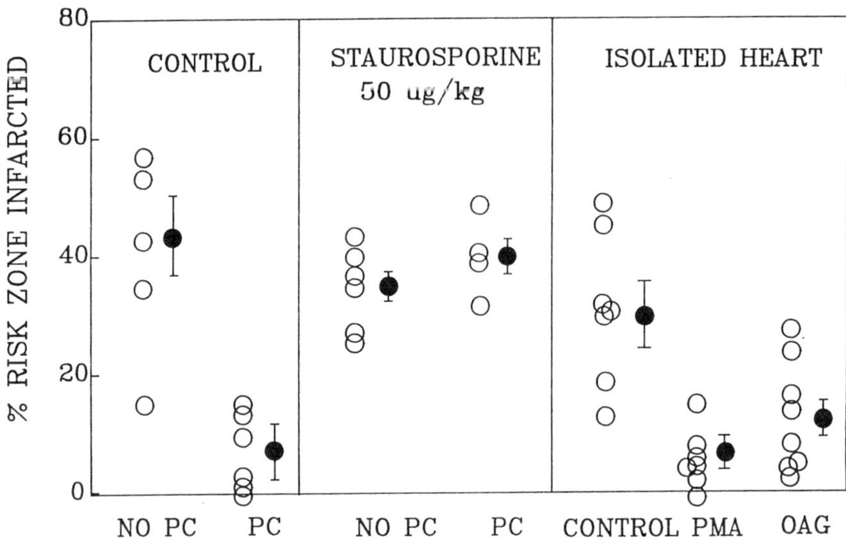

FIGURE 3. PKC's involvement in preconditioning is shown by this graph in which infarct sizes from a 30-minute regional ischemia are plotted. Control hearts did not receive any drug (left panel). Staurosporine, a potent PKC inhibitor, given 5 minutes before ischemia had no effect on non-preconditioned hearts but completely blocked protection from ischemic preconditioning (middle panel). Conversely, activation of PKC by exposure of hearts to low-dose PMA or OAG for 5 minutes prior to 30-minute ischemia mimicked preconditioning (right panel). OAG: oleoyl acetyl glycerol; PMA: phorbol myristate acetate. Based on data reported in Ytrehus et al., 1994.

preconditioning completely, as did oleoyl acetyl glycerol, a water soluble diacylglycerol that acts as a cofactor for PKC.

Staurosporine blocks PKC at the ATP-binding site of the kinase. It is potent but not very specific for PKC. Polymyxin B, which prevents phospholipid cofactors from binding to the enzyme, is highly selective for PKC among the kinases (Casnellie, 1991), but has other pharmacological effects. For example, a recent report has suggested that this antibiotic additionally closes ATP-sensitive K^+ channels (Harding et al., 1994), which have also been implicated in the mechanism of ischemic preconditioning (see below). Chelerythrine is a very specific PKC antagonist, which competitively inhibits the enzyme's phosphate acceptor and is thought to have few nonspecific effects. Both chelerytherine (Liu et al., 1994b) and polymyxin B (Ytrehus et al., 1994) selectively blocked protection from preconditioning as did staurosporine.

There has been some confirmation of the PKC hypothesis, especially in small animal models. Goto et al. (1994) were able to block PIA's protection with staurosporine in the rabbit. Speechly-Dick et al. (1994) and Mitchell et al. (1995) were able to block protection in rats with the PKC antagonist chelerytherine. Li and Kloner (1995) blocked protection in rats with calphostin C. Several others have also reported aborted protection with PKC inhibitors in the rat mode (Hu and Nattel, 1995; Cave and Apstein, 1996). Armstrong et al. (1994) found that calphostin C would prevent preconditioning protection in an isolated rabbit myocyte model of ischemic cell death.

The PKC story has yet to be confirmed in large animal models. While Kitakaze et al. (1996) report blockade of preconditioning's anti-infarct effect with polymyxin B in dogs, Przyklenk et al. (1995) failed to block protection in dogs when using the potent PKC inhibitors H7 and polymyxin B. Similarly Vogt et al. (1996), using a novel model where drugs were infused directly into the myocardium of open-chest pigs, failed to see any modulation of infarct size by the PKC activator PMA, and PKC inhibitors were actually protective. Whether or not these failures reflect a true species difference remains to be seen. Interestingly Ikonomidis et al. (1995) have noted that PKC inhibitors prevent what appears to be a preconditioning-like protection in their human cardiomyocyte model, suggesting that preconditioning in humans may be PKC-dependent.

OTHER PKC-COUPLED RECEPTORS THAT MIMIC PRECONDITIONING

One of the most convincing lines of evidence implicating PKC in ischemic preconditioning is the observation that all of the known PKC-coupled receptors in the heart can mimic preconditioning. The α_1-adrenergic, AT_1 angiotensin, and ET_1 endothelin receptors have all been shown to couple unambiguously to PKC in the cardiomyocyte. Activation of α_1 receptors has been shown to mimic preconditioning against infarction in rabbits (Tsuchida et al., 1994a), dogs (Kitakaze et al., 1994), and rats (Banerjee et al., 1993). While norepinephrine has been proposed to be an endogenous mediator of protection in the dog (Kitakaze et al., 1994) and the rat (Banerjee et al., 1993), we were unable to alter protection from ischemic preconditioning with α_1 blockade in the rabbit (Thornton et al., 1993c; Tsuchida et al., 1994a). These results suggest that rabbits rely more on adenosine while rats rely more on norepinephrine to ischemically precondition their hearts. Such quantitative differences between species are not surprising. Angiotensin II was also shown to mimic preconditioning in the rabbit heart by a PKC-dependent mechanism (Liu et al., 1995a) as did endothelin (Wang et al., 1996).

Since all of the above substances are released in some quantity by the ischemic heart it appears that the pathways for effecting preconditioning are highly redundant. Also, the predominant autacoid responsible for triggering preconditioning may differ from species to species.

COUPLING OF ADENOSINE RECEPTORS TO PKC BY PHOSPHOLIPASE D

One of the most vexing problems with the PKC hypothesis of preconditioning is the issue of whether adenosine receptors couple to PKC. For technical reasons few studies attempt measurements of PKC activity in response to receptor stimulation. Rather, most investigators measure the release of inositol phosphate, a product of phospholipase C (PLC) activation. Indeed, the amount of PLC activity resulting from adenosine receptor stimulation is modest at best (Kohl et al., 1990). We recently investigated whether phospholipase D (PLD) might be activated by A_1 adenosine receptors. Phospholipase D (PLD) hydrolyzes phosphatidylcholine rather than phosphatidylinositol and, therefore, inositol phosphate is not produced. The lipid product from PLD catalysis is phosphatidic acid which is then converted to diacylglycerol by a phosphohydrolase.

Phospholipase activity in the heart can be measured by its unique ability to effect the transphosphatidylation reaction. In that reaction, labeled alcohol such as butanol is added to the perfusate and PLD performs a base exchange with endogenous phospholipids leaving labeled phophatidylbutanol in the tissue (Moraru et al., 1992). We perfused rabbit hearts with labeled n-[1-^{14}C]-butanol and measured the amount of labeled phosphatidylbutanol formed. Preconditioning increased PLD activity by 90% over baseline. Furthermore, a 20-minute exposure to PIA caused a similar 78% increase in PLD activity (Cohen et al., 1996). PLD-derived diacylglycerol appears to be physiologically significant. At levels of 100 μM, propranolol is a potent inhibitor of the phosphohydrolase metabolizing phosphatidic acid (Moraru et al., 1992) and completely blocked protection from both traditional preconditioning and PIA pretreatment in the isolated myocyte preparation (Cohen et al., 1996). Therefore, we propose that PLD may serve to couple adenosine receptors to PKC in the heart.

THE TRANSLOCATION HYPOTHESIS OF PRECONDITIONING

In explaining preconditioning one must explain what is different between the first and second ischemic events and how the heart remembers that it is preconditioned. We have forwarded the PKC translocation hypothesis as a possible explanation (Cohen and Downey, 1993). PKC must translocate into a membrane compartment before it can phosphorylate protein. That translocation can be relatively slow and may take from 5 to 10 minutes. If a key protein must be phosphorylated in the first few minutes of ischemia in order to protect, then this would not occur with a simple coronary occlusion because of the delay associated with translocation. If, however, translocation had already occurred as a result of a previous ischemic episode, then the link between PKC-coupled adenosine receptors and phosphorylation would already have been established and substrate would be phosphorylated as soon as adenosine receptors were reoccupied. Timing studies clearly reveal that the important time for kinase activity is not during the preconditioning ischemia, but rather during the second prolonged period of ischemia (Liu et al., 1994c; Yang et al., 1997). That would rule out simple phosphorylation of a protein as the mechanism of the memory and favors the translocation hypothesis. Resolution of these issues must be the subject of future research in this area.

THE END-EFFECTOR FOR CARDIAC PROTECTION

The effector protein that PKC phosphorylates has not yet been identified. However, some investigators have suggested that the K_{ATP} channel is a likely candidate and adenosine is known to stimulate these channels (Kirsch et al., 1990). Gross and Auchampach (1992) have demonstrated that channel closure with glibenclamide can block the protection of ischemic preconditioning and opening the channel pharmacologically can mimic the salutary effects of a brief ischemic period. These effects, however, appear to be both anesthesia-dependent (Walsh et al., 1994) and species-dependent (Grover et al., 1992). Walsh et al. (1994) have also reported that the adenosine antagonist SPT can block the protective effect of pinacidil. It is not yet clear why adenosine should interact with a direct opener of the channel. However, the pharmacology of pinacidil is still far from clear. The other problem is that glibenclamide clearly does not block preconditioning in the rat (Liu and Downey, 1992; Grover et al., 1992).

There are other candidates for the end-effector besides the K_{ATP} channel. Ganote and Armstrong (1993) speculate that disruption of the cytoskeleton is a key event in the death of ischemic myocytes, and preconditioning may act to strengthen the myocyte's cytoskeleton. During ischemia the myocyte becomes increasingly fragile. Indeed, preconditioned myocytes are much more resistant to rupture in a hypotonic medium following simulated ischemia than non-preconditioned myocytes (Armstrong et al., 1994). Murry et al. (1991) argue that protection is the result of conservation of high-energy phosphates resulting from reduced energy expenditure during ischemia. That would suggest that a metabolic enzyme or ATPase is the target of this signal transduction pathway.

Finally, Linden (1994) has proposed that preconditioning may involve degranulation of mast cells in the heart. If the mast cells degranulate during a prolonged period of ischemia they would contribute to cell death. If, on the other hand, degranulation were induced by preconditioning and the cytotoxic products were washed out by a subsequent reperfusion period, no degranulation would be possible during the next prolonged period of ischemia. The mast cell theory is attractive because it explains the memory associated with preconditioning. An added advantage of the mast cell theory is that it would explain A_3 receptor involvement, since mast cells are rich in A_3 receptors, and furthermore, mast cell degranulation is a PKC-mediated event. It does not explain, however, how isolated myocytes can be preconditioned (Armstrong et al., 1994; Ikonomidis et al., 1995) since those preparations are essentially free of mast cells. Furthermore, recent investigations in our laboratory in isolated rabbit hearts have not confirmed a role for mast cells in ischemic preconditioning use (Wang et al., 1996).

POTENTIAL OF ADENOSINE AS A CARDIOPROTECTIVE DRUG

While adenosine agonists can clearly precondition the heart, one major problem stands in our way. These agents are only protective when given prior to the ischemic insult

(Thornton et al., 1992), a luxury not usually possible with patients who are experiencing acute myocardial infarction. To circumvent this problem we have attempted to maintain the heart in a preconditioned state by long-term infusion of the A_1-selective adenosine agonist CCPA (Tsuchida et al., 1994b). The negative chronotropic effect of the agonist was seen during the first few hours of infusion, but after 72 hours heart rates had returned to preocclusion levels. After 6 hours of CCPA infusion, infarcts following a 30-minute coronary occlusion were small (Figure 4), implying CCPA had maintained the preconditioned state over the 6-hour period. After 72 hours of infusion, however, 30-minute occlusions produced large infarcts similar in size to those observed in untreated control animals. Furthermore, preconditioning with a 5-minute coronary occlusion was no longer able to protect these hearts. Therefore, tolerance had developed to the adenosine agonist sometime between 6 and 72 hours of exposure.

Prolonged and repeated exposure to ischemic preconditioning itself seemed to have a similar effect. In rabbits chronically instrumented with balloon occluders around a major branch of a coronary artery, 5-minute coronary occlusions were performed every 30 minutes, 8 hours a day for 3 days (Cohen et al., 1994). S-T segment elevation

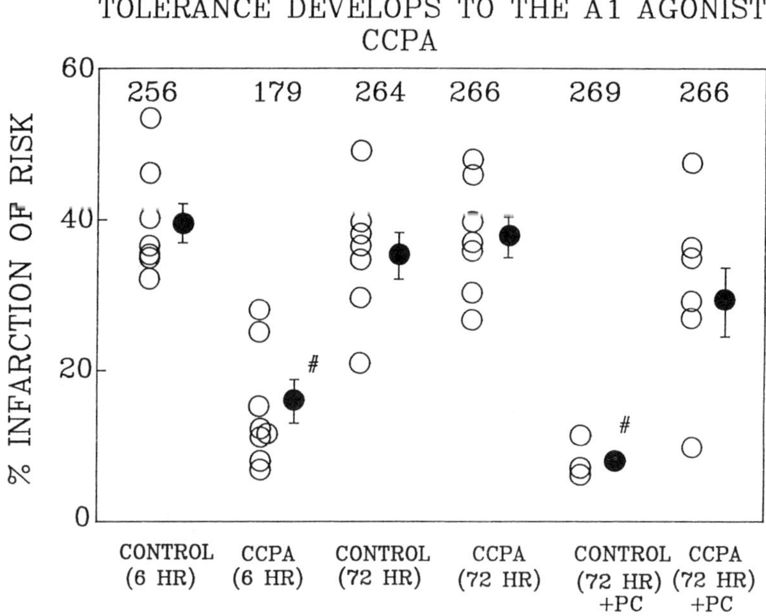

FIGURE 4. Tolerance quickly develops to a chronically infused adenosine agonist, CCPA. The plot shows infarct sizes in response to 30-minute regional ischemia in rabbit hearts. The number above each column indicates the mean heart rate in those animals. A negative chronotropic effect of CCPA was seen after 6 hours of infusion, but was lost after 72 hours of infusion. Protection against infarction was present after 6 hours of CCPA infusion but was lost after 72 hours of infusion, indicating that tolerance had developed. Note that adenosine-tolerant animals could not be protected by ischemic preconditioning either, while those receiving 72 hours of saline could be protected by ischemic preconditioning. PC: ischemic preconditioning; CCPA: 2-chloro-N^6-cyclopentyladenosine. Based on data presented in Tsuchida et al., 1994b.

attested to the adequacy of the occlusion, and normalization of the ECG following deflation of the occluder documented reflow. Following 45–65 occlusions, a 5-minute occlusion shortly before a final 30-minute occlusion no longer was protective. Hence, it seems likely that prolonged exposure to either exogenous or endogenous adenosine results in the appearance of tolerance. Obviously, these problems will have to be circumvented before preconditioning's protection can be used to benefit the patient with an acute myocardial infarction.

SUMMARY

Adenosine is an important intermediate substance in energy metabolism, as well as a powerful cardioprotective agent. Receptors for adenosine on the cardiomyocyte are capable of causing considerable resistance to infarction during ischemia if they are occupied prior to ischemia. Endogenous adenosine is thought to play a critical role in the protection of ischemic preconditioning by activating A_1AR and/or A_3AR, which couples to PKC via a PLD pathway. Before we can harness the power of adenosine's protection, however, we must find a way to administer it prior to the onset of ischemia. While this can be done in certain cases of planned ischemia such as cardiac surgery and angioplasty, it is not feasible in the setting of acute myocardial infarction. Unfortunately, the rapid development of tolerance to A_1AR-selective agonists prevents their use as a prophylactic treatment. Solving these problems and identifying the final effector remain a challenge for future research in this area.

REFERENCES

Armstrong S, Downey JM, Ganote CE (1994): Preconditioning of isolated rabbit cardiomyocytes: Induction by metabolic stress and blockade by the adenosine antagonist SPT and calphostin C, a protein kinase C inhibitor. Cardiovasc Res 28:72–77.

Armstrong S, Ganote CE (1994): Adenosine receptor specificity in preconditioning of isolated rabbit cardiomyocytes: Evidence of A_3 receptor involvement. Cardiovasc Res 28:1049–1056.

Auchampach JA, Gross GJ (1993): Adenosine A_1 receptors, K_{ATP} channels, and ischemic preconditioning in dogs. Am J Physiol 264:H1327–H1336.

Babbitt DG, Virmani R, Forman MB (1989): Intracoronary adenosine administered after reperfusion limits vascular injury after prolonged ischemia in the canine model. Circulation 80:1388–1399.

Banerjee A, Locke-Winter C, Rogers KB, Mitchell MB, Brew EC, Cairns CB, Bensard DD, Harken AH (1993): Preconditioning against myocardial dysfunction after ischemia and reperfusion by an α_1-adrenergic mechanism. Circ Res 73:656–670.

Brown JY, Lasley RD, Ely SW, Berne RM, Mentzer RM Jr (1986): Inhibition of adenosine catabolism improves postischemic myocardial function. Circulation 74 (suppl II):II–353. Abstract.

Burckhartt B, Yang X-M, Tsuchida A, Mullane KM, Downey JM, Cohen MV (1995): Acadesine extends the window of protection afforded by ischaemic preconditioning in conscious rabbits. Cardiovasc Res 29:653–657.

Casnellie JE (1991): Protein kinase inhibitors: Probes for the functions of protein phosphorylation. Adv Pharmacol 22:167–205.

Cave AC, Apstein CS (1996): Polymyxin B, a protein kinase C inhibitor, abolishes preconditioning-induced protection against contractile dysfunction in the isolated blood perfused rat heart. J Mol Cell Cardiol 28:977–987.

Charlat ML, O'Neill PG, Egan JM, Abernethy DR, Michael LH, Myers ML, Roberts R, Bolli R (1987): Evidence for a pathogenetic role of xanthine oxidase in the "stunned" myocardium. Am J Physiol 252:H566–H577.

Cohen MV, Downey JM (1993): Ischaemic preconditioning: Can the protection be bottled? Lancet 342:6.

Cohen MV, Liu Y, Liu GS, Wang P, Weinbrenner C, Cordis GA, Das DK, Downey JM (1996): Phospholipase D plays a role in ischemic preconditioning in rabbit heart. Circulation 94:1713–1718.

Cohen MV, Yang X-M, Downey JM (1994): Conscious rabbits become tolerant to multiple episodes of ischemic preconditioning. Circ Res 74:998–1004.

Conti JB, Belardinelli L, Utterback DB, Curtis AB (1995): Endogenous adenosine is an antiarrhythmic agent. Circulation 91:1761–1767.

Cronstein BN, Daguma L, Nichols D, Hutchison AJ, Williams M (1990): The adenosine/neutrophil paradox resolved: Human neutrophils possess both A_1 and A_2 receptors that promote chemotaxis and inhibit O_2 generation, respectively. J Clin Invest 85:1150–1157.

Deutsch E, Berger M, Kussmaul WG, Hirshfeld JW Jr, Herrmann HC, Laskey WK (1990): Adaptation to ischemia during percutaneous transluminal coronary angioplasty: Clinical, hemodynamic, and metabolic features. Circulation 82:2044–2051.

Downey JM, Hearse DJ, Yellon DM (1988): The role of xanthine oxidase during myocardial ischemia in several species including man. J Mol Cell Cardiol 20(suppl II):55–63.

Ely SW, Mentzer RM Jr, Lasley RD, Lee BK, Berne RM (1985): Functional and metabolic evidence of enhanced myocardial tolerance to ischemia and reperfusion with adenosine. J Thorac Cardiovasc Surg 90:549–556.

Finegan BA, Lopaschuk GD, Coulson CS, Clanachan AS (1993): Adenosine alters glucose use during ischemia and reperfusion in isolated rat hearts. Circulation 87:900–908.

Fozard JR, Carruthers AM (1993): Adenosine A_3 receptors mediate hypotension in the angiotensin II-supported circulation of the pithed rat. Br J Pharmacol 109:3–5.

Ganote C, Armstrong S (1993): Ischaemia and the myocyte cytoskeleton: Review and speculation. Cardiovasc Res 27:1387–1403.

Goto M, Miura T, Iliodoromitis EK, O'Leary EL, Ishimoto R, Yellon DM, Iimura O (1991): Adenosine infusion during early reperfusion failed to limit myocardial infarct size in a collateral deficient species. Cardiovasc Res 25:943–949.

Goto M, Miura T, Sakamoto J, Iimura O (1994): Infarct size limitation by adenosine A_1 receptor agonist was blocked by staurosporine. J Mol Cell Cardiol 26:CLII. Abstract.

Gross GJ, Auchampach JA (1992): Blockade of ATP-sensitive potassium channels prevents myocardial preconditioning in dogs. Circ Res 70:223–233.

Grover GJ, Dzwonczyk S, Sleph PG (1992): ATP-sensitive K^+ channel activation does not mediate preconditioning in isolated rat hearts. Circulation 86 (suppl I):I–341. Abstract.

Gruber HE, Hoffer ME, McAllister DR, Laikind PK, Lane TA, Schmid-Schoenbein GW, Engler RL (1989): Increased adenosine concentration in blood from ischemic myocardium by AICA riboside: Effects on flow, granulocytes, and injury. Circulation 80:1400–1411.

Harding EA, Jaggar JH, Squires PE, Dunne MJ (1994): Polymyxin B has multiple blocking actions on the ATP-sensitive potassium channel in insulin-secreting cells. Pflugers Arch 426:31–39.

Hori M, Kitakaze M, Takashima S (1990): AICA-riboside (5-amino-4-imidazole carboxamide riboside 100), a novel adenosine potentiator, attenuates myocardial stunning. Circulation 82 (suppl III):III–466. Abstract.

Hori M, Tamai J, Kitakaze M, Iwakura K, Gotoh K, Iwai K, Koretsune Y, Kagiya T, Kitabatake A, Kamada T (1989): Adenosine-induced hyperemia attenuates myocardial ischemia in coronary microembolization in dogs. Am J Physiol 257:H244–H251.

Hu K, Nattel S (1995): Mechanisms of ischemic preconditioning in rat hearts: Involvement of α_{1B}-adrenoceptors, pertussis toxin-sensitive G proteins, and protein kinase C. Circulation 92:2259–2265.

Ikonomidis JS, Shirai T, Weisel RD, Derylo B, Rao V, Whiteside CI, Mickle DAG, Li R-K (1995): "Ischemic" or adenosine preconditioning of human ventricular cardiomyocytes is protein kinase C dependent. Circulation 92 (suppl I):I–12. Abstract.

Isenberg G, Belardinelli L (1984): Ionic basis for the antagonism between adenosine and isoproterenol on isolated mammalian ventricular myocytes. Circ Res 55:309–325.

Janier MF, Vanoverschelde J-LJ, Bergmann SR (1993): Adenosine protects ischemic and reperfused myocardium by receptor-mediated mechanisms. Am J Physiol 264:H163–H170.

Jennings RB, Reimer KA (1981): Lethal myocardial ischemic injury. Am J Pathol 102:241–255.

Kirsch GE, Codina J, Birnbaumer L, Brown AM (1990): Coupling of ATP-sensitive K^+ channels to A_1 receptors by G proteins in rat ventricular myocytes. Am J Physiol 259:H820–H826.

Kitakaze M, Hori M, Morioka T, Minamino T, Takashima S, Sato H, Shinozaki Y, Chujo M, Mori H, Inoue M, Kamada T (1994): $Alpha_1$-adrenoceptor activation mediates the infarct size–limiting effect of ischemic preconditioning through augmentation of 5′-nucleotidase activity. J Clin Invest 93:2197–2205.

Kitakaze M, Node K, Minamino T, Komamura K, Funaya H, Shinozaki Y, Chujo M, Mori H, Inoue M, Hori M, Kamada T (1996): Role of activation of protein kinase C in the infarct size-limiting effect of ischemic preconditioning through activation of ecto-5′-nucleotidase. Circulation 93:781–791.

Kohl C, Linck B, Schmitz W, Scholz H, Scholz J, Tóth M (1990): Effects of carbachol and (−)-N^6-phenylisopropyladenosine on myocardial inositol phosphate content and force of contraction. Br J Pharmacol 101:829–834.

Kusuoka H, Marban E (1992): Cellular mechanisms of myocardial stunning. Annu Rev Physiol 54:243–256.

Lasley RD, Rhee JW, Van Wylen DGL, Mentzer RM Jr (1990): Adenosine A_1 receptor mediated protection of the globally ischemic isolated rat heart. J Mol Cell Cardiol 22:39–47.

Lawson CS, Downey JM (1993): Preconditioning: State of the art myocardial protection. Cardiovasc Res 27:542–550.

Li GC, Vasquez JA, Gallagher KP, Lucchesi BR (1990): Myocardial protection with preconditioning. Circulation 82:609–619.

Li Y, Kloner RA (1993): The cardioprotective effects of ischemic "preconditioning" are not mediated by adenosine receptors in rat hearts. Circulation 87:1642–1648.

Li Y, Kloner RA (1995): Does protein kinase C play a role in ischemic preconditioning in rat hearts? Am J Physiol 268:H426–H431.

Linden J (1994): Cloned adenosine A_3 receptors: Pharmacological properties, species differences and receptor functions. Trends Pharmacol Sci 15:298–306.

Linden J, Taylor HE, Robeva AS, Tucker AL, Stehle JH, Rivkees SA, Fink JS, Reppert SM (1993): Molecular cloning and functional expression of a sheep A_3 adenosine receptor with widespread tissue distribution. Mol Pharmacol 44:524–532.

Liu G-S, Jacobsen KA, Downey JM (1996): An irreversible A_1-selective adenosine agonist preconditions rabbit heart. Canadian J Cardiol 12:517–521.

Liu GS, Richards SC, Olsson RA, Mullane K, Walsh RS, Downey JM (1994a): Evidence that the adenosine A_3 receptor may mediate the protection afforded by preconditioning in the isolated rabbit heart. Cardiovasc Res 28:1057–1061.

Liu GS, Stanley AWH, Downey J (1992): Cyclooxygenase products are not involved in the protection against myocardial infarction afforded by preconditioning in rabbit. Am J Cardiovasc Path 4:56–63.

Liu GS, Thornton J, Van Winkle DM, Stanley AWH, Olsson RA, Downey JM (1991): Protection against infarction afforded by preconditioning is mediated by A_1 adenosine receptors in rabbit heart. Circulation 84:350–356.

Liu Y, Cohen MV, Downey JM (1994b): Chelerythrine, a highly selective protein kinase C inhibitor, blocks the antiinfarct effect of ischemic preconditioning in rabbit hearts. Cardiovasc Drugs Ther 8:881–882.

Liu Y, Downey JM (1992): Ischemic preconditioning protects against infarction in rat heart. Am J Physiol 263:H1107–H1112.

Liu Y, Tsuchida A, Cohen MV, Downey JM (1995a): Pretreatment with angiotensin II activates protein kinase C and limits myocardial infarction in isolated rabbit hearts. J Mol Cell Cardiol 27:883–892.

Liu Y, Ytrehus K, Downey JM (1994c): Evidence that translocation of protein kinase C is a key event during ischemic preconditioning of rabbit myocardium. J Mol Cell Cardiol 26:661–668.

Mazur C, Mullane K, Young MA (1991): Acadesine preserves cardiac function and enhances coronary blood flow in isolated, blood perfused rabbit hearts with repeated ischemia and reperfusion. J Mol Cell Cardiol 23 (suppl III):S45. Abstract.

Mitchell MB, Meng X, Ao L, Brown JM, Harken AH, Banerjee A (1995): Preconditioning of isolated rat heart is mediated by protein kinase C. Circ Res 76:73–81.

Miura T, Goto M, Urabe K, Endoh A, Shimamoto K, Iimura O (1991): Does myocardial stunning contribute to infarct size limitation by ischemic preconditioning? Circulation 84:2504–2512.

Moraru II, Popescu LM, Maulik N, Liu X, Das DK (1992): Phospholipase D signaling in ischemic heart. Biochim Biophys Acta 1139:148–154.

Mullane K (1993): Acadesine: The prototype adenosine regulating agent for reducing myocardial ischaemic injury. Cardiovasc Res 27:43–47.

Murry CE, Jennings RB, Reimer KA (1986): Preconditioning with ischemia: A delay of lethal cell injury in ischemic myocardium. Circulation 74:1124–1136.

Murry CE, Jennings RB, Reimer KA (1991): New insights into potential mechanisms of ischemic preconditioning. Circulation 84:442–445.

Murry CE, Richard VJ, Reimer KA, Jennings RB (1990): Ischemic preconditioning slows energy metabolism and delays ultrastructural damage during a sustained ischemic episode. Circ Res 66:913–931.

Norton ED, Jackson EK, Turner MB, Virmani R, Forman MB (1992): The effects of intravenous infusions of selective adenosine A_1-receptor and A_2-receptor agonists on myocardial reperfusion injury. Am Heart J 123:332–338.

Norton ED, Jackson EK, Virmani R, Forman MB (1991): Effect of intravenous adenosine on myocardial reperfusion injury in a model with low myocardial collateral blood flow. Am Heart J 122:1283–1291.

Olsson RA, Pearson JD (1990): Cardiovascular purinoceptors. Physiol Rev 70:761–845.

Parratt J (1993): Endogenous myocardial protective (antiarrhythmic) substances. Cardiovasc Res 27:693–702.

Przyklenk K, Sussman MA, Simkhovich BZ, Kloner RA (1995): Does ischemic preconditioning trigger translocation of protein kinase C in the canine model? Circulation 92:1546–1557.

Ramkumar V, Stiles GL, Beaven MA, Ali H (1993): The A_3 adenosine receptor is the unique adenosine receptor which facilitates release of allergic mediators in mast cells. J Biol Chem 268:16887–16890.

Reimer KA, Jennings RB (1979): The "wavefront phenomenon" of myocardial ischemic cell death. II. Transmural progression of necrosis within the framework of ischemic bed size (myocardium at risk) and collateral flow. Lab Invest 40:633–644.

Sajjadi FG, Firestein GS (1993): cDNA cloning and sequence analysis of the human A3 adenosine receptor. Biochim Biophys Acta 1179:105–107.

Sandhu R, Diaz RJ, Wilson GJ (1994): Raising cAMP levels with forskolin does not block ischemic preconditioning. J Mol Cell Cardiol 26:CLXXXIV. Abstract.

Schömig A, Richardt G (1990): Cardiac sympathetic activity in myocardial ischemia: Release and effects of noradrenaline. Basic Res Cardiol 85(suppl I):9–30.

Schott RJ, Rohmann S, Braun ER, Schaper W (1990): Ischemic preconditioning reduces infarct size in swine myocardium. Circ Res 66:1133–1142.

Schrier DJ, Lesch ME, Wright CD, Gilbertsen RB (1990): The antiinflammatory effects of adenosine receptor agonists on the carrageenan-induced pleural inflammatory response in rats. J Immunol 145:1874–1879.

Schulz R, Rose J, Martin C, Heusch G (1993): Activation of ATP-dependent potassium channels is involved in ischemic preconditioning in swine. Circulation 88(suppl I):I–102. Abstract.

Schwarz ER, Mohri M, Sack S, Arras M (1991): The role of adenosine and its A1-receptor in ischemic preconditioning. Circulation 84(suppl II):II–191. Abstract.

Sekili S, Jeroudi MO, Tang X-L, Zughaib M, Sun J-Z, Bolli R (1995): Effect of adenosine on myocardial "stunning" in the dog. Circ Res 76:82–94.

Silva PH, Dillon D, Van Wylen DGL (1995): Adenosine deaminase inhibition augments interstitial adenosine but does not attenuate myocardial infarction. Cardiovasc Res 29:616–623.

Smolenski RT, Suitters A, Yacoub MH (1992): Adenine nucleotide catabolism and adenosine formation in isolated human cardiomyocytes. J Mol Cell Cardiol 24:91–96.

Speechly-Dick ME, Mocanu MM, Yellon DM (1994): Protein kinase C: Its role in ischemic preconditioning in the rat. Circ Res 75:586–590.

Stein B, Mende U, Neumann J, Schmitz W, Scholz H (1993): Pertussis toxin unmasks stimulatory myocardial A_2-adenosine receptors on ventricular cardiomyocytes. J Mol Cell Cardiol 25:655–659.

Stiles GL (1992): Adenosine Receptors. J Biol Chem 267:6451–6454.

Strasser RH, Weinbrenner C, Röthele J, Kübler W, Marquetant R (1994): Blockade of A_1 adenosine receptors prevents sensitization of adenylyl cyclase and activation of protein kinase C in early myocardial ischemia. Drug Dev Res 31:325. Abstract.

Thornton JD, Thornton CS, Downey JM (1993a): Effect of adenosine receptor blockade: Preventing protective preconditioning depends on time of initiation. Am J Physiol 265:H504–H508.

Thornton JD, Liu GS, Downey JM (1993b): Pretreatment with pertussis toxin blocks the protective effects of preconditioning: Evidence for a G-protein mechanism. J Mol Cell Cardiol 25:311–320.

Thornton JD, Daly JF, Cohen MV, Yang X-M, Downey JM (1993c): Catecholamines can induce adenosine receptor-mediated protection of the myocardium but do not participate in ischemic preconditioning in the rabbit. Circ Res 73:649–655.

Thornton JD, Liu GS, Olsson RA, Downey JM (1992): Intravenous pretreatment with A_1-selective adenosine analogues protects the heart against infarction. Circulation 85:659–665.

Thornton J, Striplin S, Liu GS, Swafford A, Stanley AWH, Van Winkle DM, Downey JM (1990): Inhibition of protein synthesis does not block myocardial protection afforded by preconditioning. Am J Physiol 259:H1822–H1825.

Toombs CF, McGee DS, Johnston WE, Vinten-Johansen J (1992): Myocardial protective effects of adenosine: Infarct size reduction with pretreatment and continued receptor stimulation during ischemia. Circulation 86:986–994.

Tsuchida A, Liu GS, Mullane K, Downey JM (1993): Acadesine lowers temporal threshold for the myocardial infarct size limiting effect of preconditioning. Cardiovasc Res 27:116–120.

Tsuchida A, Liu Y, Liu GS, Cohen MV, Downey JM (1994a): α_1-Adrenergic agonists precondition rabbit ischemic myocardium independent of adenosine by direct activation of protein kinase C. Circ Res 75:576–585.

Tsuchida A, Miura T, Miki T, Shimamoto K, Iimura O (1992): Role of adenosine receptor activation in myocardial infarct size limitation by ischaemic preconditioning. Cardiovasc Res 26:456–461.

Tsuchida A, Thompson R, Olsson RA, Downey JM (1994b): The anti-infarct effect of an adenosine A_1-selective agonist is diminished after prolonged infusion as is the cardioprotective effect of ischaemic preconditioning in rabbit heart. J Mol Cell Cardiol 26:303–311.

Turrens JF, Thornton J, Barnard ML, Snyder S, Liu G, Downey JM (1992): Protection from reperfusion injury by preconditioning hearts does not involve increased antioxidant defenses. Am J Physiol 262:H585–H589.

Van Winkle DM, Chien GL, Wolff RA, Soifer BE, Kuzume K, Davis RF (1994): Cardioprotection provided by adenosine receptor activation is abolished by blockade of the K_{ATP} channel. Am J Physiol 266:H829–H839.

Van Winkle DM, Thornton JD, Downey DM, Downey JM (1991): The natural history of preconditioning: Cardioprotection depends on duration of transient ischemia and time to subsequent ischemia. Coron Artery Dis 2:613–619.

Vander Heide RS, Jennings RB, Reimer KA (1993a): Intravenous adenosine before reperfusion does not limit infarct size after 90 minutes of ischemia in dogs. FASEB J 7:A193. Abstract.

Vander Heide RS, Reimer KA, Jennings RB (1993b): Adenosine slows ischaemic metabolism in canine myocardium in vitro: Relationship to ischaemic preconditioning. Cardiovasc Res 27:669–673.

Visentin S, Wu S-N, Belardinelli L (1990): Adenosine-induced changes in atrial action potential: Contribution of Ca and K currents. Am J Physiol 258:H1070–H1078.

Vogt A, Htun P, Arras M, Podzuweit T, Schaper W (1996): Intramyocardial infusion of tool drugs for the study of molecular mechanisms in ischemic preconditioning. Basic Res Cardiol 91:389–400.

Walsh RS, Tsuchida A, Daly JJF, Thornton JD, Cohen MV, Downey JM (1994): Ketamine-xylazine anaesthesia permits a K_{ATP} channel antagonist to attenuate preconditioning in rabbit myocardium. Cardiovasc Res 28:1337–1341.

Wang P, Downey J. Cohen MV (1996): Mast cell degranulation does not contribute to ischemic preconditioning in isolated rabbit hearts. Basic Res Cardiol 92:458–467.

Wang P, Gallagher KP, Downey JM, Cohen MV (1996): Pretreatment with endothelin-1 mimics ischemic preconditioning against infarction in isolated rabbit heart. J Mol Cell Cardiol 28:579–588.

Wyatt DA, Edmunds MC, Rubio R, Berne RM, Lasley RD, Mentzer RM Jr (1989): Adenosine stimulates glycolytic flux in isolated perfused rat hearts by A_1-adenosine receptors. Am J Physiol 257:H1952–H1957.

Yang X-M, Sato H, Downey JM, Cohen MV (1997): Protection of ischemic preconditioning is dependent upon a critical timing sequence of protein kinase C activation. J Mol Cell Cardiol 29:000–000.

Yellon DM, Alkhulaifi AM, Pugsley WB (1993): Preconditioning the human myocardium. Lancet 342:276–277.

Ytrehus K, Liu Y, Downey JM (1994): Preconditioning protects ischemic rabbit heart by protein kinase C activation. Am J Physiol 266:H1145–H1152.

Zhao Z-Q, Williams MW, Sato H, Hudspeth DA, McGee DS, Vinten-Johansen J, Van Wylen DGL (1995): Acadesine reduces myocardial infarct size by an adenosine mediated mechanism. Cardiovasc Res 29:495–505.

Zhou Q-Y, Li C, Olah ME, Johnson RA, Stiles GL, Civelli O (1992): Molecular cloning and characterization of an adenosine receptor: The A3 adenosine receptor. Proc Natl Acad Sci USA 89:7432–7436.

Zughaib ME, Abd-Elfattah AS, Jeroudi MO, Sun J-Z, Sekili S, Li X-Y, Bolli R (1992): Augmentation of endogenous adenosine levels attenuates myocardial stunning. Circulation 86 (suppl I):I–612. Abstract.

CHAPTER 9

ATP as a Cotransmitter With Noradrenaline in Sympathetic Nerves—A Target for Hypertensive Therapy?

CHARLES KENNEDY, GERALD J. McLAREN, and PETER SNEDDON

Department of Physiology and Pharmacology, University of Strathclyde, Royal College, Glasgow G1 1XW, United Kingdom

INTRODUCTION

Over the past 68 years a great number of studies have shown that adenosine, adenosine 5′-triphosphate (ATP) and related compounds are active in most blood vessels and vascular beds examined. Adenosine usually causes vasodilation via P_1-purinoceptors found on the smooth muscle cells, whereas the most common effects of ATP are vasodilation via endothelial P_{2Y}-receptors and vasoconstriction via smooth muscle P_{2X}-receptors. Physiological roles have been proposed for each of these in modulating vascular tone and blood pressure (see Burnstock and Kennedy, 1986).

Until recently, noradrenaline was considered to be the sole neurotransmitter released from sympathetic nerves. Acting mainly at α_1-adrenoceptors, noradrenaline induces contraction of vascular and visceral smooth muscle via the inositol 1,4,5-trisphosphate (IP_3) second messenger system and release of intracellular calcium stores, or in some cases by depolarizing the smooth muscle cells (Bolton and Large, 1986; Walsh, 1994). However, in numerous studies a nonadrenergic component to the sympathetic responses in blood vessels and the vas deferens was reported, and it is now clear that both ATP and neuropeptide Y (NPY) are costored with noradrenaline, and that when released, ATP evokes smooth muscle contraction (Figure 1). Thus, sympathetic neurotransmission is not the simple single-neurotransmitter process previously envisaged. Rather, multiple transmitters act and interact, in varying degrees depending on the tissue, to produce an integrated control of smooth muscle tone.

Purinergic Approaches in Experimental Therapeutics, Edited by Kenneth A. Jacobson and Michael F. Jarvis
ISBN 0-471-14071-6 © 1997 Wiley-Liss, Inc.

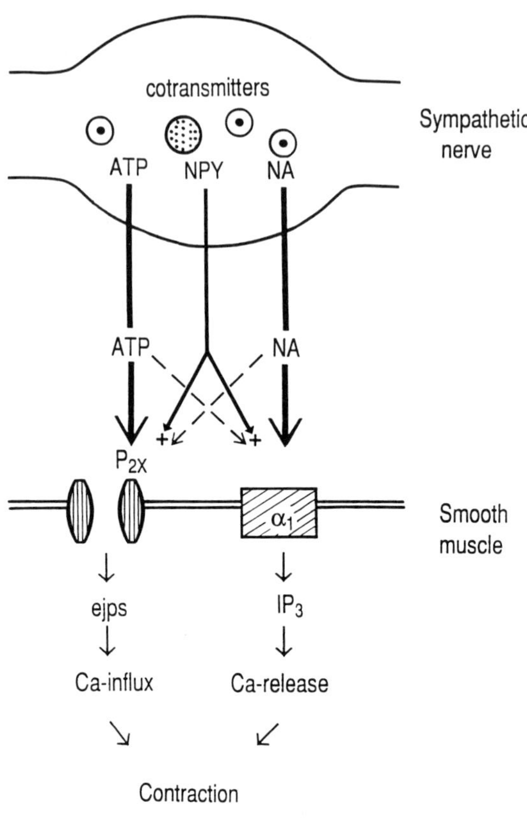

FIGURE 1. Schematic representation of cotransmission by ATP and noradrenaline in sympathetic nerves. ATP = adenosine 5′-triphosphate, NPY = neuropeptide Y,NA = noradrenaline, P_{2X} = P_{2X}-receptor, α_1 = α_1-adrenoceptor, e.j.p.s = excitatory junction potentials, IP_3 = inositol 1,4,5-trisphosphate. + indicates that ATP and noradrenaline can interact synergistically and that neuropeptide Y appears to act mainly by potentiating contractions to ATP and noradrenaline. The receptors for NPY have been omitted for clarity.

In this chapter I will discuss recent advances in our understanding of the function of ATP as an excitatory cotransmitter in sympathetic nerves innervating vascular and, where relevant, visceral smooth muscle and will consider the evidence that vascular purinergic transmission is altered during hypertension.

ATP AS A COTRANSMITTER

Neuronal Storage and Release of ATP

In support of the cotransmission model, numerous biochemical studies have shown that noradrenaline and ATP are costored in sympathetic synaptic vesicles (see Burnstock and Kennedy, 1986). However, an increasing number of reports now suggest that there may also be further populations of vesicles that store ATP and noradrenaline separately. Initially this was based on functional studies which showed that activation of prejunctional receptors could produce differential modulation of the purinergic and noradrenergic components of sympathetic contractions, but interpretation of these results is complicated by the fact that an end-organ response is not a reliable measure of neurotransmitter release. Instead, it is much preferable to measure release of the neurotransmitter per se. For example, in the guinea pig vas deferens the time courses of release for ATP and noradrenaline have been found to differ (Todorov et al., 1996). ATP overflow peaks 20 seconds after the start of sympathetic nerve stimulation and then decreases, even though the nerves are still being stimulated. In contrast, noradrenaline overflow peaks at 40 seconds and then remains constant. Also, activation of prejunctional receptors for angiotensin II and III (Ellis and Burnstock, 1989), NPY (Ellis and Burnstock, 1990), noradrenaline (Driessen et al., 1993) and adenosine (Driessen et al., 1994) can differentially modulate the release of ATP and noradrenaline in this tissue. Thus, these results are consistent with a mixed population of synaptic vesicles, which contain ATP and noradrenaline in varying proportions. Some may even store ATP and noradrenaline on their own.

Tetrodotoxin- and guanethidine-sensitive, calcium-dependent release of ATP has been demonstrated in a number of blood vessels and in the vas deferens of several species following sympathetic nerve stimulation (see Burnstock and Kennedy, 1986; Burnstock, 1990; Von Kügelgen and Starke, 1991). However, with *in vitro* preparations at least, this can be complicated by release of ATP from nonneuronal sites. For example, removal of the endothelium from blood vessels can depress electrical field stimulation–induced release of ATP by up to 90% (Sedaa et al., 1990). Also, in the guinea pig vas deferens, stimulation of postjunctional α_1-adrenoceptors and P_2-receptors by exogenous (Katsuragi et al., 1991; Vizi et al., 1992) and endogenous (Driessen et al., 1993, 1994) noradrenaline and ATP can elicit further release of ATP. However, it is unlikely that postjunctional release of ATP contributes to its neurotransmitter role, as purinergic contractions evoked by nerve stimulation or exogenous ATP reach a peak within 5 seconds (Sneddon and Westfall, 1984), whereas noradrenaline-induced release of ATP is much slower, taking 2–6 minutes to reach a peak (Katsuragi et al., 1991; Vizi et al., 1992). The cellular mechanisms underlying postjunctional release of ATP are not known and it is not clear if they can function under normal *in vivo* conditions. However, what is clear is that sufficient ATP can be released from sympathetic neurons to account for its neurotransmitter role.

The Effect of Endogenously-Released ATP on Membrane Potential

The initial postjunctional response of the vas deferens and most blood vessels to sympathetic nerve stimulation is a rapid, transient excitatory junction potential (e.j.p.). With sufficient stimulation the e.j.p.s summate and the membrane can become sufficiently depolarized to open voltage-dependent calcium channels and so initiate a calcium action potential and contraction. Although e.j.p.s are abolished by tetrodotoxin and guanethidine, they are resistant to α-adrenoceptor blockade (see Bolton and Large, 1986). The first evidence that e.j.p.s are mediated by ATP was obtained in the guinea pig vas deferens, where they were inhibited by the P_2-receptor antagonist $ANAPP_3$ (Sneddon et al., 1982; Sneddon and Westfall, 1984). Further studies in the vas deferens showed that e.j.p.s are also inhibited by other P_2-receptor antagonist, such as suramin (Sneddon, 1992) and pyridoxalphosphate-6-azophenyl-2′,4′-disulphonic acid (PPADS)

(McLaren et al., 1994) and by desensitization of the P_2-receptor by the agonist α,β-methyleneATP (α,β-meATP) (Sneddon and Burnstock, 1984b). Likewise, numerous studies, using mainly desensitization of the P_2-receptor by α,β-meATP (Sneddon and Burnstock, 1984a; Von Kügelgen & Starke, 1991) but also suramin (McLaren et al., 1995) (Figure 2), clearly demonstrate that e.j.p.s in blood vessels are also due to ATP.

In most blood vessels, noradrenaline released from sympathetic nerves evokes contraction without changing the membrane potential of smooth muscle cells. However, this is not the case in the rat tail artery, where stimulation of the sympathetic nerves with a train of pulses (10 or 20 pulses at 0.5, 1, or 2 Hz) produces an electrical response with two distinct phases (Sneddon and Burnstock, 1984a; McLaren et al., 1995). Each stimulus evokes a rapid transient e.j.p., and as the train of pulses progresses a slower depolarization also develops (Figure 2). The slow depolarization, which is smaller than the e.j.p.s reaches a peak 5–20 seconds after the last e.j.p. and decays back to resting membrane potential within 60 seconds. The magnitude of the slow depolarization increases with the frequency of stimulation and with the number of pulses in the train. The P_{2X}-receptor antagonist suramin abolishes the e.j.p.s and increases the amplitude of the slow depolarization, whereas the α-adrenoceptor antagonist phentolamine has no effect on the e.j.p.s, but abolishes the slow depolarization. Thus, in this vessel both purinergic and noradrenergic components in the electrical response to sympathetic nerve stimulation can be seen.

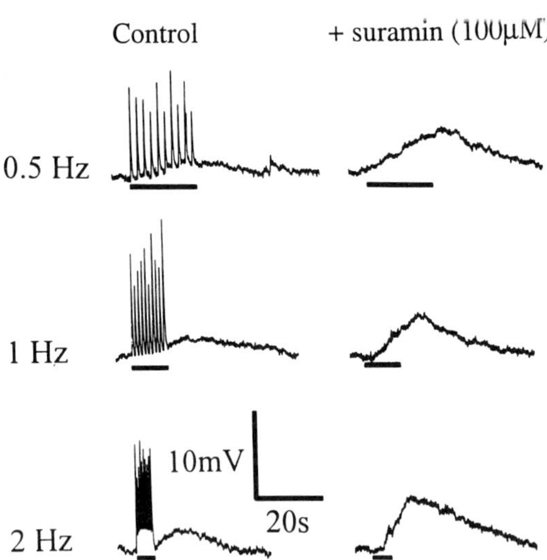

FIGURE 2. The effects of suramin on e.j.p.s and the slow depolarization in the rat isolated tail artery. The traces show the abolition of e.j.p.s and the maximum potentiation of the slow depolarization induced by suramin (100 μM). Neurogenic responses were evoked by 10 pulses at 0.5, 1, and 2 Hz. Solid bars indicate time course of stimulation. All data was obtained in a single cell of stable membrane potential. Figure reproduced from McLaren et al., 1995.

Mechanical Actions of Endogenously Released ATP

In the guinea pig vas deferens the e.j.p.s mediated by ATP initiate action potentials and calcium ion influx, whereas noradrenaline evokes the synthesis of IP_3 and release of internal calcium stores. Together ATP and noradrenaline produce a characteristic biphasic response. The initial phasic component of the contraction is predominantly purinergic, as it is inhibited by $ANAPP_3$ (Sneddon et al., 1982), α,β-meATP (Meldrum and Burnstock, 1984) and PPADS (McLaren et al., 1994). The smaller, tonic phase of the contraction is predominantly noradrenergic, because it is inhibited by α-adrenoceptor antagonists (Sneddon et al., 1982; Westfall et al., 1995).

Unlike the vas deferens, the contractile response of most blood vessels to nerve stimulation does not consist of two clearly defined components, but rather tends to be monophasic. That a component of this response was resistant to α-adrenoceptor antagonists and mediated by ATP was first shown in the mid-1980s (see Von Kügelgen and Starke, 1991). Subsequently, most blood vessels studied have been found to exhibit a component of the neurogenic contraction that is mediated by ATP (Figure 3). However, the relative contribution of ATP and noradrenaline to the total contraction varies greatly between vessels. For example, in the rat tail artery, ATP only contributes approximately 10% of the peak contraction (Bao et al., 1993). In the rabbit ear artery this increases to 20%–60% (Kennedy et al., 1986) (Figure 3), whereas in the rabbit

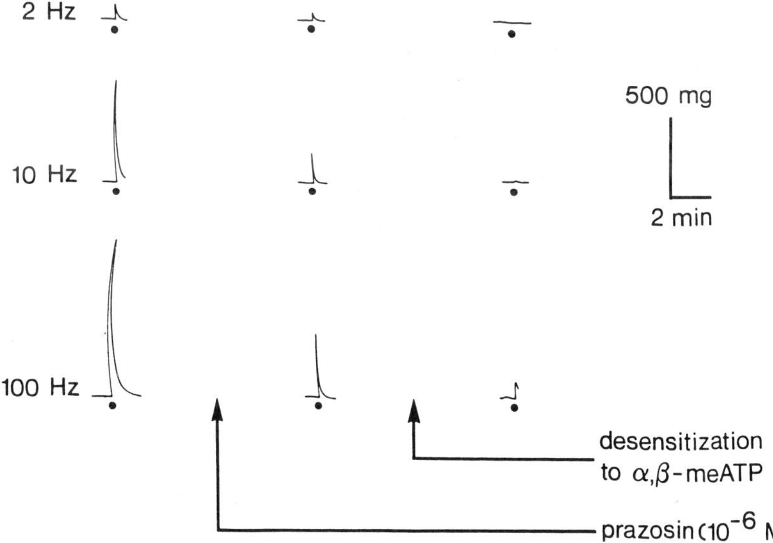

FIGURE 3. Cotransmission by ATP and noradrenaline in the rabbit isolated ear artery. Traces show contractions evoked by perivascular nerve stimulation (●) for 1 second at 2, 10, and 100 Hz, 0.1 millisecond pulse width and supramaximal voltage. Control responses (left hand traces) were inhibited, but not abolished, by prazosin (1 μM) (middle traces). Increasing the concentration of prazosin to 10 μM produced no further inhibition (not shown). Desensitization of the P_{2X}-receptor by α,β-meATP (10 μM) abolished the remaining response at low frequencies and substantially inhibited it at higher frequencies (right hand traces). Figure reproduced from Kennedy et al., 1986.

mesenteric artery ATP is the sole excitatory neurotransmitter, although noradrenaline is released and contributes to feedback inhibition (Ramme et al., 1987).

The relative contribution of ATP and noradrenaline within a given vessel can also depend on the frequency of nerve stimulation (Kennedy et al., 1986; Bao et al., 1993). Thus, in the rabbit ear artery the action of ATP is much more prominent at low frequencies of nerve stimulation and, as the frequency increases, so the contribution of ATP decreases and that of noradrenaline increases (Kennedy et al., 1986) (Figure 3). It has been suggested that this can be explained on the basis of the time-course of action of ATP and noradrenaline. Exogenous ATP evokes rapid transient contractions that fade in the continued presence of agonist, i.e., a postjunctional mechanism appears to limit the effects of ATP. Exogenous noradrenaline on the other hand evokes slower, more maintained contractions that are not subject to desensitization over the same timescale. If the postjunctional responses to endogenously released ATP and noradrenaline are similar to those of exogenously administered drug, then increasing the number of nerve stimulation pulses applied to the nerve is likely to lead to a progressively smaller action of ATP, but an increasing action of noradrenaline.

Cellular Mechanisms of Action of ATP

Fast Responses Smooth muscle cells in visceral and vascular tissues are interconnected via low-resistance gap junctions, and for many years this hampered the study of the ionic mechanisms which underlie purinergic e.j.p.s. However, the development of enzymatic dissociation procedures and of the patch clamp technique led to the demonstration that ATP evokes a rapidly developing cation current (I_{cat}) in single smooth muscle cells from the rabbit ear artery (Benham and Tsien, 1987) and portal vein (Xiong et al., 1991); and the rat tail artery (Evans and Kennedy, 1994), portal vein (Pacaud et al., 1994), and vas deferens (Nakazawa and Matsuki, 1987; Friel, 1988; Khakh et al., 1995). I_{cat} activates with a latency to onset of several milliseconds, consistent with an action at a ligand-gated channel, is inward at negative membrane potentials, so depolarizing the cell, and is not maintained during continued application of ATP, i.e., desensitization is seen (Figure 4). In the rabbit ear artery, I_{cat} displays a 3:1 selectivity for calcium ions over sodium ions, so at physiological concentrations calcium can enter the cell by two routes, (i) via I_{cat} and (ii) through dihydropyridine-sensitive L-type calcium channels that are opened by the ATP-induced depolarization. Thus, I_{cat} has many properties in common with the purinergic e.j.p.s seen in blood vessels and the vas deferens, and it is considered that I_{cat} underlies these e.j.p.s following interaction of ATP with the P_{2X}-receptor.

The P_{2X}-receptor was originally defined by a rank order of agonist potency of α,β-meATP $\gg$ 2-methylthioATP (2-meSATP) $\geq$ ATP, with α,β-meATP being approximately 1000 times more potent than ATP (Burnstock and Kennedy, 1985; Kennedy, 1990). However, it is now clear that the influence of ecto-ATPase on agonist potency is much greater than anticipated, such that the potencies of ATP and 2-meSATP are decreased 100- to 1000-fold by breakdown. When this is prevented, ATP and 2-meSATP are in fact slightly more potent agonists than α,β-meATP at the P_{2X}-receptor (Kennedy and Leff, 1995). This conclusion was reached in part following comparison of the ability of ATP, 2-meSATP, and α,β-meATP to evoke inward currents and contractions in the rat tail artery (Evans and Kennedy, 1994). In single acutely dissociated smooth muscle cells, ATP, 2-meSATP, and α,β-meATP activated I_{cat} in a concentration-dependent manner, with a rank order of potency of ATP = 2-meSATP

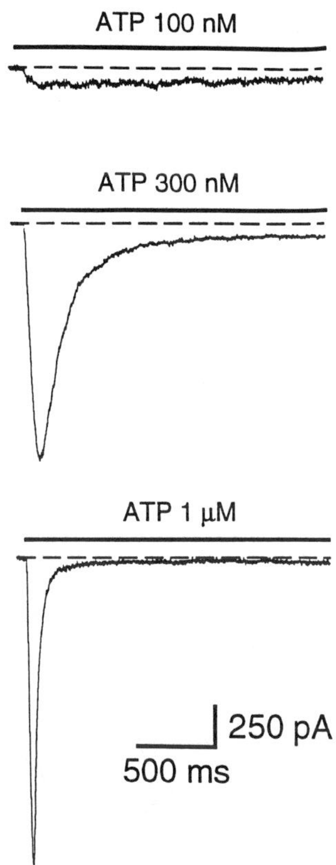

FIGURE 4. Excitatory effects of ATP in single smooth muscle cells of the rat tail artery. Traces show the inward currents evoked by ATP (100 nM–1 μM) when applied rapidly for 2 seconds using a U-tube superfusion system, as indicated by the solid bars. The dashed lines represent zero current levels. The holding potential was -60 mV. Each record is from a different cell. Figure reproduced from Evans and Kennedy, 1994.

$\geq$ α,β-meATP (Figure 4). All responses were abolished by the P_{2X}-receptor antagonist suramin. Full concentration–response curves could not be obtained, but the partial curves for ATP and 2-meSATP were less than 10-fold to the left of that for α,β-meATP. These agonists also produced suramin-sensitive contractions of isolated rat tail artery rings, with α,β-meATP $\gg$ 2-meSATP $>$ ATP. The agonist α,β-meATP was almost 1000 times more potent than ATP, which fits the "classical" profile of a P_{2X}-receptor, but is very different from that seen in single cells. ATP evoked inward currents at concentrations three orders of magnitude lower than those required to elicit contraction. Likewise, 2-meSATP was two orders of magnitude more potent in the single cells. In contrast, α,β-meATP evoked both responses at similar concentrations.

The great difference in the pharmacological profile seen in single cells and artery rings was surprising because opening of a ligand-gated cation channel is the initial step by which P_{2X}-receptor agonists contract smooth muscle. The simplest explanation is that ATP, 2-meSATP, and α,β-meATP act at the same site (the P_{2X}-receptor) to evoke I_{cat} and contraction, but their relative potencies are determined by differences in their breakdown in intact muscle and single cells. ATP and 2-meSATP are broken down by ecto-ATPase, but α,β-meATP is relatively resistant.

Slow Responses In addition to evoking the fast I_{cat} described above that underlies e.j.p.s, ATP can also act through G protein–coupled receptors to elicit currents that develop more slowly and maintained longer. In the rabbit portal vein, ATP activates a small cationic current that does not show desensitization and which is due at least in part to calcium ion influx (Xiong et al., 1991). ATP can also evoke slow, calcium-dependent chloride currents in the rabbit portal vein (Xiong et al., 1991) and pulmonary artery (Halliwell et al., 1994), and in the aorta of the rat (Von der Weid et al., 1993) and pig (Droogmans et al., 1991). The source of the calcium depends on the vessel studied. In the rabbit portal vein the chloride current is triggered by calcium influx via the slow cationic current, whereas in the rabbit pulmonary artery and pig aorta, the current is dependent upon release of internal calcium ion stores, following production of IP_3.

Each of these currents is slower than I_{cat} and shows less desensitization and so will lead to a more maintained depolarization and influx of calcium ions. However, unlike I_{cat}, it is not clear what the physiological roles of these currents are. Similarly, the subtype(s) of P_2-receptor that mediate the currents is uncertain. Both P_{2Y}- and P_{2U}-receptors are present in vascular smooth muscle cells and both have been shown to mediate release of intracellular calcium stores (Pacaud et al., 1995). A complicating factor is that, in some studies, freshly dissociated cells were used, whereas in others cultured cells were used. Pacaud et al. (1995) have recently shown that P_{2X}-receptors are downregulated during culture of rat aorta smooth muscle cells, whereas P_{2Y}-receptors, which were absent in freshly dissociated cells, are upregulated during the culturing process. In contrast, P_{2U}-receptors are present under both types of condition.

Functional Significance of Cotransmission

What is the functional significance of having more than one transmitter in sympathetic nerves? Several possibilities can be considered. First, the difference in the time-course of the postjunctional responses to ATP and noradrenaline may allow flexibility in the response to different conditions. As discussed above, ATP has a short, rapid action that could be involved in short-term (pulse–pulse or second–second) regulation of blood pressure, whereas noradrenaline elicits a slower, longer lasting vasoconstriction, which may be more important when a maintained rise in blood pressure is required. Second, ATP, noradrenaline, and NPY interact synergistically in numerous blood vessels, leading to greater vasoconstriction. In most cases, the mechanisms by which potentiation occurs are unclear, but NPY increases the calcium influx through L-type channels in single smooth muscle cells of the rat tail artery (Xiong et al., 1993) and increases the amplitude of the component of α_1-adrenoceptor-mediated contraction in rat mesenteric arterioles that is associated with these channels (Andriantsitohaina and Stoclet, 1990). Since purinergic e.j.p.s also cause these channels to open, this could represent one mechanism by which NPY potentiates purinergic contractions. Finally, because there is now evidence that ATP and noradrenaline may be stored in different synaptic vesicles, their relative rates of release may depend on the number, frequency, and pattern of nerve pulses reaching the nerve terminal. There is a precedent for this in parasympathetic nerves that innervate blood vessels to the cat submaxillary gland and which release both acetylcholine (ACh) and vasoactive intestinal polypeptide (VIP) as inhibitory transmitters (Edwards and Bloom, 1982). Maximal release of ACh is seen at low frequencies of stimulation, whereas optimal release of VIP is seen with short bursts of high frequency. Presumably, this is due to ACh and VIP being stored in different vesicles, which in turn are associated with different types of calcium channel. Such a mechanism could also lead to differential release of ATP, noradrenaline, and NPY from sympathetic nerves.

Hypertension

Hypertension is recognized as a major health problem that, if uncorrected, leads to cardiovascular complications such as increased probability of stroke, myocardial infarction, and heart failure. However, apart from a few cases, the underlying cause of hypertension is unknown. Many factors have been suggested to contribute to the genesis of the disease, but in general it is difficult to determine if these factors are caused by hypertension or are themselves the cause of hypertension. In most cases of essential hypertension the elevated arterial blood pressure is associated with an increase in peripheral vascular resistance (see Head, 1989), and it has been suggested that an enhanced activity of the sympathetic nervous system and adaptive structural changes in the vessel wall are important components in this change.

If there is greater activity of sympathetic nerves in hypertension then it would be reasonable to expect that ATP would play a greater role as a neurotransmitter. In a small number of studies the contribution of ATP to neurogenic contractions has been investigated using the spontaneously hypertensive rat (SHR), a widely used model of hypertension. In an early study in the SHR tail artery, Vidal et al. (1986) did indeed find that, compared with control animals, there was an increased contribution from neuronally released ATP to neurogenic contractions, suggesting that an increased purinergic response could be involved in hypertension. However, subsequent studies in this artery have not been able to reproduce these findings. Both Dalziel et al. (1989) and Muir and Wardle (1989) found no evidence for a greater purinergic component to neurotransmission in SHR tail artery compared with normotensive controls.

Evidence for a greater purinergic component in neurotransmission has also been found in vessels from a hypertensive rabbit model (Bulloch and McGrath, 1992). In the distal saphenous artery of normotensive animals, ATP had little or no effect as an excitatory cotransmitter, but in preparations from hypertensive animals, ATP played a significant role. Likewise, in two vessels, the ileocolic and proximal saphenous arteries, in which ATP was a transmitter in the normotensive state, an increase in its function was seen in the hypertensive state. The authors further showed that, compared with normotensive vessels, there was less negative feedback via presynaptic α_2-adrenoceptors in the hypertensive vessels and they suggested that the increase in ATP release could be due to the reduction in ongoing autoinhibition.

At present the evidence for an increase in purinergic neurotransmission in hypertensive animals is clearly thin, but what evidence there is in the hypertensive rabbit model does point to such a change. Clearly more studies, in a wider range of vessels, are required before a firm conclusion can be reached.

ACKNOWLEDGMENTS

These studies were supported by grants from the Medical Research Council and the Wellcome Trust.

REFERENCES

Andriantsitohaina R, Stoclet JC (1990): Enhancement by neuropeptide Y (NPY) of the dihydropyridine-sensitive component of the response of α_1-adrenoceptor stimulation in rat isolated mesenteric arterioles. Br J Pharmacol 99:389–395.

Bao JX, Gonon F, Stjärne L (1993): Frequency- and train length-dependent variation in the roles of postjunctional α_1- and α_2-adrenoceptors for the field stimulation-induced neurogenic contraction of the rat tail artery. Naunyn Schmiedebergs Arch Pharmacol 347:601–616.

Benham CD, Tsien RW (1987): A novel receptor-operated Ca-permeable channel activated by ATP in smooth muscle. Nature 328:275–278.

Bolton TB, Large WA (1986): Are junction potentials essential? Dual mechanism of smooth muscle cell activation by transmitter released from autonomic nerves. Q J Exp Physiol 71:1–28.

Bulloch JM, McGrath JC (1992): Evidence for increased purinergic contribution in hypertensive blood vessels exhibiting cotransmission. Br J Pharmacol 107:145P.

Burnstock G (1990): Purinergic mechanisms. Ann NY Acad Sci 603:1–17.

Burnstock G, Kennedy C (1985): Is there a basis for distinguishing two types of P_2-purinoceptor? Gen Pharmacol 5:433–440.

Burnstock G, Kennedy C (1986): A dual function for ATP in regulation of vascular tone. Circ Res 58:319–330.

Dalziel HH, Machaly M, Sneddon P (1989). Comparison of the purinergic contribution to sympathetic vascular responses in SHR and WKY rats in vitro and in vivo. Eur J Pharmacol 173:19–26.

Driessen B, Kügelgen IV, Starke K (1993): Neural ATP and its α_2-adrenoceptor-mediated modulation in guinea-pig vas deferens. Naunyn Schmiedebergs Arch Pharmacol 348:358–366.

Driessen B, Kügelgen IV, Starke K (1994): P_1-purinoceptor-mediated modulation of neural noradrenaline and ATP release in guinea-pig vas deferens. Naunyn Schmiedebergs Arch Pharmacol 350:42–48.

Droogmans G, Callewaert G, Declerck I, Casteels R (1991): ATP-induced Ca^{2+} release and Cl^- current in cultured smooth muscle cells from pig aorta. J Physiol 440:623–634.

Edwards AV, Bloom SR (1982): Recent physiological studies of the alimentary autonomic innervation. Scand J Gastroent 17(suppl):77–89.

Ellis JL, Burnstock G (1989): Angiotensin neuromodulation of adrenergic and purinergic cotransmission in the guinea-pig vas deferens. Br J Pharmacol 97:1157–1164.

Ellis JL, Burnstock G (1990): Neuropeptide Y neuromodulation of sympathetic cotransmission in the guinea-pig vas deferens. Br J Pharmacol 100:457–462.

Evans RJ, Kennedy C (1994): Characterisation of P_2-purinoceptors in the smooth muscle of the rat tail artery; a comparison between contractile and electrophysiological responses. Br J Pharmacol 113:853–860.

Friel DD (1988): An ATP-sensitive conductance in single smooth muscle cells from the rat vas deferens. J Physiol 401:361–380.

Halliwell RM, Wang Q, Hogg RC, Large WA (1994): Synergistic action of histamine and adenosine triphosphate on the response to noradrenaline in rabbit pulmonary artery smooth muscle cells. Pflügers Arch 426:433–439.

Head RJ (1989): Hypernoradrenergic innervation: Its relationship to functional and hyperplastic changes in the vasculature of the spontaneously hypertensive rat. Blood Vessels 26:1–20.

Katsuragi T, Tokunaga T, Ogawa S, Osamu S, Chiemi S, Furukawa T (1991): Existence of ATP-evoked ATP release system in smooth muscles. J Pharmacol Exp Ther 259:513–518.

Kennedy C (1990): P_1- and P_2-purinoceptor subtypes—an update. Arch Int Pharmacodyn 303:30–50.

Kennedy C, Leff P (1995): How should P_{2X}-purinoceptors be characterised pharmacologically? Trends Pharmacol Sci 16:168–174.

Kennedy C, Saville V, Burnstock G (1986): The contributions of noradrenaline and ATP to the responses of the rabbit central ear artery to sympathetic nerve stimulation depend on the parameters of stimulation. Eur J Pharmacol 122:291–300.

Khakh BS, Surprenant A, Humphrey PPA (1995): A study on P_{2X}-purinoceptors mediating the electrophysiological and contractile effects of purine nucleotides in rat vas deferens. Br J Pharmacol 115:177–185.

McLaren GJ, Kennedy C, Sneddon P (1995): The effects of suramin on purinergic and noradrenergic neurotransmission in the rat isolated tail artery. Eur J Pharmacol 277:57–61.

McLaren GJ, Lambrecht G, Mutschler E, Bäumert HG, Sneddon P, Kennedy C (1994): Investigation of the actions of PPADS, a novel P_{2X}-purinoceptor antagonist, in the guinea-pig isolated vas deferens. Br J Pharmacol 111:913–917.

Meldrum LA, Burnstock G (1984): Evidence that ATP acts as a cotransmitter with noradrenaline in sympathetic nerves supplying the guinea-pig vas deferens. Eur J Pharmacol 92:161–163.

Muir TC, Wardle KA (1989): Vascular smooth muscle responses in normo- and hypertensive rats to sympathetic nerve stimulation and putative transmitters. J Auton Pharmacol 9:23–34.

Nakazawa K, Matsuki N (1987): Adenosine triphosphate-activated inward current in isolated smooth muscle cells from rat vas deferens. Pflügers Arch 409:644–646.

Pacaud P, Grégoire G, Loirand G (1994): Release of Ca^{2+} from intracellular store in smooth muscle cells of rat portal vein by ATP-induced Ca^{2+} entry. Br J Pharmacol 113:457–462.

Pacaud P, Malam-Souley R, Loirand G, Desgranges C (1995): ATP raises $[Ca^{2+}]_i$ via different P_2-receptor subtypes in freshly isolated and cultured aortic myocytes. Am J Physiol 269:H30–H36.

Ramme D, Regenold JT, Starke K, Busse R, Illes P (1987): Identification of the neuroeffector transmitter in jejunal branches of the rabbit mesenteric artery. Naunyn Schmiedebergs Arch Pharmacol 336:267–273.

Sedaa KO, Bjur RA, Shinozuka K, Westfall DP (1990): Nerve and drug-induced release of adenine nucleosides and nucleotides from rabbit aorta. J Pharmacol Exp Ther 252:1060–1067.

Sneddon P (1992): Suramin inhibits excitatory junction potentials in guinea-pig vas deferens. Br J Pharmacol 107:101–103.

Sneddon P, Burnstock G (1984a): ATP as a cotransmitter in rat tail artery. Eur J Pharmacol 106:149–152.

Sneddon P, Burnstock G (1984b): Inhibition of excitatory junction potentials in guinea-pig vas deferens by α,β-methyleneATP: Further evidence for ATP and noradrenaline as cotransmitters. Eur J Pharmacol 100:85–90.

Sneddon P, Westfall DP (1984): Pharmacological evidence that adenosine triphosphate and noradrenaline are co-transmitters in the guinea-pig vas deferens. J Physiol 347:561–580.

Sneddon P, Westfall DP, Fedan JS (1982): Cotransmitters in the motor nerves of the guinea-pig vas deferens: Electrophysiological evidence. Science 218:693–695.

Todorov LD, Mihaylova-Todorova S, Craniso GL, Bjur RA, Westfall TD (1996): Evidence for the differential release of the cotransmitters ATP and noradrenaline from sympathetic nerves of the guinea-pig vas deferens. J Physiol 496:731–768.

Von der Weid PY, Serebryakov VN, Orallo F, Bergmann C, Snetkov VA, Takeda K (1993): Effects of ATP on cultured smooth muscle from rat aorta. Br J Pharmacol 108:638–645.

Von Kügelgen I, Starke K (1991): Noradrenaline-ATP cotransmission in the sympathetic nervous system. Trends Pharmacol Sci 12:319–324.

Vidal M. Hicks PE, Langer SZ (1986): Differential effects of α,β-methyleneATP on responses to nerve stimulation in SHR and WKY tail arteries. Naunyn Schmiedebergs Arch Pharmacol 332:384–390.

Vizi ES, Sperlagh B, Baranyi M (1992): Evidence that ATP, released from the postsynaptic site by noradrenaline, is involved in mechanical responses of guinea-pig vas deferens: Cascade transmission. Neuroscience 50:455–465.

Walsh MP (1994): Regulation of vascular smooth muscle tone. Can J Physiol Pharmacol 72:919–936.

Westfall TD, Kennedy C, Sneddon P (1996): Enhancement of sympathetic purinergic neurotransmission in the guinea-pig isolated vas deferens by the novel ecto-ATPase inhibitor ARL 67156. Br J Pharmacol 117:867–872.

Xiong Z, Bolzon BJ, Cheung DW (1993): Neuropeptide Y potentiates calcium-channel currents in single vascular smooth muscle cells. Pflügers Arch 423:504–510.

Xiong Z, Kitamura K, Kuriyama H (1991): ATP activates cationic currents and modulates the calcium current through GTP-binding protein in rabbit portal vein. J Physiol 440:143–165.

CHAPTER 10

Cardiac Electrophysiology of Adenosine: Antiarrhythmic and Proarrhythmic Actions

LUIZ BELARDINELLI, ANNE B. CURTIS, and BARRY BERTOLET

Departments of Medicine (L.B., A.B.C., B.B.) and Pharmacology (L.B), University of Florida, Gainesville, FL 32610

INTRODUCTION

It is well established that adenosine is a potent endogenous regulator of various cardiac functions whose actions are mediated by specific cell surface receptors (Belardinelli et al., 1989; Olsson and Pearson, 1990). Adenosine, in addition to being a coronary vasodilator, is an inhibitor of cardiac activity (Belardinelli et al., 1989). Adenosine slows heart rate (negative chronotropic effect), causes atrioventricular (AV) nodal conduction slowing and block (negative dromotropic effect), and antagonizes the cardiac stimulatory effects (e.g., inotropic and arrhythmogenic) of β-adrenergic agonists (anti-β-adrenergic effect) (Belardinelli et al., 1989). These cardiac actions of adenosine are mediated by at least two subtypes of cell surface receptors called A_1 and A_2. Activation of the A_1-receptor subtype is responsible for the electrophysiological, inotropic, and anti-β-adrenergic actions of adenosine, whereas the coronary vasodilation caused by this nucleoside is mediated primarily (but not exclusively) by the adenosine receptor A_{2a} subtype (Belardinelli et al., 1995; Olsson and Pearson, 1990).

Although the cardiovascular actions of adenosine including its chronotropic and dromotropic effects have been known for over 60 years (Drury and Szent-Gyorgyi, 1929), only in the last decade has the clinical relevance of this nucleoside been recognized (Belardinelli and Lerman, 1990). The use of adenosine for the treatment of paroxysmal supraventricular tachycardia (PSVT) was the first and, until recently, the only clinical application of adenosine approved by the United States Food and Drug Administration.

Purinergic Approaches in Experimental Therapeutics, Edited by Kenneth A. Jacobson and Michael F. Jarvis
ISBN 0-471-14071-6 © 1997 Wiley-Liss, Inc.

The purpose of this review is to describe the cardiac electrophysiological effects of adenosine on myocytes, in the intact heart, and on the human heart *in vivo*. Emphasis is given to the basis of the antiarrhythmic and proarrhythmic properties of this nucleoside, and to the clinical use of adenosine as a therapeutic and diagnostic agent of human cardiac arrhythmias. The potential role of adenosine antagonists in the treatment of bradyarrhythmias and tachyarrhythmias associated with conditions in which myocardial interstitial concentration of adenosine is elevated (e.g., ischemia) is also discussed.

CARDIAC ELECTROPHYSIOLOGY OF ADENOSINE

The electrophysiological responses of single cardiac myocytes and *in vitro* heart preparations to adenosine have, to a great extent, predicted the electrophysiological effects of this nucleoside on the human heart (Belardinelli and Lerman, 1990; Belardinelli et al., 1989; Lerman and Belardinelli, 1991). Although the main cardiac electrophysiological effects of adenosine, slowing of heart rate and impairment of AV nodal conduction, were reported as early as 1929 (Drury and Szent-Gyorgi, 1929), only in the last 15 years has the cellular basis for these effects been identified. Likewise, the modulators (inhibitors and potentiators) of the cardiac electrophysiological actions of adenosine have become recognized only in the last two decades. The receptor subtype (A_1) that mediates these actions of adenosine has now been identified. More detailed reviews of the cardiac electrophysiological actions of adenosine have been published in recent years (Belardinelli et al., 1995; Belardinelli et al., 1989; Lerman and Belardinelli, 1991).

Single Cardiomyocytes

Activation of the cardiac A_1 adenosine receptor leads to several electrophysiological signals that depend on the cell type and species (Belardinelli et al., 1995). In atrial sinoatrial (SA), and atrioventricular (AV) nodal cells, stimulation of the A_1 adenosine receptors activates the potassium outward current I_{KAdo} (similar to the K-current regulated by acetylcholine, I_{KACh}) via a mechanism that is independent of an effect on cAMP (Belardinelli et al., 1995). In addition, in these cells (atrial, SA, and AV nodal), adenosine attenuates the stimulatory effects of catecholamines on L-type calcium inward current $I_{ca,L}$ and pacemaker current I_F, secondary to a reduction of cellular cAMP (Belardinelli et al., 1989; Belardinelli et al., 1988; Visentin et al., 1990). In atrial and AV nodal myocytes, adenosine also causes a small (12%–18%) inhibition of basal (nonstimulated) $I_{Ca,L}$, an effect that is not seen in SA nodal or ventricular myocytes. In ventricular myocytes, the main effect of adenosine is to antagonize the stimulatory actions of catecholamines on $I_{Ca,L}$, delayed rectifier potassium current I_K, chloride current I_{Cl}, and on the transient inward current I_{Ti} (Isenberg and Belardinelli, 1984; Song et al., 1992; Song et al., 1995). The inwardly rectifying current $I_{KACh,Ado}$, which is present in supraventricular tissues, is significantly less prominent in ventricular myocytes. The low density of I_{KAdo} in guinea pig ventricular myocytes explains the lack of effect of adenosine on the ventricular action potential of this species; whereas in ferret ventricular myocytes, where this current seems to be present, adenosine causes significant shortening of the ventricular action potential (Belardinelli et al., 1995; Qu et al., 1993).

Intact Heart

The negative chronotropic (sinus bradycardia) and negative dromotropic (slowing of AV nodal conduction and AV block) effects of adenosine are mediated by the A_1 receptor subtype. They are concentration-dependent, competitively antagonized by alkylxanthines, attenuated or abolished by adenosine deaminase, and potentiated by nucleoside uptake blockers and allosteric enhancers (Belardinelli et al., 1989; Kollias-Baker et al., 1994). The magnitude and duration of these effects of adenosine depends on (i) the experimental preparation (*in vitro* vs *in vivo*), and conditions (e.g., temperature), (ii) the species, (iii) the method of administration (e.g., infusion vs bolus), and (iv) the presence of other drugs such as nucleoside uptake blockers and xanthine derivatives. Although the direct effect of adenosine is to slow heart rate and AV nodal conduction, these actions of adenosine are influenced *in vivo* by reflex activation of the autonomic nervous system (because of the reduced systemic vascular resistance) and/or by direct stimulation of chemoreceptors (Conradson et al., 1987; Biaggioni et al., 1987). For instance, when autonomic reflexes are intact, intravenous infusion of adenosine causes tachycardia instead of bradycardia (Conradson et al., 1987). These and many other conditions that modulate the negative chronotropic and dromotropic effects of adenosine have been previously reviewed (Belardinelli and Lerman, 1990; Belardinelli and Pelleg, 1990; Belardinelli et al., 1989).

In addition to a direct depressant effect on SA node function, adenosine also suppresses AV junctional and ventricular pacemakers (Belardinelli and Lerman, 1990; Belardinelli et al., 1983; Belardinelli and Pelleg, 1990). However, these pacemakers have different sensitivities to adenosine. The sinus node pacemakers are the most sensitive to adenosine, followed by AV junctional pacemakers, and the least sensitive to adenosine are the ventricular pacemakers (Belardinelli et al., 1983). This depressant effect of adenosine on AV junctional and ventricular pacemakers explains why adenosine-induced third degree AV block is often accompanied by ventricular standstill or a very slow ventricular escape rhythm. In summary, the two main cardiac electrophysiological effects of adenosine are the depression of pacemaker activity and AV nodal conduction. These effects of adenosine vary among heart preparations, are species-dependent, and are influenced by the experimental conditions, such as whether or not the autonomic reflexes are intact.

BASIS FOR THE ANTI- AND PROARRHYTHMIC ACTIONS OF ADENOSINE

The direct negative chronotropic, dromotropic, and anti-β-adrenergic actions of adenosine, and the shortening of the duration of the atrial action potential, form the basis for the anti- and proarrhythmic properties of this nucleoside. The most common types of arrhythmias that are sensitive to adenosine, and those that can be initiated (or facilitated) by this nucleoside are listed in Table 1. The efficacy of adenosine in terminating AV reentrant and AV nodal reentrant tachycardias, and catecholamine-dependent ventricular tachycardia is now well established (Lerman and Belardinelli, 1991; Camm and Garratt, 1991). Likewise, adenosine-induced sinus bradycardia and AV block in humans has been widely reported (DiMarco et al., 1983; Lerman and Belardinelli, 1991; Camm and Garratt, 1991). Atrial flutter and fibrillation, which have been found in laboratory animals (Kabell et al., 1994), are also observed in approximately 5% of humans receiving adenosine (DiMarco et al., 1990). In compari-

TABLE 1 Effects of Adenosine on Cardiac Rhythms

Antiarrhythmic	Proarrhythmic
Terminates PSVT	Causes bradyarrhythmias (bradycardia and AV block)
Terminates catecholamine-dependent VT	Facilitates the induction of atrial flutter and fibrillation

PSVT: paroxysmal supraventricular tachycardia; VT: ventricular tachycardia.

son, the role of endogenous adenosine, released in response to myocardial hypoxia and ischemia, is less well established as the mediator of bradyarrhythmias, atrial flutter, and atrial fibrillation. The evidence in support of this latter action is discussed below.

Basis for the Antiarrhythmic Actions

The basis for the AV nodal conduction delay and block caused by adenosine is the direct depressant effect of this nucleoside on the action potentials of AV nodal cells (Clemo and Belardinelli, 1986). Activation of I_{KAdo} and, to a lesser extent, inhibition of basal $I_{Ca,L}$, are the cause of the depression of AV nodal action potentials by adenosine. Thus, reentrant circuits involving the AV node, which are the underlying basis of a great number of supraventricular arrhythmias, are interrupted by adenosine because of the suppression of AV nodal action potentials. The direct depressant effect of action potentials of AV nodal cells and the AV nodal conduction delay caused by adenosine and A_1-receptor agonists is rate-dependent (Belardinelli and Shryock, 1992; Belardinelli et al., 1994; Nayebpour et al., 1993; Stark et al., 1993). In keeping with these observations, it was recently shown that in human hearts the negative dromotropic effect of adenosine is also rate-dependent (Lai et al., 1994).

The anti-β-adrenergic actions of adenosine are responsible for the termination of non-reentrant and nonautomatic ventricular tachycardia (VT) induced by isoproterenol (Lerman et al., 1986; Lerman and Belardinelli, 1991). More specifically, the decrease of adenylyl cyclase activity caused by adenosine lowers cellular cAMP and inhibits cAMP-mediated activation of $I_{Ca,L}$ and I_{Ti}, and thus explains the observation that catecholamine-dependent afterdepolarizations and triggered activity in ventricular myocytes are attenuated or abolished by adenosine, respectively. Whether catecholamine-dependent ventricular tachycardias in patients are due to cAMP-mediated triggered activity is difficult to prove. Although the inhibition of catecholamine-stimulated $I_{Ca,L}$ and I_{Ti} by adenosine is the most likely mechanism responsible for the efficacy of this nucleoside in terminating catecholamine-dependent VT, they may not be the only ionic currents responsible for the arrhythmogenic actions of isoproterenol. Regardless, adenosine is expected to antagonize all ion currents that are modulated (e.g., stimulated) by catecholamines, or any other agent or condition that is associated with cellular cAMP accumulation (Song et al., 1992). It is of interest that ventricular tachycardias (e.g., reentrant VT) that are catecholamine-independent (i.e., not sensitive to β-blockade or not due to elevated cellular cAMP) are insensitive to adenosine. The lack of effect of adenosine on these arrhythmias can be explained in part by the lack of effect of adenosine on the ventricular action potential and ion currents in the absence of catecholamine stimulation or in the absence of other agents known to stimulate the cellular production of cAMP (Belardinelli et al., 1995, and references herein).

Basis for the Proarrhythmic Actions

In atrial myocytes, adenosine causes marked shortening of the action potential. The shortening of the atrial action potential explains the decrease in atrial refractory period caused by adenosine, which in turn facilitates the initiation and maintenance of atrial flutter and atrial fibrillation (Kabell et al., 1994). In keeping with this, adenosine has been shown to facilitate induction of atrial flutter and fibrillation in laboratory animals (Kabell et al., 1994) and in humans (DiMarco et al., 1985; DiMarco et al., 1990). Similarly, the conversion of atrial flutter to transient atrial fibrillation by adenosine (Kabell et al., 1994) is likely to be due to an additional shortening of the atrial refractory period caused by adenosine. In contrast to this proarrhythmic action, adenosine also causes hyperpolarization. The hyperpolarization of atrial myocytes by adenosine should decrease excitability and may lead to an increase in conduction velocity in the atrial myocardium, an effect that would be antiarrhythmic. The ionic basis for the pro- and antiarrhythmic actions of adenosine in atrial tissue is the activation of the inward rectifying potassium current I_{KAdo}, which explains the shortening of the atrial action potential and hyperpolarization.

Results of studies in isolated and *in situ* hearts (e.g., guinea pig) have shown that adenosine fulfills the following four important criteria necessary to establish that this nucleoside is the mediator of AV nodal conduction disturbances associated with myocardial hypoxia and ischemia (Belardinelli et al., 1987).

1. Adenosine mimics the depressant effects of hypoxia and ischemia on AV nodal conduction.
2. Adenosine is released by hypoxic and ischemic myocardium in amounts sufficient to account for the AV nodal conduction delay observed during conditions of oxygen deprivation.
3. The potency (and the rank order) of adenosine-receptor antagonists in attenuating the effects of adenosine on AV nodal conduction is nearly identical to that required to attenuate the effects of hypoxia on AV nodal conduction delay.
4. Nucleoside transport blockers and the allosteric enhancer PD 81,723 potentiate the depressant effects of adenosine on AV nodal conduction, and exacerbate the depressant effects of hypoxia and ischemia on AV nodal conduction (Kollias-Baker et al., 1994).

Although these findings apply mainly to AV nodal conduction, similar evidence exists for the role of endogenous adenosine as the mediator of sinus slowing (bradycardia) associated with hypoxia and ischemia (Belardinelli et al., 1987). Nevertheless, the evidence for the role of endogenous adenosine in bradyarrhythmias in patients with inferior myocardial infarction, cardiac arrest, and sinus node dysfunction (e.g., sick sinus syndrome and bradycardia in cardiac transplants) remains indirect, and is exclusively based on the reversal of these bradyarrhythmias by the weak, receptor subtype–nonselective, and nonspecific antagonist theophylline (see below).

Adenosine, released in increased amounts during myocardial hypoxia and ischemia, could also be the mediator of atrial tachyarrhythmias (e.g., atrial fibrillation and flutter). Adenosine-induced shortening of the atrial action potential and refractory period should facilitate the induction and maintenance of atrial flutter and fibrillation. However, the evidence that such events take place during myocardial hypoxia or ischemia remains limited to a few studies. In one such study, it was shown that hypoxia caused significant shortening of the atrial action potential, and this effect of hypoxia was reversed by the adenosine antagonist aminophylline (Szentmiklosi et al., 1979). However, to our knowledge, no animal study has shown that atrial flutter or fibrillation during hypoxia or ischemia is reversed by an adenosine receptor antagonist. More evidence will be needed before this hypothesis can be widely accepted.

CLINICAL ELECTROPHYSIOLOGICAL EFFECTS OF ADENOSINE

Adenosine administered to a human patient as an intravenous bolus injection causes a biphasic effect on heart rate. The initial effect is a slowing of sinus rate that is, in general, seen within 15–30 seconds of a peripheral venous injection and lasts up to 10 seconds. This bradycardia is followed by a sinus tachycardia of variable magnitude that could be either caused by direct activation of carotid body chemoreceptors, resulting in respiratory stimulation and secondary activation of pulmonary stretch receptors (Watt and Routledge, 1986), or caused by baroreceptor activation due to reduction in systemic vascular resistance (Conradson et al., 1987; Biaggioni et al., 1987).

In addition to its negative chronotropic effect, adenosine has a negative dromotropic effect on the human AV node. It causes depression or block of AV nodal conduction for several seconds when given as an intravenous bolus injection (DiMarco et al., 1983; Favale et al., 1985). This effect is responsible for the efficacy of adenosine in terminating supraventricular tachycardias in which the AV node is part of a reentrant circuit. The negative dromotropic effect of adenosine is dependent on the atrial rate. That is, the dose of adenosine required to cause AV block is greater during normal sinus rhythm or atrial pacing at slow rates than during supraventricular tachycardia (DiMarco et al., 1983). More recently, Lai et al. (1994) showed that in humans the prolongation of the atrium-to-His bundle (A–H) interval caused by intravenous bolus injections of adenosine is more pronounced at shorter than longer atrial pacing cycle lengths. Consistent with these observations, the rate of conversion of PSVT to sinus rhythm with adenosine is higher in patients with faster than slower tachycardia (Gausche et al., 1994).

Adenosine has limited effects on other tissues in the heart. As in laboratory animals, adenosine shortens the atrial action potential and refractoriness in humans (O'Nunain et al., 1992), but has little or no direct effect on the ventricular action potential and refractoriness (Belardinelli et al., 1995). Likewise, adenosine appears to have little or no direct effect on the His-Purkinje system (Lerman et al., 1988). Most, but not all, accessory AV pathways are insensitive to adenosine. Adenosine-sensitive accessory AV pathways are typically those pathways that exhibit decremental conduction (Engelstein et al., 1994a), that is, that have electrophysiological properties similar to that of AV nodal tissue.

ANTIARRHYTHMIC EFFECTS

Effects on Specific Arrhythmias

The efficacy of adenosine in restoring sinus rhythm in patients with a variety of supraventricular tachyarrhythmias is now well established (Belardinelli and Lerman,

1990; Lerman and Belardinelli, 1991). Adenosine causes transient slowing of sinus tachycardia. There are few reports of clinical studies on the effect of adenosine on sinus node reentrant tachycardia. Although one such study concluded that adenosine had no effect on sinus node reentrant tachycardia (DiMarco et al., 1985), two more recent studies reported slowing or termination of the tachycardia in all patients given adenosine (Griffith et al., 1989; Engelstein et al., 1994b).

Early studies suggested that atrial tachycardias are insensitive to adenosine (Lerman and Belardinelli, 1991; DiMarco et al., 1985; Haines and DiMarco, 1990). However, more recent data (Engelstein et al., 1994b; Shenasa et al., 1994), and the experience in our own laboratory revealed that it is possible for atrial tachycardias to be terminated by adenosine. It is still not clear what precisely are the factors that determine whether an atrial tachycardia will terminate with adenosine. It has been proposed that sensitivity of atrial tachycardia to adenosine is determined by proximity of the focus of the arrhythmia to the crista terminalis (Shenasa et al., 1994). Adenosine terminated 95% of ectopic atrial tachycardias arising from the crista terminalis but only 29% of those arising elsewhere (Shenasa et al., 1994). On the other hand, the response to adenosine may be dictated by the mechanism of the tachycardia. In support of this latter hypothesis, adenosine was reported to terminate atrial tachycardia due to triggered activity, to transiently suppress automatic atrial tachycardia, and to have no effect on intraatrial reentrant tachycardia (Engelstein et al., 1994b). The different responses to adenosine can be explained by the cellular electrophysiological effects of this nucleoside. Adenosine does not prolong atrial refractoriness, and hence should have little or no effect on intraatrial reentrant tachycardia. Transient suppression of automatic atrial tachycardia might be due to activation of I_{KAdo}, which causes hyperpolarization, whereas termination of atrial tachycardia due to triggered activity could be explained by the anti-β-adrenergic action and/or direct activation of I_{KAdo} by adenosine.

Although adenosine has been shown to terminate atrial flutter in dogs (Kabell et al., 1994), in human patients adenosine has not been shown to consistently terminate either atrial flutter or fibrillation. However, administration of adenosine will cause transient AV block, which may unmask flutter waves, and hence help to identify the mechanism of the tachycardia (Rankin et al., 1989; Conti et al., 1995a). In addition, rather than terminating these arrhythmias, adenosine may actually precipitate atrial flutter or fibrillation because of its shortening of atrial action potential duration and refractoriness (DiMarco et al., 1985; Lerman et al., 1989). Whereas adenosine typically causes transient slowing of the ventricular response to atrial fibrillation, it may actually cause ventricular fibrillation when administered to patients with the Wolff-Parkinson-White syndrome and atrial fibrillation (Exner et al., 1995). This may be due to transient block of AV nodal conduction, which is accompanied by a reduction in concealed penetration and block in the accessory pathway. Alternatively, acceleration of the ventricular response may be due to shortening of antegrade refractoriness of the AV accessory pathway (Garratt et al., 1991).

Adenosine has an important therapeutic role in the termination of two clinically significant supraventricular tachycardias, AV nodal reentrant tachycardia and AV reentrant tachycardia. Both tachycardias involve reentrant circuits in which the AV node is a necessary link in the completion of the electrical circuit. The most common form of AV nodal reentrant tachycardia uses a "slow" AV nodal pathway (i.e., long A–H interval) in the antegrade direction and a "fast" pathway (short H–A interval) in the retrograde direction. The depressant effect of adenosine on conduction through the AV node causes termination of AV nodal reentrant tachycardia in more than 90%

of patients when given as a rapid intravenous bolus injection (DiMarco et al., 1983; DiMarco et al., 1985). Indeed, failure of the tachycardia to terminate after a 12-mg bolus of adenosine should call into question the diagnosis of the arrhythmia, unless the patient is being medicated with theophylline, which antagonizes the effects of adenosine.

The most common tachycardia that occurs in the Wolff-Parkinson-White syndrome is AV reentrant tachycardia. This tachycardia is caused by a macro-reentrant circuit, involving in the majority of cases the AV node in the antegrade direction and an accessory pathway in the retrograde direction. Impulses travel through the atrium and ventricle to complete the circuit. Termination of the tachycardia requires only transient block in any limb of the reentrant circuit. Given the vulnerability of the AV node to pharmacological agents, this is the usual site for termination of these tachycardias. Adenosine is highly effective in the termination of AV reentrant tachycardia (Lerman and Belardinelli, 1991; DiMarco et al., 1985). As the tachycardia terminates, adenosine sometimes has the added advantage of unmasking transient preexcitation, which may help to confirm a diagnosis of AV reentrant tachycardia.

Adenosine has no significant direct effect on ventricular myocardium, and thus it is not surprising that most forms of ventricular tachycardia are insensitive to this nucleoside (Griffith et al., 1990). The most common type of ventricular tachycardia is typically found in patients with structural heart disease, particularly coronary artery disease. This tachycardia is completely insensitive to adenosine. On the other hand, there is a specific form of ventricular tachycardia characterized by a left bundle branch block morphology that is most commonly seen in younger patients without structural heart disease. This tachycardia originates in the right ventricular outflow tract, commonly terminates with adenosine, and is curable with radiofrequency ablation. The underlying cellular and electrophysiological mechanism of this tachycardia has been debated, but there is strong support for the hypothesis that the arrhythmia is due to cAMP-dependent triggered activity (Lerman et al., 1986).

Minor side effects such as dyspnea, chest pain, and flushing are observed in approximately 30% of patients who receive adenosine (DiMarco et al., 1990). Adenosine should be used with caution in patients with asthma, because of the potential for bronchoconstriction. There are single case reports of more serious adverse reactions to adenosine. These include torsades de pointes in a patient with long QT syndrome (Wesley and Turnquest, 1992); prolonged sinus arrest and syncope, and syncope with bradycardia and hypotension (Reed et al., 1991); and the aforementioned acceleration of the ventricular response to atrial fibrillation in the Wolff-Parkinson-White syndrome (Exner et al., 1995). Despite occasional reports of adverse reactions to adenosine, overall the drug is quite safe and well tolerated.

The cardiac electrophysiological effects of adenosine in humans are inhibited by the adenosine receptor antagonist theophylline, whereas the nucleoside uptake blocker dipyridamole markedly potentiates the effects of adenosine (Lerman and Belardinelli, 1991; Belardinelli and Lerman, 1991). Theophylline is a nonspecific adenosine antagonist that may be used to reverse adverse effects of adenosine. Patients who are receiving theophylline may have little or no response to adenosine. The lack of termination of supraventricular tachycardia in these patients may simply be due to the antagonism by theophylline, and not because the tachycardia (or its mechanism) is insensitive to adenosine. Dipyridamole inhibits the transport of adenosine into cells, and thus increases the magnitude and duration of the actions of adenosine (Belardinelli et al., 1989; Belardinelli et al., 1983). It is therefore important to reduce the dose of adenosine in any patient taking dipyridamole, or possibly to avoid its use in these patients. Dipyridamole reduced the effective dose of adenosine required to terminate SVT by 3.6-fold from 62 ± 3 μg/kg to 17 ± 6 μg/kg after dipyridamole (Lerman et al., 1989).

Diagnostic Uses

The reproducibility and brevity of AV nodal conduction block elicited by intravenous bolus injection of adenosine has generated interest in using this nucleoside in the diagnosis of various types of supraventricular and ventricular tachycardias, including their underlying mechanism(s). Adenosine has been successfully used in the differential diagnosis of both narrow and wide complex tachycardias (Rankin et al., 1989). In fact, detailed and elaborate algorithms for the diagnosis of tachycardia with adenosine have been devised (Belardinelli and Lerman, 1991; Lerman and Belardinelli, 1991; Camm and Garratt, 1991). Misdiagnosis of cardiac arrhythmias is not uncommon and often leads to inappropriate treatment. A short-acting drug such as adenosine that can differentiate SVT from VT, and in many cases distinguish the various forms of SVT, has the potential to provide a more accurate diagnosis, and hence lead to a more specific treatment for cardiac arrhythmias. In a recent study, the accuracy of a group of internal medicine house staff in diagnosing unknown tachycardias was significantly improved when rhythm strips obtained during administration of adenosine were made available for the diagnosis (Conti et al., 1995a).

Adenosine has also been advocated as an aid in the assessment of the success of radiofrequency or surgical ablation of accessory pathways (Keim et al., 1992). If adenosine is to be used in this manner, it is important to administer the drug prior to ablation in order to ascertain that the accessory pathway itself is not sensitive to adenosine. If the accessory pathway is insensitive to adenosine, then adenosine may be given after ablation to assess the results. If all electrical pathways between the atria and the ventricles have been severed except for the AV node, then administration of adenosine should lead to transient AV block. If residual conduction in the accessory pathway remains, preexcitation or VA conduction over the accessory pathway may become manifest transiently. This technique was found helpful in achieving successful ablation (Keim et al., 1992), although there have been occasional reports of exceptions (Englestein et al., 1994a). However, in general adenosine can be a very useful tool in assessing the presence and persistence of accessory pathways during ablation procedures.

Endogenous Antiarrhythmic Effect

As discussed above, adenosine, when given as a rapid intravenous bolus, is a highly effective agent for the termination of reentrant supraventricular tachycardias. Endogenous adenosine, which is released in increased amounts in response to myocardial O_2 deprivation, is known to contribute to the increase in AV nodal conduction delay associated with myocardial ischemia and hypoxia (Belardinelli et al., 1989; Belardinelli et al., 1987). However, until recently, it has not been clear whether endogenous adenosine has physiological effects under other circumstances, specifically as an endogenous antiarrhythmic agent in an otherwise healthy patient. Because dipyridamole, an inhibitor of nucleoside transport known to potentiate the actions of adenosine (Belardinelli et al., 1989), has demonstrable electrophysiological effects on the AV node, at rapid atrial rates there is the possibility that pharmacological modulation of endogenous adenosine levels could be employed in the treatment of tachycardias in which the AV node is an integral part of a reentrant circuit (Jenkins and Belardinelli, 1988; Lerman et al., 1989). In a recent study it was found that endogenous adenosine levels are, indeed, elevated in the presence of dipyridamole, and that in many cases SVT could be terminated by this nucleoside transport blocker (Conti et al., 1995b). It is of interest that, in the presence of dipyridamole, coronary sinus adenosine levels were not significantly elevated during sinus rhythm, but they were significantly elevated during supraventricular tachycardia or during rapid atrial pacing (Conti et al., 1995b). Consistent with this finding, dipyridamole has been shown to terminate isoproterenol-induced VT in human patients (Lerman, 1993). Altogether, these observations raise the possibility that potent and more specific nucleoside transport inhibitors other than dipyridamole could be used as event-specific antiarrhythmic agents. Such drugs would have minimal effects during sinus rhythm, but they would have significant electrophysiological effects during tachycardia.

PROARRHYTHMIC EFFECTS OF ADENOSINE

Intravenous administration of adenosine may have proarrhythmic effects in addition to its aforementioned antiarrhythmic actions. There is substantial data, reviewed above, suggesting that endogenous adenosine may also play a mechanistic role in the genesis of both brady- and tachyarrhythmias, such as atrioventricular conduction delay and atrial fibrillation, respectively.

Bradyarrhythmias

Bradyarrhythmias that could be potentially associated with elevated levels of endogenous adenosine include sinus bradycardia, sinus node arrest, various degrees of AV nodal block and persistent asystole. These arrhythmias become manifest in a variety of clinical situations such as acute myocardial infarction, sick sinus syndrome, cardiac arrest, or cardiac transplant rejection.

Atrioventricular Node Dysfunction (AV Nodal Block) The prototypic clinical condition of adenosine-induced bradyarrhythmias is AV nodal block associated with acute inferior wall myocardial infarction. Bradyarrhythmias most often complicate inferior wall myocardial infarctions in humans because the sinoatrial and atrioventricular nodal arteries arise from the right coronary artery (Rotman et al., 1972).

Preliminary observations in patients with acute inferior wall myocardial infarction revealed that aminophylline can reverse atropine-resistant third degree AV block into lower degree AV block or restore 1 : 1 atrioventricular conduction (Wesley et al., 1986; Shah et al., 1986). In a larger study, in which aminophylline was administered as primary therapy to patients who developed "late" AV block (3.5 ± 1.4 days) after acute inferior myocardial infarction, it was found that this xanthine derivative restored sinus rhythm in only 4 of 15 patients (Strasberg et al., 1991). In contrast, our study in patients with clinically significant dysrhythmias occurring in the "early" (within hours) phase of acute inferior myocardial infarction, theophylline was highly effective in restoring sinus rhythm (Bertolet et al., 1995). The patients of this latter study who developed a persistent and clinically significant AV nodal conduction delay and block within 4 hours of the onset of symptoms were resistant to the initial standard therapy with atropine. In contrast to the lack of effect of atropine, normal 1:1 AV nodal conduction and sinus rhythm were restored within 1.8 ± 0.7 minutes after the intravenous injection of 250 mg of theophylline. The resumption of sinus rhythm and 1:1 AV nodal conduction was accompanied by a significant increase in systolic blood pressure from 74 ± 13 to 112 ± 6 mm Hg, and all signs and symptoms of hypoperfusion subsided. Patients who develop "early," as opposed to "late," AV nodal block following acute inferior myocardial infarction are less likely to respond to atropine, and hence, more likely to require temporary pacing, and to suffer a morbid or fatal event (Sclarvosky et al., 1984). Because of this poor prognosis and the ineffectiveness of the current medical therapy, alternative therapies, such as the use of A_1 adenosine receptor antagonists, may prove beneficial.

Another clinical condition that can be complicated by high-grade AV nodal block is moderate-to-severe cardiac transplant rejection (Haught et al., 1994). Acute cardiac transplant rejection involves myocyte injury and necrosis, which may lead to increased formation of adenosine. Furthermore, the transplanted (denervated) human heart has been shown to be "supersensitive" to the cardiac effects of adenosine (Ellenbogen et al., 1990). Therefore, it is possible that excessive adenosine production in this clinical situation may cause bradyarrhythmias. In support of this hypothesis, we recently reported a case of a patient with a transplanted heart who, during an episode of cardiac rejection, developed high-grade AV nodal block that was promptly converted to 1:1 AV nodal conduction and normal sinus rhythm after administration of theophylline (Haught et al., 1994). After the resolution of the cardiac transplant rejection, invasive electrophysiological testing revealed normal sinoatrial and atrioventricular nodal function.

Sinoatrial Node Dysfunction Excessive adenosine production by the ischemic myocardium may also affect sinoatrial nodal function. We have recently described two cases of sustained accelerated idioventricular rhythm, related to myocardial infarction, that were promptly converted to normal sinus rhythm after intravenous administration of theophylline (Bertolet, 1994). The most likely explanation for this observation is that endogenous adenosine released from the ischemic tissue led to sinus bradycardia and arrest, followed by an accelerated idioventricular rhythm that emerged as an escape rhythm. Theophylline, by reversing the negative chronotropic effect of adenosine on the sinoatrial node, allowed the resumption of normal sinus rhythm.

Sinoatrial node dysfunction frequently complicates the early postoperative period following cardiac transplantation (Miyamoto et al., 1990). Although incompletely understood, endogenously produced adenosine has been implicated in the pathogenesis of sinoatrial node dysfunction in the immediate post–cardiac transplant period (Heinz et al., 1993; Ellenbogen et al., 1988). Indirect but supportive evidence for this hypothesis includes findings that (i) the sinus nodal artery often has angiographic disease (which may lead to sinus node ischemia) in patients with sinoatrial node dysfunction, and (ii) transplanted hearts exhibit adenosine "supersensitivity" (Ellenbogen et al., 1990). Furthermore, acute cardiac allograft rejection may also contribute to an excess of adenosine formation, and hence, sinoatrial node dysfunction. Theophylline has been reported to reverse bradyarrhythmias following orthotopic heart transplantation. Heinz et al. (1993) showed that sinus node dysfunction occurring early following cardiac transplant could be corrected with aminophylline. In this study, the corrected sinus node recovery time, which was markedly prolonged (1770 msec) was significantly shortened to 440 milleseconds, a value that is within the normal range, after administration of aminophylline (Heinz et al., 1993). Theophylline was shown to restore normal sinus rhythm in patients with posttransplant bradyarrhythmias (Redmond et al., 1993; Bertolet et al., 1996), thereby reducing the need for permanent pacemaker implantation (Redmond et al., 1993). In keeping with these findings, sinus bradycardia associated with moderate to severe cardiac allograft rejection has been reported to be reversed by theophylline (Ellenbogen et al., 1988). Therefore, in the milieu of increased endogenous adenosine and clinically significant sinoatrial node dysfunction in the transplant heart, theophylline appears to be effective in restoring normal sinoatrial node function.

Less clear is the role of adenosine in the etiology of bradycardias (SA and AV nodal dysfunction) in the absence of myocardial injury, such as that found in patients with sick sinus syndrome. Patients with sick sinus syndrome may display a variety of dysrhythmias including sinus bradycardia, sinus pauses, and atrioventricular block. Frequently these patients require permanent pacemaker implantation. At least four independent studies have demonstrated that theophylline or aminophylline significantly increases sinus rate, and decreases the number and duration of sinus node pauses in patients with sick sinus syndrome (Alboni et al., 1991, 1993; Saito et al., 1993; Benditt et al., 1983). The results of these studies strongly suggest that adenosine receptor antagonists may be useful in the therapy of bradyarrhythmias, as in sick sinus syndrome.

Persistent Asystole During cardiac arrest, endogenously produced adenosine is expected to markedly increase in response to cellular hypoxia. Wesley and Belardinelli (1989), using a porcine model of ventricular fibrillation/defibrillation, observed that postdefibrillation bradyarrhythmias and hypotension (i.e., cardiovascular collapse) could be significantly prevented or reversed by BWA-1433U (a derivative of 1,3-dipropyl-8-phenylxanthine). In contrast, dipyridamole, an adenosine uptake blocker that potentiates the cardiovascular actions of adenosine, had the opposite effect. More recently, Lerman and Engelstein (1992) reported that adenosine, via its indirect anti-β-adrenergic actions, raises the defibrillation threshold during ventricular fibrillation, an effect that was reversed by the adenosine receptor blocker, 8-cyclopentyltheophylline. Viskin et al. (1993) applied these concepts clinically and demonstrated that aminophylline was efficacious in restoring a stable cardiac rhythm and blood pressure in patients with cardiac arrest refractory to atropine and epinephrine. Thus, adenosine antagonists appear to be effective in restoring postdefibrillation electrical and hemodynamic stability in situations of cardiac arrest, particularly in cases refractory to usual and standard care.

Tachyarrhythmias

Although the phenomenon is less well described, adenosine may also precipitate tachyarrhythmias. It is hypothesized that adenosine, by shortening the atrial action potential and refractory period, may facilitate the induction of atrial flutter or fibrillation (Kabel et al., 1994; DiMarco et al., 1985, 1990). Intravenously administered adenosine has been shown to induce atrial fibrillation in 2%–5% of patients (Kabell et al., 1994; DiMarco et al., 1985). In ischemic atrial tissue, a greater than threefold rise in the tissue adenosine levels has been reported (Thomas et al., 1975). Thus, it is conceivable that endogenous adenosine may also cause atrial tachyarrhythmias. In keeping with this, atrial flutter concomitant with the administration of dipyridamole, which is known to increase plasma levels of adenosine, has been reported in man (Lerman et al., 1989).

In the same study it was shown that the dipyridamole-induced atrial flutter was promptly converted to normal sinus rhythm following intravenous administration of the adenosine antagonist aminophylline (Lerman et al., 1989).

More recently, we studied two patients who developed atrial fibrillation within 2 hours of the onset of inferior myocardial infarction. Both were converted to normal sinus rhythm within 5 minutes of the administration of theophylline (unpublished data). This is the first study in man to demonstrate that atrial fibrillation, due presumably to elevated interstitial concentration of adenosine caused by myocardial ischemia, may be converted to sinus rhythm with an adenosine receptor antagonist. This interpretation is consistent with the observation that hypoxia shortens the atrial refractory period, an effect that is reversed by aminophylline (Szentimikosi et al., 1979).

In summary, endogenous adenosine may induce a variety of bradyarrhythmias, as well as tachyarrhythmias. These arrhythmias, which often are resistant to conventional therapy, appear to be responsive to the adenosine receptor antagonist theophylline. Adenosine receptor antagonists could be considered as an alternative therapeutic approach when standard frontline antiarrhythmic therapies fail. We hypothesize that more potent, specific, and selective A_1 adenosine receptor antagonists than theophylline would prove to be of clinical value in short- and long-term management of cardiac arrhythmias in conditions such as myocardial ischemia, sick sinus syndrome, cardiac arrest, cardiac transplant rejection, and following cardiac transplant, coronary artery bypass grafting, or aortic crossclamp.

CONCLUSIONS

The cardiac electrophysiological actions of adenosine described in this review are mediated by the A_1 adenosine receptor, inhibited by adenosine receptor antagonists, and potentiated by nucleoside uptake blockers and allosteric enhancers. Adenosine has important antiarrhythmic and proarrhythmic actions, and thus, agents that mimic or modulate the actions of this nucleoside may have therapeutic value (Table 2). The usefulness of adenosine for the acute management of supraventricular tachycardia and diagnosis of cardiac arrhythmias has been well established. However, the short half-life of adenosine limits its usefulness for the acute management of arrhythmias. A_1 receptor agonists with longer half-lives than adenosine have the potential to be effective and useful agents for chronic management of cardiac arrhythmias. Because the negative dromotropic effect of adenosine is frequency dependent, it is conceivable that at normal heart rates A_1 receptor agonists would cause minimal slowing of AV nodal conduction, whereas during tachycardia they should cause marked prolongation of AV nodal

TABLE 2 Adenosine-Related Antiarrhythmic Agents

	Type of Action	Agents
Tachyarrhythmias	Short-acting	Adenosine
	Long-acting	A_1-receptor agonists[a]
	Event-specific	Nucleoside uptake blockers
		Allosteric enhancers
Bradyarrhythmias	Long-acting	A_1-receptor antagonists

[a] A_1 receptor agonists could potentially act as event-specific antiarrhythmic agents (see text).

conduction and hence, terminate reentrant tachycardias involving the AV node and slow ventricular rate during atrial flutter or atrial fibrillation (Belardinelli et al., 1994). Thus, it is possible that A_1-receptor agonist–mediated slowing of AV nodal conduction would manifest itself only (or mainly) during tachycardia.

Because evidence now exists that endogenous adenosine also plays a role as an antiarrhythmic metabolite (Conti et al., 1995b), nucleoside uptake blockers and allosteric enhancers may prove useful as antiarrhythmic agents that have the added advantage of having effects only during tachycardia, i.e., act as event-specific antiarrhythmic agents (Kollias-Baker et al., 1994; Dennis et al., 1996). As to the bradyarrhythmias associated with conditions in which cardiac interstitial adenosine levels are elevated, potent and selective A_1 receptor antagonists may prove to be efficacious in restoring sinus rhythm. The results of clinical observations with the use of theophylline for treatment of bradyarrhythmias, although not uniformly positive, are encouraging. In summary, based on the evidence reviewed in this chapter, the rationale for the development of adenosine-related agents to treat cardiac arrhythmias is becoming increasingly strong.

REFERENCES

Alboni P, Paparella N, Cappato R, Pirani R, Yiannacopulu P, Antonioli GE (1993): Long-term effects of theophylline in atrial fibrillation with a slow ventricular response. Am J Cardiol 72:1142–1145.

Alboni P, Ratto B, Cappato R, Rossi P, Gatto E, Antonioli GE (1991): Clinical effects of oral theophylline in sick sinus syndrome. Am Heart J 122:1361–1367.

Belardinelli L, Lerman B (1990): Electrophysiological basis for the use of adenosine in the diagnosis and treatment of cardiac arrhythmias. Br Heart J 63:3–4.

Belardinelli L, Lerman BB (1991): Adenosine: Cardiac electrophysiology. PACE 14:1672–1680.

Belardinelli L, Pelleg A (1990): Cardiac electrophysiology and pharmacology of adenosine. J Cardiovasc Electrophysiol 1:327–339.

Belardinelli L, Shryock JC (1992): Does adenosine function as a retaliatory metabolite in the heart? NIPS 7:52–56.

Belardinelli L, Giles WR, West A (1988): Ionic mechanisms of adenosine actions in pacemaker cells from rabbit heart. J Physiol (London) 405:615–633.

Belardinelli L, Linden J, Berne RM (1989): The cardiac effects of adenosine. Prog Cardiovasc Dis 32:73–97.

Belardinelli L, Lu J, Dennis D, Martens J, Shryock JC (1994): The cardiac effects of a novel A_1-adenosine receptor agonist in guinea pig isolated heart. J Pharmacol Exp Ther 271:1371–1382.

Belardinelli L, Shryock JC, Song Y, Wang D, Srinivas M (1995): Ionic basis of the electrophysiological actions of adenosine on cardiomyocytes. FASEB J 9:359–365.

Belardinelli L, West GA, Clemo SHF (1987): Regulation of atrioventricular node function. In Gerlach E, Becker BF (eds): "Topics and Perspectives in Adenosine Research." Berlin: Springer-Verlag, pp 344–355.

Belardinelli L, West GA, Crampton R, Berne RM (1983): Chronotropic and dromotropic effects of adenosine. In Berne RM, Rall TW, Rubio R (eds): "The Regulatory Function of Adenosine." The Hague: Martinus Nijhoff pp 377–398.

Benditt DG, Benson W Jr, Kreitt J, Dunnigan A, Pritzker MR, Crouse L, Scheinman MM (1983): Electrophysiologic effects of theophylline in young patients with recurrent symptomatic bradyarrhythmias. Am J Cardiol 52:1223–1229.

Bertolet BD, Belardinelli L, Kerensky R, Hill JA (1994): Adenosine blockade as primary therapy for ischemia-induced accelerated idioventricular rhythm: Rationale and potential clinical application. Am Heart J 123:185–188.

Bertolet BD, Eagle DA, Conti JB, Mills, RM Jr, Belardinelli L (1996): Bradycardia after heart transplantation: Reversal with theophylline. J Am Cell Cardiol 28:396–399.

Bertolet BD, McMurtrie EB, Hill JA, Belardinelli L (1995): Treatment of atrioventricular block after myocardial infarction with theophylline. Ann Int Med 123:509–511.

Biaggioni I, Olafsson B, Robertson RM, Hollister AS, Robertson D (1987): Cardiovascular and respiratory effects of adenosine in conscious man. Evidence for chemoreceptor activation. Circ Res 61:779–786.

Camm AJ, Garratt CJ (1991): Adenosine and supraventricular tachycardia. N Engl J Med 325:1621–1629.

Clemo HF, Belardinelli L (1986): Effect of adenosine on atrioventricular conduction, I. Site and characterization of adenosine action in the guinea pig atrioventricular node. Circ Res 59:427–436.

Conradson TBG, Clarke B, Dixon CMS, Dalton RN, Barnes PJ (1987): Effects of adenosine on autonomic control of heart rate in man. Acta Physiol Scand 131:525–531.

Conti JB, Belardinelli L, Curtis AB (1995a): Usefulness of adenosine in diagnosis of tachyarrhythmias. Am J Cardiol 75:952–955.

Conti JB, Belardinelli L, Utterback DB, Curtis AB (1995b): Endogenous adenosine is an antiarrhythmic agent. Circulation 91:1761–1767.

Dennis DM, Pekka Reatikainen MJ, Martens JR, Belardinelli L (1996): Modulation of atrioventricular nodal function by metabolic and allosteric regulators of endogenous adenosine in guinea pig heart. Circulation 94:2551–2559.

DiMarco JP, Miles W, Akhtar M, Milstein S, Sharma AD, Platia E, McGovern B, Scheinman MM, Govier WC, and the Adenosine for PSVT Study Group (1990): Adenosine for paroxysmal supraventricular tachycardia: Dose ranging and comparison with verapamil. Ann Int Med 113:104–110.

DiMarco JP, Sellers TD, Berne RM, West GA, Belardinelli L (1983): Adenosine: Electrophysiologic effects and therapeutic use for terminating paroxysmal supraventricular tachycardia. Circulation 68:1254–1263.

DiMarco JP, Sellers TD, Lerman BB, Greenberg ML, Berne RM, Belardinelli L (1985): Diagnostic and therapeutic use of adenosine in patients with supraventricular tachyarrhythmias. J Am Coll Cardiol 6:417–425.

Drury AN, Szent-Gyorgyi A (1929): The physiological activity of adenine compounds with especial reference to their action upon the mammalian heart. J Physiol (London) 68:213–237.

Ellenbogen KA, Szentpetery S, Katz MR (1988): Reversibility of prolonged chronotropic dysfunction with theophylline following orthotopic cardiac transplantation. Am Heart J 116:202–206.

Ellenbogen KA, Thames MD, DiMarco JP, Sheehan H, Lerman BB (1990): Electrophysiological effects of adenosine in the transplanted human heart: Evidence of supersensitivity. Circulation 81:821–828.

Engelstein E, Wilber D, Wadas M, Stein K, Lippman N, Lerman B (1994a): Limitations of adenosine in assessing the efficacy of radiofrequency catheter ablation of accessory pathways. Am J Cardiol 73:774–779.

Engelstein ED, Lippman N, Stein KM, Lerman BB (1994b): Mechanism-specific effects of adenosine on atrial tachycardia. Circulation 89:2645–2654.

Exner DV, Muzyka T, Gillis AM (1995): Proarrhythmia in patients with the Wolff-Parkinson-White syndrome after standard doses of intravenous adenosine. Ann Int Med 122:351–352.

Favale S, DiBaise M, Rizzo U, Belardinelli L, Rizzo P (1985): Effect of adenosine and adenosine-5′-triphosphate on atrioventricular conduction in patients. J Am Coll Cardiol 5:1212–1219.

Garratt CJ, Griffith MJ, O'Nunain S, Ward DE, Camm AJ (1991): Effects of intravenous adenosine on antegrade refractoriness of accessory atrioventricular connections. Circulation 84:1962–1968.

Gausche M, Persse DE, Sugarman T, Shea SR, Palmer GL, Lewis RJ, Brueske PJ, Mahadevan S, Melio FR, Kuwata JH, Niemann JT (1994): Adenosine for the posthospital treatment of paroxysmal supraventricular tachycardia. Ann Emerg Med 24:2:183–189.

Griffith MJ, Garratt CJ, Ward DE, Camm AJ (1989): The effects of adenosine on sinus node reentrant tachycardia. Clin Cardiol 12:409–411.

Griffith MJ, Linker NJ, Garratt CJ, Ward DE, Camm AJ (1990): Relative efficacy and safety of intravenous drugs for termination of sustained ventricular tachycardia. Lancet 336:670–673.

Haines DE, DiMarco JP 1990: Sustained intraatrial reentrant tachycardia: Clinical, electrocardiographic and electrophysiologic characteristics and long-term follow-up. J Am Coll Cardiol 15:1345–1354.

Haught WH, Bertolet BD, Conti JB, Curtis AB, Mills RM Jr (1994): Theophylline reverses high-grade atrioventricular block resulting from cardiac transplant rejection. Am Heart J 128:1255–1257.

Heinz G, Kratochwill C, Buxbaum P, Laufer G, Kreiner G, Siostrzonek P, Gasic S, Derfler K, Gossinger H (1993): Immediate normalization of profound sinus node dysfunction by aminophylline after cardiac transplantation. Am J Cardiol 71:346–349.

Isenberg G, Belardinelli L (1984): Ionic basis for the antagonism between adenosine and isoproterenol on isolated mammalian ventricular myocytes. Circ Res 55:309–325.

Jenkins JR, Belardinelli L (1988): Atrioventricular nodal accommodation in isolated guinea pig hearts: Physiological significance and role of adenosine. Circ Res 63:97–116.

Kabell G, Buchanan LV, Gibson JK, Belardinelli L (1994): Effects of adenosine on atrial refractoriness and arrhythmias. Cardiovasc Res 28:1385–1389.

Keim S, Curtis AB, Belardinelli L, Epstein ML, Staples ED, Lerman BB (1992): Adenosine-induced atrioventricular block: A rapid and reliable method to assess surgical and radiofrequency catheter ablation of accessory atrioventricular pathways. J Am Coll Cardiol 19:1005–1012.

Kollias-Baker C, Ruble J, Dennis D, Bruns RF, Linden J, Belardinelli L (1994): Allosteric enhancer PD 81,723 acts by novel mechanism to potentiate cardiac actions of adenosine. Circ Res 75:961–971.

Lai WT, Lee CS, Wu SN (1994): Rate-dependent properties of adenosine-induced negative dromotropism in humans. Circulation 90:1832–1839.

Lerman BB (1993): Response of nonreentrant catecholamine-mediated ventricular tachycardia to endogenous adenosine and acetylcholine. Evidence for myocardial receptor-mediated effects. Circulation 87:382–390.

Lerman BB, Belardinelli L (1991): Cardiac electrophysiology of adenosine: Basic and clinical concepts. Circulation 83:1499–1509.

Lerman BB, Engelstein ED (1995): Metabolic Determinants of Defibrillation: Role of Adenosine 91:838–844.

Lerman BB, Belardinelli L, West GA, Berne RM, DiMarco JP (1986): Adenosine-sensitive ventricular tachycardia: evidence suggesting cyclic AMP-mediated triggered activity. Circulation 74:270–280.

Lerman BB, Wesley RC, Belardinelli L (1989): Electrophysiologic effects of dipyridamole on atrioventricular nodal conduction and supraventricular tachycardia: role of endogenous adenosine. Circulation 80:1536–1543.

Lerman BB, Wesley RC, DiMarco JP, Haines DE, Belardinelli L (1988): Antiadrenergic effects of adenosine on His-Purkinje automaticity. J Clin Invest 82:2127–2135.

Miyamoto Y, Curtiss EI, Kormos RL, Armitage JM, Hardesty RL, Griffith BP (1990): Bradyarrhythmia after heart transplantation: Incidence, time course, and outcome. Circulation 82:IV-313–IV-317.

Nayebpour M, Billette J, Amellal F, Nattel S (1993): Effects of adenosine on rate-dependent atrioventricular nodal function: Potential roles in tachycardia termination and physiological regulation. Circulation 88:2632–2645.

Olsson RA, Pearson JD (1990): Cardiovascular purinoceptors. Physiol Rev 70:761–845.

O'Nunain SO, Garratt C, Paul V, Debbas N, Ward DE, Camm AJ (1992): Effect of intravenous adenosine on human atrial and ventricular repolarization. Cardiovasc Res 26:939–943.

Qu Y, Campbell DL, Strauss HC (1993): Modulation of L-type Ca^{2+} current by extracellular ATP in ferret isolated right ventricular myocytes. J Physiol (Lond) 471:295–317.

Rankin AC, Oldroyd KG, Chong E, Rae AP, Cobbe SM (1989): Value and limitations of adenosine in the diagnosis and treatment of narrow and broad complex tachycardia. Br Heart J 162:195–203.

Redmond JM, Zehr KJ, Gillinov MA, Baughman KL, Augustine SM, Cameron DE, Stuart RS, Acker MA, Gardner TJ, Reitz BA, Baumgartner WA (1993): Use of theophylline for treatment of prolonged sinus node dysfunction in human orthotopic heart transplantation. J Heart Lung Transplant 12:133–139.

Reed R, Falk JL, O'Brien J (1991): Untoward reaction to adenosine therapy for supraventricular tachycardia. Am J Emerg Med 9:566–570.

Rotman M, Wagner GS, Wallace AG (1972): Bradyarrhythmias in acute myocardial infarction. Circulation 45:703–722.

Saito D, Matsubara K, Yamanari H, Obayashi N, Uchida S, Maekawa K, Sato T, Mizuo K, Kobayashi H, Haraoka S (1993): Effects of oral theophylline on sick sinus syndrome. J Am Coll Cardiol 21:1199–1204.

Sclarovsky S, Strasberg B, Hirschberg A, Arditi A, Lewin RF, Agmon J (1984): Advanced early and late atrioventricular block in acute inferior wall myocardial infarction. Am Heart J 108:19–24.

Shah PK, Nalos P, Peter T (1986): Atropine resistant post infarction complete AV block: Possible role of adenosine and improvement with aminophylline. Am Heart J 113:194–195.

Shenasa H, Cooper RA, Pressley J, Sorrentino RA, Greenfield RA, Merrill JJ, Wharton JM (1994): Site of origin of ectopic atrial tachycardia predicts its response to adenosine. J Am Coll Cardiol 23:250A. Abstract.

Song Y, Shryock J, Belardinelli L (1995): Modulation of cardiomyocyte membrane currents by A_1 adenosine receptors. In Belardinelli L, Pelleg A (eds): "Adenosine and Adenine Nucleotides: From Molecular Biology to Integrative Physiology," Norwell, MA: Kluwer Academic.

Song Y, Thedford S, Lerman BB, Belardinelli L (1992): Adenosine-sensitive after depolarizations and triggered activity in guinea pig ventricular myocytes. Circ Res 70:743–753.

Stark G, Sterz F, Stark U, Bachernegg M, Decrinis M, Tritthart HA (1993): Frequency-dependent effects of adenosine and verapamil on atrioventricular conduction of isolated guinea pig hearts. J Cardiovasc Pharmacol 21:955–959.

Strasberg B, Bassevich R, Mager A, Kusniec J, Sagie A, Sclarovsky S (1991): Effects of aminophylline on atrioventricular conduction in patients with late atrioventricular block during inferior wall acute myocardial infarction. Am J Cardiol 67:527–528.

Szentmiklosi AJ, Nemeth M, Szegi J, Papp JG, Szekeres L (1979): On the possible role of adenosine in the hypoxia-induced alterations of the electrical and mechanical activity of the atrial myocardium. Arch Int Pharmacodyn 238:283–295.

Thomas RA, Rubio R, Berne RM (1975): Comparison of the adenine nucleotide metabolism of dog atrial and ventricular myocardium. J Mol Cell Cardiol 7:115–123.

Visentin S, Wu S-N, Belardinelli L (1990): Adenosine-induced changes in atrial action potential: Contribution of Ca and K currents. Am J Physiol 258:H1070–H1078.

Viskin S, Belhassen B, Roth A, Reicher M, Averbuch M, Sheps D, Shalabye E, Laniado S (1993): Aminophylline for bradyasystolic cardiac arrest refractory to atropine and epinephrine. Ann Intern Med 118:279–281.

Watt AH, Routledge PA (1986): Transient bradycardia and subsequent sinus tachycardia produced by intravenous adenosine in healthy adult subjects. Br J Clin Pharmacol 21:533–536.

Wesley RC Jr, Belardinelli L (1989): Role of endogenous adenosine in postdefibrillation bradyarrhythmia and hemodynamic depression. Circulation 80:128–137.

Wesley RC Jr, Turnquest P (1992): Torsades de pointe after intravenous adenosine in the presence of prolonged QT syndrome. Am Heart J 123:794–796.

Wesley RC Jr, Lerman BB, DiMarco JP, Berne RM, Belardinelli L (1986): Mechanism of atropine-resistant atrioventricular block during inferior myocardial infarction: Possible role of adenosine. J Am Coll Cardiol 8:1232–1234.

CHAPTER 11

The Role of ADP in Thrombosis and the Therapeutic Potential of P_{2T}-Receptor Antagonists as Novel Antithrombotic Agents

PAUL LEFF, MARK J. ROBERTSON, and ROBERT G. HUMPHRIES

Department of Pharmacology, Astra Charnwood, Loughborough, Leicestershire LE11 5RH, United Kingdom

INTRODUCTION

The Role of Platelets in Thrombosis

Coronary vascular disease and its thrombotic sequelae are the most important causes of mortality and morbidity in the developed world (World Health Organization, 1991). Appreciation of the pivotal role of platelet adhesion and aggregation as initiating events in arterial thrombosis (Davies and Thomas, 1985; Fuster et al., 1992) and the pharmacological and molecular mechanisms involved has had two main outcomes. First, recognition and acceptance of aspirin as an effective antithrombotic agent (Antiplatelet Trialists' Collaboration, 1994), and secondly, provision of a mechanistic rationale supporting the design and development of more effective inhibitors of platelet function. Much recent research activity has focused upon identification of inhibitors of the final common pathway of platelet aggregation (Figure 1), the cross-linking of platelets following binding of dimeric fibrinogen to the platelet membrane glycoprotein complex GPIIb/IIIa (Cook et al., 1994; Weller et al., 1994). Mechanistically, such agents would be anticipated to offer superior antithrombotic properties to those of aspirin, which only inhibits thromboxane-mediated amplification of platelet activation. This theoretical consideration is not only supported by results from animal models but has been confirmed in a convincing manner by clinical experience with the GPIIb/IIIa antibody, CentoRx in "high risk" percutaneous transluminal coronary angioplasty (PTCA) patients (EPIC Investigators, 1994). However, the full potential of this class of drugs

Purinergic Approaches in Experimental Therapeutics, Edited by Kenneth A. Jacobson and Michael F. Jarvis
ISBN 0-471-14071-6 © 1997 Wiley-Liss, Inc.

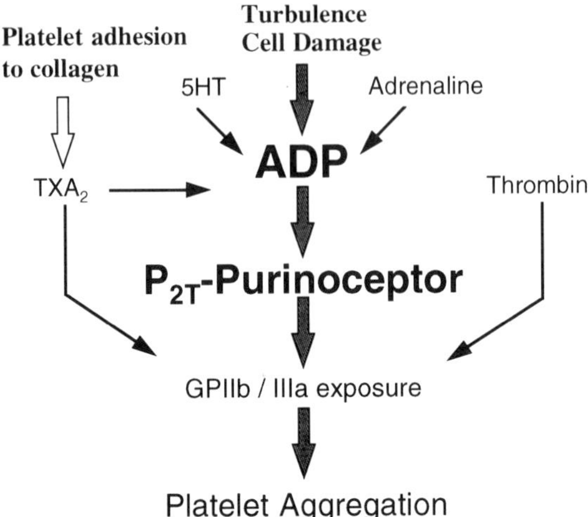

FIGURE 1. Key pathways leading to GPIIb/IIIa exposure and platelet aggregation.

may never be fully realized in the wider coronary artery disease population since, by blocking the final step of platelet aggregation, the beneficial antithrombotic effects are, in part, countered by an increased risk of major bleeding, necessitating blood or platelet transfusion. Thus, it would appear that the ideal "benefit/risk" ratio (i.e., improved efficacy over that of aspirin compared to the propensity to cause unwanted bleeding) has not been achieved.

A Role For Adenosine Diphosphate?

Clearly, the ideal antiplatelet agent would be highly effective in preventing intravascular thrombosis within the environment of a diseased coronary artery, while preserving sufficient platelet function to maintain adequate haemostasis. This might be best achieved by identifying and inhibiting a key initiating factor. Identification of agents such as thromboxane A_2 (TxA_2) and 5-hydroxytryptamine (5-HT) and their proposed roles in platelet aggregation and thrombosis have been relatively recent events. However, availability of potent, selective inhibitors or antagonists of these pathways has made preclinical and clinical hypothesis testing possible. Unfortunately, experience with inhibitors (TxA_2 synthase inhibitors) or antagonists (TxA_2 antagonists, 5-HT antagonists) of these mechanisms has, in general, proved disappointing (Lam et al., 1986; Ridogrel Versus Aspirin Patency Trial, 1994). Against this background, it is ironic that the importance of adenosine diphosphate (ADP), the first mediator postulated to be involved in thrombosis and haemostasis (Gaarder et al., 1961) and shown to produce platelet aggregation *in vitro* (Born, 1962), has remained inadequately explored due, primarily, to the lack of suitable pharmacological probes. ADP is contained at high concentrations (~1 M) in the dense granules of platelets (Ugerbil and Holmsen, 1981), is released, along with 5-HT, upon platelet activation, and achieves high concentrations (~0.1 μM) at sites of vascular damage in animal experiments and in man (Born and Kratzer, 1984). A number of studies have sought to determine whether released ADP plays a key role in aggregation responses initiated by other agents *in vitro*. However, results have been equivocal, with a pivotal role for ADP either invoked (Morinelli et al., 1983) or discounted (Huang and Detwiler, 1980) by removal of ADP from the system using the ADP scavenging enzyme, apyrase, or by conversion of any released ADP to adenosine triphosphate (ATP) by the regenerating system creatine phosphate/creatine phosphokinase. Furthermore, the potential for nonspecific effects of such enzyme systems (Nunn and Chamberlain, 1983) clearly makes them unsuitable for experiments to definitively test the hypotheses *in vitro*. Similarly, results obtained using ATP-regenerating or ADP-scavenging enzymes to invoke a role for ADP in hemostasis (Zawilska and Born, 1982) or thrombosis (Yao et al., 1992; Zawilska et al., 1975) *in vivo* are open to misinterpretation. A potent and selective antagonist of the activation of platelets by ADP would be far more desirable for this purpose.

P_{2T}-Receptor Antagonists

Macfarlane and Mills (1975) were the first to report that ATP behaves as a competitive antagonist of ADP in terms of platelet aggregation. Subsequently, this unique pharmacological profile—ADP as agonist, ATP as competitive antagonist—was used to define the platelet ADP receptor as the P_{2T} subtype of purinoceptor (Gordon, 1986), which is found only on platelets, megakaryocytes, and certain megakaryocyte-like cell lines. However, ATP is not a suitable tool for hypothesis testing *in vivo*. The nucleotide is, by definition, a nonselective P_2-receptor ligand; it has low potency and is metabolically unstable. Furthermore, during the same period, it was established that certain structural analogues of ATP can inhibit ADP-induced platelet aggregation. Notably, some of the analogs made contained substituents in the 2-position of the adenine ring (Cusack and Hourani, 1982a, 1982b) and a β,γ-methylene substituent in the triphosphate group (Hourani et al., 1986), the latter conferring relative stability to enzymatic degradation (Welford et al., 1986). However, the gain in antagonist potency over ATP (pA_2 4.6) was only modest (e.g., for 2-chloro-ATP, $pA_2 = 5.2$), and due to a lack of selectivity for the P_{2T}-receptor, those compounds, like ATP, were not suitable for use *in vivo*. Nevertheless, these observations provided the starting point for medicinal chemical exploration and some optimism that progress could be made. Indeed, a crucial early discovery in the chemical program that we undertook was that extended alkylthio substitutions at the 2-position conferred quite remarkable increases in receptor affinity, as exemplified by the prototype molecules of this series, ARL (formerly FPL) 66096 (Humphries et al., 1994a) and ARL 67085 (Humphries et al., 1995), the structures of which are shown in Figure 2.

NOVEL P_{2T}-RECEPTOR ANTAGONISTS

Pharmacological Profile *In Vitro*

In addition to possessing P_{2T}-receptor affinity in the low nanomolar range, ARL 66096 and ARL 67085 (pK_B 8.7 and 8.9, respectively) also showed unprecedented selectivity (at least 3000-fold over other P_2-receptors) *in vitro*. In addition, inclusion of a β,γ-dihalomethylene substituent meant that the compounds were substantially resistant to hydrolysis by ectonucleotidases, thus preventing breakdown of the antagonists to potential agonists (the corresponding ADP analogs). This was confirmed by the observation that the antiaggregatory potency of ARL 67085 in rat blood *in vitro* did not change significantly

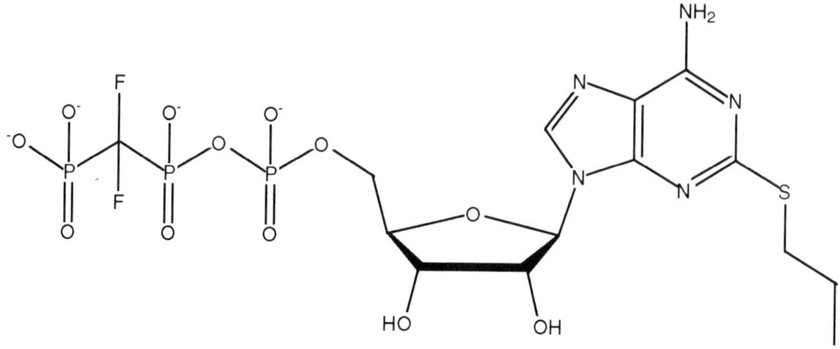

ARL 66096

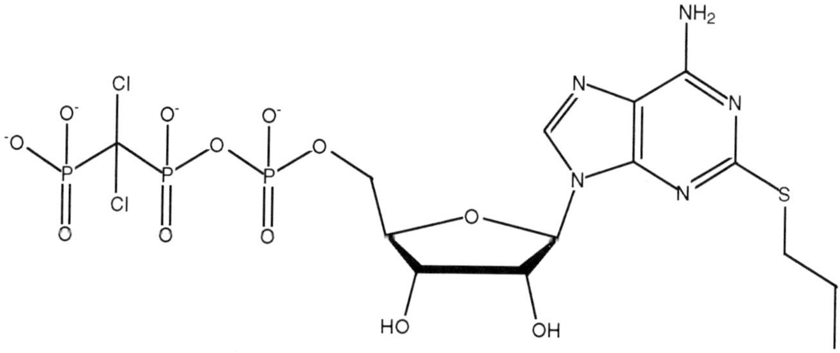

ARL 67085

FIGURE 2. Chemical structures of ARL 66096 and ARL 67085.

when the incubation time was increased from 2 to 30 minutes. This contrasts with the rapid breakdown of ATP under similar conditions: $t_{1/2}$ 10 minutes (Trams et al., 1980). These properties make ARL 66096 and ARL 67085 powerful pharmacological tools. As such, their utility in the first instance has proved to be in receptor classification, with selective antagonism by ARL 66096, with a pA_2 of 8.7, now accepted as an additional criterion defining the P_{2T}-receptor (Fredholm et al., 1994). More importantly from the present perspective, we had, in ARL 67085, discovered an agent with which the role of ADP in thrombosis could be confidently evaluated.

Hypothesis Testing *In Vivo*

We and others have demonstrated the capacity of exogenous ADP to induce platelet aggregation *in vivo* in animal studies (Humphries et al., 1990; Page et al., 1982).

However, this kind of experiment clearly does not provide evidence of a role for endogenous ADP in thrombosis. Hypothesis testing requires that the pathophysiological event of interest, in this case arterial thrombosis, can be inhibited or prevented by a selective antagonist of the agonist under investigation. To this end, we chose to evaluate the effects of ARL 67085 in a modification of a model first described by Folts (Folts et al., 1976), in which dynamic arterial thrombosis is visualized as cyclic reductions in blood flow (CFR) in a damaged, stenosed femoral artery in the anesthetized dog. The record obtained from our first experience with ARL 67085 in this model is reproduced in Figure 3 and clearly shows that intravenous (IV) infusion of this highly selective P_{2T}-receptor antagonist results in extremely potent inhibition of arterial thrombosis. In the group of experiments from which this example is taken (Humphries et al., 1994b), ARL 67085 was devoid of effects on blood pressure, heart rate, circulating cell counts, fibrinogen concentration and coagulation indices (activated partial thromboplastin time, prothrombin time) at doses up to 70-times that which abolished thrombosis (geometric mean 91 ng $\cdot$ kg^{-1} $\cdot$ min^{-1}). Predictably, the antithrombotic effect was accompanied by abolition of ADP-induced platelet aggregation measured *ex vivo,* but only a modest effect on hemostasis, as assessed by prolongation of bleeding time, was observed. These observations confirm the high degree of selectivity observed *in vitro* and provide unequivocal support for a key role for ADP-induced platelet aggregation in this model of arterial thrombosis.

Although it is stabilized against enzymatic cleavage between the β and γ positions in the triphosphate chain, ARL 67085 remains highly susceptible to metabolic cleavage at the α,β-anhydride link, which is followed by subsequent breakdown to the nucleoside. These metabolic processes provide an extremely rapid inactivation step for ARL 67085 resulting, as illustrated in Figure 3, in an "ultra-short" duration of action. Thus, although increasing the dose of ARL 67085 above that which consistently abolished thrombosis resulted in dose-related prolongation of bleeding time, this effect was fully reversed within 5 minutes of stopping the infusion (Figure 4). We perceived this pharmacokinetic profile to be of potential clinical benefit in the prevention of acute, platelet-mediated occlusion or reocclusion.

Comparative Studies

As mentioned above, a number of novel mechanistic classes of drug, originally perceived as offering an improvement in antithrombotic therapy, have failed to demonstrate adequate efficacy in the clinic. In terms of antiplatelet therapy, adequate efficacy

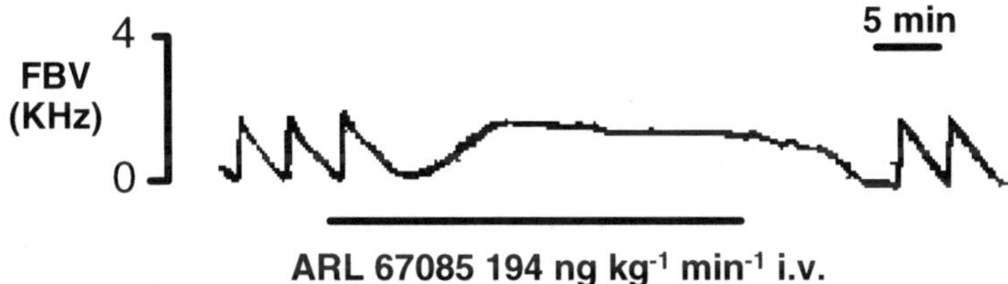

FIGURE 3. Inhibition of dynamic arterial thrombosis, visualized as cyclic reductions in blood flow in the damaged, stenosed femoral artery of the anesthetized dog, during intravenous infusion of the P_{2T}-receptor antagonist, ARL 67085.

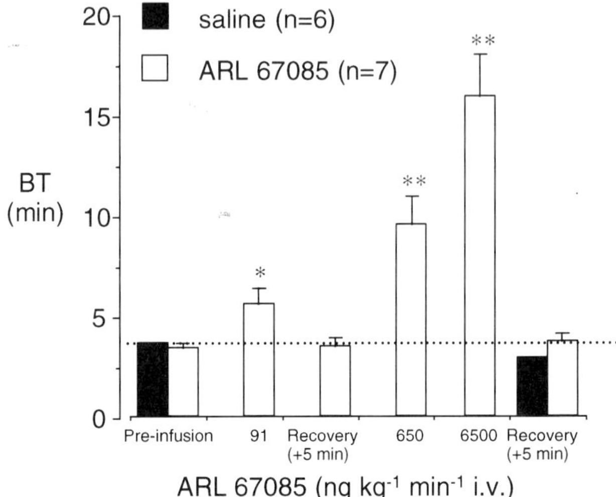

FIGURE 4. Intravenous infusion of ARL 67085 produces dose-related but rapidly reversible prolongation of bleeding time (BT) in the anesthetized dog. $*P < 0.05$ and $**P < 0.01$ *cf* preinfusion value (mean ± S.E., $n = 7$).

can be defined as demonstration of a significant improvement in outcome when the novel agent is added to the standard treatment regimen. Thus, in most cases, novel agents will be introduced as adjuncts to existing therapy and clinical trial design will reflect this. For example, standard therapy for unstable angina will include the use of aspirin, heparin, and a nitrate. It is, therefore, important to know from preclinical studies that novel antiplatelet agents not only have a superior antithrombotic efficacy compared to aspirin, but also that the combination of the novel agent with aspirin offers a significant improvement over aspirin alone. It is also important to know whether the novel mechanism of interest can offer any advantage over other novel agents in development, for instance, GPIIb/IIIa antagonists. Thus, having made the initial observation that P_{2T}-receptor antagonism results in potent and specific inhibition of arterial thrombosis *in vivo,* we embarked on a series of comparative studies. These involved comparison of the antithrombotic and antihemostatic profile of ARL 67085 with that of aspirin and two GPIIb/IIIA antagonists, Ro 449883 (Alig et al., 1992) and GR 144053 (Foster et al., 1993), and comparison of the antiaggregatory profile of ARL 67085 with that of ticlopidine (Defryn et al., 1989). The latter compound is a specific inhibitor of ADP-induced platelet aggregation, but like its close analog clopidogrel, is believed to exert its effect by interfering with the transduction pathway of P_{2T}-receptor stimulation rather than by acting as an antagonist at the receptor (Gachet et al., 1992).

Comparison with GPIIb/IIIa Antagonists In these experiments, full dose–response curves were obtained for the effects of either ARL 67085 (6.5–6500 ng · kg^{-1} · min^{-1}, IV, $n = 5$) or the GPIIb/IIIa antagonists, Ro 449883 (47–14000 ng · kg^{-1} · min^{-1}, IV, $n = 5$) and GR 144053 (30–10000 ng · kg^{-1} · min^{-1}, IV, $n = 5$) on dynamic arterial thrombosis (CFR), bleeding time, and ADP-induced platelet aggregation measured *ex vivo* in whole blood from pentobarbitone-anesthetized dogs (Robertson et al., 1994). ARL 67085 (Figure 5a), Ro 449883 (Figure 5b), and GR 144053 (Figure 5c) produced dose-related prolongation of bleeding time, and inhibition of platelet aggregation and CFR. In each case, abolition of thrombosis was associated with 100% inhibition of ADP-induced platelet aggregation and, for ARL 67085, a two-fold increase in bleeding time. Corresponding increases in bleeding time (BT) with Ro 449883 and GR 144053 were 6- and 5-fold, respectively. The dose ratio (CFR ID_{50} : BT 3.5-fold prolongation) for ARL 67085 (28) was significantly ($P < 0.01$) greater than that obtained for either Ro 449883 (2.6) or GR 144053 (1.5). Thus, for a given antithrombotic effect, a P_{2T}-receptor antagonist produces considerably less extension of bleeding time than two GPIIb/IIIa antagonists.

Comparisons with Aspirin The antithrombotic effect of ARL 67085 was compared with that of aspirin in three animal thrombosis models. First, in the experiments mentioned in the previous section, infusion of the P_{2T} antagonist abolished thrombosis in all cases (5/5), and this effect was associated with abolition of ADP-induced platelet aggregation measured *ex vivo*. However, a dosing regimen of aspirin (150 mg · day^{-1} orally, for 7 days) that abolished arachidonic acid–induced platelet aggregation, failed to abolish thrombosis in 6 out of a group of 9 dogs studied. Four of these aspirin-resistant animals subsequently received an infusion of ARL 67085, and CFR were abolished in all cases. Thus, combination of a P_{2T} antagonist with the standard antiplatelet drug aspirin resulted in an increased antithrombotic efficacy compared to the use of aspirin alone. Importantly, the ability of ARL 67085 to increase bleeding time was not enhanced in the presence of aspirin, indicating that the combination of a P_{2T} antagonist and a cyclooxygenase inhibitor does not result in an excessive compromise of normal haemostasis.

In the second model, electrical damage (300 μA) was applied to the stenosed (50% ↓ blood flow) carotid artery of chloralose/urethane-anesthetized male NZW rabbits (Nicol et al., 1995). Carotid artery blood flow velocity was monitored using a Doppler probe and the propensity for thrombosis was assessed by measurement of the time to thrombotic occlusion (zero blood flow) of the artery. The model was established in groups of 10 animals following either oral (po) pretreatment with aspirin (10 mg · kg^{-1} for days), or during infusion of ARL 67085 (3 μg · kg^{-1} · min^{-1}, IV) or vehicle (0.1 ml · min^{-1}, IV). Effectiveness of aspirin and ARL 67085 was confirmed by abolition of arachidonic acid– or ADP-induced platelet aggregation, respectively, measured *ex vivo*. Results are summarized in Figure 6a. ARL 67085 significantly increased carotid artery occlusion time (43 ± 10 min) compared to control (14 ± 4 min), while aspirin (23 ± 6 min) had no significant effect.

A similar occlusion time endpoint was used in a dog model in which a 2.5-cm length of prosthetic arterial graft (woven Dacron, i.d. 4 mm) was inserted in a femoral arterial loop in pentobarbitone-anesthetized male beagles (Nicol et al., 1995). In this case, occlusion times were assessed in groups of five dogs following either pretreatment with aspirin (150 mg · day^{-1} orally, for 7 days), or during infusion of ARL 67085 (0.2 μg kg^{-1} min^{-1}, IV) or vehicle (0.1 ml min^{-1}, IV). Results of this study are presented in Figure 6b. Dacron graft occlusion time in control animals (93 ± 53 min) was not modified by pretreatment with aspirin (78 ± 20 min) while, during infusion of ARL 67085, occlusion time was significantly extended (268 ± 32 min), with no occlusion seen within the 5-hour protocol in 4 of 5 dogs.

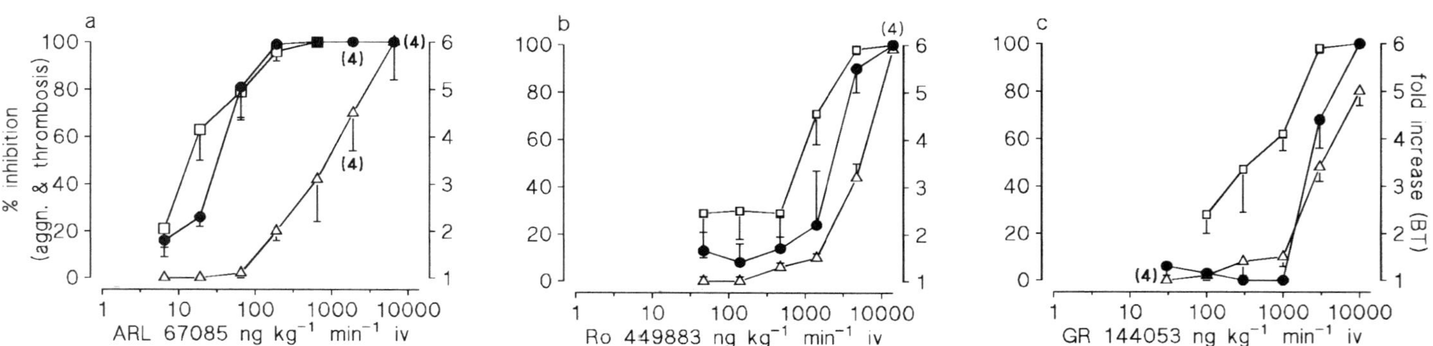

FIGURE 5. Comparison of the effects of IV infusion of (**a**) the P_{2T}-receptor antagonist, ARL 67085, or the GPIIb/IIIa antagonists (**b**) Ro 449883 and (**c**) GR 144053 on dynamic arterial thrombosis (●), ADP-induced platelet aggregation *ex vivo* (□) and bleeding time (△) in the anesthetized dog (mean ± S.E., n = 5, except where indicated)

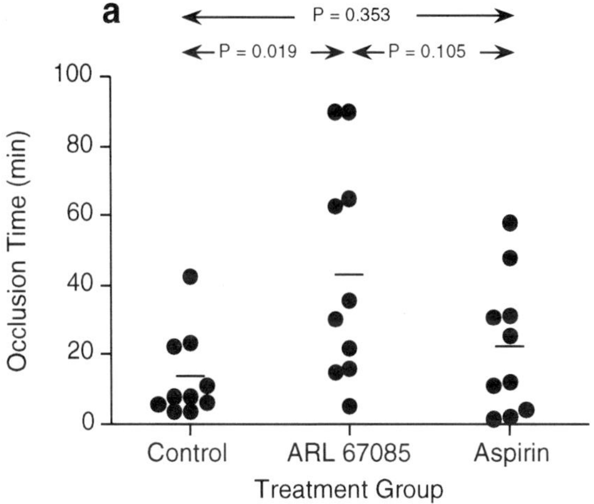

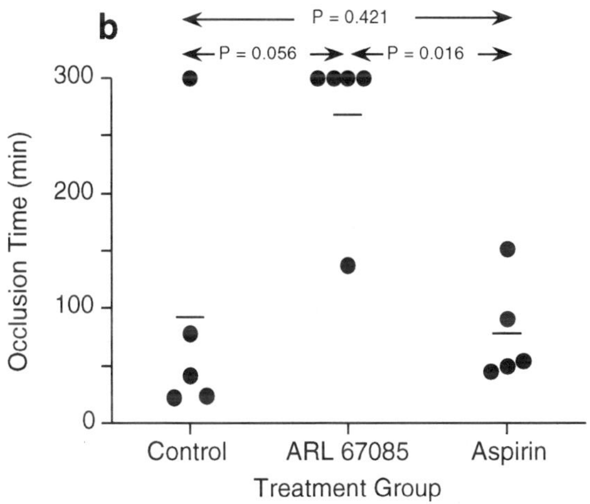

FIGURE 6. Comparison of the effect of IV infusion of ARL 67085 or oral pretreatment with aspirin on thrombotic occlusion of (**a**) the stenosed, electrically damaged carotid artery in the anesthetized rabbit or (**b**) Dacron graft segments inserted in extracorporeal femoral artery loops in the anesthetized dog.

Comparison with Ticlopidine When administered by intravenous infusion in urethane-anesthetized rats, ARL 67085 (1–100 $\mu g \cdot kg^{-1} \cdot min^{-1}$, $n = 5$–6) produced marked dose-related rightward displacement of concentration–effect curves for ADP-induced platelet aggregation measured *ex vivo* in whole blood (Figure 7) (Clegg et al., 1995). Infusion of the highest dose of ARL 67085 resulted in an approximately 500-fold increase in the concentration of ADP required to elicit an aggregation response 50% of maximal. In contrast, 5 days oral pretreatment with ticlopidine, at a maximally tolerated dose (200 $mg \cdot kg^{-1} \cdot day^{-1}$, $n = 6$), produced only a 1.6-fold rightward

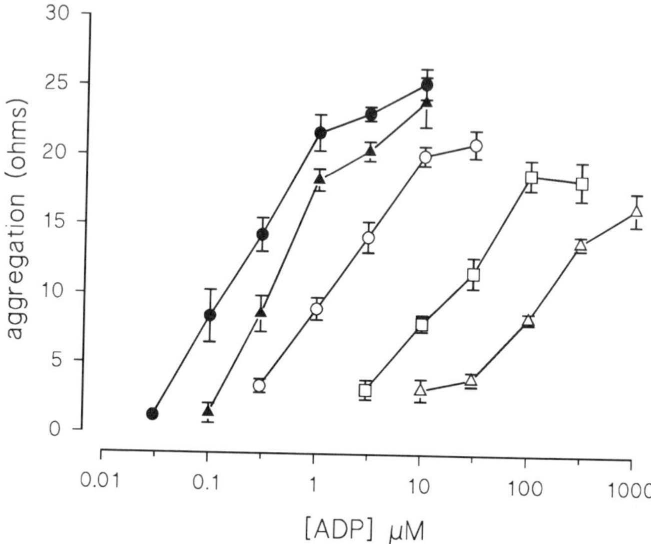

FIGURE 7. Comparison of the effect of ARL 67085 (○ 1, □ 10, △ 100 $\mu g \cdot kg^{-1} \cdot min^{-1}$ IV), vehicle (●) or ticlopidine (▲ 200 $mg \cdot kg^{-1} \cdot day^{-1}$ po, 5 days) on ADP-induced platelet aggregation measured *ex vivo* in blood from anesthetized rats (mean ± S.E., n = 5–6).

displacement in the ADP-concentration effect curve. The antithrombotic effect of ticlopidine, observed in clinical studies at a therapeutic dose of 200–500 mg day^{-1} (3–7 $mg \cdot kg^{-1}$), is attributed to inhibition of ADP-induced platelet activation by an hepatically generated metabolite. Although the exact mechanism involved remains ill defined, a number of studies have suggested that the active metabolite may act as an antagonist at the ADP receptor. Clearly, in this rat model, ticlopidine, at a dose considerably in excess of that used clinically, does not have the profile of a competitive P_{2T}-receptor antagonist and is a far less efficacious inhibitor of ADP-induced platelet aggregation than ARL 67085.

CLINICAL PHARMACOLOGY

The safety, tolerability, and activity (inhibition of ADP-induced platelet aggregation and prolongation of BT) of ARL 67085 has been assessed in healthy male human volunteers (Nassim et al., 1995). Preliminary pharmacokinetic data were also obtained. ARL 67085 (1–4000 $ng \cdot kg^{-1} \cdot min^{-1}$, IV) or placebo was administered in a series of 4-step (45 min per step) incremental (ratios 1 : 2.5 : 5 : 10) infusions over increasing, but overlapping, dose ranges to provide 8 active and 4 placebo subjects at most dose levels. Platelet aggregation, produced by an individually titrated, just submaximal, concentration of ADP, was measured *ex vivo* in whole blood using impedance aggregometry. Lancet bleeding time was determined at baseline and 15 minutes into and after the last infusion step.

Short-term infusions of ARL 67085 produced no drug-related adverse events and no clinically significant within-subject changes in blood and urine safety parameters.

ARL 67085 (10–1000 ng · kg^{-1} · min^{-1}, IV) produced dose-related inhibition of ADP-induced platelet aggregation, with complete or near-complete inhibition (98 ± 5%, mean ± SD, n = 8) achieved in the majority of subjects once an intravenous dose of 1000 ng · kg^{-1} · min^{-1} was attained (Figure 8). Reversal of inhibition upon cessation of infusion was rapid, with recovery essentially complete after 15 minutes at all infusion rates. Preliminary pharmacokinetic analysis indicated a $t_{1/2}$ for ARL 67085 of approximately 2 minutes and total clearance of approximately 20 ml · min^{-1} · kg^{-1}. There was good dose linearity as measured either by C_{max} or AUC. Bleeding time increased two-fold at an intravenous administration rate of 250 ng · kg^{-1} · min^{-1} rising to 3.5-fold at a rate of 4000 ng · kg^{-1} · min^{-1}. As with the aggregation measurements, however, the values reverted to baseline within 15 minutes of the cessation of infusion.

This first experience in man with a potent, selective P_{2T}-receptor antagonist confirms the pharmacodynamic and pharmacokinetic profile observed in preclinical studies, but clearly can provide no direct indication of clinical efficacy in terms of an antithrombotic effect of this novel mechanism. However, if, as in the anesthetized dog studies presented above, complete or near-complete inhibition of aggregation by ARL 67085 were associated with abolition of thrombosis, an intravenous dose of 1 μg · kg^{-1} · min^{-1} might be predicted to be antithrombotic in man. Significant compromise of hemostasis would not be expected until much higher doses since, as in the anesthetized dog studies, there is a marked separation between the dose–response curves for inhibition of platelet aggregation and prolongation of bleeding time.

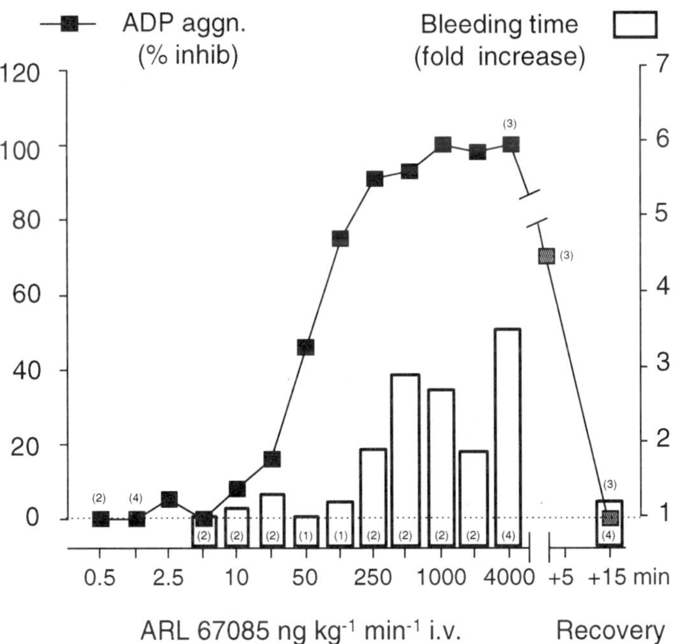

FIGURE 8. The effect of IV infusion of ARL 67085 on bleeding time and ADP-induced platelet aggregation measured *ex vivo* in blood from healthy male volunteers (n = 6–10, except where indicated).

CONCLUSION

A targeted medicinal chemical program has provided potent and selective P_{2T}-receptor antagonists and enabled us to probe the importance of ADP in platelet aggregation *in vitro* and thrombosis and hemostasis *in vivo.* More importantly, the preclinical and clinical pharmacological profile of compounds such as ARL 67085 supports the clinical development of P_{2T}-receptor antagonists as a novel class of antithrombotic drug. Results to date indicate a pivotal role for ADP in arterial thrombosis in animal models. The complex nature of the thrombotic process and the interdependence of platelet activation and coagulation require a polypharmacological approach to the therapy of arterial thrombosis, and it remains to be seen whether a P_{2T}-receptor antagonist can provide significant additional benefit in man. There are certainly grounds for optimism: Shear stress-induced platelet aggregation is believed to be particularly important in the genesis of thrombosis in stenosed arteries and arterioles, and there is accumulating evidence that ADP may play a permissive role in this process (Moake et al., 1988; Oda et al., 1995).

In animal models, the P_{2T}-receptor antagonist mechanism combines a high degree of antithrombotic efficacy with only a modest effect on hemostasis and, in this respect, offers a superior profile to that of the GPIIb/IIIa antagonists tested to date. Clinical experience with representatives of the two classes is required to determine whether this observation translates to a superior benefit–risk profile for the P_{2T}-antagonist mechanism in man. This initial experience with compounds intended for acute intravenous use in closely monitored patients will provide vital guidance for the further development of orally bioavailable representatives intended for chronic dosing, where minimizing bleeding risk will assume paramount importance.

Clinical experience with ticlopidine has set a precedent for an orally active agent interfering with ADP induced platelet aggregation and showing superior clinical efficacy to aspirin. Our results in the anesthetized rat demonstrate that direct antagonism at the P_{2T}-receptor by ARL 67085 provides a marked improvement in antiaggregatory efficacy compared to ticlopidine.

Overall, these observations now provide considerable impetus for development of the next generation of P_{2T}-receptor antagonists that are not only potent and selective but also orally bioavailable.

REFERENCES

Alig L, Edenhofer A, Hadvary P, Hurzeler M, Knopp D, Muller M, Steiner B, Trzeciak A, Weller T (1992): Low molecular weight, non-peptide fibrinogen receptor antagonists. J Med Chem 35:4393–4407.

Antiplatelet Trialists' Collaboration (1994): Collaborative overview of randomised trials of antiplatelet therapy, I. Prevention of death, myocardial infarction, and stroke by prolonged antiplatelet therapy in various categories of patients. Br Med J 308:81–106.

Born GVR (1962): Quantitative investigations into the aggregation of blood platelets. J Physiol 162:67P–68P.

Born GVR, Kratzer MAA (1984): Source and concentration of extracellular adenosine triphosphate during haemostasis in rats, rabbits and man. J Physiol 354:419–429.

Clegg JA, Fraser-Rae L, Humphries RG, Robertson MJ (1995): The effect of FPL 67085 on ADP-induced platelet aggregation *ex vivo* in the urethane-anesthetized rat: A comparison with oral aspirin and ticlopidine. Br J Pharmacol 114:102P.

Cook NS, Kottirsch G, Zerwes HG (1994): Platelet glycoprotein IIb/IIIa antagonists. Drugs of the Future 19:135–159.

Cusack NJ, Hourani SMO (1982a): Adenosine 5′-diphosphate antagonists and human platelets: No evidence that aggregation and inhibition of stimulated adenylate cyclase are mediated by different receptors. Br J Pharmacol 76:221–227.

Cusack NJ, Hourani SMO (1982b): Specific but noncompetitive inhibition by 2-alkylthio analogues of adenosine 5′-monophosphate and adenosine 5′-triphosphate of human platelet aggregation induced by adenosine 5′-diphosphate. Br J Pharmacol 75:397–400.

Davies MJ, Thomas AC (1985): Plaque fissuring—the cause of acute myocardial infarction, sudden ischaemic death, and crescendo angina. Br Heart J 53:363–373.

Defreyn G, Bernat A, Delebasse D, Maffrand JP (1989): Pharmacology of ticlopidine: A review. Semin Thromb Hemost 15:159–166.

EPIC Investigators (1994): Use of a monoclonal antibody directed against the platelet glycoprotein IIb/IIIa receptor in high-risk coronary angioplasty. N Engl J Med 330:956–961.

Folts JD, Crowell EB, Rowe GG (1976): Platelet aggregation in partially obstructed vessels and its elimination with aspirin. Circulation 54:365–370.

Foster MR, Pike NB, Hornby EJ, Porter B, Eldred CD, Ross BC, Lumley P (1993): GR 144053, a potent and specific non-peptide fibrinogen receptor antagonist on human platelets *in vitro*. Thromb Haemostas 69:559.

Fredholm BB, Abbrachio MP, Burnstock G, Daly JW, Harden TK, Jacobson KA, Leff P, Williams M (1994): Nomenclature and classification of purinoceptors. Pharmacol Rev 46:143–156.

Fuster V, Badimon L, Badimon JJ, Chesebro JH (1992): The pathogenesis of coronary artery disease and the acute coronary syndromes. N Engl J Med 326:310–318.

Gaarder A, Jonsen J, Laland S, Hellem A, Owren PA (1961): Adenosine diphosphate in red cells as a factor in the adhesiveness of human blood platelets. Nature 192:531–532.

Gachet C, Savi P, Ohlmann P, Maffrand JP, Jakobs KH, Cazenave JP (1992): ADP receptor induced activation of guanine nucleotide binding proteins in rat platelet membranes—an effect selectively blocked by the thienopyridine clopidogrel. Thromb Haemostas 68:79–83.

Gordon JL (1986): Extracellular ATP: Effects, sources and fate. Biochem J 233:309–319.

Hourani SMO, Welford LA, Cusack NJ (1986): 2-MeS-AMP-PCP and human platelets: Implications for the role of adenylate cyclase in ADP-induced aggregation? Br J Pharmacol 87:84P.

Huang EM, Detwiler TC (1980): Reassessment of the evidence for the role of secreted ADP in biphasic platelet aggregation. J Lab Clin Med 95:59–68.

Humphries RG, Tomlinson W, Clegg JA, Ingall AH, Kindon ND, Leff P (1995): Pharmacological profile of the novel P_{2T}-purinoceptor antagonist, FPL 67085 *in vitro* and in the anaesthetized rat *in vivo*. Br J Pharmacol 115:1110–1116.

Humphries RG, Tomlinson W, Ingall AH, Cage PA, Leff P (1994a): FPL 66096: A novel, highly potent and selective antagonist at human platelet P_{2T}-purinoceptors. Br J Pharmacol 113:1057–1063.

Humphries RG, Tomlinson W, Leff P, Ingall AH, Kindon ND (1994b): The P_{2T}-purinoceptor antagonist, FPL 67085, is a potent, efficacious and selective inhibitor of dynamic arterial thrombosis in the pentobarbitone-anaesthetized dog. Br J Pharmacol 113:63P.

Humphries RG, Tomlinson W, O'Connor SE, Leff P (1990): Inhibition of collagen- and ADP-induced platelet aggregation by substance P *in vivo:* Involvement of endothelium-derived relaxing factor. J Cardiovasc Pharmacol 16:292–297.

Lam JYT, Chesebro JH, Badimon L, Fuster V (1986): Serotonin and thromboxane A_2 receptor blockage decrease vasoconstriction but not platelet deposition after deep arterial injury. Circulation 7(suppl II):II–97.

Macfarlane DE, Mills DCB (1975): The effects of ATP on platelets: Evidence against the central role of released ADP in primary aggregation. Blood 46:309–320.

Moake JL, Turner NA, Stathopoulos NA, Nolasco L, Hellums JD (1988): Shear-induced platelet aggregation can be mediated by vWF released from platelets, as well as by exogenous large or unusually large vWF multimers, requires adenosine diphosphate, and is resistant to aspirin. Blood 71:1366–1374.

Morinelli TA, Niewiarowski S, Kornecki E, Figures WR, Wachtfogel Y, Colman RW (1983): Platelet aggregation and exposure of fibrinogen receptors by prostaglandin endoperoxide analogues. Blood 61:41–49.

Nassim MA, Gardner JJ, Wilkinson D, Corfield JE, Rudol L, Wyld PJ (1995): The short-acting P_{2T}-purinoceptor antagonist, FPL 67085, reliably, reversibly and safely inhibits ADP-induced platelet aggregation *ex vivo* in man. Br J Clin Pharmacol 39:98P.

Nicol AK, Smith JA, Humphries RG, Robertson MJ (1995): FPL 67085: Comparative studies in animal thrombosis models. Int Angiol 14(suppl 1):367.

Nunn B, Chamberlain PD (1983): Further evidence against the validity of using an ADP-removing enzyme system (CP/CPK) for demonstrating the role of secreted ADP in platelet activation. Thromb Res 30:19–26.

Oda A, Yokoyama K, Murata M, Tokuhira M, Nakamura K, Handa M, Watanabe K, Ikeda Y (1995): Protein tyrosine phosphorylation in human platelets during shear stress-induced platelet aggregation (SIPA) is regulated by glycoprotein (GP) Ib/IX as well as GP IIb/IIIa and requires intact cytoskeleton and endogenous ADP. Thromb Haemostas 74:736–742.

Page CP, Paul W, Morley J (1982): An *in vivo* model for studying platelet aggregation and disaggregation. Thromb Haemostas 47:210–213.

Ridogrel Versus aspirin Patency Trial (RAPT) (1994): Randomized trial of ridogrel, a combined thromboxane A_2 synthase inhibitor and thromboxane A_2/prostaglandin endoperoxide receptor antagonist, versus aspirin as adjunct to thrombolysis in patients with acute myocardial infarction. Circulation 89:588–595.

Robertson MJ, Humphries RG, Tomlinson W, Nicol AK, Willis PA (1994): The effect of FPL 67085 on thrombosis and haemostasis in the anaesthetized dog: A comparison with aspirin and the GPIIb/IIIa antagonists. Ro 449883 and GR 144053. Br J Pharmacol 113:64P.

Trams EG, Kaufmann H, Burnstock G (1980): A proposal for the role of ectoenzymes and adenylates in traumatic shock. J Theor Biol 87:609–621.

Ugurbil K, Holmsen H (1981): In Gordon JL (ed): "Platelets in Biology and Pathology, Volume 2." New York: Elsevier/North-Holland, pp 147–177.

Welford LA, Cusack NJ, Hourani SMO (1986): ATP analogues and the guinea-pig taenia coli: A comparison of the structure-activity relationships of ectonucleotidases with those of the P_2-purinoceptor. Eur J Pharmacol 129:217–224.

Weller T, Alig L, Hürzeler Müller M, Kouns WC, Steiner B (1994): Fibrinogen receptor antagonists—a novel class of promising antithrombotics. Drugs of the Future 19 (5):461–476.

World Health Organization (1992): World Health Statistics Annual 1991. Geneva: World Health Organization.

Yao SK, Ober JC, McNatt J, Benedict CR, Rosolowsky M, Anderson HV, Cui K, Maffrand JP, Campbell WB, Buja LM, Willerson JT (1992): ADP plays an important role in mediating platelet aggregation and cyclic flow variations *in vivo* in stenosed and endothelium-injured canine coronary arteries. Circ Res 70:39–48.

Zawilska KM, Born GVR, Begent NA (1982): Effect of ADP-utilising enzymes on the arterial bleeding time in rats and rabbits. Br J Haematol 50:317–325.

Zawilska KM, Izrael V, Chelloul N, Caen J (1975): Prévention de la coagulation intravasculaire disséminée au cours du Phénomenène de Schwartzmann généralisé chez le lapin par un système enzymatique consommant 1'ADP plasmatique. Actualitées Hématologiques 9:269.

CHAPTER 12

Renal Actions of Purines

EDWIN K. JACKSON

Departments of Pharmacology and Medicine, Center for Clinical Pharmacology, University of Pittsburgh Medical Center, Pittsburgh, PA 15213-2582

EFFECTS MEDIATED BY RENAL P_1 (ADENOSINE) RECEPTORS

Renal Hemodynamics

One of the most reproducible and robust effects of activating renal A_1 adenosine receptors is a marked reduction in glomerular filtration rate (GFR). Agonists of A_1 receptors diminish GFR whether infused intravenously (Churchill, 1982; Cook and Churchill, 1984; Churchill et al., 1984; Churchill and Bidani, 1987; Levins et al., 1991a; Zäll et al., 1993; Edlund and Sollevi, 1993; Yagil, 1994); administered directly into the renal artery (Tagawa and Vander, 1970; Oßwald, 1975; Osswald et al., 1975; Osswald et al., 1978b; Hall and Granger, 1986b; Nies et al., 1991; Yagil, 1994; Edlund et al., 1994; Kövér and Tost, 1994); or infused into the renal interstitium (Pawlowska et al., 1987). A_1 agonists decrease GFR whether administered acutely (Tagawa and Vander, 1970; Oßwald, 1975; Osswald et al., 1975; Osswald et al., 1978b; Churchill, 1982; Cook and Churchill, 1984; Churchill et al., 1984; Churchill and Bidani, 1987; Pawlowska et al., 1987; Nies et al., 1991; Levens et al., 1991a; Zäll et al., 1993; Edlund and Sollevi, 1993; Yagil, 1994; Edlund et al., 1994; Kövér and Tost, 1994) or chronically (Hall and Granger, 1986b). A_1 agonists diminish GFR in rats (Osswald et al., 1975; Churchill, 1982; Cook and Churchill, 1984; Churchill et al., 1984; Churchill and Bidani, 1987; Pawlowska et al., 1987; Nies et al., 1991; Yagil, 1994), dogs (Tagawa and Vander, 1970; Oßwald, 1975; Osswald et al., 1978b; Hall and Granger, 1986b; Levens et al., 1991a; Kövér and Tost, 1994), and humans (Zäll et al., 1993; Edlund and Sollevi, 1993; Edlund et al., 1994). A_1 agonists reduce GFR whether the agonist is selective for A_1 receptors (Cook and Churchill, 1984; Churchill and Bidani, 1987; Levens et al., 1991a; Nies et al., 1991); or activates both A_1 and A_2 receptors (Tagawa and Vander, 1970; Oßwald, 1975; Osswald et al., 1975; Osswald et al., 1978b; Churchill, 1982; Churchill et al., 1984; Hall and Granger, 1986b; Zäll et al., 1993; Edlund and Sollevi, 1993; Yagil, 1994; Edlund et al., 1994; Kövér and Tost, 1994; Pawlowska et al., 1987). Although selective

Purinergic Approaches in Experimental Therapeutics, Edited by Kenneth A. Jacobson and Michael F. Jarvis
ISBN 0-471-14071-6 © 1997 Wiley-Liss, Inc.

activation of renal A_2 receptors does not decrease GFR (Levens et al., 1991a,b), the reduction in filtration fraction indicates a disproportionate increase in renal blood flow (RBF).

In contrast to their effects on GFR, the effects of selective A_1 agonists and nonselective A_1/A_2 agonists on RBF are highly variable and depend on a number of conditions including species, route of administration, rate of administration, dose, experimental paradigm, and presence of modulating factors, particularly nitric oxide (NO) and angiotensin II (Ang II). The classical RBF response to an intrarenal infusion of adenosine *in vivo* is an immediate reduction in RBF that begins to wane in less than a minute, followed by a gradual return of RBF to baseline or above within a few minutes (Thurau, 1964; Tagawa and Vander, 1970; Oßwald, 1975; Osswald et al., 1978b; Spielman and Thompson, 1982). When administered *in vivo* at nonhypotensive doses, sustained intrarenal infusions of adenosine and adenosine agonists with selective A_1 or mixed A_1/A_2 activity usually cause little, if any, change in RBF (Thurau, 1964; Tagawa and Vander, 1970; Oßwald, 1975; Osswald et al., 1978b; Spielman and Thompson, 1982; Hall and Granger, 1986b; Miyamoto et al., 1988; Nies et al., 1991; Yagil, 1994; Edlund et al., 1994; Kövér and Tost, 1994). When administered *in vitro* to isolated, perfused kidneys, A_1 selective agonists decrease RBF, and A_1/A_2 mixed agonists increase RBF (Nies et al., 1991; Murray and Churchill, 1984, 1985; Rossi et al., 1987b). When injected *in vivo* as boluses into the renal artery, A_1 agonists, whether selective or not, reduce renal blood flow (Hashimoto and Kumakura, 1965; Osswald et al., 1975; Spielman and Osswald, 1979; Barrett and Droppleman, 1993; Marraccini et al., 1996). Intravenous infusions of selective A_1 or mixed A_1/A_2 agonists usually reduce RBF in part because of systemic hypotension (Churchill, 1982; Cook and Churchill, 1984; Churchill et al., 1984; Churchill and Bidani, 1987; Zäll et al., 1993). Unlike selective A_1 agonists or mixed A_1/A_2 agonists, selective A_2 agonists increase RBF (Levens et al., 1991a,b).

In order to comprehend how adenosine agonists affect the macroparameters of GFR and RBF, it is necessary to understand the interactions of adenosine receptors with various renal structures. In the renal microcirculation, activation of A_1 receptors increases resistance primarily of preglomerular, rather than postglomerular, vessels (Osswald et al., 1978b; Haas and Osswald, 1981; Murray and Churchill, 1984, 1985), with the exception that A_1 receptor–mediated vasoconstriction occurs in the outer medullary descending vasa recta (Silldorff et al., 1996). However, the degree to which A_1 receptors cause preglomerular vasoconstriction is modulated by several endogenous factors. For instance, the intensity of renal vasoconstriction caused by the A_1 selective agonist N^6-cyclopentyladenosine is markedly enhanced by blocking the biosynthesis of NO, suggesting that endogenous NO attenuates A_1 receptor–mediated preglomerular vasoconstriction (Barrett and Droppleman, 1993). Thus, there appears to be a *negative* synergy between A_1 receptor activation and NO in preglomerular vessels.

In contrast, in several studies (Osswald et al., 1975; Spielman and Osswald, 1979; Hall and Granger, 1986a; Deray et al., 1990b; Dietrich et al., 1991; Weihprecht et al., 1994; Munger and Jackson, 1994), but not all (Barrett and Droppleman, 1993; Carmines and Inscho, 1994), a *positive* synergy was observed between Ang II and A_1 receptor activation in the renal microcirculation. For example, the preglomerular vascular response to N^6-cyclohexyladenosine, a selective A_1 receptor agonist, was reduced by inhibition of angiotensin converting enzyme, volume expansion, and antagonism of Ang II receptors; conversely, low doses of exogenous Ang II potentiated preglomerular vascular responses to N^6-cyclohexyladenosine (Weihprecht et al., 1994). Since A_1 receptors also mediate inhibition of renin release (*vide infra*), with the consequence being a reduction in the rate of Ang II biosynthesis, preglomerular vasoconstriction induced by A_1 receptor activation may be somewhat self-limiting, due to a reduction in the biosynthesis of the positive modulator Ang II.

Unlike A_1 receptor activation, stimulation of A_2 receptors in the renal microcirculation leads to relaxation of postglomerular vessels in the inner cortex and medulla, but not in the outer cortex (Spielman et al., 1980; Murray and Churchill, 1984, 1985; Agmon et al., 1993). A_2 receptors also mediate vasodilation of outer medullary descending vasa recta (Silldorf et al., 1996). When both A_1 and A_2 receptors are activated in the microvasculature of deeper nephrons, A_2 receptor–mediated vasodilation overpowers A_1 receptor–mediated vasoconstriction leading to a net increase in blood flow to the renal medulla.

As described in the preceding paragraph and as illustrated in Figure 1, the net effects of adenosine agonists on GFR and RBF are the result of an interplay among numerous factors. Activation of A_1 receptors increases preglomerular resistance, thus decreasing hydrostatic pressure in the glomerular capillaries and diminishing the driving force for glomerular filtration. Inhibition of Ang II formation by A_1 receptors may limit the preglomerular effects of A_1 receptor activation and diminish the effects of A_1 receptor activation on GFR. However, since Ang II normally constricts postglomerular vessels, reducing the amount of Ang II would also decrease Ang II–mediated postglomerular vasoconstriction, which would reinforce the A_1 receptor–mediated reduction in GFR. Activation of A_2 receptors decreases postglomerular resistance, thus also decreasing the hydrostatic pressure in the glomerular capillaries and diminishing the driving force for glomerular filtration. It is little wonder, then, that administration of either an A_1-selective or a mixed A_1/A_2 agonist consistently reduces GFR. Since A_2-selective agonists should reduce glomerular filtration pressure, one might be inclined to predict that A_2-selective agonists would also decrease GFR. However, reducing postglomerular resistance without a concomitant increase in preglomerular resistance augments RBF; and increases in RBF tend to increase GFR by maintaining a lower concentration of plasma proteins in the glomerular capillaries (i.e., by maintaining filtration disequilibrium). Therefore, the net effect of A_2 receptor activation is to increase RBF, while having little overall effect on GFR.

Accounting for the net effects of A_1-selective or mixed A_1/A_2 agonists on RBF requires a more involved explanation. With regard to mixed agonists, the onset of A_1 receptor–mediated preglomerular vasoconstriction is faster than the onset of A_2 receptor–mediated postglomerular vasodilation; hence, the time-dependent effects of intrarenal infusions of adenosine (Thurau, 1964; Tagawa and Vander, 1970; Oßwald, 1975; Osswald et al., 1978b; Spielman and Thompson, 1982). The immediate preglomerular vasoconstriction caused by adenosine reduces RBF; however, the delayed decrease in postglomerular resistance gradually increases RBF to the original baseline or above. Since vasoconstriction predominates in the outer renal cortex and vasodilation predominates in the inner cortex, adenosine redistributes RBF from the outer to the inner cortex (Ueda, 1972; Spielman et al., 1980; Dinour and Brezis, 1991).

Three factors may contribute to the variable effects of selective A_1 agonists on RBF. First, suppression of renin release may diminish the preglomerular effects of A_1 agonists and attenuate Ang II–induced postglomerular vasoconstriction. Second, NO production in the preglomerular vessels may attenuate the preglomerular vasoconstriction caused by A_1 agonists. Third, preglomerular resistance, although critically important in regulating pressure in the glomerular capillaries, is only one of several resistances in series that determines RBF. Thus small changes in preglomerular resistance may

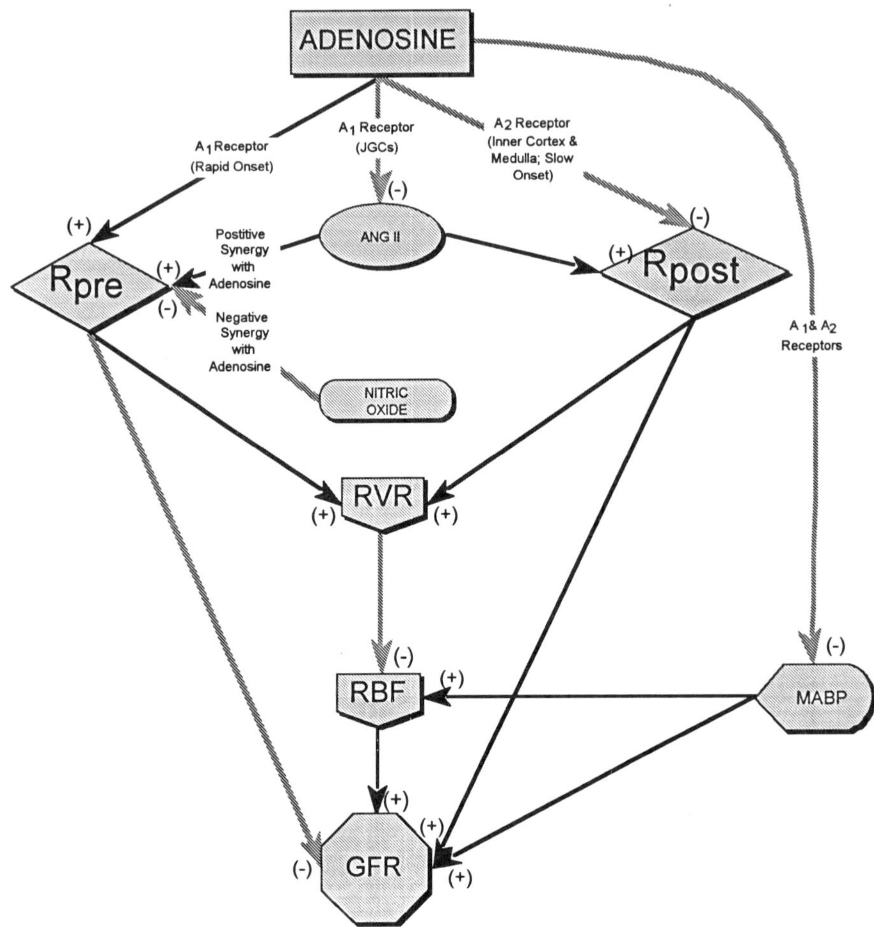

FIGURE 1. Mechanisms by which adenosine alters glomerular filtration rate (GFR) and renal blood flow (RBF). R_{pre}, resistance of preglomerular vessels; R_{post}, resistance of postglomerular vessels; JGCs, juxtaglomerular cells; ANG II, angiotensin II; RVR, total renal vascular resistance; MABP, mean arterial blood pressure. Solid lines indicate positive effect (parameter is increased) and dotted lines denote negative effect (parameter is decreased).

not be observed as significant changes in overall RBF in experimental settings where postglomerular resistances are dominant. Given the complexity of interactions, it is not surprising that the reported effects of A_1-selective and mixed A_1/A_2 agonists on RBF have been highly variable.

Tubuloglomerular Feedback

Variations in flow rate in the loop of Henle cause inverse changes in single nephron glomerular filtration rate (SNGFR). This mechanism, called tubuloglomerular feedback (TGF), assures that the capacity of the nephron to reabsorb ultrafiltrate is not exceeded by the load of ultrafiltrate. In the event that SNGFR is too rapid, proximal reabsorptive mechanisms in that particular nephron become overwhelmed so that the NaCl load is increased to the macula densa in the thick ascending limb of Henle's loop of that same nephron. The increased distal delivery of NaCl causes the macula densa to signal the afferent arteriole to constrict. Constriction of the afferent arteriole reduces glomerular capillary pressure, single nephron blood flow, and SNGFR. The reduction in SNGFR diminishes tubular load, which closes the negative feedback loop.

In 1980, Osswald and colleagues (1980) proposed that TGF was mediated by adenosine. The fundamental idea proposed by Osswald is as follows: Increased NaCl delivery to the macula densa → increased transport by the macula densa → increased ATP utilization → increased adenosine formation → diffusion of adenosine from the macula densa to the nearby afferent arteriole → adenosine-mediated afferent arteriolar vasoconstriction (Figure 2).

Several lines of evidence strongly indicate that adenosine is an important, although perhaps not the only, mediator of TGF:

1. Maneuvers that stimulate energy utilization by the kidney cause a reduction in renal tissue levels of ATP and a concomitant increase in renal levels of adenosine (Osswald et al., 1980). In this regard, recent studies have demonstrated that hypertonic saline enhances adenosine release from isolated thick ascending limbs of Henle's loop (Baudouin-Legros et al., 1995) and a high sodium diet increases renal interstitial adenosine levels approximately 18-fold compared with levels in animals on a low sodium diet (Siragy and Linden, 1996).
2. Several adenosine receptor antagonists including theophylline (Osswald et al., 1980), 1,3-dipropyl-8-(p-sulfophenyl)xanthine (Franco et al., 1989), and 1,3-dipropyl-8-cyclopentylxanthine (Schnermann et al., 1990) block the reduction in stop flow pressure (an index of preglomerular vasoconstriction) caused by increased flow rates in the loop of Henle.
3. Reducing intrarenal adenosine levels with exogenous adenosine deaminase blocks TGF responses (Osswald et al., 1980).
4. Enhancement of renal adenosine levels by inhibiting either adenosine transport with dipyridamole or inhibiting adenosine deaminase with erythro-9-(2-hydroxy-3-nonyl)adenine potentiates TGF responses (Osswald et al., 1982).
5. Intraluminal (Franco et al., 1989) or peritubular (Schnermann et al., 1990) infusions of A_1-selective agonists reduce stop flow pressure.
6. Activation of TGF by intrarenal infusions of hypertonic saline is blocked by adenosine receptor antagonists (Gerkens et al., 1983a; Gerber and Nies, 1986; Callis et al., 1989; Deray et al., 1990b).

Osswald et al. (1980) cautioned that, although adenosine is likely a mediator of TGF in the outer renal cortex, adenosine may not participate in TGF in deeper nephrons. This hesitancy to implicate adenosine as a mediator of TGF in deeper nephrons was based on the fact that intrarenal infusions of adenosine vasodilate, rather than vasoconstrict, the renal medulla. Thus, at the time Osswald et al. (1980) proposed the adenosine hypothesis for TGF, and indeed for many years thereafter, whether A_1 receptors mediate vasoconstriction in the renal medulla was an open question.

Recent studies by Agmon, Dinour, and Brezis (1993), however, indicate that adenosine may participate in TGF throughout the kidney. The highly localized release of

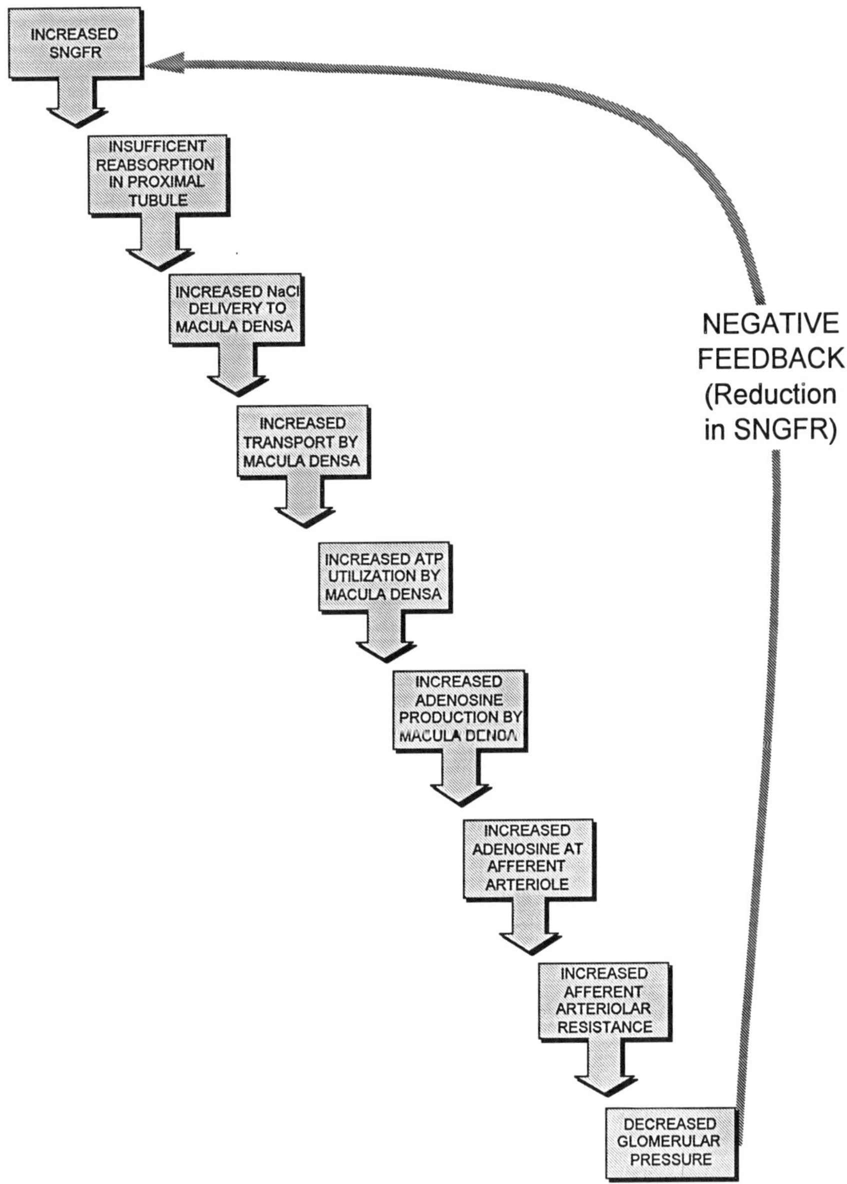

FIGURE 2. Mechanism of tubuloglomerular feedback. SNGFR, single nephron glomerular filtration rate.

adenosine by macula densa cells would provide adenosine predominantly to the A_1 receptors in the afferent arteriole, and the delivery of adenosine would be via the interstitial compartment. Intrarenal infusions of adenosine, however, deliver adenosine to both the afferent and efferent arterioles via the vascular lumen. The ideal way to mimic the interstitial, localized delivery of adenosine to the A_1 receptors of the afferent arteriole is to infuse a selective A_1 agonist into the renal interstitium. Interstitial infusion would provide the agonist from the proper orientation, and employment of a selective A_1 agonist would prevent any activation of efferent arteriolar A_2 receptors. Agmon et al. (1993) infused N^6-cyclopentyladenosine into the renal interstitium and measured both outer cortical and medullary blood flow using laser-Doppler probes. Importantly, these investigators noted that N^6-cyclopentyladenosine markedly reduced both cortical and medullary blood flow, suggesting preglomerular vasoconstriction in both superficial and deep nephrons. However, the recent study by Silldorf et al. (1996) raises the possibility that reductions in medullary blood flow induced by selective A_1 agonists could be due to a direct effect on the descending vasa recta. Nonetheless, it is possible that adenosine released locally to the afferent arteriole from the interstitial space causes afferent arteriolar vasoconstriction throughout the kidney.

Renin Release

Enhancement of renin release by many physiological and pharmacological stimuli appears to involve activation of adenylyl cyclase in juxtaglomerular cells (Jackson, 1991). Since adenosine can inhibit adenylyl cyclase via high-affinity A_1 receptors on juxtaglomerular cells, endogenous adenosine could function as a molecular brake on renin release responses that are mediated by the cyclic AMP pathway. This idea is known as the *adenosine-brake hypothesis* (Jackson, 1991).

Several lines of evidence strongly support the adenosine-brake hypothesis:

1. Exogenous adenosine and A_1-selective receptor agonists attenuate renin release in rats, (Churchill and Bidani, 1987; Osswald et al., 1978a), dogs (Tagawa and Vander, 1970; Spielman, 1984; Arend et al., 1984; Macias-Nuñez, 1985; Macias-Nuñez, 1986; Deray et al., 1987, 1989b), humans (Edlund et al., 1994), perfused rat kidneys (Murray and Churchill, 1985), rat renal cortical slices (Skøtt and Baumbach, 1985; Churchill and Churchill, 1985; Rossi et al., 1987a; Churchill et al., 1987a,b), rabbit renal cortical slices (Barchowsky et al., 1987), isolated rat glomeruli (Skøtt and Baumbach, 1985), and the rabbit isolated perfused juxtaglomerular apparatus (Lorenz et al., 1993).
2. Increasing levels of endogenous adenosine with either dipyridamole in humans (Taddei et al., 1992) or intrarenal infusion of maleic acid in dogs (Arend et al., 1986) inhibits renin release.
3. In rats, exogenous adenosine deaminase augments and blockade of adenosine deaminase with erythro-9-(2-hydroy-3-nonyl)adenine attenuates renin release responses to the vasodilator hydralazine (Kuan et al., 1990b).

A fourth line of evidence in favor of the adenosine-brake hypothesis is that antagonism of adenosine receptors enhances renin release. Theophylline stimulates renin release in dogs (Reid et al., 1972; Langård et al., 1983); isolated rabbit afferent arterioles (Cannon et al., 1989) and in isolated perfused rabbit (Viskoper et al., 1977) and rat

(Peart et al., 1975) kidneys. Caffeine enhances renin release responses to renal artery hypotension in dogs (Deray et al., 1989a); to furosemide (Paul et al., 1989), hydralazine (Tofovic et al., 1991) and salt-depletion (Tseng et al., 1993) in rats; and to diazoxide in humans (Brown et al., 1991). The extracellular adenosine receptor antagonist, 1,3-dipropyl-8-(*p*-sulfophenyl)xanthine, augments renin release responses to sodium restriction (Kuan et al., 1989), hydralazine (Kuan et al., 1990b; Tofovic et al., 1991) and renal artery clipping (Kuan et al., 1990b) in rats. The highly selective A_1 receptor antagonist FK453, but not its relatively inactive enantiomer FR113452, enhances renin release responses to isoproterenol in rats (Pfeifer et al., 1995), and FK453 increases renin release in humans (van Buren et al., 1993; Balakrishnan et al., 1993, 1996). Finally, 1,3-dipropyl-8-cyclopentylxanthine, another highly selective A_1 receptor antagonist, augments renin release responses to isoproterenol in the rat (Pfeifer et al., 1995).

The above evidence indicates that the adenosine-brake hypothesis is well confirmed and appears to be mediated by A_1, not A_2, receptors. Although activation of A_2 receptors stimulates, not inhibits, renin release (Churchill and Bidani, 1987; Churchill and Churchill, 1985), the lower affinity for adenosine of A_2 receptors, compared with A_1 receptors, makes it unlikely that A_2 receptors participate in the physiological control of renin release. This conclusion is confirmed by the recent observation that KF17837, a A_2-selective receptor antagonist, does not alter the renin release response to isoproterenol in the rat (Pfeifer et al., 1995).

The source of the endogenous adenosine that mediates the adenosine-brake on renin release may be a pathway called the *cyclic AMP-adenosine pathway.* As mentioned previously, many renin-releasing stimuli act via adenylyl cyclase. Stimulation of adenylyl cyclase is always associated with the egress of intracellular cyclic AMP (King and Mayer, 1974; Barber and Butcher, 1981, 1983), and if ectophosphodiesterase is present, extracellular cyclic AMP would be metabolized to AMP and hence to adenosine by ecto-5′-nucleotidase. Adenosine could also be formed from the intracellular metabolism of cyclic AMP to adenosine with subsequent transport of adenosine into the extracellular space. At any rate, adenosine would be synthesized and/or transported directly onto the cell surface such that modest increases in adenylyl cyclase activity could produce pharmacologically active concentrations of adenosine in the biophase near the cell membrane.

Studies from our laboratory are consistent with the hypothesis that the cyclic AMP–adenosine pathway in the kidneys is an important biosynthetic route for renal adenosine production. In this regard, we have shown that:

1. In the perfused rat kidney, exogenous cyclic AMP is converted to adenosine, and this conversion is blocked by inhibition of either phosphodiesterase or ecto-5′-nucleotidase (Mi and Jackson, 1995).
2. In the perfused rat kidney, isoproterenol increases purine release and this effect is blocked by inhibition of either β-adrenoceptors, phosphodiesterase, or ecto-5′-nucleotidase (manuscript in preparation).
3. Local administration of a phosphodiesterase inhibitor into the renal cortical interstitial space results in a significant reduction in renal cortical interstitial levels of adenosine and inosine as determined by microdialysis (Mi et al., 1994).
4. Intrarenal and intravenous infusions *in vivo* of either cyclic AMP or isoproterenol cause a concomitant increase in the urinary excretion rate of both cyclic AMP and adenosine (manuscript in preparation).

The adenosine-brake on renin release was initially conceived as a restraint that was exerted by endogenous adenosine acting directly on juxtaglomerular cells, and indeed this appears to be true. However, in addition to the *renal adenosine-brake* on renin release there may also exist a *CNS adenosine-brake* on renin release. Adenosine exerts a sympathoinhibitory effect on the CNS (Barraco et al., 1986, 1988; Mosqueda-Garcia, 1989; Laborit et al., 1990), and caffeine augments central sympathetic tone (Robertson et al., 1978; Smits et al., 1983, 1986; Mosqueda-Garcia et al., 1990; Lane et al., 1993). Therefore, endogenous adenosine may also curtail renin release by acting on the CNS to inhibit sympathetic activation of β_1-adrenoceptors on juxtaglomerular cells, a well-known stimulus for renin release.

The CNS adenosine-brake on renin release has not been adequately studied. However, the available data support the existence of this putative regulatory mechanism. For instance, an adenosine receptor antagonist that does not penetrate the blood-brain barrier augments renin release responses to hydralazine less than does an antagonist that gains access to CNS sites (Tofovic et al., 1991). Also, selective blockade of CNS adenosine receptors with intracerebroventricular infusions of caffeine enhances renin release responses to hydralazine (Tofovic et al., 1996). Further work is needed to determine whether the CNS adenosine-brake on renin release is physiologically important.

Tubular Transport

The renal tubules are exposed to endogenous adenosine from several sources. Adenosine is freely filtered by the glomerulus, and approximately one half of filtered adenosine escapes tubular uptake (Thompson et al., 1985). Therefore, filtered adenosine contributes substantially to the total amount of urinary adenosine (Thompson et al., 1985). In addition, filtered nucleotides, as well as nucleotides released by proximal tubule epithelial cells, would be converted to adenosine, since renal epithelial cells in the proximal tubule are richly endowed with ecto-ATPase, ecto-ADPase, and ecto-5′-nucleotidase (Čulić et al., 1990; Pawelczyk et al., 1992; Le Hir and Kaissling, 1993). Finally, the cyclic AMP–adenosine pathway may contribute to luminal adenosine. As reabsorption of ultrafiltrate occurs, adenosine in the luminal space would be concentrated until facilitated transport of adenosine into epithelial cells (Trimble and Coulson, 1984) is driven by a concentration gradient between the tubular lumen and the intracellular compartment. The concentration of adenosine in both the urine (Svensson and Jonzon, 1990) and renal interstitial space (Baranowski and Westenfelder, 1994) is several hundred nanomoles per liter, well within the range for activation of A_1 receptors. Therefore, if renal epithelial cells express adenosine receptors and if these receptors are coupled to epithelial transport, a physiological role for adenosine in tubular transport would be anticipated.

Only a few studies have examined the renal expression of adenosine receptors. Binding studies have detected A_1 receptors in the renal medulla (Weber et al., 1988), glomeruli (Palacios et al., 1987; Freissmuth et al., 1987a; Toya et al., 1993), renal microvessels (Freissmuth et al., 1987a), collecting tubules (Palacios et al., 1987), and a cell line derived from cortical collecting tubular cells (Spielman et al., 1992). However, A_1 receptors were not detected in whole kidney membranes (Wu and Churchill, 1985), renal cortex (Weber et al., 1988), cortical tubules (Freissmuth et al., 1987b; Palacios et al., 1987) or proximal tubular brush border membranes (Blanco et al., 1992). Weaver and Reppert (1992), using *in situ* hybridization, observed A_1 receptor mRNA expression only in the collecting tubules and juxtaglomerular apparatus. Yamaguchi et al. (1995) recently studied the distribution of A_1 receptor mRNA in rat nephron segments using the reverse transcription-polymerase chain reaction technique. In this study, A_1 receptor mRNA was detected in glomeruli, medullary collecting ducts, cortical thick ascending limbs, medullary thick ascending limbs, proximal convoluted tubules, and proximal straight tubules. Binding studies have detected A_2 receptors in whole kidney membranes (Wu and Churchill, 1985); and A_{2a} receptor mRNA, but not A_{2b} receptor mRNA, was expressed in the renal papilla and inner medulla (Weaver and Reppert, 1992) (at a location outside the collecting duct). Woodcock et al. (1984) failed to observe A_2 receptors in the renal medulla, and binding studies by Blanco et al. (1992) in brush-border membranes suggest a unique adenosine receptor that is neither A_1 nor A_2.

Functional studies demonstrate that most of the binding and molecular biological approaches mentioned above greatly underestimate the importance of epithelial adenosine receptors. Numerous studies *in vitro* indicate that adenosine receptors play an important role in regulating epithelial transport in several nephron segments. In this regard, A_1 receptors have been linked to signal transduction mechanisms in:

1. Cultured cortical collecting tubule cells [↓ cyclic AMP (Arend et al., 1987b, 1988), ↑ Ca^{2+} (Arend et al., 1988), ↑ diacylglycerol (Schwiebert et al., 1992), ↑ protein kinase C activity (Schwiebert et al., 1992), and ↑ 305 pS Cl^- channel activity (Schwiebert et al., 1992)].
2. Cultured inner medullary collecting duct cells [↓ cyclic AMP (Yagil, 1990, 1992)].
3. Cultured LLC-PK_1 renal cell line, a cell line with mixed proximal and distal phenotypes [↑ inositol phosphates and ↓ cyclic AMP (LeVier et al., 1992)].
4. Cultured opossum kidney epithelial cells, a model of proximal tubular epithelial cells [↑ protein kinase C activity and ↓ cyclic AMP (Coulson et al., 1991)].
5. Perfused rectal gland of the dogfish shark, a model of chloride transport in the thick ascending limb of Henle's loop [↓ cyclic AMP (Kelley et al., 1990)].
6. Perfused inner medullary collecting duct [↓ cyclic AMP (Edwards and Spielman, 1994)].
7. Cultured A_6 cells [↑ Ca^{2+} (Casavola et al., 1996)].

Activation of A_1 receptors also has been shown to alter tubular transport kinetics in:

1. Cultured opossum kidney epithelial cells [↑ Na^+–glucose symport and ↑ Na^+–phosphate symport (Coulson et al., 1991)].
2. Microperfused proximal convoluted tubules [↑ basolateral Na^+–$3HCO_3^-$ symport (Takeda et al., 1993)].
3. Perfused rectal gland of the dogfish shark [↓ Cl^- transport (Kelley et al., 1990)].
4. Isolated perfused medullary thick ascending limb [↓ Cl^- transport (Beach and Good, 1992)].
5. Perfused inner medullary collecting duct [↓ AVP-induced osmotic water permeability (Edwards and Spielman, 1994)].
6. Cultured A_6 cells [↑ short-circuit current (Casavola et al., 1996)].

7. Cultured inner medullary collecting duct cells [↑ transepithelial resistance and ↓ Na^+ uptake (Yagil et al., 1994)].
8. Cultured mouse inner medullary collecting duct cell line [↓ AVP-induced Cl^- secretion (Moyer et al., 1995)].

Taken together, these *in vitro* studies suggest that A_1 receptors are functional in the proximal tubule, thick ascending limb of Henle's loop, and collecting duct; and either enhance (proximal tubule) or inhibit (thick ascending limb and collecting duct) tubular transport. A stimulatory effect for A_1 receptors in the proximal tubule is further supported by the observations that KW-3902, a selective A_1 receptor antagonist, inhibits sodium-dependent phosphate transport in cultured rat proximal tubular cells (Cai et al., 1994, 1995).

Experiments have linked A_2 receptors to signal transduction mechanisms in:

1. Cultured cortical collecting tubule cells [↑ cyclic AMP (Arend et al., 1987b, 1988)].
2. Cultured LLC-PK_1 renal cell line [↑ cyclic AMP (LeVier et al., 1992)].
3. Isolated tubules of renal cortex [↑ cyclic AMP (Freissmuth et al., 1987b)].
4. A6 cells derived from *Xenopus laevis* kidney [↑ cyclic AMP (Lang et al., 1985)].

In A6 cells, A_2 receptors stimulated Na^+ transport; however, the *in vitro* effects of A_2 receptor activation on transport processes have not been adequately investigated.

Numerous *in vivo* studies have examined the effects of adenosine receptor agonists on urine volume and sodium excretion. Intravenous infusions of mixed A_1/A_2 receptor agonists (Churchill, 1982; Churchill et al., 1984; Zäll et al., 1993; Edlund and Sollevi, 1993), selective A_1 agonists (Cook and Churchill, 1984; Churchill and Bidani, 1987; Levens et al., 1991a), and selective A_2 agonists (Levens et al., 1991a) reduce urine volume and sodium excretion. Likewise, intrarenal or suprarenal infusions of mixed A_1/A_2 receptor agonists usually (Tagawa and Vander, 1970; Oßwald, 1975; Osswald et al., 1975, 1978b; Spielman, 1984; Deray et al., 1987, 1989b; Kövér and Tost, 1994), but not always (Hall and Granger, 1986b; Miyamoto et al., 1988; Edlund et al., 1994), reduce urine volume and sodium excretion. However, intrarenal infusions of CGS21680, a selective A_2 receptor agonist, do not affect urine volume or sodium excretion in the dog, whereas intrarenal infusions of N^6-cyclohexyladenosine cause a marked diuresis/natriuresis in rats (Yagil, 1994; Fransen and Koomans, 1995).

It is difficult, if not impossible, to infer from the *in vivo* studies mentioned above whether adenosine receptors have direct effects on tubular transport, and if so, whether the direct tubular effects at physiological levels of adenosine are natriuretic/diuretic or antinatriuretic/antidiuretic. In nearly all studies, systemic and intrarenal infusions of adenosine agonists altered arterial blood pressure and/or renal hemodynamic parameters (usually both, particularly during systemic infusions). Changes in blood pressure, total RBF, the intrarenal distribution of RBF, GFR, and renin release per se alter renal excretory function, thus confounding inferences regarding the direct tubular effects of infused adenosine agonists. Another experimental flaw affecting the interpretation of all *in vivo* studies published to date is that no attempts were made to remove endogenous adenosine before administering exogenous adenosine agonists. Consequently, if intrarenal levels of endogenous adenosine are high enough to cause saturation or near-saturation of A_1 receptors on tubular epithelial cells, then the effects of exogenous adenosine agonists on tubular transport would be quantitatively and qualitatively misleading. For instance, assume for the sake of discussion that endogenous adenosine levels are high in the proximal tubule, thus saturating A_1 receptors in that nephron segment, yet are low at more distal nephron sites, thereby leaving distal A_1 receptors unoccupied. Also assume that A_1 receptors stimulate transport in the proximal tubule while inhibiting transport at more distal sites, as is suggested by the studies in isolated epithelial cells and nephron segments (see above). In this situation, administration of either exogenous adenosine or an A_1 receptor antagonist would cause diuresis/natriuresis due to stimulation of distal A_1 receptors and blockade of proximal A_1 receptors, respectively. In fact, in the rat just such a paradox has emerged, i.e., adenosine infused directly into the renal artery causes diuresis/natriuresis as does administration of selective A_1 receptor antagonists (see below).

Another approach to experimentally assess the role of adenosine as a regulator of tubular transport is to administer adenosine receptor antagonists and observe the effects of blocking the actions of endogenous adenosine. In this regard, caffeine and theophylline are well-known diuretics; however, since both of these agents can affect multiple biochemical/cellular processes, any interpretation must be guarded.

More convincing than the studies with caffeine and theophylline are the growing number of reports using newer, more potent adenosine receptor antagonists. These studies provide strong support for the hypothesis that endogenous adenosine acting via A_1, but not A_2, receptors enhances tubular transport *in vivo.* Administration of the selective A_1 receptor antagonists 1,3-dipropyl-8-cyclopentylxanthine (Collis et al., 1991; Knight et al., 1993b), 8-(dicyclopropylmethyl)-1,3-dipropylxanthine (Shimada et al., 1991), and KW-3902 (Suzuki et al., 1992; Mizumoto et al., 1993) causes diuresis/natriuresis. Kuan et al. (1993) have demonstrated that administration of equipotent doses of two structurally dissimilar, highly potent and selective A_1 antagonists (FK453 and 1,3-dipropyl-8-cyclopentylxanthine) caused similar degrees of diuresis/natriuresis without influencing systemic or renal hemodynamics. Further, they found that equal doses of the less active enantiomer of FK453, i.e., FR113452, did not alter urine volume or sodium excretion. Thus, the results of Kuan et al. (1993) firmly establish that endogenous adenosine interacts with A_1 receptors in the tubules to regulate (enhance) sodium reabsorption. On the other hand, adenosine antagonists selective for the A_2 receptor do not express diuretic/natriuretic activity, suggesting that tubular A_2 receptors have little if any role in the regulation of tubular reabsorption (Suzuki et al., 1992).

Regarding the tubular site of the endogenous adenosine/A_1 receptor interaction that enhances tubular reabsorption, studies by van Buren et al. (1993) and Balakrishnan et al. (1993) indicate that blockade of A_1 receptors in humans with FK453 causes diuresis/natriuresis by interfering with tubular transport mostly in the proximal tubule. However, more distal nephron segments also appear to be involved, particularly the thick ascending limb of Henle's loop.

Erythropoietin Production

Renal hypoxia, the central stimulus for erythropoietin secretion, markedly increases the expression of erythropoietin mRNA and greatly increases the secretion rate of erythropoietin from the kidney. In addition to directly inducing renal cortical cells to produce more erythropoietin, hypoxia also amplifies its direct effects by increasing the interstitial levels of additional factors that augment erythropoietin synthesis and release (Fisher and Nakashima, 1992). Adenosine may be one such factor.

In 1988, Paul et al. and Ueno et al. reported that adenosine stimulates erythropoietin production. Paul and coworkers found that administration of adenosine increases the secretion of erythropoietin from isolated, perfused rat kidneys, and that in hypoxic rats adenosine deaminase diminished, and adenosine deaminase inhibition increased, erythropoietin levels. Ueno et al. (1988) reported that A_1 and A_2 receptor activation decreased and increased erythropoietin levels, respectively, in exhypoxic polycythemic mice. Several additional studies have confirmed that stimulation of A_2 receptors increases erythropoietin synthesis and release by a mechanism involving activation of adenylyl cyclase (Nakashima et al., 1991, 1993a, 1993b, 1994; Ohigashi et al., 1993). One study, however, failed to detect a change in renal erythropoietin mRNA levels when perfused rat kidneys were exposed to an adenosine receptor agonist (Tan and Ratcliff, 1992).

Kaissling et al. (1993) discovered that anemia increases the expression of ecto-5′-nucleotidase on fibroblasts in the renal cortical interstitium. This phenomenon could participate in the stimulation of erythropoietin synthesis/release, since increased levels of ecto-5′-nucleotidase activity would enhance adenosine biosynthesis and, consequently, facilitate the erythropoietin system.

THERAPEUTIC AND CLINICAL IMPLICATIONS OF RENAL P_1 (ADENOSINE) RECEPTORS

Acute Renal Failure

Acute renal failure (ARF), defined as an abrupt decline in GFR leading to accumulation of nitrogenous wastes, affects 5% of all hospitalized patients and is associated with a high rate of mortality (Hou et al., 1983). Since the late 1970s, many reports concluding that adenosine participates in various forms of ARF have appeared in the literature, raising hopes that adenosine receptor antagonists might be of clinical utility in ARF. An attempt will be made to summarize and objectively evaluate the available database regarding the pathophysiological role of adenosine in ARF and the possible clinical applications of adenosine receptor antagonists in this regard.

Acute Tubular Necrosis–Induced ARF Acute tubular necrosis (ATN) is responsible for two thirds of all intrinsic ARF, and the mortality rate in patients with ATN-induced ARF is approximately one half (Lunding et al., 1964). The modern view of ATN-induced ARF is that damage to tubular epithelial cells diminishes GFR by physical obstruction of tubules with epithelial cells, backleak of ultrafiltrate through a disrupted epithelial barrier, and dysfunction of the renal microcirculation.

The hypothesis that adenosine might participate in ATN-induced ARF is based on the idea that adenosine may mediate functional alterations of the renal microcirculation. In this regard, adenosine could reduce GFR following epithelial cell damage by two pathways. Epithelial cell damage may impair the ability of cells to generate energy, leading to depletion of ATP and ADP and thus to an accumulation of AMP and adenosine. Adenosine-mediated constriction of the afferent arteriole via stimulation of A_1 receptors would then result in a functional (and theoretically reversible) reduction in GFR. Epithelial cell damage also reduces reabsorption of ultrafiltrate by the proximal tubule. Thus, increased distal delivery of NaCl might diminish GFR via activation of TGF, which, as already discussed, is most likely mediated by endogenous adenosine released from macula densa cells acting on A_1 receptors in the afferent arteriole. In support of the hypothesis that adenosine participates in ATN-induced ARF, many investigators have found that blockade of adenosine receptors ameliorates various forms of experimental ATN.

One popular model of ATN-induced ARF is glycerol-induced ARF. When injected intramuscularly, glycerol damages muscle tissue and leads to a myohemoglobinuric ARF that is similar to that observed in patients with crush injuries and rhabdomyolysis. Pretreatment of animals with adenosine receptor antagonists before or soon after injection of glycerol attenuates the glycerol-induced reduction in RBF and GFR. These results have been obtained with a variety of adenosine receptor antagonists including the nonselective antagonists aminophylline (Bidani and Churchill, 1983) and theophylline (Bidani et al., 1987), the relatively nonselective antagonist 8-phenyltheophylline (Bowmer et al., 1986), and several highly selective A_1 receptor antagonists such as 8-cyclopentyl-1,3-dipropylxanthine (Kellett et al., 1989; Panjehshahin et al., 1991b), 8-(dicyclopropylmethyl)-1,3-dipropylxanthine (Shimada et al., 1991), FK453 (also called FR-113453; Ishikawa et al., 1993a; Andoh et al., 1991) and KW-3902 (Suzuki, 1992). The observations that glycerol injections increase circulating levels of adenosine in the rat (Ishikawa et al., 1993b) and increase the renal constrictor response to adenosine (Gould et al., 1995) lend further support to the hypothesis that adenosine participates in glycerol-induced ARF.

In addition to glycerol-induced renal failure, pretreatment or cotreatment with adenosine receptor antagonists attenuates ARF induced by several nephrotoxic drugs that directly damage renal epithelial cells. Cisplatin-induced nephrotoxicity was attenuated by the nonselective antagonist aminophylline (Heidemann et al., 1989) and by the highly selective A_1 antagonists 8-cyclopentyl-1,3-dipropylxanthine (Knight et al., 1991), FK453 (Andoh et al., 1991), and KW-3902 (Suzuki, 1992; Nagashima et al., 1995). In addition, maneuvers that attenuate TGF (e.g., a high salt diet and furosemide) reduced cisplatin-induced ARF (Heidemann et al., 1985). Gentamicin-induced ARF was attenuated by the selective A_1 antagonists FK453 (Andoh et al., 1991) and KW-3902 (Yao et al., 1994a), and cephaloridine-induced nephrotoxicity was reduced by KW-3902 (Nagashima et al., 1994). Amphotericin B nephrotoxicity was reduced by aminophylline (Gerkens et al., 1983b; Heidemann et al., 1983a) and theophylline (Heidemann et al., 1991), as well as by inhibiting TGF with salt repletion (Heidemann et al., 1983b; Branch et al., 1987; Ohnishi et al., 1989; Llanos et al., 1991). However, 1,3-dipropyl-8-(p-sulfophenyl)xanthine, an adenosine receptor antagonist that is restricted to the extracellular space, did not attenuate acute amphotericin B nephrotoxicity (Kuan et al., 1990a). Also, theophylline did not prevent $HgCl_2$-induced ARF (Rossi et al., 1990), and neither theophylline (Gerkens and Smith, 1985) nor 8-cyclopentyl-1,3-dipropylxanthine (Panjehshahin et al., 1991a) attenuated cyclosporine induced ARF.

There is also evidence for the involvement of adenosine in ARF caused by renal ischemia or hypoxia. The diminished renal function induced by ischemia or hypoxia was reduced by theophylline (Osswald et al., 1977; Sakai et al., 1979; Lin et al., 1986, 1987, 1988; Gouyon and Guignard, 1988) and by inhibition of ecto-5′-nucleotidase with adenosine α,β-methylene diphosphate (van Waarde et al., 1989), whereas renal function was still further reduced by inhibition of adenosine uptake with dipyridamole (Lin et al., 1987). In addition, renal ischemia or hypoxia is a potent stimulant for the renal production of adenosine (Osswald et al., 1977; Miller et al., 1978; Ramos-Salazar and Baines, 1986; Zager et al., 1990).

Doubtless, antagonists that block A_1 receptors are effective in *preventing* many forms of experimental ATN-induced ARF. However, to our knowledge no study has demonstrated that blockade of A_1 receptors can *reverse* established ATN-induced ARF. This is a critically important issue, since if A_1 receptor antagonists must be administered before ATN-induced ARF occurs, clinical use would be limited to prophylaxis of, rather than treatment of established, ARF. Whether blockade of A_1 receptors can reverse ATN-induced ARF depends upon the mechanism by which A_1 receptor antagonists are renoprotective. If A_1 receptor antagonists are renoprotective because they interfere with the functional effects of adenosine on the renal microcirculation, then A_1 receptor antagonists should be able to attenuate ATN-induced ARF whether the antagonists are administered before or after the offending nephrotoxin.

As indicated earlier, the initial rationale for investigating A_1 receptor blockers in ATN-induced ARF was that adenosine is involved in the pathophysiology of ATN-induced ARF as a mediator of functional afferent arteriolar vasoconstriction. Despite this "received view" regarding the pathophysiological role of adenosine in ATN-induced ARF, this hypothesis has not been subjected to critical experimental evaluation. Therefore, we conducted an extensive series of experiments to test the received view (Jackson et al., 1995).

In our experiments, rats were administered a dose of cisplatin or its vehicle and were studied 6 days later after ATN-induced ARF was fully developed in the cisplatin-treated rats. We found that:

1. High-dose intrarenal infusion of an A_1 agonist (N^6-cyclopentyladenosine) could not mimic the reductions in GFR and the increases in renal vascular resistance (RVR) caused by cisplatin, as would be anticipated if endogenous adenosine plays an important role with regard to altering renal hemodynamics during ARF.
2. In animals with established cisplatin-induced ARF, the GFR and RVR responses to N^6-cyclopentyladenosine were not altered compared with vehicle-treated rats. (If endogenous adenosine is reducing renal function by activating A_1 receptors, then the renal responses to exogenous A_1 agonists should be diminished since the receptors would already be occupied.)
3. Renal cortical levels of adenosine, inosine, xanthine, and hypoxanthine (as measured by renal microdialysis) were not increased in rats with established ARF (as would be expected if ARF were in part due to increased production of adenosine).
4. Administration of 1,3-dipropyl-8-cyclopentylxanthine, a highly selective A_1 receptor antagonist, did not improve renal function in established cisplatin-induced ARF (a result that is not consistent with a role for endogenous adenosine in established ARF).

Taken together, our results indicate that the received view regarding the role of adenosine in ATN-induced ARF is incorrect.

A more likely explanation for the remarkable effectiveness of A_1 receptor antagonists in the prevention of ATN-induced ARF is that endogenous adenosine facilitates reabsorption of ultrafiltrate by the proximal tubule, thus resulting in higher concentrations of nephrotoxins in the tubular lumen. Higher concentrations of nephrotoxins in the tubular lumen would result in a greater flux of nephrotoxins into the proximal epithelial cells, thus enhancing epithelial cell injury. By blocking A_1 receptors in the proximal tubule, reabsorption of ultrafiltrate is diminished and nephrotoxins are not as concentrated in proximal tubular epithelial cells. The support for this hypothesis is:

1. A_1 receptor antagonists are most effective in preventing ARF induced by glycerol, cisplatin, gentamicin, and cephaloridine; and less effective or ineffective in ARF caused by amphotericin B and cyclosporin. Glycerol, cisplatin, gentamicin and cephaloridine primarily damage the proximal tubule, whereas the damage caused by amphotericin B and cyclosporin is more widespread along the tubule.
2. A_1 receptors in the proximal tubule mediate increased transport (Coulson et al., 1991; Takeda et al., 1993), whereas A_1 receptors in more distal nephron segments have lesser effects in stimulating transport (van Buren et al., 1993; Balakrishnan et al., 1993), or may even inhibit reabsorption (Kelley et al., 1990; Beach and Good, 1992; Edwards and Spielman, 1994).
3. A_1 receptor antagonists prevent, but do not reverse, ATN-induced ARF (Jackson et al., 1995).
4. Acetazolamide, a diuretic known to inhibit transport in the proximal tubule, prevents cisplatin-induced ARF when given before, but not after, cisplatin administration (Heidemann et al., 1985).

If the hypothesis that A_1 receptor antagonists diminish tissue levels of nephrotoxins in the proximal tubule is correct, then the best clinical utilization of A_1 receptor blockers with regard to ARF would be to prophylactically treat patients receiving nephrotoxic drugs who are at high risk for developing ARF. However, additional studies are required to evaluate the suggested hypothesis.

Other Types of ARF Intrarenal or intravenous injection of hypertonic (ionic, high-osmolality) contrast media causes a rapid reduction in GFR and RBF that is usually transient, but may lead to ARF in susceptible patients (Talner and Davidson, 1968; Caldicott et al., 1970; Chou et al., 1971; Ansari and Baldwin, 1976; Katzberg et al., 1977, 1983; Larson, 1983; Reed et al., 1983). It is conceivable that the large osmotic diuresis that accompanies the administration of hypertonic contrast media triggers an adenosine-mediated TGF response that reduces GFR and RBF. In support of this hypothesis, Arend and coworkers (1987a) reported that the reductions in GFR and RBF caused by intrarenal injections of hypertonic contrast media in dogs are accompanied by a marked increase in the renal excretion rate of adenosine. Moreover, blockade of adenosine receptors with theophylline and inhibition of adenosine uptake with dipyridamole reduced and enhanced, respectively, the RBF and GFR responses to hypertonic contrast media. Similarly, Deray et al. (1990a) found that theophylline attenuated the renovascular response to hypertonic contrast media in dogs. Moreover, Arakawa et al. (1996) reported that KW-3902 attenuated the reductions in GFR and RBF induced by non-ionic radiocontrast media in dogs with renal insufficiency

Recently, these findings in animals have been extended to humans. Erley and co-workers (1994) reported that theophylline blocked the reduction in GFR induced by non-ionic low osmolar contrast media in patients. Katholi et al. (1995) observed in patients that:

1. Both low- and high-osmolality contrast media increased the urinary excretion rate of adenosine, and the increase in urinary adenosine excretion with high-osmolality contrast media was greater than that caused by low-osmolality contrast media.

2. Both low- and high-osmolality contrast media decreased creatinine clearance; however, the effect of high-osmolality contrast media on creatinine clearance was more pronounced.
3. Theophylline significantly attenuated the nephrotoxic effects of both low- and high-osmolality contrast media.

These results have three important implications. First, urinary adenosine excretion rate may be a useful clinical parameter for assessing a given patient's susceptibility to contrast media-induced renal damage. Second, theophylline could find clinical application to prevent contrast media–induced ARF. Third, additional studies should be performed both in animals and humans to determine whether selective A_1 receptor antagonists would attenuate contrast media–induced ARF more effectively and with fewer adverse effects than theophylline.

Although revascularization of a kidney with a stenotic renal artery usually improves renal function, in some instances technically successful revascularization is not accompanied by an initial improvement in renal function (Dean, 1985). A possible explanation for this phenomenon is that the ischemic kidney has adapted to a state of diminished metabolic activity such that with sudden reperfusion a significant quantity of ultrafiltrate escapes proximal reabsorption and activates an adenosine-mediated TGF response. In support of this hypothesis, Abels et al. (1992) reported that the adenosine receptor antagonist CGS 15943A blocked the increase in RVR in dogs following revascularization of a chronically stenotic renal artery. If this finding can be confirmed and extended to humans, adenosine receptor antagonists may find clinical utility as agents to improve renal function in patients undergoing renal artery reconstruction.

Endotoxemia due to gram-negative sepsis is often accompanied by ARF, and mortality in such patients is greater than 60% (Turney et al., 1990). Interestingly, incubation of endothelial cells with endotoxin releases adenosine and adenosine metabolites (Rounds et al., 1994). It is possible, therefore, that endotoxin-induced ARF may be in part mediated by adenosine. Recent studies by Knight et al. (1993a) demonstrated that administration of the selective A_1 receptor antagonist 1,3-dipropyl-8-cyclopentylxanthine attenuated to a small degree the reductions in RBF and GFR caused by systemic administration of endotoxin to rats. If these results can be confirmed and extended to humans, A_1 receptor antagonists could find clinical utility in the treatment of endotoxemia.

During some surgical procedures, it is desirable to produce a degree of controlled arterial hypotension. Adenosine is well suited for this use. Since it has an extraordinarily short half-life in humans, the targeted level of blood pressure can be precisely sustained and controlled by adjusting the rate of adenosine infusion, and once the infusion is stopped the blood pressure recovers rapidly. However, a disadvantage to the use of adenosine is that systemic infusions usually adversely affect renal hemodynamics and diminish renal excretory function. Recent studies by Barrett and Wright (1994) demonstrate in the rat that administration of N-0861, a highly selective A_1 receptor antagonist, attenuates the adverse effects of systemic infusions of adenosine on the kidneys, while having little effect on the hypotensive response. Confirmation and extension of these studies to humans may lead to a new clinical indication for A_1 receptor antagonists.

As is obvious from the discussion thus far, most work regarding the renal adenosine system is predicated on the notion that too much adenosine is detrimental to the kidney. A recent study by Vallon and Osswald (1994) challenges that simple view.

Diabetes is all too often accompanied by increased glomerular capillary pressure leading to hyperfiltration, increased urinary protein excretion, and ultimately diabetic nephropathy. Studies in animal models of diabetes suggest that these changes in renal function may be due to diminished preglomerular vascular resistance. Since adenosine constricts the afferent arteriole, it is plausible that augmenting the renal adenosine system could prevent diabetes-induced changes in renal function. To test this hypothesis, Vallon and Osswald (1994) induced diabetes in rats by administering streptozotocin, and augmented the renal adenosine system in some animals by administering dipyridamole to block uptake of adenosine. Importantly, dipyridamole attenuated the increase in urinary protein excretion induced by streptozotocin. Also, in streptozotocin-treated rats, TGF responses were diminished and GFR was increased; however, these defects were not present in diabetic rats receiving dipyridamole. These experiments raise the intriguing possibility that augmenting renal adenosine levels could prevent diabetic nephropathy, a serious, common, and costly disease.

Chronic Renal Failure

The many positive results with A_1 receptor antagonist in various models of ARF has led to optimism that A_1 receptor antagonists might be beneficial in chronic renal failure (CRF). Unfortunately, there are no published data that directly address this issue. However, in this author's opinion it is unlikely that A_1 receptor antagonists would ameliorate CRF, and in fact there are reasons to think that antagonism of A_1 receptors could worsen CRF.

Regardless of the initial insult, the pathophysiology of CRF appears to involve the following vicious cycle: initial insult → loss of nephron units → decreased afferent arteriolar resistance in remaining nephrons → increased glomerular pressure and hyperfiltration in remaining nephrons → glomerulosclerosis → further loss of nephron units with acceleration of the cycle. Any agent that decreases afferent arteriolar resistance would accelerate, not attenuate, this vicious cycle of nephron loss. Since adenosine constricts the afferent arteriole, blockade of renovascular A_1 receptors might hasten renal deterioration in CRF.

Several lines of experimental evidence suggest that blocking A_1 receptors might be disadvantageous in CRF. First, as previously mentioned, Vallon and Osswald (1994) have demonstrated that augmenting renal adenosine levels improves, not worsens, experimental diabetic nephropathy, a form of CRF. Second, blockade of A_1 receptors enhances renin release (*vide supra*). Since Ang II constricts the efferent arteriole and thereby increases glomerular capillary pressure, activation of the renin-angiotensin system in CRF is probably ill-advised in patients with CRF. This view is supported by a recent study which clearly demonstrates that blocking the renin-angiotensin system with captopril attenuates the decline in renal function in patients with diabetic nephropathy (Lewis et al., 1993). Third, in a recent study, we demonstrated that acute blockade or renal A_1 receptors causes an increase in glomerular capillary pressure (Munger and Jackson, 1994). Fourth, chronic administration of caffeine enhances the ability of Ang II to increase filtration fraction, a phenomenon suggesting that blockade of renal adenosine receptors augments Ang II–induced glomerular hypertension (Holycross and Jackson, 1992). Fifth, acute administration of the selective A_1 receptor antagonist FK453 augmented renin release, but did not improve GFR, in patients with chronic renal disease (Balakrishnan et al., 1996). *The above considerations suggest that augmenting the preglomerular actions of adenosine might diminish the rate of nephron loss in CRF.*

K^+-Sparing Diuretics

As mentioned previously (*vide supra*) blockade of renal A_1 receptors inhibits tubular transport, predominantly in the proximal tubule, and induces a brisk natriuresis/diuresis with lesser effects on potassium excretion (Coulson and Scheinman, 1989; Terai et al., 1990; Collis et al., 1991; Ibarrola et al., 1991; Shimada et al., 1991; Suzuki et al., 1992; Kuan et al., 1993; Knight et al., 1993b; Mizumoto et al., 1993; van Buren et al., 1993; Balakrishnan et al., 1993; Munger and Jackson, 1994). The diuretic and natriuretic effects of A_1 receptor antagonists are observed in animals with ARF (Yao et al., 1994b). Studies by Takeda et al. (1993) suggest that blockade of A_1 receptors inhibits basolateral Na^+-$3HCO_3^-$ symport in the proximal tubule. Whether chronic administration of A_1 receptor antagonists causes a sustained reduction in total body sodium presently is unclear. Also, it will be important to determine whether A_1 receptor antagonists augment the diuretic efficacy of loop diuretics and/or thiazide diuretics by providing simultaneous blockade of several nephron segments. If so, adding an A_1 receptor antagonist to loop and/or thiazide diuretics could be used to overcome the common problem of diuretic resistance. Animal studies suggest that the natriuretic effects of A_1 receptor blockade may lower arterial blood pressure in those forms of hypertension characterized by volume expansion (Nomura et al., 1995). Clinical studies are required to resolve these issues and to clarify the role of A_1 receptor antagonists as diuretic and antihypertensive agents.

Caffeine in High Renin States

The adenosine-brake on renin release, like any braking mechanism, is only important when there exists a stimulus to be restrained. Therefore, in patients with elevated plasma renin activity (PRA) the adenosine-brake on renin release should be of physiological significance, and interruption of the adenosine-brake in such patients would enhance renin release.

Patients with renovascular hypertension, hypertensive patients being treated with vasodilating drugs, diuretics and/or dietary salt restriction, and patients with heart failure or liver cirrhosis have an activated renin-angiotensin system. Moreover, in all such patients activation of the renin-angiotensin system has adverse consequences. For example, in hypertensive individuals, activation of the renin-angiotensin system increases blood pressure and attenuates the response to antihypertensive drugs. In patients with hypertension, heart failure, or liver cirrhosis, the more activated is the renin-angiotensin system the less optimistic is the prognosis.

Caffeine, a widely consumed psychotropic drug, is a well-known adenosine receptor antagonist (von Borstel et al., 1984; Schwabe et al., 1985) that theoretically could inactivate the adenosine-brake on renin release. If this hypothesis is correct, then consumption of caffeine by patients with an activated renin-angiotensin system could carry adverse consequences.

As an initial test of this hypothesis, we examined the effects of chronic administration of caffeine on arterial blood pressure, renal function, PRA, and plasma aldosterone levels in rats with high-renin renovascular hypertension (Ohnishi et al., 1986, 1988). These studies demonstrated that pharmacologically relevant plasma levels of caffeine have many adverse effects in rats with high-renin renovascular hypertension (2-kidney-1-clip rats) including: (i) induction of malignant hypertension; (ii) increased levels of plasma renin; (iii) increased plasma levels of aldosterone; (iv) renal failure; and (v) accelerated mortality. Our initial findings have been verified by two independent groups (Kohno et al., 1991; Choi et al., 1993), and also by a telemetry study conducted in our laboratory (Kost et al., 1994).

The adverse effects of caffeine in high-renin renovascular hypertension are mediated by activation of the renin-angiotensin system. Evidence for this conclusion is twofold. First, blocking the renin-angiotensin system with an angiotensin-converting enzyme inhibitor prevents the adverse effects of caffeine in high-renin renovascular hypertension, indicating that activation of the renin-angiotensin system is the mechanism by which caffeine worsens renovascular hypertension (Ohnishi et al., 1986). Second, caffeine does not worsen hypertension in rats with normal or low-renin hypertension including 1-kidney-1-clip renovascular hypertensive rats (Ohnishi et al., 1988), or rats with genetic hypertension (Ohnishi et al., 1986). In fact, caffeine lowers blood pressure in animals with deoxycorticosterone acetate–salt hypertension (Choi et al., 1993), a model of hypertension associated with very low levels of PRA.

Future studies are required to determine whether caffeine ingestion in other high-renin states causes adverse consequences and whether our findings in rats can be extrapolated to humans.

EFFECTS MEDIATED BY RENAL P_2 (ATP) RECEPTORS

Introduction

Although many researchers have examined the participation of P_1 (adenosine) receptors in renal physiology/pathophysiology, only a few investigators have explored the potential roles of P_2 (ATP) receptors in renal function. This relative paucity of information is because the medicinal chemistry and molecular biology of P_2 receptors has lagged far behind that of P_1 receptors, and renal physiologists interested in this research area have only scant and relatively nonselective pharmacological probes in their experimental armamentarium. Even so, enough data has been generated to warrant guarded optimism that P_2 receptors will eventually be shown to participate importantly in renal physiology, particularly with regard to modulation of the renal microcirculation and tubular transport (see references Inshco et al., 1994 and Friedlander and Amiel, 1995 for more detailed discussions regarding the renal vascular and tubular effects, respectively, of nucleotides).

Nomenclature of P_2 Receptors

The nomenclature for P_2 receptors is both confusing and evolving (Williams, 1995). Through 1992, P_2 receptors were subclassified as P_{2x}, P_{2y}, P_{2t}, P_{2Z}, $P_{2u/n}$, and P_{2D}. However, the International Union of Pharmacology Purinoceptor Committee recommends as of 1994 the use of superfamily nomenclature in which P_2 receptors are divided into three superfamilies designated P2X, P2Y, and P2Z with subtypes within each superfamily indicated by additional Arabic numerals. If the receptor has been pharmacologically characterized but not cloned subscripts are not used (e.g., P2X1), whereas if the receptor has been cloned subscripts are employed (e.g., P_{2x1}). When this chapter was written, the P2X superfamily contained three members; the P2Y superfamily, seven members; and the P2Z family, one member. Unfortunately, the ability of state-of-the-art P_2 receptor nomenclature to resolve P_2 receptor subtypes far exceeds the sophistication of most published studies in the area of renal physiology. Therefore in the subsequent discussion, few attempts are made to resolve P_2 receptor subtypes beyond the P2X and P2Y superfamily designations. (Subscripts are not used in the subsequent discussion when reference is to the superfamily class rather than to specifically cloned members of the class.)

Renal Hemodynamics

When delivered directly into the renal artery, ATP vasodilates the dog kidney (Harvey, 1964; Tagawa and Vander, 1970; Majid and Navar, 1992) yet vasoconstricts the rabbit (Needleman et al., 1974) and rat (Churchill and Ellis, 1993a) kidney. However, in the dog kidney, blockade of renal NO synthase converts ATP-induced renal vasodilation to vasoconstriction (Majid and Navar, 1992). This finding suggests that the renal vasculature contains both vasoconstrictor P2X receptors as well as vasodilator P2Y receptors. Consistent with this conclusion are the observations by Churchill and Ellis (1993a) that 2-methylthio-ATP, a P2Y receptor agonist, vasodilates the isolated, perfused rat kidney, whereas α,β-methylene-ATP, a P2X receptor agonist, vasoconstricts the same preparation. Moreover, renal vasodilation induced by 2-methylthio-ATP is nearly abolished by an NO synthase inhibitor (Churchill and Ellis, 1993a). Most likely, P2X receptors exist directly on vascular smooth muscle cells, whereas P2Y receptors reside primarily on endothelial cells and cause indirect vasodilation by stimulating the release of NO (Dalziel and Westfall, 1994).

Recent studies by Inshco and coworkers (1991, 1992) using the *in vitro* blood-perfused juxtamedullary nephron technique demonstrate that P2X receptor agonists vasoconstrict afferent, but not efferent, arterioles. Similarly, Weihprecht and colleagues (1992) using the isolated rabbit afferent arteriole preparation observed P2X receptor–mediated vasoconstriction of the afferent arteriole. Although other preglomerular vascular elements, such as arcuate and interlobular arteries, vasoconstrict when exposed to P2X receptor agonists, the response appears to be transient (Inscho et al., 1992).

Tubuloglomerular Feedback

The unique responsiveness of the afferent arteriole to P2X receptor agonists suggests that ATP might participate in TGF. Consistent with this idea are the observations by Mitchell and Navar (1993) that saturating doses of either ATP or β,γ-methylene-ATP markedly attenuate TGF as assessed by stop-flow pressure responses to late proximal perfusion. However, the inferential strength of this finding is weakened by the subsequent report by Sehic et al. (1994) that administration of α,β-methylene-ATP desensitizes the rat kidney to vasoconstriction induced by a number of substances including ATP, renal nerve stimulation, norepinephrine, Ang II, and arginine vasopressin. Therefore, blockade of TGF responses by ATP and β,γ-methylene-ATP could be due to a nonspecific desensitization of afferent arterioles to all vasoconstrictors. Confirmation or refutation of the hypothesis that ATP mediates TGF must await the availability of highly potent and selective P2X receptor antagonists.

Tubular Transport

In vitro studies with receptor agonists suggest that P_2 receptors, like P_1 receptors, may influence tubular transport. In cultured Madin-Darby canine kidney (MDCK) cells, a cell line with distal phenotype, ATP increased short circuit current due both to an enhanced transepithelial potential difference and to an increased ionic conductance (Simmons, 1981a,b). This effect of ATP was not mediated by metabolism of ATP to adenosine (Simmons, 1981a). Subsequent studies by Lang et al. (1988), Paulmichl and Lang (1988), Friedrich et al. (1989), Zegarra-Moran et al. (1995), and Post et al. (1996) in MDCK cells indicate that:

1. ATP hyperpolarizes MDCK cells by increasing potassium conductance.
2. Activation of potassium channels by ATP requires calcium from both intracellular and extracellular sources.
3. ATP increases the open probability of an inwardly rectifying potassium-selective channel.
4. ATP increases transepithelial transport of chloride.
5. The effect of ATP on short circuit current is biphasic and consists of an initial rapid and transient peak followed by a more prolonged response.
6. ATP stimulates prostaglandin synthesis.
7. UTP has the same effects as ATP (which suggests that the P_2 receptor is a member of the P2Y superfamily, perhaps P_{2y2} [formerly $P_{2u/n}$]).

In another cell line with distal phenotype (A6 cells), Middleton et al. (1993) observed nucleotide-induced changes in Na^+–K^+–$2Cl^-$ symport that (i) were not blocked by an adenosine receptor antagonist, (ii) were not associated with changes in cyclic AMP or adenylyl cyclase activity, and (iii) appeared to be mediated by release of intracellular stores of calcium leading to activation of chloride channels. The order of nucleotide potency was ATP $>$ UTP $>$ ATPγS $>$ 2-methylthio-ATP $>$ α,β-methylene-ATP, again suggesting a P_{2y2}-like receptor. Data in cultured LLC-PK_1 cells (mixed proximal/distal phenotype) also indicate P2Y receptor–mediated release of intracellular calcium and activation of protein kinase C activity (Weinberg et al., 1989; Anderson et al., 1991). Adenylyl cyclase activity was also reduced by nucleotides, but this response was not mediated by a pertussis toxin–sensitive G protein (Anderson et al., 1991). Although most studies have utilized cell lines with distal phenotype, Nanoff et al. (1990) have observed P2Y receptor–mediated stimulation of inositol phosphate formation in slices or rat renal cortex. More recently, Lederer et al. (1995) and Rouse et al. (1994) have investigated the effects of nucleotides on transport in proximal tubules and cortical collecting tubules, respectively. Lederer et al. (1995) report the existence of a P2X receptor in proximal tubules that inhibits sodium dependent phosphate transport, and Rouse et al. (1994) demonstrate the existence of a nucleotide receptor that attenuates AVP-induced osmotic water permeability. In summary, the limited available evidence suggests that a P2Y receptor (perhaps P_{2y2}) mediates stimulation of distal (and maybe proximal) tubular transport, a P2X receptor inhibits proximal transport, and AVP-induced water permeability may be modified by a nucleotide receptor.

Other Effects

Besides affecting the renal microcirculation and tubular transport, renal P_2 receptors may influence other important renal processes. Churchill and Ellis (1993b) report that

P2Y receptor activation stimulates renin release via increased production of NO, and Pfeilschifter (1990a,b) finds that P2Y receptors activate the hydrolysis of polyphosphoinositides in cultured mesangial cells. Finally, limited evidence suggests that P_2 receptors may also stimulate proliferation of mesangial cells (Schulze-Lohoff et al., 1992) and increase DNA synthesis by proximal tubular cells (Humes and Cieslinski, 1991). There is little doubt that future studies will elucidate the involvement of P_2 receptors in many aspects of renal physiology/pathophysiology.

ACKNOWLEDGMENTS

This work was supported in part by NIH grants HL40319, HL55314, and HL35909.

REFERENCES

Abels BC, Callis JT, Branch RA, Sabra R (1992): Influence of the adenosine receptor antagonist, CGS 15943A, on renal function after reconstruction of chronic renal artery stenosis in the dog. J Pharmacol Exp Ther 262:166–172.

Agmon Y, Dinour D, Brezis M (1993): Disparate effects of adenosine A_1- and A_2-receptor agonists on intrarenal blood flow. Am J Physiol 265:F802–F806.

Anderson RJ, Breckon R, Dixon BS (1991): ATP receptor regulation of adenylate cyclase and protein kinase C activity in cultured renal LLC-PK_1 cells. J Clin Invest 87:1732–1738.

Andoh TF, Terai T, Nakano K, Horiai H, Mori J, Kohsaka M (1991): Protective effects of FR113453, an adenosine A_1-receptor antagonist, on experimental acute renal failure in rats. In: Proceedings of International Symposium on Acute Renal Failure; October 2–4, 1991; University of North Carolina at Chapel Hill; p 92.

Ansari Z, Baldwin DS (1976): Acute renal failure due to radio-contrast agents. Nephron 17:28–40.

Arakawa K, Suzuki H, Naitoh M, Matsumoto A, Hayashi K, Matsuda H, Ichihara A, Kubota E, Saruta T (1996): Role of adenosine in the renal responses to contrast medium. Kidney Int 49:1199–1206.

Arend LJ, Bakris GL, Burnett JC Jr, Megerian C, Spielman WS (1987a): Role for intrarenal adenosine in the renal hemodynamic response to contrast media. J Lab Clin Med 110:406–411.

Arend LJ, Burnatowska-Hledin MA, Spielman WS (1988): Adenosine receptor–mediated calcium mobilization in cortical collecting tubule cells. Am J Physiol 255:C581–C588.

Arend LJ, Haramati A, Thompson CI, Spielman WS (1984): Adenosine-induced decrease in renin release: Dissociation from hemodynamic effects. Am J Physiol 247:F447–F452.

Arend LJ, Sonnenburgh WK, Smith WL, Spielman WS (1987b): A_1 and A_2 adenosine receptors in rabbit cortical collecting tubule cells. Modulation of hormone-stimulated cAMP. J Clin Invest 79:710–714.

Arend LJ, Thompson CI, Brandt MA, Spielman WS (1986): Elevation of intrarenal adenosine by maleic acid decreases GFR and renin release. Kidney Int 30:656–661.

Balakrishnan VS, Coles GA, Williams JD (1993): A potential role for endogenous adenosine in control of human glomerular and tubular function. Am J Physiol 265:F504–F510.

Balakrishnan VS, Coles GA, Williams JD (1996): Functional role of endogenous adenosine in human chronic renal disease. Exp Nephrol 4:26–36.

Baranowski RL, Westenfelder C (1994): Estimation of renal interstitial adenosine and purine metabolites by microdialysis. Am J Physiol 267:F174–F182.

Barber R, Butcher RW (1981): The quantitative relationship between intracellular concentration and egress of cyclic AMP from cultured cells. Mol Pharmacol 19:38–43.

Barber R, Butcher RW (1983): The egress of cyclic AMP from metazoan cells. Adv Cyclic Nucleotide Res 15:119–138.

Barchowsky A, Data JL, Whorton RA (1987): Inhibition of renin release by analogues of adenosine in rabbit renal cortical slices. Hypertension 9:619–623.

Barraco RA, Janusz CJ, Polasek PM, Parizon M, Roberts PA (1988): Cardiovascular effects of microinjection of adenosine into nucleus tractus solitarius. Brain Res Bull 20:129–132.

Barraco RA, Phillis JW, Campbell WR, Marcantonio DR, Salah RS (1986): The effects of central injections of adenosine analogs on blood pressure and heart rate in the rat. Neuropharmacology 25:675–680.

Barrett RJ, Droppleman DA (1993): Interactions of adenosine A_1 receptor–mediated renal vasoconstriction with endogenous nitric oxide and ANG II. Am J Physiol 265:F651–F659.

Barrett RJ, Wright KF (1994): A selective adenosine A_1 receptor antagonist attenuates renal dysfunction during controlled hypotension with adenosine in rats. Anesth Analg 79:460–465.

Baudouin-Legros M, Badou A, Paulais M, Hammet M, Teulon J (1995): Hypertonic NaCl enhances adenosine release and hormonal cAMP production in mouse thick ascending limb. Am J Physiol 269:F103–F109.

Beach RE, Good DW (1992): Effects of adenosine on ion transport in rat medullary thick ascending limb. Am J Physiol 263:F482–F487.

Bidani AK, Churchill PC (1983): Aminophylline ameliorates glycerol-induced acute renal failure in rats. Can J Physiol Pharmacol 61:567–571.

Bidani AK, Churchill PC, Packer W (1987): Theophylline-induced protection in myoglobinuric acute renal failure: further characterization. Can J Physiol Pharmacol 65:42–45.

Blanco J, Canela EI, Mallol J, Lluis C, Franco R (1992): Characterization of adenosine receptors in brush-border membranes from pig kidney. Br J Pharmacol 107:671–678.

Bowmer CJ, Collis MG, Yates MS (1986): Effect of the adenosine antagonist 8-phenyltheophylline on glycerol-induced acute renal failure in the rat. Br J Pharmacol 88:205–212.

Branch RA, Jackson EK, Jacqz E, Stein R, Ray WA, Ohnhaus EE, Meusers P, Heidemann H (1987): Amphotericin-B nephrotoxicity in humans decreased by sodium supplements with coadministration of ticarcillin or intravenous saline. Klin Wochenschr 65:500–506.

Brown NJ, Porter J, Ryder D, Branch RA (1991): Caffeine potentiates the renin response to diazoxide in man. Evidence for a regulatory role of endogenous adenosine. J Pharmacol Exp Ther 256:56–61.

Cai H, Batuman V, Puschett DB, Puschett JB (1994): Effect of KW-3902, a novel adenosine A_1 receptor antagonist, on sodium-dependent phosphate and glucose transport by the rat renal proximal tubular cell. Life Sci 55:839–845.

Cai H, Puschett DB, Guan S, Batuman V, Puschett JB (1995): Phosphate transport inhibition by KW-3902, an adenosine A_1 receptor antagonist, is mediated by cyclic adenosine monophosphate. Am J Kidney Dis 26:825–830.

Caldicott WJ, Hollenberg NK, Abrams HL (1970): Characteristics of responses of the renal vascular bed to contrast media: Evidence of vasoconstriction induced by renin-angiotensin system. Invest Radiol 5:539–547.

Callis JT, Kuan CJ, Branch KR, Abels, BC, Sabra R, Jackson EK, Branch RA (1989): Inhibition of renal vasoconstriction induced by intrarenal hypertonic saline by the nonxanthine adenosine antagonist CGS 15943A. J Pharmacol Exp Ther 248:1123–1129.

Cannon ME, Twu BM, Yang CS, Hsu CH (1989): The effect of theophylline and cyclic adenosine 3′,5′-monophosphate on renin release by afferent arterioles. J Hypertens 7:569–576.

Carmines PK, Inscho EW (1994): Renal arteriolar angiotensin responses during varied adenosine receptor activation. Hypertension 23:I-114–I-119.

Casavola V, Guerra L, Reshkin SJ, Jacobson KA, Verrey F, Murer H (1996): Effect of adenosine on Na+ and Cl− currents in A6 monolayers. Receptor localization and messenger involvement. J Membr Biol 151:237–245.

Choi KC, Lee J, Moon KH, Park KK, Kim SW, Kim NH (1993): Chronic caffeine ingestion exacerbates 2-kidney, 1-clip hypertension and ameliorates deoxycorticosterone acetate-salt hypertension in rats. Nephron 65:619–622.

Chou CC, Hook JB, Hsieh CP, Burns TD, Dabney JM (1971): Effects of radiopaque dyes on renal vascular resistance. J Lab Clin Med 78:705–712.

Churchill PC (1982): Renal effects of 2-chloroadenosine and their antagonism by aminophylline in anesthetized rats. J Pharmacol Exp Ther 222:319–323.

Churchill PC, Bidani A (1987): Renal effects of selective adenosine receptor agonists in anesthetized rats. Am J Physiol 252:F299–F303.

Churchill PC, Churchill MC (1985): A_1 and A_2 adenosine receptor activation inhibits and stimulates renin secretion of rat renal cortical slices. J Pharmacol Exp Ther 232:589–594.

Churchill PC, Ellis VR (1993a): Pharmacological characterization of the renovascular P_2 purinergic receptors. J Pharmacol Exp Ther 265:334–338.

Churchill PC, Ellis VR (1993b): Purinergic P_{2Y} receptors stimulate renin secretion by rat renal cortical slices. J Pharmacol Exp Ther 266:160–163.

Churchill, PC, Bidani AK, Churchill MC, Prada J (1984): Renal effects of 2-chloroadenosine in the two-kidney goldblatt rat. J Pharmacol Exp Ther 230:302–306.

Churchill PC, Jacobson KA, Churchill MC (1987a): XAC, a functionalized congener of 1,3-dialkylxanthine, antagonizes A_1 adenosine receptor-mediated inhibition of renin secretion in vitro. Arch Int Pharmacodyn 290:293–301.

Churchill PC, Rossi NF, Churchill MC (1987b): Renin secretory effects of N^6-cyclohexyladenosine: Effects of dietary sodium. Am J Physiol 252:F872–F876.

Collis MG, Shaw G, Keddie JR (1991): Diuretic and saliuretic effects of 1,3-dipropyl-8-cyclopentylxanthine, a selective A_1-adenosine receptor antagonist. J Pharm Pharmacol 43:138–139.

Cook, CB, Churchill PC (1984): Effects of renal denervation on the renal responses of anesthetized rats to cyclohexyladenosine. Can J Physiol Pharmacol 62:934–938.

Coulson R, Scheinman SJ (1989): Xanthine effects on renal proximal tubular function and cyclic AMP metabolism. J Pharmacol Exp Ther 248:589–595.

Coulson R, Johnson RA, Olsson RA, Cooper DR, Scheinman SJ (1991): Adenosine stimulates phosphate and glucose transport in opossum kidney epithelial cells. Am J Physiol 260:F921–F928.

Čulić O, Sabolić I, Žanić-Grubišić T (1990): The stepwise hydrolysis of adenine nucleotides by ectoenzymes of rat renal brush-border membranes. Biochim Biophys Acta 1030:143–151.

Dalziel HH, Westfall DP (1994): Receptors for adenine nucleotides and nucleosides: Subclassification, distribution, and molecular characterization. Pharmacol Rev 46:449–466.

Dean RH (1985): Complications of renal revascularization. In Bernhard VM, Towne JB (eds): "Complications of Vascular Surgery," Orlando, FL: Grune & Stratton, pp 239–241.

Deray G, Branch RA, Herzer WA, Ohnishi A, Jackson EK (1987): Adenosine inhibits β-adrenoceptor but not DBcAMP-induced renin release. Am J Physiol 252:F46–F52.

Deray G, Branch RA, Jackson EK (1989a): Methylxanthines augment the renin response to suprarenal-aortic constriction. Naunyn Schmiedebergs Arch Pharmacol 339:690–696.

Deray G, Branch RA, Ohnishi A, Jackson EK (1989b): Adenosine inhibits renin release induced by suprarenal-aortic constriction and prostacyclin. Naunyn Schmiedebergs Arch Pharmacol 339:590–595.

Deray G, Martinez F, Cacoub P, Baumelou B, Baumelou A, Jacobs C (1990a): A role for adenosine, calcium and ischemia in radiocontrast-induced intrarenal vasoconstriction. Am J Nephrol 10:316–322.

Deray G, Sabra R, Herzer WA, Jackson EK, Branch RA (1990b): Interaction between angiotensin II and adenosine in mediating the vasoconstrictor response to intrarenal hypertonic saline infusions in the dog. J Pharmacol Exp Ther 252:631–635.

Dietrich MS, Endlich K, Parekh N, Seinhausen M (1991): Interaction between adenosine and angiotensin II in renal microcirculation. Microvasc Res 41:275–288.

Dinour D, Brezis M (1991): Effects of adenosine on intrarenal oxygenation. Am J Physiol 261: F787–F791.

Edlund A, Sollevi A (1993): Renal effects of i.v. adenosine infusion in humans. Clin Physiol 13:361–371.

Edlund A, Ohlsén H, Sollevi A (1994): Renal effects of local infusion of adenosine in man. Clin Sci 87:143–149.

Edwards RM, Spielman WS (1994): Adenosine A_1 receptor-mediated inhibition of vasopressin action in inner medullary collecting duct. Am J Physiol 266:F791–F796.

Erley CM, Duda SH, Schlepckow S, Koehler J, Huppert PE, Strohmaier WL, Bohle A, Risler T, Osswald H (1994): Adenosine antagonist theophylline prevents the reduction of glomerular filtration rate after contrast media application. Kidney Int 45:1425–1431.

Fisher JW, Nakashima J (1992): The role of hypoxia in renal production of erythropoietin. Cancer 70:928–939.

Franco M, Bell PD, Navar LG (1989): Effect of adenosine A_1 analogue on tubuloglomerular feedback mechanism. Am J Physiol 257:F231–F236.

Fransen R, Koomans HA (1995): Adenosine and renal sodium handling: Direct natiuresis and renal nerve-mediated antinatriuresis. J Am Soc Nephrol 6:1491–1497.

Freissmuth M, Hausleithner V, Tuisl E, Nanoff C, Schütz W (1987a): Glomeruli and microvessels of the rabbit kidney contain both A_1- and A_2-adenosine receptors. Naunyn Schmiedebergs Arch Pharmacol 335:438–444.

Freissmuth M, Nanoff C, Tuisl E, Schütz W (1987b): Stimulation of adenylate cyclase activity via A_2-adenosine receptors in isolated tubules of the rabbit renal cortex. Eur J Pharmacol 138:137–140.

Friedlander G, Amiel C (1995): Extracellular nucleotides as modulators of renal tubular transport. Kidney Int 47:1500–1506.

Friedrich F, Weiss H, Paulmichl M, Lang F (1989): Activation of potassium channels in renal epithelioid cells (MDCK) by extracellular ATP. Am J Physiol 256:C1016–C1021.

Gerber JG, Nies AS (1986): Renal vasoconstrictor response to hypertonic saline in the dog: Effects of prostaglandins, indomethacin and theophylline. J Physiol 380:35–43.

Gerkens, JF, Smith AJ (1985): Captopril and theophylline treatment on cyclosporine-induced nephrotoxicity in rats. Transplantation 40:213–214.

Gerkens JF, Heidemann HT, Jackson EK, Branch RA (1983a): Aminophylline inhibits renal vasoconstriction produced by intrarenal hypertonic saline. J Pharmacol Exp Ther 225:611–615.

Gerkens JF, Heidemann HT, Jackson EK, Branch RA (1983b): Effect of aminophylline on amphotericin B nephrotoxicity in the dog. J Pharmacol Exp Ther 224:609–613.

Gould J, Bowmer CJ, Yates MS (1995): Renal haemodynamic responses to adenosine in acute renal failure. Nephron 71:184–189.

Gouyon JB, Guignard JP (1988): Theophylline prevents the hypoxemia-induced renal hemodynamic changes in rabbits. Kidney Int 33:1078–1083.

Haas JA, Osswald H (1981): Adenosine induced fall in glomerular capillary pressure. Effect of ureteral obstruction and aortic constriction in the Munich-Wistar rat kidney. Naunyn Schmiedebergs Arch Pharmacol 317:86–89.

Hall JE, Granger JP (1986a): Adenosine alters glomerular filtration control by angiotensin II. Am J Physiol 250:F917–F923.

Hall JE, Granger JP (1986b): Renal hemodynamics and arterial pressure during chronic intrarenal adenosine infusion in conscious dogs. Am J Physiol 250:F32–F39.

Harvey RB (1964): Effects of adenosine triphosphate on autoregulation of renal blood flow and glomerular filtration rate. Circ Res 14:I178–I182.

Hashimoto K, Kumakura S (1965): The pharmacological features of the coronary, renal, mesenteric and femoral arteries. Jpn J Physiol 15:540–551.

Heidemann HT, Bolten M, Inselmann G (1991): Effect of chronic theophylline administration on amphotericin B nephrotoxicity in rats. Nephron 59:294–298.

Heidemann HT, Gerkens JF, Jackson EK, Branch RA (1983a): Effect of aminophylline on renal vasoconstriction produced by amphotericin B in the rat. Naunyn Schmiedebergs Arch Pharmacol 324:148–152.

Heidemann HT, Gerkens JF, Jackson EK, Branch RA (1985): Attenuation of cisplatinum-induced nephrotoxicity in the rat by high salt diet, furosemide and acetazolamide. Naunyn Schmiedebergs Arch Pharmacol 329:201–205.

Heidemann HT, Gerkens JF, Spickard WA, Jackson EK, Branch RA (1983b): Amphotericin B nephrotoxicity in humans decreased by salt depletion. Am J Med 75:476–481.

Heidemann HT, Müller S, Mertins L, Stepan G, Hoffmann K, Ohnhaus EE (1989): Effect of aminophylline on cisplatin nephrotoxicity in the rat. Br J Pharmacol 97:313–318.

Holycross BJ, Jackson EK (1992): Effects of chronic treatment with caffeine on kidneys responses to angiotensin II. Eur J Pharmacol 219:361–367.

Hou SH, Bushinky DA, Wish JB, Cohen JJ, Harrington JT (1983): Hospital-acquired renal insufficiency: A prospective study. Am J Med 74:243–248.

Humes HD, Cieslinski DA (1991): Adenosine triphosphate stimulates thymidine incorporation but does not promote cell growth in primary cultures or renal proximal tubular cells. Renal Physiol Biochem 14:253–258.

Ibarrola AM, Inscho EW, Vari RC, Navar LG (1991): Influence of adenosine receptor blockade on renal function and renal autoregulation. J Am Soc Nephrol 2:991–999.

Inscho EW, Carmines PK, Navar LG (1991): Juxtamedullary afferent arteriolar responses to P_1 and P_2 purinergic stimulation. Hypertension 17:1033–1037.

Inscho EW, Mitchell KD, Navar LG (1994): Extracellular ATP in the regulation of renal microvascular function. FASEB J 8:319–328.

Inscho EW, Ohishi K, Navar LG (1992): Effects of ATP on pre- and postglomerular juxtaglomedullary microvasculature. Am J Physiol 263:F886–F893.

Ishikawa I, Shikura N, Takada K (1993a): Amelioration of glycerol-induced acute renal failure in rats by an adenosine A_1 receptor antagonist (FR-113453). Renal Failure 15:1–5.

Ishikawa I, Shikura N, Takada K, Sato Y (1993b): Changes of adenosine levels in the carotid artery, renal vein and inferior vena cava after glycerol or mercury injection in the rat. Nephron 64:605–608.

Jackson EK (1991): Adenosine: A physiological brake on renin release. Annu Rev Pharmacol Toxicol 31:1–35.

Jackson, EK, Mi Z, Herzer WA (1995): Studies on the mechanism by which adenosine receptor antagonists attenuate acute renal failure. In Belardinelli L, Pelleg A (eds): "Adenosine and Adenine Nucleotides: From Molecular Biology to Integrative Physiology," Boston: Kluwer, pp 415–423.

Kaissling B, Spiess S, Rinne B, LeHir M (1993): Effects of anemia on morphology of rat renal cortex. Am J Physiol 264:F608–F617.

Katholi RE, Taylor GJ, McCann WP, Woods WT Jr, Womach KA, McCoy CD, Katholi CR, Moses HW, Mishkel GJ, Lucore CL, Holloway RM, Miller BD, Woodruff RC, Dove JT, Mikell FL, Schneider JA (1995): Nephrotoxicity from contrast media: Attenuation with theophylline. Radiology 195:17–22.

Katzberg RW, Morris TW, Burgener FA, Kamm DE, Fischer HW (1977): Renal renin and hemodynamic responses to selective renal artery catheterization and angiography. Invest Radiol 12:381–388.

Katzberg RW, Schulman G, Meggs LG, Caldicott WJ, Damiano MM, Hollenberg NK (1983): Mechanism of renal response to contrast media in dogs: Decrease in renal function due to hypertonicity. Invest Radiol 18:74–80.

Kellett R, Bowmer CJ, Collis MG, Yates MS (1989): Amelioration of glycerol-induced acute renal failure in the rat with 8-cyclopentyl-1,3-dipropylxanthine. Br J Pharmacol 98:1066–1074.

Kelley GG, Poeschla EM, Barron HV, Forrest JN Jr (1990): A_1 adenosine receptors inhibit chloride transport in the shark rectal gland: Dissociation of inhibition and cyclic AMP. J Clin Invest 85:1629–1636.

King CD, Mayer SE (1974): Inhibition of egress of adenosine 3′,5′-monophosphate from pigeon erythrocytes. Mol Pharmacol 10:941–953.

Knight RJ, Bowmer CJ, Yates MS (1993a): Effect of the selective A_1 adenosine antagonist 8-cyclopentyl-1,3-dipropylxanthine on acute renal dysfunction induced by *Escherichia Coli* endotoxin in rats. J Pharm Pharmacol 45:979–984.

Knight RJ, Bowmer CJ, Yates MS (1993b): The diuretic action of 8-cyclopentyl-1,3-dipropylxanthine, a selective A_1 adenosine receptor antagonist. Br J Pharmacol 109:271–277.

Knight RJ, Collis MG, Yates MS, Bowmer CJ (1991): Amelioration of cisplatin-induced acute renal failure with 8-cyclopentyl-1,3-dipropylxanthine. Br J Pharmacol 104:1062–1068.

Kohno M, Murakawa K, Horio T, Yokokawa K, Yasunari K, Fukui T, Takeda T (1991): Plasma immunoreactive endothelin-1 in experimental malignant hypertension. Hypertension 18:93–100.

Kost CK Jr, Li P, Pfeifer CA, Jackson EK (1994): Telemetric blood pressure monitoring in benign 2-kidney, 1-clip renovascular hypertension: Effect of chronic caffeine ingestion. J Pharmacol Exp Ther 270:1063–1070.

Kövér G, Tost H (1994): The renal actions of adenosine. Acta Physiologica Hungarica 82:215–228.

Kuan CJ, Branch RA, Jackson EK (1990a): Effect of an adenosine receptor antagonist on acute amphotericin B nephrotoxicity. Eur J Pharmacol 178:285–291.

Kuan, CJ, Herzer WA, Jackson EK (1993): Cardiovascular and renal effects of blocking A_1 adenosine receptors. J Cardiovasc Pharm 21:822–828.

Kuan CJ, Wells JN, Jackson EK (1989): Endogenous adenosine restrains renin release during sodium restriction. J Pharmacol Exp Ther 249:110–116.

Kuan CJ, Wells JN, Jackson EK (1990b): Endogenous adenosine restrains renin release in conscious rats. Circ Res 66:637–646.

Laborit H, Manzo-Fry G, Baron C, Hasni H (1990): Changes in plasma catecholamine levels after the intraperitoneal and intracerebroventricular administration of adenosine analogues and of clonidine in conscious rats. Res Commun Chem Pathol Pharmacol 68:307–327.

Lane JD, Adcock RA, Williams RB, Kuhn CM (1990): Caffeine effects on cardiovascular and neuroendocrine responses to acute psychosocial stress and their relationship to level of habitual caffeine consumption. Psychosom Med 52:320–336.

Lang F, Plöckinger B, Häussinger D, Paulmichl M (1988): Effects of extracellular nucleotides on electrical properties of subconfluent Madin Darby canine kidney cells. Biochim Biophys Acta 943:471–476.

Lang MA, Preston AS, Handler JS, Forrest JN Jr (1985): Adenosine stimulates sodium transport in kidney A6 epithelia in culture. Am J Physiol 249:C330–C336.

Langård O, Holdaas H, Eide I, Kiil F (1983): Conditions for augmentation of renin release by theophylline. Scand J Clin Lab Invest 43:9–14.

Larson TS, Hudson K, Mertz JI, Romero JC, Knox FG (1983): Renal vasoconstrictive response to contrast medium: The role of sodium balance and the renin-angiotensin system. J Lab Clin Med 101:385–391.

Lederer ED, McLeish KR (1995): P2 purinoceptor stimulation attenuates PTH inhibition of phosphate uptake by a G protein-dependent mechanism. Am J Physiol 269:F309–F316.

Le Hir M, Kaissling B (1993): Distribution and regulation of renal ecto-5′-nucleotidase: Implications for physiological functions of adenosine. Am J Physiol 264:F377–F387.

Levens N, Beil M, Jarvis M (1991a): Renal actions of a new adenosine agonist, CGS 21680A selective for the A_2 receptor. J Pharmacol Exp Ther 257:1005–1012.

Levens N, Beil M, Schulz R (1991b): Intrarenal actions of the new adenosine agonist CGS 21680A, selective for the A_2 receptor. J Pharmacol Exp Ther 257:1013–1019.

LeVier DG, McCoy DE, Spielman WS (1992): Functional localization of adenosine receptor-mediated pathways in the LLC-PK_1 renal cell line. Am J Physiol 263:C729–C735.

Lewis EJ, Hunsicker LG, Bain RP, Rohde RD (1993): The effect of angiotensin-converting-enzyme inhibition on diabetic nephropathy. The collaborative study group. N Engl J Med 329:1456–1462.

Lin JJ, Churchill PC, Bidani AK (1986): Effect of theophylline on the initiation phase of postischemic acute renal failure in rats. J Lab Clin Med 108:150–154.

Lin JJ, Churchill PC, Bidani AK (1987): The effect of dipyridamole on the initiation phase of postischemic renal failure in rats. Can J Physiol Pharmacol 65:1491–1495.

Lin JJ, Churchill PC, Bidani AK (1988): Theophylline in rats during maintenance phase of post-ischemic acute renal failure. Kidney Int 33:24–28.

Llanos A, Cieza J, Bernardo J, Echevarria J, Biaggioni I, Sabra R, Branch RA (1991): Effect of salt supplementation on amphotericin B nephrotoxicity. Kidney Int 40:302–308.

Lorenz JN, Weihprecht H, He XR, Skøtt O, Briggs JP, Schnermann J (1993): Effects of adenosine and angiotensin on macula densa–stimulated renin secretion. Am J Physiol 265:F187–F194.

Lunding M, Steiness I, Thaysen JH (1964): Acute renal failure due to tubular necrosis. Immediate prognosis and complications. Acta Med Scand 176:103–119.

Macias-Nuñez JF, García-Iglesias C, Santos JC, Sanz E, López-Novoa JM (1985): Influence of plasma renin content, intrarenal angiotensin II, captopril, and calcium channel blockers on the vasoconstriction and renin release promoted by adenosine in the kidney. J Lab Clin Med 106:562–567.

Macias-Nunez JF, Revert M, Fiksen-Olsen M, Knox FG, Romero JC (1986): Effect of allopurinol on the renovascular responses to adenosine. J Lab Clin Med 108:30–36.

Majid DS, Navar LG (1992): Suppression of blood flow autoregulation plateau during nitric oxide blockade in canine kidney. Am J Physiol 262:F40–F46.

Marraccini P, Fedele S, Marzilli M, Orsini E, Dukic G, Serasini L, L'Abbate A (1996): Adenosine-induced renal vasoconstriction in man. Cardiovasc Res 32:949–953.

Mi Z, Jackson EK (1995): Metabolism of exogenous cyclic AMP to adenosine in the rat kidney. J Pharmacol Exp Ther 273:728–733.

Mi Z, Herzer WA, Zhang Y, Jackson EK (1994): 3-Isobutyl-1-methylxanthine decreases renal cortical interstitial levels of adenosine and inosine. Life Sci 54:PL277–PL282.

Middleton JP, Mangel AW, Basavappa S, Fitz JG (1993): Nucleotide receptors regulate membrane ion transport in renal epithelial cells. Am J Physiol 264:F867–F873.

Miller WL, Thomas RA, Berne RM, Rubio R (1978): Adenosine production in the ischemic kidney. Circ Res 43:390–397.

Mitchell KD, Navar LG (1993): Modulation of tubuloglomerular feedback responses by extracellular ATP. Am J Physiol 264:F458–F466.

Miyamoto M, Yagil Y, Larson T, Robertson C, Jamison RL (1988): Effects of intrarenal adenosine on renal function and medullary blood flow in the rat. Am J Physiol 255:F1230–F1234.

Mizumoto H, Karasawa A, Kubo K (1993): Diuretic and renal protective effects of 8-(noradamantan-3-yl)-1,3-dipropylxanthine (KW-3902), a novel adenosine A_1-receptor antagonist, via pertussis toxin insensitive mechanism. J Pharmacol Exp Ther 266:200–266.

Mosqueda-Garcia R, Killian TJ, Haile V, Tseng CJ, Robertson RM, Robertson D (1990): Effects of caffeine on plasma catecholamines and muscle sympathetic nerve activity in man. Circulation 82:III–335.

Mosqueda-Garcia R, Tseng CJ, Appalsamy M, Robertson D (1989): Modulatory effects of adenosine on baroreflex activation in the brainstem of normotensive rats. Eur J Pharmacol 174:119–122.

Moyer BD, McCoy DE, Lee B, Kizer N, Stanton BA (1995): Adenosine inhibits arginine vasopressin-stimulated chloride secretion in a mouse IMCD cell line (mIMCD-K2). Am J Physiol 269:F884–F891.

Munger KA, Jackson EK (1994): Effects of selective A_1 receptor blockade on glomerular hemodynamics: Involvement of renin-angiotensin system. Am J Physiol 267:F783–F790.

Murray RD, Churchill PC (1984): Effects of adenosine receptor agonists in the isolated, perfused rat kidney. Am J Physiol 247:H343–H348.

Murray RD, Churchill PC (1985): Concentration dependency of the renal vascular and renin secretory responses to adenosine receptor agonists. J Pharmacol Exp Ther 232:189–193.

Nagashima K, Kusaka H, Karasawa A (1995): Protective effects of KW-3902, an adenosine A_1-receptor antagonist, against cisplatin-induced acute renal failure in rats. Jpn J Pharmacol 67:349–357.

Nagashima K, Kusaka H, Sato K, Karasawa A (1994): Effects of KW-3902, a novel adenosine A_1-receptor antagonist, on cephaloridine-induced acute renal failure in rats. Jpn J Pharmacol 64:9–17.

Nakashima J, Brookins J, Beckman B, Fisher JW (1991): Increased erythropoietin secretion in human hepatoma cells by N^6-cyclohexyladenosine. Am J Physiol 261:C455–C460.

Nakashima J, Brookins J, Ohigashi T, Fisher JW (1994): Adenosine A_2 receptor modulation of erythropoietin secretion in hepatocellular carcinoma cells. Life Sci 54:109–117.

Nakashima J, Ohigashi T, Brookins JW, Bechman BS, Agrawal KC, Fisher JW (1993a): Effects of 5′-N-ethylcarboxamideadenosine (NECA) on erythropoietin production. Kidney Int 44:734–740.

Nakashima J, Tazak H, Fisher JW (1993b): Increase in erythropoietin secretion mediated by adenosine A_2 receptors. Human Cell 6:36–46.

Nanoff C, Freissmuth M, Tuisl E, Schutz W (1990): P_2-, but not P_1-purinoreceptors mediate formation of 1,4,5-inositol trisphosphate and its metabolites via a pertussis toxin-insensitive pathway in the rat renal cortex. Br J Pharmacol 100:62–68.

Needleman P, Minkes MS, Douglas JR Jr (1974): Stimulation of prostaglandin biosynthesis by adenine nucleotides. Profile of prostaglandin release by perfused organs. Circ Res 34:455–460.

Nies AS, Beckmann ML, Gerber JG (1991): Contrasting effects of changes in salt balance on the renovascular response to A_1-adenosine receptor stimulation *in vivo* and *in vitro* in the rat. J Pharmacol Exp Ther 256:542–546.

Nomura H, Nagashima K, Kusaka H, Karasawa A (1995): Antihypertensive effects of KW-3902, an adenosine A_1-receptor antagonist, in Dahl salt-sensitive rats. Jpn J Pharmacol 68:389–396.

Oßwald H (1975): Renal effects of adenosine and their inhibition by theophylline in dogs. Naunyn Schmiedebergs Arch Pharmacol 288:79–86.

Ohigashi T, Brookins J, Fisher JW (1993): Adenosine A_1 receptors and erythropoietin production. Am J Physiol 265:C934–C938.

Ohnishi A, Branch RA, Jackson K, Hamilton R, Biaggioni I, Deray G, Jackson EK (1986): Chronic caffeine administration exacerbates renovascular, but not genetic, hypertension in rats. J Clin Invest 78:1045–1050.

Ohnishi A, Li P, Branch RA, Biaggioni I, Jackson EK (1988): Adenosine in renin-dependent renovascular hypertension. Hypertension 12:152–161.

Ohnishi A, Ohnishi T, Stevenhead W, Robinson RD, Glick A, O'Day DM, Sabra R, Jackson EK, Branch RA (1989): Sodium status influences chronic amphotericin B nephrotoxicity in rats. Antimicrob Agents Chemother 33:1222–1227.

Osswald H, Hermes HH, Nabakowski G (1982): Role of adenosine in signal transmission of tubuloglomerular feedback. Kidney Int 22:S-136–S-142.

Osswald H, Nabakowski G, Hermes H (1980): Adenosine as a possible mediator of metabolic control of glomerular filtration rate. Int J Biochem 12:263–267.

Osswald H, Schmitz HJ, Heidenreich O (1975): Adenosine response of the rat kidney after saline loading, sodium restriction and hemorrhagia. Pflügers Arch 357:323–333.

Osswald H, Schmitz HJ, Kemper R (1977): Tissue content of adenosine, inosine and hypoxanthine in the rat kidney after ischemia and postischemic recirculation. Pflügers Arch 371:45–49.

Osswald H, Schmitz HJ, Kemper R (1978a): Renal action of adenosine: Effect on renin secretion in the rat. Naunyn Schmiedebergs Arch Pharmacol 303:95–99.

Osswald H, Spielman WS, Knox FG (1978b): Mechanism of adenosine-mediated decreases in glomerular filtration rate in dogs. Circ Res 43:465–469.

Palacios JM, Fastbom J, Wiederhold KH, Probst A (1987): Visualization of adenosine A_1 receptors in the human and the guinea-pig kidney. Eur J Pharmacol 138:273–276.

Panjehshahin MR, Chahil RS, Collis MG, Bowmer CJ, Yates MS (1991a): The effect of 8-cyclopentyl-1,3-dipropylxanthine on the development of cyclosporin-induced acute renal failure. J Pharm Pharmacol 43:525–528.

Panjehshahin MR, Munsey TS, Collis MG, Bowmer CJ, Yates MS (1991b): Further characterization of the protective effect of 8-cyclopentyl-1,3-dipropylxanthine on glycerol-induced acute renal failure in the rat. J Pharm Pharmacol 44:109–113.

Paul S, Jackson EK, Robertson D, Branch RA, Biaggioni I (1989): Caffeine potentiates the renin response to furosemide in rats. Evidence for a regulatory role of endogenous adenosine. J Pharmacol Exp Ther 251:183–187.

Paul P, Rothmann SA, Meagher RC (1988): Modulation of erythropoietin production by adenosine. J Lab Clin Med 112:168–173.

Paulmichl M, Lang F (1988): Enhancement of intracellular calcium concentration by extracellular ATP and UTP in Madin Darby canine kidney cells. Biochem Biophys Res Commun 156:1139–1143.

Pawelczyk T, Bizon D, Angielski S (1992): The distribution of enzymes involved in purine metabolism in rat kidney. Biochim Biophys Acta 1116:309–314.

Pawlowska D, Granger JP, Knox FG (1987): Effects of adenosine infusion into renal interstitium on renal hemodynamics. Am J Physiol 252:F678–F682.

Peart WS, Quesada T, Tenyl I (1975): The effects of cyclic adenosine 3′,5′-monophosphate and guanosine 3′,5′-monophosphate and theophylline on renin secretion in the isolated perfused kidney of the rat. Br J Pharmac 54:55–60.

Pfeifer CA, Suzuki F, Jackson EK (1995): Selective A_1 receptor antagonism augments β adrenergic-induced renin release *in vivo.* Am J Physiol 269:F469–F479.

Pfeilschifter J (1990a): Comparison of extracellular ATP and UTP signalling in rat renal mesangial cells. Biochem J 272:469–472.

Pfeilschifter J (1990b): Extracellular ATP stimulates polyphosphoinositide hydrolysis and prostaglandin synthesis in the renal mesangial cells. Involvement of a pertussis toxin-sensitive guanine nucleotide binding protein and feedback inhibition by protein kinase C. Cell Signal 2:129–138.

Post SR, Jacobson JP, Insel PA (1996): P_2 purinergic receptor agonists enhance cAMP production in Madin-Darby canine kidney epithelial cells via an autocrine/paracrine mechanism. J Biol Chem 271:2029–2032.

Ramos-Salazar A, Baines AD (1986): Role of 5′-nucleotidase in adenosine-mediated renal vasoconstriction during hypoxia. J Pharmacol Exp Ther 236:494–499.

Reed JR, Williams RH, Luke RG (1983): The renal hemodynamic response to diatrizoate in normal and diabetic rats. Invest Radiol 18:536–540.

Reid IA, Stockigt JR, Goldfien A, Ganong WF (1972): Stimulation of renin secretion in dogs by theophylline. Eur J Pharmacol 17:325–332.

Robertson D, Frolich JC, Carr RK, Watson JT, Holifield JW, Shand DG, Oates JA (1978): Effects of caffeine on plasma renin activity, catecholamines and blood pressure. N Engl J Med 298:181–186.

Rossi N, Ellis V, Kontry T, Gunther S, Churchill P, Bidani A (1990): The role of adenosine in $HgCl_2$-induced acute renal failure in rats. Am J Physiol 258:F1554–F1560.

Rossi NF, Churchill PC, Churchill MC (1987a): Pertussis toxin reverses adenosine receptor-mediated inhibition of renin secretion in rat renal cortical slices. Life Sci 40:481–487.

Rossi NF, Churchill PC, Jacobson KA, Leahy AE (1987b): Further characterization of the renovascular effects of N^6-cyclohexyladenosine in the isolated perfused rat kidney. J Pharmacol Exp Ther 240:911–915.

Rounds S, Hsieh L, Agarwal KC (1994): Effects of endotoxin injury on endothelial cell adenosine metabolism. J Lab Clin Med 123:309–317.

Rouse D, Leite M, Suki WN (1994): ATP inhibits the hydrosmotic effect of AVP in rabbit CCT: Evidence for a nucleotide P_{2u} receptor. Am J Physiol 267:F289–F295.

Sakai K, Akima M, Nabata H (1979): A possible purinergic mechanism for reactive ischemia in isolated, cross-circulated rat kidney. Jpn J Pharmacol 29:235–242.

Schnermann J, Weihprecht H, Briggs JP (1990): Inhibition of tubuloglomerular feedback during $adenosine_1$ receptor blockade. Am J Physiol 258:F553–F561.

Schulze-Lohoff E, Zanner S, Ogilvie A, Sterzel RB (1992): Extracellular ATP stimulates proliferation of cultured mesangial cells via P_2-purinergic receptors. Am J Physiol 263:F374–F383.

Schwabe U, Ukena D, Lohse MJ (1985): Xanthine derivates as antagonists at A_1 and A_2 adenosine receptors. Naunyn Schmiedebergs Arch Pharmacol 330:212–221.

Schwiebert EM, Karlson KH, Friedman PA, Dietl P, Spielman WS, Stanton BA (1992): Adenosine regulates a chloride channel via protein kinase C and a G protein in a rabbit cortical collecting duct cell line. J Clin Invest 89:834–841.

Sehic E, Ruan Y, Malik KU (1994): Attenuation by α,β-methylenadenosine-5′-triphosphate of periarterial nerve stimulation-induced renal vasoconstriction is not due to desensitization of purinergic receptors. J Pharmacol Exp Ther 271:983–992.

Shimada J, Suzuki F, Nonaka H, Karasawa A, Mizumoto H, Ohno T, Kubo K, Ishii A (1991): 8-(Dicyclopropylmethyl)-1,3-dipropylxanthine: A potent and selective adenosine A_1 antagonist with renal protective and diuretic activities. J Med Chem 34:466–469.

Silldorff EP, Kreisberg MS, Pallone TL (1996): Adenosine modulates vasomotor tone in outer medullary descending vasa recta of the rat. J Clin Invest 98:18–23.

Simmons NL (1981a): Identification of a purine (P_2) receptor linked to ion transport in cultured renal (MDCK) epithelium. Br J Pharmacol 73:379–384.

Simmons NL (1981b): Stimulation of Cl^- secretion by exogenous ATP in cultured MDCK epithelial monolayers. Biochim Biophys Acta 646:231–242.

Siragy HM, Linden J (1996): Sodium intake markedly alters renal interstitial fluid adenosine. Hypertension 27:404–407.

Skøtt O, Baumbach L (1985): Effects of adenosine on renin release from isolated rat glomeruli and kidney slices. Pflügers Arch 404:232–237.

Smits P, Hoffmann H, Thien T, Houben H, van't Laar A (1983): Hemodynamic and humoral effects of coffee after β_1-selective and nonselective β-blockade. Clin Pharmacol Ther 34:153–158.

Smits P, Pieters G, Thien T (1986): The role of epinephrine in the circulatory effects of coffee. Clin Pharmacol Ther 40:431–437.

Spielman WS (1984): Antagonistic effect of theophylline on the adenosine-induced decrease in renin release. Am J Physiol 247:F246–F251.

Spielman WS, Osswald H (1979): Blockade of postocclusive renal vasoconstriction by angiotensin II antagonist: Evidence for an angiotensin-adenosine interaction. Am J Physiol 237:F463–F467.

Spielman WS, Thompson CI (1982): A proposed role for adenosine in the regulation of renal hemodynamics and renin release. Am J Physiol 242:F423–F435.

Spielman WS, Britton SL, Fiken-Olsen MJ (1980): Effect of adenosine on the distribution of renal blood flow in dogs. Circ Res 46:449–456.

Spielman WS, Klotz KN, Arend LJ, Olson BA, LeVier DG, Schwabe U (1992): Characterization of adenosine A_1 receptor in a cell line (28A) derived from rabbit collecting tubule. Am J Physiol 263:C502–C508.

Suzuki F (1992): KW-3902. Drugs of the Future 17:876–878.

Suzuki F, Shimada J, Mizumoto H, Karasawa A, Kubo K, Nonaka H, Ishii A, Kawakita T (1992): Adenosine A_1 antagonists, 2. Structure-activity relationships on diuretic activities and protective effects against acute renal failure. J Med Chem 35:3066–3075.

Svensson JO, Jonzon B (1990): Determination of adenosine and cyclic adenosine monophosphate in urine using solid-phase extraction and high-performance liquid chromatography with fluorimetric detection. J Chromatography 529:437–441.

Taddei S, Arzilli F, Arrighi P, Salvetti A (1992): Dipyridamole decreases circulating renin-angiotensin system activity in hypertensive patients. Am J Hypertension 5:29–31.

Tagawa H, Vander AJ (1970): Effects of adenosine compounds on renal function and renin secretion in dogs. Circ Res 26:327–338.

Takeda M, Yoshitomi K, Imai M (1993): Regulation of Na^+–$3HCO^-_3$ cotransport in rabbit proximal convoluted tubule via adenosine A_1 receptor. Am J Physiol 265:F511–F519.

Talner LB, Davidson AJ (1968): Renal hemodynamic effects of contrast media. Invest Radiol 3:310–317.

Tan CC, Ratcliff PJ (1992): Rapid oxygen-dependent changes in erythropoietin mRNA in perfused rat kidneys: Evidence against mediation by cAMP. Kidney Int 41:1581–1587.

Terai T, Kita Y, Kusunoki T, Ando T, Shimazaki T, Deguchi Y, Akahane A, Shiokawa Y, Yoshida K (1990): The renal effects of FR-113453, a potent non-xanthine adenosine antagonist. Eur J Pharmacol 183:1057–1058.

Thompson CI, Sparks HV, Spielman WS (1985): Renal handling and production of plasma and urinary adenosine. Am J Physiol 248:F545–F551.

Thurau K (1964): Renal hemodynamics. Am J Med 36:698–719.

Tofovic SP, Branch KR, Oliver RD, Magee WD, Jackson EK (1991): Caffeine potentiates vasodilator-induced renin release. J Pharmacol Exp Ther 256:850–860.

Tofovic SP, Kusaka H, Pfeifer CA, Jackson EK (1996): Central effects of caffeine on renal renin secretion and norepinephrone spillover. J Cardiovasc Pharm 28:302–313.

Toya Y, Umemura S, Iwamoto T, Hirawa N, Kihara M, Takagi N, Ishii M (1993): Identification and characterization of adenosine A_1 receptor-cAMP system in human glomeruli. Kidney Int 43:928–932.

Trimble ME, Coulson R (1984): Adenosine transport in perfused rat kidney and renal cortical membrane vesicles. Am J Physiol 246:F794–F803.

Tseng CJ, Kuan CJ, Chu H, Tung CS (1993): Effect of caffeine treatment on plasma renin activity and angiotensin I concentrations in rats on a low sodium diet. Life Sci 52:883–890.

Turney JH, Marshall DH, Brownjohn AM, Ellis CM, Parsons FM (1990): The evolution of acute renal failure 1956–1988. Q J Med 74:83–104.

Ueda J (1972): Adenine nucleotides and renal function: Special reference with intrarenal distribution of blood flow. Jpn J Pharmacol 22:5.

Ueno M, Brookins J, Beckman B, Fisher JW (1988): A_1 and A_2 adenosine receptor regulation of erythropoietin production. Life Sci 43:229–237.

Vallon V, Osswald H (1994): Dipyridamole prevents diabetes-induced alterations of kidney function in rats. Naunyn Schmiedebergs Arch Pharmacol 349:217–222.

van Buren M, Bijlsma JA, Boer P, van Rijn HJM, Koomans HA (1993): Natriuretic and hypotensive effect of adenosine-1 blockade in essential hypertension. Hypertension 22:728–734.

van Waarde A, Stromski ME, Thulin G, Gaudio KM, Kashgarian M, Shulman RG, Siegel NJ (1989): Protection of the kidney against ischemic injury by inhibition of 5′-nucleotidase. Am J Physiol 256:F298–F305.

Viskoper RJ, Maxwell MH, Lupu AN, Rosenfeld S (1977): Renin stimulation by isoproterenol and theophylline in the isolated perfused kidney. Am J Physiol 232:F248–F253.

von Borstel WR, Wurtman RJ (1984): Caffeine and the cardiovascular effects of physiological levels of adenosine. In Dews PB (ed): "Caffeine," Berlin: Springer-Verlag, pp. 142–150.

Weaver DR, Reppert SM (1992): Adenosine receptor gene expression in rat kidney. Am J Physiol 263:F991–F995.

Weber RG, Jones CR, Palacios JM, Lohse MJ (1988): Autoradiographic visualization of A_1-adenosine receptors in brain and peripheral tissues from rat and guinea-pig using ^{125}I-HPIA. Neurosci Lett 87:215–220.

Weihprecht H, Lorenz JN, Briggs JP, Schnermann J (1992): Vasomotor effects of purinergic agonists in isolated rabbit afferent arterioles. Am J Physiol 263:F1026–F1033.

Weihprecht H, Lorenz JN, Briggs JP, Schnermann J (1994): Synergistic effects of angiotensin and adenosine in the renal microvasculature. Am J Physiol 266:F227–F239.

Weinberg JM, Davis JA, Shayman JA, Knight PR (1989): Alterations of cytosolic calcium in LLC-PK_1 cells induced by vasopressin and exogenous purines. Am J Physiol 256:C967–C976.

Williams M (1995): Purioceptor nomenclature: Challenges for the future. In Belardinelli L, Pelleg A (eds): "Adenosine and Adenine Nucleotides: from Molecular Biology to Integrative Physiology," Boston: Kluwer, pp. 39–48.

Woodcock EA, Loxlely R, Leung E, Johnston CI (1984): Demonstration of R_A-adenosine receptors in rat renal papillae. Biochem Biophys Res Commun 121:434–440.

Wu PH, Churchill PC (1985): 2-chloro-[^{3}H]-adenosine binding in isolated rat kidney membranes. Arch Int Pharmacodyn 273:83–87.

Yagil Y (1990): Interaction of adenosine with vasopressin in the inner medullary collecting duct. Am J Physiol 259:F679–F687.

Yagil Y (1992): Differential effect of basolateral and apical adenosine on AVP-stimulated cAMP formation in primary culture of IMCD. Am J Physiol 263:F268–F276.

Yagil Y (1994): The effects of adenosine on water and sodium excretion. J Pharmacol Exp Ther 268:826–835.

Yagil C, Katni G, Yagil Y (1994): The effects of adenosine on transepithelial resistance and sodium uptake in the inner medullary collecting duct. Pflügers Arch 427:225–232.

Yamaguchi S, Umemura S, Tamura K, Iwamoto T, Nyui N, Ishigami T, Ishii M (1995): Adenosine A_1 receptor mRNA in microdissected rat nephron segments. Hypertension 26:1181–1185.

Yao K, Kusaka H, Sato K, Karasawa A (1994a): Protective effects of KW-3902, a novel adenosine A_1-receptor antagonist, against gentamicin-induced acute renal failure in rats. Jpn J Pharmacol 65:276–170.

Yao K, Kusaka H, Sano J, Sato K, Karasawa A (1994b): Diuretic effects of KW-3902, a novel adenosine A_1-receptor antagonist, in various models of acute renal failure in rats. Jpn J Pharmacol 64:281–288.

Zager RA, Gmur DJ, Bredl CR, Eng MJ, Fisher L (1990): Regional responses within the kidney to ischemia: Assessment of adenine nucleotide and catabolite profiles. Biochim Biophys Acta 1035:29–36.

Zäll S, Milocco I, Ricksten SE (1993): Effects of adenosine on renal function and central hemodynamics after coronary artery bypass surgery. Anesth Analg 76:493–497.

Zegarra-Moran O, Romeo G, Galietta LJ (1995): Regulation of transepithelial ion transport by two different purinoceptors in the apical membrane of canine kidney (MDCK) cells. Br J Pharmacol 114:1052–1056.

2. Metabolism

CHAPTER 13

P_2 Purinoceptor Agonists: New Insulin Secretagogues Potentially Useful in the Treatment of Non-Insulin-Dependent Diabetes Mellitus

MARIE-MADELEINE LOUBATIERES-MARIANI, DOMINIQUE HILLAIRE-BUYS, JEANNIE CHAPAL, GYSLAINE BERTRAND, and PIERRE PETIT

Laboratoire de Pharmacologie, Faculté de Médecine, Montpellier, France

INTRODUCTION

Diabetes is related to a dysfunction in insulin secretion. In insulin-dependent diabetes mellitus (IDDM) or type I diabetes, pancreatic B cells are destroyed, and treatment by exogenous insulin is absolutely necessary. In non-insulin-dependent diabetes mellitus (NIDDM) or type II diabetes, a relative deficiency in insulin secretion is associated with a resistance of target tissues to the action of this hormone. In clinical practice, drugs aimed at stimulating insulin release are often used in the treatment of NIDDM. These are mainly represented by the hypoglycemic sulfonamides, which have been known since the 1940s, and the compounds currently used belong to the closely related family of sulfonylureas. The search for alternative insulin secretagogues has been underway for several years.

In this paper, we present the current state of knowledge concerning the stimulating effects of P_2 purinoceptor agonists on insulin secretion. After a brief summary of the role of ATP in insulin-secreting pancreatic B cells, we present the pharmacological evidence for the existence of P_2 purinoceptors on the B cell, the proposed cellular events involved in their coupling mechanisms, and their possible physiological relevance. Finally we report results suggesting that B cell P_2 purinoceptors may constitute a target for new insulin secretagogues.

ROLE OF ATP IN INSULIN-SECRETING B CELLS

The pancreatic B cell is able to release insulin in response to an increase in the concentration of glucose, which is its major nutrient stimulus. This response may be

Purinergic Approaches in Experimental Therapeutics, Edited by Kenneth A. Jacobson and Michael F. Jarvis
ISBN 0-471-14071-6 © 1997 Wiley-Liss, Inc.

modulated by a variety of other nutrients, by hormones, and by neurotransmitters. It is well known that glucose has to be metabolized to initiate insulin secretion, and that adenosine triphosphate (ATP) plays an important intracellular role. For many years, ATP was considered only as an intracellular fuel, but more recently its key role in the regulation of insulin secretion through ATP-dependent K^+ channels has been recognized and is now well documented (see review by Petit and Loubatières-Mariani, 1992). The ATP generated by glucose metabolism leads to closure of ATP-dependent K^+ channels; the resulting decrease in K^+ permeability induces depolarization of the plasma membrane and opening of voltage-activated calcium channels. The subsequent increase in the cytoplasmic calcium concentration triggers the exocytosis of insulin granules.

However, in addition to these intracellular effects, ATP is also able to act on the extracellular side of the B cell membrane to increase insulin secretion. A systematic pharmacological study of the extracellular effects of ATP and structural analogs on insulin secretion was performed in our laboratory and permitted us to characterize P_2 purinoceptors on the pancreatic B cell.

STIMULATING EFFECT OF ATP AND STRUCTURAL ANALOGS ON INSULIN SECRETION; CHARACTERIZATION OF A P_2 PURINOCEPTOR ON THE PANCREATIC B CELL

In 1963, Candela et al. were the first to report that external ATP stimulated insulin release *in vitro* by pieces of rabbit pancreas. In 1972, Loubatières et al., using the isolated perfused rat pancreas, reported that the administration of ATP induced a stimulation of insulin secretion higher than that elicited by cAMP. Subsequently, ATP was shown to be ineffective on insulin secretion in the absence of glucose. In the presence of a non-stimulating glucose concentration, it induced a weak and transient stimulation, whereas in the presence of a slightly stimulating glucose concentration, ATP elicited a marked and long-lasting stimulation of insulin secretion (Loubatières-Mariani et al., 1976). Therefore ATP is able to potentiate insulin secretion in the presence of glucose but, unlike this sugar, it cannot initiate the secretion. The insulin response to ATP displayed a biphasic pattern: the first phase displayed a peak and was followed by a long-lasting second phase (Figure 1). This insulin-stimulating effect was concentration-dependent between 10^{-6} and 10^{-4} M (Loubatières-Mariani et al., 1977, 1979).

The study of natural adenine derivatives showed that ATP and ADP displayed similar insulin-stimulating effects, while AMP and adenosine were much less effective (Loubatières-Mariani et al., 1979). Concerning the triphosphate nucleotides other than ATP, only purine nucleotides were effective (ATP > GTP > ITP), whereas pyrimidine nucleotides (CTP and UTP) showed little or no activity. Thus, there was a structure–activity relationship for nucleotide-induced insulin secretion: the purine base and the presence of a di- or triphosphate chain seemed to be essential. These results suggested the presence of a purinoceptor of the P_2 type on the B cell membrane. When analogs modified on the polyphosphate chain between α phosphorus and β phosphorus (α,β-methylene ATP and α,β-methylene ADP) were used, the insulin-secreting activity was not altered, and these substances were both as potent as ATP. When analogs modified on the polyphosphate chain between β phosphorus and γ phosphorus (adenyl-imido-diphosphate or AMP-PNP, and β,γ methylene ATP or AMP-PCP) were tested, the

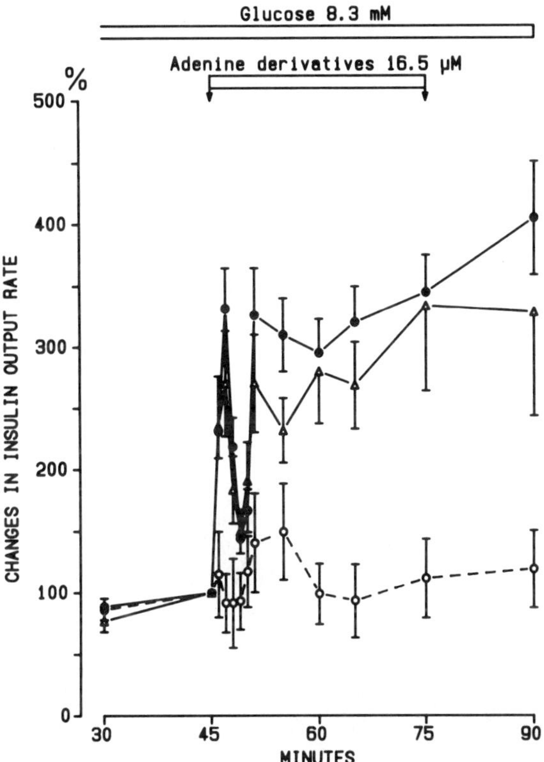

FIGURE 1. Effects of ATP (●——●), ADP (△——△), or adenosine (○---○) at the same concentration on the insulin secretion from isolated rat pancreas perfused with a Krebs-Ringer solution containing 8.3 mM glucose.

response was different: the former was ten times less potent than ATP, and the latter was unable to stimulate insulin release (Chapal and Loubatières-Mariani, 1981a). This discrepancy in the insulin-stimulating effects of these analogs may be related to the differences in electronic and steric characteristics of the polyphosphate chain, particularly between ATP and AMP-PCP. Indeed, the structural configurations of the P-C-P and the P-O-P bonds differ substantially. Furthermore, the ionization constant of the last diphosphonate hydrogen of β,γ-methylene ATP is much lower than that of the comparable hydrogen for ATP. Thus the net charge of β,γ-methylene ATP may be quite different from ATP at pH 7 (Yount, 1975). All these results supported the view of an extracellular nonmetabolic effect of ATP and emphasized the importance of the steric and electronic characteristics of the polyphosphate chain for the binding to B cell purinoceptors (Chapal and Loubatières-Mariani, 1981a). The use of a purinoceptor antagonist, 2-2′-pyridylisatogene tosylate (PIT), gave another argument in favor of an extracellular effect of ATP by counteracting the insulin-stimulating effect of ATP (Chapal and Loubatières-Mariani, 1981b).

When P_2 purinoceptors were subdivided into P_{2x} and P_{2y}, the use of an ATP analog more specific for P_{2y} receptors, 2-methylthio ATP, permitted the characterization of the P_2 purinoceptor of the B cell as being of the P_{2y} subtype (Bertrand et al., 1987). More recently, another P_{2y} purinoceptor agonist, adenosine 5′-O-(2-thiodiphosphate) (ADPβS), was found to be 100 times more potent than ATP or ADP in stimulating insulin secretion (Bertrand et al., 1991). Thus, the pancreatic B cell is provided with P_2 purinoceptors of the P_{2y} subtype, the activation of which stimulates insulin secretion. Recently a cDNA clone encoding a rat P_{2y} purinoceptor was isolated from an insulinoma cDNA library (Tokuyama et al., 1995).

CELLULAR EVENTS INVOLVED IN THE INSULIN SECRETORY EFFECT OF P_2-PURINOCEPTOR AGONISTS

The way in which P_2 receptors operate in the B cell has not been fully elucidated. In rat isolated islets, the insulin-stimulating effect of P_2 agonist α,β-methylene ADP was accompanied by an increase in calcium uptake (Petit et al., 1987). These results, as well as those reported by Geschwind et al. (1989) in the pancreatic B cell line HIT, suggest the involvement of an influx of extracellular calcium. On the other hand, Gylfe and Hellman (1987), reported a mobilization of intracellular calcium in mice pancreatic islets in the presence of α,β-methylene ATP or α,β-methylene ADP. Concerning the contribution of inositol phosphate production in the coupling mechanisms of B cell purinoceptors, conflicting results have also been reported, depending on the experimental preparation or the substance used and its concentration. Under our conditions, membrane phosphoinositide hydrolysis did not play a prominent role in rat islets, since α,β-methylene ADP at a concentration of 100 μmol/l was able to stimulate insulin release without affecting inositol phosphate accumulation (Petit et al., 1988). In experiments by Blachier and Malaisse (1988), a similar concentration of ATP failed to significantly enhance the production of inositol phosphates in rat islets, only a 10 fold higher concentration was effective; however, the low concentration was effective when tested on tumoral islet cells (RINm5F line). Most investigators now agree that purinergic agonists elevate cytoplasmic free calcium by mechanisms that involve both calcium influx and calcium mobilization from intracellular stores. Kindmark et al. (1991) have shown a rapid increase in cytosolic calcium in response to extracellular ATP in human pancreatic islets. On the other hand, it has been shown by Li et al. (1991) that externally applied ATP decreased the relative open state probability of the ATP-sensitive potassium channel on an excised outside-out patch of membrane from tumoral insulin-secreting cells.

THE P_2 PURINOCEPTORS MAY BE OF PHYSIOLOGICAL RELEVANCE

As regards physiological implications, P_2 purinoceptors may modulate insulin secretion. Indeed, there is evidence that ATP is stored and coreleased with acetylcholine from nerve terminals. Accordingly, ATP released with acetylcholine may potentiate the effect of this neurotransmitter on insulin secretion. Indeed ATP or ADP and acetylcholine act in potentiating synergism on B cells in the presence of a physiological glucose concentration (Bertrand et al., 1986). This potentiation may play a role in the parasympathetic control of insulin secretion, particularly during the early prandial period (cephalic phase). Perhaps more importantly, ATP and ADP are present in high concentrations in B cell secretory granules and are released during glucose-induced insulin secretion. These nucleotides may exert a positive feedback and amplify glucose-induced insulin secretion; in fact, ATP and ADP, via the activation of P_{2y} purinoceptors, potentiate the biphasic response to a physiological glucose increment (Bertrand et al., 1989). Therefore, the activation of B cell P_2 purinoceptors may participate in the mechanisms of glycemia control by stimulating insulin release.

P_2 PURINOCEPTOR AGONISTS AS POTENTIAL ANTIDIABETIC DRUGS

In this section are considered the preclinical data suggesting that in NIDDM B cell P_{2y} purinoceptors may be a potential target for new antidiabetic drugs with the ability to stimulate insulin release. Indeed, B cell dysfunction, particularly an impairment of insulin response to glucose, is involved in the pathophysiology of NIDDM; therefore, it was necessary to establish that P_{2y} agonists administered *in vivo* were effective in increasing insulin secretion and reducing hyperglycemia. For these studies it was necessary to have a structural analog sufficiently stable to be administered to animals. The P_{2y}-selective agonist ADPβS was used for these investigations in rats and dogs. Experiments were performed under basal conditions, as well as during glucose tolerance tests (Hillaire-Buys et al., 1993). In anesthetized rats, ADPβS injected into a peripheral vein evoked an increase in plasma insulin levels. The insulin-stimulating effect of ADPβS appeared to be dependent on the nutritional state of the animals. In rats that had fasted overnight, ADPβS elicited only a transient insulin response without significant modification in glycemia. In contrast, in fed rats, in which the basal glycemia was slightly but significantly higher, the same dose of ADPβS evoked a sustained insulin response and decreased glycemia to a level similar to the baseline in fasting animals. Thus, the lack of ADPβS effect under fasting conditions may limit the risk of hypoglycemia. When an intravenous glucose tolerance test was performed in anesthetized rats, the simultaneous administration of ADPβS with glucose strongly increased insulin secretion and accelerated the return of glycemia to baseline values. As it is known that purine nucleotides are metabolized in the gastrointestinal tract, it was necessary to investigate the effects of ADPβS after oral administration. *In vivo* experiments in conscious dogs that had fasted overnight showed that this substance was effective after oral administration, transiently increasing plasma insulin. Moreover, during an oral glucose tolerance test, ADPβS also markedly enhanced insulin secretion and reduced hyperglycemia. From these results, it appears that the P_{2y}-purinoceptor agonist ADPβS is a potent insulin secretagogue *in vivo,* improves glucose tolerance, and is effective after oral administration.

In addition, the effect of a P_{2y} agonist on insulin secretion was investigated in the perfused pancreas isolated from rats with streptozotocin-induced diabetes, in which the most of the B cells are destroyed. The typical biphasic insulin response to a stimulating glucose concentration observed in normal rats was totally suppressed in these diabetic animals (Figure 2). In the same way, the response to the hypoglycemic sulfonylurea tolbutamide was also drastically reduced (Hillaire-Buys et al., 1992). In contrast, the insulin response of remnant B cells to the P_{2y} agonist ADPβS was preserved. Recently, similar results were observed in the perfused pancreas isolated from Zucker diabetic fatty rats, a genetic model of NIDDM; insulin secretory responses mediated by P_{2y} purinoceptors are preserved whereas B cells are unresponsive to glucose (Tang et al., 1996).

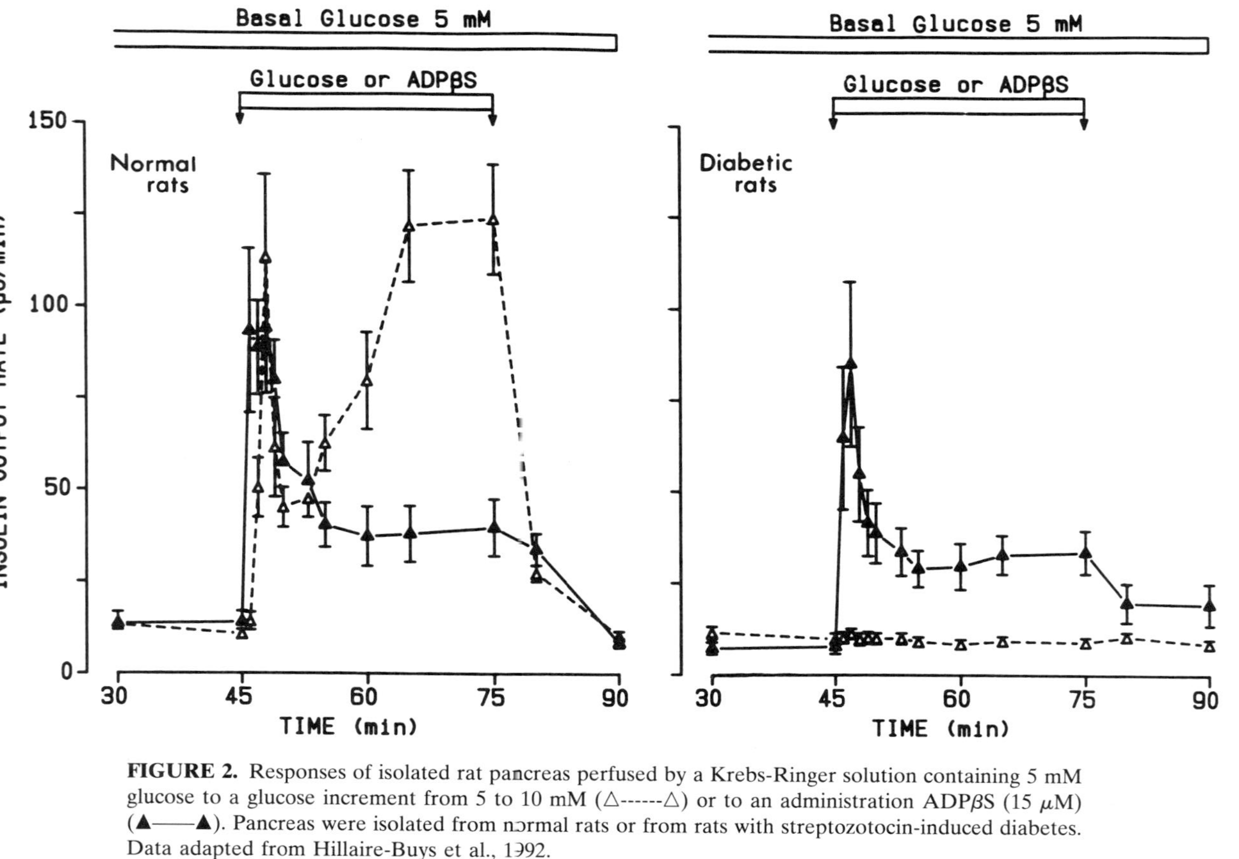

FIGURE 2. Responses of isolated rat pancreas perfused by a Krebs-Ringer solution containing 5 mM glucose to a glucose increment from 5 to 10 mM (△------△) or to an administration ADPβS (15 μM) (▲——▲). Pancreas were isolated from normal rats or from rats with streptozotocin-induced diabetes. Data adapted from Hillaire-Buys et al., 1992.

CONCLUSION

From these *in vitro* and *in vivo* pharmacological investigations, it can be suggested that due to their ability to stimulate insulin release the P_{2y} agonists may be considered as potential new substances in the treatment of non-insulin-dependent diabetes. A specific goal for future investigation would be to obtain P_{2y} agonists more selective for B cell purinoceptors and to study the effects of a long-term treatment with such drugs in normal and diabetic animals. However, if P_2 purinoceptor agonists open new therapeutic perspectives in NIDDM, it should be kept in mind that much experimental research will have to be performed before their clinical use can be considered.

REFERENCES

Bertrand G, Chapal J, Loubatières-Mariani MM (1986): Potentiating synergism between adenosine diphosphate or triphosphate and acetylcholine on insulin secretion. Am J Physiol 251:E416–E421.

Bertrand G, Chapal J, Loubatières-Mariani MM, Roye M (1987): Evidence for two different P_2-purinoceptors on β cell and pancreatic vascular bed. Br J Pharmacol 102:783–787.

Bertrand G, Gross R, Chapal J, Loubatières-Mariani MM (1989): Difference in the potentiating effect of adenosine triphosphate and α,β-methylene ATP on the biphasic insulin response to glucose. Br J Pharmacol 98:998–1004.

Bertrand G, Chapal J, Puech R, Loubatières-Mariani MM (1991): Adenosine-5′-O-(2-thiodiphosphate) is a potent agonist at P_2 purinoceptors mediating insulin secretion from perfused rat pancreas. Br J Pharmacol 102:627–630.

Blachier F, Malaisse WJ (1988): Effect of exogenous ATP upon inositol phosphate production, cationic fluxes and insulin release in pancreatic islet cells. Biochim Biophys Acta 970:222–229.

Candela JLR, Martin-Hernandez D, Castilla Cortazar T (1963): Stimulation of insulin secretion in vitro by adenosine triphosphate. Nature 197:A1304.

Chapal J, Loubatières-Mariani MM (1981a): Effects of phosphate-modified adenine nucleotide analogues on insulin secretion from perfused rat pancreas. Br J Pharmacol 73:105–110.

Chapal J, Loubatières-Mariani MM (1981b): Attempt to antagonize the stimulatory effect of ATP on insulin secretion. Eur J Pharmacol 74:127–134.

Geschwind JF, Hiriart M, Glennon MC, Najafi H, Corkey BE, Matschinsky FM, Prentki M (1989): Selective activation of Ca^{2+} influx by extracellular ATP in a pancreatic β-cell line (HIT). Biochim Biophys Acta 1012:107–115.

Gylfe E, Hellman B (1987): External ATP mimics carbachol in initiating calcium mobilization from pancreatic β-cells conditioned by previous exposure to glucose. Br J Pharmacol 92:281–289.

Hillaire-Buys D, Bertrand G, Chapal J, Puech R, Ribes G, Loubatières-Mariani MM (1993): Stimulation of insulin secretion and improvement of glucose tolerance in rat and dog by the P_{2y}-purinoceptor agonist, adenosine-5′-O-(2-thiodiphosphate). Br J Pharmacol 109:183–187.

Hillaire-Buys D, Gross R, Chapal J, Ribes G, Loubatières-Mariani MM (1992): P_{2y} purinoceptor response of β-cells and vascular bed are preserved in diabetic rat pancreas. Br J Pharmacol 106:610–615.

Kindmark H, Köhler M, Nilsson T, Arkhammar P, Wiechel KL, Rorsman P, Efendic S, Berggren PO (1991): Measurements of cytoplasmic free Ca^{2+} concentration in human pancreatic islets and insulinoma cells. FEBS 291:310–314.

Li G, Milani D, Dunne MJ, Pralong WF, Theler JM, Petersen OH, Wollheim CB (1991): Extracellular ATP causes Ca^{2+}-dependent and -independent insulin secretion in RINm5F cells. J Biol Chem 266:3449–3457.

Loubatières-Mariani MM, Chapal J, Lignon F, Valette G (1979): Structural specificity of nucleotides for insulin secretory action from the isolated perfused rat pancreas. Eur J Pharmacol 59:277–286.

Loubatières-Mariani MM, Chapal J, Valette G (1977): Adénosine triphosphate (ATP) et insulinosécrétion: Effet de différentes concentrations et influence de la température. C R Soc Biol 171:864–869.

Loubatières-Mariani MM, Loubatières A, Chapal J, Valette G (1976): Adénosine triphosphate (ATP) et glucose. Action sur les sécrétions d'insuline et de glucagon. C R Soc Biol 170:833–836.

Loubatières A, Loubatières-Mariani MM, Chapal J (1972): Adénosine triphosphate (ATP), adénosine 3′5′ monophosphate cyclique (3′5′ AMPc) et sécrétion d'insuline. C R Soc Biol 166:1742–1746.

Petit P, Loubatières-Mariani MM (1992): Potassium channels of the insulin-secreting B cell. Fundam Clin Pharmacol 6:123–134.

Petit P, Manteghetti M, Loubatières-Mariani MM (1988): Differential effects of purinergic and cholinergic activation on the hydrolysis of membrane polyphosphoinositides in rat pancreatic islets. Biochem Pharmacol 37:1213–1217.

Petit P, Manteghetti M, Puech R, Loubatières-Mariani MM (1987): ATP and phosphate-modified adenine nucleotide analogues: Effects on insulin secretion and calcium uptake. Biochem Pharmacol 36:377–380.

Tang J, Pugh W, Polonsky KS, Zhang H (1996): Preservation of insulin secretory responses to P_2 purinoceptor agonists in Zucker diabetic fatty rats. Am J Physiol 270:E504–E512.

Tokuyama Y, Hara M, Jones EMC, Fan Z, Bell GI (1995): Cloning of rat and mouse P_{2Y} purinoceptors. Biochem Biophys Res Commun 211:211–218.

Yount RG (1975): ATP analogs. Adv Enzymol 43:1–56.

CHAPTER 14

Purinergic Modulation of Gastrointestinal Function

FEDIAS L. CHRISTOFI and MICHAEL A. COOK

Department of Anesthesiology, College of Medicine, Ohio State University, Columbus, OH, 43210 (F.L.C.); Department of Pharmacology and Toxicology, Faculty of Medicine, University of Western Ontario, London, Ontario, N6A 5C1, Canada (M.A.C.)

INTRODUCTION

The primary physiological functions of the gastrointestinal (GI) tract, motility and secretion, are capable of modulation by purines. Endogenous and exogenous purinergic agonists act at P_1 and/or P_2 receptors in mediating such actions and are implicated in both the physiological regulation of gastrointestinal function, and in its therapeutic modulation. This brief treatment will review the current state of the field and will focus on purinergic modulation of motor and secretory functions. Although we have attempted to emphasize more recent contributions to the field, we have included all appropriate citations wherever necessary. However, we have dealt only fleetingly with the actions of purines on vascular targets in the GI tract, and the reader is referred to Chapters 8 and 9, this volume, for relevant information. Similarly, we have not addressed extensively the actions of purines on endocrine functions of the pancreas. This area was recently reviewed and the reader is referred to Loubatières-Mariani et al., (1995) and the citations therein for relevant information, as well as to Chapter 13, this volume.

The physiological actions of purines on motor functions of the GI tract are largely mediated by receptors on the nerves located within the ganglionated plexuses that comprise the enteric nervous system (ENS). The first section deals with these actions and addresses the influence of purines on enteric neurophysiological function. The second section addresses the ability of purines to modulate secretory functions, primarily at intestinal epithelial cells, although neural and vascular targets are also relevant.

Purinergic Approaches in Experimental Therapeutics, Edited by Kenneth A. Jacobson and Michael F. Jarvis
ISBN 0-471-14071-6 © 1997 Wiley-Liss, Inc.

PURINERGIC REGULATION OF GASTROINTESTINAL MOTOR FUNCTION

Regulation at Postsynaptic Sites on AH/Type 2 and S/Type 1 Neurons in Enteric Ganglia

Enteric neurons of the gastrointestinal tract are classified according to their morphology (Furness et al., 1988) and subclassified into AH/type 2 and S/type 1 neurons, according to their electrophysiological properties (Hirst et al., 1974; Nishi and North, 1973). Slow excitatory postsynaptic potentials (EPSPs) occur in both S and AH cells, while fast EPSPs occur mainly in S cells. The probable mediators of slow EPSPs in myenteric ganglia are substance P (SP) (Johnson et al., 1980; Bornstein et al., 1984), acetylcholine (ACh, muscarinic) (North and Tokimasa, 1982), and serotonin (5-HT); while fast EPSPs are mediated by ACh (Hirst et al., 1974; Nishi and North, 1973). Recent progress in our understanding of neurogastroenterology, as it applies to enteric circuits and gastrointestinal motility, has been reviewed by Wood (1992, 1994). Available evidence suggests significant involvement of adenosine and ATP as distinct neuromodulators within the microcircuits of the ENS. Actions are at both pre- and postsynaptic sites in enteric ganglia. Such actions will be described in some detail in this chapter, since no comparable reviews have appeared recently.

Electrophysiological data from the ENS show that adenosine hyperpolarizes the cell body, decreases input resistance, enhances postspike hyperpolarizing potentials, and suppresses excitability in myenteric AH/type 2 neurons (Zafirov et al., 1985a; Palmer et al., 1987a). It also blocks the excitatory actions of forskolin in AH/type 2 neurons, but has no effect on the excitatory actions of intracellularly injected cyclic AMP, membrane permeant analogs of cyclic AMP, inhibitors of phosphodiesterase, or elevation of magnesium and reduction of calcium in the bathing medium (Zafirov et al., 1985b; Palmer et al., 1987b). These findings suggest that adenosine inhibits the response to forskolin by suppression of the catalytic activity of adenylyl cyclase and consequent reduction of intraneuronal levels of cAMP. The actions of some, but not all, slow EPSP mimetics are suppressed by adenosine in AH/type 2 neurons (Palmer et al., 1987b). Thus, excitatory actions of gastrin-releasing peptide (GRP) and vasoactive intestinal peptide (VIP) are blocked while those of calcitonin gene-related peptide (CGRP), SP, or 5-HT are enhanced by adenosine. The data are consistent with A_1 inhibition of adenylyl cyclase and A_2 stimulation of adenylyl cyclase. Alternatively, stimulation could involve another cAMP-independent signaling pathway linked to a distinct P_1 purinoceptor. A recent report described high-affinity A_2 excitatory receptors that are coupled to adenylyl cyclase and the elevation of cAMP in a minority subset of AH/type 2 myenteric neurons. Furthermore, inhibitory A_1 and excitatory A_2 receptors are colocalized on some AH/type 2 neurons (Christofi et al., 1994). Adenosine was shown to interact with A_1 purinoceptors to suppress the slow EPSP-mimetic response to the pituitary adenylyl cyclase activating peptide (PACAP) in AH neurons with Dogiel type II morphology (Christofi and Wood, 1993a). Other electrophysiological findings (Katayama and Morita, 1989) indicate that ATP, like adenosine, also causes membrane hyperpolarization in the majority of AH neurons, whereas it causes a membrane depolarization in the majority of S neurons.

The pharmacology of P_1 and P_2 purinoceptors with respect to myenteric neurons is complex. Conventional intracellular microelectrode studies subclassified P_1 purinoceptors on AH/type 2 neurons (Christofi and Wood, 1994). Thus, neuronal excitability of AH/type 2 neurons is suppressed by activation of high-affinity A_1 receptors that may be linked to a cAMP-dependent pathway, leading to an increase in calcium-dependent potassium conductance and enhancement of the afterhyperpolarizing potential. Activation of lower affinity, non-A_1/P_1 purinoceptors linked to a cAMP-independent pathway reduces excitability and leads mainly to a steady-state hyperpolarization. Interestingly, adenosine produces a smaller hyperpolarization in AH/type 2 neurons than ATP and is less potent than the nucleotide (Katayama and Morita, 1989). Furthermore, methylxanthines only partially block the ATP-mediated effect in AH/type 2 neurons and do not significantly affect the ATP-induced depolarization in S/type 1 neurons. These data are consistent with the presence of P_2 purinoceptors on both subtypes of neurons. ATP modulates a potassium conductance in myenteric neurons; it increases the conductance in AH neurons and suppresses the conductance in S neurons. Another effect of ATP in these neurons is to enhance the afterhyperpolarization (AHP) at nanomolar concentrations (Katayama and Morita, 1989). The P_2 purinoceptor subtypes on myenteric neurons were not identified.

Purinergic Inhibition of Synaptic Transmission to AH/type 2 and S/type 1 Neurons in Enteric Ganglia

The presynaptic actions of adenosine and ATP in electrophysiologically identified myenteric neurons were recently studied in detail (Christofi and Wood, 1993a, 1994; Kamiji et al., 1994). Adenosine acts are presynaptic A_1 sites to suppress fast and slow EPSPs in all neurons (Christofi and Wood, 1993b, 1994; Kamiji et al., 1994). Adenosine also suppresses nicotinic cholinergic transmission in S/type 1 neurons of the guinea pig small intestine by activating presynaptic high affinity A_1 receptors (Christofi and Wood, 1993b, 1994). Suppression of nicotinic synaptic transmission also occurs in myenteric ganglia of the gastric antrum via P_1 purinoceptors, likely representing A_1 sites at cholinergic release sites (Christofi et al., 1992). With respect to excitatory transmission mediated by the tachykinins, Christofi et al., (1990) have described adenosine A_1-mediated inhibition of substance P-like immunoreactivity (SP-LI) and neurokinin A-like immunoreactivity (NKA-LI) release from perifused enteric nerve endings prepared from the guinea pig myenteric plexus. Broad et al., (1992) provided further evidence for the presence of adenosine A_1 purinoceptors on enteric nerve endings and later showed that they may represent a low-affinity A_{1b} subtype linked to inhibition of transmitter release (Broad and Cook, 1993). The latter study demonstrated the anomalous nature of the interaction of adenosine analogs bearing the 5′-uronamide substituent with the receptor on enteric neurons and suggested that such interactions may arise as a consequence of the unique structure of such compounds. The data did not support the presence of an A_2 receptor on enteric nerves. Kwok et al. (1990) used the vascularly perfused rat stomach to study purinergic modulation of gastric somatostatin-like immunoreactivity (SLI). They found that the action of ATP in augmenting the release of SLI was likely to be a result of metabolism to adenosine. Such neurotransmitter release studies on SLI and other neuropeptides in different regions of the gastrointestinal tract are lacking.

The inhibitory actions of ATP on fast EPSPs in S neurons, and slow EPSPs in both S and AH neurons, were prevented by pretreatment with theophylline, caffeine, quinidine, and 8-phenyltheophylline (8-PT). Furthermore, the ATP analogs ATPγS and α-β-methylene ATP also depressed the synaptic potentials from both types of neurons, and the inhibitory effect of ATP was 10 times more potent than that of adenosine. Moreover, at concentrations exceeding 1 μM, ATP augmented nicotinic fast depolarizations of S neurons produced by ACh, but inhibited muscarinic and SP-mediated depolarizations in both AH and S neurons (Kamiji et al., 1994). These data suggest that P_2 purinoceptors are also involved in presynaptic inhibition of EPSPs in enteric neurons. An earlier study by Sperlagh and Vizi (1991) also provided evidence for the presence of P_{2x} purinoceptors on varicose nerve endings of myenteric neurons, through which stimulation-evoked release of transmitter is modulated by ATP in a *positive manner.* Thus, as suggested by the authors, if ATP is coreleased with another transmitter or released from another source such as smooth muscle (Takeshi et al., 1993), it might enhance the transmitter release evoked by axonal firing.

The effect of the P_2 purinoceptor agonists α,β-methylene ATP and β,γ-methylene ATP on endogenous ACh release from ileal longitudinal nerve-muscle strips was also described recently (Takeshi et al., 1993). Their findings suggested the intriguing possibility that, in intact myenteric plexus, ATP acts at P_2 purinoceptors on smooth muscle to release ATP, which is then metabolized to adenosine in the synapse and acts at P_1 purinoceptors to inhibit cholinergic transmission. Christofi and Wood (1993b) also provided quantitative evidence for accumulation of endogenous adenosine within the ganglia to levels sufficient for suppression of excitatory neurotransmission and neuronal excitability. Thus application of selective P_1 purinoceptor antagonists, appeared to release the neurons from ongoing inhibition as reflected by enhancement of fast EPSPs, noncholinergic slow EPSPs, as well as neuronal excitability. Other agents, such as dipyridamole and adenosine deaminase, had the anticipated effects (Christofi and Wood, 1989). Whether such effects of accumulation of endogenous adenosine are an artifact of the *in vitro* experimental conditions is unclear. Regardless of this possibility, the results are suggestive of effects to be anticipated in the functional bowel in case of accumulation of adenosine caused by pathological states such as ischemia or hypoxia (Milusheva et al., 1990), or physiological conditions such as intense contraction of the musculature.

Until recently, ACh was believed to be the sole mediator of enteric fast EPSPs. However, other mediators in the ENS, such as ATP and 5-HT, also cause fast depolarization responses similar to nicotinic responses that are due to an increase in a cation conductance (Mawe et al., 1986; Surprenant and Crist, 1988). ATP causes fast inward currents and activates cationic channels in cell-free patches of enteric nerves (Barajas-Lopez et al., 1993, 1994). A recent study using conventional intracellular electrophysiological methods has provided some evidence for ATP as a mediator of noncholinergic fast EPSPs in the enteric nervous system (Galligan and Bertrand, 1994). The reported inhibitory effects of suramin on GABA- and glutamate-gated channels in hippocampal neurons (Nakazawa et al., 1995) could, however, explain the actions of the compound on noncholinergic fast EPSPs in myenteric neurons, since suramin was used to reveal purinergic transmission. A further complicating factor in the interpretation of these data is the documented potentiating effect of ATP on fast GABA responses at primary afferent neurons (Morita et al., 1984) and fast nicotinic responses at myenteric neurons (Kamiji et al., 1994). Suramin was also shown to potentiate the ATP-evoked Ca^{2+} current in cultured myenteric neurons of the guinea pig ileum (Barajas-Lopez et al., 1993). This effect is not unique to enteric P_2 purinoceptors, since suramin also facilitates hippocampal potentials (i.e., EPSPs and population spike; Wieraszko, 1995). The failure of suramin to antagonize the ATP-evoked Ca^{2+} current in cultured myenteric neurons (Barajas-Lopez et al., 1993) suggests an interaction at atypical P_2 purinoceptors, since suramin alone does not appear to have nonspecific effects. Thus, in enteric neurons, suramin did not change the holding current, voltage-dependent Ca^{2+} currents or fast

EPSPs (Barajas-Lopez et al., 1993), the cationic inward current induced by ACh, nicotinic EPSPs, or noradrenergic inhibitory post-synaptic potentials (IPSPs) (Evans et al., 1992).

Suramin is known to modulate the activity of enzymes using ATP as a substrate (Mahoney et al., 1990). Therefore, suramin could increase the efficacy of ATP in these neurons by suppressing ecto-nucleotidase activity. A recent review by Kennedy and Leff (1995) however, provides evidence to support the view that, in dissociated cells such as cultured myenteric neurons, which contain no diffusion barrier to ATP or other hydrolyzable nucleotides, ectonucleotidases are unlikely to influence agonist potencies. Thus, suramin's effects on these enzymes could not explain the potentiation of the ATP response in myenteric neurons. Our preliminary data (Christofi et al., 1995) indicate that activation of several intracellular enzymes, including protein kinase C, contribute to ATP-induced elevation in $[Ca^{2+}]_i$ in myenteric neurons. It is conceivable then, that suramin potentiates ATP Ca^{2+} responses by modulating the activity of protein kinase C as previously reported (Mahoney et al., 1990), or that of other enzymes in a subpopulation of neurons with P_2 purinoceptors that are distinct from those blocked by suramin.

Synaptic inhibition is presumably an important function in the integrated operations of enteric microcircuits (Wood, 1994). In spite of this, robust stimulus-evoked IPSPs have not been shown consistently in the myenteric plexus (Wood, 1989). Recent findings (Christofi and Wood, 1993b) suggest that this may be due to masking of the slow IPSP by the slow excitatory synaptic inputs that are activated simultaneously by electrical stimulation of either individual interganglionic fiber tracts or the ganglionic surface. Masking of such IPSPs by simultaneous activation of excitatory inputs would be analogous to the experimental situation in intestinal circular muscle, where electrical field stimulation simultaneously evokes both excitatory and inhibitory junction potentials in the muscle cells. Atropine blockade of the excitatory junction potentials unmasks robust inhibitory junction potentials. Thus adenosine A_1 inhibition of slow EPSPs unmasked robust slow IPSPs in AH/type 2 neurons (Christofi and Wood, 1993b). Enteric microcircuits are apparently organized for adenosine to suppress excitatory synaptic functions without interfering with inhibitory synaptic transmission. Adenosine's ability to facilitate inhibitory transmission would complement its ability to shut down the microcircuitry through its dual inhibitory pre- and postsynaptic actions. Adenosine antagonists offset the unmasking action of A_1 agonists without suppressing slow IPSPs alone, arguing against the involvement of adenosine as a slow IPSP mediator. However, ATP remains a potential candidate.

Slow IPSPs in the submucous plexus are suppressed rather than facilitated by adenosine acting at P_1 purinoceptors (Zafirov et al., 1993), suggesting that adenosine's actions on inhibitory transmission may differ according to the inhibitory transmitters involved and the specialized functions of the two nerve plexuses. In the submucous plexus, adenosine inhibits noradrenergic slow IPSPs and cholinergic fast EPSPs by acting at A_1 presynaptic receptors, whereas it depolarizes S neurons by acting at A_2 receptors (Barajas-Lopez et al., 1991). Effects on submucous AH neurons are qualitatively the same as those in myenteric AH neurons (F. L. Christofi, unpublished observations). In contrast to myenteric S neurons, the adenylyl cyclase activator forskolin depolarizes submucosal S neurons by reducing the membrane potassium conductance (Mihara et al., 1987). Recent electrophysiological findings indicate that adenosine reduces this conductance by activating protein kinase A in submucosal neurons (Barajas-Lopez, 1993).

In submucosal neurons, ATP closes a K^+ ion channel and opens a ligand-gated cation conductance through distinct P_2 purinoceptors (Barajas-Lopez et al., 1994). In AH/type 2 neurons, ATP induces a fast depolarization (latency 30 msec; duration, 5 sec), a fast depolarization followed by slow hyperpolarization, or just slow hyperpolarization. In S/type 1 neurons, ATP elicits a biphasic depolarization. The ATP-induced depolarization is likely to be mediated by the opening of ligand-gated cation channels, and Na^+, Ca^{2+}, K^+, and Cs^+ permeate this channel. These electrophysiological effects of ATP on submucosal neurons are mediated by two distinct purinoceptors, similar to those described as P_{2y} purinoceptors (Barajas-Lopez et al., 1994). It is likely that ATP plays a key role in synaptic communication within the neuronal circuits of the submucous plexus. The differential purinergic modulation of S and AH neurons is likely to be of pharmacological, neurophysiological, and perhaps even pathophysiological significance in the way the enteric microcircuits drive and coordinate peristalsis with secretomotor functions in the gastrointestinal tract (see later discussion on putative functions of these neuronal subtypes).

Implications for Neurophysiology and Gastroenterology

Understanding the function of purines in the enteric nervous system requires a detailed analysis and knowledge of the anatomical projections, chemical coding, as well as the integrated function of AH and S cells within enteric circuits and intestinal reflexes. Although a great deal of progress has been made in this regard (see reviews by Wood, 1992, 1994), we have a long way to go in understanding the complex neuronal mechanisms underlying one of the most basic functions of gastrointestinal motor control, the peristaltic reflex. Our discussion will only highlight some of the putative functions of those cell types that are responsive to purines.

Neurons of the AH type with Dogiel type II morphology possess numerous long varicose fibers, most of which terminate around nerve cell bodies, either in the same ganglion or in adjacent ganglia circumferential to the AH cell body (Iyer et al., 1988; Hendriks et al., 1990; Bornstein et al., 1991). The circumferential axons of Dogiel type II neurons send branches to the submucosa and mucosa (Brookes et al., 1995), and approximately 70%–85% of Dogiel type II neurons are immunoreactive for the calcium-binding protein calbindin (Furness et al., 1988; Iyer et al., 1988; Furness et al., 1990; Song et al., 1994). Dogiel type II neurons account for 30%–40% of all myenteric neurons. All calbindin-immunoreactive Dogiel type II neurons (~80%) and non-immunoreactive Dogiel II neurons (~20%) project to the mucosa from the myenteric plexus (Song et al., 1994). *It has been proposed that AH/Dogiel type II neurons may function as intrinsic sensory neurons* (Furness et al., 1988) *or primarily as interneurons in the guinea pig small intestine* (Wood, 1994). A subtype of Dogiel type II neurons with AH cell characteristics also has long, aborally directed projections up to 100 mm long and exhibits dendritic morphology. This subtype accounts for about 10% of all Dogiel type II neurons identified either by random impalement (Bornstein et al., 1992) or by the proportion of all calbindin neurons that can be labeled with the retrogradely transported fluorescent dye 1,1′-didodecyl-3,3,3′,3′-tetramethylindocarbocyanine perchlorate ($DiIC_{12}$ (3), "*DiI*"). The aboral fiber usually arises from a circumferentially directed axon (Brookes et al., 1995). *These neurons are suggested to play an important role in aborally directed reflexes in the small intestine.* Myenteric AH neurons were shown to transmit *only via slow EPSPs to both AH and S neurons in enteric ganglia* (Kunze et al., 1993). This was revealed in studies utilizing simultaneous intracellular recordings from pairs of neurons in adjacent circumferential ganglia. Excitatory motor neurons have Dogiel type I morphology (Brookes et al., 1991a). Furthermore, calretinin-immunoreactive Dogiel type I neurons could be distinguished by their projections and by their neurochemical coding. Celretinin immunocytochemical studies revealed three distinct functional classes of Dogiel type I neurons: cholinergic motor neurons to longitudinal muscle, myenteric cholinergic interneurons, and submucosal vasomotor neurons (Brookes et al., 1991b).

Since both ATP and adenosine have effects on most AH/type 2 neurons, they may convey sensory and interneuronal signals within the "little brain of the gut" or even participate in aborally directed reflexes. Furthermore, the putative role of ATP in slow synaptic transmission in enteric ganglia remains to be elucidated. Detailed studies utilizing intraneuronal staining techniques, in combination with electrophysiology and immunohistochemical localization of neurotransmitters with P_1 and P_2 purinoceptor agonists in intact myenteric plexus preparations or organotypic cultures, are necessary to resolve these possibilities.

Future Studies on Intracellular Signaling Pathways Linked to P_1 and P_2 Purinoceptors in Enteric Ganglia

Although considerable progress has been made in the last 5 years in our understanding of purinergic neuroregulation in the digestive tract, further studies are needed to elucidate the cascade of intracellular signaling pathways that may be linked to each of the P_1 and P_2 purinoceptors described in this chapter, i.e., presynaptic A_1, postsynaptic/somal A_1, postsynaptic A_2, pre- and postsynaptic P_{2x}, and postsynaptic P_{2y} subtypes, as well as others that will probably be described soon.

Ongoing research in the laboratory of one of us (F.L.C.) uses several complimentary neurophysiological approaches to study purinergic signaling pathways in enteric ganglia. These include (i) photometry of intracellular free calcium ion concentration $[Ca^{2+}]_i$, $[Na^+]_i$, and $[H^+]_i$, (ii) high-resolution fluorescence imaging at video rates of 30 frames/sec to study the spatial and temporal distribution of purinergic signals in enteric circuits, (iii) imaging, for the first time of $[cAMP]_i$ in individual myenteric AH/type 2 and submucosal S neurons microinjected with the new cAMP fluorosensor FlCRhR (Adams et al., 1991, 1993), and (iv) simultaneous imaging of $[Ca^{2+}]_i$ and $[cAMP]_i$, or $[Ca^{2+}]$ and $[H^+]_i$ in enteric ganglia. These approaches should facilitate further characterization and subclassification of the multitude of P_1 and P_2 purinoceptors in enteric ganglia. Preliminary findings, shown in Figure 1, demonstrate that P_2 purinoceptor activation in calbindin-immunoreactive cultured multipolar neurons leads to an increase in $[Ca^{2+}]_i$ ($n = 15$), while P_1 purinoceptor activation inhibits electrically-evoked elevation of $[Ca^{2+}]_i$ ($n = 7$) in these neurons. Adenosine (0.5 mM) does not elevate $[Ca^{2+}]_i$ in the neurons ($n = 12$). Both receptors are linked to an N-ethyl maleimide (NEM)-sensitive G protein ($n = 7$ for ATP, $n = 3$ for N^6-cyclopentyl adenosine (CPA)). Calbindin-immunoreactive neurons (Christofi and Wood, 1993b) display AH/Dogiel type II morphology, and their response to ATP strongly supports the hypothesis that ATP activation of P_2 purinoceptors on these neurons elevates $[Ca^{2+}]_i$, which in turn activates calcium-dependent potassium channels and leads to hyperpolarization of the neurons and a decrease in their excitability.

Motility and Contractility as They Relate to Function and Digestive Disorders

In a study in man, adenosine injection that provokes transient chest pain did not affect swallowing-induced peristaltic contractions of the esophagus and lower esophageal

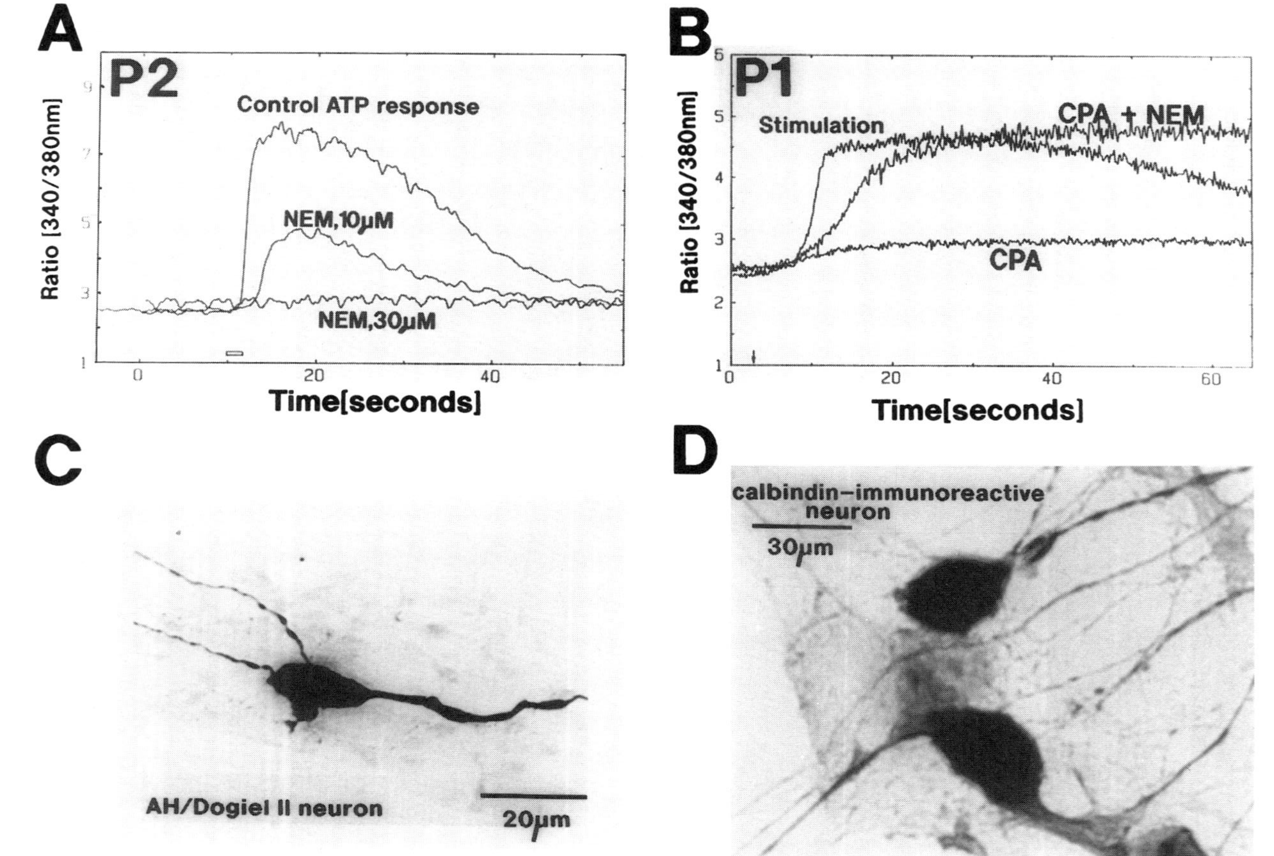
A
P2
Ratio [340/380nm]
Control ATP response
NEM,10μM
NEM,30μM
Time[seconds]
B
P1
Ratio [340/380nm]
Stimulation
CPA + NEM
CPA
Time[seconds]
C
AH/Dogiel II neuron
20μm
D
calbindin–immunoreactive neuron
30μm

sphincter, although the peristaltic wave was delayed significantly (Melcher and Sylven, 1989). Adenosine also had a relaxing effect on the lower esophageal sphincter, but did not block its normal reactions to swallowing. Interference with esophageal impulse propagation is putatively via P_1-mediated inhibition of local reflexes involving postsynaptic cholinergic and nonadrenergic noncholinergic (NANC) nerves to the esophagus. This study also showed that chest pain caused by adenosine cannot be caused by spastic esophageal contractions.

Introduction of predigested food into the jejunal lumen of an anesthetized dog significantly increases adenosine release into the local venous blood during the initial several minutes after food placement. The increased production and release of adenosine may be of importance in postprandial intestinal hyperemia (Sawmiller and Chou, 1990).

Adenosine was shown to be a mediator of ethanol-induced gastric vasodilation in dogs (Wood et al., 1993). Functional P_{2x} purinoceptors are present on both arterial and venous blood vessels of the cat intestinal circulation (Taylor and Parsons, 1991). The potential of adenosine to regulate organ blood flow in the intact rabbit upper alimentary tract was evaluated using the radioactive microsphere technique by Pennanen et al. (1994). Their results indicated that, in rabbit alimentary organs, adenosine preferentially increases blood flow to the esophageal mucosa, antral mucosa, and small intestine, and its effects are mediated by A_2/P_1 purinoceptors.

Adenosine and adenine nucleotides have been shown to influence gastric acid secretion in several models; this topic is discussed in some detail later (see section on purinergic regulation of gastrointestinal secretory function, below). The relevance of such purinergic modulation to peptic ulcer disease is not presently understood, although the central attenuation of stress-induced ulceration by adenosine *in vivo* has been described (Geiger and Glavin, 1985).

Adenosine has been implicated as a potential anti-inflammatory or cytoprotective agent in postischemic intestine (Kaminski and Proctor, 1989, 1990; Grisham et al., 1989). More recent findings by Kaminski and Proctor (1992) suggest that (i), endogenous adenosine modulates the inflammatory response evoked by intestinal reperfusion following a brief ischemia, and (ii), exogenously applied adenosine arrests most of the inflammatory changes associated with reperfusion by mechanisms that include both extracellular receptor-mediated vasodilation and granulocyte inhibition, and intracellular restoration of ATP. The authors point out that the effectiveness of adnenosine compounds, even when administered after ischemia, attests to the practicality of salvag-

FIGURE 1. Activation of P_1 and P_2 purinoceptors linked to G proteins leads to changes in $[Ca^{2+}]_i$ in cultured myenteric neurons of the guinea pig small intestine. **(a)** The ATP-induced Ca^{2+} transient is blocked by N-ethylmaleimide (NEM) in a dose-dependent manner. **(b)** The electrically-stimulated increase in $[Ca^{2+}]_i$ in another cultured myenteric neuron is suppressed by the A_1 agonist CPA, and reversed by NEM (30 μM). **(c)** Photomicrograph showing the morphology of a typical neuron, which responds to both adenosine and ATP with a sustained membrane hyperpolarization in the intact, microdissected myenteric plexus. The recorded neuron was injected with biocytin and later fixed and reacted with avidin-HRP and carried through a diaminobenzidine color-developing reaction. **(d)** Photomicrograph of calbindin-immunoreactive cultured myenteric neurons, which respond to ATP with a Ca^{2+} transient. Adenosine does not increase $[Ca^{2+}]_i$ in these cells.

ing ischemic bowel, at least under some conditions. Ischemia/reperfusion injury, particularly to splanchnic organs, can contribute to multiple-system organ failure (Deitch, 1990). The potential beneficial role of purines in such clinical emergencies is at present unclear, although prospects for useful therapeutic intervention may be emerging.

Studies on the actions of purinergic compounds on intestinal motility are very rare in the literature. An early report by Tonini et al. (1982) showed that ATP inhibited the propulsive contractions in the isolated rabbit colon. Adenosine inhibited vagally stimulated gastric motility in rabbits (Baccari et al., 1990), as well as suppressing the interdigestive mirgrating myoelectric complex (MMC) in rat small intestine (Fargeas et al., 1990).

Feit and Roche (1988) carried out experiments in dogs to test the hypothesis that adenosine exerts beneficial effects on intestinal motility after experimental mesenteric ischemia. It was concluded that, under physiological conditions, adenosine modifies jejunal motility directly via A_1 receptors on smooth muscle and indirectly via A_2 receptors on mesenteric vessels. Furthermore, after experimental ischemia, adenosine improves the postischemic restoration via vessel A_2 receptors only. Adenosine is also likely to have a presynaptic effect in potentiating the contractile action of direct vagal stimulation on gastric motility (Cho et al., 1993). In contrast, adenosine lacks an effect on the contractile cholinoceptors located on the smooth muscles of stomachs and blood vessels.

Propagated giant migrating contractions (GMCs) of the colon with high amplitude and long duration, have been reported to be closely associated with the provocation of urgency and defecation (Karaus and Sarna, 1987; Bazzocchi et al., 1991). The glycerol enema used to elicit GMCs is used clinically in colonic lavage procedure prior to colonoscopy and for treating chronic constipation. New evidence indicates that adenosine A_1 agonists suppress glycerol enema–induced GMCs via peripheral adenosine receptors. Karaus and Sarna (1987) hypothesized that occurrence of GMCs with abnormally high frequency may result in rapid propulsion and diarrhea. A later clinical study provided support for this hypothesis, in that patients with functional diarrhea had an increase in the frequency of GMCs (Bazzocchi et al., 1991). An earlier study also found that adenosine A_1 activation suppressed morphine withdrawal diarrhea in mice (Tucker et al., 1984). Clinical use of selective adenosine A_1 agonists as antidiarrheal drugs, or to inhibit diarrhea in patients with conspicuous GMCs, is likely to be limited due to undesirable side effects on other organ systems, unless, of course, its administration could be somehow restricted to the site of therapeutic action. No information is available on anticipated actions of P_2 purinoceptor agonists during the digestive and interdigestive motility states, especially with respect to GMCs and MMCs.

Stimulation of lumbar sympathetic nerves contracts the circular muscle in cat colon via corelease of noradrenaline and ATP or a related purine nucleotide from sympathetic fibers. The neurogenic responses are likely to be mediated through both excitatory suramin-sensitive and suramin-insensitive P_{2x} purinoceptors (Venkova and Krier, 1993).

The diversity of potential therapeutic applications for ATP is illustrated by the recent finding that exogenous ATP provides a radioprotective effect following damaging neutron radiation of the small intestine (Szeinfeld and Villiers, 1993). ATP is believed to exert a number of metabolic adaptations as a defense mechanism in which the small intestinal smooth muscle cells exposed to neutron radiation could remain viable because the injury is potentially repairable.

Nonadrenergic, Noncholinergic Inhibitory Neurons

While the coordinated firing of excitatory enteric neurons leading to stimulation of smooth muscle provides for organized contractile activity, inhibitory transmission similarly contributes to this fundamental activity. The relaxation of smooth muscle and of sphincters is critical to normal peristaltic function as well as to the MMC and other motor patterns. The putative role of ATP in NANC neurotransmission in the proximal stomach was recently reviewed by Lefebvre (1993) and will not be dealt with at length here. The available evidence suggests that ATP, vasoactive intestinal polypeptide (VIP), and nitric oxide (NO) contribute to NANC inhibitory neurotransmission in the gastrointestinal tract (Goyal et al., 1980; Burnstock, 1981, 1993; Sanders and Ward, 1992; Saffrey et al., 1992; Maggi and Giuliani, 1993; Boeckxstaens et al., 1993). The fast, apamin-sensitive component of NANC inhibitory responses is best mimicked by ATP, while VIP mediates the slowest, most prolonged component; NO is of intermediate duration. The role of NO in NANC inhibitory transmission was reviewed recently by Brookes (1993). A recent report provided evidence for the coexistence of ATP and NO in NANC inhibitory neurons in the rat ileum, colon, and anococygeus muscle (Belai and Burnstock, 1994). However, clear differences were observed in the number of NANC inhibitory neurons that copackaged these neurotransmitters in different regions of the gastrointestinal tract. NANC inhibitory neurotransmission in rat pyloric sphincter also involves ATP acting directly at P_{2y} purinoceptors. NO contributed to the later component of NANC relaxation (Soediono and Burnstock, 1994). Although both P_1 and P_2 purinoceptors exist on rat duodenal smooth muscle, neither ATP nor adenosine is involved in NANC relaxation of the muscle.

Zagorodnyuk et al. (1989) carried out a detailed analysis of the role of ATP in NANC inhibitory neuromuscular transmission in normal human gut, as well as tissue from patients with Hirschsprung's disease. Their results are consistent with the purinergic hypothesis of nonadrenergic inhibition. In addition to inhibitory receptors, excitatory purinoceptors were identified in the longitudinal muscle of human gut specimens, and these receptors may play a role in the pathophysiology of Hirschsprung's disease.

PURINERGIC REGULATION OF GASTROINTESTINAL SECRETORY FUNCTION

Although the actions of adenosine and adenine nucleotides on secretory functions of the gastrointestinal tract have undergone less overall investigation than their motor actions, some progress has occurred since one of us previously reviewed the area (Cook, 1991). Other relevant reviews have also appeared (Westerberg and Geiger, 1988; Daniel, 1989). There are two principal areas in which such progress has occurred: actions of nucleosides and nucleotides on gastric acid secretion, and their actions on intestinal electrolyte (primarily chloride) secretion. These topics will be examined, together with documented progress in relevant related areas.

Gastric Acid Secretion

Following the long-standing observation that caffeine and theophylline are gastric secretagogues (Krasnow and Grossman, 1849; Foster et al., 1979), several groups examined the ability of adenosine and its analogs to modify gastric H^+ production.

Different approaches, models and species have been used with predictably conflicting results. For example, inhibition of acid secretion (Gerber et al., 1984, 1985; Scarpignato et al., 1987; Glavin et al., 1987; Westerberg and Geiger, 1989), stimulation of acid secretion (Puurunen et al., 1986; Gil-Rodrigo et al., 1990; Ainz et al., 1993), and lack of effect (Puurunen et al., 1987) have all been reported as actions of the nucleoside (for additional references see Cook, 1991). The actions of adenine nucleotides on acid secretion have received less attention, although some intriguing data have been reported.

Inhibition of Acid Secretion Studies in conscious rats prepared with chronic indwelling gastric cannulae (Westerberg and Geiger, 1989) demonstrated that the adenosine analogs 5′-N-ethylcarboxamido adenosine (NECA), *R*- and *S*-phenylisopropyladenosine (PIA), and 2-chloroadenosine (CADO), compounds that are metabolically stable and resistant to nucleoside transport, decreased basal gastric H^+ output. The rank order of potency of these analogs, as well as the stereospecificity of PIA, indicated mediation of an A_1 receptor. These investigators also noted that NECA and CADO reduced secretion volume while R-PIA did not. The site of action of adenosine could not be determined from these functional data and the authors suggested that central, as well as peripheral receptors may be involved and that the parietal cell may be a target. They also suggested that unmasking of bicarbonate secretion may also contribute. It is noteworthy that this study reports inhibition of *basal* (unstimulated) secretion (confirming the initial report from Geiger's group (Glavin et al., 1987)). Other studies demonstrating inhibitory actions have been obtained in models using *stimulated* output of acid. For example, in anesthetized dogs with gastric fistulas, Gerber et al. (1984) noted that intra-arterial adenosine inhibited both histamine- and methacholine-stimulated acid secretion. Using a more reductionist approach, the same group (Gerber et al., 1985) reported the inhibitory action of R-PIA on histamine- or carbachol-stimulated acid secretion (measured by the [^{14}C]aminopyrine uptake technique, (see Sack and Spenney, 1982) in isolated canine parietal cells. Interestingly, they speculated that mediation via A_1 receptors, which would inhibit histamine-stimulated cAMP generation, was probably occurring at low concentrations of R-PIA, while at higher concentrations activation of stimulatory A_2 receptors by the analog would occur. The response to receptor blockade, transport inhibition, and deamination by adenosine deaminase provided evidence for the presence of endogenous adenosine in this preparation (Gerber and Payne, 1988). Another report of inhibitory actions was that of Scarpignato et al. (1987). These investigators used the Shay rat (in which gastric secretion is stimulated by the reflex response to pyloric ligation (Shay et al., 1954) to demonstrate inhibitory responses to both adenosine and R-PIA. The response to R-PIA was inhibitory at all concentrations used. However, it is interesting to note that the response to adenosine itself was biphasic, with the highest concentration giving values not significantly different from control. It is possible that the higher concentration activated A_2 receptors, although it is not clear whether such receptors are located on parietal cells or if indirect actions are involved. A brief report of the ability of adenosine to inhibit histamine-stimulated acid production by isolated human parietal cells has also appeared (England et al., 1990).

Stimulation of Acid Secretion Stimulation of gastric acid secretion in intact, urethane-anesthetized rats was reported by Puurunen et al. (1986) following intravenous administration of NECA, R-PIA, N^6-cyclohexyl adenosine (CHA), S-PIA,

CADO, or adenosine, the response exhibiting that rank order of potency. The response was abolished by vagotomy, and the authors suggested mediation occurred via afferent vagal pathways. However, probable systemic vascular effects of the analogs used were not addressed. Using isolated parietal cells prepared from rabbit fundic mucosa, Ota et al. (1989) showed that NECA, CADO, and adenosine, in that rank order, stimulated both histamine-mediated acid secretion, as measured by [^{14}C]aminopyrine uptake, and cAMP accumulation. They postulated the presence of an A_2 receptor on parietal cells. Gil-Rodrigo et al. (1990) used gastric glands, isolated enzymatically from a mince of rabbit gastric mucosa, to examine not only the actions of adenosine and R-PIA, but also the adenine nucleotides ATP, ADP, AMP, and the ATP analog β,γ-methylene ATP, on resting and histamine-stimulated acid secretion, as measured by [^{14}C]aminopyrine uptake. They showed that adenosine and R-PIA increased basal acid secretion, while the nucleotides gave no response. Histamine-stimulated acid secretion was also increased by adenosine, R-PIA, and AMP, while ATP, ADP, and the analog β,γ-methylene ATP inhibited. Theophylline at low concentrations inhibited the ability of adenosine to stimulate acid secretion, but at higher concentrations (>10 μM) augmented acid secretion. The methylxanthine also augmented histamine-stimulated acid secretion at concentrations greater than 100 μM, while lower concentrations had no effect. The inhibitory actions of ATP, which were ascribed to activation of a P_2 receptor, were sensitive to indomethacin, suggesting involvement of prostaglandin synthesis. While hydrolysis of AMP to adenosine provides a basis for the actions of AMP, the distinct responses to the other nucleotides provide evidence for the presence of functional P_2 receptors at this locus. These investigators extended their findings in a study using isolated rabbit parietal cells (Ainz et al., 1993), which yielded additional evidence for the presence of A_2 receptors on parietal cells. In this study, resting acid secretion (by [^{14}C]aminopyrine uptake) was stimulated by NECA, CADO, and adenosine, in that rank order. Secretion stimulated by either histamine or dibutyryl-cAMP, both secretagogues in this system, was similarly augmented. It is noteworthy that the studies using the more isolated preparations—gastric glands or parietal cells (*canine,* Gerber et al., 1985; *human,* England et al., 1990)—had shown inhibitory actions, while those obtained from rabbit preparations showed stimulation. Clearly, species differences may be important in interpreting nucleoside responses at this site of action. Parietal cells may express both A_1 and A_2 receptors, responding according to relative receptor density as well as endogenous adenosine concentration.

Interpretation of the findings reported in the more intact models is complicated by the existence of multiple potential targets for the actions of adenosine and adenine nucleotides. In addition to vascular effects, which are well known to influence acid secretory function (for refs. see Guth et al., 1989), other factors involved in the regulation of gastric acid secretion may be influenced by the purines.

Release of Gastrin The actions of adenosine on release of gastrin was examined by Schepp et al. (1990), using dispersed canine antral cells. Their study, which followed a brief report on data from rat antral tissue by Harty and Franklin (1984), showed that adenosine is able to modulate immunoreactive gastrin release. Forskolin-stimulated gastrin release was inhibited by adenosine and R-PIA in a pertussis toxin–sensitive manner, such release also being sensitive to the A_1 antagonist 1,3-dipropyl-8-cyclopentylxanthine (DPCPX). Mediation by an A_1 receptor was inferred. In contrast, bombesin-stimulated gastrin release was augmented by adenosine and by the modestly A_2-selective analog 2-phenyaminoadenosine (CV1808) but not by R-PIA. These findings are compatible with mediation by an A_2 receptor putatively located on antral G cells. Such augmentation was insensitive to pertussis toxin and may not involve signaling by alteration of cAMP levels but rather by altered Ca^{2+}/phosphatidylinositol levels. The specificity of the immunoassay system used in this study allowed actions of the nucleosides on gastrin release to be measured despite the heterogeneous nature of the preparation (approximately 12% enrichment with G cells). This was recognized by the authors who contemplated possible paracrine effects, which may have contributed to their findings. Two mediators that are undoubtedly able to influence acid secretion are histamine and somatostatin, and the ability of adenosine to affect their release may then indirectly alter acid secretion.

Release of Histamine It has been known for some time that adenosine can enhance the release of histamine from rat peritoneal mast cells stimulated by several secretagogues (Marquardt et al., 1978). Studies on the ability of the nucleotides to alter histamine release from gastric mast cells, the cell population most relevant to altered acid secretion, were carried out by Soll et al. (1988). This group used mast cells isolated from canine fundic mucosa by enzymatic dispersion and elutriation, and subjected to short-term culture. Adenosine and several analogs augmented histamine release stimulated by concanavalin A, with R-PIA being the most potent. The rank order of potency, R-PIA > adenosine > CADO > NECA, was suggestive of mediation by an A_1 receptor, although this ranking may be species specific. It is not clear from these studies whether adenosine is involved in the physiological regulation of histamine release, although the possibility arises that endogenous adenosine may function as a potential stimulator. As the authors point out, many of the cell types in their preparation are potential sources for adenosine, including the mast cells themselves (Marquardt et al., 1984), and autocrine mechanisms may therefore also contribute. Mast cells also express receptors for gastrin (Ekblad, 1985; Nylander et al., 1985), increasing the levels of regulation available; furthermore, mast cells are not the sole source of histamine available to the fundic mucosa. A histamine-containing neuron capable of releasing histamine at this locus in response to gastrin has also been described (Hakanson et al., 1986). Another paracrine influence that may be relevant is somatostatin.

Release of Somatostatin Somatostatin is an inhibitor of gastric acid secretion (Creutzfeldt and Arnold, 1978). Somatostatin is an inhibitor of gastric acid secretion (Creutzfeldt and Arnold, 1978). The ability of adenosine and several nucleotides to alter somatostatin release was examined by Kwok et al. (1990) using an arterially perfused rat stomach model. They demonstrated the ability of adenosine to stimulate somatostatin release under basal conditions and showed that vascular actions were probably not involved in this response. ATP and the analog β,γ-methylene ATP were similarity shown to stimulate somatostatin release, while both α,β-methylene ATP and α,β-methylene ADP did not. It is possible, as the authors speculate, that β,γ-methylene ATP was metabolized to adenosine, while the α,β-methylene isosteres were resistant. However, mediation via a P_{2x} receptor has not been ruled out. The release of somatostatin would provide an inhibitory influence on acid secretion, as well as on multiple endocrine or paracrine mediators, and the complexity of these possibilities, together with the potency of this peptide mediator, suggest that further studies would be useful.

Intestinal Electrolyte Secretion Evidence for the presence of purine receptors on secretory cells from intestinal epithelium was obtained by Grasl and Turnheim (1984) using rabbit colonic mucosal strips mounted in Ussing chambers. They demonstrated that the addition of adenosine to the serosal, but not luminal, side increased a short-circuit current and conductance resulting from the electrogenic secretion of chloride ions. The addition of AMP elicited a similar response and hydrolysis to adenosine appears likely. Active transport of sodium was not altered by adenosine and the presence of an adenosine receptor on the basolateral membrane coupled to chloride secretion was implied. A human colonic epithelial cell line, designated T_{84} and usually grown in monolayers, has provided additional information. Barrett et al. (1989) showed that adenosine, R-PIA, and NECA produce sustained increases in chloride secretion when added to either apical or basolateral surfaces of these cells, addition to the basolateral surface being more potent. Similarly, NECA applied to the basolateral surface caused greater increase in cAMP than did apical addition. The expression of an adenosine receptor on the basolateral surface of these cells seems highly probable.

More recently, Madara et al. (1993) investigated the paracrine factor released by neutrophils activated on exposure to endotoxin and so on, and responsible for stimulation of chloride secretion when applied to the apical surface of T_{84} cells. The factor was identified as 5′-AMP and, since inhibition of ectonucleotidase activity by α,β-methylene-ADP inhibited the response to AMP but not to adenosine, the mechanism was suggested to involve conversion of AMP to adenosine at the epithelial surface. The presence of ectonucleotidase activity on the apical, but not basolateral, surface may explain the differences in AMP potency obtained in these studies. The authors suggest that AMP may function as a paracrine mediator of intestinal secretion under conditions of active intestinal inflammation. A similar mechanism was invoked to explain the ability of activated eosinophils to elicit chloride secretion from T_{84} cells on release of AMP (Resnick et al., 1993). Release of AMP from neutrophils in the cyst abcesses that characterize active intestinal inflammation would directly access the luminal surface of crypt epithelial cells and provide a stimulus for the activation of chloride channels leading to the substantial secretion observed. Interestingly, ATP may serve to block chloride channel activation at this locus (Venglarik et al., 1993).

The adenosine receptor on T_{84} cells mediating chloride secretion has been characterized as an A_{2b} subtype (Strohmeier et al., 1995). These investigators showed that the potency rank order for agonists at this receptor was NECA > adenosine > CGS-21680, while the hierarchy of antagonist affinity was xanthine amine congener (XAC) > 1,3-diethyl-8-phenylxanthine > aminophylline—profiles compatible with the presence of the A_{2b} subtype. Moreover, Northern blotting of RNA extracted from T_{84} cells revealed no expression of A_1 or A_{2a} receptors, but did show strong expression of A_{2b} receptors. Similar data was obtained for RNA extracted from human intestine, which the strongest expression of A_{2b} receptors occurring in the colon and appendix, and less in the ileum. Positive coupling of this receptor to adenylate cyclase was also demonstrated. Since the A_{2b} receptor can contribute significantly to the development of secretory diarrhea, it represents an important potential target for therapeutic innovation.

In a distinct cell line derived from human intestinal goblet cells and designated HT29-C1.16E, both ATP and the analog ATPγS were shown to increase mucin and chloride secretion when applied to the apical surface of monolayers (Merlin et al., 1994). Mediation via an adenosine receptor following hydrolysis of nucleotides was discounted as a mechanism because the ATP analog was as effective as ATP, while adenosine was far less effective. It is unfortunate that adenosine analogs were not examined concurrently since transport or degradation of adenosine may have masked the presence of a functional P_1 receptor. The presence of a P_2 receptor coupled to signaling pathways that initiate granule fusion and mucin release, as well as chloride channel activation, was also suggested by these authors. Clearly, selective expression of P_1 and/or P_2 receptors on intestinal epithelial cells is to be expected, and considerable additional study is required to illuminate this area.

CONCLUSION

Progress in defining the actions of purines at the GI tract continues to be made and the physiological and pathophysiological roles of endogenous purines are beginning to emerge. Advances in our overall understanding of the molecular biology of purinergic receptors are being applied to the GI tract, although clearer definition of receptor subtypes and the transduction mechanisms involved are still required. Opportunities for useful therapeutic intervention are also becoming clearer, and it is suggested that these will accelerate with the development and application of our understanding.

ACKNOWLEDGMENTS

The authors' research quoted in this chapter was supported by grants from the National Institutes of Health, NIH R29 DK44179 (F.L.C.) and from the Medical Research Council of Canada, MT 10167 (M.A.C.).

REFERENCES

Adams SR, Bacskai BJ, Taylor SS, Tsien RY (1993): Optical probes for cyclic AMP. In Mason WT, Relf, G (eds): "Fluorescent Probes for Biological Activity of Living Cells—A Practical Guide." London: Academic Press, pp 133–149.

Adams SR, Harootunian AT, Buechler YJ, Taylor SS, Tsien RY (1991): Fluorescence ratio imaging of cyclic AMP in single cells. Nature 349:694–697.

Ainz LF, Salgado C, Gandarias JM, Gomez R, Vallejo A, Gil-Rodrigo CE (1993): P_1 (A_2/R_a)-purinoceptors may mediate the stimulatory effect of adenosine and adenosine analogs on acid formation in isolated rabbit parietal cells. Pharmacol Res 27:319–334.

Baccari MC, Calamai F, Staderini G (1990): Effects of arterial infusions of adenosine 5′-triphosphate (ATP) and vasoactive intestinal polypeptide (VIP) on vagal excitatory motor responses in the rabbit stomach in vivo. J Auton Nerv Syst 30:S15–S18.

Barajas-Lopez C (1993): Adenosine reduces the potassium conductance of guniea pig submucosal plexus neurons by activating protein kinase A. Pflugers Arch 424:410–415.

Barajas-Lopez C, Barrientos M, Espinosa-Luna R (1993): Suramin increases the efficacy of ATP to activate an inward current in myenteric neurons from guinea pig ileum. Eur J Pharmacol 250:141–145.

Barajas-Lopez C, Espinosa-Luna R, Gerzanich V (1994): ATP closes potassium and opens a cationic conductance through different receptors in neurons of guinea pig submucous plexus. J Pharmacol Exp Ther 268:1396–1402.

Barajas-Lopez C, Surprenant A, North RA (1991): Adenosine A_1 and A_2 receptors mediate presynaptic inhibition and postsynaptic excitation in guinea pig submucosal neurons. J Pharmacol Exp Ther 258:490–495.

Barrett KE, Huott PA, Shah SS, Dharmsathaphorn K, Wasserman, SI (1989): Differing effects of apical and basolateral adenosine on colonic epithelial cell line T_{84}. Am J Physiol 256:C197–C203.

Bazzocchi G, Ellis J, Villanueva-Meyer J, Reddy SN, Mena I, Snape WJ Jr (1991): Effect of eating on colonic motility and transit in patients with functional diarrhea: Simultaneous scintigraphic and manometric evaluations. Gastroenterology 101:1298–1306.

Belai A, Burnstock G (1994): Evidence for coexistence of ATP and nitric oxide in non-adrenergic, non-cholinergic (NANC) inhibitory neurones in the rat ilium, colon and anococcygeus muscle. Cell Tissue Res 278:197–200.

Boeckxstaens GE, Pelckmans PA, Herman AG, Van Maercke YM (1993): Involvement of nitric oxide in the inhibitory innervation of the human isolated colon. Gastroenterology 104:690–697.

Bornstein JC, Kunze AA, Furness JB (1992): Calbindin immunoreactivity in morphologically distinct classes of putative enteric sensory neurons in guinea pig ileum. Proc Aust Physiol Pharmacol Soc 23:44P Abstract.

Bornstein JC, Hendricks R, Furness JB, Trussell DC (1991): Ramifications of the axons of AH-neurons injected with the intracellular marker biocytin in the myenteric plexus of the guinea-pig small intestine. J Comp Neurol 314:437–451.

Bornstein JC, North RA, Costa M, Furness JB (1984): Excitatory synaptic potentials due to activation of neurones with short projections in the myenteric plexus. Neuroscience 11:723–731.

Broad RM, Cook MA (1993): Inhibition of neurotransmitter release from enteric nerve endings by 2-[*p*-(carboxyethyl)-phenethylamino]-5′-N-ethylcarboxamido-adenosine (CGS-21680) and related adenosine analogs: Lack of simple competition by antagonists. J Pharmacol Exp Ther 266:634–641.

Broad RM, McDonald TJ, Brodin E, Cook MA (1992): Adenosine A_1 receptors mediate inhibition of tachykinin release from perfused enteric nerve endings. Am J Physiol 262:G525–G531.

Brookes SJH (1993): Neuronal nitric oxide in the gut. J Gastroenterol Hepat 8:590–603.

Brookes SJH, Song Z.-M, Ramsey GA, Costa M (1995): Long aboral projections of Dogiel type II, AH neurons within the myenteric plexus of the guinea pig small intestine. J Neurosci 15:4013–4022.

Brookes SJH, Steele PA, Costa M (1991a): Identification and immunohistochemistry of cholinergic and non-cholinergic circular muscle motor neurons in the guinea-pig small intestine. Neuroscience 42:863–878.

Brookes SJH, Steele PA, Costa M (1991b): Calretinin immunoreactivity in cholinergic motor neurones, interneurones and vasomotor neurones in the guinea-pig small intestine. Cell Tissue Res 263:471–481.

Burnstock G (1993): Physiological and pathological roles of purines: An update. Drug Dev Res 28:195–206.

Burnstock G (1981): Neurotransmitter and trophic factors in the autonomic nervous system. J Physiol (London) 313:1–35.

Cho CH, Qiu BS, Ballard HJ, Ogle CW (1993): Adenosine and cholinergic-induced gastric contraction in rats. Digestion 54:98–104.

Christofi FL, Wood JD (1989): Release of endogenous adenosine modulates the synaptic and electrical behavior of myenteric neurons of the guinea pig distal ileum. J Gastrointest Mot 1:77.

Christofi FL, Wood JD (1993a): Effects of PACAP on morphologically identified myenteric neurons in guinea pig small bowel. Am J Physiol 264:G414–G421.

Christofi FL, Wood JD (1993b): Presynaptic inhibition by adenosine A_1 receptors on guinea pig small intestinal myenteric neurons. Gastroenterology 104:1420–1429.

Christofi FL, Wood JD (1994): Electrophysiological subtypes of inhibitory P_1 purinoceptors on myenteric neurones of guinea-pig small bowel. Br J Pharmacol 113:703–710.

Christofi FL, Baidan LV, Fertel RH, Wood JD (1994): Adenosine A_2 receptor-mediated excitation of a subset of AH/Type 2 neurons and elevation of cAMP levels in myenteric ganglia of guinea-pig ileum. Neurogastroenterol Mot 6:67–78.

Christofi FL, Baidan LV, Wood JD, Stokes BT (1995): Multiple intracellular signaling mechanisms are linked to the elevation of cytosolic free calcium in myenteric multipolar neurons of the guinea-pig small bowel. Gastroenterology 108:A921.

Christofi FL, McDonald TJ, Cook MA (1990): Adenosine receptors are coupled negatively to release of tachykinin(s) from enteric nerve endings. J Pharmacol Exp Ther 253:290–295.

Cook MA (1991): Purinergic regulation of gasrointestinal motility and secretion. In Phillis JW (ed): "Adenosine and Adenine Nucleotides as Regulators of Cellular Function. Boston: CRC Press, pp 267–282.

Creutzfeldt W, Arnold R (1978): Somatostatin and the stomach: Exocrine and endocrine aspects. Metabolism 27(suppl 1):1309–1315.

Daniel EE, Collins SM, Fox JET, Huizinga JD (1989): Pharmacology of drugs acting on gastrointestinal motility. In Schultz SG (section ed), Wood JD, (volume ed): "Handbook of Physiology, Volume 1, Part 1." Bethesda MD: Americal Physiological Society, pp 715–757.

Deitch EA (1990): Role of intestinal barrier failure and bacterial translocation in development of systemic infection and multiple organ failure. Arch Surg 125:403–404.

Ekblad EBM (1985): Histamine—the sole mediator of pentagastrin stimulated acid secretion. Acta Physiol Scand 125:135–143.

England S, Daly MJ, Jacob S, Bardhan KD (1990): Adenosine inhibits acid production by human isolated parietal cells. (Abstract). Br J Pharmacol 101:604P.

Evans RJ, Derkach V, Surprenant A (1992): ATP mediates fast synaptic transmission in mammalian neurons. Nature 357:503–505.

Fargeas MJ, Fioramonti J, Bueno L (1990): Central and peripheral actions of adenosine and its analogues on intestinal myoelectric activity and propulsion in rats. J Gastrointest Mot 2:121–127.

Feit C, Roche M (1988): Action of adenosine on intestinal motility after experimental mesenteric ischaemia in the dog. Gastroenterol Clin Biol 12:803–809.

Foster LJ, Trudeau WL, Goldman AL (1979): Bronchodilator effects on gastric acid secretion. J Am Med Assoc 241:2613–2615.

Furness JB, Bornstein JC, Trussell DC (1988): Shapes of nerve cells in the myenteric plexus of the guinea pig small intestine revealed by the intracellular injection of dye. Cell Tissue Res 254:561–571.

Furness JB, Trussell DC, Pompolo S, Bornstein JC, Smith TK (1990): Calbindin neurons of the guinea-pig small intestine: Quantative analysis of their numbers and projections. Cell Tissue Res 260:261–272.

Galligan JJ, Bertrand PP (1994): ATP mediates fast synaptic potentials in enteric neurons. J Neurosci 14:7563–7571.

Geiger JG, Glavin GB (1985): Adenosine receptor activation in brain reduces stress-induced ulcer formation. Eur J Pharmacol 115:185–190.

Gerber JG, Payne NA (1988): Endogenous adenosine modulates gastric acid secretion to histamine in canine parietal cells. J Pharmacol Exp Ther 244:190–194.

Gerber JG, Nies AS, Payne NA (1985): Adenosine receptors on canine parietal cells modulate gastric acid secretion to histamine. J Pharmacol Exp Ther 233:623–627.

Gerber JG, Fadul S, Payne NA, Nies AS (1984): Adenosine: A modulator of gastric acid secretion in vivo. J Pharmacol Exp Ther 231:109–113.

Gil-Rodrigo CE, Galdiz JM, Gandarias JM, Gomez R, Ainz LF (1990): Characterization of the effects of adenosine, adenosine 5′-triphosphate and related purines on gastric acid secretion in isolated rabbit gastric glands. Pharmacol Res 22:103–113.

Glavin GB, Westerberg VS, Geiger JD (1987): Modulation of gastric acid secretion by adenosine in conscious rats. Can J Physiol Pharmacol 65:1182–1185.

Goyal RK, Rattan S, Said SI (1980): VIP as a possible neurotransmitter of non-cholinergic, non-adrenergic inhibitory neurones. Nature 288:378–380.

Grasl M, Turnheim K (1984): Stimulation of electrolyte secretion in rabbit colon by adenosine. J Physiol (London) 346:93–110.

Grisham MB, Hernandez LA, Granger DN (1989): Adenosine inhibits ischemia-reperfusion-induced leukocyte adherence and extravasation. Am J Physiol 257:H1334–H1339.

Guth PH, Leung FW, Kaffman GL (1989): Physiology of gastric circulation. In Schultz SG (section ed), Wood JD (volume ed): "Handbook of Physiology, Volume 1, Part 2." Bethesda, MD: Americal Physiological Society, pp 1371–1404.

Hakanson R, Bottcher G, Ekblad P, Panula M, Simonsson M, Dohlsed M Hallberg T, Sundler F (1986): Histamine in endocrine cells in the stomach: A survey of several species using a panel of histamine antibodies. Histochemistry 86:5–17.

Harty RF, Franklin PA (1984): Effects of exogenous and endogenous adenosine on gastrin release from rat antral mucosa. Gastroenterology 86:1107 Abstract.

Hendriks R, Bornstein JC, Furness JB (1990): An electrophysiological study of projections of putative sensory neurons within the myenteric plexus of the guinea-pig ileum. Neurosci Lett 110:286–290.

Hirst GDS, Holman ME, Spence I (1974): Two types of neurones in the myenteric plexus of duodenum in the guinea-pig. J Physiol (London) 236:303–326.

Iyer V, Bornstein JC, Costa M, Furness JB, Takahashi Y, Iwanaga T (1988): Electrophysiology of guinea-pig myenteric neurons correlated with immunoreactivity for calcium binding proteins. J Auton Nerv Syst 22:141–150.

Johnson SM, Katayama Y, North RA (1980): Slow synaptic potentials in the neurones of the myenteric plexus. J Physiol (London) 301:505–516.

Kamiji T, Morita K, Katayama Y (1994): ATP regulates synaptic transmission by pre- and postsynaptic mechanisms in guinea-pig myenteric neurons. Neuroscience 59:165–174.

Kaminski PM, Proctor KG (1989): Attenuation of no-flow phenomenon, neutrophil activation, and perfusion injury in intestinal microcirculation by topical adenosine. Circ Res 65:426–435.

Kaminski PM, Proctor KG (1990): Actions of adenosine on nitro blue tetrazolium deposition and surface pH during intestinal reperfusion injury. Circ Res 66:1713–1719.

Kaminski PM, Proctor KG (1992): Extracellular and intracellular actions of adenosine and related compounds in the reperfused rat intestine. Circulation Res 71:720–731.

Karaus M, Sarna SK (1987): Giant migrating contractions during defecation in the dog colon. Gastroenterology 92:925–933.

Katayama Y, Morita K (1989): Adenosine 5′-triphosphate modulates membrane potassium conductance in guinea-pig myenteric neurones. J Physiol (London) 408:373–390.

Kennedy C, Leff P (1995): How should P_{2x} purinoceptors be classified pharmacologically? Trends Pharmacol Sci 16:168–174.

Krasnow S, Grossman MI (1949): Stimulation of gastric secretion in man by theophylline ethylenediamine. Proc Soc Exp Biol Med 71:335–336.

Kunze WAA, Furness JB, Bornstein JC (1993): Simultaneous intracellular recordings from enteric neurons reveal that myenteric AH neurons transmit via slow excitatory postsynaptic potentials. Neuroscience 55:685–694.

Kwok YN, McIntosh C, Brown J (1990): Augmentation of release of gastric somatostatin-like immunoreactivity by adenosine, adenosine triphosphate and their analogs. J Pharmacol Exp Ther 255:781–788.

Lefebvre RA (1993): Non-adrenergic non-cholinergic neurotransmission in the proximal stomach. Gen Pharmacol 24:257–266.

Loubatieres-Mariani MM, Petit P, Chapal J, Hillaire-Buys D, Bertrand G, Ribes G (1995): Effects of purinoceptor agonists on insulin secretion. In Belardinelli L, Pelleg A (eds): "Adenosine and Adenine Nucleotides: From Molecular Biology to Integrative Physiology." Boston: Kluwer, pp 337–345.

Madara JL, Patapoff TW, Gillece-Castro B, Colgan SP, Parkos CA, Delp C, Mrsny RJ (1993): 5′-Adenosine monophosphate is the neutrophil-derived paracrine factor that elicits chloride secretion from T_{84} intestinal epithelial cell monolayers. J Clin Invest 91:2320–2325.

Maggi CA, Guiliani S (1993): Multiple inhibitory mechanisms mediate non-adrenergic relaxation in the circular muscle of the guinea-pig colon. Naunyn Schmiedebergs Arch Pharmacol 347:630–632.

Mahoney CW, Azzi A, Huang KP (1990): Effects of suramin, an anti-human immunodeficiency virus reverse transcriptase agent, on protein kinase C. J Biol Chem 265:5424–5428.

Marquardt DL, Walker LL, Wasserman SI (1984): Adenosine receptors on mouse bone marrow-derived mast cells: Functional significance and regulation by aminophylline. J Immunol 133:932–937.

Marquardt DL, Parker CW, Sullivan TJ (1978): Potentiation of mast cell mediator release by adenosine. J Immunol 120:871–878.

Mawe G, Branchek TA, Gershon MD (1986): Peripheral neural serotonin receptors: Identification and characterization with specific agonists and antagonists. Proc Natl Acad Sci USA 83:9799–9803.

Melcher A, Sylven C (1989): Chest pain and oesophageal pressure relationships in man following an intravenous bolus of adenosine. Clin Physiol 9:441–448.

Merlin D, Augeron C, Tien X-Y, Guo X, Laboisse CL, Hopfer U (1994): ATP-stimulated electrolyte and mucin secretion in the human intestinal goblet cell line HT29-C1.16E. J Membr Biol 137:137–149.

Mihara S, North RA, Surprenant A (1987): Somatostatin increases an inwardly rectifying potassium conductance in guinea-pig submucous plexus neurones. J Physiol (London) 390:335–355

Milusheva E, Sperlagh B, Kiss B, Szporny L, Pastztov E, Papasova M, Visi ES (1990): Inhibitory effect of hypoxic condition on acetylcholine release is partly due to the effect of adenosine released from the tissue. Brain Res Bull 24:369–373.

Morita K, Katayama, Y, Koketsu K, Akasu T (1984): Actions of ATP on the soma of bullfrog primary afferent neurones and its modulating action of the GABA-induced response. Brain Res 293:360–363.

Nakazawa K, Inoue K, Ito K, Koizumi S, Inoue K (1995): Inhibition by suramin and reactive blue 2 of GABA and glutamate receptor channels in rat hippocampal neurons. Naunyn Schmiedebergs Arch Pharmacol 351:202–208.

Nishi S, North RA (1973): Intracellular recording from the myenteric plexus of the guinea-pig ileum. J Physiol (London) 231:471–491.

North RA, Tokimasa T (1982): Muscarinic synaptic potentials in guinea-pig mesenteric plexus neurones. J Physiol (London) 33:151–156.

Nylander OE, Berqvist E, Obrink KJ (1985): Dual inhibitory actions of somatostatin on isolated gastric glands. Acta Physiol Scand 125:111–119.

Ota S, Hiraishi H, Terano A, Mutoh H, Kurachi Y, Shimada T, Ivey KJ, Sugimoto T (1989): Effects of adenosine and adenosine analogs on [^{14}C]aminopyrine accumulation by rabbit parietal cells. Dig Dis Sci 34:1882–1889.

Palmer JM, Wood JD, Zafirov DH (1987a): Purinergic inhibition in the small intestine myenteric plexus of the guinea-pig. J Physiol (London) 387:357–369.

Palmer JM, Wood JD, Zafirov DH (1987b): Transduction of aminergic and peptidergic signals in enteric-neurones of the guinea-pig. J Physiol (London) 387:371–383.

Pennanen MF, Bass BL, Dziki AJ, Harmon JW (1994): Adenosine: Differential effect on blood flow to subregions of the upper gastrointestinal tract. J Surg Res 56:461–465.

Puurunen J, Aittakumpu R, Tanskanen T (1986): Vagally mediated stimulation of gastric acid secretion by intravenously administered adenosine derivatives in anaesthetized rats. Acta Pharmacol Toxicol (Copenh) 58:265–271.

Puurunen J, Ruoff H-J, Schwabe U (1987): Lack of direct effect of adenosine on the parietal cell function in the rat. Pharmacol Toxicol 60:315–317.

Resnick MB, Colgan SP, Patapoff TW, Mrsny RJ, Awtrey CS, Delp-Archer C, Weller PF, Madara JL (1993): Activated eosinophils evoke chloride secretion in model intestinal epithelia primarily via regulated release of 5′-AMP. J Immunol 151:5716–5723.

Sack J, Spenney JG (1982): Aminopyrine accumulation by mammalian gastric glands: An analysis of the technique. Am J Physiol 243:G313–G319.

Saffrey MJ, Hassall CJS, Hoyle CHV, Belai A, Moss J, Schmidt HH, Forstermann U, Murad F, Burnstock G (1992): Colocalization of nitric oxide synthase and NADPH-diaphorase in cultured myenteric neurones. Neuroreport 3:333–336.

Sanders KKM, Ward SM (1992): Nitric oxide as a mediator of non-adrenergic non-cholinergic neurotransmission. Am J Physiol 262:G379–G392.

Sawmiller DR, Chou CC (1990): Jejunal adenosine increases during food-induced jejunal hyperemia. Am J Physiol 258:G370–G376.

Scarpignato C, Tramacere R, Zappia L, Del Soldato P (1987): Inhibition of gastric acid secretion by adenosine receptor stimulation in the rat. Pharmacology 34:264–268.

Schepp W, Soll AH, Walsh JH (1990): Dual modulation by adenosine of gastrin release from canine G-cells in primary culture. Am J Physiol 259:G556–G563.

Shay H, Sun CH, Gruenstein M (1954): A quantitative method for measuring spontaneous gastric secretion in the rat. Gastroenterology 26:906–913.

Soediono P, Burnstock G (1994): Contribution of ATP and nitric oxide to NANC inhibitory transmission in rat pyloric sphincter. Br J Pharmacol 113:681–686.

Soll AH, Toomey M, Culp D, Shanahan F, Beaven MA (1988): Modulation of histamine release from canine fundic mucosal mast cells. Am J Physiol 254:G40–G48.

Song Z.-M, Brooks SJH, Costa M (1994): All calbindin-immunoreactive myenteric neurons project to the mucosa of the guinea-pig small intestine. Neurosci Lett 180:219–222.

Sperlagh B, Vizi ES (1991): Effect of presynaptic P_2 receptor stimulation on transmitter release. J Neurochem 56:1466–1470.

Strohmeier GR, Reppert SM, Lencer WI, Madara JL (1995): The A_{2b} adenosine receptor mediates cAMP responses to adenosine receptor agonists in human intestinal epithelia. J Biol Chem 270:2387–2394.

Surprenant A, Crist J (1988). Electrophysiological characterization of functionally distinct 5-hydroxytryptamine receptors on guinea pig submucous plexus neurones. Neuroscience 24:283–295.

Szeinfeld D, de Villiers N (1993): Cholinesterase response in the rhabdomyosarcoma tumor and small intestine of the BALB/c mice and the radioprotective actions of exogenous ATP after lethal dose of neutron radiation. Strahlenther Onkol 169:311–316.

Takeshi K, Shirakabe K, Soejima O, Tokunaga T, Matsuo K, Sato C, Furukawa T (1993): Possible transsynaptic cholinergic neuromodulation by ATP released from ileal longitudinal muscles of guinea pigs. Life Sci 53:911–918.

Taylor EM, Parsons ME (1991): Effects of α,β-methylene ATP on resistance and capacitance blood vessels of the cat intestinal circulation; a comparison with other vasoconstrictor agents and sympathetic nerve stimulation. Eur J Pharmacol 205:35–41.

Tonini M, Onori L, Lecchini S, Frigo G, Perucca E, Crema A (1982): Mode of action of ATP on propulsive activity in rabbit colon. Eur J Pharmacol 82:21–28.

Tucker JF, Plant NT, von Uexkull A, Collier HOJ (1984): Inhibition by adenosine analogs of opiate withdrawal effects. NIDA Res Monogr 49:85–91.

Venglarik CJ, Singh AK, Wang R, Bridges RJ (1993): Trinitrophenyl-ATP blocks colonic Cl^- channels in planar phospholipid bilayers. Evidence for two nucleotide binding sites. J Gen Physiol 101:545–569.

Venkova K, Krier J (1993): Stimulation of lumbar sympathetic nerves evokes contractions of cat colon circular muscle mediated by ATP and noradrenaline. Br J Pharmacol 110:1260–1270.

Westerberg VS, Geiger JG (1988): Adenosine and gastric function. Trends Pharmacol Sci 9:345–347.

Westerberg VS, Geiger JD (1989): Adenosine analogs inhibit gastric secretion. Eur J Pharmacol 160:275–281.

Wieraszko A (1995): Facilitation of hippocampal potentials by suramin. J Neurochem 64:1097–1101.

Wood JD (1989): Electrical and synaptic behaviour of enteric neurons. In Schultz SG (section ed), Wood JD (volume ed): "Handbook of Physiology, Volume 1, Part 1." Bethesda, MD: Americal Physiological Society, pp 465–517.

Wood JD (1992): Integrated circuits: The basis of gastrointestinal-motility programs. In Heading RC, Wood JD (eds): "Gastrointestinal Dysmotility: Focus on Cisapride." New York: Raven Press, pp 15–39.

Wood JD (1994): Application of classification schemes to the enteric nervous system. J Auton Nerv Syst 48:17–29.

Wood JG, Darnell MP, Cheung LY (1993): Adenosine is a mediator of ethanol-induced gastric vasodilation in dogs. Am J Physiol 264:G664–G670.

Zafirov DH, Cooke HJ, Wood JD (1993): Elevation of cAMP facilitates noradrenergic transmission in submucous neurons of guinea pig ileum. Am J Physiol 264:G442–G446.

Zafirov DH, Palmer JM, Nemeth PR, Wood JD (1985a): Cyclic 3′,5′-adenosine monophosphate mimics slow synaptic excitation in myenteric plexus neurons. Brain Res 347:368–371.

Zafirov DH, Palmer JM, Wood JD (1985b): Adenosine inhibits forskolin-induced excitation in myenteric neurons. Eur J Pharmacol 113:143–144.

Zagorodnyuk VP, Vladimirova IA, Vovk EV, Shuba MF (1989): Studies of the inhibitory non-adrenergic neuromuscular transmission in the smooth muscle of the normal human intestine and from a case of Hirschsprung's disease. J Auton Nerv Syst 26:51–60.

3. Immunology

CHAPTER 15

Adenosine Regulation of Neutrophil Function and Inhibition of Inflammation via Adenosine Receptors

BRUCE N. CRONSTEIN

New York University Medical Center, New York, NY 10016

NEUTROPHILS AND INFLAMMATION

Neutrophils comprise a larger percentage of circulating leukocytes than any other cell type. They are the first white blood cells to arrive at an injured or infected site and histological evidence of their presence is required to demonstrate acute inflammation in tissue specimens. The dissection of signal transduction mechanisms in the neutrophil and the characterization of the proteins and lipids involved in neutrophil function over the past 30 years has provided a framework within which the molecular interactions that govern neutrophil function can be understood. Moreover, the description of the functional effects of a variety of endogenous and exogenous agents—most notably adenosine—that modulate neutrophil function has given rise to a growing recognition of their utility in the treatment of such inflammatory conditions as rheumatoid arthritis and reperfusion injury. This review will discuss recent information on the effects of adenosine on the function of neutrophils and how adenosine is currently used in the treatment of inflammatory diseases.

NEUTROPHIL FUNCTION

Neutrophils arise in the bone marrow and then must traverse the vasculature to arrive at injured or inflamed sites. Under normal circumstances neutrophils spend approximately 6 hours in the circulation before emigrating into the extravascular space. It is interesting to note that, in general, neutrophils leave the circulation only from postcapillary venules. Although it was thought that rheological or other mechanisms

Purinergic Approaches in Experimental Therapeutics, Edited by Kenneth A. Jacobson and Michael F. Jarvis
ISBN 0-471-14071-6 © 1997 Wiley-Liss, Inc.

governed the location at which neutrophils emigrated from the vasculature, it is now known that the endothelium that lines the postcapillary venules expresses, after appropriate stimulation, several "sticky" molecules that interact with counterligands on the neutrophil. The neutrophil is not passive in its interaction with the endothelium and is also capable, after stimulation, of adhering more tightly to the microvascular endothelium. It is through the specific interactions of "sticky" molecules that neutrophils and other cells find their way out of the vasculature at the appropriate location (reviewed in Cronstein and Weissmann, 1993; Springer, 1994).

The endothelial molecules that govern neutrophil recruitment include P-selectin (expressed rapidly after stimulation by activated complement components, histamine or thrombin), E-selectin (expressed more slowly after stimulation with endotoxin, tumor necrosis factor α or interleukin-1) and intercellular adhesion molecule (ICAM-1) (expressed constitutively but increased after stimulation with endotoxin, tumor necrosis factor α or interleukin-1). The molecules on the neutrophil that "stick" to the endothelium are L-selectin, which mediates the initial rolling of neutrophils along the vascular wall and which is shed after stimulation of the neutrophil by chemoattractants, and the β_2 integrins CD11b,c/CD18 (heterodimers which must be "activated" after stimulation of the cell to become adhesive but which are also increased in number after stimulation). The appropriate sequential activation or expression of these molecules on the neutrophil and the endothelium lining the vasculature permits pinpoint accuracy in recruitment of neutrophils to inflamed or injured sites. Indeed, the molecules that govern neutrophil interactions with the vessel wall prior to emigration from the vasculature have become the target for anti-inflammatory pharmacological interventions (Cronstein and Weissmann, 1993; Springer, 1994).

Once in the extravascular space, neutrophils follow a trail of chemoattractants such as activated complement components (C5a, C3a), cytokines (interleukin-8, etc.), lipids (platelet activating factor, leukotrienes), or bacterial products (e.g., N-formylated peptides) by means of specific cell surface receptors that are members of the seven transmembrane spanning, G protein–binding family of receptors. Neutrophils avidly ingest foreign organisms and cellular debris by means of several other cell surface receptors; these receptors include receptors for the Fc portion of immunoglobulin (FcRII and FcRIII) and activated complement components bound to particles (C1q, C3b, and C3bi). Neutrophils fuse their lysosomal granules to the phagosome containing the ingested organism or debris and empty their lysosomal contents into the newly formed phagolysosome. The contents of the phagolysosome include a variety of bactericidal peptides (e.g., lysozyme, defensins, and bacterial permeability inducing protein (BPI)) and acid hydrolases (e.g., cathepsins and collagenase) that can kill ingested organisms and break down proteinaceous debris. In addition, the neutrophil acidifies the phagolysosome and generates a number of toxic metabolites of molecular oxygen (superoxide anion, H_2O_2, HOCl) into the phagolysosome. Although the primary role of the neutrophil is to rid the body of injurious organisms and clean up the debris after tissue injury, the extracellular release of any of the contents of the phagolysosome or the generation of toxic oxygen metabolites into the extracellular space can lead to destruction of normal, uninjured cells surrounding the injured or infected site. It is the destruction of the surrounding tissue by overactive neutrophils that adds so greatly to tissue destruction in the setting of reperfusion injury.

THE EFFECT OF ADENOSINE ON NEUTROPHIL FUNCTION

In 1980, Marone et al. hypothesized that since β-adrenergic agents inhibited neutrophil degranulation, presumably via increasing intracellular cAMP, adenosine, which was known to increase cAMP in other cell types, ought to inhibit neutrophil degranulation as well. They therefore tested the effect of adenosine on chemoattractant N-formyl-methionyl-leucyl-phenylalanine (FMLP)-stimulated neutrophil degranulation but were disappointed to find that adenosine did not affect neutrophil degranulation. No further work in this area was performed until 1983, when Cronstein and coworkers (Cronstein et al., 1983) confirmed the negative findings of Marone et al. (1980), but went on to demonstrate a marked inhibitory effect of adenosine on stimulated generation of superoxide anion, the molecular precursor of H_2O_2 and HOCl. Just as striking was the demonstration that neutrophils released adenosine and that the adenosine so released acted as an endogenous inhibitor of stimulated neutrophil superoxide anion generation (Cronstein et al., 1983). Moreover, it was extracellular adenosine that was responsible for inhibition of superoxide anion generation since dipyridamole enhanced the effect of adenosine on superoxide anion generation. The effects of adenosine on neutrophil function were found to be both stimulus- and function-specific; adenosine inhibited superoxide anion generation stimulated by the chemoattractant FMLP, the lectin concanavalin A and the Ca^{2+} ionophore A23187 but not the similar action generated by phorbol myristate acetate, a direct activator of protein kinase C. Results of later studies demonstrated that adenosine inhibited superoxide anion generation by occupying specific receptors (A_2, now thought to be A_{2A} receptors) on the neutrophil surface (Cronstein et al., 1985; Roberts et al., 1985; Cronstein, 1994). Subsequent studies have confirmed these initial results (de la Harpe and Nathan, 1989; Currie et al., 1990, 1991; Nielson and Vestal, 1989; Zheng et al., 1991a,b; Burkey and Webster, 1993; Sipka et al., 1989; McGarrity et al., 1989; Thiel and Bardenheuer, 1992).

The effects of adenosine on stimulated degranulation and aggregation are more controversial. Marone et al. (1980) were unable to demonstrate any biologically significant effect of adenosine on chemoattractant-stimulated degranulation, an observation confirmed by Cronstein et al. (1983). Moreover, Cronstein and coworkers (1983) originally reported that adenosine did not significantly affect stimulated neutrophil aggregation. Some investigators have subsequently reported that adenosine inhibits degranulation and aggregation (Schmeichel and Thomas, 1987; Skubitz et al., 1988; Richter, 1992), whereas other laboratories have confirmed the original reports (Grinstein and Furuya, 1986; Becker et al., 1985; McGarrity et al., 1989; Walker et al., 1989). It is not at all clear why there is such variation in the results obtained by different laboratories using similar functional assays.

Acting at extracellular receptors, adenosine significantly affects phagocytosis of immunoglobulin-coated particles via Fc receptors on neutrophils (Salmon and Cronstein, 1990; Zalavary et al., 1994). At low concentrations adenosine promotes phagocytosis, whereas at higher concentrations adenosine inhibits phagocytosis (Salmon and Cronstein, 1990). This difference in effects can be explained by the interaction of adenosine with different receptors on the surface of the neutrophil. Occupancy of adenosine A_1 receptors enhances phagocytosis dramatically, whereas occupancy of adenosine A_2 receptors inhibits phagocytosis (Salmon and Cronstein, 1990).

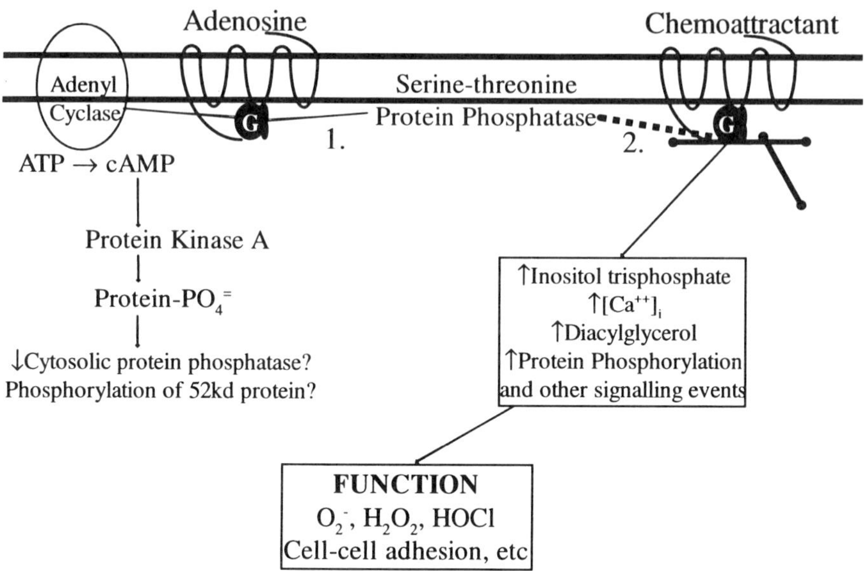

FIGURE 1. Signal transduction at neutrophil adenosine A_2 receptors. Adenosine A_2 receptors activate adenyl cyclase and, in a cAMP-independent fashion, activate a serine threonine protein phosphatase (1.). Enhanced serine-threonine phosphatase activity desensitizes chemoattractant receptors by increasing their association with the cytoskeleton (2.).

Similar to the controversy over adenosine's effects on degranulation and aggregation is the disagreement over the effect of adenosine on neutrophil chemotaxis. Rose et al. (1988) first demonstrated that concentrations of adenosine receptor agonists orders of magnitude lower than those which inhibit superoxide anion generation enhance the rate at which neutrophils migrate in a chemotactic gradient formed by either FMLP or C5a. In a later study using adenosine receptor–specific agonists it was reported that enhancement of chemotaxis resulted from occupancy of adenosine A_1 receptors (Cronstein et al., 1990). This work has been confirmed by Sullivan and coworkers (Sullivan et al., 1990). Other laboratories have reported that adenosine inhibits chemotaxis (Garcia-Castro et al., 1983). Different methods for studying chemotaxis probably account for the variation in experimental results.

Adenosine also modulates stimulated neutrophil adhesion to vascular endothelial cells and other surfaces. Initial reports indicated that the poorly metabolized adenosine receptor agonist 2-chloroadenosine inhibits stimulated neutrophil adhesion to cultured vascular endothelial cells (Cronstein et al., 1986); however, subsequent reports indicated that the effects of adenosine on stimulated adhesion are more complex (Cronstein et al., 1990; Derian et al., 1995; Felsch et al., 1995; Bazzoni et al., 1991). As with phagocytosis, occupancy of adenosine A_1 receptors promotes, and occupancy of A_2 (most likely A_{2A}) receptors inhibits, stimulated neutrophil adhesion to vascular endothelium. Most likely these effects are mediated by modulation of the function of different adhesive molecules on the neutrophil, since A_1 receptor occupancy promotes stimulated neutrophil adhesion to surfaces coated with gelatin (denatured collagen), but A_2 receptor occupancy inhibits adhesion to fibrinogen-coated surfaces (Cronstein et al., 1986). Activation of the β_2 integrins of the neutrophil mediate stimulated adhesion to fibrinogen-coated surfaces, whereas the neutrophil surface proteins involved in adhesion to denatured collagen have not been completely delineated. Adenosine, acting at A_2 receptors, also inhibits neutrophil adhesion to cultured cardiac myocytes, an interaction mediated by β_2 integrins on the neutrophil adhering to ICAM-1 on the myocyte (Bullough et al., 1995). Thus, adenosine acting at A_2 receptors either inhibits "activation" of β_2 integrins or promotes the "deactivation" of these receptors to inhibit stimulated neutrophil adhesion to vascular endothelium. Firestein et al. (1995) have recently reported that adenosine also modulates L-selectin-mediated adhesion of neutrophils to cultured vascular endothelium, and this may explain the observation *in vivo* of the rapid inhibition by adenosine of neutrophil rolling along the endothelium (Asako et al., 1993). The mechanism by which adenosine A_1 receptor occupation promotes stimulated neutrophil adhesion is totally unexplained.

After exposure to bacterial products (lipopolysaccharide), lipid products (platelet activating factor [PAF]), or cytokines (TNF-α), neutrophils become "primed" to generate greater quantities of superoxide anion and other oxygen metabolites and adhere to surfaces more tightly. Adenosine, acting at A_2 receptors, inhibits superoxide anion generation by primed neutrophils as well as unprimed neutrophils (Dianzani et al., 1994). Adenosine inhibits TNF-α-mediated priming of adherent neutrophils but not of neutrophils in suspension, although adenosine does inhibit PAF-mediated priming of neutrophils in suspension (de la Harpe and Nathan, 1989; Stewart and Harris, 1993). After TNA-α-mediated priming, neutrophils migrate less well in a chemotactic gradient; adenosine reverses the effect of TNF-α on neutrophil migration (Sullivan et al., 1990). Thus, as with other neutrophil functions, adenosine inhibition of neutrophil priming is both stimulus- and function-specific.

The ability of neutrophils to injure or kill bacteria depends upon the capacity of the neutrophils to phagocytose the bacteria, transport them into the cell, release the bactericidal contents of their granules, and generate toxic oxygen metabolites into the phagolysosome. Hadart and coworkers (1991) have reported that adenosine inhibits the capacity of neutrophils to kill *Staphylococcus aureus* and that the diminished capacity of neutrophils to kill the bacteria most likely results from inhibition of stimulated generation of toxic oxygen metabolites. Even at extremely high concentrations (10 mM), adenosine was a far less effective inhibitor of bacterial killing than a poorly metabolized A_2 agonist (2-phenylaminoadenosine, 10–100 μM), most likely due to the rapid metabolism of adenosine by both the neutrophils and the bacteria. Some microorganisms have even adapted this mechanism to enhance their own survival; *Candida albicans,* in its invasive form, releases adenosine into the extracellular milieu and thereby modulates neutrophil function *in vitro* (Smail et al., 1992). The capacity of adenosine and its analogs to inhibit neutrophil microbicidal activity may diminish the ability of neutrophils to control or eliminate potentially serious infections.

Although inhibition of bacterial killing may be deleterious, inhibition of the capacity of neutrophils to injure healthy tissue surrounding injured or infected sites may be beneficial. The poorly metabolized adenosine analog 2-chloroadenosine was first reported to diminish neutrophil-mediated injury of cultured vascular endothelium (Cronstein et al., 1986). The diminished capacity of neutrophils to injury the cultured endothelium most likely results from inhibition of stimulated neutrophil adhesion to endothelium and diminished generation of toxic metabolites by the stimulated neutrophils (Cronstein et al., 1986). Thus, it can be inferred from these studies that adenosine diminishes neutrophil-mediated endothelial injury by occupying A_2 receptors on the neutrophils. More important than the capacity of 2-chloroadenosine to inhibit neutrophil-mediated injury to cultured vascular endothelium was the observation that adenosine, released by either the neutrophils or the vascular endothelium, is a potent regulator of neutrophil-mediated injury (Cronstein et al., 1986). This latter observation suggested that adenosine is a potent, endogenous feedback inhibitor of neutrophil-mediated tissue injury *in vitro.* This hypothesis has been validated *in vivo,* and several anti-inflammatory agents have been demonstrated to take advantage of the capacity of endogenously released adenosine to inhibit inflammation (*vide infra*).

A more recently discovered function of the neutrophil is the regulation of the synthesis of inflammatory cytokines such as TNF-α. High concentrations of TNF-α can help to promote further accumulation of neutrophils by stimulating endothelial expression of adhesion molecules and by priming neutrophils to increase their destructive and bactericidal potential. Thiel and Chouker (1995) have reported that adenosine, acting at an A_2 receptor, inhibits endotoxin-stimulated production of TNF-α. Indeed, increased concentrations of endogenous adenosine have been shown to diminish TNF-α concentrations in the circulation and at inflamed sites *in vivo* (Firestein et al., 1994; Cronstein et al., 1995). The effect of adenosine on stimulated neutrophil TNF-α synthesis and secretion is similar to the effect of adenosine on TNF-α secretion by stimulated macrophages (Parmely et al., 1993; Bouma et al., 1994).

SIGNAL TRANSDUCTION AT NEUTROPHIL ADENOSINE RECEPTORS

It is now clear that adenosine modulates neutrophil function via interaction with specific adenosine receptors on the surface of the neutrophil. Of the four different subtypes of adenosine receptor that have been described, three have been implicated as playing a role in modulating neutrophil function (A_1, A_{2A}, A_{2D}) Pharmacological data (agonist and antagonist specificity) underlies the assertion that specific receptors are involved in adenosine-mediated modulation of neutrophil function. Results of radioligand binding studies reveal only a single receptor type, although the interpretation of these studies has been hampered by the high background binding and rapid degradation of neutrophil proteins by proteases contained within neutrophil granules (Cronstein et al., 1985; Martini et al., 1991). Nonetheless, the pharmacological data clearly demonstrate the presence of more than one adenosine receptor on the neutrophil (Cronstein et al., 1990). It is interesting to note that A_1 receptors, which are active at lower concentrations of adenosine, promote neutrophil function (phagocytosis and chemotaxis), whereas A_2 receptors, which are active at the higher concentrations that may prevail at sites of tissue injury, inhibit stimulated neutrophil function (Cronstein et al., 1990; Salmon and Cronstein, 1990).

Signal transduction at neutrophil adenosine A_1 receptors appears to be similar to signal transduction at A_1 receptors on other cell types. Adenosine A_1 receptors signal via pertussis toxin–sensitive G proteins in most cell types, and neutrophils are no exception. At concentrations that do not inferfere with chemotaxis to FMLP, pertussis toxin completely abrogates the effect of adenosine on neutrophil chemotaxis (Cronstein et al., 1990). Agents that disrupt microtubules, such as colchicine and vinblastine, reverse the effects of A_1 receptor occupancy on neutrophil function, although it is not clear whether microtubule disruption alters adenosine receptor expression or signal transduction (Cronstein et al., 1990). It is tempting to speculate that since adenosine A_1 receptors signal through the same G proteins as chemoattractants, adenosine receptor occupancy acts to enhance the magnitude or duration of those signals (inositol trisphosphate generation, Ca^{2+} fluxes, etc.) sent after occupancy of chemoattractant receptors.

Adenosine A_2 receptors were first described on the basis of their capacity to stimulate adenylyl cyclase; the intracellular messenger for adenosine A_2 receptors is thought, therefore, to be cAMP. In support of this hypothesis are studies that show parallel effects of adenosine and cAMP analogs, or agents that directly increase intracellular cAMP concentrations (forskolin, phosphodiesterase inhibitors). This sort of observation has been advanced in support of the hypothesis that cAMP mediates the effect of adenosine A_2 receptor occupancy on neutrophil function (Nielson and Vestal, 1989). However, on closer examination it has been demonstrated that, although the effects of increased intracellular cAMP concentrations parallel those of adenosine A_2 receptor occupancy, they cannot mediate all of the effects of adenosine receptor occupancy in the neutrophil. This is best shown in experiments using a nonmethylxanthine phosphodiesterase inhibitor (RO-20-1724) that increases intracellular cAMP and inhibits superoxide anion generation (Cronstein et al., 1988). RO-20-1724 and adenosine inhibit superoxide anion generation in an additive fashion whereas they increase cAMP concentration synergistically (Cronstein et al., 1988; Zalavary et al., 1994). Thus, although the IC_{50} of 5′N-ethylcarboxamidoadenosine (NECA) for superoxide anion generation is identical in the presence and absence of RO-20-1724, the magnitude of the inhibition is equal to the sum of the effects of NECA plus RO-20-1274 on superoxide anion generation at all concentrations of NECA tested. Moreover, an agent that inhibits protein kinase A (KT5720), the downstream signaling element for cAMP, reverses the effects of dibutyryl cAMP but not of NECA on superoxide anion generation (Cronstein et al., 1992). These results indicate that cAMP probably does not mediate inhibition of superoxide anion generation by adenosine A_2 receptor occupancy, although other functional effects of adenosine may be mediated by cAMP.

Occupancy of neutrophil chemoattractant receptors leads to increased phospholipid turnover (generation of diacylglycerol and inositol 1,4,5-trisphosphate [IP_3]) via activation of G_i proteins. Adenosine receptor occupancy affects neither the early wave in diacylglycerol production nor the stimulated increment in IP_3 (Walker et al., 1990; Cronstein et al., 1992). These intracellular signals are most tightly linked to degranulation, a function usually not affected by adenosine (*vide supra*). However, adenosine clearly inhibits the late, sustained increase in diacylglycerol generation (Cronstein et al., 1992). Similarly, as would be predicted from its lack of effect on IP_3 generation, adenosine receptor occupancy does not affect the early and rapid increment in $[Ca^{2+}]_i$ but markedly inhibits the sustained increase in $[Ca^{2+}]_i$ that occurs after stimulation with chemoattractants (Cronstein et al., 1988; Ward et al., 1988a,b; Walker et al., 1990). This latter observation, plus the demonstration that adenosine inhibits superoxide anion generation in response to the calcium ionophore A23187, has suggested that occupancy of adenosine receptors acts to block calcium channels, diminish influx of Ca^{2+}, and thereby inhibit superoxide anion generation (Pasini et al., 1985; Di Perri et al., 1989). That adenosine inhibits superoxide anion generation even in the absence of extracellular Ca^{2+} does not support this hypothesis, however (Cronstein et al., 1988). Iannone et al. (1991) have reported that the calcium ionophore A23187, in the presence of adenosine, increases adenyl cyclase activity in a Ca^{2+}- and calmodulin-dependent manner, but that the adenyl cyclase activity induced by adenosine in chemoattractant-stimulated neutrophils is independent of $[Ca^{2+}]_i$ and calmodulin.

It has been suggested that adenosine receptor occupancy modulates neutrophil function via phosphorylation or dephosphorylation of various proteins. Bengis-Garber and Gruener (1992) have reported that adenosine, presumably acting via cAMP, cooperates with chemoattractant (FMLP) to promote phosphorylation of a 52-kDa protein in neutrophils, the function of which is unknown. Phosphorylation of this protein can be reproduced by a combination of the calcium ionophore A23187 and dibutyryl cAMP. More recently, Revan et al. (1996) have reported that adenosine receptor occupancy inhibits generation of superoxide anion by a mechanism dependent upon a calyculin A–sensitive protein phosphate. This group went on to show that adenosine receptor occupancy increases the activity of a plasma membrane–associated protein phosphatase in a cAMP-independent fashion, while decreasing the activity of a cytosolic protein phosphatase (Revan et al., 1996). These studies suggest that it is likely that the plasma membrane protein phosphatase activity stimulated by adenosine receptor occupancy is responsible for the adenosine receptor–mediated inhibition of superoxide anion generation. Moreover, these studies are consistent with the hypothesis that the role of the adenosine receptor in the phosphorylation of the 52-kDa protein is to inhibit, by diminishing cytosolic protein phosphatase activity, the dephosphorylation of this protein that would otherwise occur and permit the detection of the phosphorylated form of this protein.

Although the molecular mechanism is not well understood, adenosine receptor occupancy appears to interfere with signal transduction at chemoattractant receptors at a proximal step. Direct activation of G proteins with NaF leads to superoxide generation in the neutrophil, and adenosine receptor occupancy does not inhibit superoxide anion generation stimulated by NaF (Cronstein et al., 1990). Moreover, adenosine receptor occupancy inhibits chemoattractant-mediated increments in plasma membrane GTPase activity, a measure of chemoattractant-stimulated G protein activation (Cronstein et al., 1990; Burkey and Webster, 1993). One potential explanation for the adenosine receptor–mediated uncoupling of chemoattractant receptors is that adenosine receptor occupancy promotes desensitization of chemoattractant receptors. In the neutrophil, chemoattractant receptor desensitization is associated with and possibly due to association with the cytoskeleton (Jesaitis et al., 1984, 1986, 1988, 1989; Zigmond and Tranquillo, 1986); adenosine receptor occupancy promotes the association of chemoattractant receptors with the cytoskeleton in a protein phosphatase–dependent fashion (Cronstein et al., 1990; Cronstein and Haines, 1992; Revan et al., 1995). Thus, these results suggest that adenosine receptors uncouple chemoattractant receptors from their signal transduction mechanisms by promoting their premature desensitization.

THE EFFECT OF ENDOGENOUS AND EXOGENOUS ADENOSINE ON INFLAMMATION

Neutrophils, endothelial cells, and other cells release adenosine into the extracellular space *in vitro.* Adenosine is detectable in whole blood (~300 nM; Hirschhorn et al., 1981) and in higher concentrations in inflammatory exudates or ischemic tissue (600–1200 nM; Cronstein et al., 1993, 1995; Gadangi et al., 1995). The precise molecular source of the adenosine in blood, tissue, or exudates is not known, although it has been suggested that, since inhibition of ecto-5′-nucleotidase activity inhibits apparent adenosine release, adenosine is generated extracellularly from nucleotides (Kitakaze et al., 1993; Poelstra et al., 1992). Regardless of the mechanism by which adenosine concentrations are maintained in the extracellular space, it is clear that adenosine inhibits inflammation *in vitro* and *in vivo*. In a model of neutrophil-mediated vascular injury it was demonstrated that cultured endothelial cells release adenosine that inhibits the capacity of stimulated neutrophils to adhere to and injure the endothelium (Cronstein et al., 1986). Adenosine deaminase, released from dead and dying cells, adheres to phagocytic particles and serves as a counterregulatory inflammatory agent by metabolizing adenosine to inosine, thereby promoting phagocytosis and superoxide anion generation by activated leukocytes (Kubersky et al., 1989).

When studied *in vivo* endogenously released adenosine also appears to be an important regulator of inflammation. Rosengren et al. (1991) reported that endogenously released adenosine diminishes inflammation since an adenosine receptor antagonist, 8-phenyltheophylline, enhances inflammation in the hamster cheek pouch model of inflammation (Rosengren et al., 1991). Exogenous adenosine and adenosine receptor agonists are also potent anti-inflammatory agents when studied *in vivo*. Asako and colleagues (1992, 1993) have demonstrated that, as predicted from *in vitro* studies, high concentrations of exogenous adenosine prevent neutrophil adhesion to microvascular endothelium in the rat mesenteric circulation. Green et al. (1991) reported that daily intraperitoneal injection of adenosine reduced the severity of experimental adjuvant arthritis in rats, a finding that is quite surprising considering the extremely rapid metabolism of adenosine (half-life in whole blood is measured in seconds; Moser et al., 1989). Exogenous adenosine and adenine nucleotides inhibit the development of experimental glomerulonephritis in rats (Poelstra et al., 1992, 1993) and neutrophil-mediated pulmonary injury (Adkins et al., 1995; Marts et al., 1993). Schrier and cowork-

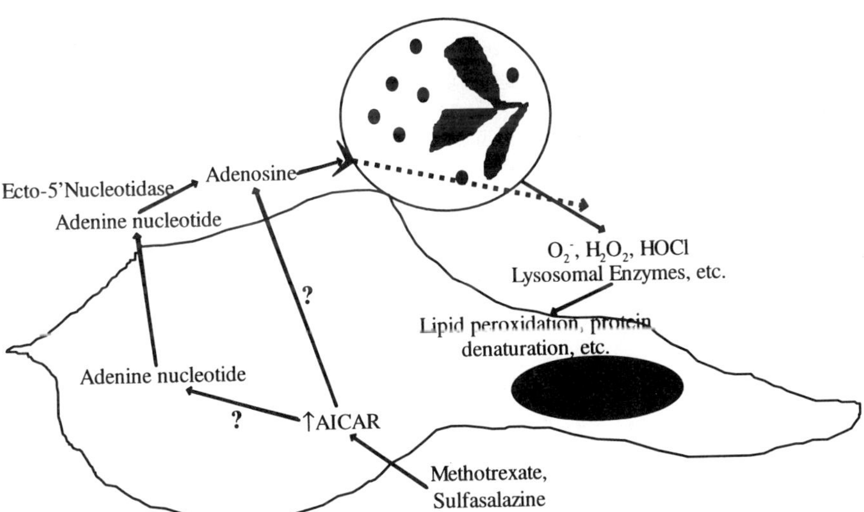

FIGURE 2. Endogenously-released adenosine diminishes the capacity of inflammatory cells to injure tissue at inflamed sites. Leukocytes adhere to cells at inflamed sites and direct a barrage of potentially injurious oxygen metabolites and proteolytic enzymes at target cells. Adenosine, either endogenously released or released in larger quantities after treatment with drugs like methotrexate, sulfasalazine or adenosine kinase inhibitors, diminishes leukocyte adherence and release of toxic products. The end result is diminished leukocyte-mediated tissue injury and inflammation.

ers have reported that adenosine A_1 receptor agonists are better inhibitors of pleural and peritoneal inflammation than A_2 receptor agonists (Schrier and Imre, 1986; Schrier et al., 1990; Lesch et al., 1991), although based upon *in vitro* (*vide supra*) and *in vivo* studies (Cronstein et al., 1993, 1995; Gadangi et al., 1995; Adkins et al., 1995; Marts et al., 1993) A_2 receptor agonists tend to be anti-inflammatory, whereas A_1 receptor occupancy promotes inflammatory functions of neutrophils. Unfortunately, exogenous adenosine analogs may, by virtue of their myriad effects in virtually every tissue and organ studied, not be useful pharmacological agents for the treatment of inflammatory conditions.

Although the clinical utility of adenosine or its agonists in the treatment of inflammatory diseases may be limited by unwanted effects on other organs and tissues it has been suggested that agents that enhance endogenous adenosine release may be useful. Indeed, recent studies indicate that enhanced release of adenosine mediates the anti-inflammatory effects of both methotrexate and sulfasalazine, the two most widely used second-line agents for the treatment of rheumatoid arthritis (Cronstein et al., 1993; Gadangi et al., 1995). Both methotrexate and sulfasalazine share the capacity to inhibit 5-aminoimidazole-4-carboxamide ribonucleotide (AICAR) transformylase *in vitro* (Baggott et al. 1986, 1992, 1993) leading to intracellular accumulation of AICAR *in vivo* (Cronstein et al., 1993; Gadangi et al., 1995). The intracellular accumulation of AICAR has previously been associated with an increase in adenosine release, both *in vitro* (Barankiewicz et al., 1990a,b) and *in vivo* (Gruber et al., 1989). Treatment of mice with either methotrexate or sulfasalazine leads to an increase in the adenosine concentration in an inflammatory exudate (carrageenan-induced inflammation of murine air pouches), which is associated with diminished leukocyte accumulation. The effect of either drug on leukocyte accumulation is reversed by either an adenosine A_2 receptor antagonist or adenosine deaminase, but not by an adenosine A_1 receptor antagonist. These studies indicated that the salutary effects of these two widely used anti-inflammatory drugs are mediated via adenosine and its occupancy of A_2 receptors on inflammatory cells. Other agents that induce adenosine release by different mechanisms may also be anti-inflammatory. AICAR itself, in high concentrations, prevents reperfusion injury in dogs (Gruber et al., 1989), and adenosine kinase inhibitors inhibit inflammation in a number of different *in vivo* models by an adenosine A_2 receptor–mediated mechanism (Firestein et al., 1994; Rosengren et al., 1995; Cronstein et al., 1995).

CONCLUSION

Adenosine, acting at specific surface receptors, is a potent, specific regulator of neutrophil function. Based on this initial observation it is now abundantly clear that adenosine is a potent endogenous regulator of inflammation. Moreover, pharmacological manipulation of adenosine release underlies the therapeutic efficacy of two widely used anti-inflammatory agents. Recognition of the potential clinical utility of adenosine, either endogenously released or as exogenous analogs, has led to the development of new agents that may be useful in the treatment of inflammatory diseases.

ACKNOWLEDGMENTS

This work was supported by grants from the Public Health Service (AR/AI41911, AR11949, HL19721) and the S.L.E. Foundation, Inc.

REFERENCES

Adkins WK, Barnard JW, Moore TM, Allison RC, Prasad VR, Taylor AE (1995): Adenosine prevents PMA-induced lung injury via an A2 receptor mechanism. J Appl Physiol 74:982–988.

Asako H, Kubes P, Baethge B, Wolf R, Granger DN (1992): Colchicine and methotrexate reduce leukocyte adherence and emigration in rat mesenteric venules. Inflammation 16:45–56.

Asako H, Wolf RE, Granger DN (1993): Leukocyte adherence in rat mesenteric venules: Effects of adenosine and methotrexate. Gastroenterology 104:31–37.

Baggott JE, Morgan SL, Ha T, Alarcon GS, Koopman WJ, Krumdieck CL (1993): Antifolates in rheumatoid arthritis: A hypothetical mechanism of action. Clin Exp Rheumatol 11(suppl. 8):S101–S105.

Baggott JE, Morgan SL, Ha T, Vaughn WH, Hine RJ (1992): Inhibition of folate-dependent enzymes by non-steroid anti-inflammatory drugs. Biochem J 282:197–202.

Baggott JE, Vaughn WH, Hudson BB (1986): Inhibition of 5-aminoimidazole-4-carboxamide ribotide transformylase, adenosine deaminase and 5′-adenylate deaminase by polyglutamates of methotrexate and oxidized folates and by 5′-aminoimidazole-4-carboxamide riboside and ribotide. Biochem J 236:193–200.

Barankiewicz J, Jimenez R, Ronlov G, Magill M, Gruber HE (1990a): Alteration of purine metabolism by AICA-riboside in human B lymphoblasts. Arch Biochem Biophysics 282:377–385.

Barankiewicz J, Ronlov G, Jimenez R, Gruber HE (1990b): Selective adenosine release from human B but not T lymphoid cell line. J Biol Chem 265:15738–15743.

Bazzoni G, Dejana E, Del Maschio A (1991): Adrenergic modulation of human polymorphonuclear leukocyte activation. Potentiating effect of adenosine. Blood 77:2042–2048.

Becker EL, Kermode JC, Naccache PH, Yassin R, Marsh ML, Munoz JJ, Sha'afi RI (1985): The inhibition of neutrophil granule enzyme secretion and chemotaxis by pertussis toxin. J Cell Biol 100:1641–1646.

Bengis-Garber C, Gruener N (1992): Cross-talk between cAMP and formylmet-leu-phe in human neutrophils: Phosphorylation of a 52,000 molecular weight protein. Cell Signal 4:247–260.

Bouma MG, Stad RK, van den Wildenberg FAJM, Buurman WA (1994): Differential regulatory effects of adenosine on cytokine release by activated human monocytes. J Immunol 153:4159–4168.

Bullough DA, Magill MJ, Firestein GS, Mullane KM (1995): Adenosine activates A_2 receptors to inhibit neutrophil adhesion and injury to isolated cardiac myocytes. J Immunol 155:2579–2586.

Burkey TH, Webster RO (1993): Adenosine inhibits FMLP-stimulated adherence and superoxide anion generation by human neutrophils at an early step in signal transduction. Biochim Biophys Acta 1175:312–318.

Cronstein BN (1994): Adenosine, an endogenous anti-inflammatory agent. J Appl Physiol 76:5–13.

Cronstein BN, Haines KA (1992): Adenosine A2 receptor occupancy does not affect "triggering" but inhibits "activation" of human neutrophils by a mechanism independent of actin filament formation. Biochem J 281:631–635.

Cronstein BN, Weissmann G (1993): The adhesion molecules of inflammation. Arthritis Rheum 36:147–157.

Cronstein BN, Duguma L, Nicholls D, Hutchison A, Williams M (1990): The adenosine/neutrophil paradox resolved. Human neutrophils possess both A1 and A2 receptors which promote chemotaxis and inhibit O_2-generation, respectively. J Clin Invest 85:1150–1157.

Cronstein BN, Haines KA, Kolasinski SL, Reibman J (1992): Occupancy of $G\alpha 5$-linked receptors uncouples chemoattractant receptors from their stimulus-transduction mechanisms in the neutrophil. Blood 80:1052–1057.

Cronstein BN, Kramer SB, Rosenstein ED, Korchak HM, Weissmann G, Hirschhorn R (1988): Occupancy of adenosine receptors raises cyclic AMP alone and in synergy with occupancy of chemoattractant receptors and inhibits membrane depolarization. Biochem J 252:709–715.

Cronstein BN, Kramer SB, Weissmann G, Hirschhorn R (1983): Adenosine: A physiological modulator of superoxide anion generation by human neutrophils. J Exp Med 158:1160–1177.

Cronstein BN, Levin RI, Belanoff J, Weissmann G, Hirschhorn R (1986): Adenosine: An endogenous inhibitor of neutrophil-mediated injury to endothelial cells. J Clin Invest 78:760–770.

Cronstein BN, Naime D, Firestein GS (1995): The antiinflammatory effects of an adenosine kinase inhibitor are mediated by adenosine. Arthritis Rheum 38:1040–1045.

Cronstein BN, Naime D, Ostad E (1993): The antiinflammatory mechanism of methotrexate: Increased adenosine release at inflamed sites diminishes leukocyte accumulation in an in vivo model of inflammation. J Clin Invest 92:2675–2682.

Cronstein BN, Rosenstein ED, Kramer SB, Weissmann G, Hirschhorn R (1985): Adenosine: A physiologic modulator of superoxide anion generation by human neutrophils. Adenosine acts via an A2 receptor on human neutrophils. J Immunol 135:1366–1371.

Currie MS, Rao KM, Padmanabhan J, Jones A, Crawford J, Cohen HJ (1990): Stimulus-specific effects of pentoxifylline on neutrophil CR3 expression, degranulation, and superoxide production. J Leukoc Biol 47:244–250.

Currie MS, Simel DL, Christenson RH, Holmes C, Crawford J, Cohen HJ, Rao KM (1991): Anti-inflammatory effects of pentoxifylline in claudication. Am J Med Sci 301:85–90.

de la Harpe J, Nathan CF (1989): Adenosine regulates the respiratory burst of cytokine-triggered human neutrophils adherent to biologic surfaces. J Immunol 143:596–602.

Derian CK, Santulli RJ, Rao PE, Solomon HF, Barrett JA (1995): Inhibition of chemotactic peptide-induced neutrophil adhesion to vascular endothelium by cAMP modulators. J Immunol 154:308–317.

Di Perri T, Pasini FL, Pecchi S, De Franco V, Damiani P, Pasqui AL, Capecchi PL, Orrico A, Materazzi M, Domini L, et al (1989): In vivo and in vitro evidence of an adenosine-mediated mechanism of calcium entry blocker activities. Angiology 40:190–198.

Dianzani C, Brunelleschi S, Viano I, Fantozzi R (1994): Adenosine modulation of primed human neutrophils. Eur J Pharmacol 263:223–226.

Felsch A, Stocker K, Borchard U (1995): Phorbol ester-stimulated adherence of neutrophils to endothelial cells is reduced by adenosine A_2 receptor agonists. J Immunol 155:333–338.

Firestein GS, Boyle D, Bullough DA, Gruber HE, Sajjadi FG, Montag A, Sambol B, Mullane K (1994): Protective effect of an adenosine kinase inhibitor in septic shock. J Immunol 152:5853–5859.

Firestein GS, Bullough DA, Erion MD, Jimenez R, Ramirez-Weinhouse M, Barankiewicz J, Smith CW, Gruber HE, Mullane KM (1995): Inhibition of neutrophil adhesion by adenosine and an adenosine kinase inhibitor: the role of selectins. J Immunol 154:326–334.

Gadangi P, Longaker M, Naime D, Levin RI, Recht PA, Carlin G, Cronstein BN (1996): The antiinflammatory mechanism of sulfasalazine is related to adenosine release at inflamed sites. J Immunol 156:1937–1941.

Garcia-Castro I, Mato JM, Vasanthakumar G, Wiesmann WP, Schiffmann E, Chiang PK (1983): Paradoxical effects of adenosine on neutrophil chemotaxis. J Biol Chem 258:4345–4349.

Green PG, Basbaum AI, Helms C, Levine JD (1991): Purinergic regulation of bradykinin-induced plasma extravasation and adjuvant-induced arthritis in the rat. Proc Natl Acad Sci USA 88:4162–4165.

Grinstein S, Furuya W (1986): Cytoplasmic pH regulation in activated human neutrophils: Effects of adenosine and pertussis toxin on Na^+/H^+ exchange and metabolic acidification. Biochim Biophys Acta 889:301–309.

Gruber HE, Hoffer ME, McAllister DR, Laikind PK, Lane TA, Schmid-Schoenbein GW, Engler RL (1989): Increased adenosine concentration in blood from ischemic myocardium by AICA riboside: Effects on flow, granulocytes and injury. Circulation 80:1400–1411.

Hardart GE, Sullivan GW, Carper HT, Mandell GL (1991): Adenosine and 2-phenylaminoadenosine (CV-1808) inhibit human neutrophil bactericidal function. Infect Immun 59:885–889.

Hirschhorn R, Roegner-Maniscalco V, Kuritsky L, Rosen FS (1981): Bone marrow transplantation only partially restores purine metabolites to normal in adenosine deaminase-deficient patients. J Clin Invest 68:1387–1393.

Iannone MA, Wolberg G, Zimmerman TP (1991): Ca^{2+} ionophore-induced cyclic adenosine-3′,5′-monophosphate elevation in human neutrophils. A calmodulin-dependent potentiation of adenylate cyclase response to endogenously produced adenosine: Comparison to chemotactic agents. Biochem Pharmacol 42(suppl):S105–S111.

Jesaitis AJ, Bokoch GM, Tolley JO, Allen RA (1988): Lateral segregation of neutrophil chemotactic receptors into actin- and fodrin-rich plasma membrane microdomains depleted in guanyl nucleotide regulatory proteins. J Cell Biol 107:921–928.

Jesaitis AJ, Naemura JR, Sklar LA, Cochrane CG, Painter RG (1984): Rapid modulation of N-formyl chemotactic peptide receptors on the surface of human granulocytes: Formation of high affinity ligand-receptor complexes in transient association with cytoskeleton. J Cell Biol 98:1378–1387.

Jesaitis AJ, Tolley JO, Allen RA (1986): Receptor-cytoskeleton interactions and membrane traffic may regulate chemoattractant-induced superoxide production in human granulocytes. J Biol Chem 261:13662–13669.

Jesaitis AJ, Tolley JO, Bokoch GM, Allen RA (1989): Regulation of chemoattractant receptor interaction with transducing proteins by organizational control in the plasma membrane of human neutrophils. J Cell Biol 109:2783–2790.

Kitakaze M, Hori M, Morioka T, Takashima S, Minamino T, Sato H, Inoue M, Kamada T (1993): Attenuation of ecto-5′-nucleotidase activity and adenosine release in activated human polymorphonuclear leukocytes. Circ Res 73:524–533.

Kubersky SM, Hirschhorn R, Broekman MJ, Cronstein BN (1989): Occupancy of adenosine receptors on human neutrophils inhibits the respiratory burst stimulated by ingestion of complement coated particles and occupancy of chemoattractant but not Fc receptors. Inflammation 13:591–599.

Lesch ME, Ferin MA, Wright CD, Schrier DJ (1991): The effects of (R)-N-(1-methyl-2-phenylethyl) adenosine (L-PIA), a standard A1-selective adenosine agonist on rat acute models of inflammation and neutrophil function. Agents Actions 34:25–27.

Marone G, Thomas L, Lichtenstein L (1980): The role of agonists that activate adenylate cyclase in the control of cAMP metabolism and enzyme release by human polymorphonuclear leukocytes. J Immunol 125:2277–2283.

Martini C, Di Sacco S, Tacchi P, Bazzichi L, Soletti A, Bondi F, Ciompi ML, Lucacchini A (1991): A2 adenosine receptors in neutrophils from healthy volunteers and patients with rheumatic disease. In Harkness RA (ed): "Purine and Pyrimidine Metabolism in Man, Volume 7, Part A" New York: Plenum Press, pp 459–462.

Marts BC, Baudendistel LJ, Naunheim KS, Dahms TE (1993): Protective effect of 2-chloroadenosine on lung ischemia reperfusion injury. J Surg Res 54:523–529.

McGarrity ST, Stephenson AH, Webster RO (1989): Regulation of human neutrophil functions by adenine nucleotides. J Immunol 142:1986–1994.

Moser GH, Schrader J, Deussen A (1989): Turnover of adenosine in plasma of human and dog blood. Am J Physiol 256:C799–C806.

Nielson CP, Vestal RE (1989): Effects of adenosine on polymorphonuclear leukocyte function, cyclic 3′,5′-adenosine monophosphate, and intracellular calcium. Br J Pharmacol 97:882–888.

Parmely MJ, Zhou W-W, Edwards CK, III, Borcherding DR, Silverstein R, Morrison DC (1993): Adenosine and a related carbocyclic nucleoside analogue selectively inhibit tumor necrosis factor-alpha production and protect mice against endotoxin challenge. J Immunol 151:389–396.

Pasini FL, Capecchi PL, Orrico A, Ceccatelli L, DiPierri T (1985): Adenosine inhibits polymorphonuclear leukocyte in vitro activation: A possible role as an endogenous calcium entry blocker. Int J Immunopharmol 7:203–215.

Poelstra K, Brouwer E, Baller JFW, Hardonk MJ, Bakker WW (1993): Attenuation of anti-Thy 1 glomerulonephritis in the rat by anti-inflammatory platelet-inhibiting agents. Am J Pathol 142:441–450.

Poelstra K, Heynen ER, Baller JF, Hardonk MJ, Bakker WW (1992): Modulation of anti-Thy 1 nephritis in the rat by adenine nucleotides. Evidence for an anti-inflammatory role of nucleotidases. Lab Invest 66:555–563.

Revan S, Montesinos MC, Naime D, Landau S, Cronstein BN (1996): Adenosine A_2 receptor occupancy regulates stimulated neutrophil function via activation of a serine/threonin protein phosphate. J Biol Chem 271:17114–17118.

Richter J (1992): Effect of adenosine analogues and cAMP-raising agents on TNF-, GM-CSF-, and chemotactic peptide-induced degranulation in single adherent neutrophils. J of Leukoc Biol 51:270–275.

Roberts PA, Newby AC, Hallett MB, Campbell AK (1985): Inhibition by adenosine of reactive oxygen metabolite production by human polymorphonuclear leucocytes. Biochem J 227:669–674.

Rose FR, Hirschhorn R, Weissmann G, Cronstein BN (1988): Adenosine promotes neutrophil chemotaxis. J Exp Med 167:1186–1194.

Rosengren S, Arfors KE, Proctor KG (1991): Potentiation of leukotriene B4-mediated inflammatory response by the adenosine antagonist, 8-phenyl theophylline. Intl J Microcirc Clin Exp 10:345–357.

Rosengren S, Bong GW, Firestein GS (1995): Anti-inflammatory effects of an adenosine kinase inhibitor: Decreased neutrophil accumulation and vascular leakage. J Immunol 154:5444–5451.

Salmon JE, Cronstein BN (1990): Fcgamma receptor-mediated functions in neutrophils are modulated by adenosine receptor occupancy: A1 receptors are stimulatory and A2 receptors are inhibitory. J Immunol 145:2235–2240.

Schmeichel CJ, Thomas LL (1987): Methylxanthine bronchodilators potentiate multiple human neutrophil functions. J Immunol 138:1896–1903.

Schrier DJ, Imre KM (1986): The effects of adenosine agonists on human neutrophil function. J Immunol 137:3284–3289.

Schrier DJ, Lesch ME, Wright CD, Gilbertsen RB (1990): The antiinflammatory effects of adenosine receptor agonists on the carrageenan-induced pleural inflammatory response in rats. J Immunol 145:1874–1879.

Sipka S, Szentmiklosi AJ, Nagy A, Taskov V, Szegedi G (1989): Inhibition of the zymosan-induced chemiluminescence of human phagocytes by adenosine, polyadenylic acid and agents influencing adenosine metabolism. Allergologia Et Immunopathologia 17:209–212.

Skubitz KM, Wickham NW, Hammerschmidt DE (1988): Endogenous and exogenous adenosine inhibit granulocyte aggregation without altering the associated rise in intracellular calcium concentration. Blood 72:29–33.

Smail EH, Cronstein BN, Meshulam T, Esposito AL, Ruggeri RW, Diamond RD (1992): In vitro, Candida albicans releases the immune modulator adenosine and a second, high-molecular weight agent that blocks neutrophil killing. J Immunol 148:3588–3595.

Springer TA (1994): Traffic signals for lymphocyte recirculation and leukocyte emigration: The multistep paradigm. Cell 76:301–314.

Stewart AG, Harris T (1993): Adenosine inhibits platelet-activating factor, but not tumour necrosis factor-alpha-induced priming of human neutrophils. Immunol 78:152–158.

Sullivan GW, Linden J, Hewlett EL, Carper HT, Hylton JB, Mandell GL (1990): Adenosine and related compounds counteract tumor necrosis factor-alpha inhibition of neutrophil migration: Implication of a novel cyclic AMP-independent action on the cell surface. J Immunol 145:1537–1544.

Thiel M, Bardenheuer H (1992): Regulation of oxygen radical production of human polymorphonuclear leukocytes by adenosine: The role of calcium. Pflugers Arch 420:522–528.

Thiel M, Chouker A (1995): Acting via A2 receptors, adenosine inhibits production of tumor necrosis factor-α of endotoxin-stimulated polymorphonuclear leukocytes. J Lab Clin Med 126:275–282.

Walker BA, Cunningham TW, Freyer DR, Todd RF, Johnson KJ, Ward PA (1989): Regulation of superoxide responses of human neutrophils by adenine compounds. Independence of requirement for cytoplasmic granules. Lab Invest 61:515–521.

Walker BAM, Hagenlocker BE, Douglas VK, Ward PA (1990): Effects of adenosine on inositol 1,4,5-trisphosphate formation and intracellular calcium changes in formyl-met-leu-phe-stimulated human neutrophils. J Leuk Biol 48:281–283.

Ward PA, Cunningham TW, McCulloch KK, Johnson KJ (1988a): Regulatory effects of adenosine and adenine nucleotides on oxygen radical responses of neutrophils. Lab Invest 58:438–447.

Ward PA, Cunningham TW, Walker BAM, Johnson KJ (1988b): Differing calcium requirements for regulatory effects of ATP, ATPgS and adenosine on O_2-responses of human neutrophils. Biochem Biophys Res Commun 154:746–751.

Zalavary S, Stendahl O, Bengtsson T (1994): The role of cyclic AMP, calcium and filamentous actin in adenosine modulation of Fc receptor-mediated phagocytosis in human neutrophils. Biochim Biophys Acta 1222:249–256.

Zheng H, Crowley JJ, Chan JC, Raffin TA (1991a): Attenuation of LPS-induced neutrophil thromboxane b2 release and chemiluminescence. J Cell Physiol 146:264–269.

Zheng H, Crowley JJ, Chan JC, Raffin TA (1991b): Attenuation of LPS-induced neutrophil thromboxane b2 release and chemiluminescence. J Cell Physiol 146:264–269.

Zigmond SH, Tranquillo AW (1986): Chemotactic peptide binding by rabbit polymorphonuclear leukocytes: Presence of two compartments having similar affinities but different kinetics. J Biol Chem 261:5283–5288.

CHAPTER 16

Purinergic Modulation of Septic Shock and Systemic Inflammation Response Syndrome

SANNA ROSENGREN and GARY S. FIRESTEIN

Gensia, Inc., San Diego, CA 92121 (S.R.), and UCSD School of Medicine, LaJolla, CA, 92093 (G.S.F.)

INTRODUCTION

Septic shock and its complications constitute the most common cause of death in noncoronary intensive care units [1]. Commonly, the cause is a gram-negative bacterial infection, although sepsis and septic shock can arise from infections by gram-positive bacteria and fungi. Identical symptoms are also seen in some patients without infection, including those with extensive burns, ischemia, trauma, pancreatitis, and hemorrhagic shock. This has lead to the adoption of a new term, SIRS or systemic inflammation response syndrome [2], which defines a constellation of clinical findings independent of etiology.

Systemic inflammation response syndrome is characterized by tachycardia, tachypnea, severe hypotension, fever, and leukocytosis. Frequently, this syndrome leads to the development of organ system dysfunction and failure, including myocardial dysfunction, acute renal failure, hepatic failure, and multiple organ dysfunction syndrome. The adult respiratory distress syndrome (ARDS), which is seen in 20%–40% of SIRS patients, is distinguished by noncardiogenic pulmonary edema resulting from increased capillary permeability [3]. Taken together, this set of conditions constitute a significant clinical problem, and the therapy to date is mostly supportive. Despite the development of new antibiotics and sophisticated critical care procedures, the morbidity and mortality in severe SIRS have remained essentially unchanged for the last 10 to 15 years.

The progression of SIRS evolves from the inflammatory process that occurs in response to infection or trauma, rather than as a direct result from the insult itself.

Purinergic Approaches in Experimental Therapeutics, Edited by Kenneth A. Jacobson and Michael F. Jarvis
ISBN 0-471-14071-6 © 1997 Wiley-Liss, Inc.

In other words, SIRS can be regarded as the result of an inflammatory response gone awry. Novel approaches to SIRS treatment include anti-inflammatory therapies of various kinds. In this context, both cyclooxygenase inhibitors [4] and corticosteroids [5] have been evaluated as possible anti-SIRS therapies. The outcomes of these trials have been largely disappointing. Therefore, many pharmaceutical companies are seeking to identify new ways of interfering with the aggressive inflammatory response. During the last decade, adenosine and its analogs have been described to inhibit several aspects of inflammation through actions on purinergic receptors. This chapter seeks to give an overview of these studies as well as to describe some recent advances towards purinergic SIRS therapy.

THE INFLAMMATORY REACTION IN SEPSIS AND SIRS

In gram-negative bacteremic sepsis, the main agent responsible for the deleterious inflammatory response is the lipopolysaccharide (LPS) of the bacterial cell wall, also called endotoxin. Monocytes and macrophages have receptors for endotoxin and produce several pro-inflammatory cytokines in response to stimulation of these receptors. One of the first cytokines to be produced after LPS stimulation is tumor necrosis factor alpha (TNF-α), which is capable of initiating a sepsis-like condition when injected into experimental animals [6]. The crucial role of TNF-α in the progression of sepsis is supported by studies in which antibodies to TNF-α protected animals from shock, organ dysfunction, and death resulting from a bacterial infusion [7]. Other cytokines released in response to endotoxin include interleukins-1,6, and 10. Interleukin-1 (IL-1) potentiates the lethal effects of TNF-α [8], and IL-1 receptor blockade improves survival in models of septic shock [9]. The role of IL-6 remains controversial, in that either the cytokine itself or neutralizing antibodies can be protective in endotoxic shock, depending on the model. On the other hand, IL-10 reduces inflammatory responses to endotoxin [10]. Apart from the cytokine response, the arachidonic acid pathway is also activated leading to the production of prostaglandins, thromboxane, and leukotrienes. These potent mediators have profound effects on vascular tone, as well as on the permeability of the vascular wall to plasma proteins. In addition, some leukotrienes as well as TNF-α and IL-1 are capable of activating polymorphonuclear leukocytes (PMN).

Activated neutrophils are capable of binding to vascular walls using adhesion molecules like L-selectin and the CD11/CD18 complex, and migrating out into the tissue. Their granules contain proteases, lysozyme, and an oxygen radical–producing apparatus normally used to kill ingested microorganisms. However, when overstimulated, neutrophils can release these products to the extracellular space. This is thought to be among the chief causes of endothelial damage in SIRS and ARDS, and indeed neutrophils have proven crucial for the progression of disease in animal models of sepsis [11]. Intravascular aggregation of activated neutrophils and capillary plugging have also been suggested as causes of hypoperfusion of certain organs. The destructive role of the neutrophil in SIRS has been recently reviewed [12]. However, neutrophils also constitute a line of defense against infection, and complete eradication of neutrophil adhesion could actually worsen septic shock [13].

The final important player in SIRS consists of the vascular endothelium, which normally maintains the integrity of the vessel wall by means of barrier junctions between adjacent endothelial cells. Many inflammatory mediators act directly on endothelial cells, causing them to retract from each other and allow plasma to leak into the interstitium [14]. Activated neutrophils can also destroy the endothelial layer [15]. The end result is edema and, if this is widespread, hypotension, shock, and ARDS. The endothelium is also capable of producing various inflammatory mediators itself, as well as expressing adhesion proteins such as E- and P-selectin and intercellular adhesion molecule-1 (ICAM-1), which enhance neutrophil adhesion. Once the SIRS reaction is fully expressed, the various cell types involved are capable of activating and recruiting each other through a variety of mediators and adhesion proteins. It may not be enough to target just one of these entities in the full-blown syndrome, and it may be desirable to identify therapies that interfere with several pathways. For this reason, adenosinergic therapy is especially attractive as it confers protection on several levels.

THE ANTI-INFLAMMATORY ACTIONS OF ADENOSINE

Stimulation of adenosine receptors has a multitude of effects on various inflammatory cells (Table 1). Adenosine attenuates production of cytokines, inhibits many aspects of neutrophil function, and reduces endothelial permeability.

Effects on Cytokine Production

Recently, it was discovered that adenosine and some of its analogs are capable of reducing TNF-α production from macrophages and macrophage-like cell lines *in vitro* [16–18]. This effect was not simply due to a general inhibition of cell function, since

TABLE 1. Potentially Beneficial Effects of Adenosinergic Stimulation in Sepsis/SIRS

Macrophage effects
Reduced TNF-α production
Decreased interferon-induced priming
Reduced major histocompatibility complex (MHC) expression
Decreased nitric oxide production
Neutrophil effects
Decreased free radical formation
Reduced degranulation
Prevention of upregulation of surface CD11/CD18
Decreased rolling
Decreased adhesion to endothelium
Reduced destruction of endothelium
Reduced migration into inflammatory sites
Endothelium effects
Reduced baseline permeability
Decreased leakage response to inflammatory mediators
Cardiovascular effects
Negative inotropic, chronotropic, and dromotropic action
Vasodilation
Metabolic effects
Alleviation of LPS-induced insulin resistance

the production of other cytokines such as IL-1 [17] and IL-6 [16] remained intact (Figure 1). On the other hand, adenosine decreased IL-1, IL-6, IL-8, and TNF-α in LPS-stimulated human monocytes [19]. The effect of adenosine on TNF-α production appeared to be pre-translational since TNF-α mRNA levels were reduced as well [18]. Reversal of the inhibition by a nonspecific adenosine antagonist [16] indicated that adenosine indeed exerted this action through binding to its receptors. However, the exact subtype of receptor involved is still under investigation, although an A2-mediated effect of adenosine on cytokine production has been suggested based on the lack of effect by an A1 agonist [19]. In reality, the rank order of agonist potency is characteristic neither A1 nor A2 receptors [16], and similar results have been obtained studying the effects of adenosine on other aspects of macrophage function, e.g., interferon-induced priming and expression of major histocompatibility complex antigens [20]. Recent data from our laboratory implicate the A3 receptor [20a].

In vivo experiments utilizing R-PIA, a nonmetabolizable adenosine analog, confirmed these *in vitro* findings, in that elevation of plasma TNF-α by an endotoxin injection into rats was reduced by R-PIA, whereas IL-6 levels remained unchanged [16]. These results may be difficult to interpret due to the unavoidable effects on cardiovascular parameters by adenosine analogs. On the other hand, treatment with an adenosine kinase inhibitor that elevates endogenous tissue adenosine concentrations (see below) reduced TNF-α plasma levels in endotoxic mice, an effect that was reversed by an adenosine antagonist [18].

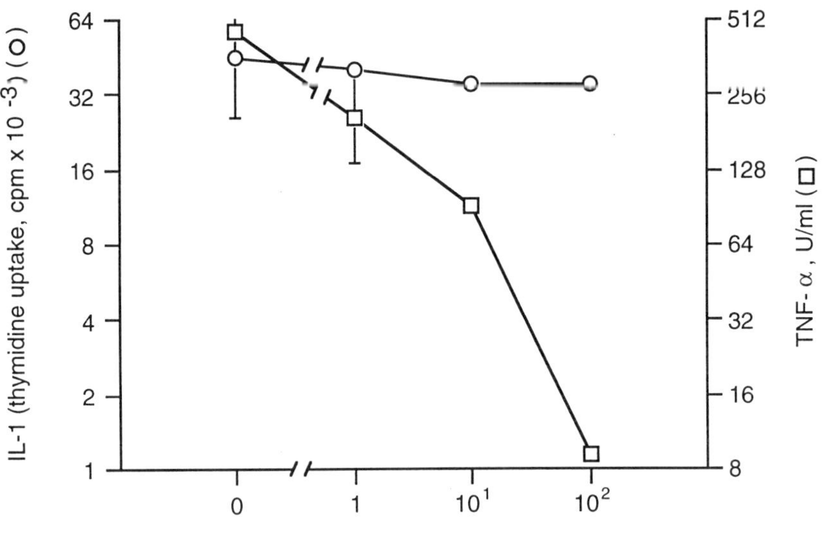

FIGURE 1. Adenosine reduces LPS-induced production of TNF-α, but not of IL-1, in mouse peritoneal macrophages. Culture supernatant fluids were harvested 18 hours after stimulation with LPS. TNF-α was measured with a modified L cell cytotoxicity bioassay, using mouse recombinant TNF-α as a standard. IL-1 was measured by a thymocyte comitogenesis assay, wherein thymidine uptake indicates IL-1 activity. Modified from Parmely et al. [17], reprinted with permission.

Effects on Neutrophil Function

The majority of reports regarding anti-inflammatory actions of adenosine concern themselves with effects on neutrophils. In 1983, Cronstein and others [21] described the first such action, namely inhibition of superoxide production of neutrophils stimulated with a chemotactic factor. This has since been ascribed to binding of the A2 receptor [22,23]. Cytokine stimulation of the neutrophil respiratory burst is also reduced by adenosine, a finding highly relevant to sepsis [24]. Additionally, many other aspects of neutrophil function have been found to be attenuated by adenosine A2 receptor stimulation, including actin polymerization [25], degranulation [26], and phagocytosis [27], as well as adhesion to and injury of endothelium [28–30]. A2 receptor ligation in neutrophils increases intracellular cyclic AMP (cAMP) [31,32] and also has effects on intracellular calcium [32] and G protein function [33]. The exact intracellular signal transduction step responsible for the inhibition of neutrophil function by adenosine remains controversial. An interesting observation, however, is the fact that adenosine is a much more effective inhibitor of superoxide production and phagocytosis in adherent neutrophils than in neutrophils in suspension [24,27].

A crucial step in neutrophil adhesion to the vascular wall is the rolling phenomenon, a behavior exhibited by neutrophils when traveling in the bloodstream through small venules of various microvascular beds. The neutrophils are seen in close contact with the vessel wall, but retain their spheric shape and roll along the endothelial surface at a steady speed. Rolling is dependent on L-selectin, an adhesion protein localized to microvilli-like protrusions on the neutrophil surface. This localization appears to depend on an association between L-selectin and the underlying cytoskeleton [34]. Adenosine interferes with L-selectin-dependent neutrophil adhesion to endothelium *in vitro* [30], possibly by interrupting the cytoskeletal connection to L-selectin. The resulting inhibition of the rolling phenomenon by adenosine receptor stimulation is evident both *in vitro* [30] and *in vivo* [35,36]. Other adhesion proteins on the neutrophil surface, such as the CD11b/CD18 complex, are upregulated on the cell surface when the neutrophil is activated and are essential to firm adhesion and migration. This upregulation is reduced by adenosine [37,38], as is CD11b/CD18-mediated neutrophil adhesion to fibrinogen [29]. Given these effects on the function of adhesion proteins, adenosine should reduce neutrophil migration into an inflammatory site, and this has been observed in animal models [39,40]. Likewise, elevation of the endogenous levels of adenosine results in significant inhibition of dermal neutrophil accumulation [41].

Effect on Endothelial Permeability

Apart from its well-documented effect on neutrophil function, adenosine also exerts a direct effect on endothelial permeability. Several studies of endothelial monolayers *in vitro* have documented a lowering of baseline permeability in the presence of adenosine [42–44]. In one study the effect of adenosine depended on the source of the endothelial cells, in that coronary microvascular endothelial cells increased their permeability in response to adenosine, whereas aortic endothelial cells showed the opposite effect [43]. The adenosine receptor subtype responsible for changes in permeability is not yet clear, as both the A1 [42] and A2 [44] receptors have been reported as solely responsible. Most likely, however, adenosine exerts the effects through an elevation of intracellular cAMP in endothelial cells [45]. Supporting this notion, increasing cAMP in endothelial monolayers through other means such as cholera toxin, forskolin, or stable prostacyclin analogs results in a tightening of the monolayer as well [46,47]. When stimulated by certain inflammatory mediators, the endothelium becomes leaky due to retraction of individual endothelial cells from each other, but less so under conditions that elevate intracellular cAMP levels [47,48]. Adenosine might have the same attenuating effect on inflammatory permeability changes, although this remains to be demonstrated *in vitro*.

In vivo studies of adenosine's effect on inflammation-associated permeability changes appear at first contradictory. In the isolated perfused lung, phorbol ester-induced lung injury is prevented by the inclusion of adenosine in the perfusate [49], but at least part of this inhibition can be ascribed to an effect of adenosine in leukocytes in the perfusate [50]. On the other hand, adenosine potentiates bradykinin-induced plasma extravasation when coinjected with the mediator [51]. This latter finding, however, is probably related to the potent vasodilatory actions of adenosine, which will increase blood flow and perfusion pressure, thereby increasing fluid flux across the vascular wall.

THE EFFECT OF ADENOSINE IN DISEASE MODELS

Given these anti-inflammatory activities of adenosine, it is understandable that adenosine receptor stimulation confers beneficial effects in a multitude of animal models of human disease. In a rat pleurisy model, both leukocyte accumulation and exudate formation was inhibited by adenosine analogs [40]. Adenosine significantly reduced the extent of the lung injury, as measured by permeability increases, in PMA-infused isolated lungs [49]. In a model of reperfusion-induced intestinal injury, local infusion of adenosine diminished vascular injury and leukocyte accumulation [39]; and in myocardial infarct studies intracoronary adenosine improved the outcome both in terms of infarct size and myocardial function [52]. The beneficial effects of adenosine in these reperfusion injury models may be related to inhibition of leukocyte accumulation and vascular leakage, but also to the vasodilation induced by the autacoid. Since the organ dysfunction seen in SIRS is related to vasoconstriction and hypoperfusion, this activity might have a direct bearing on SIRS as well.

In a model of kidney dysfunction in sepsis, an A1 receptor antagonist partially counteracted the detrimental effect of endotoxin on renal function [53]. This could be interpreted as evidence that adenosine contributes to endotoxin-related pathology. However, in the kidney, A1 receptor stimulation induces preglomerular vascular constriction, whereas A2 stimulation dilates both pre- and postglomerular vessels [54]. The A1 receptor antagonist might unmask the action of endogenous adenosine on A2 receptors, leading to a reduction of tissue injury. In support of this, the relatively nonspecific adenosine analog 2-chloroadenosine significantly decreased proteinuria in an animal model of glomerulonephritis [55]. Interestingly, in this latter study intraglomerular production of superoxide by neutrophils was also found to be significantly inhibited by the analog, in line with the observed inhibition of superoxide production *in vitro* [21].

Despite these beneficial effects of adenosine in various disease models, there is a remarkable paucity of literature regarding the direct effect of adenosine on survival in endotoxic shock. In the only such study to date, adenosine was given as a bolus injection 1 hour before LPS challenge [17] and was ineffective, probably due to rapid metabolism of the nucleoside *in vivo*. Interestingly, however, a carbocyclic nucleoside analog with longer half-life than adenosine was protective in this model [17]. This analog did not bind to A1 or A2 receptors, but A3 activity has not been ruled out. Additionally, adenosine agonists reduced nitric oxide release in mice in response to an endotoxin injection [56] with a rank order of potency characteristic of neither A1 or A2 receptors. Adenosine also plays a role in glucose metabolism during endotoxic shock. During the shock response, the myocardium was found to display resistance to insulin. In the presence of adenosine, however, the normal glucose-uptake response to insulin was restored [57], an effect demonstrating yet another possibly advantageous action of adenosine in SIRS.

CARDIOVASCULAR EFFECTS OF ADENOSINE

Direct studies of the effect of adenosine on endotoxic shock are very difficult to perform since adenosine exhibits potent cardiodepressant and vasodilatory properties. The negative inotropic, chronotropic, and dromotropic actions of adenosine on cardiac muscle, as well as the relaxing effects on vascular tone have been known for over 50 years and are reviewed by Olsson and Pearson [58]. In shock, such as observed in serious SIRS, life-threatening hypotension could be further exacerbated by the depressing effect of adenosine on blood pressure. Hence, the infusion of large doses of adenosine or an analog in shock would not be a viable therapeutic option.

ADENOSINE REGULATORY AGENTS: ADENOSINE KINASE INHIBITION

An alternative to a direct administration of adenosine is provided by agents that enhance endogenous adenosine release through the inhibition of certain enzymes involved in adenosine metabolism. In this context, adenosine kinase is one potential target, since this enzyme catalyzes the phosphorylation of adenosine to AMP. For instance, an inhibitor of adenosine kinase, GP-1-515 (also known as GP515), inhibited adenosine phosphorylation in cultured endothelial cells, increased adenosine production, and decreased neutrophil adhesion to endothelial monolayers in an adenosine-dependent fashion [30]. Like direct adenosine agonists, GP-1-515 interfered with L-selectin function on neutrophils. In two dermal inflammation models, GP-1-515 reduced neutrophil infiltration through the action of adenosine on the A2 receptor [41,59]. The adenosine kinase inhibitor also significantly diminished vascular leakage induced by histamine or bradykinin in a manner reversible by adenosine antagonists [41]. This constitutes the first unequivocal *in vivo* demonstration of adenosine's protective action on vascular leakage in a model independent of neutrophil activation. GP-1-515 treatment had no effect on blood pressure or heart rate in conscious rats [41], although local adenosine concentrations in inflammatory sites were elevated in a mouse model [59]. This demonstrates the potential site-specific effect of adenosine-regulating agents on adenosine levels, which is possibly related to the extremely short half-life of adenosine in plasma [60].

The relative absence of cardiovascular side effects enabled testing of GP-1-515 in models of septic shock. In endotoxin-injected mice, mortality was reduced by up to 80% by a single GP-1-515 injection (Figure 2) [18], and an adenosine antagonist reversed this effect. Plasma levels of TNF-α in treated mice were significantly lower than in control mice, while IL-1 and IL-6 levels remained unchanged. These results

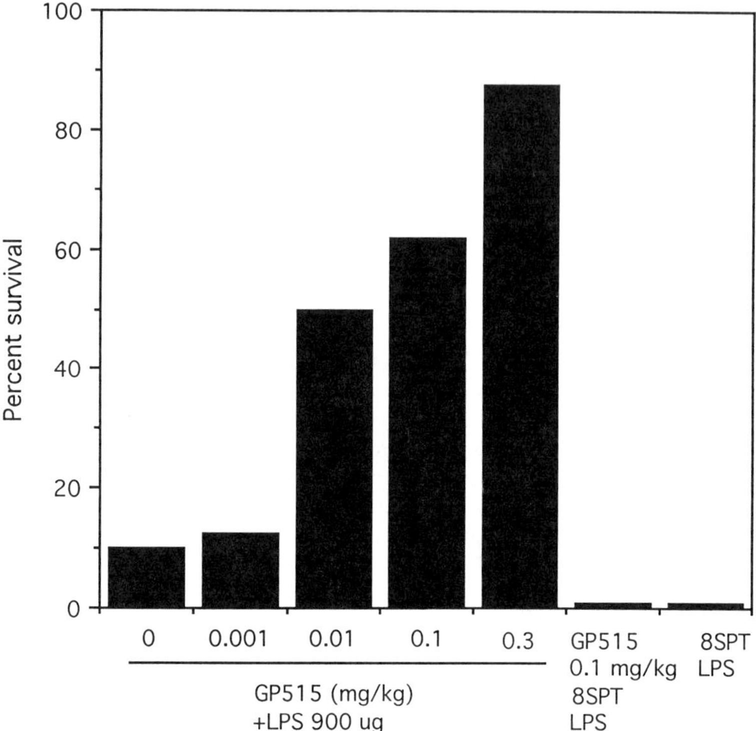

FIGURE 2. An adenosine kinase inhibitor, GP-1-515 (also known as GP515), decreases mortality from endotoxic shock in mice through an adenosine receptor-mediated mechanism. Mice received endotoxin and various doses of GP-1-515 by intravenous injection. In addition, some animals were treated with 8SPT. Mortality was assessed 72 hrs after endotoxin injection. 8SPT, 8-(p-sulfophenyl)theophylline, a nonspecific adenosine receptor antagonist. Modified from Firestein et al. [18], reprinted with permission.

parallel *in vitro* findings regarding the effect of adenosine on LPS-induced cytokine production [16,17]. The lung histology from the two groups demonstrated a marked protection by the adenosine kinase inhibitor. GP-1-515 was also protective in a cecal ligation and puncture model in rats, which is characterized by bacterial peritonitis and high mortality [18]. Hence, adenosine kinase inhibition confers protection even in the presence of bacteremia, despite the fact that adenosine inhibits neutrophil bactericidal function *in vitro* [61].

OTHER PURINERGIC APPROACHES

Apart from the vast literature on adenosine as an anti-inflammatory compound, there are few reports regarding alternative approaches to purinergic therapy in septic shock or SIRS in general. Several classes of P2 purinergic receptors, which bind ADP and ATP, are present on macrophages [62], and ATP promotes several features of macrophage activation. For example, in macrophage cell lines ATP potentiates LPS-induced nitric oxide production [63] and acts synergistically with LPS in stimulating GTPase activity [64]. 2-Methylthio-ATP, which binds to certain classes of P2 receptors, inhibits this synergistic effect [65]. It also protects mice from endotoxin-associated mortality [65]. This may be partly due to hydrolysis of 2-methylthio-ATP to the corresponding adenosine analog, a notion supported by the fact that a nonspecific adenosine antagonist partially reverses the protective action of 2-methylthio-ATP on mortality (G. S. Firestein, unpublished observation.) Alternatively, the protective effect might be associated with the beneficial cardiovascular effects of purinoreceptor ligation [58].

On the other hand, ATP-$MgCl_2$ infusion has proven beneficial in various shock situations including septic shock [66]. This has been attributed to a repletion of the depressed tissue adenine nucleotide levels normally observed in late sepsis, although this seems unlikely, since ATP is not efficiently transported into cells. Alternatively, ATP could function as an adenosine source due to rapid hydrolysis, and this might contribute to its protective effect in shock. ATP-$MgCl_2$ treatment has been recommended for clinical trials [67], although in some models of endotoxic shock no apparent effect on survival was seen [68].

There are also several other compounds with purine structures that are currently under consideration for sepsis therapy. One such compound is the xanthine oxidase inhibitor allopurinol. Xanthine oxidase catalyzes the metabolism of xanthine to hypoxanthine and uric acid, with superoxide formed as a byproduct. A substantial part of the free radical damage seen after ischemia and reperfusion has been attributed to this reaction (reviewed in ref 69). Also, allopurinol is protective against lung injury induced by severe burns [70]. Another purine with potential is pentoxifylline, which elevates intracellular cAMP, inhibits formation of TNF-α, and is protective in animal models of sepsis (reviewed in ref. 12). However, neither of these act through binding to purinergic receptors.

CONCLUDING REMARKS

Systemic inflammation response syndrome, with its multifaceted inflammatory response, remains a significant clinical problem. The purinergic approach has the advantage that it affects several cell types and mediators involved in septic shock, rather than inhibiting only one aspect of inflammation, as is the case with many of the other alternative therapy strategies proposed for SIRS treatment. Adenosine can cause adverse cardiovascular side effects upon infusion, but novel strategies designed to elevate endogenous adenosine levels could potentially circumvent this problem.

REFERENCES

1. Parrillo, JE (1993): Pathogenetic mechanisms of septic shock. N Engl J Med 328:1471–1477.
2. Bone RC, Balk RA, Cerra FB, Dellinger RP, Fein AM, Knaus WA, Schein RM, Sibbald WJ (1992): Definitions for sepsis and organ failure and guidelines for the use of innovative therapies in sepsis. The ACCP/SCCM Conference Committee. Chest 101:1644–1655.
3. Kollef MH, Schuster DP (1995): The acute respiratory distress syndrome. N Engl J Med 332:27–37.

4. Haupt MT, Jastremski MS, Clemmer TP, Metz CA, Goris GB (1991): Effect of ibuprofen in patients with severe sepsis: A randomized, double-blind, multicenter study. Crit Care Med 19;1339–1347.
5. Bone RC, Fischer CJ, Clemmer TP, Slotman GJ, Metz CA, Balk RA (1987): A controlled clinical trial of high-dose methylprednisolone in the treatment of severe sepsis and septic shock. N Engl J Med 317:653–658.
6. Tracey KJ, Beutler B, Lowry SF, Merryweather J, Wolpe S, Milsark, IW, Hairi RJ, Fahey TJ, Zentella A, Albert JD, et al. (1986): Shock and tissue injury induced by recombinant human cachectin. Science 234:470–474.
7. Tracey KJ, Fong Y, Hesse DG, Manogue RK, Lee AT, Kuo GC, Lowry SF, Cerami A (1987): Anti-cachectin/TNF monoclonal antibodies prevent septic shock during lethal bacteremia. Nature 330:662–664.
8. Waage A, Espevik T, (1988): Interleukin 1 potentiates the lethal effect of tumor necrosis factor alpha/cachectin in mice. J Exp Med 167:1987–1992.
9. Ohlsson K, Björk P, Bergenfeld M, Hageman R, Thompson RC (1990): Interleukin-1 receptor antagonist reduces mortality from endotoxin shock. Nature 348:550–552.
10. Marchant A, Bruyns C, Vandenabeele P, Abramowicz D, Gerard C, Delvaux A, Ghezzi P, Velu T, Goldman M (1994): The protective role of interleukin-10 in endotoxin shock. Progr Clin Biol Res 388:417–423.
11. Heflin AC, Brigham KL (1981): Prevention by granulocyte depletion of increased vascular permeability of sheep lung following endotoxemia. J Clin Invest 68:1253–1260.
12. Bone RC (1992): Inhibitors of complement and neutrophils: A critical evaluation of their role in the treatment of sepsis. Crit Care Med 20:891–898.
13. Eichacker PQ, Hoffman WD, Farese A, Danner RL, Suffredini AF, Waisman Y, Banks SM, Mouginis T, Wilson L, Rothlein R et al. (1993): Leukocyte CD18 monoclonal antibody worsens endotoxemia and cardiovascular injury in canines with septic shock. J Appl Physiol 74:1885–1892.
14. Clough G (1991): Relationship between microvascular permeability and ultrastructure. Prog Biophys Mole Biol 55:47–69.
15. Weiss SJ (1989): Tissue destruction by neutrophils. N Engl J Med 320:365–376.
16. LeVraux V, Chen YL, Masson I, DeSousa M, Giroud JP, Floretin I, Chauvelot-Moachon L (1993): Inhibition of human monocyte TNF production by adenosine receptor agonists. Life Sci 52:1917–1924.
17. Parmely MJ, Zhou WW, Edwards CK, Borcherding DR, Silverstein R, Morrison DC (1993): Adenosine and a related carbocyclic nucleoside analogue selectively inhibit tumor necrosis factor-alpha production and protect mice against endotoxin challenge. J Immunol 151:389–396.
18. Firestein GS, Boyle D, Bullough DA, Gruber HE, Sajjadi FG, Montag A, Sambol B, Mullane KM (1994): Protective effect of an adenosine kinase inhibitor in septic shock. J Immunol 152:5853–5859.
19. Bouma MG, Stad RK, van der Wildenberg FAJM, Buurman WA (1994): Differential regulatory effects of adenosine on cytokine release by activated human monocytes. J Immunol 153:4159–4168.
20. Edwards CK, Watts LM, Parmely MJ, Linnik MD, Long RE, Borcherding DR (1994): Effect of the carbocyclic nucleoside analogue MDL 201,112 on inhibition of interferon-γ-induced priming of Lewis (LEW/N) rat macrophages for enhanced respiratory burst and MHC class II Ia^+ antigen expression. J Leukoc Biol 56:133–144.
20a. Sajjadi FG, Takabayashi K, Foster A, Domingo R, Firestein GS (1996): Inhibition of TNF-alpha gene expression by adenosine: The role of A3 adenosine receptors. J Immunol 156:3435–3442.

21. Cronstein BN, Kramer SB, Weissmann G, Hirschhorn R (1983): Adenosine: A physiological modulator of superoxide anion generation by human neutrophils. J Exp Med 158:1160–1173.

22. Cronstein BN, Rosenstein ED, Kramer SB, Weissmann G, Hirschhorn R (1985): Adenosine: A physiologic modulator of superoxide anion generation by human neutrophils. Adenosine acts via an A2 receptor on human neutrophils. J Immunol 135:1366–1371.

23. Roberts PA, Newby AC, Hallett MB, Campbell AK (1985): Inhibition by adenosine of reactive oxygen metabolite production by human polymorphonuclear leucocytes. Biochem J 227:669–674.

24. de la Harpe J, Nathan CF (1989): Adenosine regulates the respiratory burst of cytokine-triggered human neutrophils adherent to biologic surfaces. J Immunol 143:596–602.

25. Tsuruta S, Ito S, Mikawa H (1993): Effects of adenosine and its analogues on actin polymerization in human polymorphonuclear leucocytes. Clin Exp Pharmacol Physiol 20:89–94.

26. Schrier DJ, Imre KM (1986): The effects of adenosine agonists on human neutrophil function. J Immunol. 137:3284.

27. Salmon JE, Cronstein BN (1990): Fc gamma receptor-mediated functions in neutrophils are modulated by adenosine receptor occupancy. A1 receptors are stimulatory and A2 receptors are inhibitory. J Immunol 145:2235–2240.

28. Cronstein BN, Levin RI, Belanoff J, Weissmann G, Hirschhorn R (1986): Adenosine: An endogenous inhibitor of neutrophil-mediated injury to endothelial cells. J Clin Invest 78:760.

29. Cronstein BN, Levin RI, Philips M, Hirschhorn R, Abramson SB, Weissmann G (1992): Neutrophil adherence to endothelium is enhanced via adenosine A1 receptors and inhibited via adenosine A2 receptors. J Immunol 148:2201–2206.

30. Firestein GS, Bullough DA, Erion MD, Jimenez R, Ramirez-Weinhouse M, Barankiewicz J, Smith CW, Gruber HE, Mullane KM (1995): Inhibition of neutrophil adhesion by adenosine and an adenosine kinase inhibitor: The role of selectins. J Immunol 154:326–334.

31. Cronstein BN, Kramer SB, Rosenstein ED, Korchak HM, Weissmann G, Hirschhorn R (1988): Occupancy of adenosine receptors raises cyclic AMP alone and in synergy with occupancy of chemoattractant receptors and inhibits membrane depolarization. Biochem J 252:709–715.

32. Nielson CP, Vestal RE (1989): Effects of adenosine on polymorphonuclear leukocyte function, cyclic 3′ : 5′-adenosine monophosphate, and intracellular calcium. Br J Pharmacol 97:882–888.

33. Burkey TH, Webster RO (1993): Adenosine inhibits fMLP-stimulated adherence and superoxide anion generation by human neutrophils at an early step in signal transduction. Biochim Biophys Acta 1175:312–318.

34. Kansas GS, Ley K, Munro JM, Tedder TF (1993): Regulation of leukocyte rolling and adhesion to high endothelial venules through the cytoplasmic domain of L-selectin. J Exp Med 177:833–838

35. Nolte D, Lehr H-A, Messmer K (1991): Adenosine inhibits postischemic leukocyte-endothelium interaction in postcapillary venules of the hamster. Am J Physiol 261:H651–H655.

36. Asako H, Wolf RE, Granger DN (1993): Leukocyte adherence in rat mesenteric venules: Effects of adenosine and methotrexate. Gastroenterology 104:31–37.

37. Wollner A, Wollner S, Smith JB (1993): Acting via A2 receptors, aadenosine inhibits the upregulation of Mac-1 (CD11b/CD18) expression on fMLP-stimulated neutrophils. Am J Respir Cell Mol Biol 9:179–185.

38. Mathew JP, Rinder CS, Tracey JB, Auszura LA, O'Connor T, Davis E, Smith BR (1995): Acadesine inhibits neutrophil CD11b up-regulation in vitro and during in vivo cardiopulmonary bypass. J Thorac Cardiovasc Surg 109:448–456.

39. Grisham MB, Hernandez LA, Granger DN (1989): Adenosine inhibits ischemia-reperfusion-induced leukocyte adherence and extravasation. Am J Physiol 257:H1334–H1339.

40. Schrier DJ, Lesch ME, Wright CD, Gilbertsen RB (1990): The antiinflammatory effects of adenosine receptor agonists on the carrageenan-induced pleural inflammatory response in rats. J Immunol 145:1874–1879.
41. Rosengren S, Bong GW, Firestein GS (1995): Anti-inflammatory effects of an adenosine kinase inhibitor: Decreased neutrophil accumulation and vascular leakage. J Immunol 154:5444–5451.
42. Paty PSK, Sherman PF, Shepard JM, Malik AB, Kaplan JE (1992): Role of adenosine in platelet-mediated reduction in pulmonary vascular permeability. Am J Physiol 262:H771–H777.
43. Watanabe H, Kuhne W, Schwartz P, Piper HM (1992): A2-adenosine receptor stimulation increases macromolecule permeability of coronary endothelial cells. Am J Physiol 262:H1174–H1181.
44. Haselton FR, Alexander JS, Mueller SN (1993): Adenosine decreases permeability of in vitro endothelial monolayers. J Appl Physiol 74:1993.
45. Legrand AB, Narayanan TK, Ryan US, Aronstam RS, Catravas JD (1989): Modulation of adenylate cyclase activity in cultured bovine pulmonary arterial endothelial cells. Effects of adenosine and derivatives. Biochem Pharmacol 38:423–430.
46. Stelzner TJ, Weil JV, O'Brien RF (1989): Role of cyclic adenosine monophosphate in the induction of endothelial barrier properties. J Cell Physiol 139:157–166.
47. Casnocha SA, Eskin SG, Hall ER, McIntire LV (1989): Permeability of human endothelial monolayers: effect of vasoactive agonists and cAMP. J Appl Physiol 67:1997–2005.
48. Carson MR, Shasby SS, Shasby DM (1989): Histamine and inositol phosphate accumulation in endothelium: cAMP and a G protein. Am J Physiol 257:L259–L264.
49. Allison RC, Hernandez EM, Prasad VR, Grisham MB, Taylor AE (1988): Protective effects of O_2 radical scavengers and adenosine in PMA-induced lung injury. J Appl Physiol 64:2175–2182.
50. Perry M, Taylor AE (1988): Phorbol myristate acetate-induced injury of isolated perfused rat lungs: Neutrophil dependence. J Appl Physiol 65:2164–2169.
51. Sugio K, and Daly JW (1984): Adenosine analogs: Potentiation of bradykinin-induced plasma exudation and prevention by caffeine and theophylline. Life Sci 35:1575–1583.
52. Olafsson B, Forman MB, Puett DW, Pou A, Cates CU, Friesinger GC, Virmani R (1987): Reduction of reperfusion injury in the canine preparation by intracoronary adenosine: Importance of the endothelium and the no-reflow phenomenon. Circulation 76:1135–1145.
53. Knight RJ, Bowner CJ, Yates MS (1993): Effect of the selective A1 adenosine antagonist 8-cyclopentyl- 1,2-dipropylxanthine on acute renal dysfunction induced by Escherichia coli endotoxin in rats. J Pharm Pharmacol 45:979–984.
54. Holz FG, Steinhausen M (1987): Renovascular effects of adenosine receptor agonists. Renal Physiol 10:272–282.
55. Poelstra K, Brouwer E, Baller JFW, Hardonk MJ, Bakker WW (1993): Attenuation of anti-Thy 1 glomerulonephritis in the rat by anti-inflammatory platelet-inhibiting agents. Am J Pathol 142:441–450.
56. Hon WM, Khoo HE, Ngoi SS, Moochhala S (1995): Effects of adenosine receptor agonists on nitric oxide release in mouse during endotoxemia. Biochem Pharmacol 50:45–47.
57. Law WR, McLane MP, Raymond RM (1989): Adenosine restores myocardial responsiveness to insulin during acute endotoxin shock in vivo. Cir Shock 28:333–345.
58. Olsson RA, Pearson JD (1990): Cardiovascular purinoceptors. Physiol Rev 70:761–845.
59. Cronstein BN, Naime D, Firestein GS (1995): The antiinflammatory effects of an adenosine kinase inhibitor are mediated by adenosine. Arthritis Rheum 38:1040–1045.
60. Moser GH, Schrader J, Deussen A (1989): Turnover of adenosine is plasma of human and dog blood. Am J Physiol 256:C799–C806.

61. Hardart GE, Sullivan GW, Carper HT, Mandell GL (1991): Adenosine and 2-phenylaminoadenosine (CV-1808) inhibit human neutrophil bactericidal function. Infect Immun 59:885–889.

62. Dubyak GR, el-Moatassim C (1993): Signal transduction via P2-purinergic receptors for extracellular ATP and other nucleotides. Am J Physiol 265:C577–C606.

63. Tonetti M, Sturla L, Bistolfi T, Benatti U, De Flora A (1994): Extracellular ATP potentiates nitric oxide synthase expression induced by lipopolysaccharide in RAW 264.7 murine macrophages. Biochem Biophys Res Commun 203:430–435.

64. Tanke T, van de Loo JW, Rhim H, Leventhal PS, Proctor RA, Bertics PJ (1991): Bacterial lipopolysaccharide-stimulated GTPase activity in RAW 264.7 macrophage membranes. Biochem J 277:379–385.

65. Proctor RA, Denlinger LC, Leventhal PS, Daughtery SK, van de Loo JW, Tanke T, Firestein GS, Bertics PJ (1994): Protection of mice from endotoxic death by 2-methylthio-ATP. Proc Natl Acad Sci USA 91:6127–6020.

66. Chaudry IH, Baue AE (1980): The use of substrates and energy in the treatment of shock. Adv Shock Res 3:27–46.

67. Harkema JM, Chaudry IH (1992): Magnesium-adenosine triphosphate in the treatment of shock, ischemia and sepsis. Crit Care Med 20;263–275.

68. van der Meer C, Snijders PM, Valkenburg PW (1982): The effect of ATP on survival in intestinal ischemia shock, hemorrhagic shock, and endotoxin shock in rats. Circ Shock 9:619–628.

69. Emerit I, Fabiani JN (1990): Allopurinol in ischemia-reperfusion injury of heart. Adv Exp Med Biol 264:367–372.

70. Demling RH, LaLonde C (1990): Early postburn lipid peroxidation: Effect of ibuprofen and allopurinol. Surgery 107:85–93.

CHAPTER 17

The Role of Adenosine in Asthma

MARLENE A. JACOBSON and TONY R. BAI

Department of Pharmacology, Merck Research Laboratories, West Point, PA 19486 (M.A.J.); University of British Columbia Pulmonary Research Laboratory, St. Paul's Hospital, Vancouver, British Columbia V6Z 1Y6, Canada (T.R.B.)

INTRODUCTION

Asthma is an etiologically heterogeneous disorder, causal factors being identifiable in some cases, but not in others. Three separate components can be recognized: episodic airway obstruction that resolves spontaneously or as a result of treatment (i.e., "the asthma attack"); hyperresponsiveness of the airways to a variety of provocative stimuli; and persistent, although poorly defined, inflammation of the airways.

A linear relationship exists between the risk of developing asthma and serum IgE level (Burrows et al., 1989), and between the presence of airways hyperresponsiveness and serum IgE level, irrespective of skin test responses (Sears et al., 1991). IgE responsiveness and the production of eosinophils are considered to be largely determined by the interaction of B cells with either Th2 cells or cells bearing high-affinity IgE receptors such as mast cells, basophils, and some other cell types. Atopic ("allergic") individuals differ from non-atopic individuals in that they have increased production of cytokines, such as IL-4, 5, 10, and 13 (classically but not exclusively derived from Th2 lymphocytes), increased levels of IgE, altered mast cell responsiveness, and increased numbers of mast cells and eosinophils in tissues near mucosal surfaces (Barnes et al., 1995). Genetic linkage studies have demonstrated associations between risk for atopy, asthma, and/or high IgE and several other markers including the IL-4 cytokine gene cluster and the high-affinity IgE receptor.

Products of activated eosinophils and mast cells are thought to produce many of the pathologic abnormalities in the airway wall in asthma. Although it is widely conceded that the inflammatory process is orchestrated by T lymphocytes, a pivotal role for other cell types in the airway injury, such as epithelial cells, fibroblasts, macrophages, and neuronal structures has by no means been excluded. These data and the results of numerous other epidemiological and immunopathological studies in asthmatics

Purinergic Approaches in Experimental Therapeutics, Edited by Kenneth A. Jacobson and Michael F. Jarvis
ISBN 0-471-14071-6 © 1997 Wiley-Liss, Inc.

suggest that allergic airway inflammation underlies most features of asthma, including airways hyperresponsiveness, in the majority of patients, although they do not exclude separate contributing factors such as smooth muscle phenotype.

Interest in the role of adenosine as a modulator of asthma and airway inflammatory reactions developed from several independent observations. Atopic and non-atopic asthmatics, but not normal individuals exhibit bronchoconstriction to inhaled adenosine or adenosine 5′-monophosphate (Cushley et al., 1983). Adenosine potentiates mediator release from antigen-stimulated mast cells (Marquardt et al., 1978). Lung tissue and plasma levels of adenosine are increased by stimuli such as hypoxia, energy deficit, and antigen challenge (Mann et al., 1986). Recently, adenosine levels were found to be markedly increased in bronchoalveolar lavage fluid from asthmatics in comparison to normal individuals (Driver et al., 1993). Evidence for the clinical relevance of increases in endogenous adenosine concentrations in asthmatics is derived from reports of adverse effects of dipyridamole. This agent is used intravenously as a coronary vasodilator in myocardial imaging for suspected coronary artery disease. Dipyridamole results in increased extracellular adenosine concentrations and can cause bronchospasm as a side effect in both asthma and chronic obstructive pulmonary disease (COPD) (Eagle and Boucher, 1989; Lette et al., 1995).

Development of a Link Between Adenosine and Asthma: Chronology

The effects of adenosine on airway tone were first reported by Coleman (1976), and Farmer and Farrar (1976). These investigators reported that micromolar concentrations of adenosine had a direct effect on isolated normal guinea pig airway smooth muscle and caused a weak contraction followed by more sustained relaxation. Subsequently Marquardt and coworkers (1978) reported that adenosine enhanced histamine release from calcium ionophore stimulated rat peritoneal mast cells. In a series of seminal reports, Holgate, Church, and coworkers, demonstrated that inhalation of adenosine by both atopic and non-atopic asthmatics caused airway narrowing (bronchoconstriction) (Cushley and Holgate, 1983; Cushley et al., 1983, 1984; Mann et al., 1986). In contrast, normal individuals were unaffected. Inosine, the deaminated produce of adenosine, and the closely related nucleoside guanosine had no measurable effect. Adenosine mono- and diphosphates (AMP, ADP), which are rapidly metabolized to adenosine under physiological conditions, also caused bronchoconstriction. Subjects who demonstrated atopy (enhanced skin or nose reactivity to common inhaled aeroallergens), but without current asthma, also developed exaggerated airway narrowing following inhalation of adenosine. However as non-atopic asthmatics and smokers with airway obstruction also show heightened responses to adenosine (Pin et al., 1991; Oosterhoff et al., 1993), abnormal responses to inhaled adenosine are not specific to the atopic state.

Inhaled or orally administered theophylline protected against the adenosine-induced bronchoconstriction in asthmatics (Cushley et al., 1984; Mann and Holgate, 1985). In these studies, theophylline was found to be less effective against histamine-induced versus adenosine-induced bronchoconstriction by a factor of three. These results provided evidence that the bronchoconstrictor response to adenosine was receptor-specific, and it was proposed that theophylline acted selectively, via antagonism of adenosine receptors (Cushley et al., 1984; Mann and Holgate, 1985).

Mechanisms of Adenosine Action in Asthma

The mechanism underlying the airway response to adenosine in asthma is not clearly defined, nor is the significance or origin of the elevated levels of adenosine in the asthmatic individual well understood. The multitude of actions of adenosine are mediated by G protein–coupled receptors classified into A_1, A_{2a}, A_{2b}, and A_3 subtypes (Collis and Hourani, 1993). These receptors have a wide tissue distribution and are found on many cell types. The effect of adenosine in asthma could be initially considered to be the result of a direct action on smooth muscle to produce contraction, or alternatively, the result of an indirect effect via stimulation of inflammatory cells or neuronal reflexes. This review will examine the possible mechanisms by which adenosine may contribute to the pathophysiology of asthma and discuss which specific adenosine receptor subtypes participate in this process.

Indirect Mechanism of Adenosine in Asthma Sodium cromoglycate and nedocromil, which reduce mediator release from mast cells (among other actions such as inhibiting activation of sensory nerves, eosinophils, and epithelial cells; see review by Barnes et al., 1995), were also shown to reduce adenosine-induced airway narrowing in asthmatics (Phillips et al., 1989). Furthermore, the bronchoconstrictor response to inhaled AMP was reduced by pretreatment with selective histamine H_1-receptor antagonists (Rafferty et al., 1987; Phillips and Holgate, 1989). These data suggested that adenosine functions as an indirect mediator of asthma in that the mast cell probably plays a central role in initial airway responses to inhaled allergen. Other important evidence that adenosine may play a role in the pathogenesis of asthma was the observation by Bjorck and coworkers (1992) that isolated bronchi removed from asthmatics display an enhanced responsiveness to the contractile effects of adenosine compared with that in nonasthmatics. The contractile effect was demonstrated to be indirect, and secondary to the release of the known contractile mediators, leukotrienes and histamine, in that leukotriene antagonists and an antihistamine together completely abolished the response. Clearly, the mast cell is one cell type that could produce both contractile mediators in response to adenosine, although other cell types could contribute to the leukotriene release. Further evidence that the mechanism of action in asthma was indirect came from the observations that, coincident with bronchoconstriction, inhalation of adenosine caused release of histamine and tryptase (Phillips et al., 1990; Polosa et al., 1995), prostanoids (Phillips and Holgate, 1989), thromboxane B_2 (Narushima et al., 1990), and a high molecular weight neutrophil chemotactic factor (Driver et al., 1991).

Adenosine Modulation of Lung Mast Cell Function The mast cell is an inflammatory cell that plays a key role in mediating the acute response in asthmatic and allergic states, as discussed above. The time-course of mast cell activation and histamine release is similar to the immediate bronchoconstrictor response observed in asthmatics provoked by adenosine or AMP inhalation (Cushley et al., 1983). Supporting evidence for the involvement of mast cell mediator release as a predominant mechanism for adenosine-induced bronchoconstriction has been generated *in vitro* on isolated mast cell preparations and *in vivo* in asthmatic and atopic nonasthmatics. Adenosine and adenosine analogs *in vitro* have been shown to enhance histamine release from stimulated rat peritoneal mast cells (Marquardt et al., 1978; Church and Hughes, 1985), mouse bone marrow–derived mast cells (Marquardt et al., 1994), rat basophilic leukemia RBL-2H3 cells (Ramkumar et al., 1993), human peripheral blood basophils (Church et al., 1983), and human lung mast cells (Hughes et al., 1984), in response to challenge with a variety of stimuli such as anti-IgE, conconavalin A, compound 48/80, and calcium ionophore A23187. Adenosine and adenosine analogs have also been reported to inhibit histamine release from immunologically stimulated human lung mast cells and basophils (Nishibori et al., 1983; Hughes et al., 1984; Marone et al., 1989; Peachell et al., 1991). In these studies, the inhibitory response to adenosine was dependent upon experimental conditions, such as incubation with adenosine prior to immunological challenge, the level of immunological stimulation, and adenosine concentration.

Adenosine alone, at micromolar concentrations, has no effect on mediator release from mast cells in the absence of antigen stimulation or nonimmunological challenge (Marquardt et al., 1978). The release of adenosine from stimulated rat serosal and mouse bone marrow–derived mast cells has been measured in response to antigen, calcium ionophore A23187, or compound 48/80 challenge (Marquardt et al., 1978; Fredholm, 1981). The increase in extracellular adenosine was correlated with a decrease in ATP levels (Marquardt et al., 1984). It has been postulated that the local release of endogenous adenosine from stimulated mast cells modulates a positive feedback mechanism in promoting mediator release and may contribute to hyperresponsiveness and immediate hypersensitivity.

In addition to the potentiation of histamine release, adenosine at micromolar concentrations has been demonstrated to enhance the generation of leukotriene C_4 (LTC_4) in dispersed human lung mast cells (Peachell et al., 1988). This is of significance since LTC_4 is a newly formed lipid mediator, arising from arachidonic acid metabolism, whereas histamine is preformed and stored in cytoplasmic granules of mast cells. Furthermore, LTC_4 is a more potent bronchoconstrictive agent than histamine, with a wider range of stimulant activities.

The contractile response to adenosine observed *in vitro* on bronchi isolated from asthmatics can be completely abolished by histamine antagonists (mepyramine, H_1 and metiamide, H_2) in combination with either a leukotriene antagonist or synthesis inhibitor (Bjorck et al., 1992). In clinical studies, pretreatment with the selective histamine H_1-receptor antagonists terfenadine and astermizole reduced the bronchoconstriction to AMP inhalation in asthmatics and atopic individuals by up to 80% (Phillips et al., 1987; Rafferty et al., 1987). Agents that function as inhibitors of mast cell mediator release, including sodium cromoglycate and nedocromil sodium, have been shown to be highly effective in attenuating bronchoconstriction induced in asthmatics by inhaled adenosine and AMP (Phillips et al., 1989). The evidence that adenosine-induced responses are initiated by acute release of mast cell–derived mediators has been strengthened by a recent report of measurements of mediators in bronchoalveolar fluid following local segmental adenosine challenge in mild atopic asthmatics (Polosa et al., 1995). In this study, significant elevations were noted in prostaglandin D_2, tryptase, and histamine concentrations, but not in the macrophage product, β-glucuronidase concentrations.

The rat basophilic leukemic cell line, RBL-2H3, has been used as a model system to study adenosine receptor–mediated secretion and transduction systems linked to mast cell degranulation. Adenosine potentiates antigen-induced mediator secretion and degranulation in RBL-2H3 cells. Stimulation of adenylate cyclase and phospholipase C have both been postulated to couple to adenosine modulation of mast cell mediator release (Ali et al., 1990; Abbracchio et al., 1992). Ali et al. (1990) suggested that the adenosine receptor subtype mediating mast cell degranulation was atypical (non-A_1, non-A_2) in that methylxanthine antagonists were ineffective in inhibiting the enhancement of antigen-induced secretory responses by adenosine analogs, such as 5′-N-ethylcarboxamidoadenosine (NECA). The receptor was not found to be coupled to adenylyl cyclase but was instead linked to stimulation of phospholipase C activity and induction of transient increases in inositol phosphates and intracellular calcium. The response to NECA in the RBL-2H3 cells was pertussis toxin–sensitive suggesting coupling to Gi-like proteins.

A_3 Adenosine Receptors and Mast Cells Subsequent to the cloning of the rat A_3 adenosine receptor subtype (Zhou et al., 1992), the enhancement of antigen-induced secretion from RBL-2H3 was correlated with an affinity order profile for adenosine receptor agonists similar to that established for the cloned A_3 adenosine receptor subtype (Ramkumar et al., 1993). In addition, RBL-2H3 cells specifically bound ^{125}I-labeled aminophenylethyladenosine (^{125}I-APNEA) with an affinity comparable to the rat A_3 receptor expressed in CHO cells, and adenosine analogs competed for binding with an order potency characteristic of the cloned rat A_3 receptor (Zhou et al., 1992). From these studies, it was concluded that the A_3 adenosine receptor was the specific adenosine receptor subtype that mediated the secretion from antigen-stimulated mast cells. The enhanced response of RBL-2H3 cells to adenosine after dexamethasone exposure has been attributed to an increase in A_3 adenosine receptor expression (Ramkumar et al., 1995).

Early studies of adenosine potentiation of mediator release from immunologically activated human mast cells suggested the involvement of an A_2 receptor (Church et al., 1983; Hughes et al., 1984; Peachell et al., 1988). A rank order potency for the adenosine analogs NECA > R-PIA was established for potentiation of LTC_4 and histamine secretion. Theophylline was ineffective, even at 100 μM, in preventing the potentiation of histamine release; however, it partially reduced the enhancement of LTC_4 generation. These studies were performed prior to the identification and characterization of the A_3 adenosine receptor subtype and conclusions implicating A_2 receptors were drawn without the knowledge of the A_3 receptor pharmacological profile for adenosine analogs or affinity for xanthine antagonists. Preference for NECA over R-PIA and insensitivity to methylxanthine antagonism is also characteristic of the A_3 adenosine receptor (Zhou et al., 1992; Salvatore et al., 1993); therefore, it may be possible to alternatively interpret these results, as activation of A_3 adenosine receptors could be responsible for the observed mast cell mediator release. The differential effects of theophylline on histamine versus LTC_4 generation could reflect either involvement of more than one adenosine receptor subtype or a mechanism of inhibitory action not involving adenosine receptor antagonism as discussed above. The functional response of A_3 adenosine receptor activation in human mast cells has not yet been established and is of great interest in light of the observations in RBL-2H3 cells. The A_3 adenosine receptor subtype has also been identified in a canine BR mastocytoma cell line by molecular cloning and saturation binding of the A_3 agonist, N^6-amino[^{125}I]iodoaminobenzyladenosine; however, in functional studies of mast cell degranulation, the affinity order profile established for adenosine analogs did not correlate with A_3 adenosine receptor activation (see below) (Auchampach et al., 1995).

A correlation of A_3 adenosine receptor activation and nonimmunologically dependent mast cell mediator release has recently been demonstrated *in vivo*. The hypotension induced in anesthetized rats by the intravenous administration of the agonist APNEA is qualitatively similar to that induced by the mast cell degranulating agent compound 48/80 (Hannon et al., 1995). In contrast to the response observed with compound 48/80, which increased airway resistance when high concentrations were administered, no changes in airway resistance was noted concurrent with the APNEA-induced hypotension. The hypotensive response to APNEA was blocked by the A_3 antagonist, 3-(3-iodo-4-aminobenzyl)-8-(4-oxyacetate)-1-propylxanthine (I-ABOPX) and partially protected by prior treatment with either sodium cromoglycate or lodoxamide. A marked increase in plasma histamine levels was associated with the APNEA-induced hypotension. Depletion of mast cell mediators with compound 48/80 decreased the effectiveness of APNEA-induced hypotension, providing evidence for the involvement of mast cells.

In isolated perfused hamster cheek pouch, a mast cell–dependent vasoconstrictor response of arterioles to adenosine and adenosine analogs has also been observed and correlated with A_3 adenosine receptor activation (Doyle et al., 1993; Klindt et al., 1994).

A_{2b} Adenosine Receptors and Mast Cells The presence of A_{2a} and A_{2b} adenosine receptors, in addition to A_3 adenosine receptors, has been confirmed on RBL-2H3 cells by saturation binding and detection of mRNA transcripts (Marquardt et al., 1994; Ramkumar et al., 1995). Since adenosine potentiation of antigen-induced degranulation of RBL-2H3 cells is coupled to phospholipase C activation, and recombinant A_{2b} adenosine receptors expressed in *Xenopus* oocytes have been shown to couple to phospholipase C activation (Yakel et al., 1993), it has been postulated that the A_{2b} receptor in RBL-2H3 cells may be responsible for modulation of mediator secretion (Marquardt et al., 1994). However, direct evidence for a functional role of A_{2b} adenosine receptor subtypes in RBL-2H3 cells has not been established.

Recent evidence from studies of mouse, canine, and human mast cell lines has also suggested that the A_{2b} adenosine receptor subtype may be involved in mediating mast cell degranulation. In canine BR mastocytoma cells, an affinity order profile of NECA > R-PIA > I-ABA >> CGS21680 for degranulation measured by β-hexosaminidase release was established (Auchampach et al., 1995). This pharmacological profile is more consistent with activation of an A_{2b} adenosine receptor subtype. In antigen or calcium ionophore stimulated mouse bone marrow–derived mast cells, activation of A_{2b} receptors has been correlated with mast cell mediator release (Marquardt et al., 1994). A_{2b} adenosine receptors have also been implicated in mediating the release of interleukin-8 (IL-8) in a human mast cell line, HMC-1 (Feoktistov and Biaggioni, 1995). The expression of A_{2b} receptors on HMC-1 cells has recently been confirmed by immunoblotting and fluorescent immunostaining with an antibody specific for the A_{2b} receptor (Feoktistov et al., 1996). The characterization of A_{2b} adenosine receptor subtypes is limited by the lack of high-affinity, selective agonists and antagonists. Further evaluation of the role of A_{2b} receptors in modulating mast cell degranulation would be facilitated by the availability of selective ligands.

In Vitro *Effects on Airway Smooth Muscle* The response of normal guinea pig airway smooth muscle to adenosine *in vitro* is relaxation, with only occasional reports of preceding small contractions (Fredholm et al., 1979; Farmer et al., 1988; Thorne and Broadley, 1992). In normal guinea pigs, the dominant response to inhaled adenosine *in vivo* is also bronchodilation, indicative of smooth muscle relaxation. The relaxant effect has been shown to be mediated through stimulation of A_2 adenosine receptors with induction of increased cyclic AMP levels (Brown and Collis, 1982). The relaxant effect is potentiated by the adenosine transporter inhibitor dipyridamole, inhibited by low concentrations of methylxanthines, and reproduced more effectively by the nonselective A_2 agonist NECA than by the A_1 agonist R-PIA. In contrast, the contractile response of isolated guinea pig tracheal muscle to adenosine analogs exhibits a potency order profile consistent with activation of A_1 adenosine receptor subtypes (Ghai et al., 1987; Farmer et al., 1988). The guinea pig airway smooth muscle contractile response is inhibited by indomethacin (Advenir et al., 1982), and is thus mediated through the release of contractile prostaglandins. In contrast to normal animals, adenosine produced more marked contractions in lung strips and trachea from ovalbumin-sensitized guinea pigs (Thorne and Broadley, 1992). Relaxation was observed when high doses of adenosine ($\geq$1 mg) were employed. However, in this model, the adenosine receptor antagonist, 8-phenyltheophylline was ineffective in blocking the contractile response. Contractile responses to adenosine are observed in airway smooth muscle preparations from ragweed-sensitized rabbits but not in those from normal rabbits, (Ali et al., 1994b). In these studies, peripheral airway preparations were more responsive than central (tracheal) preparations, a feature indicating preferential sensitivity to adenosine in small airways. The nonxanthine adenosine receptor antagonist CGS-15943 inhibited the adenosine-induced contraction. *In vitro,* in trachea from nonsensitized rats, adenosine-induced contractions have been reported to be modulated by 5-HT_2 receptors and therefore are probably secondary to mast cell mediator release (Nordstrom and Delbro, 1986).

In normal isolated human airway smooth muscle preparations, adenosine either does not affect baseline tone (Finney et al., 1985; Bai et al., 1989), or induces inconsistent contractions. As previously discussed, Bjorck et al. (1992) demonstrated that bronchi isolated from asthmatics were more sensitive to adenosine than nonasthmatic tissue, confirming *in vitro* what has been observed in asthmatics in response to inhaled adenosine. In bronchi isolated from asthmatics, adenosine induced dose-dependent contraction in the concentration range of 10–100 μM. That these concentrations of adenosine have been demonstrated in bronchoalveolar lavage fluid (Driver et al., 1993) suggests pathophysiological relevance. The contractile effect was inhibited by an A_1 adenosine receptor antagonist, 2-thio-[1,3-dipropyl)-8-cyclopentyl]-xanthine (1 μM) or an A_1/A_2 receptor antagonist, 8-(*p*-sulfo)phenyltheophylline (30 μM). Pretreatment with histamine antagonists (mepyramine, H_1; metiamide, H_2) in combination with either a leukotriene antagonist or synthesis inhibitor completely abolished the contractile response to adenosine. These results provided supporting evidence for an indirect mechanism of adenosine-induced contraction involving mediator release.

Adenosine Effect on Airway Vasculature Of particular interest to airway disease is the role of adenosine in the control of bronchovascular tone via the bronchial circulation. Adenosine has been shown to relax bovine bronchial arteries (Alexander and Eyre, 1985), so that by increasing bronchial blood flow, airway lumenal diameter can decrease, thus increasing airflow resistance.

In Vivo *Models of Adenosine-Induced Bronchoconstriction* *In vivo* adenosine-induced bronchoconstriction has been studied in a variety of animal models, including normal BDE rats, ovalbumin-sensitized guinea pigs, and ragweed- or dust mite–sensitized rabbits. In general, an animal model of adenosine-induced airway narrowing should mimic as many features of the human phenomenon as possible. Thus it should occur in an animal with airway structure, neural innervation, and immune function similar to those of humans. In addition, the phenomenon should not occur in the normal host but rather should be a function of the allergic or inflammatory state. The distribution and pharmacological profile of adenosine receptor subtypes should be similar to those of the human receptors. Clearly, no current model fulfills all of those criteria.

Adenosine induces bronchoconstriction in normal, anesthetized BDE rats (Pauwels and Van Der Straeten, 1987; Pauwels and Joos, 1995). The response follows an affinity order profile of NECA = CPA > APNEA > CHA > R-PIA > CGS21680.* An A_{2a} selective antagonist, KF17837 had no significant effect, and A_1 antagonists partially blocked the NECA-induced bronchoconstriction. This affinity order profile is consistent with activation of an A_{2b} adenosine receptor subtype (Rivkees and Reppert, 1992). NECA-induced bronchoconstriction was markedly inhibited by methysergide, cromoglycate, nedocromil, and high concentrations of atropine, suggesting a multicomponent mechanism involving both mast cell activation and mast cell–derived mediators stimulating vagal nerves. Some inbred rat strains including the BDE strain are innately atopic, producing increased levels of IgE. This could explain the contractile response to adenosine in this nonallergic, normal rat model. Interestingly, a difference in the sensitivity to the A_3 agonist APNEA was observed between rat strains, with the BDE strain having significantly greater sensitivity to APNEA-induced bronchoconstriction.

Both anesthetized and conscious guinea pigs sensitized to ovalbumin, but not normal animals, develop airway constriction following inhaled and intravenous adenosine (Manzini and Ballanti, 1990; Thorne and Broadley, 1994). *In vivo* models of adenosine-induced bronchoconstriction have also been developed in rabbits neonatally sensitized to house dust mite (Ali et al., 1994a) and ragweed allergen (Ali et al., 1994b). Although this is known to be a difficult model to reproduce in terms of significant changes in pulmonary function, these workers have reported clear evidence of adenosine-induced reductions in dynamic compliance, which could be blocked by the A_1 antagonist DPCPX and the nonxanthine antagonist CGS15943. The agonist potency profile for the bronchoconstriction was consistent with A_1 receptor mediated responses. The results obtained in the *in vivo* rabbit model are similar to those in isolated guinea pig tracheal muscle (Farmer et al., 1988). To our knowledge adenosine responses have not been tested in primate models of allergic airway hyperresponsiveness.

Neural Mediation of Adenosine Effects on Pulmonary Function Adenosine or adenosine analogs act presynaptically in some animal species to modulate release of acetylcholine from cholinergic nerves. In most nonpulmonary tissues an inhibitory effect has been reported. However, in rabbit bronchi, adenosine has been reported to enhance neural release of acetylcholine and hence airway smooth muscle contraction, although this effect has not been demonstrated in human airway smooth muscle (Bai et al., 1989). Muscarinic antagonists oppose adenosine-induced airway constriction *in vivo,* suggesting the involvement of cholinergic neural reflexes (Pauwels et al., 1990; Okayama et al., 1986; Polosa et al., 1991). The threshold for activation of afferent C fibers of these neural pathways may have been lessened by mediators such as leukotrienes released by adenosine from inflammatory cells. Whether adenosine directly effects sensory nerve activation is unresolved. Morimoto and coworkers (1993) have reported that, in normal anesthetized guinea pigs, A_2 receptor activation by intravenous analogs reduces capsaicin-induced bronchoconstriction, suggesting negative regulation of C fiber activation. The vagally mediated, atropine-resistant, but capsaicin-sensitive, bronchoconstriction following intravenous 2-chloroadenosine reported in anesthetized sensitized guinea pigs is consistent with a peptidergic neural response in allergic airways

* CPA, N^6-cyclopentyladenosine; CHA, N^6-cyclohexyladenosine; CGS21680, 2-p-(2-carboxyethyl) phenylamino 5′-N-ethylcarboxamidoadenosine; NECA, 5′-N-ethylcarboxamidoadenosine; APNEA, N^6-2-(4-aminophenyl) ethyladenosine; R-PIA, R-N^6-(2-phenylisopropyl) adenosine.

(Manzini and Ballanti, 1990). The vagally mediated component of the bronchoconstriction induced by intravenous adenosine in "normal" inbred BDE rats may be secondary to stimulation of afferent neurons by mast cell–derived products (Pauwels and Van Der Straeten, 1987). Polosa and coworkers (1992) reported a reduced airway constriction upon AMP challenge in asthmatics after induction of tachyphylaxis to bradykinin. One interpretation of these findings is that the adenosine airway response involve local peptidergic (e.g., tachykinergic) neural pathways, since the response to bradykinin, at least in rodents, involves capsaicin-sensitive neural pathways.

Adenosine Effects on Airway Epithelial Cells The function of the adenosine receptors identified on human airway epithelial cells (Nilsen et al., 1995) are unknown. Increased mucus production by specialized epithelia is an important abnormality in asthma and related inflammatory airway disorders, particularly in more severe cases. Adenosine has been shown to be a potent and rapid acting secretagogue in a canine *in vivo* tracheal mucus model (Johnson and McNee, 1985). In the guinea pig trachea, when smooth muscle effects are examined in the presence of epithelium, the airway epithelium seems to be involved only in the uptake and metabolism of adenosine (Advenier et al., 1988). Changes in the phenotype of epithelial cells in asthma are potentially of great importance, as these cells are the line of first contact for inhaled antigens, gases, and infectious organisms. Epithelial cells may play an important role in airway inflammation, first by releasing chemokines (for example IL-8) that attract mononuclear cells and granulocytes to the airway, and second by releasing eicosanoids and other lipid mediators. The similarities between adenosine challenge responses and exercise responses are also of interest with regard to the epithelium. The primary stimulus in exercise is the airway drying effect of increased ventilation. The role of adenosine receptors in modulating these responses is unexplored.

Other Effects Adenosine is reported to increase the uncoupling and/or downregulation of β_2 adrenoceptors in guinea pig trachea (Matran et al., 1989). Whether the high concentrations of adenosine likely to be present in acute asthma contribute to the uncoupling of β_2 adrenoceptors demonstrated in airways in fatal asthma (Bai et al., 1992) is unknown.

A_3 Adenosine Receptors and Asthma

The discovery of the A_3 adenosine receptor on RBL-2H3 cells and establishment of its functional role in mast cell degranulation (see above) has prompted speculation that the A_3 receptor could be the "missing link" between adenosine and the pathogenesis of asthma (Beaven et al., 1994; Linden, 1994). The differential sensitivity of asthmatics to adenosine or AMP challenge in comparison to normal individuals may be related to an increased number of mast cells possessing A_3 adenosine receptors in bronchial airways in the diseased state. Analysis of inflammatory cell populations in bronchoalveolar lavage fluid (BAL) from asthmatic and atopic nonasthmatic individuals revealed a significant elevation in mast cells in comparison to normal nonasthmatic controls (Tomioka et al., 1984; Wardlaw et al., 1988). Levels of histamine measured in the BAL fluid of asthmatics are elevated in comparison to those of normal individuals (Driver et al., 1993). The histamine is most likely the result of increased releasibility from airway mast cells. The utility of A_3 adenosine receptor antagonists as antiasthma therapeutics has been proposed on the basis of the correlation between mast cell mediator release and A_3 receptor activation in rat RBL-2H3 cells (Beaven et al., 1994; Linden, 1994). However, this relationship has not yet been established on human mast cells, and therefore the potential efficacy of A_3 antagonists in the treatment of asthma is currently unknown.

In humans, the most abundant expression of A_3 transcripts is detected in the lung and liver (Salvatore et al., 1993). The cellular expression of A_3 transcripts in the lung was localized by *in situ* hybridization on eosinophils and mesenchymal cells in the lamina propria of nonasthmatic airways and adventia of blood vessels (Bai et al., 1994). In contrast to RBL-2H3 cells, where A_3 transcripts were detected by Northern blot analysis and polymerase chain reaction (Ramkumar et al., 1993), no hybridization was detected on toluidine blue–positive mast cells *in situ*. This discrepancy may reflect a species difference in tissue and cellular distribution of A_3 transcript expression (Linden, 1994), or alternatively, it may be the result of differences in the sensitivity of *in situ* hybridization of tissue sections versus amplification or hybridization of mRNA isolated from clonal cell lines. In isolated peripheral eosinophils, relatively high levels of A_3 transcripts are expressed (Knight et al., 1996), and a high density of A_3 receptors has been confirmed by radioligand binding (Kohno et al., 1996). Selective activation of A_3 receptors on peripheral cosinophils produced a dose-dependent inhibition of platelet-activating factor induced chemotaxis *in vitro* (Walker et al., 1997; Knight et al., 1996). These results suggested that A_3 receptors may play a role in eosinophilic airway inflammation. To determine whether A_3 adenosine receptor expression was altered in inflammatory airway disorders, A_3 transcript expression was quantified in peripheral lung and small airways from nonsmokers and smokers with and without chronic obstructive pulmonary disease. The mean transcript abundance was greater in lung tissue from subjects with airway inflammation (0.33 ± 0.04 pg/μg total RNA) than in normal lung (0.24 ± 0.03 pg/μg total RNA). Three-fold higher levels of A_3 transcripts were detected in parenchyma compared to large airways (Walker et al., 1997).

Eosinophils are important effector cells in inflammatory, allergic, and asthmatic diseases. They are a source of potent inflammatory mediators, including leukotrienes and platelet activating factor. The recruitment of activated eosinophils after allergen challenge to airways and their accumulation in the bronchial mucosa has been correlated with the development of the late-phase response and bronchial hyperresponsiveness in asthma (DeMonchy et al., 1985; Wardlaw et al., 1988). Activated eosinophils release major basic protein, cationic proteins, and collagenases, which damage respiratory epithelium and can induce airway hyperresponsiveness. Prior to identification and function of A_3 receptors on eosinophils, the effect of adenosine on eosinophils had not been studied as extensively as on other inflammatory cells such as mast cells and neutrophils. Adenosine at micromolar concentrations has been shown to exhibit anti-inflammatory effects on activated guinea pig peritoneal eosinophils and human peripheral blood eosinophils (Yukawa et al., 1989). NECA was more potent than R-PIA and it was suggested that an A_2 adenosine receptor was involved in the inhibitory response. Theophylline exhibited biphasic responses on the stimulated eosinophils. At subtherapeutic concentrations (0.1–10 μM), theophylline enhanced superoxide anion generation. At higher concentrations (10 μM–1 mM), theophylline inhibited superoxide anion release. In comparison, the adenosine receptor antagonist 8-phenyltheophylline potentiated superoxide anion generation from the stimulated eosinophils. From these results, it was concluded that at low concentrations, theophylline was acting as an adenosine receptor antagonist and competed with the anti-inflammatory effects of endogenous adenosine. At higher concentrations, theophylline acting as a phosphodiesterase inhibitor, preventing the release of anions through increasing intracellular cAMP levels. This mechanism is consistent with the observation that agents such as β-adrenoceptor agonists and PGE_2 increase intracellular cAMP and inhibit eosinophil activation (Kita et al., 1991).

Mechanism of Theophylline Action

Theophylline, a methylxanthine, is commonly used in the treatment of asthma and COPD. The therapeutic action of theophylline has been primarily attributed to its bronchodilator properties; however the specific mechanism of action is unknown and remains a controversial topic (Church et al., 1986; Howell, 1990; Persson et al., 1982; Barnes and Pauwels, 1994). The ability of theophylline to protect against adenosine-induced bronchoconstriction initially suggested that the therapeutic effectiveness was attributed to adenosine receptor antagonism. The affinity of theophylline for adenosine receptor subtypes is relatively weak, with K_i values reported on rat brain A_1 and A_{2a} receptors as 8.5 and 25 μM, respectively (Bruns et al., 1986), and on cloned sheep and human A_3 receptors as 424 and 56 μM, respectively (Linden et al., 1994; M. Jacobson, unpublished). Enprophylline (3-propylxanthine), a xanthine with negligible adenosine receptor antagonist activity, is more potent than theophylline in protecting against antigen-induced bronchoconstriction in asthmatics (Pauwels et al., 1985). Because of the similar activities of enprophylline and theophylline, it has been suggested that the bronchodilator activity of alkylxanthines is not attributed to adenosine receptor antagonism.

An alternative mechanism for the action of theophylline is its ability to inhibit cyclic nucleotide phosphodiesterase, thereby producing increases in intracellular cyclic AMP that result in bronchial smooth muscle relaxation and reduction of mast cell mediator secretion. There is poor correlation, however, between therapeutic concentrations of theophylline and the concentrations required for effective inhibition of phosphodiesterase activity (Polson et al., 1978). The therapeutic serum concentration range of theophylline used clinically is 10–20 μg/ml, corresponding to a free concentration of 28–56 μM (Weinberger, 1984). In comparison, the inhibitory constants of theophylline on phophodiesterase isozymes I–V measured *in vitro* are reported as 155–630 μM (Ukena et al., 1993), values that significantly exceed therapeutically relevant concentrations of theophylline.

There is significant interest in the anti-inflammatory and immunomodulatory properties of theophylline, including inhibition of cytokine synthesis and release from monocytes and T lymphocytes, inhibition of inflammatory cell activation, and microvascular leakage (Barnes and Pauwels, 1994). The effectiveness of theophylline is more pronounced on the late asthmatic reaction than on the immediate response after allergen challenge, which would suggest an anti-inflammatory mechanism of action (Pauwels et al., 1985). Lower doses of theophylline than those required for bronchodilator effects (plasma concentrations 5–10 μg/mL) have been found to be clinically effective as anti-inflammatory agents in asthma.

Adenosine as a Diagnostic Agent in Asthma

Inhaled adenosine is increasingly recognized as a safe, practical, and very useful diagnostic test for asthma (particularly exercise-induced asthma). A recent study in children showed inhaled adenosine distinguished asthmatics from normals with a sensitivity of

84% and specificity of 98%, a result not significantly different from the diagnostic value of an exercise test (Avital et al., 1995). Indeed exercise challenge is cross-tachyphylactic with adenosine challenge, suggesting a common pathway of action (Finnerty, 1990). Current tobacco smokers with COPD show greater responsiveness to inhaled adenosine than ex-smokers with COPD (Oosterhoff et al., 1993). The responsiveness of smokers with COPD approximates that of asthmatics, whereas smokers without COPD show negligible adenosine responsiveness. These results suggest that the indirect stimulus of adenosine, when the potential for increased airway narrowing is present (as judged by methacholine responsiveness), identifies an inflammatory state in the airways.

CONCLUSION

The differences between normal and asthmatic subjects in responses to adenosine may result from a number of different mechanisms, or, in all likelihood, a combination of mechanisms. First, the coupling of increased numbers of activated resident mast cells and other inflammatory cells in the airway wall with increased adenosine concentrations may augment mediator release. Second, central and local reflex neuronal activity may be increased by the high concentrations of adenosine found in inflamed airways. Third, the predominant adenosine receptor subtype may change in airway inflammation, thus augmenting pro-inflammatory effects of adenosine. Conceivably, genetic polymorphisms in adenosine receptors may also alter responses to ligands, as has been demonstrated for $\beta 2$ adrenergic receptors (Hall et al., 1995), although molecular characterization of adenosine receptor genes from asthmatic patients has not been explored.

The evidence presented to date suggests that inhaled adenosine identifies a mast cell–derived pathway leading to increased airway responses in asthmatics. However the role of the mast cell in *chronic* asthma, in controversial. Thus if the role of adenosine is restricted purely to potentiation of mast cell mediator release in asthma, its overall role in the pathophysiology of the chronic phase of the disorder may be minimal. However, the presence and effects of adenosine receptors on other cell types involved in asthma may be of greater relevance in assessing the overall airway response to the alterations in adenosine concentrations that occur in symptomatic phases of the illness.

REFERENCES

Abbracchio MP, Paoletti AM, Luini A, Cattabeni F, De Matteis MA (1992): Adenosine receptors in rat basophilic leukaemia cells: Transductional mechanisms and effects of 5-hydroxytryptamine-release. Br J Pharmacol 105:405–411.

Advenier C, Bidet D, Floch-Saint-Aubin A, Renier A (1982): Contribution of prostaglandins and thromboxanes to the adenosine and ATP-induced contraction of guinea-pig isolated trachea. Br J Pharmacol 77:39–44.

Advenier C, Devillier P, Matran R, Naline E (1988): Influence of epithelium on the responsiveness of guinea pig isolated trachea to adenosine. Br J Pharmacol 93:295–302.

Alexander I, Eyre P (1985): P1-purinoceptors mediate relaxation of bovine bronchial artery. Eur J Pharmacol 107:359–362.

Ali H, Cunha-Melo JR, Saul WF, Beaven MA (1990): Activation of phospholipase C via adenosine receptors provides synergistic signals for secretion in antigen-stimulated. RBL-2H3 cells: Evidence for a novel adenosine receptor. J Biol Chem 265:745–753.

Ali S, Mustafa SJ, Metzger WJ (1994a): Adenosine receptor–mediated bronchoconstriction and broncial hyperresponsiveness in allergic rabbit model. Am J Physiol 266:L271–L277.

Ali S, Mustafa SJ, Metzger WJ (1994b): Adenosine-induced bronchoconstriction and contraction of airway smooth muscle from allergic rabbits with late phase airway obstruction: Evidence for an inducible adenosine A1 receptor. J Pharmacol Exp Ther 268:1328–1334.

Auchampach JA, Caughey GH, Linden J (1995): Molecular cloning and pharmacological characterization of the canine A_3 adenosine receptor from BR mastocytoma cells. FASEB J 9:A122.

Avital A, Springer C, Bar-Yishay E, Godfrey S (1995): Adenosine, methacholine, and exercise challenges in children with asthma or pediatric chronic obstructive pulmonary disease. Thorax 50:511–516.

Bai TR, Lam R, Prasad YF (1989): Effects of adrenergic agonists and adenosine on cholinergic neurotransmission in human tracheal smooth muscle. Pulmon Pharmacol 1;193–199.

Bai TR, Mak JC, Barnes PJ (1992): A comparison of beta-adrenergic receptors and in vitro relaxant responses to isoproterenol in asthmatic airway smooth muscle. Am J Respir Cell Mol Biol 6:647–651.

Bai TR, Weir T, Walker BAM, Salvatore CA, Johnson RG, Jacobson MA (1994): Comparison and localization of adenosine A_3 receptor expression in normal and asthmatic lung. Drug Dev Res 31:244.

Barnes PJ, Pauwels RA (1994): Theophylline in the management of asthma: Time for reappraisal. Eur Resp J 7:579–591.

Barnes PJ, Holgate ST, Laitinen LA, Pauwels R (1995): Asthma mechanisms, determinants of severity and treatment: The role of nedocromil sodium. Clin Exp Allergy 25:771–787.

Beaven MA, Ramkumar V, Ali H (1994): Adenosine A_3 receptors and mast cells. Trends Pharmacol Sci 15:13–14.

Bjorck T, Gustafsson LE, Dahlen SE (1992): Isolated bronchi from asthmatics are hyperresponsive to adenosine which apparently acts indirectly by liberation of leukotrienes and histamine. Am Rev Respir Dis 145:1087–1091.

Brown CM, Collis MG (1982): Evidence for an A_2/Ra adenosine receptor in the guinea-pig trachea. Br J Pharmacol 76:381.

Bruns RF, Lu GH, Pugsley TA (1986): Characterization of the A_2 adenosine receptor labeled by [^{3}H]NECA in rat striatal membranes. Mol Pharmacol 29:331–346.

Burrows B, Martinez FD, Haonen M, Baebee RA, Cline MG (1989): Association of asthma with serum IgE levels and skin test reactivity to allergens. N Engl J Med 320:271–277.

Church MK, Hughes PJ (1985): Adenosine potentiates immunological histamine release from rat mast cells by a novel cyclic AMP independent cell-surface action. Br J PHarmacol 85:3–5.

Church MK, Featherstone RL, Cushley MJ, Mann JS, Holgate ST (1986): Relationships between adenosine, cyclic nucleotides and xanthines in asthma. J Allergy Clin Immunol 78:670–675.

Church MK, Holgate ST, Hughes PJ (1983): Adeonsine inhibits and potentiates IgE-dependent histamine release from human basophils from an A2-receptor mediated mechanism. Br J Pharmacol 80:719–726.

Coleman RA (1976): Effects of some purine derivatives on the guinea pig trachea and their interaction with drugs that block adenosine uptake. Br J Pharmacol 57:51–57.

Collis MG, Hourani SMO (1993): Adenosine receptor subtypes. Trends Pharmacol Sci 14:360–366.

Cushley MJ, Holgate ST (1983): Adenosine-induced bronchoconstriction in asthma: Specificity and relationship to airway reactivity. Thorax 38:705.

Cushley MJ, Tattersfield AE, Holgate ST (1983): Inhaled adenosine and guanosine on airway resistance in normal and asthmatic subjects. Br J Clin Pharmacol 15:161.

Cushley MJ, Tattersfield AE, Holgate ST (1984): Adenosine-induced bronchoconstriction in asthma: Antagonism by inhaled theophylline. Am Rev Respir Dis 129:380–384.

DeMonchy JGR, Kauffman HF, Venge P, Koeter GH, Jansen HM, Sluiter HJ, De Vires K (1985): Bronchoalveolar eosinophillia during allergen-induced late phase asthmatic reactions. Am Rev Respir Dis 131:373–376.

Doyle MP, Linden J, Duling BR (1993): Nucleoside-induced arteriolar constriction: A mast cell dependent response. Am J Physiol 266:H2042–H2050.

Driver AG, Kukoly CA, Ali S, Mustafa SJ (1993): Adenosine in bronchoalveolar lavage fluid in asthma. Am Rev Respir Dis 148:91–97.

Driver AG, Kukoly CA, Metzger WJ, Mustafa SJ (1991): Bronchial challenge with adenosine causes the release of serum neutrophil chemotactic factor in asthma. Am Rev Respir Dis 143:1002–1007.

Eagle KA, Boucher CA (1989): Intravenous dipyridamole infusion causes severe bronchospasm in asthmatic patients. Chest 95:258–259.

Farmer JB, Farrar DG (1976): Pharmacological studies with adenine, adenosine and some phosphorylated derivatives on guinea-pig tracheal muscle. J Pharm Pharmacol 28:748–752.

Farmer SG, Canning BJ, Wilkins DE (1988): Adenosine receptor–mediated contraction and relaxation of guinea pig isolated tracheal smooth muscle: Effect of adenosine antagonists. Br J Pharmacol 95:371–378.

Feoktistov I, Biaggioni I (1995): Adenosine A2b receptors activate interleukin-8 (IL-8) production in human mast cells. FASEB J 9:A1381.

Feoktistov I, Sheller JR, Vallejo V, Biaggioni I (1996): Immunological identification of adenosine A_{2b} receptors in human lung mast cells. Drug Dev Res 37:146.

Finnerty JP, Polosa R, Holgate ST (1990): Repeated exposure of asthmatic airways to inhaled adenosine 5′-monophosphate attenuates bronchoconstriction provoked by exercise. J Allergy Clin Immunol 86:353–359.

Finney MJB, Karlsson JA, Persson CGA (1985): Effects of bronchoconstrictors and bronchodilators on a novel human small airway preparation. Br J Pharmacol 85:29–36.

Fredholm BB (1981): Release of adenosine from rat lung by antigen and compound 48/80. Acta Physiol Scand 111:507–508.

Fredholm BB, Brodin K, Strandberg K (1979): On the mechanism of relaxation of tracheal muscle by theophylline and other cyclic nucleotide phosphodiesterase inhibitors. Acta Pharmacol 45:336–344.

Ghai G, Zimmerman MB, Hopkins MF (1987): Evidence for A1 and A2 adenosine receptors in guinea pig trachea. Life Sci 41:1215–1224.

Hall IP, Wheatly A, Wilding P, Liggett SB (1995): Association of Glu 27 beta-2-adrenoceptor polymorphism with lower airway reactivity in asthmatic subjects. Lancet 345:1213–1214.

Hannon JP, Pfannkuche HJ, Fozard JR (1995): A role of mast cells in adenosine A_3 receptor mediated hypotension in the rat. Br J Pharmacol 115:945–952.

Howell RE (1990): Multiple mechanisms of xanthine actions on airway reactivity. J Pharmacol Exp Ther 255:1008–1011.

Hughes PJ, Holgate ST, Church MK (1984): Adenosine inhibits and potentiates IgE-dependent histamine release from human lung mast cells by an A_2-purinoceptor mediated mechanism. Biochem Pharmacol 33:3847–3852.

Johnson HG, McNee ML (1985): Adenosine-induced secretion in the canine trachea: Modification by methylxanthines and adenosine derivatives. Br J Pharmacol 86:63–67.

Kita H, Abu-ghazaleh RI, Gleich GJ, Abraham RT (1991): Regulation of Ig-induced eosinophil degranulation by adenosine 3′,5′-cyclic monophosphate. J Immunol 146:2712–2718.

Klindt RL, Linden J, Duling BR (1994): Inosine causes vasoconstriction of *in vivo* hamster cheek pouch arterioles via an interaction with mast cells. FASEB J 8:A1046.

Knight DA, Bai TR, Weir T, Salvator CA, Jacobson MA, Walker BAM (1996): Expression and function of adenosine A_3 receptors (A_3-AR) in human lung tissues. Am J Respir Crit Care Med 153:A645.

Kohno Y, Ji K, Mawhorter SD, Koshiba M, Jacobson KA (1996): Activation of A_3 adenosine receptors on human, eosinophils elevates intracellular calcium. Blood 88:3569–3574.

Lette J, Taturn JL, Fraser S, Miller DD, Waters DD, Heller G, Stanton EB, Born HS, Leppo J, Nattel S (1995): Safety of dipyridamole testing in 73,806 patients: The Multicenter Dipyridamole Safety Study. J Nuclear Cardiol 2:3–17.

Linden J (1994): Cloned adenosine A_3 receptors: Pharmacological properties, species differences and receptor functions. Trends Pharmacol Sci 15:298–306.

Mann JS, Holgate ST (1985): Specific antagonism of adenosine-induced bronchoconstriction in asthma by oral theophylline. Br J Clin Pharmacol 19:685–692.

Mann JS, Holgate ST, Renwick AG, Cushley MJ (1986): Airway effects of purine nucleosides and nucleotides and release with bronchial provocation in asthma. J Appl Physiol 61:1667–1676.

Manzini S, Ballati L (1990): 2-Chloroadenosine induction of vagally mediated and atropine-resistant bronchomotor responses in anaesthetised guinea-pigs. Br J Pharmacol 100:251–256.

Marone G, Cirillo R, Genovese A, Marino O, Quattrin S (1989): Human basophil/mast cell releasability, VII. Heterogeneity of the effect of adenosine on mediator secretion. Life Sci 45:1745–1754.

Marquardt DL, Gruber HE, Wasserman SI (1984): Adenosine release from stimulated mast cells. Proc Natl Acad Sci USA. 81:6192–6196.

Marquardt DL, Parker CW, Sullivan TJ (1978): Potentiation of mast cell relese by adenosine. J Immunol 120:871–878.

Marquardt DL, Walker LL, Heinemann S (1994): Cloning of two adenosine receptor subtypes from mouse bone marrow-derived mast cells. J Immunol 152:4508–4515.

Matran R, Naline E, Advenier C, Duroux P (1989). In vitro desensitization of β-adrenoceptors in guinea pig trachea: Interaction between β-adrenoceptor agonists and influence of adenosine and other drugs. Fundam Clin Pharmacol 3:103–113.

Morimoto H, Yamashita M, Imazumi K, Matsuda A, Ochi T, Seki N, Mizuhara H, Fuji T, Senoh H (1993): Effects of adenosine A2 receptor agonists on the excitation of capsaicin-sensitive afferent sensory nerves in airway tissues. Eur J Pharmacol 240:121–126.

Narushima MK, Akisawa K, Tanaka K, Nakagami H, Noguchi E (1990): Studies of adenosine inhalation in asthmatic patients. Arerugi 9:322–329.

Nilsen D, Hirschhorn B, Cronstein B, Talbot A, Aston C, Reibman J (1995): Expression of adenosine receptors in human alveolar macrophages and airway epithelial cells. Am J Resp Crit Care Med 151:A190.

Nishibori M, Shimamura K, Yokoyama H, Tsutsumi K, Saeki K (1983): Differential effects of adenosine on histamine secretion induced by antigen and chemical stimuli. Arch Pharmacodyn 265:17–28.

Nordstrom O, Delbro D (1986): Adenosine-induced contractions of the isolated rat trachea are mediated by 5-hydroxytryptamine. Acta Physiol Scand 127:557–558.

Okayama M, Ma JY, Mataoka I, Kimura K, Miura M, Iifuna H, Inoue H, Takishima T (1986): Role of vagal nerve activity on adenosine-induced bronchoconstriction in asthma. Am Rev Respir Dis 133(suppl):A93. Abstract.

Oosterhoff Y, De Jong JW, Jansen AM, Koeter GH, Postma DS (1993): Airway responsiveness to adenosine 5′-monophosphate in chronic obstructive pulmonary disease is determined by smoking. Am Rev Respir Dis 147:553–558.

Pauwels R, Joos GF (1995): Characterization of the adenosine receptors in airways. Arch Int Pharmacodyn 329:151–160.

Pauwels RA, Van Der Straeten ME (1987): An animal model for adenosine-induced bronchoconstriction. Am Rev Respir Dis 136:374–378.

Pauwels R, Joos G, Kips J, Van der Straeten M (1990): Synergistic mechanisms in the adenosine and neuropeptide-induced bronchoconstriction. Arch Int Pharmacodyn 303:113–121.

Pauwels R, Van Renterghem D, Van Der Straren M, Johannesson N, Persson CGA (1985): The effect of theophylline and enprophylline on allergen-induced bronchoconstriction. J Allergy Clin Immunol 76:583–590.

Peachell PT, Columbo M, Kagey-Sobotka A, Lichtenstein LM, Marone G (1988): Adenosine potentiates mediator release from human lung mast cells. Am Rev Respir Dis 138:1143–1151.

Peachell PT, Lichtenstein LM, Schleimer RP (1991): Differential regulation of human basophil and lung mast cell function by adenosine. J Pharmacol Exp Ther 256:717–726.

Persson CGA, Karlsson J-A, Erjefalt I (1982): Differentiation between bronchodilation and universal adenosine antagonism among xanthine derivatives. LIfe Sci 30;2181–2189.

Phillips GD, Holgate ST (1989): The effect of oral terfenadine alone and in combination with flurbiprofen on the bronchoconstrictor response to inhaled adenosine 5′-monophosphate in nonatopic asthma. Am Rev Respir Dis 139:463–469.

Phillips GD, Ng WH, Church MK, Holgate ST (1990): The response of plasma histamine to bronchoprovocation with methacholine, adenosine 5′-monophoshpate, and allergen in atopic nonasthmatic subjects. Am Rev Respir Dis 141:9–13.

Phillips GD, Rafferty P, Beasley CRW, Holgate ST (1987): The effect of oral terfenadine on bronchoconstriction response to inhaled histamine and adenosine 5′-monophosphate in non-atopic asthma. Thorax 42:939–945.

Phillips GD, Scott VL, Richards R, Holgate ST (1989): Nedocromil sodium is more potent than sodium cromoglycate against AMP-induced bronchonconstriction in atopic asthmatic subjects. Clin Exp Allergy 19:285–291.

Pin I, Hepperle MJ, Wong BJO, Ramsdale EH, Hargreave FE (1991): Methacholine and adenosine monophosphate (AMP) airway hyperresponsiveness in asthmatics and smokers with chronic airflow limitation. Am Rev Respir Dis 143:A413.

Polosa R, Ng WH, Crimi N, Vanchereri C, Holgate ST, Church MK, Mistretta A (1995): Release of mast-cell-derived mediators after endobronchial adenosine challenge in asthma. Am J Respir Crit Care Med 151:624–629.

Polosa R, Phillips GD, Rajakulasingam K, Holgate ST (1991): The effect of inhaled ipratropium bromide alone and in combination with oral terfenadine on bronchoconstriction provoked by adenosine 5′-monophosphate and histamine in asthma. J Allergy Clin Immun 87:939–947.

Polosa R, Rajakulasingam K, Church MK, Holgate ST (1992): Repeated inhalation of bradykinin attenuates adenosine 5′-monophosphate (AMP) induced bronchoconstriction in asthmatic airways. Eur Respir J 5:700–706.

Polson JB, Krazanowski JJ, Goldman AL, Szentivanyi A (1978): Inhibition of human pulmonary phosphodiesterase activity by therapeutic levels of theophylline. Clin Exp Pharmacol Physiol 5:535–539.

Rafferty P, Beasley CR, Holgate ST (1987): The contribution of histamine to bronchoconstriction produced by inhaled allergen and adenosine 5′-monophosphate in asthma. Am Rev Respir Dis 136:369–373.

Ramkumar V, Stiles GL, Beaven MA, Ali H (1993): The A_3 adenosine receptor is the unique adenosine receptor which facilitates release of allergic mediators in mast cells. J Biol Chem 268:16887–16890.

Ramkumar V, Wilson M, Dhanraj DN, Gettys TW, Ali H (1995): Dexamethasone up-regulates A_3 adenosine receptors in rat basophilic leukemia (RBL-2H3) cells. J Immunol 154:5436–5443.

Rivkees SA, Reppert SM (1992): RFL9 encodes an A_{2b}-adenosine receptor. Mol Endocrinol 6:1598–1604.

Salvatore CA, Jacobson MA, Taylor HE, Linden J, Johnson RG (1993): Molecular cloning and characterization of the human A_3 adenosine receptor. Proc Natl Acad Sci USA 90:10365–10369.

Sears MR, Burrows B, Flannery EM, Herbison GP, Hewitt CJ, Holdaway MD (1991): Relation between airway responsiveness and serum IgE in children with asthma and in apparently normal children. N Engl J Med 325:1067–1071.

Thorne JR, Broadley KJ (1992): Adenosine-induced bronchoconstriction of isolated lung and trachea from sensitized guinea pigs. Br J Pharmacol 106:978–985.

Thorne JR, Broadley KJ (1994): Adenosine-induced bronchoconstriction in conscious hyperresponsive and sensitized guinea pigs. Am J Respir Crit Care Med 149:392–399.

Tomioka M, Ida S, Shindoh Y, Ishihara T, Takishima T (1984): Mast cell in bronchoalvelor lumen of patients with bronchial asthma. Am Rev Respir Dis 129:1000–1005.

Ukena D, Schudt C, Sybrecht GW (1993): Adenosine receptor-blocking xanthines as inhibitors of phosphodiesterase isozymes. Biochem Pharmacol 45:847–851.

Walker BAM, Jacobson MA, Knight DA, Salvatore CA, Weir T, Zhou D, Bai TR (1997): Adenosine A_3 receptor expression and functions in eosinophils. Am J Respir Cell Mol Biol 16:1–7.

Wardlaw AJ, Dunnette S, Gleich GJ, Collins JV, Kay AB (1988): Eosinophils and mast cells in bronchoalveolar lavage in subjects with mild asthma. Am Rev Respir Dis 137:62–69.

Weinberger M (1984): The pharmacology and therapeutic use of theophylline. J Allergy Clin Immunol 73:525–543.

Yakel JL, Warren RA, Reppert SM, North RA (1993): Functional expression of adenosine A_{2b} receptor in *Xenopus* oocytes. Mol Pharmacol 43:277–280.

Yukawa T, Kroegel C, Chanez P, Dent G, Ukena D, Chung KF, Barnes PJ (1989): Effect of theophylline and adenosine on eosinophil function. Am Rev Respir Dis 140:327–333.

Zhou Q-Y, Li C, Olah ME, Johnson RA, Stiles GA, Civelli O (1992): Molecular cloning and characterization of an adenosine receptor: The A_3 receptor. Proc Natl Acad Sci USA 89:7432–7436.

CHAPTER 18

Nucleotides and Nucleosides in Pulmonary Function

JEFFREY S. FEDAN

Pathology and Physiology Research Branch, Health Effects Laboratory Division, National Institute for Occupational Safety and Health, Morgantown, WV 26505; Department of Pharmacology and Toxicology, Robert C. Byrd Health Sciences Center, West Virginia University, Morgantown, WV 26505

INTRODUCTION

The lung is a complex organ composed of many cell types that are involved in gas exchange, secretion, fluid balance, the regulation of airway diameter, and immune defense. A discussion of the effects of purine and pyrimidine compounds in the lungs must take into account the fact that there is cross-talk between the cell types, and that the composition of the lung changes in several pulmonary diseases due to transmigration of inflammatory cells out of the blood and into the air spaces and airway walls of the lung. This chapter will attempt to summarize the main effects of nucleotides and nucleosides in the lungs with two caveats: first, there may exist critical cellular effects of these compounds that have yet to be described; and second, the precise roles of endogenous nucleotides and nucleosides in pulmonary function *in toto* are not well understood. Most studies of nucleotide and nucleoside effects on cells present in the lungs have been obtained from *in vitro* experiments. The unique effects of nucleotides on ion transport in airway epithelia, which are being explored for novel therapy of cystic fibrosis, and responses to inhaled adenosine that are exaggerated in asthmatics (discussed later in this chapter), indicate that the pharmacological actions of these substances have direct medical relevance.

AIRWAY SMOOTH MUSCLE: INTERACTIONS WITH EPITHELIUM

The contractile and relaxant effects of nucleosides and nucleotides on the airways that have been described from *in vitro* and *in vivo* studies have been contradictory. Among

Purinergic Approaches in Experimental Therapeutics, Edited by Kenneth A. Jacobson and Michael F. Jarvis
ISBN 0-471-14071-6 © 1997 Wiley-Liss, Inc.

several factors affecting the outcome of *in vitro* experiments, the method of tissue preparation (i.e., transverse strips, spiral strips, rings, etc.) has a significant, qualitative influence. Second, it is becoming increasingly apparent that the respiratory epithelium may mediate or modulate responses to contractile and relaxant agonists (Flavahan et al., 1985; Fedan et al., 1988; Farmer and Hay, 1991). Responses of airway preparations will involve the direct interaction of the agonist on the airway smooth muscle as well as the indirect effect(s) of the agonist on the epithelium. Thus, knowledge of the role of epithelium in the response is needed to interpret the *in vitro* effects of nucleotides on airway preparations. Third, heterogeneity of airway responsiveness to nucleosides and nucleotides may involve interspecies variability. Fourth, differences in the orientation of the smooth muscle in isolated airway preparations, relative to the axis of circumferentially and longitudinally arranged smooth muscle cells, may contribute to the type of response obtained, i.e., contraction versus relaxation, at least in guinea pigs (Satchell and Smith, 1984).

Adenosine Agonists

The rank order of adenosine analog potencies indicates that relaxation responses of airway smooth muscle to adenosine and its synthetic congeners is mediated via methylxanthine-sensitive A_2 adenosine receptors, i.e., NECA > R-PIA > adenosine (Brown and Collis, 1982; Ghai et al., 1987; Farmer et al., 1988; Brackett and Daly, 1991).

In guinea pig isolated airways, adenosine and its congeners may also induce contraction; the concentrations inducing contraction are lower than those which cause relaxation. Contractile responses generally are smaller in magnitude than those elicited by agonists such as methacholine and histamine. Contraction in response to adenosine occurs in human airways *in vitro* (Finney et al., 1985), but a contractile effect may not be seen in all species (i.e., dog, Krzanowski et al., 1987). On the basis of relative potencies, the contractile response is thought to be mediated by A_1 receptors (Caparrotta et al., 1984; Ghai et al., 1987; Farmer et al., 1988). While adenosine-induced relaxation responses are inhibited by methylxanthines such as 8-phenyltheophylline (8-PT), the ineffectiveness of methylxanthines against adenosine-induced contractile responses, observed by some investigators, has led to the suggestion that A_3 receptors might mediate the responses (Pauwels and Joos, 1995), at least in some cases. For example, while Farmer et al. (1988) and Matera et al. (1995) observed that contractile responses to R-PIA were inhibited by methylxanthines, several investigators (Caparrotta et al., 1984; Ghai et al., 1987; Thorne and Broadley, 1992) did not observe such antagonism. Furthermore, Matera et al. (1995) observed that R-PIA-induced contractions of guinea pig tracheal rings were antagonized in the absence or presence of epithelium by ketanserin, a 5-HT2-receptor blocker, but not by atropine and diphenhydramine, the muscarinic and histamine H_1 receptor blockers, respectively, suggesting that 5-HT had mediated the responses.

In vivo experiments involving inhalation of aerosolized adenosine analogs by rats demonstrated that bronchoconstriction was elicited by N^6-2-(4-aminophenyl)ethyladenosine (APNEA), an A_3-selective compound, and NECA, an A_{2B}-selective agonist, but not by the A_{2A}-selective agonist CGS 21680. Additionally, bronchoconstriction to NECA was inhibited by A_1-selective antagonists (Pauwels and Joos, 1995). Further evidence will be needed to evaluate whether such *in vivo* responses involved signaling by A_3 receptors on the airway smooth muscle per se. An A_3 receptor transcript has been reported to be abundant in the human lung (Salvatore et al., 1993); more recently, it has been determined that the greatest expression of the receptor in inflamed airways is in eosinophils (Knight et al., 1996).

The respiratory epithelium modulates the reactivity of airway smooth muscle to adenosine receptor agonists and antagonists; much of the evidence revealing this effect has been obtained from guinea pigs. Compared to intact airways, adenosine concentration–response curves for relaxation responses were shifted to the left in the absence of the epithelium (Holroyde, 1986; Advenier et al., 1988; Lundblad and Persson, 1988). Farmer et al. (1986) observed that the potentiating effect of epithelium removal in guinea pig tracheal strips was observed only in the presence of the adenosine transporter inhibitor, dipyridamole, and an adenosine deaminase inhibitor, EHNA. It has been suggested that the epithelium is involved in uptake and metabolism of adenosine (Advenier et al., 1988).

The epithelium also affects reactivity to the contractile effects of adenosine. Devillier et al. (1992) observed that contractile responses to adenosine were potentiated in the absence of the epithelium.

While it is generally assumed that relaxation and contraction of airways in response to adenosine agonists *in vitro* are mediated directly via G protein–coupled receptors linked to adenylate cyclase and inositol phosphate turnover, respectively, evidence has also been obtained to suggest that indirect mechanisms may be involved. For example, Yamamoto et al. (1991) and Candenas et al. (1992) observed that contraction of guinea pig trachea to R-PIA was inhibited by the cyclooxygenase inhibitor, indomethacin; it is possible that the source of the prostanoids was the epithelium. Björck et al. (1992) obtained evidence that adenosine caused the release of histamine and leukotrienes, which are potent bronchoconstrictors, from isolated airways—possibly from mast cells. On the other hand, Kroll et al. (1990) concluded that adenosine did not release the neuropeptide calcitonin gene related peptide (CGRP) to cause contraction of guinea pig perfused lungs.

Nucleotide Agonists

Generally speaking, contractile responses to nucleotides are produced in unstimulated isolated tracheal preparations, whereas relaxant responses are produced after addition of the compounds to contracted tissues (Advenier et al., 1982). Biphasic responses to ATP, consisting of an initial contraction followed by relaxation, also have been seen in guinea pig airways *in vitro* (Kamikawa and Shimo, 1976). In transverse strip preparations of *guinea pig* trachea, or variants of this preparation, adenosine tri-, di-, and monophosphates (ATP, ADP, and AMP) have been reported to evoke relaxation responses of the smooth muscle. These responses were suggested to involve adenosine formed from the degradation of the nucleotides, inasmuch as they were potentiated by dipyridamole and inhibited by adenosine deaminase (Coleman, 1976; Christie and Satchell, 1980; Abe et al., 1983; Satchell, 1984). Studies using 5′-anhydride-substituted ATP analogs to ascertain whether P_2 purinoceptors are involved in relaxation of guinea pig airway smooth muscle have given divergent results. In one laboratory (Christie and Satchell, 1980) APPCP possessed relaxant activity (less potent than adenosine and claimed to be due to enzymatic formation of adenosine), while APCPP was without activity. In another laboratory (Brown and Burnstock, 1981) these two compounds were not relaxant but caused weak contractions. Welford and Anderson (1988), on the other hand, observed that APPNP was not broken down enzymatically, but that it was a more potent relaxant than adenosine, and responses were potentiated by dipyridamole. The authors concluded that a P_2 purinoceptor mediated the relaxant responses. Candenas et al. (1992) observed that APCPP induced contractile responses that were unaffected by 8-phenyltheophylline (8-PT) or indomethacin, but were inhibited by dipyridamole. The authors suggested that the contractile responses were mediated by P_{2X} purinoceptors. It is very difficult to reconcile the diversity of responses to adenine nucleotides that have been reported in guinea pig isolated tracheal strip preparations.

The interactions of nucleotides with airway epithelium may give rise to an indirect component of the smooth muscle response. Aksoy and Kelsen (1994) observed in *rabbit* isolated tracheal strip preparations that ATP, ADP, APPCP, APCPP, APPNP, and adenosine elicited relaxation responses (APPNP, APPCP, and APCPP > adenosine). Neither ATP nor APCPP elicited contraction of uncontracted strips containing or lacking the epithelium. However, epithelium removal attenuated relaxation responses to ATP, ADP, APPCP, and adenosine. Relaxation to ATP but not adenosine was inhibited by reactive blue 2 and indomethacin, suggesting that inhibitory prostanoids had been released from the epithelium in response to activation of a P_2 receptor. Subsequent studies (Aksoy et al., 1995) have shown that ATP stimulated the release of the inhibitory prostaglandin E_2 and, to a lesser extent, the excitatory prostaglandin $F_{2\alpha}$, predominantly from an indomethacin-sensitive epithelial source. The order of potency of a series of analogs indicated that the receptor involved in phospholipase A_2 activation and prostaglandin release by cultured rabbit epithelial explants had attributes associated with P_{2U} purinoceptors (P2Y_2 receptors), i.e., UTP $\geq$ ATP > 2MeSATP $\gg$ APCPP $\geq$ adenosine.

The studies discussed thus far were done using strips prepared from airways. It is not possible in most reports to ascertain whether the preparations contained or lacked the epithelium, or whether the epithelium had been damaged.

Strip preparations of airways contain an epithelial layer on one of the four exposed surfaces. Applied agents have access to both sides of the epithelium, to the smooth muscle, and to other elements in the wall, where they can initiate responses indirectly and/or directly. To examine polarity in the effects of purinoceptor stimulation on the airways, the isolated, perfused trachea of the guinea pig (Fedan et al., 1990a,b; Fedan and Frazer, 1992) has been employed to apply agents separately to the mucosal or serosal surfaces (Fedan et al., 1993a,b, 1994a,b). In this preparation, methacholine and histamine are more active as contractile agonists when applied to the outside, serosal bath (where there is no hindrance to the access of the agents to the smooth muscle) than when applied to the perfusate, mucosal compartment (where the agents must diffuse across the epithelium to reach the smooth muscle) (Fedan and Frazer, 1992). In contrast to methacholine and histamine, ATP and UTP were more potent contractile agonists when they were applied to the *mucosal* surface of the trachea than when they were applied to the serosal surface. This is a reflection of the involvement of epithelial mediators in the development of the contractile responses. Three lines of evidence suggested that ATP and UTP interact with separate receptors to cause contraction of the perfused trachea. First, responses to ATP were abolished by cyclooxygenase inhibition with indomethacin, while those to UTP were resistant to indomethacin. Second, contractions to ATP were inhibited by epithelium removal while responses to UTP were potentiated. Third, when APPCP was used to cause desensitization, contractile responses to ATP were inhibited while those to UTP were not. In addition to contraction, some nucleotides caused relaxation of the trachea when used in concentrations higher than those that caused contraction. While APPCP itself did not cause contraction, as was the case in rabbit tracheal strips (Aksoy et al., 1994), it was a more potent relaxant than ATP (Fedan et al., 1993b); UTP did not cause relaxation. In smooth muscle, APPCP is typically a more potent P_{2X} receptor contractile agonist than ATP, and a less active P_{2Y} receptor relaxant agonist than ATP (Abbracchio et al., 1993). Relaxation of the perfused trachea in response to ATP and APPCP was unaffected by indomethacin or 8-PT. Therefore, the relaxant effects of the nucleotides in the guinea pig perfused trachea differ in many respects from those obtained in rabbit tracheal strips. It is possible that nitric oxide may have mediated the relaxation responses (Hassséssian and Burnstock, 1995); however, while ATP increased mRNA levels for inducible nitric oxide synthase in murine cultured macrophages (Tonetti et al., 1995), recent evidence has suggested that ATP does not cause the release of nitric oxide from cultured tracheal epithelium (Tamaoki et al., 1995).

The relative contractile activities of purine and pyrimidine compounds in the guinea pig isolated perfused trachea were assessed in detail (Fedan et al., 1994a). When applied to the serosal surface of the preparation, the order of activity (2MeSATP = ADP > ADPβS = ATP = ATPγS > APPNP = APCPP > UTP = UDP = ITP > XTP $\gg$ APPCP) resembled some characteristics of P_{2Y} purinoceptors, perhaps of the cloned $P2Y_1$ subtype.[1] On the other hand, the order of activity for the agents applied to the mucosal surface (ATP = UTP = ITP > ATPγS = ADP = APPNP = 2MeSATP > UDP = ADPβS = XTP > APCPP) resembled that for P_{2U} purinoceptors, or cloned $P2Y_2$ receptors. The presence of other receptors on the apical and/or basolateral membranes may have given rise to the deviations of the potency series from the predicted ones for $P2Y_1$ and $P2Y_2$ receptors. For example, the high potency of ADP when added to the serosal surface could indicate that a population of $P2Y_5$ receptors participated in the responses (Fedan et al., 1994a). Differences in the metabolism of the compounds and in their diffusion rates across the epithelium could have contributed to deviations from predicted potency order. Metabolism probably does not explain the differences between the 5′-anhydride-substituted analogs and ATP, because ATP was more potent than the substituted analogs on the mucosal and serosal surfaces.

Nevertheless, there is a strong similarity between the orders of potency for the effects of serosally applied nucleotides on short-circuit current responses of human nasal epithelium and the contractile responses of guinea pig perfused trachea; likewise, the orders of potency for the mucosal application of nucleotides for short-circuit current responses of nasal epithelium and perfused trachea contraction were also in very good agreement (Mason et al., 1991; Fedan et al., 1994a). This suggests the possibility that the population of receptors for nucleotides in human and *perfused* guinea pig airways, but not strips of guinea pig airways, are similar.

Responses of the guinea-pig perfused trachea to nucleotides involve Na^+ and Cl^- channels. Responses to both ATP and UTP were inhibited by the Na^+ channel blocker amiloride and the Cl^- channel blocker, 4,4′-diisothiocyano-stilbene-2,2′-disulfonic acid (DIDS; Fedan et al., 1993a). While relaxation responses to UTP were not observed, those to ATP were unaffected by DIDS or indomethacin (Fedan et al., 1993b). Taken together, the sensitivity of contractile responses to the ion channel blockers and to epithelium removal would suggest that ATP initiates cross-talk between ion channel and prostanoid production pathways. Ion channels, but not prostanoids, appear to

[1] ADPBS = adenosine 5′-*O*-(2-thiodiphosphate); ITP = inosine 5′-triphosphate; XTP = xanthosine 5′-triphosphate; ATPγS = adenosine 5′-*O*-(3-thiotriphosphate).

mediate contractile responses to UTP, which are resistant to indomethacin and potentiated in the absence of the epithelium.

Using selective application of amiloride and DIDS to the mucosal or serosal surfaces of perfused trachea, evidence was obtained that apical Na^+ channels mediate responses to ATP applied to the mucosal or serosal surfaces. DIDS applied mucosally inhibited responses to mucosally applied ATP, while serosally applied DIDS did not. (Amiloride inhibited responses of the smooth muscle in epithelium-denuded tracheas to ATP, whereas DIDS resulted in potentiation of the contractions). Thus, while responses to ATP reflect the complex interactions of several cell types and polarity in the effects of amiloride and DIDS, many of which are not described here, the results suggest that apical Na^+ and Cl^- channels in epithelium are involved in contraction of the guinea pig perfused trachea to ATP.

EPITHELIUM

Many functions of the respiratory epithelium are affected by nucleosides and nucleotides. These include ion transport and fluid balance, the production of mucus and surfactant, and mucociliary clearance. The respiratory epithelium, a heterogenous cell layer throughout the lungs, undergoes functional as well as morphological transitions from the nasal cavity to the alveoli related to the types and frequencies of the cells found in different levels of the airways. Other than studies involving cultured epithelial cells of a single type, much of what is known regarding nucleoside and nucleotide effects on signaling in epithelium has been obtained on epithelial preparations containing multiple cell types.

Ion Transport

Application of ATP and UTP to the apical surface of human, rabbit, dog, and mouse respiratory epithelium stimulated apical Cl^- secretion, which was measurable as an increase in short-circuit current (I_{sc}) across the epithelium (Mason et al., 1991; Clarke et al., 1992; Koslowsky et al., 1994; Benali et al., 1994; Haas and McBrayer, 1994; Van Scott et al., 1995; Satoh et al., 1995; Crack et al., 1995). This response is mediated via P_{2U} receptors. In human nasal epithelium, application of the nucleotides to the basolateral membrane also activated I_{sc}, via a receptor with a rank order of potency different from that of the P_{2U} series seen after apical membrane application (Mason et al., 1991).

Addition of ATP to the basolateral surface of rabbit cultured Clara cells did not activate Cl^- conductance (Van Scott et al., 1995). Apical P_{2U} receptor stimulation of Clara cells stimulated HCO_3^- secretion (Van Scott et al., 1995). Such responses may reflect the endogenous autocrine effects on apical P_{2U} receptors of cellular ATP, which has been transported out of the cell via cystic fibrosis transmembrane regulator (CFTR) (Schwiebert et al., 1995).

The activation of G protein–coupled P_2 receptors in respiratory epithelium results in an activation of phospholipase A_2 and phospholipase C, inositol phosphate (IP_3) formation, and the release of Ca^{2+} from intracellular storage sites (human nasal epithelium, Mason et al., 1991; human CF/T43 cells, Brown et al., 1991; human BEAS39 cells, Lazarowski et al., 1994). Differences in coupling to G proteins or other aspects of cell signaling after apical versus basolateral application of the nucleotides, which is a possibility in view of the observed differences in receptor population(s) in nasal epithelial cells (Mason et al., 1991), have not been explored to date.

Activation with ATP of purinoceptors localized on both apical and basolateral membranes activates phospholipase A_2 associated with that membrane and elevates intracellular Ca^{2+} in localized, polarized domains of the cell confined to apical or basolateral plasma membrane regions of the cell and associated with that receptor (Paradiso et al., 1995); greater efficacy of response was associated with basolateral receptor activation. On the other hand, application of ATP to the apical membrane also has been reported to lead to an elevation of intracellular Ca^{2+}, $[Ca^{2+}]_i$, that spreads as a "wave" of increasing $[Ca^{2+}]_i$ to adjacent cells, perhaps mediated via the movement of IP_3 through gap junctions (Hansen et al., 1993).

Activation of apical P_{2U} receptors with ATP and UTP, and the resulting Cl^- efflux from dog tracheal epithelial cells, leads to stimulation of basolateral membrane Na^+–K^+–Cl^- cotransport, as evidenced by increased binding of bumetanide, an inhibitor of this cotransporter (Haas et al., 1993; Haas and McBrayer, 1994).

With regard to adenosine's effects on airway epithelial ion channel activity, Mason et al. (1991) observed that the nucleoside induced small (5%–10%) increases in I_{sc} in human nasal epithelium. Consonant with this observation was the finding that adenosine had no effect on phospholipase A_2 activation (Lazarowski et al., 1994). On the other hand, Lazarowski et al. (1992) demonstrated that adenosine acting via apical and basolateral A_2 receptors elevated cyclic AMP levels in human airway epithelium, which resulted in Cl^- secretion. Chao et al. (1994) also suggested, on the basis of experiments using a cystic fibrosis airway epithelial cell line (CFPEo−), that Cl^- secretion is promoted by A_2 receptor activation accompanying elevation in $[Ca^{2+}]_i$; curiously, adenosine was more potent than ATP in their studies. A_1 receptors have been localized in human airway epithelium with molecular biology techniques, and patch clamp experiments in 9HTEo− tracheal epithelial cells have suggested, based on the effects of the antagonist DPCPX on a membrane-permeable cyclic AMP analog, that antagonism of A_1 receptors increases cyclic AMP-associated Cl^- conductance (McCoy et al., 1995). These latter findings suggest that adenosine arising from the degradation of ATP by ectonucleotidases could modulate the nucleotide's secretory effect or epithelial secretory function.

Mucus Secretion

Mucins are secreted by mucous cells of submucosal glands in the airways and goblet cells in the respiratory epithelium. The mucus lining of the airways and its movement by the ciliary escalator are important defense mechanisms for clearing foreign materials from the lungs.

Adenosine triphosphate and UTP stimulate the secretion of mucus from human (Lethem et al., 1993), dog (Davis et al., 1992), and hamster (Kim and Lee, 1991; Kim et al., 1993) airway epithelium and human cultured tracheobronchial glands (Merten et al., 1993). The receptor(s) involved have not been identified unequivocally, but it appears that P_{2U} purinoceptors ($P2Y_2$ receptors) may be involved, at least in human cells, based on comparable secretory potencies of ATP and UTP and the lesser potency of 2MeSATP (Kim and Lee, 1991; Lethem et al., 1993; Merten et al., 1993). Receptors for the nucleotides may be present on both apical and basolateral membranes of goblet cells; however, it has been suggested that different receptors are present on each side of goblet cells or that multiple receptor types are present (Kim et al., 1994). For example, Davis et al.

(1992) observed that application of ATP to the apical surface of the cells stimulated a biphasic burst of secretion, and that ADP was ineffective. Basolateral application of ATP gave complex secretory responses, ADP was effective, and reactivity was variable between cells for both nucleotides. Evidence has been obtained suggesting that the secretory response to ATP and UTP is coupled to Ca^{2+} entry, $[Ca^{2+}]_i$ release and G protein–coupled phospholipase C activation (Merten et al., 1993; Kim et al., 1993).

Kim et al. (1994) observed saturable, reversible binding of $[^{35}S]ATP\gamma S$ to two sites on cultured hamster tracheal epithelial cells. They observed good correlation between competition for binding versus mucin secretory activity among a panel of analogs (ATP > ADP > APCPP > APPCP) but not with 2MeSATP. They noted further that UTP, while similar in secretory potency to ATP, was weaker than ATP at competing for $[^{35}S]ATP\gamma S$ binding. While these findings support the authors' conclusion that multiple receptors may mediate the secretory response, an alternative explanation for the failure of UTP to interfere with $[^{35}S]ATP\gamma S$ binding is that one of the sites labelled with radioactive sulfur $[^{35}S]$thiophosphate is an ectokinase (Steinberg et al., 1990), analogous to one which uses $ATP\gamma S$ as a substrate in transducing contractile responses of smooth muscle (Fedan and Lamport, 1990; Lamport-Vrana et al., 1991).

Adenosine does not appear to play an important role in mucin secretion. The nucleoside was inactive in hamster goblet cells (Kim and Lee, 1991). In canine epithelium, adenosine was inactive when applied to the apical side of the cells but stimulated secretion when added to the basolateral surface (Davis et al., 1992). It is not clear from any of these studies whether separate adenosine receptors were involved in these responses, or if adenosine acted on P_2 receptors. Nevertheless, the cells appear to be polarized with respect to responsiveness to adenosine.

Surfactant Secretion

Surfactant is synthesized, stored and secreted by type II alveolar epithelium and is a mixture largely of lipids (phosphatidylcholine) with smaller amounts of protein (surfactant protein A). Surfactant reduces surface tension at the air–fluid interface in the alveolus, and therefore, plays an important role in the compliance of the lung during inspiration and expiration.

Multiple regulated mechanisms may be involved in surfactant synthesis and secretion. For nucleotides, these may include stimulation of protein kinase C and its translocation from cytosol to membrane (Chander et al., 1995), stimulation of phospholipase D, activation of adenylate cyclase, elevation of $[Ca^{2+}]_i$ and incompletely understood calmodulin-dependent mechanisms (Rice, 1990; Rice et al., 1990; Griese et al., 1991a, 1993b; Rooney and Gobran, 1993; Wali et al., 1993). The potent stimulatory effect of ATP on surfactant production by type II cells has been known for at least a decade (Gilfillan et al., 1986; Rice and Singleton, 1986; Gilfillan and Rooney, 1987). It was suggested on the basis of an antagonistic effect of reactive blue 2 and on the potency series in early studies (Rice and Singleton, 1987, 1989; Rice, 1990) that the phospholipid secretory response to ATP was mediated by P_{2Y} receptors. Interestingly, cibacron blue, a putative P_{2Y} receptor antagonist in some systems, did not antagonize ATP-induced secretory responses (Griese et al., 1992). A number of investigators have more recently attempted to characterize the receptor(s) that mediate the surfactant secretion response in type II cells. The characterization is not yet complete, as evidenced by some occasional inconsistencies in the results when compared to the accepted receptor criteria. Gobran et al. (1994) observed recently that, with respect to phospholipid secretion by rat isolated type II cells, ATP and UTP were equipotent (both more potent than 2MeSATP and ADP) and that responses to both nucleotides were coupled to phospholipase C, elevation of IP_3, and activation of phospholipase D; these agents were not coupled to adenylate cyclase, in contrast to β adrenoceptor–stimulated responses. These results suggested that both nucleotides activate secretion via a P_{2U} purinoceptor, a view that was supported by the evidence (Gobran et al., 1994) that P_{2U} receptor mRNA was expressed in rat type II cells. Of interest is the observation (Warburton et al., 1989; Griese et al., 1991b; Gobran et al., 1994) that the adenosine antagonists xanthine amine congener (XAC) and 8-PT inhibited phosphatidylcholine secretion and phospholipase D activation in response to ATP, UTP, and other ATP analogs. These antagonisms were not considered to result from activation of adenylate cyclase by the two nucleotides. These inhibitory effects of adenosine receptor antagonists on nucleotide-induced responses are reminiscent of the attributes of a putative P_3 receptor (Dalziel and Westfall, 1994).

While it is generally agreed that nucleotides stimulate surfactant phospholipid synthesis and secretion, the effects of ATP on the synthesis and secretion of surfactant protein A are not clear. On the one hand, it was demonstrated that surfactant protein A synthesis is stimulated in rat lung type II cells by ATP (Wali et al., 1993) and in human adenocarcinoma cells by ATPγS (Pryhuber et al., 1990). On the other hand, Rooney et al. (1993) did not observe secretion of surfactant protein A after rat type II cells were stimulated with ATP in combination with other agonists. Further work will be needed to answer the question of whether the synthesis of surfactant phospholipid and protein are regulated independently or identically.

Adenosine receptors appear to influence surfactant secretion (Rooney, 1990) via coupling to adenylate cyclase, although the nucleoside has been observed generally to be less potent as a secretagogue than ATP, UTP, and related analogs. Stimulation of A_2 receptors with NECA increased phospholipid secretion in concert with elevation in cAMP (cyclic AMP) level (Griese et al., 1993a). Cibacron blue, the putative P_{2Y} receptor antagonist, which did not inhibit ATP-induced phospholipid secretion responses (*vide supra*), did potentiate responses to NECA (Griese et al., 1992). The significance of this finding is unclear. In contrast, activation of adenylate cyclase–coupled A_1 receptors leads to inhibition of surfactant secretion by type II cells (Gobran and Rooney, 1990). A_2 receptor–stimulated secretory responses appear not to involve IP_3 but are inhibited by the standard adenosine receptor antagonists (Warburton et al., 1989; Griese et al., 1991b). The further characterization of adenosine receptor subtypes in phospholipid and surfactant protein production are fertile areas for exploration. However, the finding that 8-Br-cAMP stimulated surfactant protein A synthesis could suggest that A_2 receptors regulate this mechanism (Wali et al., 1993).

Cilia

The synchronous, wavelike beating of cilia on the apical surface of respiratory epithelial cells is required to move mucus and debris up the mucociliary escalator. Nucleotides and nucleosides affect ciliary beat frequency (CBF) and ciliary beat amplitude (CBA) of upper and lower airway epithelial cells. In combination with their effects on mucus secretion, modification of cilia function by these substances would constitute a dual and coordinated approach for modification of the mucociliary escalator in diseases such as primary ciliary dyskinesia (de Iongh et al., 1995; Rayner et al., 1995) and cystic fibrosis (Knowles et al., 1995). This area of research is in its relative infancy.

Forrest et al. (1979) and Lewis et al. (1983) observed that human nasal epithelial cilia from patients with "immobile" cilia could be activated by ATP. Levrier et al. (1989) reported later that topical (i.e., luminal or apical) application of ATP to guinea pig nasal epithelium *in vivo* stimulated CBF. ATP produced a mucociliary stimulation both in tracheal explants *in vitro* and *in vivo* after intravenous (IV) injection (Saano et al., 1990, 1992). The fact that cilia were stimulated after IV administration of ATP could suggest that receptors coupled to ciliary activity may exist on the basolateral surface of the cells. Comparable findings have been made using human nasal and tracheal ciliated cells in culture (Clary-Meinesz et al., 1992; Yoshitsugu et al., 1993), and also after cultured human nasal epithelium were exposed to the CBF- and CBA-depressing effects of bacterial endotoxin (Rautiainen et al., 1994).

Insight into the receptors involved in the effects of purine and pyrimidine agonists on ciliary activity is beginning to evolve. Rautiainen et al. (1994) observed that stimulation of human airway cilia by ATP was not antagonized by 8-PT. In a novel series of experiments involving laser light scattering analysis along with inhalation delivery of a series of compounds, Wong and Yeates (1992) observed that ATP, GTP, APPNP, and GPPNP stimulated CBF of canine tracheal epithelium. The nature of this preparation does not lend itself to rigorous comparison of relative potency. However, the authors' results suggest that all four nucleotides had comparable potency. While it is not necessarily surprising that the adenine compounds behaved similarly, the effectiveness of GTP and GPPNP in this system should prompt further study of the effects of UTP. Inasmuch as it is quite rare for GTP activity to equal that of ATP, it is possible that GTP acting through a novel receptor could be a relatively specific, cilia-stimulating agonist.

A P_2 purinoceptor mediating the effects of nucleotides on *nonrespiratory* mucociliary epithelium (Gheber et al., 1995) could provide clues into the nature of the receptor(s) linked to ciliary activity in airway epithelium. For stimulation of CBF, the order of potency was: ADPβS, UTP $\geq$ 2MeSATP, GTPγS, APPNP, ATP $\geq$ ADP $>$ APPCP, this order would tend to rule out the involvement of $P2Y_1$ receptors, but could suggest the involvement of $P2Y_2$ or $P2Y_6$ receptors. In their ability to inhibit the photolabelling of two proteins (46 and 96 kDa) by 3′-*O*-(benzoyl)benzoyl adenosine 5′-triphosphate (BzATP), the nucleotides displayed differing orders of potency: for p46 the order was ATP $\approx$ ADP $>$ GTPγS $>$ ADPβS, UTP, 2MeSATP, APPNP $>$ APPCP; for p96 the order was ADP $\approx$ ADPβS $\geq$ ATP $\gg$ APPCP, APPNP $>$ GTPγS $\geq$ 2MeSATP, UTP. These findings suggested that interactions with p46 might reflect receptor interactions associated with ciliary stimulation activity, whereas p96 could represent a different receptor ($P2_{Y5}$?) or an ectonucleotidase. It should be noted, again, that the GTP analog, GTPγS, had high potency at p46 protein. The authors probed the divalent cation–dependence of these interactions and concluded that interactions with p46 required bidentate-complex agonists, whereas at p96 nucleotides were active as uncomplexed species. This is consistent with nucleotides acting as tetrabasic agonists at P_{2X} receptors (Fedan et al., 1990a), as well as at P_{2U} receptors (Lustig et al., 1992) and P_{2Z} receptors (Steinberg and Silverstein, 1987). At least superficially, the potency order for cilia stimulatory activity bears some resemblance to the activity of the compounds as mucus secretagogues (*vide supra*).

There are large gaps in our understanding of the signaling mechanisms by which extracellular nucleotides stimulate ciliary activity in respiratory epithelium. Stimulation of CBA in rabbit tracheal epithelium by ATP was reported to be accompanied by entry of Ca^{2+} from extracellular stores (Korngreen and Priel, 1994).

Little is known about the role of adenosine receptors in the regulation of ciliary activity. In cultured rabbit tracheal epithelium, ciliary activity and intracellular cyclic AMP were depressed by adenosine, the effects were inhibited by 8-PT, and ciliary responses were characterized by a potency order (CHA > PIA > adenosine > NECA), all of which indicated that A_1 receptors mediated the response (Tamaoki et al., 1989).[2] In contrast, Wong and Yeates (1992) observed, in *in vivo* experiments on canine trachea, that while aerosolized adenosine itself caused a modest activation of cilia, adenosine also inhibited responses to aerosolized ATP, APPNP, GTP, and GPPNP. Interestingly, guanosine itself also provoked a mild ciliary response, but in the presence of guanosine the response to GPPNP was blunted. It is not clear whether the effects of adenosine and guanosine were mediated via adenosine receptors. These two studies suggest that there may be species differences in the effects of adenosine on ciliary activity. In nonrespiratory ciliated epithelium (Gheber et al., 1995), adenosine had the least or penultimate potency in the three rankings of a large series of analogs for ciliary stimulation and interaction with two putative receptors for nucleotides (*vide supra*).

METABOLISM

Most, if not all, mammalian cells are capable of degrading extracellular nucleotides through the actions of divalent cation–dependent ecto-5′-phosphohydrolases, including ecto-ATPase and ecto-5′-nucleotidase, which give rise ultimately to adenosine (see Plesner, 1995, for review). Very few studies have characterized biochemically these enzyme systems in the various cell types present in the airways and lungs, although their existence has been implied on the basis of the actions of inhibitors, and on the basis of comparisons of the effects of nucleotides that are or are not substrates for the enzymes, i.e., ATP versus APPCP.

After perfusion through the lungs, ATP is degraded rapidly into adenosine (see Plesner, 1995), and it is likely based on the efficiency and rapidity with which degradation takes place that endothelial ecto-ATPase and ecto-5′-nucleotidase enzymes metabolize ATP sequentially into ADP, AMP, and adenosine through a series of regulated enzyme steps (Ryan et al., 1985; Slakey et al., 1990). An ADPase also has been reported in rat lungs (Crutchley et al., 1978), and this degrades perfused ADP into AMP, adenosine, and inosine (Bakhle and Chelliah, 1983a,b).

Recently an ATP-diphosphohydrolase in bovine lung microsomes was described that degrades both ATP and ADP (Picher et al., 1993). Picher et al. (1994) have determined subsequently that the ATPDase is a plasma membrane ectoenzyme in airway smooth muscle. Ecto-ATPDase could play a significant role in regulating the effects of nucleotides in the lung, because it is relatively nonselective with regard to substrate, and catabolizes ATP (relative activity, 1.0) and UTP (0.8), as well as ADP (0.7) and UDP (0.5). Whether this enzyme is localized on other cell types, such as epithelium, mast cells, blood vessels, inflammatory cells, and so on, has yet to be determined.

The Cl^- secretory response of rabbit tracheal epithelial cells to UTP was potentiated by an ATP analog (FPL 67156) that inhibited the ecto-ATPase activity of human blood; in contrast, the secretory response to ATPγS, which is a substrate for kinases but not ATPases, was not potentiated (Crack et al., 1995). These findings suggested that epithe-

[2] CHA = N^6-cyclohexyladenosine; PIA = R(-)N^6-(2-phenylisopropyl)adenosine.

lial ecto-ATPase activity had been inhibited, and that this enzyme can affect nucleotide potency, especially after inhalation. With the availability of ecto-phosphohydrolase inhibitors such as FPL 67156 and periodate-oxidized ATP (Murgia et al., 1993; Fedan and Lamport, 1990; Lamport-Vrana et al., 1991; Fedan and Grant, 1995), the role of these enzymes in the responses of pulmonary cells to nucleotides will be better defined.

ASTHMA

Asthma is a disease characterized by complex pathophysiological alterations in the airways that lead to pulmonary obstruction. The alterations intensify with the severity of the disease and include, but are not limited to infiltration into the airways of several types of inflammatory cells; bronchoconstriction due to the contraction of airway smooth muscle; nonspecific hyperreactivity of the airways to inhaled provocative agents such as methacholine, histamine, leukotrienes, and so on; a widespread sloughing of epithelial cells; and mucus hypersecretion and the formation of mucus plugs. Fifty or more cell-derived mediator substances may be involved in the etiology of these changes.

Theophylline, which is a weak adenosine receptor antagonist, has been a mainstay in the therapeutic management of asthma. The precise mechanisms of its therapeutic action are not fully understood, but could involve two potential mechanisms leading to its bronchodilatory and anti-inflammatory effects: inhibition of cyclic AMP phosphodiesterase, and antagonism of adenosine receptors (Cushley et al., 1983, 1984; Holgate et al., 1993; Persson, 1986; Feoktistov and Biaggioni, 1995; Bowden and McDonald, 1995). Which of these effects predominates at therapeutic blood levels is unresolved, although adenosine receptor antagonism in patients is probably greater than is phosphodiesterase inhibition. Alternatively, phosphodiesterase may be elevated in inflammatory cells of asthmatics, and the effect of theophylline in asthma could involve some anti-inflammatory effects (Banner and Page, 1996).

Aside from these uncertainties, studies on the mechanism of action of theophylline in asthma have led to the intriguing discovery that asthmatics, but not normal individuals, experience pulmonary obstruction upon inhaling adenosine (Cushley et al., 1983; Polosa et al., 1989; Holgate et al., 1993). In addition to adenosine, inhaled AMP and ADP also provoke obstruction in asthmatics, to the same degree as adenosine, and it has been assumed that the response is due to adenosine formed from the degradation of the nucleotides (Mann et al., 1986).

Several pharmacological agents inhibit obstructive responses of asthmatics to inhaled adenosine and AMP (AMP is used alternatively to adenosine because it is more water soluble). These include H_1 histamine receptor antagonists (Phillips et al., 1987, 1989a); the "mast cell stabilizing" agents sodium cromoglycate and nedocromil (Crimi et al., 1988a; Phillips et al., 1989b); the loop diuretics furosemide and bumetanide (Polosa et al., 1993; Rajakulasingam et al., 1994); and the 5-HT2 receptor antagonist, ketanserin (Cazzola et al., 1992). The muscarinic receptor antagonist, atropine, has been reported both to inhibit (Polosa et al., 1991) and not to inhibit (Kung and Diamond, 1990) pulmonary responses to inhaled adenosine. These effects suggest that obstruction after inhalation of adenosine in humans involves the release of histamine, 5-HT, and leukotrienes, which are bronchoconstrictor agonists; and the release of acetylcholine from vagal efferent nerves and neuropeptides from sensory neuronal pathways (Ng et al., 1990; Björck et al., 1992). Inhalation of dipyridamole potentiates the obstructive response to adenosine (Crimi et al., 1988b), while cyclooxygenase inhibition with indomethacin has been suggested to inhibit (Ng et al., 1990) or not to inhibit (Kung and Diamond, 1990) the obstructive response. Many of these effects have been reproduced in laboratory animal models of asthma based on antigen sensitization and challenge (Pauwels et al., 1993; Thorne and Broadley, 1994; Ali et al., 1994). On the other hand, adenosine-induced obstruction also has been induced in guinea pigs after a 1-hour exposure to ozone and this obstruction was accompanied by airway hyperreactivity to inhaled carbachol, a muscarinic agonist (Thorne and Broadley, 1994). Hyperreactivity to bronchoconstrictor agents after short exposure to ozone is associated with neutrophil influx into the lungs and may not involve a substantial mast cell degranulation component (Krishna et al., 1995).

That the obstructive effects of inhaled adenosine probably involve an interaction with mast cells has received support from *in vitro* studies. A number of investigators have suggested that adenosine is capable of both promoting and inhibiting histamine release in a concentration dependent manner (Gilfillan et al., 1990; Peachell et al., 1991; Ott et al., 1992). In addition to histamine, mast cells are a source of cytokines and chemoattractants, which participate in the inflammation that is a central feature of asthma. Activation of A_2 receptors with adenosine in mast cells of the airways leads to activation of adenylate cyclase and phospholipase C; potentiation of mediator release, including interleukin-8 and bronchoactive mediators; and pro-inflammatory and chemotactic signals (Ali et al., 1990; Peachell et al., 1991; Mita et al., 1993; Marquardt and Walker, 1994; Feoktistov and Biaggioni, 1995), perhaps via A_{2B} receptors (Marquardt et al., 1994). An A_3 receptor has also been found on mast cells (RBL-243 cells) (Ji et al., 1994), but its role in asthma has not been clarified. Likewise, a putative A_{2A} receptor with uncertain function has been identified in murine mast cells (Marquardt et al., 1994). Mast cells themselves release adenosine (Marquardt et al., 1993), which could increase reactivity to inhaled methacholine (Rosati et al., 1989). It has been suggested, based on studies using a rabbit antigen model of asthma, that the hyperreactivity of asthmatics to inhaled adenosine reflects the upregulation of A_1 receptors in the lungs (Ali et al., 1994), perhaps on mast cells, which could heighten the potentiating effect of adenosine on mast cell mediator release.

The inhibitory effect of antihistamine and other blocking drugs on adenosine-induced obstruction in asthmatics has been taken to suggest that the response is due to *bronchoconstriction.* However, if the response is thought of as *obstruction,* rather than bronchoconstriction per se, then it is possible that there are other mechanisms by which adenosine could cause a reduction in airway diameter. The inflamed airways of asthmatics contain several cells that are affected by adenosine. If inhaled adenosine is capable of diffusing into and though the mucosa to affect mast cells, it is reasonable to suspect that adenosine also is capable of exerting an effect on other cells in regions proximate to the mast cells. For example, a large number of cytokines and chemoattractants from mononuclear cells and granulocytes release histamine from basophils and mast cells through the action of "histamine releasing factor" (Alam et al., 1994; Kuna et al., 1993). Could adenosine also release such factors from inflammatory cells as intermediate mediators?

The effects of adenosine on nonactivated and activated inflammatory cells have been studied *in vitro* and are primarily inhibitory in nature; this would tend to rule out the possibility that inflammatory cells are involved in the response to inhaled adenosine. For example, rabbit alveolar macrophages contain A_2 receptors (Hasday and Sitrin, 1987). Activation of adenosine receptors inhibits superoxide production by alveolar macrophages (Zeidler and Conley, 1986) and neutrophils (Cronstein et al., 1985). However, stimulation of adenosine receptors on human alveolar macrophages had little effect on opsonized zymosan-stimulated mediator release (Baker and Fuller, 1992).

The effects of adenosine on T cells have not been well characterized. In Jurkat cells, a human T-cell line, activation of A_{2B} receptors elevated cyclic AMP levels (Nonaka et al., 1994).

Activated eosinophils have been reported to release AMP, which, after its conversion to adenosine and action at adenosine receptors, evokes Cl^- secretion from intestinal epithelium (Resnick et al., 1993). It is tempting to speculate that AMP, released from eosinophils in asthmatic airways near smooth muscle, could add to the effects of inhaled adenosine. The release of superoxide anion by human eosinophils is inhibited by activation of A_2 adenosine receptors (Yukawa et al., 1989).

The chronic airway inflammation of asthmatics is not typically accompanied by pronounced neutrophil infiltration. Nevertheless, activation of A_2 receptors inhibits neutrophil activation, the respiratory burst, degranulation, and cytokine secretion (Dianzani et al., 1994; Zhang and Fredholm, 1994; Thiel and Chouker, 1995; Bullough et al., 1995).

In contrast to the inhibitory effects of adenosine, nucleotides generally activate inflammatory cells via P_2 receptors characteristic of the different cells (Di Virgilio, 1995). Examples include macrophages (Hickman et al., 1994; Nakanishi et al., 1994; Falzoni et al., 1995); neutrophils (Yu et al., 1991; Walker et al., 1991; O'Flaherty and Cordes, 1994); T cells (Pizzo et al., 1991; Apasov et al., 1995); and eosinophils (Saito et al., 1991). ATP, via P_{2Z} receptors, releases histamine (Tatham et al., 1988; Gomperts, 1990). The fact that inhaled ADP was equipotent with adenosine in asthmatics (Mann et al., 1986) has been taken as evidence that ADP is broken down to adenosine, which then acts at adenosine receptors to induce obstruction. In inflammatory cells, an examination of the efficacy of ADP at P_2 receptors other than P_{2Z} receptors, which have an obligate preference for ATP, and of the ability of inflammatory cells to generate ATP from AMP via ectoadenylate kinase, are areas of investigation that could better define whether nucleotides play a role in the obstructive response to inhaled adenosine. For example, Biffen and Alexander (1994) have demonstrated that ADP elevated $[Ca^{2+}]_i$ in T cells, and have suggested that a novel receptor is involved, one at which ADP is more potent than ATP.

Lastly, it is well to consider the possibility that obstruction induced by adenosine could involve vasodilation of blood vessels in the airways. Engorgement of vessels in the lamina propria, submucosa, and adventitia, and vascular leakage and edema such as that caused by histamine, can reduce the lumen diameter of airways to give rise to obstruction in the absence of bronchoconstriction (Boucher et al., 1995). Adenosine is a potent relaxant of blood vessels in the lung (McCormack et al., 1989; Szentmiklosi et al., 1995; Pearl, 1994; Neely and Matot, 1996). It is tempting to speculate that adenosine receptors in blood vessels in the lung are among those thought to be upregulated in the asthmatic, but no studies have been done to examine this possibility. For example, the study of Ali et al. (1994) demonstrated the appearance of A_1 receptors in membranes from the lungs of allergic rabbits in conjunction with the development of obstructive responsiveness to inhaled adenosine. Inasmuch as the investigators' lung membranes would have contained elements from vascular smooth muscles as well as other cell types from the lungs, the possibility that upregulation of adenosine receptors had occurred in blood vessels would seem to remain a possibility.

ACKNOWLEDGMENTS

The author thanks Kathleen Fedan for her comments on the manuscript.

REFERENCES

Abbracchio MP, Cattabeni F, Fredholm BB, Williams M (1993): Purinoceptor nomenclature: A status report. Drug Dev Res 28:207–213.

Abe M, Katsuragi T, Furukawa T (1983): Discrimination by dipyridamole of two types of responses to adenosine and ATP of the carbachol-induced contraction of guinea-pig trachea. Eur J Pharmacol 90:29–34.

Advenier C, Bidet D, Floch-Saint-Aubin A, Renier A (1982): Contribution of prostaglandins and thromboxanes to the adenosine and ATP-induced contraction of guinea-pig isolated trachea. Br J Pharmacol 77:39–44.

Advenier C, Devillier P, Matran R, Naline E (1988): Influence of epithelium on the responsiveness of guinea-pig isolated trachea to adenosine. Br J Pharmacol 93:295–302.

Aksoy MO, Borenstein M, Li XX, Kelsen SG (1995): Eicosanoid production in rabbit tracheal epithelium by adenine nucleotides: Mediation by P_2-purinoceptors. Am J Respir Cell Mol Biol 13:410–417.

Aksoy MO, Kelsen SG (1994): Relaxation of rabbit tracheal smooth muscle by adenine nucleotides: Mediation by P_2-purinoceptors. Am J Respir Cell Mol Biol 10:230–236.

Alam R, Kumar D, Anderson-Walters D, Forsythe PA (1994): Macrophage inflammatory protein-1 and monocyte chemoattractant peptide-1 elicit immediate and late cutaneous reactions and activate murine mast cells in vivo. J Immunol 152:1298–1303.

Ali S, Mustafa SJ, Metzger WJ (1994): Adenosine-induced bronchoconstriction and contraction of airway smooth muscle from allergic rabbits with late-phase airway obstruction: Evidence for an inducible A_1 receptor. J Pharmacol Exp Ther 268:1328–1334.

Ali H, Cunha-Melo JR, Saul WF, Beaven MA (1990): Activation of phospholipase C via adenosine receptors provides synergistic signals for secretion in antigen-stimulated RBL-2H3 cells. Evidence for a novel adenosine receptor. J Biol Chem 265:745–753.

Apasov S, Koshiba M, Redegeld F, Sitkovsky MV (1995): Role of extracellular ATP and P1 and P2 classes of purinergic receptors in T-cell development and cytotoxic T lymphocyte effector functions. Immunol Rev 146:5–19.

Baker AJ, Fuller RW (1992): Effect of cyclic adenosine monophosphate, 5′-(N-ethylcarboxyamido)-adenosine and methylxanthines on the release of thromboxane and lysosomal enzymes from human alveolar macrophages and peripheral blood monocytes in vitro. Eur J Pharmacol 211:157–661.

Bakhle YS, Chelliah R (1983a): Metabolism and uptake of adenosine in rat isolated lung and its inhibition. Br J Pharmacol 79:509–515.

Bakhle YS, Chelliah R (1983b): Effect of streptozotocin-induced diabetes on the metabolism of ADP, AMP and adenosine in the pulmonary circulation of rat isolated lung. Diabetologia 24:455–459.

Benali R, Pierrot D, Zahm, de Bentzmann S, Puchelle E (1994): Effect of extracellular ATP and UTP on fluid transport by human basal epithelial cells in culture. Am J Respir Cell Mol Biol 10:363–368.

Banner KH, Page CP (1996): Anti-inflammatory effects of theophylline and selective phosphodiesterase inhibitors. Clin Exp Allergy 26(suppl 2):2–9.

Biffen M, Alexander DR (1994): Mobilization of intracellular Ca^{2+} by adenine nucleotides in human T-leukaemia cells: Evidence for ADP-specific and P_{2Y}-purinergic receptors. Biochem J 304:769–774.

Björck T, Gustafsson LE, Dahlen SE (1992): Isolated bronchi from asthmatics are hyperresponsive to adenosine, which apparently acts by liberation of leukotrienes and histamine. Am Rev Resp Dis 145:1087–1091.

Boucher RC, Gilbert IA, Hogg JC, King M, Knowles MR, Nadel J (1995): Nonmuscular airway obstruction and asthma. Am J Respir Crit Care Med 152:408–410.

Bowden JJ, McDonald DM (1995): 8. The microvasculature as a participant in inflammation. In Holgate ST (ed): "Immunopharmacology of the Respiratory System." New York: Academic Press, pp 147–168.

Brackett LE, Daly JW (1991): Relaxant effects of adenosine analogs on guinea pig trachea in vitro: Xanthine-sensitive and xanthine-insensitive mechanisms. J Pharmacol Exp Ther 257:205–213.

Brown CM, Burnstock G (1981): The structural confirmation of the polyphosphate chain of the ATP molecule is critical for its promotion of prostaglandin biosynthesis. Eur J Pharmacol 69:81–86.

Brown CM, Collis MG (1982): Evidence for an A_2/R_A adenosine receptor in the guinea-pig trachea. Br J Pharmacol 76:381–387.

Brown HA, Lazarowski, Boucher RC, Harden TK (1991): Evidence that UTP and ATP regulates phospholipase C through a common extracellular 5′-nucleotide receptor in human airway epithelial cells. Mol Pharmacol 40:648–655.

Bullough DA, Magill MJ, Firestein GS, Mullane KM (1995): Adenosine activates A_2 receptors to inhibit neutrophil adhesion and injury to isolated cardiac myocytes. J Immunol 155:2579–2586.

Candenas ML, Devillier P, Naline E, Advenier C (1992): Contractile effect of α,β-methylene ATP on the guinea-pig isolated trachea. Fundam Clin Pharmacol 6:135–144.

Caparrotta L, Cillo F, Fassina G, Gaion RM (1984): Dual effect of (−)-N^6-phenylisopropyladenosine on guinea-pig trachea. Br J Pharmacol 83:23–29.

Cazzola M, Matera MG, Santangelo G, Assogna G, D'Amato G, Rossi F, Girbino G (1992): Effect of the selective 5-HT2 antagonist ketanserin on adenosine-induced bronchoconstriction in asthmatic subjects. Immunopharmacology 23:21–28.

Chander A, Sen N, Wu AM, Spitzer AR (1995): Protein kinase C in ATP regulation of lung surfactant secretion in type II cells. Am J Physiol 268:L108–L116.

Chao AC, Zifferblatt JB, Wagner JA, Dong YJ, Gruenert DC, Gardner P (1994): Stimulation of chloride secretion by P_1 purinoceptor agonists in cystic fibrosis phenotype airway epithelial cell line CDPEo-. Br J Pharmacol 112:169–175.

Christie J, Satchell DG (1980): Purine receptors in the trachea: is there a receptor for ATP? Br J Pharmacol 70:512–514.

Clarke LL, Burns KA, Bayle J-Y, Boucher RC, Van Scott MR (1992): Sodium- and chloride-conductive pathways in cultured mouse tracheal epithelium. Am J Physiol 263:L519–L525.

Clary-Meinesz CF, Cosson J, Huitorel P, Blaive B (1992): Temperature effect on the ciliary beat frequency of human nasal and tracheal ciliated cells. Biol Cell 76:335–338.

Coleman RA (1976): Effects of some purine derivatives on the guinea-pig trachea and their interaction with drugs that block adenosine uptake. Br J Pharmacol 57:51–57.

Crack BE, Pollard CE, Beukers MW, Roberts SM, Hunt SF, Ingall AH, McKechnie KC, IJzerman AP, Leff P (1995): Pharmacological and biochemical analysis of FPL 67156, a novel, selective inhibitor of ecto-ATPase. Br J Pharmacol 114:475–481.

Crimi N, Palermo F, Vancheri C, Oliveri R, Distefano SM, Polosa R, Mistretta A (1988a): Effect of sodium cromoglycate and nifedipine on adenosine-induced bronchoconstriction. Respiration 53:74–80.

Crimi N, Palermo F, Oliveri R, Maccarrone C, Palermo B, Vancheri C, Polosa R, Mistretta A (1988b): Enhancing effect of dipyridamole inhalation on adenosine-induced bronchospasm in asthmatic patients. Allergy 43:179–183.

Cronstein BN, Kramer SB, Rosenstein ED, Weissmann G, Hirschhorn R (1985): Adenosine modulates the generation of superoxide by stimulated human neutrophils via interaction with a specific cell surface receptor. Ann NY Acad Sci 451:291–301.

Crutchley DJ, Eling TE, Anderson MW (1978): ADPase activity of isolated perfused lung. Life Sci 22:1413–1420.

Cushley MJ, Tattersfield AE, Holgate ST (1983): Inhaled adenosine and guanosine on airway resistance in normal and asthmatic subjects. Br J Clin Pharmacol 15:161–165.

Cushley MJ, Tattersfield AE, Holgate ST (1984): Adenosine-induced bronchoconstriction in asthma. Antagonism by inhaled theophylline. Am Rev Resp Dis 129:380–384.

Dalziel HH, Westfall DP (1994): Receptors for adenine nucleotides and nucleosides: Subclassification, distribution, and molecular characterization. Pharmacol Rev 46:449–466.

Davis CW, Dowell ML, Lethem M, Van Scott M (1992): Goblet cell degranulation in isolated canine tracheal epithelium: Response to exogenous ATP, ADP, and adenosine. Am J Physiol 262:C1313–C1323.

de Iongh RU, Rutland J (1995): Ciliary defects in healthy subjects, bronchiectasis, and primary ciliary dyskinesia. Am J Respir Crit Care Med 151:1559–1567.

Devillier P, Candenas ML, Naline E, Advenier C (1992): Influence of benzodiazepines on the response of the guinea-pig isolated trachea to the contractile action of adenosine. Eur J Pharmacol 214:67–74.

Dianzani C, Brunelleschi S, Viano, I, Fantoozzi R (1994): Adenosine modulation of primed human neutrophils. Eur J Pharmacol 263:223–226.

Di Virgilio F (1995): The P2Z purinoceptor: an intriguing role in immunity, inflammation and cell death. Immunol Today 16:524–528.

Falzoni S, Munerati M, Ferrari D, Spisani S, Moretti S, Di Virgilio F (1995): The purinergic P_{2Z} receptor of human macrophage cells. Characterization and possible physiological role. J Clin Invest 95:1207–1216.

Farmer SG, Hay DWP (eds) (1991): "The Airway Epithelium. Physiology, Pathophysiology, and Pharmacology." New York: Marcel Dekker.

Farmer SG, Canning BJ, Wilkins DE (1988): Adenosine receptor-mediated contraction and relaxation of guinea-pig isolated tracheal smooth muscle: Effects of adenosine antagonists. Br J Pharmacol 95:371–378.

Farmer SG, Fedan JS, Hay DWP, Raeburn D (1986): The effects of epithelium removal on the sensitivity of guinea-pig isolated trachealis to bronchodilator drugs. Br J Pharmacol 89:407–414.

Fedan JS, Frazer DG (1992): Influence of epithelium on the reactivity of guinea-pig isolated, perfused trachea to bronchoactive drugs. J Pharmacol Exp Ther 262:741–750.

Fedan JS, Grant LJR (1995): Potentiating and inhibitory effects of periodate-oxidized ATP analogs on ATP-induced contractions of guinea-pig isolated vas deferens. Eur J Pharmacol 281:213–217.

Fedan JS, Lamport SJ (1990): Two dissociable phases in the contractile response of the smooth muscle of the guinea pig isolated vas deferens to adenosine triphosphate. J Pharmacol Exp Ther 253:993–1001.

Fedan JS, Belt JJ, Yuan L-X, Frazer DG (1993a): Contractile effects of nucleotides in guinea-pig isolated, perfused trachea: Involvement of respiratory epithelium, prostanoids, and Na^+ and Cl^- channels. J Pharmacol Exp Ther 264:210–216.

Fedan JS, Belt JJ, Yuan L-X, Frazer DG (1993b): Relaxant effects of nucleotides in guinea-pig isolated, perfused trachea: Lack of involvement of prostanoids, Cl^- channels and adenosine. J Pharmacol Exp Ther 264:217–220.

Fedan JS, Dagirmanjian JP, Attfield MD, Chideckel EW (1990a): Evidence that the P_{2X}-purinoceptor of the smooth muscle of the guinea pig vas deferens is an ATP^4-receptor. J Pharmacol Exp Ther 255:46–51.

Fedan JS, Hay DWP, Farmer SG, Raeburn D (1988): Modulation of airway smooth muscle reactivity by epithelial cells. In Rodger IW, Barnes PJ, Thomson NC (eds) (1988): "Asthma: Basic Mechanisms and Clinical Management." New York: Academic Press, pp 143–162.

Fedan JS, Nutt ME, Frazer DG (1990b): Reactivity of guinea-pig isolated trachea to methacholine, histamine and isoproterenol applied serosally vs. mucosally. Eur J Pharmacol 190:337–345.

Fedan JS, Stem JL, Day B (1994a): Contraction of the guinea-pig isolated, perfused trachea to purine and pyrimidine agonists. J Pharmacol Exp Ther 268:1321–1327.

Fedan JS, Yuan L-X, Belt JJ, Frazer DG (1994b): Polarized effects of amiloride and 4,4′-diisothiocyanostilbene-2,2′-disulfonic acid on ATP-induced contraction of trachea. Eur J Pharmacol 256:51–56.

Feoktistov I, Biaggoni I (1995): Adenosine A_{2b} receptors evoke interleukin-8 secretion in human mast cells. An enprofylline-sensitive mechanism with implications for asthma. J Clin Invest 96:1979–1986.

Finney MJ, Karlsson JA, Persson CG (1985): Effects of bronchodilators on a novel human small airway preparation. Br J Pharmacol 85:29–36.

Flavahan NA, Aarhus LL, Rimele TJ, Vanhoutte PM (1985): Respiratory epithelium inhibits bronchial smooth muscle tone. J Appl Physiol 58:834–838.

Forrest JB, Rossman CM, Newhouse MJ, Ruffin R (1979): Activation of nasal cilia in immotile cilia syndrome. Am Rev Resp Dis 120:511–515.

Ghai G, Zimmerman MB, Hopkins MF (1987): Evidence for A1 and A2 adenosine receptors in guinea pig trachea. Life Sci 41:1215–1224.

Gheber L, Priel Z, Aflalo C, Shoshan-Barmatz V (1995): Extracellular ATP binding proteins as potential receptors in mucociliary epithelium: characterization using [^{32}P]3′-O-(benzoyl)benzoyl ATP, a photoaffinity label. J Membr Biol 147:83–93.

Gilfillan AM, Rooney SA (1987): Purinoceptor agonists stimulate phosphatidylcholine secretion in primary cultures of adult rat type II pneumocytes. Biochim Biophys Acta 917:18–23.

Gilfillan AM, Hollingsworth M, Jones AW (1986): The pharmacological modulation of [^{3}H]-disaturated phosphatidylcholine overflow from perfused lung slices of adult rats: A new method for the study of lung surfactant secretion. Br J Pharmacol 79:363–371.

Gilfillan AM, Wiggan GA, Welton AF (1990): Pertussis toxin pretreatment reveals differential effects of adenosine analogs on IgE-dependent histamine and peptidoleukotriene release from RBL-2H3 cells. Biochim Biophys Acta 1052:467–474.

Gobran LI, Rooney SA (1990): Adenosine A_1 receptor-mediated inhibition of surfactant secretion in rat type II pneumocytes. Am J Physiol 258:L45–L51.

Gobran LI, Xu ZX, Lu Z, Rooney SA (1994): P_{2U} purinoceptor stimulation of surfactant secretion coupled to phosphatidyl choline hydrolysis in type II cells. Am J Physiol 267:L625–L633.

Gomperts, BD (1990): Exocytosis: The role of Ca^{2+}, GTP and ATP as regulators and modulators in the rat mast cell model. J Exp Pathol 71:423–431.

Griese M, Gobran LI, Rooney SA (1991a): ATP-stimulated inositol phospholipid metabolism and surfactant secretion in rat type II pneumocytes. Am J Physiol 260:L586–L593.

Griese M, Gobran LI, Rooney SA (1991b): A_2 and P_2 purine receptor interactions and surfactant secretion in primary cultures of type II cells. Am J Physiol 261:L140–L147.

Griese M, Gobran LI, Rooney SA (1992): Ontogeny of surfactant secretion in type II pneumocytes from fetal, newborn, and adult rats. Am J Physiol 262:L337–L343.

Griese M, Gobran LI, Rooney SA (1993a): Potentiation of A_2 purinoceptor-stimulated surfactant phospholipid secretion in primary cultures of rat type II pneumocytes. Lung 171:75–86.

Griese M, Gobran LI, Rooney SA (1993b): Signal transduction mechanisms of ATP-stimulated phosphatidylcholine secretion in rat type II pneumocytes: Interactions between ATP and other surfactant secretagogues. Biochim Biophys Acta 1167:85–93.

Haas M, McBrayer DG (1994): Na-K-Cl cotransport in nystatin-treated cells: Regulation by isoproterenol, apical UTP, and $[Cl]_i$. Am J Physiol 266:C1440–C142.

Haas M, McBrayer DG, Yankaskas JR (1993): Dual mechanisms for Na–K–Cl cotransport regulation in airway epithelial cells. Am J Physiol 264:C189–C200.

Hansen M, Boitano S, Dirksen ER, Sanderson MJ (1993): Intercellular calcium signaling induced by extracellular adenosine 5′-triphosphate and mechanical stimulation in airway epithelial cells. J Cell Sci 106:995–1004.

Hasday JD, Sitrin RG (1987): Adenosine receptors on rabbit alveolar macrophages: binding characteristics and effects on cellular function. J Lab Clin Med 110:264–273.

Hassésian H, Burnstock G (1995): Interacting roles of nitric oxide and ATP in the pulmonary circulation of the rat. Br J Pharmacol 114:846–850.

Hickman SE, el Khoury J, Greenberg S, Schieren I, Silverstein SC (1994): The P_{2Z} adenosine triphosphate receptor activity in cultured human monocyte-derived macrophages. Blood 84:2452–2456.

Holgate ST, Polosa R, Phillips, GD (1993): Role of adenosine and adenosine receptors. In Weiss EB, Stein M (eds): "Bronchial Asthma. Mechanisms and Therapeutics." Boston: Little Brown, pp 253–257.

Holroyde MC (1986): The influence of epithelium on the responsiveness of guinea-pig isolated trachea. Br J Pharmcol 87:501–507.

Ji X-D, Gallo-Rodriguez C, Jacobson KA (1994): A selective agonist affinity label for A_3 adenosine receptors. Biochem Biophys Res Commun 203:570–576.

Kamikawa Y, Shimo Y (1976): Mediation of prostaglandin E_2 in the biphasic response to ATP of the isolated tracheal muscle of guinea-pigs. Br J Pharmacol 28:294–297.

Kim KC, Lee BC (1991): P_2 purinoceptor regulation of mucin release by airway goblet cells in primary culture. Br J Pharmacol 103:1053–1056.

Kim KC, Zheng QX, Wilson AK, Lee BC, Berman JS (1994): Binding kinetics of ATPγS^{35} on cultured primary tracheal surface epithelial cells. Am J Respir Cell Mol Biol 10:154–159.

Kim KC, Zheng QX, Van-Seuningen I (1993): Involvement of a signal transduction mechanism in ATP-induced mucin release from cultured airway goblet cells. Am J Respir Cell Mol Biol 8:121–125.

Knight DA, Bai TR, Weir T, Salvatore CA, Jacobson MA, Walker BAM (1996): Expression and function of A3 adenosine receptors (A3-AR) in human lung tissue. Am J Respir Crit Care Med 153:A645.

Knowles MR, Olivier KN, Hohneker KW, Robinson J, Bennett WD, Boucher RC (1995): Pharmacologic treatment of abnormal ion transport in the airway epithelium in cystic fibrosis. Chest 107:71S–76S.

Korngreen A, Priel Z (1994): Simultaneous measurement of ciliary beating and intracellular calcium. Biophys J 67:377–380.

Koslowsky T, Hug T, Ecke D, Klein P, Greger R, Gruenert DC, Kunzelmann K (1994): Ca^{2+}-and swelling-induced activation of ion conductances in bronchial epithelial cells. Pflugers Arch 428:597–603.

Krishna MT, Mudway I, Kelly FJ, Frew AJ, Holgate ST (1995): Ozone, airways and allergic airways disease. Clin Exp Allergy 25:1150–1158.

Kroll F, Karlsson JA, Lundberg JM, Persson CG (1990): Capsaicin-induced bronchoconstriction and neuropeptide release in guinea pig perfused lungs. J Appl Physiol 68:1679–1687.

Krzanowski JJ, Urdaneta-Bohorquez A, Polson JB, Sakamoto Y, Szentivanyi A (1987): Effects of adenosine on canine tracheal smooth muscle tone. Arch Int Pharmacodyn Ther 287:224–236, 1987.

Kuna P, Reddigari SR, Schall TJ, Rucinski D, Sadick M, Kaplan AP (1993): Characterization of the human basophil response to cytokines, growth factors, and histamine releasing factors of the intecrine/chemokine family. J Immunol 150:932–943.

Kung M, Diamond L (1990): Adenosine-induced bronchoconstriction in human asthmatics: Effects of pretreatment with indomethacine or atropine sulfate. Pulmon Pharmacol 3:17–23.

Lamport-Vrana SJ, Vrana KE, Fedan JS (1991): Involvement of ecto-phosphoryl transfer in contractions of the smooth muscle of the guinea-pig vas deferens to ATP. J Pharmacol Exp Ther 258:339–348.

Lazarowski ER, Boucher RC, Harden TK (1994): Calcium-dependent release of arachidonic acid in response to purinergic receptor activation in airway epithelium. Am J Physiol 266:C406–C415.

Lazarowski ER, Mason SJ, Clarke L, Harden TK, Boucher RC (1992): Adenosine receptors on human airway epithelia and their relationship to chloride secretion. Br J Pharmacol 106:774–782.

Lethem MI, Dowell ML, Van Scott M, Yankaskas JR, Egan T, Boucher RC, Davis CW (1993): Nucleotide regulation of goblet cells in human airway epithelial explants: Normal exocytosis in cystic fibrosis. Am J Respir Cell Mol Biol 9:315–322.

Levrier J, Molon-Noblot S, Duval D, Lloyd KG (1989): A new ex vivo method for the study of nasal drops on ciliary epithelium. Fundam Clin Pharmacol 3:471–482.

Lewis FH, Beals TF, Carey TE, Baker SR, Mathews KP (1983): Ultrastructural and functional studies of cilia from patients with asthma, aspirin intolerance, and nasal polyps. Chest 83:487–490.

Lundblad LKA, Persson CGA (1988): The epithelium and the pharmacology of guinea-pig tracheal tone in vitro. Br J Pharmacol 93:909–917.

Lustig KD, Sportiello MG, Erb L, Weisman GA (1992): A nucleotide receptor in vascular endothelial cells is specifically activated by the fully ionized forms of ATP and UTP. Biochem J 284:733–739.

Mann JS, Holgate ST, Renwick AG, Cushley MJ (1986): Airway effects of purine nucleosides in nucleotides and release with bronchial provocation in asthma. J Appl Physiol 61:1667–176.

Marquardt DL, Walker LL (1994): Inhibition of protein kinase A fails to alter mast cell adenosine responsiveness. Agents Actions 43:7–12.

Marquardt DL, Walker LL, Heinemann S (1994): Cloning of two adenosine receptor subtypes from mouse bone marrow-derived mast cells. J Immunol 152:4508–4515.

Marquardt DL, Lwin A, Walker LL (1993): Mast cell desensitization to IgE fails to induce a parallel adenosine receptor desensitization. Agents Actions 40:11–17.

Mason SJ, Paradiso AM, Boucher RC (1991): Regulation of transepithelial ion transport and intracellular calcium by extracellular ATP in normal and cystic fibrosis airway epithelium. Br J Pharmacol 103:1649–1656.

Matera MG, De Santis D, D'Agostino B, Pallotta M, Vacca C, Cazzola M, Rossi F (1995): Role of 5-hydroxytryptamine in mediating adenosine-induced airway constriction. Immunopharmacology 29:73–78.

McCormack DG, Clarke B, Barnes PJ (1989): Characterization of adenosine receptors in human pulmonary arteries. Am J Physiol 256:H41–H46.

McCoy DE, Schwiebert EM, Karlson KH, Spielman WS, Stanton BA (1995): Identification and function of A_1 adenosine receptors in normal and cystic fibrosis human epithelial cells. Am J Physiol 268:C1520–1527.

Merten MD, Breittmayer J-P, Figarella C, Frelin C (1993): ATP and UTP increase secretion of bronchial inhibitor by human tracheal gland cells in culture. Am J Physiol 265:L479–L484.

Mita H, Ishii T, Yamada T, Akiyama K, Shida T (1993): Further characterization of dispersed human sinus mast cells. Life Sci 53:775–782.

Murgia M, Hanau S, Pizzo P, Rippa M, Di Virgilio F (1993): Oxidized ATP. An irreversible inhibitor of the macrophage purinergic P_{2Z} receptor. J Biol Chem 268:8199–8203.

Nakanishi M, Kawasaki M, Ogino H, Yoshida M, Yagawa K (1994): Extracellular ATP regulates the proliferation of alveolar macrophages. Am J Respir Cell Mol Biol 10:560–564.

Neely CF, Matot I (1996): Pharmacological probes for A_1 and A_2 adenosine receptors in vivo in feline pulmonary vascular bed. Am J Physiol 270:H610–H619.

Ng WH, Polosa R, Church MK (1990): Adenosine bronchoconstriction in asthma: Investigations into its possible mechanism of action. Br J Clin Pharmacol 30(suppl 1):89S–98S.

Nonaka H, Ichimura M, Takeda M, Nonaka Y, Shimada J, Suzuki F, Yamaguchi K, Kase H (1994): KF17837 ((E)-8-(3,4-dimethoxystyryl)-1,3-dipropyl-7-methylxanthine), a potent and selective adenosine A_2 receptor antagonist. Eur J Pharmacol 267:335–341.

O'Flaherty JT, Cordes, JF (1994): Human neutrophil degranulation responses to nucleotides. Lab Invest 70:816–821.

Ott I, Lohse MJ, Klotz KN, Vogt-Moykopf I, Schwabe U (1992): Effects of adenosine on histamine release from human lung fragments. Int Arch Allergy Immunol 98:50–56.

Paradiso AM, Mason SJ, Lazarowski ER, Boucher RC (1995): Membrane-restricted regulation of Ca^{2+} release and influx in polarized epithelia. Nature 377:643–646.

Pauwels RA, Joos GF (1995): Characterization of the adenosine receptors in the airways. Arch Int Pharmacodyn Ther 329:151–160.

Pauwels R, Joos GF, Kips JC, Peleman RA (1993): Mechanisms of adenosine-induced bronchoconstriction: An animal model. Drug Dev Res 28:318–321.

Peachell PT, Lichtenstein LM, Schleimer RP (1991): Differential regulation of human basophil and lung mast cell function by adenosine. J Pharmacol Exp Ther 256:717–726.

Pearl RG (1994): Adenosine produces pulmonary vasodilation in the perfused rabbit lung via an adenosine A2 receptor. Anesth Analg 79:46–51.

Persson CGA (1986): Overview of the effects of theophylline. J Allergy Clin Immunol 78:780–787.

Phillips GD, Polosa R, Holgate ST (1989a): The effect of histamine-H1 receptor antagonism with terfenadine on concentration-related AMP-induced in asthma. Clin Exp Allergy 19: 405–409.

Phillips GD, Rafferty P, Beasley R, Holgate ST (1987): Effect of oral terfenadine on the bronchoconstrictor response to inhaled histamine and adenosine 5′-monophosphate in non-atopic asthma. Thorax 42:939–945.

Phillips GD, Scott VL, Richards R, Holgate ST (1989b): Effect of nedocromil and sodium cromoglycate against bronchoconstriction induced by inhaled adenosine 5′-monophosphate. Eur Resp J 2:210–217.

Picher M, Béliveau R, Potier M, Savaria D, Rousseau E, Beaudoin AR (1994): Demonstration of an ectoATP-diphosphohydrolase (E.C.3.6.1.5.) in non-vascular smooth muscles of the bovine trachea. Biochim Biophys Acta 1200:167–174.

Picher M, Côté YP, Béliveau R, Potier M, Beaudoin AR (1993): Demonstration of a novel type of ATP-diphosphohydrolase (EC 3.6.1.5) in the bovine lung. J Biol Chem 268:4699–4703.

Pizzo P, Zanovello P, Bronte V, Di Virgilio F (1991): Extracellular ATP causes lysis of mouse thymocytes and activates a plasma membrane ion channel. Biochem J 274:139–144.

Plesner L (1995): Ecto-ATPases: Identities and functions. Int Rev Cytol 158:141–214.

Polosa R, Holgate ST, Church MK (1989): Adenosine as a pro-inflammatory mediator in asthma. Pulmon Pharmacol 2:21–26.

Polosa R, Phillips GD, Rajakulasingam K, Holgate ST (1991): The effect of inhaled ipratropium bromide alone and in combination with oral terfenadine on bronchoconstriction provoked by adenosine 5′-monophosphate and histamine in asthma. J Allergy Clin Immunol 87:939–947.

Polosa R, Rajakulasingam K, Propsperini G, Church MK, Holgate ST (1993): Relative potencies and time course changes in adenosine 5′-monophosphate airway responsiveness with inhaled furosemide and bumetanide in asthma. J Allergy Clin Immunol 92:288–297.

Pryhuber GS, O'Reilly MA, Clark JC, Hull WM, Fink I, Whitsett JA (1990): Phorbol ester inhibits surfactant protein SP-A and SP-B expression. J Biol Chem 265:20822–20828.

Rajakulasingam K, Polosa R, Church MK, Howarth PH, Holgate ST (1994): Effect of inhaled frusemide on responses of airways to bradykinin and adenosine 5′-monophosphate in asthma. Thorax 49:485–491.

Rautiainen M, Yoshitsugu M, Matsune S, Nuutinen J, Happonen P, Ohyama M (1994): Effect of exogenous ATP and physical stimulation on ciliary function impaired by bacterial endotoxin. Acta Otolaryngol (Stockh) 114:337–340.

Rayner CF, Rutman A, Dewar A, Cole PJ, Wilson R (1995): Ciliary disorientation in patients with chronic upper respiratory tract inflammation. Am J Respir Crit Care Med 151:800–804.

Resnick MB, Colgan SP, Patapoff TW, Mrsny RJ, Awtrey CS, Delp-Archer C, Weller PF, Madara JL (1993): Activated eosinophils evoke chloride secretion in model intestinal epithelia primarily via regulated release of 5′-AMP. J Immunol 151:5716–5723.

Rice WR (1990): Effects of extracellular ATP on surfactant secretion. Ann NY Acad Sci 603:64–74.

Rice WR, Singleton FM (1986): P_2-purinoceptors regulate surfactant secretion from rat isolated alveolar type II cells. Br J Pharmacol 89:485–491.

Rice WR, Singleton FM (1987): P_{2Y}-purinoceptor regulation of surfactant secretion from rat isolated type II cells is associated with mobilization of intracellular calcium. Br J Pharmacol 91:833–838.

Rice WR, Singleton FM (1989): Reactive blue 2 selectively inhibits P_{2Y}-purinoceptor-stimulated surfactant phospholipid secretion from rat isolated type II cells. Br J Pharmacol 97:158–162.

Rice WR, Dorn CC, Singleton FM (1990): P_2-purinoceptor regulation of surfactant phosphatidylcholine secretion. Relative roles of calcium and protein kinase C. Biochem J 266: 407–413.

Rooney SA (1990): Type II cell purinoceptors and surfactant secretion. Prog Respir Res 25:127–135.

Rooney SA, Gobran LI (1993): Activation of phospholipase D in rat type II pneumocytes by ATP and other surfactant secretagogues. Am J Physiol 264:L133–L140.

Rooney SA, Gobran LI, Umstead TM, Phelps DS (1993): Secretion of surfactant protein A from type II pneumocytes. Am J Physiol 265:L586–L590.

Rosati G, Hargreave FE, Ramsdale EH (1989): Inhalation of adenosine 5′-monophosphate increases methacholine airway responsiveness. J Appl Physiol 67:792–796.

Ryan US, Ryan JW, Crutchley DJ (1985): The pulmonary endothelial surface. Fed Proc 44:2603–2609.

Saano V, Muranen A, Nuutinen J, Karttunen P, Silvasti M (1990): Effects of intravenous ATP on tracheal ciliary activity, airway resistance and haemodynamics in rats. Pharmacol Toxicol 66:303–306.

Saano V, Virta P, Joki S, Nuutinen J, Karttunen P, Silvasti M (1992): ATP induces respiratory ciliostimulation in rat and guinea pig in vitro and in vivo. Rhinology 30:33–40.

Salvatore CA, Jacobson MA, Taylor HE, Linden J, Johnson RG (1993): Molecular cloning and characterization of the human A_3 adenosine receptor. Proc Natl Acad Sci USA 90:103655–10369.

Saito H, Ebisawa M, Reason DC, Ohno K, Kurihara K, Sakaguchi N, Ohgimi A, Saito E, Akasawa A, Skimoto K (1991): Extracellular ATP stimulated interleukin-dependent cultured mast cells and eosinophils through calcium mobilization. Int Arch Allergy Appl Immunol 94:68–70.

Satchell D (1984): Purine receptors in trachea: Studies with adenosine deaminase and dipyridamole. Eur J Pharmacol 102:545–548.

Satchell D, Smith R (1984): Adenosine causes contractions in spiral strips and relaxations in transverse strips of guinea-pig trachea: Studies on mechanism of action. Eur J Pharmacol 101:243–247.

Satoh M, Yamaya M, Shimura S, Masuda T, Yamamoto M, Sasaki T, Shirato K (1995): Isoproterenol augments ATP-evoked Cl^- secretion across canine tracheal epithelium. Respir Physiol 99:13–18.

Schwiebert EM, Egan ME, Hwang TH, Fulmer SB, Allen SS, Cutting GR, Guggino WB (1995): CFTR regulates rectifying chloride channels through an autocrine mechanism involving ATP. Cell 81:1063–1073.

Slakey LL, Gordon EL, Pearson JD (1990): A comparison of ectonucleotidase activities on vascular endothelial and smooth muscle cells. Ann NY Acad Sci 603:366–379.

Steinberg TH, Buisman HP, Greenberg S, Di Virgilio F, Silverstein SC (1990): Effects of extracellular ATP on mononuclear phagocytes. Ann NY Acad Sci 603:120–129.

Steinberg TH, Silverstein SC (1987): Extracellular ATP^{4-} promotes cation fluxes in the J774 mouse macrophage cell line. J Biol Chem 262:3118–3122.

Szentmiklosi AJ, Ujfalusi A, Cseppento A, Nosztray K, Kovacs P, Szabo JZ (1995): Adenosine receptors mediate both contractile and relaxant effects of adenosine in main pulmonary artery of guinea pigs. Naunyn Schmidebergs Arch Pharmacol 351:417–425.

Tamaoki J, Kondo M, Takemura H, Chiyotani A, Yamawaki, I, Konno K (1995): Cyclic adenosine monophosphate-mediated release of nitric oxide from cultured canine cultured tracheal epithelium. Am J Respir Crit Care Med 152:1325–1330.

Tamaoki J, Kondo M, Takizawa T (1989): Adenosine-mediated cycle AMP-dependent inhibition of ciliary activity in rabbit tracheal epithelium. Am Rev Respir Dis 139:441–445.

Tatham PE, Cusack NJ, Gomperts BD (1988): Characterization of the ATP^{4-} receptor that mediates permeabilisation of rat mast cells. Eur J Pharmacol 147:13–21.

Thiel M, Chouker A (1995): Acting via A_2 receptors, adenosine inhibits the production of tumor necrosis factor-α of endotoxin-stimulated human polymorphonuclear leukocytes. J Lab Clin Med 126:275–282.

Thorne JR, Broadley KJ (1992): Adenosine-induced bronchoconstriction of isolated lung and trachea from sensitized guinea-pigs. Br J Pharmacol 106:978–985.

Thorne JR, Broadley KJ (1994): Adenosine-induced bronchoconstriction in conscious hyperresponsive and sensitized guinea pigs. Am J Respir Crit Care Med 149:392–299.

Tonetti M, Sturla L, Giovine M, Benatti U, De Flora A (1995): Extracellular ATP enhances mRNA levels of nitric oxide synthase and TNF-α in lipopolysaccharide-treated RAW 264.7 murine macrophages. Biochem Biophys Res Commun 214:125–130.

Van Scott MR, Chinet TC, Burnette AD, Paradiso AM (1995): Purinergic regulation of ion transport across nonciliated bronchiolar epithelial (Clara) cells. Am J Physiol 269:L30–L37.

Wali A, Beers MF, Dodia C, Feinstein SI, Fisher AB (1993): ATP and adenosine 3′,5′-cyclic monophosphate stimulate the synthesis of surfactant protein A in rat lung. Am J Physiol 264:L431–L437.

Walker BA, Hagenlocker BE, Douglas VK, Tarapchuk SJ, Ward PA (1991): Nucleotide responses of human neutrophils. Lab Invest 64:105–112.

Warburton D, Buckley S, Cosico L (1989): P_1 and P_2 purinergic receptor signal transduction in rat type II pneumocytes. J Appl Physiol 66:901–905.

Welford LA, Anderson WH (1988): Purine receptors and guinea-pig trachea: Evidence for a direct effect of ATP. Br J Pharmacol 95:689–694.

Wong LB, Yeates DB (1992): Luminal purinergic regulatory mechanisms of tracheal ciliary beat frequency. Am J Respir Cell Mol Biol 7:447–454.

Yamamoto A, Shikada K, Iwama T, Sakashita M, Hibi M, Tanaka S (1991): Effects of NZ-107 on tracheal responses to adenosine in the guinea pig. Jpn J Pharmacol 56:79–84.

Yoshitsugu M, Rautiainen M, Matsune S, Nuutinen J, Ohyama M (1993): Effect of exogenous ATP on ciliary beat of human ciliated cells studied with differential interference microscope equipped with high speed video. Acta Otolaryngol (Stockh) 113:655–659.

Yu GH, Tarapchak SJ, Walker BA, Ward PA (1991): Adenosine 5′-O-(3-thiotriphosphate) binding to human neutrophils. Evidence for a common nucleotide receptor. Lab Invest 65:316–323.

Yukawa T, Kroegel C, Chanez P, Dent G, Ukena D, Chung KF, Barnes PJ (1989): Effect of theophylline and adenosine on eosinophil function. Am Rev Respir Dis 140:327–333.

Zeidler RB, Conley NS (1986): Superoxide generation by pig alveolar macrophages. Comp Biochem Physiol [B] 85:101–104.

Zhang Y, Fredholm BB (1994): Propentofylline enhancement of the actions of adenosine on neutrophil leukocytes. Biochem Pharmacol 48:2025–2032.

4. Neurology

CHAPTER 19

Adenosine Neuromodulation

THOMAS V. DUNWIDDIE and BERTIL B. FREDHOLM

Department of Pharmacology/Program in Neuroscience, University of Colorado Health Sciences Center, Denver, and Veterans Administration Medical Research Service, Denver, CO 80262 (T.V.D.); Department of Physiology and Pharmacology, Section of Molecular Neuropharmacology, Karolinska Institutet, S-171 77 Stockholm, Sweden (B.B.F.)

INTRODUCTION

The discovery that adenosine and adenine nucleotides modulate the release of neurotransmitters occurred more than 20 years ago. Following the initial discoveries using electrophysiological methods in the frog neuromuscular preparation (Ginsborg and Hirst, 1972; Ribeiro and Walker, 1975), or the release of sympathetic (Fredholm and Hedqvist, 1980; Fredholm, 1974) and parasympathetic neurotransmitters (Vizi and Knoll, 1976), the generality of the phenomenon was soon realized (Fredholm and Hedqvist, 1980). Since it was also found that nerve stimulation could release adenosine (Fredholm, 1976) the possibility that adenosine could act as a transsynaptic or retrograde modulator of transmitter release was suggested (Fredholm and Hedqvist, 1979). Thus, it was clear at the outset that the effect of adenosine was an inhibition of neurotransmitter release via heteroreceptors, i.e., presynaptic receptors responding to a factor different from the transmitter whose release was regulated (see Fredholm, 1995b).

Although most synapses are modulated by adenosine, there is considerable variability between species (Luchelli-Fortis et al., 1981; Fredholm and Hedqvist, 1980), between neurotransmitters (Fredholm and Dunwiddie, 1988), and even between different neuronal populations using the same transmitter (see Jin et al., 1993). The effect of adenosine is most prominent in excitatory neuronal pathways (Fredholm and Dunwiddie, 1988; Snyder, 1985). For example, in the hippocampus, excitatory synaptic transmission mediated by glutamate is strongly inhibited at the perforant path to the dentate gyrus, in the mossy fibers to the CA3 region, and the commissural and associational inputs to the CA1 region, but inhibitory transmission mediated by $GABA_A$ (γ-amino butyric acid) receptors is adenosine-insensitive (Thompson et al., 1992; Yoon and

Purinergic Approaches in Experimental Therapeutics, Edited by Kenneth A. Jacobson and Michael F. Jarvis
ISBN 0-471-14071-6 © 1997 Wiley-Liss, Inc.

Rothman, 1991; Lambert and Teyler, 1991). It also seems clear that adenosine most often produces an inhibition of neurotransmitter release, even though an increase is sometimes observed (see below). Since adenosine formation is markedly enhanced by excitatory stimuli inducing ATP consumption, a general role for presynaptic adenosine receptors could be to prevent overstimulation of a given neuronal effector.

Since there are several previous reviews (e.g., Dunwiddie, 1985; Stone, 1991; Dunwiddie, 1990; Greene and Haas, 1991), to which the reader is referred for more extensive coverage of the above and related issues, we will concentrate in this chapter on a few issues that are still uncertain and where there is much current interest.

FORMATION OF ADENOSINE DURING SYNAPTIC TRANSMISSION

Several lines of evidence have demonstrated that there is always a background level of adenosine in all body fluids. This results from the fact that there is a basic equilibrium between intracellular and extracellular adenosine levels that is maintained by efficient equilibrative transporters that can shuttle adenosine between the intracellular and extracellular compartments. Adenosine is a key metabolite whose intracellular concentration is kept at a level around 10^{-8}–10^{-7} M because of the kinetic characteristics of the basic enzymes involved in purine metabolism (S-adenosine homocysteine hydrolase, 5′-nucleotidases, adenosine deaminase, adenosine kinase). Because of the presence of the equilibrative transporters, the extracellular concentration of adenosine is usually within this range as well. The situation is diagrammatically illustrated in Figure 1.

In addition to this "basal" adenosine, there are several mechanisms that can result in increases in extracellular adenosine, which can then influence neurotransmitter release. Extracellular adenosine can derive from intracellular hydrolysis of adenine nucleotides and subsequent transport of intracellular adenosine into the extracellular space. There is no compelling evidence that adenosine can act as a neurotransmitter in the central nervous system, i.e., that it can be released via a calcium-dependent exocytotic release process as with other neurotransmitters. However, there is good evidence that extracellular adenosine levels increase as a consequence of synaptic activity. In brain, increases in extracellular adenosine following various stimuli (e.g., seizures, ischemia, etc.) are quite large and fairly rapid (seconds to minutes), and in the rat hippocampus there is evidence that electrical stimulation can result in very rapid (<1 sec) increases in the adenosine level in the biophase controlling transmitter release (Mitchell et al., 1993). On the other hand, in many peripheral tissues such as the heart, the adenosine level increases only sluggishly in response to nerve stimulation (Fredholm et al., 1982b; Fredholm and Hedqvist, 1978; Fredholm, 1976). In this case, the increase in adenosine appears to correspond to a decrease in tissue ATP levels, and it may therefore be a consequence of nerve stimulation of such a magnitude that a severe imbalance between ATP consumption and formation is achieved. With the rapid release of adenosine in the hippocampus, an overall compromised energy metabolism cannot account for the increase in synaptic adenosine levels. It is, however, possible that in a small compartment, such as a dendrite, a single depolarizing stimulus may be sufficient to raise ATP consumption well above the capacity for rapid ATP resynthesis. In such a limited structure, intracellular ATP breakdown all the way to adenosine may occur as a consequence of a limited number of stimuli. The intracellular adenosine

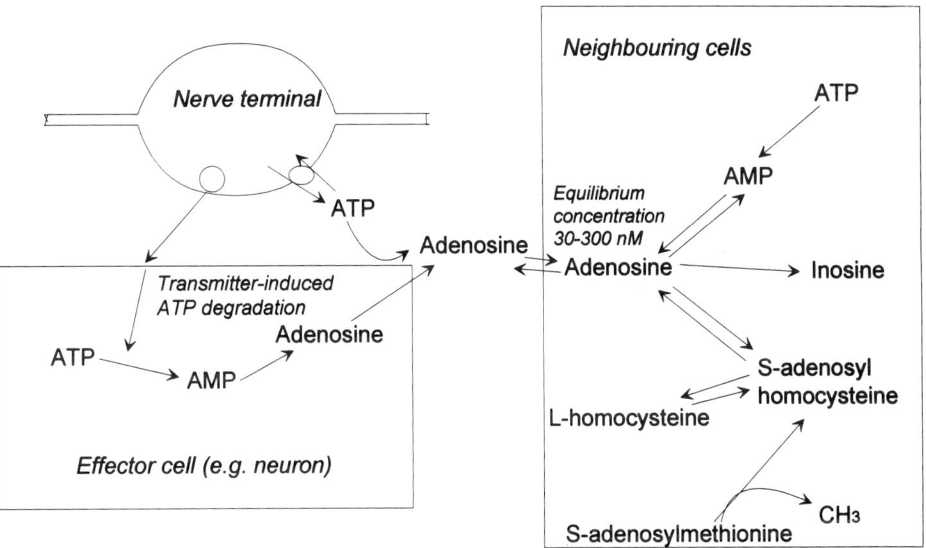

FIGURE 1. Schematic representation of the processes that regulate the extracellular level of adenosine. Note that the basal metabolism tends to give a minimal level of adenosine close to 100 nM and that increases are due predominantly to intra- or extracellular ATP breakdown.

would then be transported out into the extracellular space via equilibrative transporters (see Fredholm, 1995a; Fredholm et al., 1988).

Another possibility is that there is a rapid release of ATP, which is subsequently hydrolyzed to adenosine in the extracellular space. The ATP may derive both from pre- and postsynaptic sites. ATP is colocalized with a number of neurotransmitters such as acetylcholine (ACh), norepinephrine (NE), and glutamate, and is released together with these transmitters. The relative amount of ATP released from vesicular and from other sources appears to vary considerably from synapse to synapse. At the rat neuromuscular junction, it has been estimated that ACh and ATP are released in roughly equimolar amounts, and that approximately half of all the purines released by electrical stimulation come from the presynaptic nerve terminal (Smith, 1991). On the other hand, experiments in the rat vas deferens have suggested that the ratio of vesicular ATP : NE is quite low, that only a minute fraction of the total purine release evoked by nerve stimulation is from presynaptic ATP (Fredholm et al., 1982a), and that a major part of the total release of ATP following nerve activation appears to come not from the nerve but from the muscle (Kurz et al., 1994). Thus, the fraction of ATP that comes from vesicular release may be a function of the ATP : transmitter ratio, which may vary depending upon the synapse. At other synapses where higher concentrations of ATP are found in the vesicle, ATP derived from this source may have much greater physiological significance. Finally, at synapses such as in the vas deferens, where vesicular ATP content is quite low, it seems likely that quantitatively much more ATP is present in the nerve terminal cytosol than in vesicles,[1] so release of cytosolic ATP must be considered as a source of presynaptic ATP as well.

It is an intriguing possibility that some of the release of ATP from the nerves is also non-exocytotic. One mechanism to account for rapid, non-exocytotic release of ATP has recently been demonstrated. Certain P glycoproteins such as the multidrug resistance gene product (MDR) can transport ATP along the concentration gradient from the interior to the exterior (Abraham et al., 1993). Recently it was suggested that the cystic fibrosis transmembrane conductance (CFTR) regulator can mediate efflux of ATP through epithelial cell membranes, and this release of ATP may in fact be an important part of the chloride efflux effected by these proteins (Schwiebert et al., 1995). Since MDR and CFTR are members of a large family of traffic ATPases or ATP-binding cassette transporters (Gunderson and Kopito, 1995; Al-Awqati, 1995), it seems possible that such ATP efflux can be effected by other proteins in this family. It is very important to obtain further data on such ATP release, not least because such ATP transporters are potential targets for drugs. In fact, there is recent evidence that the sulfonylurea receptor is a member of this gene family. Thus, it will be very interesting to determine whether sulfonylurea compounds affect ATP release and, secondarily, various types of adenosine receptor–mediated responses.

ADENOSINE RECEPTORS: COUPLING MECHANISMS AND SECOND MESSENGER SYSTEMS

As outlined in other chapters in this volume there are four recognized adenosine receptors A_1, A_{2A}, A_{2B}, and A_3 (Fredholm et al., 1994). Although there may be additional receptor forms (see e.g., Cunha et al., 1996; Johansson and Fredholm, 1995; Johansson et al., 1993b), the evidence for these is not yet compelling, as additional forms have not been cloned. Of the four established receptor types the first two are expressed in high abundance in neurons. The A_{2A} receptors are found predominantly in the dopamine-rich regions of the brain, where they are found on a subset of GABAergic neurons (Johansson et al., 1993a; Fink et al., 1992; Schiffmann et al., 1990). It is not yet known if they are expressed also in the terminal regions of these neurons. A_{2A} receptors are also expressed at low levels in other regions such as the cortex and hippocampus, where there is good evidence for their presence in cholinergic neurons (Cunha et al., 1994a).

The distribution of A_1 receptors is much more widespread (Fastbom et al., 1987). It is, however, interesting that there is a slight mismatch between the distribution of A_1 receptor binding sites and A_1 receptor mRNA (Johansson et al., 1993a). One probable explanation for this is that a large proportion of the A_1 receptors are present on nerve terminals, whereas the mRNA is found only in the soma. As summarized

[1] There may be between 100 and 1000 vesicles per terminal, but stimulation of the vas deferens releases at most 1 or 2 vesicles per impulse, as opposed to the neuromuscular junction, where several hundred vesicles can be released by a single stimulus. The synaptic vesicle has a diameter of about 0.035 μm and thus its volume is about 1×10^{-5} μm^3. The terminal has a volume of between 0.1 and 0.3 μm^3 and has thus a 10,000 times larger volume. The ATP concentration in cells is close to 3 mM. The concentration of transmitters in a vesicle has been assumed to approach 1 M (5,000–10,000 molecules (0.8–1.6×10^{-20}moles) in a vesicle with a volume of about 10^{-20} l), but the ATP content is some 5 to 100 times lower. Thus, the ratio of ATP in the vesicle to that in the terminal ranges from 0.0003 to 0.01.

below, there is abundant evidence that there are presynaptic adenosine A_1 receptors on a variety of different nerves.

There is good evidence that adenosine A_{2A} receptors activate G_s proteins, but it is not known if they activate both splice variants of the G protein equally, or if they can activate other G proteins as well. Because these adenosine receptors activate G_s they activate adenylyl cyclase, and they can also activate L-type Ca^{2+} channels under some circumstances (Birnbaumer, 1992). Several types of ion channels, including P-type Ca^{2+} channels, can be modified via cAMP-dependent protein kinases. These types of actions are likely explanations for the effects of A_{2A} receptors on neurotransmitter release.

It is known that A_1 receptors activate all members of the G_i/G_o class of pertussis toxin–sensitive G proteins (Munshi et al., 1991; Freissmuth et al., 1991). The α_i subunits thus activated can reduce adenylyl cyclase activity. Liberated α_o subunits can inactivate some types of Ca^{2+} channels, and finally the liberated $\beta\gamma$ subunits can activate some types of phospholipase C, open several types of K^+ channels, influence adenylyl cyclase, and stimulate phospholipase A_2 (Fredholm et al., 1993; Birnbaumer, 1992). Thus, A_1 receptors have the ability to influence a host of different relevant signaling pathways via pertussis toxin–sensitive G proteins. In addition, adenosine A_1 receptors may be able to influence pertussis toxin–insensitive pathways, which may be quite important in explaining actions on transmitter release (Thompson et al., 1992; Fredholm et al., 1989).

CELLULAR MECHANISMS OF NEUROTRANSMITTER RELEASE

Before we turn to a specific discussion of the mechanism(s) underlying the presynaptic actions of adenosine it will be helpful to briefly summarize what is known about vesicular transmitter release. Transmitter release is now considered to be a special case of membrane fusion with features conserved from yeast to man—and is believed to be part of a cycle involving nine different steps (Sudhof, 1995):

1. *Docking* of synaptic vesicles filled with transmitter at active zones at the cell membrane.
2. *Priming* of the docked vesicles making them capable of fast Ca^{2+}-stimulated release.
3. *Exocytosis* triggered by a Ca^{2+} signal in less than 0.3 milliseconds.
4. *Endocytosis,* by which the empty vesicle membranes are rapidly internalized, probably by way of clathrin-coated pits that turn into coated vesicles.
5. *Translocation* of the vesicles involving shedding of the coat and acidification followed by step 6.
6. *Fusion with endosomes.*
7. *Budding of synaptic vesicles from endosomes.*
8. *Vesicular uptake of neurotransmitter* by transporters driven by active transport of protons.
9. *Translocation* of the neurotransmitter-filled vesicles back to the area of the active zones, a process that probably involves interactions with cytoskeletal elements.

The entire process is believed to be quite rapid, taking only about a minute (Sudhof, 1995).

Steps 1, 2, 3, and 9 are the ones most likely regulated by agents that decrease (or increase) neurotransmitter release via presynaptic receptors. Electrophysiological experiments in which the release of single quanta of transmitter can be studied probably are focused on step 3, with effects at step 2 contributing; whereas studies that use overflow of transmitter may also pick up significant contributions of other steps (predominantly 9 and 1) in the cycle.

It is known that the vesicle fusion is triggered by a very localized calcium signal generated by entry through voltage-dependent calcium channels (Sudhof, 1995; Schweizer et al., 1995; Jessell and Kandel, 1993), and that calcium channels aggregate in clusters, with a spacing of about 40 nm between channels (Haydon et al., 1994). The extreme proximity explains why the release machinery senses the calcium entry so rapidly, and why the magnitude of the release is determined by the third to fourth power of the extracellular calcium concentration. The calcium transients measured in deeper compartments of the cell show a much weaker relationship to the extracellular level. The fact that the release machinery responds to calcium at the mouth of the calcium channel also has the important consequence that the responsible Ca^{2+}-sensing element must react to high concentrations of the ion (about 10^{-4} M).

Several lines of evidence indicate that synaptotagmin I may be this Ca^{2+} sensor (Sudhof, 1995; Geppert et al., 1994). Mice lacking this protein are unable to sustain Ca^{2+}-induced transmitter release, whereas other transmitter release is unaffected. Furthermore the interaction of synaptotagmin I with another protein in the docking complex, syntaxin, is influenced by Ca^{2+} in the appropriate concentration range. There are, however, many different synaptotagmins whose function(s) are unknown (Sudhof, 1995), and little, if anything, is known about the possibility of directly affecting synaptotagmin function via cell membrane receptors. If the important Ca^{2+}-regulated event is the interaction between synaptotagmin, which is associated with the vesicle, and syntaxin, which is a plasma membrane protein, then the regulated entity may be syntaxin. It has been shown that there is a very intimate relation between syntaxin and another membrane protein called n-secl (Pevsner et al., 1994), and that it can block transmitter release. Another group of proteins that interact closely with syntaxin are $\alpha\beta\gamma$-SNAP (Sudhof, 1995). Direct interactions of receptors with either of these proteins may regulate transmitter release.

Other possible players in the release process that must be considered are the rab-proteins. They comprise a family of small GTP-binding proteins known to be involved in several aspects of intracellular membrane trafficking (Fischer von Mollard et al., 1994). Two of these proteins, Rab 3a and Rab 3c, are important constituents of nerve terminals, and there seem to be differences between neurons in the type they express. These proteins may act in concert with yet another protein, rabphilin. Thus, presynaptic inhibitory receptors could also modulate release by influencing the function of a rab–rabphilin complex. As mentioned above, some test systems for detecting regulation of transmitter release could also respond to changes in the mechanisms for translocating vesicles to the docking region and docking. Synapsins, long known to be associated with synaptic vesicles, may be important in this process (Pierbone et al., 1995). It is known that synapsins can be phosphorylated by cAMP-dependent protein kinases and by calcium-calmodulin-dependent kinases. The latter process is influenced by changes in Ca^{2+} in the terminal that are much smaller than those required for the actual release. Therefore it makes sense that this process, which can be regulated by adenosine (B.

Fredholm, E. Lindgren and P. Greengard, unpublished data), occurs at a distance away from the docking region where the mouths of the calcium channels are positioned. In addition the translocation and docking might be regulated by rab-proteins.

INHIBITION OF NEUROTRANSMITTER RELEASE BY ADENOSINE

With this by way of background, there would appear to be at least four distinct ways in which adenosine could inhibit the release of neurotransmitter. First, by altering the steps in the release process that occur prior to Ca^{2+} influx, i.e., mobilization and docking of vesicles, adenosine could inhibit subsequent transmitter release. A second and perhaps more direct mechanism for inhibiting release would be to regulate the Ca^{2+} influx via the voltage-dependent calcium channels in the neighborhood of the docking complex. It is known that $G_{o\alpha}$ subunits can negatively regulate N-type (Offermanns and Schultz, 1994; Shapiro et al., 1994), and possibly Q-type (Wheeler et al., 1994) calcium channels, that $G_{s\alpha}$ subunits can directly facilitate opening of some L-type channels (Hamilton et al., 1991), and indirectly (perhaps via cAMP-dependent phosphorylation) facilitate entry of Ca^{2+} through P-type channels (Mogul et al., 1993). A third possibility is that adenosine could change the Ca^{2+} dependency of the release process itself, perhaps by changing the affinity of a Ca^{2+}-binding site on a protein such as synaptotagmin for Ca^{2+}. Finally, by hyperpolarizing the presynaptic nerve terminal (Bennett and Ho, 1991), or by preventing the invasion of nerve impulses into nerve terminals (Wall, 1995), adenosine may indirectly inhibit the voltage-dependent opening of presynaptic Ca^{2+} channels. These alternative mechanisms have been explored in a number of different ways involving direct electrophysiological measurements of channel activity, pharmacological modulation of Ca^{2+} channel activity, measurements of presynaptic Ca^{2+} concentrations using Ca^{2+}-sensitive dyes, and determinations of spontaneous release of transmitter occurring via non-Ca^{2+}-dependent mechanisms. Nerve terminals are difficult to study directly, and consequently it has been difficult to obtain evidence in nerve terminals concerning these mechanisms and to determine if they apply to the wide range of synapses that are modulated by adenosine. Perhaps for these reasons, the mechanism by which neurotransmitter release is modulated by adenosine is currently a topic of debate (Scholz, 1993; Thompson et al., 1993), and the issue is far from resolved.

BIOCHEMICAL MODIFICATION OF TRANSMITTER RELEASE MECHANISMS

The first type of proposed mechanism for adenosine neuromodulation (disruption of mobilization/docking/priming) is one that could occur in the absence of any Ca^{2+}-dependent transmitter release. The strongest electrophysiological support for an action of adenosine directly on the transmitter release mechanism consists of evidence that the spontaneous release of individual quanta of neurotransmitter is inhibited by adenosine. Since this can occur in the complete absence of extracellular Ca^{2+}, or when voltage-dependent Ca^{2+} channels are completely blocked, adenosine can apparently act in ways that are independent of any changes in Ca^{2+} influx. In most synaptic systems, this type of spontaneous release of neurotransmitter is independent of changes in intracellular Ca^{2+}, in that blocking Ca^{2+} channels, or buffering intracellular Ca^{2+} to prevent changes related to release of Ca^{2+} from intracellular stores, has little effect on spontaneous release. However, the rate of individual release events is, in many cases, modulated by agents such as adenosine. These types of actions of adenosine have been described in many systems, ranging from the neuromuscular junction (Ginsborg and Hirst, 1972; Silinsky, 1984), to organotypic hippocampal slices cultures (Scanziani et al., 1992), and hippocampal cells in culture (Scholz and Miller, 1992). At the neuromuscular junction, Silinsky and colleagues have demonstrated that the increased spontaneous release of transmitter induced by a wide variety of stimuli that do not depend upon Ca^{2+} channel activity, such as lanthanum (Silinsky, 1984), Ca^{2+}-filled liposomes (Silinsky, 1984), and ionomycin (Hunt and Silinsky, 1993), are all inhibited by adenosine. Furthermore, buffering Ca^{2+} concentrations in the nerve terminal does not affect inhibition by adenosine (Hunt et al., 1994). Silinsky and Solsona (1992) have also demonstrated that under conditions where extracellular Ca^{2+} concentrations are buffered to very low levels, opening Ca^{2+} channels *reduces* the rate of spontaneous release (presumably because of efflux of Ca^{2+} from the nerve terminal). Under these conditions, Ca^{2+} channel blockers such as Co^{2+} enhance release (by blocking Ca^{2+} efflux via the channel), whereas adenosine still reduces release under these conditions (Silinsky and Solsona, 1992). These observations demonstrate that this type of modulation of transmitter release occurs independently of any effects on Ca^{2+} channel activity.

Similar studies have been conducted in other preparations, such as cultured hippocampal neurons and brain slices, with somewhat similar results. For example, Scholz and Miller (1992) demonstrated that in hippocampal pyramidal neurons in culture, spontaneous release of glutamate can be modulated by A_1 receptors. This observation has been confirmed by others (Thompson et al., 1993), and it has been shown that adenosine inhibits release evoked by a variety of agents that work independently of Ca^{2+} channel action (e.g., α-latrotoxin, which acts directly on the release apparatus to evoke release; Capogna et al., 1994). Based upon these types of studies, it has been concluded that the mechanism underlying adenosine neuromodulation is most likely to be a direct action on the release mechanism, such as a reduction in the affinity of the Ca^{2+}-dependent component of the release apparatus for Ca^{2+} (Thompson et al., 1993; Silinsky, 1986). Taken together, these results provide convincing evidence that at least some types of inhibition of transmitter release by adenosine can occur by mechanisms that do not involve Ca^{2+} channels. In mechanistic terms, adenosine appears to inhibit release by reducing the ability of Ca^{2+} to induce release, rather than by changing the amount of Ca^{2+} available to stimulate release.

What is still a somewhat contentious issue is whether effects of adenosine on the release apparatus can account for the modulation of *evoked* release (Scholz, 1993; Thompson et al., 1993). Despite the fact that adenosine can inhibit spontaneous release of transmitter, there is little concrete evidence that this phenomenon bears any direct relationship to inhibition of Ca^{2+}-dependent transmitter release, even though this is commonly assumed to be the case. Unlike the situation at the frog neuromuscular junction, in hippocampus (Wu and Saggau, 1994) and in the avian ciliary ganglion (Yawo and Chuhma, 1993), there is convincing evidence that presynaptic Ca^{2+} currents *are* modulated by adenosine, and that the reduction is quantitatively sufficient to account for the inhibition of release. Thus, one could hypothesize that these two actions of adenosine (inhibition of spontaneous and evoked release) are completely unrelated effects that occur via different mechanisms. For example, adenosine might inhibit evoked release via modulation of Ca^{2+} channel activity, and modulate spontaneous release by altering some aspect of the release process (e.g., the number of vesicles that are docked at release sites but not primed for release) that would modify spontaneous release without affecting evoked release. This latter type of effect might not directly affect evoked transmitter release, but could modify other properties of the system, such as the ability to respond to repeated stimulation. When studying transmitter release by biochemical methods one can also measure a basal release of transmitter. Like spontaneous quantal release, this basal transmitter release is not markedly affected by changes in extracellular Ca^{2+} levels, indicating that it does not require Ca^{2+} influx. It is, however, strongly affected by decreasing intracellular Ca^{2+} by BAPTA (1,2-bis(2-aminophenoxy)ethane-N,N,N′,N′-tetraacetic acid) (Fredholm and Hu, 1993), indicating that its magnitude depends on the level of intracellular Ca^{2+}.

One other line of evidence that might support actions of adenosine on the release apparatus is the observation that in many systems it has been difficult to demonstrate that pertussis toxin, which inactivates certain types of G proteins, disrupts the modulation of transmission. The postsynaptic modulation of Ca^{2+} and K^{+} channel function by adenosine is readily blocked by pertussis toxin, but the presynaptic effects of adenosine on excitatory transmission have been reported to be insensitive to pertussis toxin in hippocampal slice cultures (Thompson et al., 1992), in hippocampal brain slices (Fredholm et al., 1989), but not in cultured hippocampal neurons (Scholz and Miller, 1992). The pertussis toxin insensitivity of presynaptic modulation suggests that mechanisms other than ion channel modulation mediate these effects. An alternative possibility is that there are factors that limit the accessibility of pertussis toxin to presynaptic nerve terminals, but not to postsynaptic neurons. This issue could be resolved if there were positive controls, i.e., if pertussis toxin successfully blocked the actions of one presynaptic modulator but not another, but so far this has not been observed; when pertussis toxin fails to block the action of one modulator, it also fails to block the actions of other modulators at the same synapse.

As a final note, it should be stressed that conclusions based on experimental observations in a single system are unlikely to generalize to other systems, despite the fact that they are both modulated by adenosine. There clearly are fundamental differences in the way in which transmitter release is regulated at different synapses. Even within a structure such as the hippocampus, a single modulator may work via different mechanisms at different synapses (e.g., the modulatory actions of baclofen are blocked by Ba^{2+} inhibitory synapses, but not at excitatory ones; Thompson and Gahwiler, 1992). Finally, there is much that remains to be learned about the cellular processes that underlie release, and certain experimental observations (e.g., that synchronous, *stimulated* release of transmitter can occur in the absence of Ca^{2+} influx; Silinsky et al., 1995) suggest that substantial revisions of the classical models of neurotransmitter release may need to be made.

EVIDENCE *AGAINST* INHIBITION OF EVOKED RELEASE BY ADENOSINE ACTIONS ON RELEASE MACHINERY

As discussed in the preceding section, there is direct evidence that adenosine inhibits the spontaneous release of transmitter that occurs independently of Ca^{2+} influx through voltage-dependent Ca^{2+} channels. On this basis it has been suggested that it is not necessary to invoke modulation of presynaptic ionic conductances as a mechanism to underlie synaptic modulation (Thompson et al., 1993). However, the fact that in some systems the inhibition of Ca^{2+} influx is sufficient to account quantitatively for the reduction in release of transmitter suggests the opposite, i.e., that it is unnecessary to invoke mechanisms that *do not* involve Ca^{2+} channels. Perhaps one of the best ways to differentiate between these possibilities would be to compare neuromodulation of release mediated by different populations of Ca^{2+} channels. At many synapses there are multiple populations of Ca^{2+} channels on nerve terminals that cooperatively mediate the Ca^{2+} influx necessary for neurotransmitter release (Wheeler et al., 1994; Takahashi and Momiyama, 1993), and which can often be differentiated on the basis of inhibition by selective Ca^{2+} channel blockers. If release mediated by one population of presynaptic Ca^{2+} channels is sensitive to adenosine, while another is not, as often appears to be the case (see below), this would argue *against* an action of adenosine on a common substrate that lies beyond Ca^{2+} entry into the nerve terminal.

Another source of information concerning mechanisms of adenosine action might be quantal analysis of transmitter release at synapses where such analysis is appropriate. For example, mean quantal content can be expressed as $N \times p$, where N is usually considered to be the number of release sites, and p is the probability of release from a given site. Changes in intraterminal Ca^{2+} would be expected to be reflected in changes in p, whereas changes in N would most likely correspond to biochemical events (e.g., phosphorylation of a release site) that would render them incapable of releasing transmitter. The interpretation of quantal analysis in the central nervous system is complicated, however, and postsynaptic changes may influence quantal content (Kullmann and Siegelbaum, 1995). Further studies may help to explore some of these possibilities.

TRANSMITTER RELEASE INHIBITION VIA INHIBITION OF CALCIUM CHANNEL ACTIVITY

Modulation of Ca^{2+} channel activity is a very direct means by which activation of adenosine receptors could affect neurotransmitter release, and numerous studies have shown that adenosine inhibits the activity of Ca^{2+} channels, including the N and Q type channels that are thought to be primarily involved in release (Wheeler et al., 1994; Mogul et al., 1993; MacDonald et al., 1986; Scholz and Miller, 1991; Dolphin et al., 1986). Because of the difficulty in recording Ca^{2+} currents from nerve terminals, three general approaches have been used to address this issue. The first approach has been to measure calcium currents directly at a few unusual synapses where this can be done by direct means. The chick ciliary ganglion calyciform synapse is one such synapse where direct recordings can be made from the presynaptic nerve terminal, and at this synapse adenosine has been reported to activate K^+ channels *and* inhibit Ca^{2+}-dependent action potentials recorded under conditions where K^+ currents are largely blocked (Bennett and Ho, 1991). When Ca^{2+} influx into the nerve terminal is measured at this synapse with a Ca^{2+}-sensitive dye, a direct inhibition of Ca^{2+} influx can be demonstrated, and the magnitude of this inhibition accounts quantitatively for the inhibition of transmitter release (Yawo and Chuhma, 1993). This suggests that it is unlikely that other mechanisms are acting in parallel at this synapse to inhibit release. Because the latter experiments only indicate the endpoint of Ca^{2+} channel action (i.e., increases in intraterminal Ca^{2+}), they do not rule out a possible role of K^+ channels in this effect. However, it was found that when the N-type of Ca^{2+} channel is blocked with ω-conotoxin, the remaining presynaptic Ca^{2+} signal was no longer modulated by adenosine (Yawo and Chuhma, 1993). This suggests that at this synapse, adenosine, acting via A_1 receptors and a G protein–dependent pathway can specifically inhibit N channel function.

A second approach to this general issue is to use indirect (extracellular) measurements of putative Ca^{2+} currents at synapses where Ca^{2+} currents cannot be measured directly in voltage-clamped nerve terminals. Silinsky and Solsona (1992) have used this approach at the frog neuromuscular junction to characterize ω-conotoxin-sensitive Ca^{2+} currents at this synapse, and have reported them to be completely insensitive to adenosine, under conditions where release is markedly inhibited by adenosine. On the other hand, at the rat neuromuscular junction presynaptic Ca^{2+}currents *are* inhibited by adenosine (Hamilton and Smith, 1991), emphasizing once again that adenosine may act via entirely different mechanisms at different synapses to antagonize release.

In preparations where Ca^{2+} currents cannot be recorded electrophysiologically, fluorescent indicators of intracellular Ca^{2+} concentrations have been used to measure the Ca^{2+} influx associated with stimulated release. This has been accomplished at excitatory synapses in the CA1 region of the rat hippocampus, and the magnitude of the changes in Ca^{2+} influx has been compared with parallel changes in the magnitude of the evoked response. In this system, activation of A_1 receptors inhibits both the synaptic response (Dunwiddie and Hoffer, 1980) and the transient increase in intraterminal Ca^{2+} associated with synaptic stimulation (Wu and Saggau, 1994). Furthermore, the decrease in Ca^{2+} influx is sufficient to account quantitatively for the decrease in transmitter release, making it unnecessary to invoke other parallel mechanisms for release inhibition. On the basis of experiments with selective Ca^{2+} channel blockers, these authors concluded that in the CA1 region of hippocampus, at least three different types of Ca^{2+} channels contribute to the presynaptic Ca^{2+} influx that supports transmitter release. N channels were estimated to contribute about 40% of the Ca^{2+} influx, and could be inhibited by about 40% by adenosine; an ω-agatoxin-sensitive channel(s) contributed about 27% of the Ca^{2+} and was essentially adenosine-insensitive, and the remaining current, contributed by unknown Ca^{2+} channel(s), could be inhibited by 73% by adenosine. Somewhat similar conclusions were reported by Wheeler et al. (1994), who suggested that synaptic transmission at this synapse is primarily supported by N and Q channels, and that both channels could be inhibited by adenosine. Although there are still aspects of these studies that need to be resolved, several issues seem clear; multiple Ca^{2+} channels support transmission at single synapses, adenosine can modulate multiple subtypes of Ca^{2+} channels, and there are also Ca^{2+} channels on nerve terminals that are insensitive to adenosine.

The evidence concerning adenosine actions on nerve terminal Ca^{2+} channels suggests that there may be considerable diversity in the way in which adenosine acts at different synapses. It appears that there are specific systems where Ca^{2+} channels are affected by adenosine, and others where this is not the case. Although it is possible that multiple mechanisms (i.e., changes in channel function, changes in release mechanisms) may act in parallel to inhibit release, it does not seem to be necessary to invoke this type of model to explain most of the existing results. Where there is evidence for an inhibition of presynaptic Ca^{2+} currents, the magnitude of the inhibition is in most cases adequate to account for the reduction in release.

K^+ CHANNEL–MEDIATED MECHANISMS OF TRANSMITTER RELEASE INHIBITION

The activation of K^+ channels by adenosine acting via A_1 receptors occurs in a number of tissues, including heart and brain, and in all cases a pertussis toxin–sensitive G protein seems to be involved in the response. The cellular hyperpolarization that results will inhibit release in semi-intact preparations such as brain slices by reducing the rate of cell firing. Although this effect would be significant in terms of the activity of the intact brain, it would not provide a mechanistic explanation for inhibition of transmitter release in more reduced preparations such as synaptosomes, or where transmitter release is evoked by electrical stimulation rather than by the spontaneous activity of neurons. Adenosine can also hyperpolarize nerve terminals by this same mechanism, which could result in a failure of action potentials to invade the nerve terminal, or could reduce the activation of voltage-gated Ca^{2+} channels in the nerve terminal that are directly responsible for the Ca^{2+} influx that leads to release. Although there may be synapses where this type of action is important, several lines of evidence suggest that this is generally not the case. First, direct measurements suggest that in most instances, adenosine does not affect K^+ currents in the nerve terminal (Silinsky et al., 1990), or the ability of the action potential to invade the nerve terminal (Dunwiddie and Miller, 1993; Silinsky and Solsona, 1992). Second, Ba^{2+} and tetrahydroaminoacridine are relatively potent blockers of the inwardly rectifying K^+ conductance that is activated by A_1 receptors, but do not have any effect on adenosine modulation of transmitter release (Akhondzadeh and Stone, 1994; Thompson et al., 1992; Dunwiddie and Proctor, 1988; Silinsky, 1984; Scholfield and Steel, 1989; Dunwiddie et al., 1991; but see Birnstiel et al., 1992). Third, the activation of K^+ channels by adenosine can be blocked by pertussis toxin, while adenosine receptor–mediated inhibition of transmitter release in the same preparation is unaffected (Fredholm et al., 1989). Finally, even in systems where adenosine clearly activates K^+ channels in the nerve terminal (e.g., the ciliary ganglion of the chick, where adenosine hyperpolarizes nerve terminals by almost 20 mV; Bennett and Ho, 1991), the primary mechanism underlying modulation of transmitter release still appears to be a direct modulation of Ca^{2+} currents

In addition to providing insights into the role of presynaptic K^+ channels in release, experiments with K^+ channel blockers have also established that there are important differences between modulators that act at the same synapse. For example, 4-aminopyridine (4-AP), which blocks several kinds of K^+ channels and enhances neurotransmitter release (primarily by prolonging the presynaptic spike and thus enhancing Ca^{2+} influx), completely blocks the effects of certain presynaptic modulators of [^{3}H]ACh release in hippocampus (muscarinic and opiate), while others (adenosine A_1) are unaffected (Fredholm, 1990b). One possible interpretation of this is that some modulators act upon 4-AP-sensitive K^+ channels, but this seems somewhat unlikely based upon the evidence presented above. A more likely alternative is suggested by the observation that multiple populations of Ca^{2+} channels appear to support transmitter release, and that not all are sensitive to modulators such as adenosine. On this basis, we can hypothesize that in the previous example, 4-AP increases Ca^{2+} influx into the terminal through muscarinic- and opiate-*insensitive* Ca^{2+} channels to the point where they are by themselves able to support maximal levels of transmitter release; this type of situation is illustrated schematically in Figure 2a. At this point, inhibition of other channels by agonists at these receptors would have no effect on transmitter release. On the other hand, this would imply that adenosine inhibits a somewhat different set of presynaptic Ca^{2+} channels, and that even in the presence of 4-AP, adenosine-insensitive channels are unable to support maximal transmitter release.

In this context, a number of studies that have characterized the effects of 4-AP on presynaptic modulation have provided an additional control for the effects of 4-AP, i.e., they have reduced the concentration of Ca^{2+} in the extracellular buffer (or increased

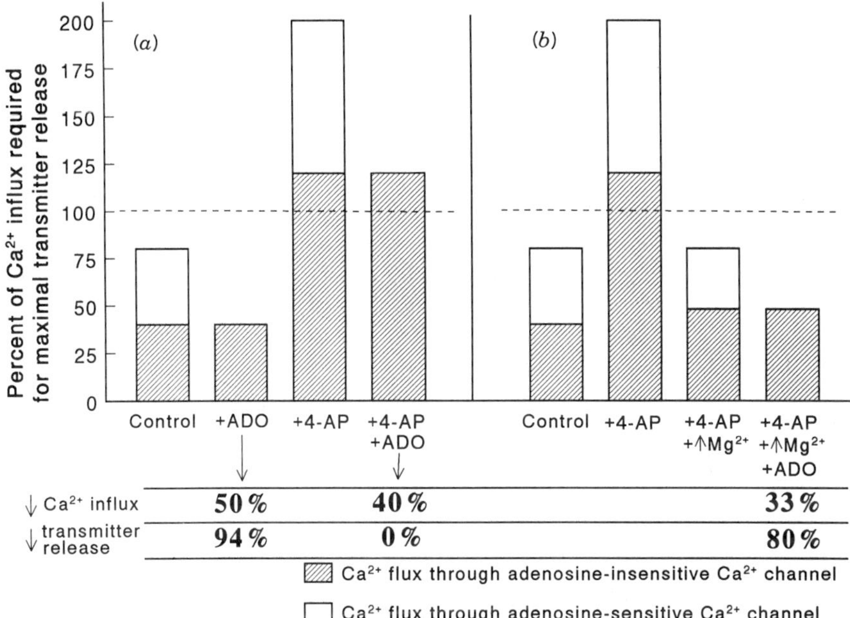

FIGURE 2. This figure illustrates some of the expected effects of 4-aminopyridine (4-AP) at a hypothetical synapse where some of the presynaptic Ca^{2+} channels are adenosine-sensitive (open bars) and some are adenosine-insensitive (hatched bars). Under control conditions, the Ca^{2+} influx into the nerve terminal is 80% of that which would be required to saturate transmitter release; of that Ca^{2+} current, half is carried by adenosine-sensitive and half by adenosine-insensitive channels. Under these conditions, adenosine inhibits 50% of the presynaptic Ca^{2+} influx, which reduces transmitter release by 94% (assuming that release is proportional to the fourth power of the Ca^{2+} influx). (**a**), 4-AP has increased the Ca^{2+} flux through both types of channels, but has increased the current through the adenosine-insensitive channel to a greater extent (reflecting a somewhat different voltage- and time-dependence for the activation of this type of channel). Adenosine now has no effect on release, because Ca^{2+} flux through the adenosine-insensitive channel is able to evoke maximal levels of release. This type of loss of adenosine sensitivity with 4-AP is often observed (Dunwiddie et al., 1991; Klapstein and Colmers, 1992; Fredholm, 1990b). (**b**) The effect of increasing Mg^{2+} to reduce Ca^{2+} influx (and hence transmitter release) back down to basal levels. If 4-AP differentially enhances Ca^{2+} currents through the two types of channels, the relative proportions of adenosine-sensitive and adenosine-insensitive channels supporting release have changed, and the inhibitory effect of adenosine is changed. In the example shown, the normal inhibitory effect of adenosine (94% reduction in release) has been reduced to 80% because of a shift in the relative numbers of adenosine-sensitive channels.

Mg^{2+}) to compensate for the additional release induced by 4-AP. Under these conditions, the ability of 4-AP to block the modulatory effects of various agonists is often reduced or completely reversed (Dunwiddie et al., 1991; Klapstein and Colmers, 1992; Hu and Fredholm, 1989). If the effects of 4-AP are not reversed completely under these conditions, several conclusions are possible. As before, it is possible that the modulator is acting via a 4-AP-sensitive K^+ channel, and thus the 4-AP occludes the effect of the modulator, but based on other evidence this seems somewhat unlikely.

An alternative possibility is that because 4-AP markedly changes the presynaptic action potential, it may shift the relative contributions of different kinds of Ca^{2+} channels to the total Ca^{2+} current in the nerve terminal; this of course could result in a situation where the effects of some modulators would be reduced, and others might be unaffected or even enhanced (see Figure 2b).

If we accept the hypothesis that the primary effects of 4-AP are to shift the relative dependence upon various types of presynaptic Ca^{2+} currents, where does adenosine fit into this picture? We have found that at cholinergic (Fredholm, 1990b) and noradrenergic (Hu and Fredholm, 1989) nerve terminals in hippocampus, 4-AP does not affect the inhibitory effects of A_1 agonists, suggesting that at these synapses, most of the Ca^{2+} influx into the nerve terminal is through adenosine-sensitive Ca^{2+} channels. At hippocampal glutamatergic synapses, our lab as well as others (Klapstein and Colmers, 1992; Dunwiddie et al., 1991) have observed that 4-AP by itself significantly blocks A_1-mediated inhibition, and that these effects can be largely, although not completely, reversed by changing divalent cation concentrations to reverse the stimulatory effect of 4-AP on transmission. These results would suggest that at this synapse, there is a significant contribution to presynaptic Ca^{2+} currents from adenosine-insensitive channels, which is the same conclusion reached by Wu and Saggau (1994) based on direct measurements of Ca^{2+} influx into these same nerve terminals. The incomplete reversal of the effects of 4-AP on adenosine sensitivity suggests that in the presence of 4-AP, these adenosine insensitive channel(s) make a greater relative contribution to the total presynaptic Ca^{2+} flux than under control conditions. Finally, in the olfactory cortex, the effects of adenosine are completely blocked by 3,4-diaminopyridine (which blocks the same channels as 4-AP) and are not reinstated by reducing transmitter release with increased Mg^{2+} (Scholfield and Steel, 1989). This result would be expected if 3,4-diaminopyridine shifted presynaptic Ca^{2+} influx in such a way that an adenosine-insensitive Ca^{2+} channel completely supports synaptic transmission.

Thus, the observation that multiple Ca^{2+} channels support transmission at many of the synapses that have been tested has forced us to reevaluate the earlier evidence that suggested that adenosine effects on K^+ channels played a significant role in the mechanism of synaptic modulation. Most, if not all, of the evidence that has been used to implicate adenosine actions on K^+ channels in synaptic modulation can also be explained by a model that does not involve effects on K^+ channels, provided there are adenosine-sensitive and adenosine-insensitive populations of Ca^{2+} channels on nerve terminals. To determine which of these hypotheses is correct, it will be necessary to isolate each of the channels involved in transmitter release and identify its adenosine sensitivity, as well as to determine the relative contributions of each of these channels to presynaptic Ca^{2+} flux, under control conditions and in the presence of drugs such as 4-AP.

THE ROLE OF A_{2A} RECEPTORS IN REGULATING TRANSMITTER RELEASE

Whereas the concept that A_1 adenosine receptors are present on nerve terminals and mediate inhibition of transmitter release is well established, the role of A_2 receptors is more controversial. The possible existence of A_2 receptors that stimulate ACh release from cholinergic synaptosomes isolated from the striatum has been reported (Kirk and Richardson, 1994; Brown et al., 1990; Richardson et al., 1987). These putative receptors were exquisitely sensitive to adenosine, which suggested that they might be of the A_{2A} rather than the A_{2B} subtype. Indeed, the same group of researchers later found that an adenosine analog, CGS 21680, that is quite selective for A_{2A} receptors is able to stimulate striatal ACh release (Kirk and Richardson, 1994). At first sight this finding is entirely compatible with the distribution of A_{2A} receptors, which are highly enriched in the striatum of rats (Johansson et al., 1993a; Martinez-Mir et al., 1991; Parkinson and Fredholm, 1990; Jarvis and Williams, 1989; Alexander and Reddington, 1989). However, the mRNA coding for the adenosine A_{2A} receptors is found not in the large aspiny cholinergic neurons, but in a subgroup of the medium-sized spiny GABAergic neurons (Johansson et al., 1994; Johansson et al., 1993a; Fink et al., 1992; Schiffmann et al., 1990). Furthermore, in a previous study using the A_{2A}-selective agonist CGS 21680 we did not observe any stimulation of acetylcholine release, but instead we noted an inhibition (Jin et al., 1993). This compound also caused inhibition of the release of dopamine, as reported earlier (Lupica et al., 1990). Thus, both stimulatory and inhibitory effects of A_{2A} selective agonists have been reported in striatum. There is also some evidence that the noradrenaline release in the nucleus tractus solitarius is stimulated by CGS 21680 (Barraco et al., 1995), even though there is scant evidence for control of catecholamine release elsewhere by A_{2A} receptors (Fredholm and Dunwiddie, 1988).

There is good evidence that stimulation of A_{2A} receptors on cholinergic neurons innervating the hippocampus enhances electrically evoked release of the transmitter (Cunha et al., 1995; Cunha et al., 1994a), even though inhibitory effects mediated by adenosine A_1 receptors are easier to detect (Fredholm, 1990a). Thus, release of acetylcholine from these neurons appears to be regulated in an opposite manner by inhibitory adenosine A_1 and stimulatory A_{2A} receptors (Cunha et al., 1994a).

The evidence for the selective presence of adenosine A_{2A} receptors on hippocampal, but not on striatal, cholinergic neurons largely rests on data using the agonist CGS 21680. Recent data suggest that this compound might not be quite as selective as previously assumed. Thus, in cortical (including hippocampal) tissue, a major part of CGS 21680 binding is to a site that shows different pharmacological properties than the A_{2A} receptors present in striatum (Cunha et al., 1996; Johansson and Fredholm, 1995; Johansson et al., 1993b).

We recently used the nonselective adenosine receptor agonist 2-chloroadenosine to examine the role of A_1 and A_{2A} receptors in regulating catecholamine and ACh release from striatal and hippocampal slices (Jin and Fredholm, 1997). The results from these experiments are shown in Figure 3. It is clear that in all instances 2-chloroadenosine produced a clear, concentration-dependent inhibition of transmitter release that was competitively antagonized by the A_1 receptor antagonist 8-cyclopentyl-1,3-diprophyl xanthine (DPCPX). In one instance, namely the release of ACh from hippocampus, there was also a stimulatory effect of 2-chloroadenosine that was revealed in the presence of DPCPX. This was blocked by a selective A_{2A} receptor antagonist KF 17387 (not shown). However, such a stimulatory A_{2A} component was not observed in the case of ACh release from the striatum or in the case of catecholamine release from either slice preparation.

It is unclear why these results differ from those of others who have detected stimulatory effects in the striatum (Kurokawa et al., 1994; Kirk and Richardson, 1994; Brown et al., 1990), but there are some potential explanations. It is probably significant that an increase in striatal ACh release due to adenosine acting at A_{2A} receptors has been observed when K^+ stimulated synaptosomes have been investigated, but not when electrical field stimulation was used. In a fashion analogous to that outlined

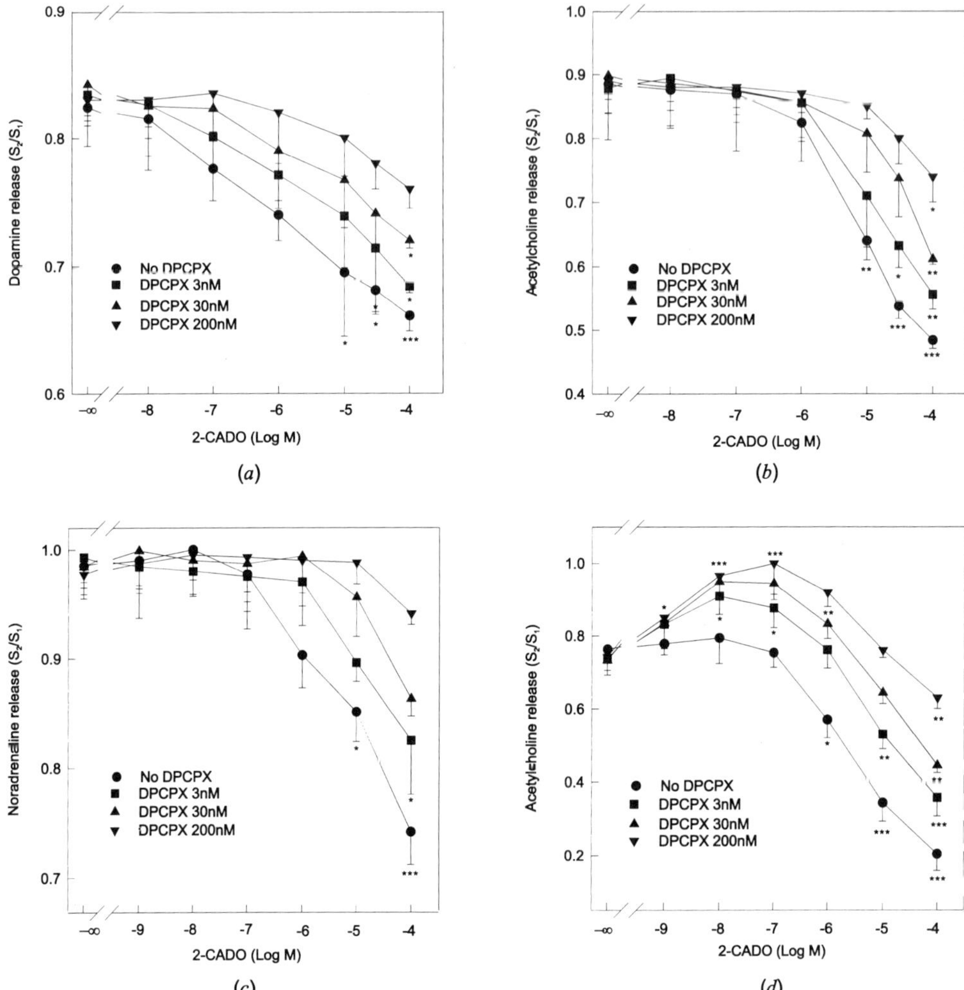

FIGURE 3. The effect of increasing concentrations of 2-chloroadenosine (2-CADO) on the electrically evoked release of (**a**) [^{3}H] dopamine and (**b**) [^{14}C] ACh from rat striatal slices, and the electrically evoked release of (**c**) [^{3}H] noradrenaline and (**d**) [^{14}C] ACh from rat hippocampal slices. Slices were stimulated twice with 10-V pulses delivered at the rate of 1 Hz for 5 minutes. Concentration–response curves to 2-CADO were studied in the absence and presence of DPCPX (3, 30, and 200 nM). DPCPX was present from the start of superfusion and 2-CADO was added to the superfusion buffer from 15 minutes prior to the second period of stimulation (S2) until the end of the experiment. Responses were expressed as the ratio of radioactivity overflow during the control period (S1) and S2. Each point is the mean ± S.E.M. of 4 to 18 observations. A significant difference is represented by: $*P < 0.05$; $**P < 0.01$; $***P < 0.001$.

previously with respect to the inhibition of transmitter release, adenosine stimulation of release may rely upon the enhancement of the activity of only certain kinds of Ca^{2+} channels, and different sets of these channels may support electrically and K^+-stimulated transmitter release. Alternatively the site of action may be completely different in the two preparations. In the slice preparation, the release of ACh from the short cholinergic neurons may be influenced by adenosine mainly at the somatodendritic receptors, whereas adenosine obviously must act at the level of the terminal in the synaptosomal preparation. There is one more, rather remote, possibility: in striatal slices the inhibitory effect of the D_2 agonist quinpirole can be antagonized by an A_{2A} agonist (Jin et al., 1993). Perhaps endogenous dopamine released from the dopaminergic synaptosomes exerts an influence on the cholinergic synaptosomes that is stronger than that found in slices. Thus, the effect of A_{2A} agonists may be to reduce the inhibitory actions of dopamine, rather than to directly enhance release.

The differential effect of A_{2A} receptors in regulating striatal and hippocampal ACh release agrees with what is known about the distribution of the corresponding mRNA. The cholinergic neurons in the striatum are interneurons, with short processes (Gerfen, 1992), whereas the cholinergic nerve terminals in the hippocampus belong to long neurons originating in the septum (Amaral and Kurz, 1985; Wainer et al., 1985). The latter have been shown to express mRNA for A_{2A} receptors (Cunha et al., 1995), whereas the former have not (Johansson et al., 1994; Fink et al., 1992; Schiffmann et al., 1990). Thus, the expression of A_{2A} receptors appears to be linked not to the type of main transmitter used in a neuron, but to other characteristics. This conclusion is also borne out by the fact that mRNA for A_{2A} receptors is only found in one subset of GABAergic neurons in the caudate putamen, namely those that also express enkephalin (Johansson et al., 1994; Fink et al., 1992). There is even some evidence that A_{2A} receptors may modulate ACh release only in a part of the hippocampal cholinergic nerve supply (Cunha et al., 1994b).

The hippocampal cholinergic neurons thus appear to possess both A_1 and A_{2A} receptors which modulate ACh in opposite directions, in agreement with a previous suggestion (Cunha et al., 1994a). In fact, the two types of adenosine effects seem to be so balanced that when low concentrations of a nonselective agonist are used, there is no net effect on release, despite the fact that these same low concentrations can modulate release when one of the receptors is blocked. The fact that the A_{2A}-selective antagonist KF 17387, when given alone, reduced ACh release from the hippocampus suggests that even the background level of adenosine found in such a slice preparation (about 0.2–1 μM; Dunwiddie et al., 1981) is sufficient to activate A_{2A} and A_1 receptors to influence release. A basal, tonic influence of adenosine on ACh release from the hippocampus may occur also *in vivo,* as suggested by recent microdialysis data (Carter et al., 1995). However, the latter interpretation is complicated by the fact that the adenosine antagonist was given systemically and may thus have affected ACh release in the hippocampus, not only via presynaptic receptors at the level of the nerve terminal, but also at the somatodendritic level. There is, in fact, good evidence that the rate of firing of some cholinergic neurons, and thus the release of ACh from them, can be regulated by somatodendritic A_1 receptors (Rainnie et al., 1994).

Thus, A_1 receptors appear to be widely present, whereas A_{2A} receptors appear to play a more circumscribed role, as might also have been predicted from their more limited distribution in the nervous system.

SUMMARY

Studies of neurotransmitter release have clearly established the widespread importance of adenosine as a modulator of synaptic transmission, and in the brain at least, it appears that the majority of synapses that have been studied are modulated in some way by adenosine. Recent advances in the analysis of adenosine actions at a number of synapses suggest that there is not a *single* common mechanism by which adenosine modulates release, but rather that there are *multiple* mechanisms, which in some cases may be acting in parallel at the same synapse to achieve similar ends. Thus, all of the proposed mechanisms of adenosine action (modulation of vesicle docking and priming, inhibition of Ca^{2+} channel function, reduction in the Ca^{2+} sensitivity of the release mechanism, and alteration of electrical events in the presynaptic nerve terminal) may occur at different synapses, and at some, several of these mechanisms may act in a concerted fashion to modulate release. The relatively recent observation that multiple Ca^{2+} channel subtypes (adenosine-sensitive as well as adenosine-insensitive) contribute to the Ca^{2+} influx that supports transmitter release from a single terminal has suggested that some of the earlier studies concerning mechanisms of adenosine modulation need reevaluation. As our knowledge of the cellular mechanisms that underlie transmitter release becomes more sophisticated, our hypotheses concerning how adenosine acts will no doubt become more detailed and specific.

REFERENCES

Abraham EH, Prat AG, Gerweck L, Seneveratne T, Arceci RJ, Kramer R, Guidotti G, Cantiello HF (1993): The multidrug resistance (mdr 1) gene product functions as an ATP channel. Proc Natl Acad Sci USA 90:312–316.

Akhondzadeh S, Stone TW (1994): Interaction between adenosine and $GABA_A$ receptors on hippocampal neurones. Brain Res 665:229–236.

Al-Awqati Q (1995): Regulation of ion channels by ABC transporters that secrete ATP. Science 269:805–806.

Alexander SP, Reddington M (1989): The cellular localization of adenosine receptors in rat neostriatum. Neuroscience 28:645–651.

Amaral DJ, Kurz J (1985): An analysis of the origin of the cholinergic and noncholinergic septal projections to the hippocampal formation of the rat. J Comp Neurol 240:37–59.

Barraco RA, Clough-Helfman C, Goodwin BP, Anderson GF (1995): Evidence for presynaptic adenosine A_{2A} receptors associated with norepinephrine release and their desensitization in the rat nucleus tractus solitarius. J Neurochem 65:1604–1611.

Bennett MR, Ho S (1991): Probabilistic secretion of quanta from nerve terminals in avian ciliary ganglia modulated by adenosine. J Physiol (London) 440:513–527.

Birnbaumer L (1992): Receptor-to-effector signaling through G proteins: Roles of beta-gamma dimers as well as alpha subunits. Cell 71:1069–1072.

Birnstiel S, Gerber U, Greene RW (1992): Adenosine-mediated synaptic inhibition: Partial blockade by barium does not prevent anti-epileptiform activity. Synapse 11:191–196.

Broad RM, Fredholm BB (1996): A_1, but not A_{2A}, adenosine receptors modulate electrically-stimulated [^{14}C] acetylcholine release from rat cortex. J Pharmacol Exp Ther 277:193–197.

Brown SJ, James S, Reddington M, Richardson PJ (1990): Both A_1 and A_{2A} purine receptors regulate striatal acetylcholine release. J Neurochem 55:31–38.

Capogna M, Gahwiler BH, Thompson SM (1994): alpha-Latrotoxin enhances synaptic transmission but does not prevent presynaptic inhibition induced by adenosine, baclofen, and opioid peptides in hippocampus. Soc Neurosci Abstr 20:1120.

Carter AJ, O'Connor WT, Carter MJ, Ungerstedt U (1995): Caffeine enhances acetylcholine release in the hippocampus in vivo by a selective interaction with adenosine A_1 receptors. J Pharmacol Exp Ther 273:637–642.

Cunha RA, Johansson B, Van der Ploeg I, Sebastiáo AM, Ribeiro JA, Fredholm BB (1994a): Evidence for functionally important adenosine A_{2A} receptors in the rat hippocampus. Brain Res 649:208–216.

Cunha RA, Johansson B, Constantino D, Sebastiáo AM, Fredholm BB (1996): Evidence for high affinity binding sites for the adenosine A_{2A} receptor agonist [^{3}H]CGS21680 in the rat hippocampus and cerebral cortex that are different from striatal A_{2A} receptors. Naunyn Schmiedebergs Arch Pharmacol 353:261–271.

Cunha RA, Johansson B, Fredholm BB, Ribeiro JA, Sebastiáo AM (1995): Adenosine A_{2A} receptors stimulate acetylcholine release from nerve terminals of the rat hippocampus. Neurosci Lett 196:41–44.

Cunha RA, Milusheva E, Vizi ES, Ribeiro JA, Sebastiáo AM (1994b): Excitatory and inhibitory effects of A_1 and A_{2A} adenosine receptor activation on the electrically evoked [^{3}H]acetylcholine release from different areas of the rat hippocampus. J Neurochem 63:207–214.

Dolphin AC, Forda SR, Scott RH (1986): Calcium-dependent currents in cultured rat dorsal root ganglion neurones are inhibited by an adenosine analogue. J Physiol (London) 373:47–61.

Dunwiddie TV (1985): The physiological role of adenosine in the central nervous system. Int Rev Neurobiol 27:63–139.

Dunwiddie TV: (1990): Electrophysiological aspects of adenosine receptor function. In Williams M (ed): "Adenosine and Adenosine Receptors." Clifton, NJ: Humana Press, pp 143–172.

Dunwiddie TV, Hoffer BJ (1980): Adenine nucleotides and synaptic transmission in the in vitro rat hippocampus. Br J Pharmacol 69:59–68.

Dunwiddie TV, Miller KK (1993): Effects of adenosine and cadmium on presynaptic fiber spikes in the CA1 region of rat hippocampus in vitro. Neuropharmacology 32:1061–1068.

Dunwiddie TV, Proctor WR (1988): Mechanisms underlying adenosine modulation of hippocampal excitatory transmission. In Paton DM (ed): "Adenosine and Adenine Nucleotides: Physiology and Pharmacology." London: Taylor & Francis, pp 111–120.

Dunwiddie TV, Hoffer BJ, Fredholm (1981): Alkylxanthines elevate hippocampal excitability: Evidence for a role of endogenous adenosine. Naunyn Schmiedebergs Arch Pharmacol 316:326–330.

Dunwiddie TV, Taylor M, Lupica CR, Cass WA, Zahniser NR, Fredholm BB (1991): Adenosine modulation of excitatory transmission in the rat hippocampus. In Imai S, Nakazawa M (eds): "Role of Adenosine and Adenine Nucleotides in the Biological System." Amsterdam: Elsevier, pp 631–644.

Fastbom J, Pazos A, Palacios JM (1987): The distribution of adenosine A_1 receptors and 5′-nucleotidase in the brain of some commonly used experimental animals. Neuroscience 22:813–826.

Fink JS, Weaver DR, Rivkees SC, Peterfreund RA, Pollack AE, Adler EM, Reppert SM (1992): Molecular cloning of the rat A_2 adenosine receptor: Selective co-expression with D_2 dopamine receptors in rat striatum. Mol Brain Res 14:186–195.

Fischer von Mollard G, Stahl B, Khokhlatchev A, Südhof TC, Jahn R (1994): Rab3C is a synaptic vesicle protein that dissociates from synaptic vesicles after stimulation of exocytosis. J Biol Chem 269:10971–10974.

Fredholm BB (1974): Vascular and metabolic effects of theophylline, dibutyryl cyclic AMP, and dibutyryl cyclic GMP in canine subcutaneous adipose tissue in situ. Acta Physiol Scand 90:226–236.

Fredholm BB (1976): Release of adenosine-like material from isolated perfused dog adipose tissue following sympathetic nerve stimulation and its inhibition by adrenergic alpha-receptor blockade. Acta Physiol Scand 96:422–430.

Fredholm BB (1990a): Adenosine A_1-receptor-mediated inhibition of evoked acetylcholine release in the rat hippocampus does not depend on protein kinase C. Acta Physiol Scand 140:245–255.

Fredholm BB (1990b): Differential sensitivity to blockade by 4-aminopyridine of presynaptic receptors regulating [^{3}H]acetylcholine release from rat hippocampus. J Neurochem 54:1386–1390.

Fredholm BB (1995a): Purinoceptors in the nervous system. Pharmacol Toxicol 76:228–239.

Fredholm BB: (1995b): Modulation of neurotransmitter release by heteroreceptors. In Powis DA, Bunn SJ (eds): "Neurotransmitter Release and Its Modulation: Biochemical Mechanisms, Physiological Function and Clinical Relevance." Cambridge: Cambridge University Press, pp 104–121.

Fredholm BB, Dunwiddie TV (1988): How does adenosine inhibit transmitter release? Trends Pharmacol Sci 9:130–134.

Fredholm BB, Hedqvist P (1978): Release of [^{3}H]-purines from [^{3}H]-adenine labelled rabbit kidney following sympathetic nerve stimulation, and its inhibition by alpha-adrenoceptor blockade. Br J Pharmacol 64:239–245.

Fredholm BB, Hedqvist P: (1979): Adenosine—a transsynaptic modulator of norepinephrine release? In E Usdin, IJ Kopin, JD Barchas (eds): "Catecholamines: Basic and Clinical Frontiers." New York: Pergamon Press, pp 1146–1148.

Fredholm BB, Hedqvist P (1980): Modulation of neurotransmission by purine nucleotides and nucleosides. Biochem Pharmacol 29:1635–1643.

Fredholm BB, Hu P-S (1993): Effect of an intracellular calcium chelator on the regulation of electrically evoked [^{3}H]-noradrenaline release from rat hippocampal slices. Br J Pharmacol 108:126–131.

Fredholm BB, Abbracchio MP, Burnstock G, Daly JW, Harden TK, Jacobson KA, Leff P, Williams M (1994): Nomenclature and classification of purinoceptors. Pharmacol Rev 46:143–156.

Fredholm BB, Dunér-Engström M, Fastbom J, Jonzon B, Lindgren E, Nordstedt C (1988): Formation and actions of adenosine in the rat hippocampus, with special reference to the interactions with classical transmitters. In M Avoli, TA Reader, RW Dykes, P Gloor (eds): "Molecules to Mind: Neurotransmitters and Cortical Function." New York: Plenum Press, pp 437–451.

Fredholm BB, Fried G, Hedqvist P (1982a): Origin of adenosine released from rat vas deferens by nerve stimulation. Eur J Pharmacol 79:233–243.

Fredholm BB, Gerwins P, Johansson B, Parkinson FE, Van der Ploeg I (1993): Signalling via G-protein coupled receptors—using adenosine receptors as an example. Drug Des Discov 9:189–197.

Fredholm BB, Hedqvist P, Lindström K, Wennmalm R (1982b): Release of nucleosides and nucleotides from the rabbit heart by sympathetic nerve stimulation. Acta Physiol Scand 116:285–295.

Fredholm BB, Proctor WR, Van der Ploeg I, Dunwiddie TV (1989): In vivo pertussis toxin treatment attenuates some, but not all, adenosine A_1 effects in slices of the rat hippocampus. Eur J Pharmacol 172:249–262.

Freissmuth M, Schutz W, Linder ME (1991): Interactions of the bovine brain A_1-adenosine receptor with recombinant G protein alpha-subunits. Selectivity for rG_i alpha-3. J Biol Chem 266:17778–17783.

Geppert M, Goda Y, Hammer RE, Li C, Rosahl TW, Stevens CF, Sudhof TC (1994): Synaptotagmin I: A major Ca^{2+} sensor for transmitter release at central synapses. Cell 79:717–727.

Gerfen CR (1992): The neostriatal mosaic: Multiple levels of compartmental organization. Trends Neurosci 15:133–139.

Ginsborg BL, Hirst GDS (1972): The effect of adenosine on the release of the transmitter from the phrenic nerve of the rat. J Physiol (London) 224:629–645.

Greene RW, Haas HL (1991): The electrophysiology of adenosine in the mammalian central nervous system. Prog Neurobiol 36:329–341.

Gunderson KL, Kopito RR (1995): Conformational states of CFTR associated with channel gating: The role of ATP binding and hydrolysis. Cell 82:231–239.

Hamilton BR, Smith DO (1991): Autoreceptor-mediated purinergic and cholinergic inhibition of motor nerve terminal calcium currents in the rat. J Physiol (London) 432:327–341.

Hamilton SL, Codina J, Hawkes MJ, Yatani A, Sawada T, Strickland FM, Froehner SC, Spiegel AM, Toro L, Stefani E, Birnbaumer L, Brown AM (1991): Evidence for direct interaction of $G_{s\alpha}$ with the Ca^{2+} channel of skeletal muscle. J Biol Chem 266:19528–19535.

Haydon PG, Henderson E, Stanley EF (1994): Localization of individual calcium channels at the release face of a presynaptic nerve terminal. Neuron 13;1275–1280.

Hu PS, Fredholm BB (1989): Alpha 2-adrenoceptor agonist-mediated inhibition of [^{3}H]noradrenaline release from rat hippocampus is reduced by 4-aminopyridine, but that caused by an adenosine analogue or omega-conotoxin is not. Acta Physiol Scand 136:347–353.

Hunt JM, Silinsky EM (1993): Ionomycin-induced acetylcholine release and its inhibition by adenosine at frog motor nerve endings. Br J Pharmacol 110:828–832.

Hunt JM, Redman RS, Silinsky EM (1994): Reduction by intracellular calcium chelation of acetylcholine secretion without occluding the effects of adenosine at frog motor nerve endings. Br J Pharmacol 111:753–758.

Jarvis MF, Williams M (1989): Direct autoradiographic localization of adenosine A_2 receptors in the rat brain using the A_2-selective agonist [^{3}H]CGS 21680. Eur J Pharmacol 168:243–246.

Jessell TM, Kandel ER (1993): Synaptic transmission: A bidirectional and self-modifiable form of cell–cell communication. Neuron 10:1–30.

Jin S, Fredholm BB (1997): Adenosine A_{2A} receptor stimulation increases release of acetylcholine from rat hippocampus but not striatum, and does not affect catecholamine release. J Neurochem 355:48–56.

Jin S, Johansson B, Fredholm BB (1993): Effects of adenosine A_1 and A_2 receptor activation on electrically evoked dopamine and acetylcholine release from rat striatal slices. J Pharmacol Exp Ther 267:801–808.

Johansson B, Fredholm BB (1995): Further characterization of the binding of the adenosine receptor agonist [^{3}H]CGS 21680 to rat brain using autoradiography. Neuropharmacology 34:393–403.

Johansson B, Ahlberg S, Van der Ploeg I, Brene S, Lindefors N, Person H, Fredholm BB (1993a): Effect of long-term caffeine treatment on A_1 and A_2 adenosine receptor binding and on mRNA levels. Naunyn Schmiedebergs Arch Pharmacol 347:407–414.

Johansson B, Georgiev V, Parkinson FE, Fredholm BB (1993b): The binding of the adenosine A_2 selective agonist [^{3}H]CGS 21680 to rat cortex differs from its binding to rat striatum. Eur J Pharmacol 247:103–110.

Johansson B, Lindstrom K, Fredholm BB (1994): Differences in the regional and cellular localization of c-fos messenger RNA induced by amphetamine, cocaine and caffeine in the rat. Neuroscience 59:837–849.

Kirk IP, Richardson PJ (1994): Adenosine A_{2A} receptor-mediated modulation of striatal [^{3}H]GABA and [^{3}H]acetylcholine release. J Neurochem 62:960–966.

Klapstein GJ, Colmers WF (1992): 4-Aminopyridine and low Ca^{2+} differentiate presynaptic inhibition mediated by neuropeptide Y, baclofen and 2-chloroadenosine in rat hippocampal CA1 in vitro. Br J Pharmacol 105:470–474.

Kullmann DM, Siegelbaum SA (1995): The site of expression of NMDA receptor-dependent LTP: New fuel for the fire. Neuron 15:997–1002.

Kurokawa M, Kirk IP, Kirkpatrick KA, Kase H, Richardson PJ (1994): Inhibition by KF17837 of adenosine A_{2A} receptor-mediated modulation of striatal GABA and ACh release. Br J Pharmacol 113:43–48.

Kuz AK, Bültmann R, Driessen B, von Kügelgen I, Starke K (1994): Release of ATP in rat vas deferens: Origin and role of calcium. Naunyn Schmiedebergs Arch Pharmacol 350:491–498.

Lambert NA, Teyler TJ (1991): Adenosine selectively depresses excitatory synaptic transmission in area CA1 of the rat hippocampus. Neurosci Lett 122:50–52.

Luchelli-Fortis MA, Fredholm BB, Langer SZ (1981): Evidence against the presence of presynaptic inhibitory adenosine receptors in the cat nictitating membrane. J Pharmacol Exp Ther 219:235–242.

Lupica CR, Cass WA, Zahniser NR, Dunwiddie TV (1990): Effects of the selective adenosine A_2 receptor agonist CGS 21680 on in vitro electrophysiology, cyclic AMP formation and dopamine release in the rat CNS. J Pharmacol Exp Ther 252:1134–1141.

MacDonald RL, Skerritt JH, Werz MA (1986): Adenosine agonists reduce voltage-dependent calcium conductance of mouse sensory neurones in cell culture. J Physiol (London) 370:75–90.

Martinez-Mir MI, Probst A, Palacios JM (1991): Adenosine A_2 receptors: Selective localization in the human basal ganglia and alterations with disease. Neuroscience 42:697–706.

Mitchell JB, Lupica CR, Dunwiddie TV (1993): Activity-dependent release of endogenous adenosine modulates synaptic responses in the rat hippocampus. J Neurosci 13:3439–3447.

Mogul DJ, Adams ME, Fox AP (1993): Differential activation of adenosine receptors decreases N-type but potentiates P-type Ca^{2+} current in hippocampal CA3 neurons. Neuron 10:327–334.

Munshi R, Pang I-H, Sternweis PC, Linden J (1991): A_1 adenosine receptors of bovine brain couple to guanine nucleotide-binding proteins G_{i1}, G_{i2} and G_o. J Biol Chem 266:22285–22289.

Offermanns S, Schultz G (1994): Complex information processing by the transmembrane signaling system involving G proteins. Naunyn Schmiedebergs Arch Pharmacol 350:329–338.

Parkinson FE, Fredholm BB (1990): Autoradiographic evidence for G-protein coupled A_2-receptors in rat neostriatum using [^{3}H]-CGS 21680 as a ligand. Naunyn Schmiedebergs Arch Pharmacol 342:85–89.

Pevsner J, Hsu S-C, Braun JEA, Calakos N, Ting AE, Bennett MK, Scheller RH (1994): Specificity and regulation of a synaptic vesicle docking complex. Neuron 13:353–361.

Pierbone VA, Shupliakov O, Brodin L, Hilfiker-Rothenfluh S, Czernik AJ, Greengard P (1995): Distinct pools of synaptic vesicles in neurotransmitter release. Nature 375:493–497.

Rainnie DG, Grunze HC, McCarley RW, Greene RW (1994): Adenosine inhibition of mesopontine cholinergic neurons: Implications for EEG arousal. Science 263:689–692.

Ribeiro JA, Walker J (1975): The effects of adenosine triphosphate and adenosine diphosphate on transmission at the rat and frog neuromuscular junctions. Br J Pharmacol 54:213–218.

Richardson PJ, Brown SJ, Bailyes EM, Luzio JP (1987): Ectoenzymes control adenosine modulation of immunoisolated cholinergic synapses. Nature 327:232–234.

Scanziani M, Capogna M, Gahwiler BH, Thompson SM (1992): Presynaptic inhibition of miniature excitatory synaptic currents by baclofen and adenosine in the hippocampus. Neuron 9:919–927.

Schiffmann SN, Libert F, Vassart G, Dumont JE, Vanderhaeghen J-J (1990): A cloned G protein-coupled protein with a distribution restricted to striatal medium-sized neurons. Possible relationship with D_1 dopamine receptor. Brain Res 519:333–337.

Scholfield CN, Steel L (1989): Presynaptic K-channel blockade counteracts the depressant effect of adenosine in olfactory cortex. Neuroscience 24:81–91.

Scholz KP (1993): Presynaptic inhibition in the hippocampus. Trends Neurosci 16:395–397.

Scholz KP, Miller RJ (1991): Analysis of adenosine actions on calcium currents and synaptic transmission in cultured rat hippocampal pyramidal neurones. J Physiol (London) 435:373–393.

Scholz KP, Miller RJ (1992): Inhibition of quantal transmitter release in the absence of calcium influx by a G protein-linked adenosine receptor at hippocampal synapses. Neuron 8:1139–1150.

Schweizer FE, Betz H, Augustine GJ (1995): From vesicle docking to endocytosis: Intermediate reactions of exocytosis. Neuron 14:689–696.

Schwiebert EM, Egen ME, Hwang T-H, Fulmer SB, Allen SS, Cutting GR, Guggino WB (1995): CFTR regulates outward rectifying chloride channels through an autocrine mechanism involving ATP. Cell 81:1063–1073.

Shapiro MS, Wollmuth LP, Hille B (1994): Modulation of Ca^{2+} channels by PTX-sensitive G-proteins is blocked by N-ethylmaleimide in rat sympathetic neurons. J Neurosci 14:7109–7116.

Silinsky EM (1984): On the mechanism by which adenosine receptor activation inhibits the release of acetylcholine from motor nerve endings. J Physiol (London) 346:243–256.

Silinsky EM (1986): Inhibition of transmitter release by adenosine: Are calcium currents depressed or are the intracellular effects of calcium impaired? Trends Pharmacol Sci 7:180–185.

Silinsky EM, Hunt JM, Solsona CS, Hirsh JK (1990): Prejunctional adenosine and ATP receptors. Ann NY Acad Sci 603:324–333.

Silinsky EM, Solsona CS (1992): Calcium currents at motor nerve endings: Absence of effects of adenosine receptor agonists in the frog. J Physiol (London) 457:315–328.

Silinsky EM, Watanabe M, Redman RS, Qiu R, Hirsh JK, Hunt JM, Solsona CS, Alford S, MacDonald RC (1995): Neurotransmitter release evoked by nerve impulses without Ca^{2+} entry through Ca^{2+} channels in frog motor nerve endings. J Physiol (London) 482:511–520.

Smith DO (1991): Sources of adenosine released during neuromuscular transmission in the rat. J Physiol (London) 432:343–354.

Snyder SH (1985): Adenosine as a neuromodulator. Ann Rev Neurosci 8:103–124.

Stone TW (1991): "Adenosine in the Nervous System." San Diego: Academic Press.

Sudhof TC (1995): The synaptic vesicle cycle: A cascade of protein–protein interactions. Nature 375:645–653.

Takahashi T, Momiyama A (1993): Different types of calcium channels mediate central synaptic transmission. Nature 366:156–158.

Thompson SM, Gahwiler BH (1992): Comparison of the actions of baclofen at pre- and postsynaptic receptors in the rat hippocampus in vitro. J Physiol (London) 451:329–345.

Thompson SM, Capogna M, Scanziani M (1993): Presynaptic inhibition in the hippocampus. Trends Neurosci 16:222–227.

Thompson SM, Haas HL, Gahwiler BH (1992): Comparison of the actions of adenosine at pre- and postsynaptic receptors in the rat hippocampus in vitro. J Physiol (London) 451:347–363.

Vizi ES, Knoll J (1975): The inhibitory effect of adenosine and related nucleotides on the release of acetylcholine. Neuroscience 1:391–398.

Wainer BH, Levey AI, Rye DB, Mesulam MM, Mufson EJ (1985): Cholinergic and noncholinergic septohippocampal pathways. Neurosci Lett 54:45–52.

Wall PD (1995): Do nerve impulses penetrate terminal arborizations? A pre-presynaptic control mechanism. Trends Neurosci 18:99–103.

Wheeler DB, Randall A, Tsien RW (1994): Roles of N-type and Q-type Ca^{2+} channels in supporting hippocampal synaptic transmission. Science 264:107–111.

Wu LG, Saggau P (1994): Adenosine inhibits evoked synaptic transmission primarily by reducing presynaptic calcium influx in area CA1 of hippocampus. Neuron 12:1139–1148.

Yawo H, Chuhma N (1993): Preferential inhibition of omega-conotoxin-sensitive presynaptic Ca^{2+} channels by adenosine autoreceptors. Nature 365:256–258.

Yoon KW, Rothman SM (1991): Adenosine inhibits excitatory but not inhibitory synaptic transmission in the hippocampus. J Neurosci 11:1375–1380.

CHAPTER 20

ATP in Brain Function

MARIA P. ABBRACCHIO
Institute of Pharmacological Sciences, University of Milan, 20133 Milan, Italy

INTRODUCTION

The roles of ATP as a neurotransmitter and neuromodulator in the peripheral nervous system are well established (Burnstock, 1986). The possibility that ATP could also act as a neurotransmitter in the central nervous system was suggested in the early literature by a few scattered studies (see Burnstock, 1977). Release of ATP was demonstrated from rat brain synaptosomes following electrical or chemical stimulation (Pull and MacIlwain, 1975; White, 1977); moreover, excitatory responses to this nucleotide were reported for rat cortical neurons (Phillis et al., 1975). However, for a long time any attempts to demonstrate specific effects of ATP on central neurotransmission have been hampered by the rapid degradation of this nucleotide to adenosine, which itself is a potent neural signaling substance (Scefner and Chiu, 1986). A role for ATP in brain function was also indirectly suggested by the demonstration that both astrocytes and microglial cells express various types of P2-purinoceptors, the receptors specifically responsive to ATP, which may play a role in the activation of these cells during brain repair following trauma and ischemia (Abbracchio et al., 1994, 1995b; for review, see Abbracchio et al., 1995a; Neary et al., 1996; see also Functional Roles of ATP in Brain, below).

It has been only recently that, thanks to both elegant electrophysiological demonstrations of ATP-induced currents in the brain and to the cloning of several members of the large P2 purinoceptor family, ATP has been unanimously recognized as a neurotransmitter in mammalian brain. I shall review here the available knowledge on ATP and related compounds as neural signaling substances in the brain, by providing evidence for nucleotide storage and release from cerebral nerve terminals, for specific mechanisms of inactivation by ecto-ATPases, and for potent and specific actions of ATP mediated by extracellular purinoceptors located on both neuronal and nonneuronal brain cells. Moreover, the initial exciting results on the possible functional role(s)

Purinergic Approaches in Experimental Therapeutics, Edited by Kenneth A. Jacobson and Michael F. Jarvis
ISBN 0-471-14071-6 © 1997 Wiley-Liss, Inc.

of ATP in brain function under both physiological and pathological conditions will be discussed.

STORAGE AND RELEASE OF ATP AND OTHER NUCLEOTIDES BY BRAIN SYNAPTIC TERMINALS

Growing evidence suggests that ATP and other nucleotides are stored in brain nerve endings, from which they can be released by presynaptic action potentials. Costorage of ATP and "classic" neurotransmitters, such as acetylcholine and noradrenaline, has been demonstrated in various peripheral and, more recently, central synaptic vesicles (for reviews, see Burnstock, 1976; Burnstock, 1990; Zimmermann, 1994). For example, ATP is costored in the millimolar range with acetylcholine in rat caudate nucleus and guinea pig cerebral cortex. In general, ATP is outnumbered by its neurotransmitter (in striatal cholinergic terminals the ratio acethylcholine : ATP is 7 : 1). The molecular characterization of the vesicular transporter for ATP is still lacking: future studies will show whether there is any similarity between this transporter and those packaging ubiquitous amino acids, such as glutamate and glycine, in presynaptic vesicles. Although further studies are needed in this respect, it is highly likely that, besides cholinergic and noradrenergic cells, other neurons can also store and release ATP (see also Functional Roles of ATP in Brain, below).

Although concentrations are higher for ATP, brain synaptic vesicles also contain other nucleotides in great amounts, in particular, the diadenosine polyphosphates AP_4A and AP_5A. These compounds have been previously shown to be stored in the chromaffin granules from the adrenal medulla (Pintor, et al., 1992a) and in cholinergic synaptic vesicles from the electric organ of the electric ray (Pintor et al., 1992b), and have been postulated to play a role in neurotransmission (Pintor and Miras-Portugal, 1993). Ca^{2+}-dependent release of both ATP and diadenosine polyphosphates has been demonstrated directly from rat brain synaptosomes (Richardson and Brown, 1987; Pintor and Miras-Portugal, 1993). Interestingly, certain neurotoxins were shown to dissociate release of the cotransmitter by selectively blocking release of either ATP or costored acetylcholine and noradrenaline (Farinas et al., 1992; von Kugelgen and Starke, 1991). Whether this reflects contribution of nonvesicular ATP release or effects of toxins on ecto-ATPases still remains to be established.

Adenosine triphosphate can also be released from cellular sites other than storage presynaptic vesicles. It has long been known that ATP and other nucleotides are massively released from the cytosolic metabolic pool of various cell types under conditions of hypoxia. The mechanism responsible for this nonexocytotic release of ATP is unknown, although involvement of either a specific carrier or a channel has been postulated (Zimmermann, 1994). In brain, following a number of insults to the brain, such as stroke and trauma, release of nucleotides can occur from glial and endothelial cells (Neary et al., 1996; see also Functional Roles of ATP in Brain, below). Under such pathological conditions, nucleotides are also produced from degradation of nucleic acids of dying cells. This exposes cells at the site of damage to elevated levels of these compounds for prolonged time periods. On these bases, a role for nucleotides and nucleosides in the activation of brain repair mechanisms, including astrocytic and microglial activation and neuritogenesis, has been proposed (Abbracchio et al., 1994, 1995a,b; Neary et al., 1996; see also Functional Roles of ATP in Brain, below).

REGULATION OF BRAIN CONCENTRATIONS OF NUCLEOTIDES BY ECTO-ATPASES

Following its release from nervc endings, ATP is inactivated by ubiquitous membrane-bound hydrolyzing enzymes, whose presence has been demonstrated in neuronal, glial, and endothelial cells (James and Richardson, 1993; Hohmann et al., 1993, Zimmermann, 1994). The end-product of ATP breakdown is generally adenosine (which acts itself as a neurotransmitter at P1 purinoceptors; Burnstock, 1978), as shown by the sequential appearance of ADP, AMP, and adenosine in the medium upon incubation of intact synaptosomes with ATP. Inactivation of ATP, therefore, procedes through the sequential involvement of ecto-ATPase, ecto-ADPase, (whose activity is generally lower with respect to ATPases) and ecto-5′-nucleotidases, which hydrolyse extracellular AMP to adenosine. The presence of an ectoapyrase, an ectodiphosphohydrolase that breaks down both ATP and ADP, has also been demonstrated. A common feature of these enzymes is a relatively broad substrate specificity for purine and pyrimidine trinucleotides. This enables them to also hydrolyze UTP and GTP. Although far less is known about the functional significance of the enzymatic degradation of these other nucleotides, the relatively broad substrate specificity of these enzymes would imply that elevated local concentrations of uridine and guanosine derivatives are likely produced in brain from degradation of nucleic acids, especially under pathological conditions characterized by massive cell death (see also Functional Roles of ATP in Brain, below).

The kinetics of ATP enzymatic degradation is of great functional importance, since, in neural tissues, not only does ecto-5′-nucleotidase (which mediates the last enzymatic step leading to generation of adenosine) have a much lower activity with respect to ecto-ATPase and ecto-ADPase, but this enzyme has been also shown to be potently inhibited by ATP and ADP (Zimmermann, 1994). This feedforward inhibition of 5′-nucleotidase results in a delay in the generation of extracellular adenosine following ATP release, and therefore in a physiologically significant prolongation of ATP and ADP actions.

Enzymatic breakdown of diadenosine polyphosphates (AP_xA) is also beginning to be elucidated. For AP_4A, hydrolysis by ectodinucleoside polyphosphate hydrolase leads to generation of ATP and, subsequently, of adenosine. Interestingly, in brain, diadenosine polyphosphates appear to be significantly more resistant to hydrolysis with respect to ATP, which would enable them to exert potent and prolonged effects on cerebral neurotransmission.

ACTIVATION OF BOTH LIGAND-GATED ION CHANNELS AND G PROTEIN–COUPLED RECEPTORS IN BRAIN BY ATP AND RELATED NUCLEOTIDES

Responses to ATP are due to ATP itself or to one of its metabolic products. Back in 1978, in a now seminal review, Burnstock proposed that the actions of ATP and related nucleotides and nucleosides were mediated by P1 purinoceptors (preferentially activated by adenosine, and, to a lesser extent, by AMP and ADP), and P2 purinoceptors, at which the rank order of potency for these compounds was ATP > ADP > AMP (adenosine being totally ineffective). Other criteria were proposed by Burnstock to distinguish between P1 and P2 purinoceptors, such as specific antagonism of P1 purinoceptor–mediated effects by xanthines, and coupling to adenylate cyclase for P1 purinoceptors, and to prostaglandin synthesis for P2 purinoceptors (Burnstock, 1978). Although some of the original criteria utilized for this receptor classification have inevitably been updated on the basis of new data reported in the literature, and despite the fact that several subclasses of both the P1 and P2 purinoceptor families have now been identified, this main subdivision is still acknowledged by the International Union of Pharmacologists (IUPHAR) Subcommittee for Purinoceptor Nomenclature and Classification (Fredholm et al., 1994).

It is now clear that ATP activates both ligand-gated cation channels (mainly for Ca^{2+}, K^+, and Na^+) and G protein–coupled receptors linked to activation of phospholipases C and A2. Ligand-gated channels are likely to be responsible for the fast responses induced by ATP, whereas G protein–coupled receptors mediate the slower modulatory responses to the nucleotide (Abbracchio and Burnstock, 1994). At least six different purinoceptor subtypes have been identified so far on the basis of pharmacological data (Fredholm et al., 1994): the P_{2X} (ion channel), the P_{2Y} (G protein–coupled), the P_{2U} (G protein–coupled and responsive to both ATP and UTP), the P_{2T} (the G protein–coupled receptor for ADP in platelets), the P_{2D} (the G protein–coupled receptor specifically responsive to diadenosine polyphosphates), and the P_{2Z} (relatively nonselective ion channel in mast cells). Moreover, a large heterogeneity within the P_{2X} and P_{2Y} subtypes has been hypothesized on the basis of differential responses to novel ATP analogs in different experimental models (Fischer et al., 1993; Jacobson et al., 1995). Expression cloning data are now available for the P_{2X}, P_{2Y}, and P_{2U} purinoceptors and confirm the existance of various P2 purinoceptor subtypes (Barnard et al., 1994; Surprenant et al., 1995; Williams and Jacobson, 1995), with six different P_{2X}- and six different P_{2Y}-like subtypes cloned so far (Burnstock et al., 1996; see also Tables 1 and 2). A reorganization of this nomenclature has been recently proposed, in which all the G protein–coupled receptors are grouped in a P2Y superfamily (which therefore also includes the P_{2U}, P_{2T}, and P_{2D} subtypes), and all the ion channel purinoceptors are to be considered subtypes within a P2X superfamily (Abbracchio and Burnstock, 1994). Within the two families, numbers are assigned simply based on date of cloning. On this basis, the already cloned ligand-gated receptors have been named $P2X_{1-6}$ and the cloned G protein–coupled receptors have been named $P2Y_{1-6}$. No cloning data are so far available for the P_{2D} and the P_{2z} purinoceptors.

Several P2X and P2Y purinoceptors are expressed in the mammalian brain (summarized in Tables 1 and 2), a finding that suggests their importance in a variety of central nervous system functions.

The P2X Purinoceptor

Discovery of ATP-Evoked Currents The discovery of ATP-dependent excitatory potentials at various peripheral effector junctions dates back to the early 1960s. Stimulation of the sympathetic innervation to various organs elicited rapidly rising depolarizations that were not abolished by antagonists at cholinergic and noradrenergic receptors (NANC transmission; Burnstock, 1972). These effects were indeed due to electrically evoked corelease of ATP. A number of subsequent reports confirmed that ATP acted as a fast transmitter at sympathetic and parasympathetic ganglia, and at enteric neurons (for review, see Illes and Norenberg, 1993). The first direct evidence that ATP mediates excitatory transmission at neuro–neuronal synapses came from studies performed on neurons of the guinea pig celiac ganglion (Silinsky et al., 1992; Evans et al., 1992),

TABLE 1 Brain Expression of Cloned P2X Purinoceptors

P2 Purinoceptor Subtype	Tissue (Source for the Cloning Studies)	Characteristics	Expression in Brain	Possible Function
$P2X_1$[a]	Vas deferens (rat)	Channel, $I_{Na/K/Ca}$	Cerebellum; lower levels in striatum hippocampus, cortex[b]	Role in onset and maintenance of LTP[e]; regulation of motor function[b,f]; role in apoptosis[g]? Specific roles of the various subtypes unknown
$P2X_2$[c]	PC 12 cells (rat)	Channel, $I_{Na/K}$	High, especially in olfactory tubercles, striatum hypothalamus, hippocampus, amygdala, cortex, cerebellum[d]	
$P2X_3$[h]	DRG cells (rat)	Channel, $I_{Na/K}$	No expression	
$P2X_4$	Hippocampus[i] (rat) DRG cells[j] (rat)	Channel, $I_{Na/K}$	Cerebellum (Purkinje cells), cortex, thalamus, hippocampus and spinal cord[i,j]	
$P2X_5$[j]	Celiac ganglia (rat)	Channel, $I_{Na/K/Ca}$	Very low[k]	
$P2X_6$[j]	Superior cervical ganglion (rat)	Channel, $I_{Na/K/Ca}$	Heavily expressed in cerebellum, ependyma, olfactory bulb, cortex, hippocampus, and thalamus[l]	

[a] Valera et al., 1994.
[b] Kidd et al., 1995.
[c] Brake et al., 1994.
[d] Kidd et al., 1995; see also text.
[e] Motin and Bennett, 1995.
[f] Bo and Burnstock, 1994.
[g] Hypothesized on the bases of structure identity with RP-2 cDNA, which encodes for a protein highly expressed in thymus during immunocompetent T-cells selection; see also Surprenant et al., 1995.
[h] Chen et al., 1995; Lewis et al., 1995.
[i] Bo et al., 1995.
[j] Collo et al., 1996. This study also demonstrates a remarkable overlapping of $P2X_4$ and $P2X_6$ purinoceptor mRNA distribution in rat brain which clearly suggests expression in the same cells.
[k] $P2X_5$ mRNA is detected only in the mesencephalic nucleus of the trigeminal nerve (Collo et al., 1996).
[l] With the exception of the ependyma, the brain expression of $P2X_6$ closely mirrors that of $P2X_4$, see also note 10.

The pharmacological characteristics of the various P2X subtypes are described in the text. See also Note Added in Proof.

TABLE 2 Brain Expression of Cloned P2Y Purinoceptors

P2 Purinoceptor Subtype	Tissue (Source for the Cloning Studies)	Characteristics	Expression in Brain	Possible Function
$P2Y_1$	Brain[a]	G protein–coupled ($PLC\beta/IP_3/Ca^{2+}$)	Telencephalon, diencephalon, mesencephalon, cerebellum[b]	
	Insulinoma cells[c]	G protein–coupled ($PLC\beta/IP_3/Ca^{2+}$)	Present in brain	Role in visual and auditory signal processing? Role in reactive astrogliosis Role in microglia activation?
	Placenta[d] (human)	G protein–coupled ($PLC\beta/IP_3/Ca^{2+}$)		
	Endothelium[e] (bovine)	G protein–coupled ($PLC\beta/IP_3/Ca^{2+}$)	N.D.	
$P2Y_2$	NG-108-15 cells[f] (mouse)		Present in brain (mouse)	
	CT/43 cells[g] (human)	G protein–coupled ($PLC\beta/IP_3/Ca^{2+}$)	Undetected in human brain[h]	Role in reactive astrogliosis Role in microglia activation?
	Lung[i] (rat)		Undetected in rat brain[i]	
	Bone[j] (human)			
	Pituitary[k] (rat)		N.D.	
$P2Y_3$	Brain[l] (chick)	G protein–coupled ($PLC\beta/IP_3/Ca^{2+}$)	Expressed in adult chicken brain[l]	
$P2Y_4$	Placenta[m] (human)	G protein–coupled ($PLC\beta/IP_3/Ca^{2+}$)	No apparent expression	
$P2Y_5$	HEL cells[n] (human)	N.D.	N.D.	
	Lymphocytes (chicken)[o]	N.D.	N.D.[o]	
$P2Y_6$	Aortic smooth muscle[p] (rat)	G protein–coupled ($PLC\beta/IP_3/Ca^{2+}$)	Undetected in cerebrum and cerebellum	

TABLE 2. (*Continued*)

[a] Chick brain (Webb et al., 1993) and turkey brain (Filtz et al., 1994).
[b] Webb et al., 1994.
[c] Mouse and rat (Tokuyama et al., 1995).
[d] Leon et al., 1996.
[e] Henderson et al., 1995.
[f] Lustig et al., 1993.
[g] Parr et al., 1994.
[h] Despite its substantial similarity (89% identity) to the mouse receptor sequence reported by Lustig et al. (1993), this receptor does not seem to be highly expressed in brain (as assessed by Northern blot analysis). See text for further discussion on this point.
[i] Rice et al., 1995. This receptor has been cloned from rat alveolar type II cells; in the same study. Northern blot analysis did not reveal any apparent expression in *rat* brain, despite the presence of this subtype in *mouse* brain (Lustig, et al., 1993). This apparent discrepancy, together with that described in 8, is further discussed in the text.
[j] Bowler et al., 1995.
[k] Chen et al., 1996.
[l] Webb et al., 1996a.
[m] Communi, et al., 1996; this is the cloning of the first "uridine" nucleotide receptor. At this receptor, UTP and UDP are equipotent and behave as full agonists, whereas ATP behaves as a partial agonist (ADP inactive). There are rumors that this receptor has been also cloned from rat brain, but no information is available yet.
[n] Kunapuli et al., 1995.
[o] Webb et al., 1996b. No data available yet on brain expression of $P2Y_5$; however, the $P2Y_5$ seems related to a human sequence derived from infant brain.
[p] Chang et al., 1995.

The pharmacological characteristics of the various P2X subtypes are described in the text.
N.D.-Not Determined.

where the P2 purinoceptor antagonist suramin was shown to depress the ATP-induced currents but not those produced by acetylcholine. As mentioned before, the direct demonstration of a role for ATP as a fast neurotransmitter in the brain has been delayed considerably by its rapid degradation by ecto-ATPases, and by the subsequent generation of adenosine, which instead potently inhibits synaptic transmission (Scefner and Chiu, 1986). Until the 1990s, only a few observations had been produced in favor of excitatory effects of ATP on central neurons. Salt and Hill (1983) showed that sensory neurons in the rat caudal trigeminal nucleus responded to ATP with an increased firing. Studies performed by Illes and coworkers starting in 1990 clearly demonstrated that, besides possessing inhibitory adenosine receptors (Scefner and Chiu, 1986; Regenold and Illes, 1990), noradrenergic neurons of the rat locus coeruleus nucleus also responded to bath-applied enzymatically stable analogs of ATP with an increased firing rate (Tschopl et al., 1992). ATP caused no significant changes when given alone, but increased the firing rate when applied in the presence of an adenosine receptor antagonist, suggesting a balance between an excitatory P2 purinoceptor–mediated effect and an inhibitory adenosine receptor–mediated effect due to the degradation of ATP to adenosine (Tschopl et al., 1992). Subsequent studies employing intracellular recordings (Harms et al., 1992) and patch-clamp techniques (Shen and North, 1993) confirmed that P2 purinoceptor activation in these neurons elicits inward currents and membrane depolarization. Simultaneously and independently, Ueno et al. (1992) provided evidence for an ATP-gated cation channel with high Ca^{2+} permeability in dissociated rat nucleus tracti solitarii neurons. While studying evoked and miniature synaptic currents in rat medial habenula, Edwards et al. (1992) observed that these currents were unaffected by blockers of glutamate, γ-aminobutyric acid, or acetylcholine receptors but were blocked by suramin and the desensitizing ATP receptor agonist $\alpha\beta$-methylene-ATP. This was the first evidence identifying synaptic currents in the brain mediated directly by ATP receptors. The fast rise-time of the suramin-blocked currents showed that they were mediated by a ligand-activated ion channel rather than a second-messenger system and classified this purinoceptor as belonging to the P2X subtype (Abbracchio and Burnstock, 1994).

Currents evoked by ATP have clear similarities with those mediated by other excitatory transmitters at neuro-neuronal or neuro-muscular synapses (for example, glutamate, acetylcholine, and serotonin). Studies with rapid perfusion methods have indicated that the ATP-gated channel opens within a few milliseconds upon addition of ATP, whereas experiments with excised outside-out patches have shown no need for diffusible second messengers. Similarities with previously characterized ligand-gated channels include rapid kinetics, a cation-selective ionic pore and, at least in some cases, strong desensitization. The high permeability of the P2X purinoceptor to Ca^{2+} is similar to that seen in homomeric $\alpha7$ nicotinic channels, in AMPA-type glutamate receptors, and in NMDA-type glutamate receptors (for review, see Surprenant et al., 1995).

Discovery of a New Structure for Ligand-Gated Channels by Cloning of P2X Purinoceptors On the basis of these similarities, it had been expected that the P2X purinoceptors would bear overall structural resemblance to the other ligand-gated ion channels, such as the nicotinic and NMDA receptors, which are characterized by an extracellular N-terminus and three to five hydrophobic membrane-spanning regions (Surprenant et al., 1995).

So far, six different P2X purinoceptors, $P2X_{1-6}$, have been cloned (however, see Note Added in Proof); all of them (with the single exception of the $P2X_3$ purinoceptor) are expressed in mammalian brain, although to different extents (see Table 1 and below). Isolation of the first two cDNAs encoding P2X purinoceptors in the smooth muscle of vas deferens (Valera et al., 1994) and in pheochromocytoma cells (Brake et al., 1994) showed that, unexpectedly, these receptors are not related to any other ligand-gated channels. The overall structure of the six cloned P2X purinoceptors (which are 36% to 48% identical) resembles more closely that of the amiloride-sensitive epithelial Na^+ channels and of the inward-rectifier family of K^+ channels than any other ligand-gated channels, suggesting a disposition with short intracellular N- and C-terminus regions, two membrane-spanning domains, and most protein located extracellularly (Surprenant et al., 1995). There is a striking conservation of 10 cysteine residues in the extracellular portion of the receptor, suggesting the presence of disulphide bonds, which may stabilize the terniary structure and therefore contribute to the ATP-binding site.

Pharmacological Characterization of the Various P2X Purinoceptors and Their Expression in Brain Although the various P2X purinoceptors are similar in amino acid sequence (and show a relatedness comparable to that seen between AMPA and kainate subclasses of ligand-gated channels), a number of differences emerged when the pharmacological profiles and the distribution of the first two cloned P2X receptors were compared. At the rat vas deferens receptor ($P2X_1$), ATP, 2meSATP, and $\alpha\beta$-meATP are effective at concentrations of 0.5–5 μM, and progressively smaller responses are evoked following repeated agonist applications. This type of receptor is found in many peripheral neurons and in enteric ganglia; this is also the pharmacological profile of the receptor identified by Edwards and coworkers in the medial habenula, although initial Northern blotting studies could not detect any significant expression of this transcript in mammalian brain (Valera et al., 1994, but see also below). At the PC12 receptor ($P2X_2$), ATP and 2meSATP are effective only at 10–20 times higher concentrations and $\alpha\beta$meATP is ineffective. However, more recent data suggest that full sensitivity to ATP can be revealed by acidifying the bathing solution (to pH 6.5), whereas affinity is further diminished in an alkaline solution (pH 8.0; King et al., 1996). A rapid acidic pH shift at the synaptic cleft was previously demonstrated in rat hippocampal slices (Krishtal et al., 1987). One may speculate that the sensitivity of P2X purinoceptors to ATP is enhanced as the local concentration of acid (as coreleased aminoacids or vesicle protons) builds up during neurotransmitter release.

Unlike the $P2X_1$ receptor, the $P2X_2$ subtype undergoes little or no desensitization following repeated agonist applications. Moreover, tissue distribution of the PC12 receptor mRNA showed a significantly higher expression of this receptor in the brain with respect to $P2X_1$ (Table 1), suggesting that this P2X purinoceptor subtype may be highly involved in ATP-mediated communication within brain cells. This is the ATP-gated channel identified in many ganglion cells (Cloues et al., 1993; Fieber and Adams, 1991), tubero-mammillary neurons (Furukawa et al., 1994), outer hair cells of the cochlea (Nakagawa et al., 1990), and in neurons of rat nucleus tracti solitarii (Ueno et al., 1992). Extraordinary high levels of expression of the PC12 purinoceptor mRNA were also found in the pituitary gland (Brake et al., 1994), suggesting a neuroendocrine role for ATP.

More recently, transcript-specific oligonucleotides have been utilized to examine the localization in the rat nervous system of the corresponding mRNAs of the $P2X_1$ and $P2X_2$ purinoceptors (Kidd et al., 1995; Table 1). The presence of these two transcripts in the brain was confirmed by polymerase chain reaction (PCR). The distribution of the $P2X_2$ purinoceptor matched well with the previously reported Northern blotting results (Brake et al., 1994), with significant levels in the olfactory tubercles, striatum, hypothalamus, hippocampus, dentate girus, amygdala, cortex, and cerebellum; on the other hand, data on the $P2X_1$ purinoceptor showing the presence of this transcript in the cerebellum and, at lower levels, in striatum, hippocampus, and cortex were not in keeping with the findings of Valera and coworkers (1994) who did not find any evidence for this purinoceptor mRNA in adult rat brain by Northern blot analysis. This difference can be attributed to the greater sensitivity of the methods utilized by Kidd and coworkers (1995; *in situ* analysis and PCR techniques) with respect to the previous study.

A third P2X purinoceptor subtype (the $P2X_3$) has been recently cloned independently by two groups (Chen et al., 1995; Lewis et al., 1995; Table 1). However, there is no evidence for a possible role of this purinoceptor in brain, since $P2X_3$ has a highly restricted distribution, being selectively expressed at high levels in sensory C fibers running into the spinal cord, which suggests a potential role for this receptor in the central transmission of pain signals.

The cloning of a fourth P2X purinoceptor has also been reported and this does appear to have intriguing implications for ATP signaling in the brain (Bo et al., 1995; Collo et al., 1996; Table 1). A main feature of this receptor is that it is not blocked by any of the currently used purinoceptor blockers [e.g., suramin, reactive blue 2 and pyridoxalphosphate-6-azophenyl-2′,4′-disulfonic acid (PPADS)]; in contrast, their application resulted in a potentiation of the agonist response, similar to the atypical response mediated by purinoceptors in guinea pig intracardiac neurons (Allen and Burnstock, 1990). The $P2X_4$ receptor therefore represents a new pharmacological subtype of P2X purinoceptor, and some of the previously uninterpretable data may be now classified as genuine P2X purinoceptor–mediated events. Moreover, its peculiar responses to classic P2 purinoceptor antagonists suggests that this receptor may be different enough from the other subtypes to be successfully targeted with new ligands. $P2X_4$ mRNA has a widespread distribution in the brain, with particularly high concentrations in cerebellar Purkinje cells, cortex, thalamus, hippocampus, and spinal cord—a striking overlap with the cerebral distribution of the $P2X_6$ purinoceptor mRNA (Collo et al., 1996; see also below).

Two new P2X purinoceptor cDNAs ($P2X_5$ and $P2X_6$) have been recently reported (Collo et al., 1996). Functionally, these two clones most resemble $P2X_2$ and $P2X_4$, since they desensitize slowly and do not respond to $\alpha\beta$meATP. In the study performed by Collo and coworkers (1996), *in situ* hybridization analysis with oligoprobes selective for $P2X_{1-6}$ have confirmed that there is no expression of the $P2X_3$ purinoceptor in the brain, that the $P2X_5$ is only expressed in the mesencephalic nucleus of the trigeminal nerve, and that $P2X_4$ and $P2X_6$ are the predominant brain P2X-purinoceptors (Table 1). The most dense staining for the $P2X_6$ purinoceptor was found in the cerebellar Purkinje cells and the ependyma; strong hybridization signals were also detected in the cortex, olfactory bulb, hippocampus, hypothalamic and thalamic nuclei, the mesencephalic nucleus of the trigeminal nerve, and the cranial nerve motor nuclei (Table 1). Apart from the ependyma, this pattern of staining strikingly mirrors that of the $P2X_4$ purinoceptor. This overlapping, demonstrated for more than 100 discrete brain regions, suggests that the two receptors are expressed by the same neurons (Collo et al., 1996). To further confirm this colocalization, there was a marked loss of the hybridizing signals for both the $P2X_4$ and $P2X_6$ in the hippocampus after treatment with kainic acid, a phenomenon that also supports a neuronal rather than glial localization.

The wide distribution of P2X purinoceptors suggested by the hybrydization studies with oligoprobes for the different P2X subtypes is also consistent with autoradiographic studies obtained with $[^3H]\alpha\beta$meATP binding in rat brain and spinal cord (Bo and Burnstock, 1994). The distribution of $[^3H]\alpha\beta$meATP binding sites showed that many structures in the CNS were densely labelled, and that binding was displaced by both $\beta\gamma$meATP and the P2X purinoceptor antagonist PPADS, indicating that binding occurred at P2X purinoceptors. Image analysis revealed that the nuclei of the thalamus had the highest density. Other densely labelled structures included the cerebral cortex, amygdaloid, substantia nigra, caudate putamen, geniculate nuclei, hypothalamus and medial habenula. A high density of $[^3H]\alpha\beta$meATP binding sites was also reported in the cortex and granular layer of the cerebellum, and, to a smaller extent, in the cerebellar molecular layer. So far, evidence for the presence of P2X purinoceptors in these structures is lacking, but the results of Bo and Burnstock (1994) would suggest a role for ATP as a transmitter also in these brain areas (see also below).

Another autoradiographic study employing $[^3H]\alpha\beta$meATP to detect P2X purinoceptors in rat brain has been recently published (Balcar et al., 1995). Consistent with the results of Bo and Burnstock (1994), in this study, binding of $[^3H]\alpha\beta$meATP was shown to be reversible and inhibited by suramin. Moreover, association/dissociation parameters demonstrated that it consisted of two saturable components. As in the previous study, high levels of relative binding were found in thalamus, cerebral cortex, neostriatum, and hippocampus; however, a strikingly higher density of P2X purinoceptors was detected in cerebellum. In the study of Balcar et al. (1995) the cerebellar cortex was indeed the most densely labelled structure in the brain. The apparent discrepancy between these results and those of Bo and Burnstock may be related to differences in the experimental protocols utilized in the two laboratories (in this respect, differences in the exposure times of autoradiograms may be crucial).

In any case, the high levels of binding found in both studies reinforce the view that P2X purinoceptors are highly involved in brain neurotransmission. Moreover, the lack of a complete correlation between binding studies and the results of *in situ* hybridization analysis (where specific probes for the various subtypes were utilized) suggests that the P2X purinoceptor family currently identified may be destined to further growth.

The P2Y Purinoceptor

Cloning of the G Protein–Coupled P2 Purinoceptors So far, 11 different sequences encoding for P2Y purinoceptors have been reported (for update, see Burnstock, 1996). Some of these are indeed the species homologs of the same P2Y purinoceptor subtype (Table 2). The first P2Y purinoceptor subtype was cloned from chick brain by Webb et al. (1993). Advantage was taken of the fact that an unusual burst of expression of many receptors is known to occur in chick brain at the time of hatching. Moreover, binding studies with a high-affinity P2Y purinoceptor ligand in membranes from newly hatched chick brain had indeed shown that, at this time of development, concentration of specific binding sites was 37 pmol/mg of protein, a result that suggested an exceptionally high expression of P2Y purinoceptors with respect to any other known neurotransmitter receptor (for example, in rat brain, the B_{max} value for muscimol for the ubiquitous $GABA_A$ receptor is 1–2 pmol/mg protein at its maximum; see Barnard et al., 1994). To clone the P2Y purinoceptor, a previously isolated guinea pig partial complementary DNA (cDNA) was used to screen a late embryonic chick whole brain cDNA library under conditions of low stringency, and the resultant clones were tested with adenine nucleotides in the oocyte system. Two quite different full-length cDNAs were eventually isolated that encode for receptors responsive in one case to ATP (Webb et al. 1993) and in the other to ADP (Webb et al., 1996a). The first one was named $P2Y_1$ (being the first P2Y purinoceptor cloned), whereas the ADP-preferring receptor was named $P2Y_3$ (based on the fact that, in the meantime, cloning of another G protein–coupled purinoceptor subtype had been reported, see also below and Table 2). Both have the typical molecular features of G protein–coupled receptors (seven stretches of increased hydrophobicity—probably corresponding to transmembrane domains—linked by regions of higher hydrophilicity) but share only 39% amino acid sequence identity. Simultaneously, Lustig et al. (1993) reported the cloning of the DNA encoding another P2 purinoceptor from a neuroblastoma-glioma cell line. This receptor, which was activated by both ATP and UTP (P2u or $P2Y_2$ according to Abbracchio and Burnstock, 1994) shared only about 40% amino acid sequence homology with the two other cloned receptors, but clearly belonged to the same family (Barnard et al., 1994). All these receptors also show low amino acid sequence identity with any other member of the main G protein–coupled receptor superfamily, suggesting that they may represent a separate and unusual new family of their own within this superfamily (Barnard et al., 1994). Since these original reports, three additional P2Y purinoceptors have been cloned. Their characteristics and brain expression, together with those of $P2Y_{1-3}$, are summarized in Table 2.

Pharmacological Characterization of the Various P2Y Purinoceptors and Their Brain Expression Functional expression of the isolated clones in *Xenopus* oocytes resulted in the appearance of ATP-dependent inward currents, blocked by suramin and showing differential responses to a variety of ATP analogs. For the $P2Y_1$ purinoceptor, the pharmacological characterization of the recombinant receptor revealed an agonist potency order 2meSATP > ATP > ADP ≫ $\alpha\beta$meATP, with UTP being totally ineffective. This was consistent with previously characterized P2Y purinoceptor–mediated responses in mammalian brain (Abbracchio and Burnstock, 1994). Responses to adenine nucleotides were highly specific, with EC_{50} values in the nanomolar range (Webb et al., 1993). As anticipated by the binding studies, *in situ* hybridization analysis of chick brain showed that the $P2Y_1$ mRNA has a widespread but rather specific distribution, being particularly rich in various nuclei of the telencephalon, diencephalon, and mesencephalon, as well as in the external granule, Purkinje, and internal granule cells of the cerebellum (Webb et al., 1994).

The recombinant receptor cloned by Lustig et al. (1993; $P2Y_2$ in Table 2) responded to agonists with the potency order ATP = UTP > ATPγS ≫ 2meSATP = $\beta\gamma$meATP = ADP, a pharmacological profile resembling that of the previously characterized purinoceptor responsive to both ATP and UTP. This receptor is also highly expressed in mouse brain (Table 2). The cloning of the human homolog of this receptor was later reported by Parr et al. (1994). Surprisingly, despite its substantial similarity to the mouse receptor sequence, this receptor did not seem to be significantly expressed in human brain, as detected by Northern blot analysis (Table 2). Similarly, Rice et al. (1995) have found no apparent brain expression of the rat $P2Y_2$ purinoceptor. It is highly unlikely that this reflects real species differences in the expression of this purinoceptor subtype. The inability to detect any $P2Y_2$ mRNA in human (or rat) brain may simply reflect a highly discrete expression of this receptor in the various brain areas, and/or lower levels of expression that are not detected by standard Northern blot techniques. Future studies specifically aimed at evaluating the regional expression of this receptor, with particular focus on the human brain, will clarify this important point.

Expression cloning studies of the $P2Y_3$ purinoceptor (the ADP-preferring clone) confirmed the unusual pharmacological properties of this receptor, at which UDP, UTP and ADP resulted far more potent than other adenine nucleotides, including ATP itself. Also, since this purinoceptor subtype is widely expressed in spinal cord and brain, especially during development, it may play a role in central nervous system formation and maturation (Webb et al., 1996a).

A fourth P2Y purinoceptor has been cloned from human placenta (the $P2Y_4$ purinoceptor, Communi et al., 1996). At this receptor, UTP and UDP act as full agonists, whereas ATP behaves as a partial agonist (ADP being practically inactive). For this reason, this receptor was named the "uridine nucleotide" receptor (Communi et al., 1996); however, its structure clearly indicated that also this receptor belongs to the P2Y family, which therefore encompasses selective purinoceptors (e.g., $P2Y_1$), receptors responsive to both adenine and uridine nucleotides ($P2Y_2$), and UTP-preferring receptors (the newly discovered $P2Y_4$; Communi et al., 1996). A human probe designed on the basis of the $P2Y_4$ purinoceptor sequence gave no hybridization signal in a number of rat tissues, including brain. Again, whether this is due to real species differences in the brain expression of this receptor or to lower expression levels, and/or to the use of different technologies still has to be determined.

The previously cloned G protein–coupled orphan receptor 6H1 has been recently identified as a P2Y purinoceptor, preferentially responding to ATP and, to a lower extent, to UTP (the $P2Y_5$ purinoceptor; Webb et al., 1996b). This receptor is the chicken homolog of the $P2Y_5$ purinoceptor preliminarily communicated by Kunapuli et al. (1995). Although no distribution studies are yet available for this receptor subtype, the chicken $P2Y_5$ sequence is highly related to a human sequence isolated from a cDNA library derived from infant brain (Webb et al., 1996b), suggesting that this receptor may be highly expressed in developing CNS.

The cloning of an additional P2Y purinoceptor ($P2Y_6$) has been also reported from a rat aortic smooth muscle cell line (Chang et al., 1995). Northern blot analysis did not reveal significant expression of this receptor in either cerebrum or cerebellum.

In summary, on the basis of the data available so far, the P2Y purinoceptors expressed in mammalian brain seem to mainly belong to the $P2Y_{1-3}$ subtypes (Table 2). In contrast to neurons, which seem to express both P2X and P2Y purinoceptors, cells of the glial lineage (especially astroglial cells) seem to mainly express metabotropic P2Y purinoceptors (Abbracchio et al., 1995a,b; Bolego et al., 1997; Neary et al., 1996; see also below).

Transduction Mechanisms of P2Y Purinoceptors in the Brain CNS G protein–coupled P2Y purinoceptors have been associated to a number of transduction mechanisms. Activation of P2Y purinoceptors on brain astroglial cells leads to both (i) increases of cytoplasmic Ca^{2+} due to mobilization from inositol phosphate–sensitive intracellular stores, suggesting coupling to phospholipase C, and (ii) arachidonate mobilization and prostanoid release, indicating coupling to phospholipase A_2 (Neary et al., 1988; Neary et al., 1991; Pearce, et al., 1989; Bruner and Murphy, 1990, 1993; Salter and Hicks, 1995; Bolego et al., 1997). P2Y purinoceptor–mediated responses are often, but not always, blocked by pertussis toxin, demonstrating involvement of G_i/G_o proteins. In some cases, partial blockade by pertussis toxin indicates involvement of G_q/G_{11} proteins as well (Bruner and Murphy, 1993). Expression of both the $P2Y_1$ (Filtz et al., 1994) and the $P2Y_2$ purinoceptor subtypes (Lazarowski et al., 1995) in cell lines has shown that both purinoceptors can activate phospholipase C (PLC). A third PLC-activating purinoceptor has been also reported in C6-2B rat glioma cells, and this receptor seems to exclusively respond to UTP, ATP being completely inactive (Lazarowski and Harden, 1994). Although this receptor has not yet been cloned, this report initially reinforced the hypothesis that there may exist "pyrimidinic receptors" specifically activated by UTP and unresponsive to ATP. However, the C6 glioma cells, from which the C6-2B subclone was isolated, express a typical $P2Y_2$ response. A mutation of that receptor, with selective loss of the adenine nucleotides responsiveness, may explain the peculiar behavior of the C6-2B subclone (Communi et al., 1996); therefore, the significance (and even the existence) of this ATP-resistant receptor in mammalian brain is still highly speculative. C6 rat glioma cells also express another unique purinoceptor, different from those described above, but does not affect any of the above-described signaling pathways, but instead markedly inhibits adenylyl cyclase (Boyer et al., 1993, 1994). This response is fully pertussis toxin–sensitive, suggesting the involvement of a typical G_i protein. The cyclase-inhibiting receptor also can be distinguished from the PLC-activating P2Y purinoceptor by pharmacological studies. For example, the P2X antagonist PPADS was shown to block the PLC-activating purinoceptor but had no effect on the adenylyl cylase inhibiting receptor (Boyer et al., 1994). Two other widely used P2 purinoceptor antagonists (suramin and reactive blue 2) effectively blocked both receptors, although antagonism was competitive for the cyclase-linked receptor and apparently noncompetitive for the PLC-stimulatory one (Boyer et al., 1994). However, as for the receptor isolated by Lazarowski and Harden (1994), the presence and functional significance of this receptor in brain still remain to be established.

FUNCTIONAL ROLES OF ATP IN BRAIN: HYPOTHESES AND PATHOPHYSIOLOGICAL IMPLICATIONS

The functional and molecular results produced in the last 5 years and summarized above have enormously increased our knowledge concerning ATP and its receptors in the brain. Several pathophysiological roles for this nucleotide can now be hypothesized. It is now unanimously accepted that ATP acts as a fast synaptic transmitter at central synapses, thereby extending the concept of purinergic transmission (Burnstock, 1972) to the brain. The wide distribution of both P2X and P2Y purinoceptors in the brain, as suggested by both radioligand and *in situ* hybridization studies (Tables 1 and 2) supports a crucial role for ATP in the regulation of brain function.

Presynaptic inhibition of neurotransmitter release through P1 purinoceptors is widespread in the brain and relatively well characterized (for review, see Rudolphi et al., 1992); much recent data has shown that both cerebrocortical noradrenergic (von Kugelgen et al., 1994) and cholinergic (Cunha et al., 1994) axons also possess presynaptic ATP receptors inhibitory of either noradrenaline or acetylcholine release, respectively. These results are consistent with the colocalization of ATP with "classic" transmitters in central synapses (Zimmermann, 1994) and implicate this nucleotide as a "tonic" regulator of neurotransmitter release in mammalian cortex, a role that has so far been considered as exclusive to adenosine. The functional importance of the presynaptic actions of ATP is also supported by the finding that ecto-ATPase (the enzyme responsible for ATP catabolism, see Storage and Release of ATP and Other Nucleotides by Brain Synaptic Terminals, below) is closely associated with cerebrocortical cholinergic axons (Cunha and Sebastiao, 1992). Instead, ecto-5′-nucleotidase, the enzyme responsible for the formation of adenosine, is lacking in these nerve terminals (Cuhna et al., 1992). Whether the presynaptic P2 purinoceptor modulatory on neurotransmitter release belongs to the P2X or P2Y subtype still remains to be defined.

More recently, both presynaptic and postsynaptic P2 purinoceptors have been suggested to modulate the release and action of endogenous glutamate in rat hippocampus (Motin and Bennett, 1995). Effects of P2 purinoceptor antagonists on excitatory postsynaptic currents in CA1 and CA3 pyramidal neurons suggested that ATP can facilitate glutamate release, probably via a P2X purinoceptor modulatory on calcium entry (which is consistent with the high levels of this receptor subtype in hippocampus, Table 1). ATP can also enhance the postsynaptic actions of glutamate, either via a metabotropic P2Y purinoceptor or through an ectoprotein kinase that in turn phosphorylates membrane proteins coupled to the glutamate receptor (Chen et al., 1994). On this basis, it may be speculated that ATP is costored with glutamate in hippocampal presynaptic terminals and is released together with this amino acid during physiological neurotransmission. Cotransmitter synergistic actions of ATP on the functional effects of glutamate may play a crucial role in the induction of long-term potentiation (LTP) and, consequently, in learning and memory (Wieraszko and Ehrlich, 1994). In this respect, a key role for ATP, and, in particular ATP-activated ectoprotein kinases, is supported by the observation that specific antagonists of these ectoenzymes block the maintenance phase of LTP (Fujii et al., 1994). These findings suggest that brain purinoceptors may represent a novel target for the design and development of new therapeutic entities modulatory of both cognitive functions and ischemia-associated neurodegeneration, an event characterized by a pathological increase of glutamatergic neurotransmission (Rudolphi et al., 1992; Williams, 1995).

Other specific roles for ATP are suggested by the discrete distribution of P2 purinoceptors in the brain. High expression levels in caudate-putamen, substantia nigra, and cerebellum (P2X; Bo and Burnstock, 1994: Kidd et al., 1995) suggest a role in the regulation of motor functions, while the presence of P2 purinoceptors in the pineal gland (Ferreira et al., 1994) suggests a role in the regulation of the light-dark cycle. The *in situ* hybridization studies have shown that many of the brain nuclei involved in visual and/or auditory functions express high amounts of P2 purinoceptors (e.g., the $P2Y_1$ receptor; Webb et al., 1993), and actions of ATP on nucleus solitarii neurons suggest a possible role for the nucleotide in processing afferent signals from visceral sensory receptors, vagus nerve, cardiopulmonary receptors, arterial baroreceptors, and chemoceptors (Ueno et al., 1992).

Besides its involvement in fast neurotransmission, recent data also implicate ATP (and other nucleotides and nucleosides) in long-term communication between cells. Nucleotides and nucleosides seem to act in concert with polypeptidic growth factors as novel trophic factors to induce long-term phenotypic changes of brain cells, including both neurons, and glial and endothelial cells. Indeed, P2 purinoceptors are not only found on neurons, but also on astroglia (mainly PLA_2- and PLC-associated P2Y purinoceptors, for review see Abbracchio et al., 1995a; Neary et al., 1996), microglia (both P2X and P2Y purinoceptors, Waltz et al., 1993; Norenberg et al., 1994; Banati et al., 1993), and oligodendroglia (Kastritsis et al., 1992; Kirischuk et al., 1995). Activation of these receptors has been associated to strikingly different effects, ranging from cell differentiation and apoptosis, mitogenetic and morphogenetic changes, and modulation of cytokine and trophic factor synthesis and release. A detailed review on the variety of effects evoked by ATP and other nucleotides and nucleosides on brain cells has been recently published (Neary et al., 1996). Here I shall briefly summarize them, with special reference to those implicating ATP in brain repair mechanisms and plasticity.

Following brain trauma and ischemia, extracellular levels of nucleotides and nucleosides are greatly increased, due to release of these compounds both from the intracellular soluble pool and from degradation of the DNA and RNA of dying cells. Therefore, at the site of damage, cells are exposed for long periods of time to significantly elevated concentrations of both ATP and guanosine and uridine derivatives. These compounds can therefore serve as signals to initiate brain repair mechanisms. ATP stimulates DNA synthesis and proliferation in microglia, and endothelial and astroglial cells, where it seems to act synergistically with polypeptidic growth factors, and hence to contribute to their plastic changes under such pathological conditions. Astrocytes and microglia activated by ATP release cytokines, neurotrophins, and pleiotrophins, which in turn are neuroprotective and minimize neuronal loss. Moreover, nucleotides and nucleosides directly participate to the remodelling of damaged brain circuitry by promoting axonal regrowth and therefore contribute to the formation of new synapses in order to replace those lost as a result of injury and diseases. Most of these actions are mediated by P2Y-like purinoceptors, as with other examples of G protein–coupled receptors that have been shown to interact with mitogenic pathways via mechanisms independent of tyrosine kinase receptor (Clapham and Neer, 1993; Bolego et al., 1996). In astroglial cells, the trophic actions of nucleotides seem to specifically require the activation of PLA_2-linked P2Y purinoceptors, and to be largely independent of Ca^{2+} mobilization (Bolego et al., 1997; Centemeri et al., 1997).

A role for P2 purinoceptors in induction of apoptosis has been also suggested. When comparing the sequence of the cloned $P2X_1$ and $P2X_2$ with those of previously cloned proteins, a striking similarity emerged between the ATP receptors and RP-2, a gene activated in thymocytes by desamethasone, which induces them to undergo apoptotic cell death (Surprenant et al., 1995). RP-2 may therefore encode a receptor for ATP or another metabolite released during induction of apoptosis. On the basis of these observations $P2X_1$ has been suggested to play a role in the massive apoptosis that accompanies the selection of immunocompetent T cells of the thymus (Surprenant et al., 1995). Since $P2X_1$ is widely expressed in the brain, a role for this receptor subtype can be hypothesized both in naturally occurring apoptosis during brain development, and in the remodelling of brain circuitries following trauma and ischemia (Neary et al., 1996). Purine-induced apoptosis may serve as a means of eliminating irreversibly damaged cells, in order to spare energy and space for recovering ones. In addition, it has been proposed that ATP-induced apoptosis might contribute to limiting the spread of cell death in infarcted brain (Neary et al., 1996). In fact, in contrast to necrosis (where cellular contents, including neurotoxic mediators, are released massively in the extracellular space) cell death by apoptosis occurs with no inflammation.

Globally, these data suggest that ATP participates to the initiation and maintainance of brain repair mechanisms following a variety of insults both by direct actions on astroglia, microglia, and neurons, and by acting in concert with polypeptidic growth factors in promoting changes of target cells that are instrumental to functional recovery and regeneration. In this respect, there is obvious therapeutic potential for purine derivatives, such as AIT082, that readily enter the CNS and stimulate locally the production of neurotrophins and pleiotrophins (Neary et al., 1996).

These results also imply that ATP is an endogenous regulator of brain plasticity both during development and adulthood; moreover, the presence of P2 purinoceptors on oligodendroglia supports a role for ATP as an endogenous regulator of these cells as well. Thus, P2 purinoceptors could represent novel targets in developing therapeutic strategies for a wide range of neurological disorders, including ischemia, seizures, and trauma-associated neurodegeneration, aging-associated cognitive defects, and demyelinating diseases.

FUTURE PERSPECTIVES

Among the numerous exciting developments related to the elucidation of ATP function in the brain, two issues in particular seem to deserve specific attention. The first concerns the role of purinoceptors in evolution. The role of ATP in intercellular communication is universal and phylogenetically ancient in plants and animals (Neary et al., 1996); responses to ATP have been demonstrated in very simple organisms, such as leech or amoeba (Burnstock, 1996). ATP also seems to play a key role in early development, since it is obligatory to sperm movement and is involved in fertilization and gastrulation (Neary et al., 1996). The striking homology of the cloned P2X purinoceptors with epithelial Na^+ channels (Surprenant et al., 1995) suggests that primordial membrane voltage-sensitive channels may have evolved at an early stage to detect ATP as the first chemical signal of life. The exceptionally high expression of some purinoceptor subtypes in developing brain (see Activation of Both Ligand-Gated Ion Channels and G Protein–Coupled Receptors in Brain by ATP and Related Nucleotides, above) suggests a crucial role for ATP in the early organization of the CNS. Few attempts have been made to characterize the purinoceptor subtypes specifically involved in early development; their elucidation will be of great help in understanding the role of ATP in CNS formation and maturation (with particular reference to the role of these receptors in naturally occurring apoptosis during develoment) and may disclose new exciting findings in the pathophysiology of purines and pyrimidines.

A second important perspective concerns the differential role of the various purinoceptor subtypes in brain function. The discrete presence of purinoceptors in brain areas involved in learning and memory and in the control of motor and sensory functions suggests that selective purinoceptor ligands may be developed as modulators of behavioral disorders (depression, epilepsy, and psychosis) and aging-associated neurodegenerative diseases (Alzheimer's, Huntington's, and Parkinson's diseases). The different pharmacological profiles of the various P2 purinoceptor subtypes (for example, the recently discovered differences in sensitivity to antagonists) suggest that these receptors may be different enough to be selectively targeted with specific ligands. This further underlines the urgent need for the development of specific agonists and antagonists for the various subtypes and their characterization *in vivo.*

NOTE ADDED IN PROOF

When this paper was already in press, the cloning of a new ligand-gated P2X receptor ($P2X_7$) was reported (Surprenant et al., Science 272:735–738, 1996). Although initially cloned from brain, this 595 aminoacid receptor is mostly expressed in macrophages. A number of characteristics (i.e., activation by benzoylATP, cytolytic activity through the activation of a large permeabilizing pore) suggest that this is the previously named P_{2Z} receptor. The recent detection of this receptor on microglial cells (Ferrari et al., J Immunol 156:1531–1539, 1996) further underlines its possible functional importance in the remodeling and functional recovery of brain circuitries following trauma and ischemia.

ACKNOWLEDGMENTS

The author is deeply grateful to Professors G. Burnstock (University College London), and F. Cattabeni (University of Milan) for continuous support and invaluable suggestions. The author also wishes to thank Dr. G. Collo (Glaxo Institute of Molecular Biology, Geneva), Dr. B. King (University College London) and Dr. G. Boeynaems (Erasmus Hospital Brussels) for sharing unpublished material; Ms. Carla Chieregato for typing assistance; and Ms. Annie Evans and Dr. D. Christie for help in editing. The author is a partner in the shared-cost project "Nucleotides, a Novel Class of Extracellular Signalling Substances in the Nervous System," supported by the European Union within the BIOMED 2 programme.

REFERENCES

Abbracchio MP, Burnstock G (1994): Purinoceptors: Are there families of P2X and P2Y purinoceptors? Pharmacol Ther 64:445–475.

Abbracchio MP, Ceruti S, Burnstock G, Cattabeni F (1995a): Purinoceptors on glial cells of the central nervous system: functional and pathological implications. In Belardinelli L, Pelleg A (eds): "Adenosine and Adenine Nucleotides: From Molecular Biology to Integrative Physiology." Norwell, MA: Kluwer Academic, p 271–280.

Abbracchio MP, Ceruti S, Langfelder R, Cattabeni F, Saffrey MJ, Burnstock G (1995b): Effects of ATP analogues and basic fibroblast growth factor on astroglial cell differentiation in primary cultures of rat striatum. Int J Dev Neurosci 13:658–693.

Abbracchio MP, Saffrey MJ, Hopker V, Burnstock G (1994): Modulation of astroglial cell proliferation by analogues of adenosine and ATP in primary cultures of rat striatum. Neuroscience 59:67–76.

Allen TGJ, Burnstock G (1990): The actions of adenosine 5′-triphosphate on guinea-pig intracardiac neurones in culture. Br J Pharmacol 100:269–276.

Balcar VJ, Li Y, Killinger S, Bennett MR (1995): Autoradiography of P_{2X} ATP receptors in the rat brain. Br J Pharmacol 115:302–306.

Banati RB, Gehrmann J, Schubert P, Kreutzberg W (1993): Cytotoxicity of microglia. Glia 7: 111–118.

Barnard EA, Burnstock G, Webb TE (1994): G-protein coupled receptors for ATP and other nucleotides: A new receptor family. Trends Pharmacol Sci 15:67–70.

Bo X, Burnstock G (1994): Distribution of [^{3}H] α, β-methylene ATP binding sites in rat brain and spinal cord. Neuroreport 5:1601–1604.

Bo X, Zhang Y, Nassar M, Burnstock G, Schoepfer R (1995): A P2X purinoceptor cDNA conferring a novel pharmacological profile. FEBS Lett 375:129–133.

Bolego C, Ceruti S, Brambilla R, Puglisi L, Cattabeni F, Burnstock G, Abbracchio MP (1997): Characterization of the signalling pathways involved in adenosine-tri-phosphate-and basic fibroblast growth factor-induced astrogliosis. Submitted Br J Pharmacol.

Bowler WB, Birch MA, Gallagher JA, Bilbe G (1995): Identification and cloning of human P_{2U} purinoceptor present in osteoblastoma, bone, and osteoblasts. J Bone Miner Res 10:1137–1145.

Boyer JL, Lazarowski ER, Chen K-H, Harden TK (1993): Identification of a P_{2Y} purinergic receptor that inhibits adenylyl cyclase. J Pharmacol Exp Ther 267:1140–1146.

Boyer JL, Zohn IE, Jacobson KA, Harden TK (1994): Differential effects of P_2-purinoceptor antagonists on phospholipase C- and adenylyl cyclase-coupled P_{2Y}-purinoceptors. Br J Pharmacol 113:614–620.

Brake AJ, Wagenbach MJ, Julius D (1994): New structural motif for ligand-gated ion channels defined by an ionotropic ATP receptor. Nature 371:519–523.

Bruner G, Murphy S (1990): ATP-evoked arachidonic acid mobilization in astrocytes is via a P_{2Y}-purinergic receptor. J Neurochem 55:1569–1575.

Bruner G, Murphy S (1993): UTP activates multiple second messenger systems in cultured rat astrocytes. Neurosci Lett 162:105–108.

Burnstock G (1972): Purinergic nerves. Pharmacol Rev 24:509–581.

Burnstock G (1976): Do some nerve cells release more than one transmitter? Neuroscience 1:239–248.

Burnstock G (1977): Purine nucleotides and nucleosides as neurotransmitters or neuromodulators in the central nervous system. In Usdin E, Humburg DA, Barshas JS (eds) "Neuroregulators and Psychiatric Disorders." New York: Oxford University Press, pp 470–477.

Burnstock G (1978): A basis for distinguishing two types of purinergic receptor. In Straub RW, Bolis L (eds): "Cell Membrane Receptors for Drugs and Hormones: A Multidisciplinary Approach." New York: Raven Press, pp 107–118.

Burnstock G (1986): Purines as cotransmitters in adrenergic and cholinergic neurones. In Hokfelt T, Fuxe K, Pernow B (eds): "Coexistence of Neuronal Messengers: A New Principle in Chemical Transmission." Progress in Brain Research. Amsterdam:Elsevier, Vol 68, pp 193–203.

Burnstock G (1990): Co-transmission. Arch Int Pharmacodyn Ther 304:7–33.

Burnstock G (1996): P2 purinoceptors: Historical perspective and classification. In DJ Chadwick and JA Goode (eds): "Proceedings of the Ciba Foundation Symposium 198 on P2 Purinoceptors: Localization, Function and Transduction Mechanisms." Chichester: Wiley, pp 1–33.

Centemeri C, Bolego C, Abbracchio MP, Cattabeni F, Puglisi L, Burnstock G, Nicosia S (1997): Characterization of the Ca^{2+} responses evoked by ATP and other nucleotides in mammalian brain astrocytes. Br J Pharmacol (in press).

Chang K, Hanaoka K, Kumada M, Takuwa Y (1995): Molecular cloning and functional analysis of a novel P2 nucleotide receptor. J Biol Chem 270: 26152–26158.

Chen C-C, Akoplan AN, Sivilotti L, Colquhoun D, Burnstock G, Wood JN (1995): A P2X purinoceptor expressed by a subset of sensory neurons. Nature 377:428–431.

Chen W, Hogan MV, Wieraszko A, Pawlowska Z, Soifer D, Ehlich YH (1994): NMDA-regulated ectoprotein kinase in hippocampal neurones: Role in LTP. Soc Neurosci Abstr 20:263.

Chen ZP, Krull N, Xu S, Levy A, Lightman SL (1996): Molecular cloning and functional characterization of a rat pituitary G protein-coupled ATP receptor. Endocrinology (in press).

Clapham DE, Neer EJ (1993): New roles for G protein $\beta\gamma$ dimers in transmembrane signalling. Nature 365:403–406.

Cloues R, Jones S, Brown DA (1993): Zn^{2+} potentiates ATP-activated currents in rat sympathetic neurons. Pflugers Arch 424:152–158.

Collo G, North RA, Kawashima E, Merlo-Pich E, Neidhart S, Surprenant A, Buell G (1996): Cloning of $P2X_5$ and $P2X_6$ receptors, and the distribution and properties of an extended family of ATP-gated ion channels. J Neurosci 16:2495–2507.

Communi D, Pirotton S, Parmentier M, Boeynaems J-M (1996): Cloning and functional expression of a human uridine nucleotide receptor. J Biol Chem 270:30849–30852.

Cunha RA, Sebastiao AM (1992): Ecto-ATPase activity in cholinergic nerve terminals of the hippocampus and of the cerebral cortex of the rat. Neurochem Int 21:A12.

Cunha RA, Ribeiro JA, Sebastiao AM (1994): Purinergic modulation of the evoked release of [^{3}H]acetylcholine from the hippocampus and cerebral cortex of the rat: Role of the ectonucleotidases. Eur J Neurosci 6:33–42.

Cunha RA, Sebastiao AM, Ribeiro JA (1992): Ecto-5′-nucleotidase is associated with cholinergic nerve terminals in the hippocampus but not in the cerebral cortex of the rat. J Neurochem 59:657–666.

Edwards FA, Gibb AJ, Colquhoun D (1992): ATP receptor-mediated synaptic currents in the central nervous system. Nature 359:144–147.

Evans RJ, Derkach V, Surprenant A (1992): ATP mediates fast synaptic transmission in mammalian neurons. Nature 357:503–505.

Farinas I, Solsona C, Marsal J (1992): Omega-conotoxin blocks acetylcholine and adenosine triphosphate releases from Torpedo synaptosomes. Neuroscience 47:641–648.

Ferreira ZS, Cipolla-Neto J, Markus RP (1994): Presence of P2-purinoceptors in the rat pineal gland. J Pharmacol 112:107–110.

Fischer B, Boyer JL, Hoyle CHV, Ziganshin AU, Brizzolara AL, Knight GE, Zimmet J, Burnstock G, Harden TK, Jacobson KA (1993): Identification of potent, selective P2Y-purinoceptor agonists: Structure-activity relationships for 2-thioether derivatives of adenosine-5′-triphosphate. J Med Chem 36:3937–3946.

Fieber LA, Adams DJ (1991): Adenosine triphosphate-evoked currents in cultured neurones dissociated from rat parasympathetic cardiac ganglia. J Physiol 434:239–256.

Filtz TM, Q Li, Boyer JL, Nicholas RA, Harden TK (1994): Expression of a cloned P_{2Y} purinergic receptor that couples to phospholipase C. Mol Pharmacol 46:8–14.

Fredholm BB, Abbracchio MP, Burnstock G, Daly JW, Harden KT, Jacobson KA, Leff P, Williams M (1994): VI. Nomenclature and classification of purinoceptors. Pharmacol Rev 46:143–156.

Fujii S, Kuroda Y, Ito K, Miyakawa H, Kato H (1994): Possible involvement of extracellular ATP and ecto-protein kinase in the maintenance (late) phase of hippocampal long term potentiation. Soc Neurosci Abstr 20:264.

Furukawa K, Ishibashi H, Akaike N (1994): ATP-induced inward current in neurons freshly dissociated from the tuberomammillary nucleus. J Neurophys 71:868–873.

Harms L, Finta EP, Tschopl M, Illes P (1992): Depolarization of rat locus coeruleus neurones by adenosine 5′-triphosphate. Neuroscience 48:941–952.

Henderson DJ, Elliot DG, Smith GM, Webb TE, Dainty IA (1995): Cloning and characterization of a bovine P_{2Y} receptor. Biochem Biophys Res Commun 212:648–656.

Hohmann J, Kowalewski H, Vogel M, Zimmerman H (1993): Isolation of a Ca^{2+} or Mg^{2+}-activated ATPase (ecto-ATPase) from bovine brain synaptic membranes. Biochim Biophys Acta 1152:146–154.

Illes P, Norenberg W (1993): Neuronal ATP receptors and their mechanisms of action. Trends Pharmacol Sci 14:50–54.

Jacobson KA, Fischer B, Malliard M, Boyer JL, Hoyle GHV, Harden TK, Burnstock G (1995): Novel ATP agonists reveal receptor heterogeneity within P2X and P2Y subtypes. In Belardinelli L, Pelleg A (eds): "Adenosine and Adenine Nucleotides: From Molecular Biology to Integrative Physiology." Norwell, MA: Kluwer Academic, pp 149–156.

James S, Richardson PJ (1993): Production of adenosine from extracellular ATP at the striatal cholinergic synapse. J Neurochem 60:219–227.

Kastritsis CHX, Salm AK, McCarthy K (1992): Stimulation of the P2Y purinergic receptor on type I astroglia results in inositol phosphate formation and calcium mobilization. J Neurochem 58:1277–1284.

Kidd EJ, Grahames CBA, Simon J, Michel AD, Barnard EA, Humphrey PPA (1995): Localization of P_{2X} purinoceptor transcripts in the rat nervous system. Mol Pharmacol 48:569–573.

King BF, Ziganshina LE, Pintor J, Burnstock G (1996): Full sensitivity of $P2X_2$ purinoceptor to ATP revealed by changing extracellular pH. Br J Pharmacol 117:1371–1373.

Kirischuk S, Scherer J, Kettenmann H, Verkhratsky A (1995): Activation of P2-purinoceptors triggered Ca^{2+} release from InsP3-sensitive stores in mammalian oligodendrocytes. J Physiol (London) 483:41–57.

Krishtal OA, Osipchuk YV, Shelest TN, Smirnoff SV (1987): Rapid extracellular pH transients related to synaptic transmission in rat hippocampal slices. Brain Res 436:352–356.

Kunapuli SP, Akbar GKM, Webb T, Matsumoto M, Mills DCB, Barnard EA (1995): Cloning and characterization of novel P2 purinoceptor from human erythroleukemia cells. In "Structure and Function of P2-Purinoceptors." Satellite Meeting of Experimental Biology. Atlanta, GA, April 7–9, 1995. Abstract.

Lazarowski ER, Harden TK (1994): Identification of a uridine nucleotide-selective G-protein-linked receptor that activates phospholipase C. J Biol Chem 269:11830–11836.

Lazarowski ER, Watt WC, Stutts MJ, Boucher RC, Harden TK (1995): Pharmacological selectivity of the cloned human P2u-purinoceptor: Potent activation by diadenosine tetraphosphate. Br J Pharmacol 116:1619–1627.

Leon C, Vial C, Cazenave J, Gachet C (1996): Cloning and sequencing of a human endothelial $P2Y_1$ purinoceptor. Gene (in press).

Lewis C, Neldhart S, Holy C, North RA, Buell G, Surprenant A (1995): Coexpression of $P2X_2$ and $P2X_3$ receptor subunits can account for ATP-gated currents in sensory neurons. Nature 377:432–435.

Lustig KD, Shiau AK, Brake AK, Julius D (1993): Expression cloning of an ATP receptor from mouse neuroblastoma cells. Proc Natl Acad Sci USA 90:5113–5117.

Motin L, Bennett MR (1995): Effect of P_2-purinoceptor antagonists on glutamatergic transmission in the rat hippocampus. Br J Pharmacol 115:1276–1280.

Nakagawa T, Akajke N, Kimitsuki T, Komune S, Arima T (1990): ATP-induced current in isolated outer hair cells of guinea pig cochlea. J Neurophysiol 63:1068–1074.

Neary JT, Laskey R, van Breemen C, Blicharska J, Norenberg LOB, Norenberg MD (1991): ATP-evoked calcium signal stimulates protein phosphorylation/dephosphorylation in astrocytes. Brain Res 566:89–94.

Neary JT, Rathbone MP, Cattabeni F, Abbracchio MP, Burnstock G (1996): Trophic actions of extracellular nucleotides and nucleosides on glial and neuronal cells. Trends Neurosci 19:13–18.

Neary JT, van Breemen C, Forster E, Norenberg LOB, Norenberg MD (1988): ATP stimulates calcium influx in primary astrocyte cultures. Biochem Biophys Res Commun 157:1410–1416.

Norenberg W, Langosch JM, Gebicke-Haerter PJ, Illes P (1994): Characterization and possible function of adenosine 5′-triphosphate receptors in activated rat microglia. Br J Pharmacol 111:942–950.

Parr CE, Sullivan DM, Paradiso AM, Lazarowski ER, Burch LH, Olsen JC, Erb L, Weisman GA, Boucher RC, Turner JT (1994): Cloning and expression of a human P_{2U} nucleotide receptor, a target for cystic fibrosis pharmacotherapy. Proc Natl Acad Sci USA 91:3275–3279.

Pearce B, Murphy J, Morrow C, Dandona P (1989): ATP-evoked Ca^{2+} mobilization and prostanoid release from astrocytes: P_2-purinergic receptors linked to phosphoinositide hydrolysis. J Neurochem 52:971–977.

Phillis JW, Kostopoulos GK, Limacher JJ (1975): A potent depressant action of adenine derivatives on cerebral cortical neurons. Eur J Pharmacol 30:125–129.

Pintor J, Miras-Portugal MT (1993): Diadenosine polyphosphates—ApnA—as new neurotransmitters. Drug Dev Res 28:259–262.

Pintor J, Rotllan P, Torres M, Miras-Portugal MT (1992a): Characterization and quantification of diadenosine hexaphosphate in chromaffin cells: Granular storage and secretagogue-induced release. Anal Biochem 200:296–300.

Pintor J, Kowalewski HJ, Torres M, Miras-Portugal MT, Zimmermann H (1992b): Synaptic vesicle storage of diadenosine polyphosphates in the Torpedo electric organ. Neurosci Res Commun 10:9–14.

Pull I, McIlwain H (1975): Actions of neurohumoral agents and cerebral metabolites on the output of adenine derivatives from superfused tissues of the brain. J Neurochem 24:695–700.

Regenold JT, Illes P (1990): Inhibitory adenosine A_1-receptors on rat locus coeruleus neurones. An intracellular electrophysiological study. Naunyn Schmiedebergs Arch Pharmacol 341:225–231.

Rice WR, Burton FM, Fiedeldey DT (1995): Cloning and expression of the alveolar type II cell P2U-purinergic receptor. Am J Respir Cell Mol Biol 12:27–32.

Richardson PJ, Brown SJ (1987): ATP release from affinity-purified rat cholinergic nerve terminals. J Neurochem 48:622–630.

Rudolphi KA, Schubert P, Parkinson FE, Fredholm BB (1992): Neuroprotective role of adenosine in cerebral ischaemia. Trends Pharmacol Sci 13:439–445.

Salt TE, Hill RG (1983): Excitation of single sensory neurons in the rat caudal trigeminal nucleus by iontophoretically applied adenosine 5′-triphosphate. Neurosci Lett 35:53–57.

Salter MW, Hicks JL (1995): ATP causes release of intracellular Ca^{2+} via the phospholipase $C\beta/IP_3$ pathway in astrocytes from the dorsal spinal cord. J Neurosci 15:2961–2971.

Scefner SA, Chiu TH (1986): Adenosine inhibits locus coeruleus neurons: An intracellular study in a rat brain slice preparation. Brain Res 366:364–368.

Shen K-Z, North RA (1993): Excitation of rat locus coeruleus neurons by adenosine 5′-triphosphate: Ionic mechanism and receptor characterization. J Neurosci 13:894–899.

Silinsky EM, Gerzanich V, Vanner SM (1992): ATP mediates excitatory synaptic transmission in mammalian neurones. Br J Pharmacol 106:762–763.

Surprenant A, Buell G, North AR (1995): P_{2X} receptors bring new structure to ligand-gated ion channels. Trends Neurosci 18:224–229.

Tokuyama Y, Hara M, Jones EMC, Fan Z, Bell GI (1995): Cloning of rat and mouse P_{2Y} purinoceptors. Biochem Biophys Res Commun 211:211–218.

Tschopl M, Harms L, Norenberg W, Illes P (1992): Excitatory effects of adenosine 5′-triphosphate on rat locus coeruleus neurones. Eur J Pharmacol 213:71–77.

Ueno S, Harata N, Inoue K, Akaike N (1992): ATP-gated current in dissociated rat nucleus solitarii neurons. J Neurophysiol 68:778–785.

Valera S, Hussy N, Evans RJ, Adami N, North RA, Surprenant A, Buell G (1994): A new class of ligand-gated ion channel defined by P_{2X} receptor for extracellular ATP. Nature 371:516–519.

von Kugelgen I, Starke K (1991): Release of noradrenaline and ATP by electrical stimulation and nicotine in guinea-pig vas deferens. Arch Pharmacol 344:419–429.

von Kugelgen I, Spath L, Starke K (1994): Evidence for P_2-purinoceptor-mediated inhibition of noradrenaline release in rat brain cortex. Br J Pharmacol 113:815–822.

Walz W, Ilschner S, Ohlemeyer C, Banati R, Kettenmann H (1993): Extracellular ATP activates a cation conductance and a K^+ conductance in cultured microglial cells from mouse brain. J Neurosci 13:4403–4411.

Webb TE, Henderson D, King BF, Wang S, Simon J, Bateson AN, Burnstock G, Barnard EA (1996a): A novel G protein-coupled P_2 purinoceptor ($P2Y_3$) activated preferentially by nucleoside diphosphates: Mol Pharmacol 50:258–265.

Webb TE, Kaplan MG, Barnard EA (1996b): Identification of 6H1 as a P_{2Y} Purinoceptor: $P2Y_5$. Biochem Biophys Res Commun 219:105–110.

Webb TE, Simon J, Bateson AN, Barnard EA (1994): Transient expression of the recombinant chick brain P_{2y1} purinoceptor and localization of the corresponding mRNA. Cell Mol Biol 40:437–442.

Webb TE, Simon J, Krishek BJ, Bateson AN, Smart TG, King BJ, Burnstock G, Barnard EA (1993): Cloning and functional expression of a cDNA encoding a brain G-protein-coupled ATP receptor. FEBS Lett 324:219–225.

White TD (1977): Direct detection of depolarisation-induced release of ATP from a synaptosomal preparation. Nature 267:67–68.

Wieraszko A, Ehrlich YH (1994): On the role of extracellular ATP in the induction of long-term potentiation in the hippocampus. J Neurochem 63:1731–1737.

Williams M (1995): Purinoceptors in central nervous system function. Targets for therapeutic intervention. In Bloom FE, Kupfer DJ (eds): "Psychopharmacology: the Fourth Generation of Progress." New York: Raven Press, pp 643–655.

Williams M, Jacobson KA (1995): P2-purinoceptors: Advances and therapeutic opportunities. Exp Opin Invest Drugs 4:925–934.

Zimmerman H (1994): Signalling via ATP in the nervous system. Trends Neurosci 17:420–426.

CHAPTER 21

Psychomotor Aspects of Adenosine Receptor Activation

MICHAEL F. JARVIS

Neurological and Urological Diseases Research, Pharmaceutical Products Division, Abbott Laboratories, Abbott Park, IL 60064-3500

INTRODUCTION

The study of the functional effects of purines both in the context of overt ambulatory behavior and in animal models of neuropathology has contributed greatly to our understanding of inhibitory aspects of adenosine neuromodulation. During the last 6 years the field of purine pharmacology has benefited greatly from significant advancements in the molecular cloning of purine receptor subtypes. However, a pharmacological analysis of purine effects in the behaving organism remains fundamental to our understanding of purine physiology and in the discovery of useful therapeutics (Williams, 1995). The behavioral pharmacology of adenosine has also provided significant focus for new insights into the molecular interactions of purines with other classical neurotransmitters.

This point is best illustrated by the recent work in the area of adenosine/dopamine interactions (Ferre et al., 1992). For approximately 20 years it has been appreciated that adenosine antagonists and dopamine agonists produce similar behavioral effects in laboratory animals (Fuxe and Ungerstedt, 1974; Fredholm et al., 1976). These observations were subsequently supplemented by demonstrations that adenosine agonists produce behavioral effects opposite to those of dopamine agonists (Green et al., 1982), and that adenosine A_{2a} and dopamine D_2 receptors are highly concentrated in the neostriatum (Jarvis et al, 1989; Jarvis and Williams, 1989; Parkinson and Fredholm 1990; Martinez-Mir et al., 1991). These observations coupled with the discovery of A_{2a}-selective agonist compounds (Hutchison et al., 1990; Jarvis et al., 1989) lead to the direct evaluation of adenosine A_{2a}/dopamine D_2 receptor interactions and the demonstration that adenosine A_{2a} receptor activation reduces dopamine D_2 receptor affinity (Ferre et al., 1991a; Ferre et al., 1992). These advances in the understanding

Purinergic Approaches in Experimental Therapeutics, Edited by Kenneth A. Jacobson and Michael F. Jarvis
ISBN 0-471-14071-6 © 1997 Wiley-Liss, Inc.

of the nature of purine neuromodulation have provided new perspectives in evaluating the behavioral actions of purinergic agonists, as well as new mechanisms for the development of purinergically based therapeutics.

Adenosine represents an endogenous inhibitory neuromodulator that is widely distributed in the mammalian central nervous system at concentrations ranging from 20 to 1000 μM (Zetterstrom et al., 1982). Specificity in the actions of adenosine appears to be mediated by the discrete distributions of adenosine receptor subtypes in the central nervous system (Jarvis, 1988; Jacobson et al., 1993; Ji et al., 1994). While there has been an abundance of behavioral data to indicate functional interactions between the purinergic and other neurotransmitter systems (Jarvis and Williams, 1987, 1990), the distinct neuroregulatory processes between adenosine and dopamine (Ferre et al., 1992) have stimulated a great deal of experimental work examining the functional relationship between these neurotransmitters and their role in the pathophysiology of the basal ganglia, such as in Parkinson's disease, Huntington's chorea, and schizophrenia. Motor activity continues to be a popular behavioral endpoint in investigating the effects of purinergic ligands. However, other behavioral paradigms involving schedule-controlled operant procedures and striatal lesion–induced rotational behavior have also provided useful insights into the neuropsychopharmacology of adenosine. The remainder of this chapter will be devoted to an overview and integration of the behavioral pharmacology of adenosine ligands as they relate to psychomotor function and therapeutic significance.

MOTOR ACTIVITY

The hallmark behavioral effect observed in laboratory animals following adenosine receptor activation is the suppression of exploratory motor activity (Snyder et al., 1981; Katims et al., 1983). Conversely, the prototypic effect of adenosine receptor activation is a pronounced increase in motor activity (Glowa et al., 1985; Jarvis and Williams, 1990). The ability to suppress motor activity appears to be shared by all known adenosine agonists regardless of their specificity for the known adenosine receptor subtypes A_1, A_{2a}, A_{2b}, or A_3 (Jacobson et al., 1993, 1995; Nikodijevic et al., 1991). This property is also shared by agents that enhance endogenous adenosine levels via an inhibition of adenosine metabolism (Wotring et al., 1979; Kowaluk et al., 1996).

Centrally administered adenosine receptor agonists can produce significant alterations in cardiovascular function (Barraco et al., 1983, 1984, 1986). However, the ability of adenosine agonists to inhibit motor activity appears to be centrally mediated, since peripherally selective adenosine antagonists are ineffective in blocking the suppression of locomotor activity (Durcan and Morgan, 1989). Adenosine agonists administered either centrally (Barraco et al., 1984; Phillis et al., 1986) or systemically (Snyder et al., 1981; Carney, 1982; Katims et al., 1983; Buckholtz and Middaugh, 1987) have been shown to cause dose-dependent reductions in the motor activity of rodents. The peripherally acting adenosine antagonist (8-PST) can effectively block the hypothermic effects of adenosine agonists without significantly altering adenosine-induced behavioral depression (Seale et al., 1988; Durcan and Morgan, 1989).

The classical adenosine antagonist caffeine causes biphasic and dose-dependent effects on locomotor behavior; low doses increase and high doses decrease motor activity (Thithapandha et al., 1972; Waldeck, 1975; Snyder et al., 1981; Fredholm et al., 1983; Choi et al., 1988). The motor-increasing effects of caffeine appear to result from a competitive interaction at adenosine receptors, because caffeine can block the behavioral depression induced by adenosine agonists in a dose-dependent manner (Snyder et al., 1981; Coffin et al., 1984; Phillis et al., 1986; Seale et al., 1986). In addition, compounds that are relatively selective phosphodiesterase inhibitors (e.g., IBMX and papaverine) have been found to produce only decreases in motor activity (Coffin et al., 1984; Phillis et al., 1986; Choi et al., 1988). Support for a central locus in the locomotor-stimulating effects of caffeine can be found in the observation that the relative activity of various methylxanthine analogs in producing increased motor activity correlates well with their ability to inhibit [^{3}H]CHA binding in brain (Snyder et al., 1981; Murphy and Snyder, 1982).

Alkylxanthines can also potentiate the locomotor-stimulating effects of other psychomotor stimulants. In a number of studies, caffeine has been shown to potentiate psychomotor stimulant-induced hyperactivity and stereotypy by such agents as amphetamine, cocaine, methylphenidate, and apomorphine (Klawans et al., 1974; Waldeck, 1975; White and Keller, 1984; Fredholm et al., 1983). Caffeine can also potentiate the increase in motor activity brought about by the dopamine precursor, L-dopa (Waldeck, 1975). While caffeine is ineffective alone, it has been shown to potentiate the reversal of reserpine-induced suppression of locomotor activity by the dopaminergic precursor, L-dopa (Waldeck, 1975). Additionally, the dopaminergic receptor anatgonists, pimozide and haloperidol, can effectively block the locomotor-stimulating effects of caffeine, whereas the adrenergic receptor blockers, propanolol and phenoxybenzamine, are ineffective (Waldeck, 1975).

Interestingly, treatment with the dopamine neurotoxin, 6-hydroxydopamine (6-OHDA), has been found to either attenuate (Erinoff and Snodgrass, 1986) or potentiate (Criswell et al., 1988) the locomotor-stimulating effects of caffeine. The differential effects of 6-OHDA treatment on xanthine-stimulated motor activity may be dependent upon test parameters, age of the animals, and/or extent of the dopaminergic receptor supersensitivity (Criswell et al., 1988; Erinoff and Snodgrass, 1986). Consistent with a decreased sensitivity to methylxanthines following 6-OHDA treatment α-methyltyrosine administration has been found to attenuate caffeine-stimulated motor activity (White et al., 1978; Criswell et al., 1988).

Analysis of the pharmacological profile of both adenosine agonists and antagonists to alter locomotor activity has led to the notion that the suppression of motor activity is primarily mediated by an activation of striatal adenosine A_{2a} receptors (Barraco et al., 1984; Coffin et al., 1984; Griebel et al., 1991; Phillis et al., 1986). This idea has been supported by demonstrations that A_2 agonists are generally more potent than A_1 agonists in reducing locomotion (Coffin et al., 1984; Phillis et al., 1986; Durcan and Morgan, 1989; Ferre et al., 1991b,c). However, A_1 agonists have also been shown to effectively reduce motor activity (Snyder et al., 1981; Seale et al., 1988; Heffner et al., 1989). This action is behaviorally consistent with the ability of adenosine A_1 agonists to inhibit striatal dopamine release (Wood et al., 1989). Additionally, the novel adenosine A_3 agonist, N^6-(3-iodobenzyl)-5′-N-methylcarboxamido-adenosine (3-IB-MECA), can also specifically reduce locomotor activity in mice (Jacobson et al., 1993). It is interesting to note that NECA, which has nonselective actions at all four adenosine receptor subtypes (Williams, 1995), is typically the most potent agonist in suppressing motor activity (Durcan and Morgan, 1989; Nikodijevic et al., 1990, 1991). Taken together, these experimental findings suggest that all four adenosine receptor subtypes may contribute to the ability of adenosine to reduce motor activity. The possibility also exists that the activation of different adenosine receptor subtypes can synergistically contribute to locomotor suppression. In a systematic examination of various A_1 and A_{2a} selective agonists and antagonists, Nikodijevic et al. (1991) provided direct evidence for such a synergistic interaction between the A_1-selective agonist CHA and the A_2-selective agonist 2-[4-(2-aminoethyl)propionamido]phenethylamino-adenosine (APEC) in reducing motor activity in mice. Further, the comparatively greater potency of the nonselective agonist (NECA), in reducing locomotor activity, relative to that found for A_1- or A_{2a}-selective agonists, may also be indicative of a synergistic participation of various adenosine receptor subtypes (Nikodijevic et al., 1991).

SCHEDULE-CONTROLLED BEHAVIOR

The effects of adenosine agonists on response rates are similar to their effects on overt motor activity, i.e., they decrease response rates under a variety of schedule-controlled operant paradigms in both rodents and primates (Glowa and Spealman, 1984; Coffin and Spealman, 1987). The alkylxanthines, in contrast, produce both response rate–dependent and dose-dependent effects, increasing low rates of responding and decreasing high rates of responding (Glowa and Spealman, 1984; Carney, et al., 1985b; Coffin and Spealman, 1987, 1989; Howell, 1993; Howell and Byrd, 1993). Since the behavioral depressant effects of adenosine agonists can be attenuated by the subsequent administration of methylxanthines, there appears to be a competitive interaction between the behavioral effects of purines and methylxanthines (Glowa and Spealman, 1984; Spealman, 1988). Further, the potency of methylxanthines in increasing response rates is positively correlated with their ability to attenuate the rate-suppressing effects of the adenosine agonists (Spealman, 1988).

The ability of adenosine agonists to disrupt fixed-ratio operant responding in monkeys exhibits a pharmacological profile (NECA > R-PIA > CHA) that is generally consistent with an involvement of adenosine A_{2a} receptors (Howell and Byrd, 1993; Coffin and Spealman, 1987). The same relative potency order was also obtained for the ability of these compounds to reduce blood pressure in these animals. Further, the relative potency of these compounds in reducing heart rate was indicative of activity at A_1 receptors (Coffin and Spealman, 1987), which is also consistent with data obtained *in vitro* for these cardiovascular responses (Evans et al., 1982; Hamilton et al., 1987; Oei et al., 1989). The possibility also exists of a synergistic interaction between various adenosine receptor subtypes in modulating operant responding, as has been demonstrated for locomotor activity (Nikodijevic et al., 1991). Thus, the greater relative potency of NECA in suppressing operant responding (Coffin and Spealman, 1987) could very well reflect the ability of this agonist to activate various adenosine receptor subtypes (Nikodijevic et al., 1991).

There are also several reports that the behavioral actions of the methylxanthines cannot be fully explained via interactions at adenosine receptor subtypes (Glowa et al., 1985; Goldberg et al., 1985). Caffeine has been shown to effectively restore operant responding that has been reduced by R-PIA or NECA in monkeys (Spealman, 1988), rats (Carney et al., 1985b), and mice (Glowa et al., 1985). However, the topography of operant responding obtained when both of these compounds were present in the animal has been reported to be different from that obtained with either compound alone (Goldberg et al., 1985). This discrepancy may be accounted for by the fact that methylxanthines can also produce behavioral effects that are more indicative of phosphodiesterase inhibition than of adenosine receptor antagonism. The rate-increasing effects of caffeine may reflect its ability to block central adenosine receptors, while the rate-decreasing effects, which require higher doses, may be mediated through an action at peripheral adenosine receptors, as well as an additional action on phosphodiesterase inhibition (Glowa and Spealman, 1984). In this regard, it should be noted that not all alkylxanthines produced rate-dependent effects on operant responding. CGS 15943 is a potent, but nonselective, adenosine receptor antagonist (Jarvis et al., 1991) that lacks phosphodiesterase activity (Williams et al., 1987) and has been demonstrated to produce only stimulation of operant responding (Howell, 1993; Howell and Byrd, 1993). However, compounds with significant phosphodiesterase activity such as enprofylline (Spealman, 1988), rolipram (Howell and Byrd, 1993), and isobutylmethylxanthine (Kleven and Sparber, 1987), have been shown to produce only decreases in operant responding. This may be attributed to their potency as phosphodiesterase inhibitors.

The drug discrimination paradigm is another operant procedure that has been used to elucidate the neuromodulatory actions of adenosine (Jarvis and Williams, 1990). Both adenosine agonist and antagonist ligands can serve as discriminative stimuli (Carney and Christensen, 1980; Modrow et al., 1981; Holtzman, 1986, 1987; Holloway et al., 1985b; Spealman and Coffin, 1988). The stimulus properties of adenosine agonists appear to be centrally mediated, because the peripherally acting adenosine antagonist 8-PST was ineffective in blocking R-PIA discriminations (Holloway et al., 1985b). Caffeine has been shown to effectively block a stimulus cue produced by R-PIA, but this agonist was relatively ineffective in antagonizing the stimulus cues engendered by high doses of caffeine and theophylline (Holloway et al., 1985a). This result indicates that at least some component of the caffeine discriminative cue may not be caused by an antagonist interaction at A_1 receptors (Holloway et al., 1985a). Interestingly, the cyclic nucleotide phosphodiesterase inhibitors IBMX and papaverine can also produce stimulus generalization in caffeine-trained animals (Holtzman, 1986; Holloway et al., 1985a).

The ability of other psychomotor stimulants to produce generalizable stimuli in caffeine-trained animals remains equivocal. Despite early failures to observe caffeine-induced stimulus generalization with compounds such as amphetamine, methylphenidate, or nicotine (Modrow et al., 1981; Carney et al., 1985b), there is some data indicating that psychomotor stimulants, including *d*-amphetamine, cocaine, and methylphenidate, can generalize to caffeine-engendered interoceptive cues in avoidance discrimination tests (Holtzman, 1986, 1987). Additional data indicative of a functional link between monoaminergic and purinergic systems comes from the observation that the administration of caffeine in combination with phenylpropanolamine and/or ephedrine afforded complete stimulus generalization to *d*-amphetamine in *d*-amphetamine-trained animals, whereas when any of these compounds were given alone, they produced only partial stimulus generalization (Holloway et al., 1985a). In addition, caffeine has been shown to potentiate the effects of a low dose of *d*-amphetamine (0.5 mg/kg) in producing stimulus generalization in animals trained to discriminate a higher dose of *d*-amphetamine (0.8 mg/kg) from saline (Schechter, 1977).

These findings offer additional support to the notion that both adenosine agonists and antagonists can produce specific stimulus cues and can antagonize their respective interoceptive cues. It also appears that additional neurochemical mechanisms contribute to the stimulus properties of purinergic ligands. Thus, purinergic receptor interactions, inhibition of cyclic nucleotide phosphodiesterase activity, and purinergic modulation of other neurotransmitter systems can contribute to the unique stimulus properties of adenosine ligands.

EFFECTS UPON REPEATED ADMINISTRATION

Following chronic exposure, tolerance to the acute behavioral effects of various adenosine antagonists develops (Carney, 1982; Ahiljanian and Takemori, 1986; Finn and Holtzman, 1987; 1988; Jacobson et al., 1996). Such tolerance has been demonstrated in a variety of behavioral paradigms, including schedule-controlled operant responding (Carney, 1982), locomotor activity (Finn and Holtzman, 1987, 1988; File et al., 1988; Nikodijevic et al., 1992), and drug discrimination procedures (Holtzman, 1987). In all of these procedures, drug tolerance is typically defined as a rightward shift in drug dose–response curves (surmountable tolerance) or a downward shift in drug dose–response curves (insurmountable tolerance) (Finn and Holtzman, 1987).

Using locomotor activity as the behavioral measure, chronic caffeine exposure has been shown to result in complete tolerance to the biphasic effects of caffeine and in symmetrical cross-tolerance to the stimulant effects of other alkylxanthines (Holtzman et al., 1991; Finn and Holtzman, 1988). This caffeine tolerance also appears to be pharmacologically specific, since similar cross-tolerance was not observed with other psychomotor stimulants, including *d*-amphetamine, methylphenidate, and cocaine (Finn and Holtzman, 1987). Interestingly, for rats that displayed complete tolerance to caffeine, the adenosine agonists NECA and R-PIA were found to be only 10 times less active compared to vehicle-treated controls in producing decreases in locomotor activity (Finn and Holtzman, 1987, 1988). This last observation stands in contrast to a previously reported leftward shift in adenosine agonist dose–response curves in caffeine-tolerant mice (Fisher and Hughes, 1996; Ahiljanian and Takemori, 1986) and to several demonstrations of brain adenosine receptor upregulation following chronic exposure to methylxanthines (Johansson et al., 1993; Lupica et al., 1991; Boulenger et al., 1983; Green and Stiles, 1986; Sanders and Murray, 1988; Murray, 1982).

Using drug discrimination procedures, chronic caffeine treatment has been found to produce rightward shifts in caffeine stimulus generalization curves (Holtzman, 1987). Additionally, chronic treatment with either caffeine or methylphenidate resulted in symmetrical cross-tolerance between both drugs. This demonstration of symmetrical cross-tolerance indicates that caffeine and methylphenidate produce salient interoceptive stimuli by some common mechanism (Holtzman, 1987). The mechanistic similarity between these psychomotor stimulants appears to be limited to their stimulant properties, since similar cross-tolerance in their locomotor-stimulating effects does not occur (Finn and Holtzman, 1987, 1988). One possible explanation for these observations is that the dosing regimen required to produce behavioral tolerance in locomotor activity is much greater than that required to obtain tolerance to caffeine-induced interoceptive cues (Holtzman, 1987; Finn and Holtzman, 1987). Thus, chronic caffeine-induced inhibition in nucleotide phosphodiesterase activity, as well as an upregulation of central adenosine receptors, may serve as a contributory mediator of tolerance to the locomotor-stimulating effects of caffeine. The observation that prenatal exposure to caffeine can result in long-lasting alterations in central cyclic AMP activity is consistent with this idea (Concannon et al., 1983).

It has been well documented that chronic exposure to adenosine receptor antagonists can result in increased numbers of central adenosine A_1 receptors (Johansson et al., 1993; Lupica et al., 1991; Murray, 1982; Boulenger et al., 1983; Chou et al., 1985; Green and Stiles, 1986; Szot et al., 1987). In the majority of these studies, chronic methylxanthine treatment results in an approximately 10%–20% increase in brain A_1 receptors without a significant alteration in receptor affinity. These caffeine-induced increases in A_1 receptors appear to be greatest in the cerebral cortex and cerebellum, with more moderate increases occurring in the hippocampus (Szot et al., 1987). Chronic caffeine treatment can also produce a shift in the relative proportion of high- and low-affinity states of the cortical A_1 receptor, which is manifested as an apparent increase in agonist binding (Green and Stiles, 1986). Chronic exposure to nonselective adenosine antagonists, however, does not produce a general upregulation of all adenosine receptor subtypes. There is an apparent differential sensitivity of adenosine receptor subtypes to chronic antagonist exposure, such that treatments that upregulate the densities of adenosine A_1 receptors do not produce similar alterations in adenosine A_{2a} receptors (Lupica et al., 1991) or A_{2a} mRNA (Johansson et al., 1993).

From the present discussion it appears that chronic exposure to adenosine receptor antagonists clearly results in a decreased sensitivity to the acute effects of both adenosine antagonists and agonists. Chronic caffeine treatment can produce complete tolerance to the locomotor-stimulating effects of the compound, as well as symmetrical cross-tolerance to other methylxanthines; yet, adenosine agonists still retain full efficacy in their ability to reduce locomotor activity (Finn and Holtzman, 1987). From these observations, it has been argued that tolerance to the locomotor-stimulating effects of caffeine cannot be explained solely on the basis of a caffeine-induced upregulation of central adenosine receptors (Finn and Holtzman, 1987). Given that the rate-decreasing effects of the methylxanthines on operant responding and locomotor activity may involve an inhibition of nucleotide phosphodiesterase activity (Howell and Byrd, 1993; Choi et al., 1988), as well as adenosine receptor antagonism, it is quite conceivable that the observed tolerance to caffeine in some behavioral paradigms may also involve a reduced sensitivity to caffeine-induced phosphodiesterase inhibition. In other behavioral paradigms, which may be more sensitive to drug tolerance effects (i.e., drug discrimination procedures, Holtzman, 1987), an upregulation of central adenosine receptors may be sufficient for tolerance phenomena to occur. Interestingly, tolerance to caffeine-induced interoceptive stimuli also appears to be less pharmacologically strict than that for locomotor activity, since caffeine-induced cross-tolerance to other classes of psychomotor stimulants can be obtained in drug discrimination procedures.

ADENOSINE MODULATION OF ROTATIONAL BEHAVIOR

Early support for a functional interaction between purinergic and dopaminergic systems in striatum came from demonstrations that methylxanthines could stimulate rotational behavior in rats with unilateral striatal lesions (Ungerstedt et al., 1971; Fuxe and Ungerstedt, 1974; Fredholm et al., 1976). The analysis of rotational behavior following unilateral lesions of the basal ganglia has provided great insight into the nature of dopamine–purine interactions in motor function (Jarvis and Williams, 1987; Nehlig et al., 1992; Ferre et al., 1992).

While it is clear that methylxanthines can potentiate the effects of dopaminergic agonists, it is also the case that methylxanthines can mimic the effects of dopamine receptor agonists in stimulating rotation (Garrett and Holtzman, 1994, 1996; Jiang et al., 1993; Herrera-Marschitz et al., 1988). Theophylline and caffeine administration alone have been shown to induce rotation in rats with unilateral striatal lesions (Fredholm et al., 1976; Watanabe et al., 1981; Garrett and Holtzman, 1995). While both caffeine and dopamine agonists can stimulate rotational behavior in this model, the effects of these agents can be differentiated. Apomorphine has been shown to increase rotational behavior at lower doses than are required to significantly increase motor activity (Garrett and Holtzman, 1996). Caffeine, however, increases both rotational behavior and motor activity at similar doses (Garrett and Holtzman, 1996). Interestingly, contralateral rotation to apomorphine has been demonstrated in rats following the unilateral intrastriatal administration of NECA (Green et al., 1982). In these animals, the intraperitoneal administration of theophylline blocked this apomorphine-induced rotation (Green et al., 1982). The intrastriatal administration of NECA into the unilaterally lesioned striatum also has been shown to reduce apomorphine-induced rotation (Brown et al., 1991). Similarly, the intrastriatal administration of adenosine agonists and antagonists into normal rats can also produce competitive ipsilateral and contralateral rotation, respectively (Josselyn and Beninger, 1991).

The intrastriatal administration of the A_{2a} agonist CGS 21680 can also attenuate rotational behavior to both direct and indirect dopamine agonists in unilaterally lesioned rats (Morelli et al., 1994). The apparent increased sensitivity to CGS 21680 in animals with supersensitive dopamine receptors (Morelli et al., 1994) is entirely consistent with radioligand binding data and indicates an increased efficacy of CGS 21680 in reducing the binding affinity of supersensitive D_2 receptors (Ferre and Fuxe, 1992; Ferre et al., 1992). The repeated administration of the dopamine antagonist haloperidol has also been shown to upregulate the density of dopamine D_2 and adenosine A_2 receptors in rat striatum (Parsons et al., 1995). The evidence discussed above indicates that purinergic manipulations of basal ganglia function can result in profound behavioral effects that appear to be mediated through adenosine-induced alterations in dopamine function. Further, the modulatory influence of adenosine on dopaminergic neurotransmission may be particularly enhanced in situations of increased dopamine receptor sensitivity (Casas et al., 1989a,b; Ferre et al., 1992).

THERAPEUTIC IMPLICATIONS OF PURINE–DOPAMINE INTERACTIONS

Both the specific colocalization of high-affinity adenosine A_{2a} receptor and dopamine D_2 receptors (Jarvis and Williams, 1989; Parkinson and Fredholm, 1990) and the newly appreciated functional interaction between dopamine and adenosine receptor regulation (Ferre et al., 1992, 1993, 1994) have provided additional focus in the search for novel therapeutic interventions in disorders of the basal ganglia.

The pathological consequences of a disruption in this balance in basal ganglia function is illustrated in the Lesch-Nyhan syndrome, which is characterized by motor dysfunction, mental retardation, and self-mutilatory behavior (Nyhan, 1973). The Lesch-Nyhan syndrome is a X-linked recessive disorder of purine metabolism in which the lack of hypoxanthine-guanine phosphoribosyl transferase (HGPRT) activity results in decreased levels of adenosine in the brain as well as an overproduction of uric acid (Nyhan, 1973). This disorder is also characterized by marked decreases in striatal dopamine concentrations (Lloyd et al., 1981; Kopin, 1981). Self-mutilatory behavior appears to be mediated by supersensitive dopamine receptors (Goldstein et al., 1986; Criswell et al., 1988; Sivam, 1996) rather than to be a direct result of purine dysfunction (Bitler and Howard, 1986). Interestingly, the dopaminergic dysfunction in the Lesch-

Nyhan syndrome, characterized by a selective destruction of dopamine terminals, is remarkably similar to that found in methamphetamine-induced neurotoxicity (Lloyd et al., 1981; Ricaurte et al., 1982). Methylxanthines, when administered to food-deprived rats (Ferrer et al., 1982) or in very large doses (Boyd et al., 1965), can induce self-mutilatory behavior. Similar behavior can also be produced by other psychomotor stimulants, including L-dopa, pemoline, amphetamine, apomorphine, and SKF-38393 (Muller and Nyhan, 1982; Goldstein et al., 1986; Criswell et al., 1988). This self-mutilatory behavior can be reduced by the intrastriatal administration of adenosine analogs, with NECA being more potent that 2-CADO or CPA, a finding that indicates that self-mutilatory behavior may be modulated by agonist interactions at striatal A_2 receptors (Criswell et al., 1988). Taken together these findings suggest that the behavioral consequences of Lesch-Nyhan syndrome seem to be mediated by hyperactive striatal dopamine receptor populations in a context of diminished inhibitory adenosine neuromodulation via activation of adenosine A_{2a} receptors.

The neuromodulatory relationship between adenosine and dopamine is not solely confined to A_{2a}–D_2 interactions. Activation of adenosine A_1 receptors can also reduce the high-affinity state of the dopamine D_1 receptor in the rat striatum (Ferre et al., 1994). Functionally, the A_1-selective agonist CPA can effectively block dopamine D_1 receptor–mediated locomotor activation in mice receiving reserpine (Ferre et al., 1994). NECA has also been shown to attenuate perioral dyskinesias induced by selective dopamine D_1 activation in rabbits (Caporali et al., 1992). It appears, then, that adenosine acting at the striatal A_{2a} and A_1 receptors can directly modulate dopaminergic function mediated by dopamine D_2 receptors on striatal GABA-enkephalinergic neurons and by dopamine A_1 receptors on striatal GABA-substance P neurons (Ferre et al., 1992; Williams, 1995).

This dynamic interrelationship between dopaminergic and purinergic systems in the neurochemistry of psychomotor function offers new possibilities for the amelioration of dopaminergic dysfunction via adenosine receptor modulation. Thus, selective adenosine A_2 receptor antagonists have been proposed as novel treatments for Parkinson's disease and advanced Huntington's disease (Jarvis and Williams, 1987; Ferre et al., 1992; Williams, 1995). KW 17837 (Shimada et al., 1992) represents an A_{2a} selective antagonist that is currently being evaluated in these situations of dopaminergic hypofunction (Williams, 1995).

Given that the effects of adenosine agonists on motor activity can be opposite to the effects observed for dopamine agonists, and that antagonist ligands for each receptor system also produce opposite behavioral effects (Jarvis and Williams, 1987), there has been growing interest in examining the efficacy of adenosine neuromodulation as a therapeutic intervention in psychotic disorders. Much of the recent experimental work in this area has been devoted to the evaluation of adenosine receptor agonists in animal models of dopaminergic activity that offer some degree of predictive validity for clinical efficacy as antipsychotic agents. In addition to their ability to decrease spontaneous motor activity, adenosine agonists with significant A_2 receptor activity are also effective in reducing amphetamine-induced hyperactivity at doses that do not cause ataxia, a profile that is shared with dopamine receptor antagonists (Heffner et al., 1989).

The ability of adenosine agonists to suppress hyperactive dopaminergic function without concomitantly producing ataxia remains equivocal. Unlike dopamine antagonists, adenosine agonists appear to attenuate apomorphine-induced wall climbing in mice only at ataxic doses (Heffner et al., 1989). Examinations of agonist potency profiles has lead to some suggestions that ataxia in rodents may be more closely related to activation of adenosine A_1 receptors rather than the A_{2a} receptor subtype (Heffner et al., 1989; Martin et al., 1993). However, both the central and systemic administration of the selective A_{2a} agonist CGS 21680 have been shown to produce ataxia in rodents (Ferre et al., 1993; Morelli et al., 1994; Martin et al., 1993). Yet, CGS 21680 does not produce ataxia in the same dose-dependent fashion that has been shown for adenosine A_1-selective agonists (Martin et al., 1993).

The clinical utility of adenosine agonists as antipsychotic agents has been tested more directly in the conditioned avoidance response (CAR) paradigm (Martin et al., 1993). This behavioral assay examines the relative ability of drugs to selectively disrupt lever pressing by rats to avoid a signaled foot shock. Selective disruption of avoidance responding is a property shared by all known dopamine antagonists, and activity in this behavioral task is highly correlated with clinical efficacy in treating schizophrenia in man (Arnt, 1982; Kuribara and Tadokoro, 1981; Martin et al., 1989). The conditioned avoidance response paradigm also contains a measure of nonspecific behavioral disruption of sedation (i.e., the drug-induced disruption of lever pressing following a shock presentation, termed escape responding). The CAR paradigm has been used to identify a number of novel antipsychotic agents (Martin et al., 1993).

Adenosine agonists exhibit a behavioral profile very much like dopamine antagonists in the CAR procedure, in that they potently disrupt avoidance responding without significantly impairing escape behavior (Table 1). The pharmacological profile of adenosine agonists to inhibit CAR was not indicative of a selective interaction with either adenosine A_1 or A_2 receptors. Interestingly, NECA was found to be significantly more potent than either CGS 21680 or CPA in blocking CAR. This observation may reflect a NECA-induced synergistic interaction of both A_1 and A_2 receptor stimulation similar to that found in locomotor activity studies (Nikodijevic et al., 1991). Despite the fact that NECA was approximately 5 to 10 times more potent than prototypic neuroleptic agents (e.g., chlorpromazine), neither NECA nor any of the other adenosine agonists examined were found to fully block CAR to the same degree as dopamine antagonists (Martin et al., 1993). Adenosine agonists can also produce catalepsy at the same general

TABLE 1 Comparitive Efficacy of Adenosine Agonist and Dopamine Antagonist Compounds in Blocking Conditioned Avoidance Responding in the Rat

Compound	Receptor	Conditioned $Block_{MAX}$	Avoidance ED_{50} (mg/kg)	Escape %Loss	Catalepsy %CAR ED_{50}
Chloropromazine	D_2	90%	4.9	7	N.D.
Thioridazine	D_2	80%	12.5	2	N.D.
Trifluroperazine	D_2	100%	0.9	3	N.D.
Haloperidol	D_2	60%	0.2	3	87%
Metoclopramide	D_2	98%	7.8	8	N.D.
CPA	A_1	70%	1.5	2	80%
R-PIA	A_1	80%	0.3	1	25%
CV-1808	A_2/A_1	60%	1.3	4	N.D.
NECA	A_1/A_2	60%	0.07	4	30%
CGS 21680	A_2	40%	1.1	1	30%

N.D. = not determined. $Block_{MAX}$ indicates disruption of lever pressing before signaled shock presentation. %Escape Loss indicates disruption of lever pressing during shock presentation. %CAR indicates disruption of conditioned avoidance. Data from Martin et al. (1993).

dose levels that are effective in attenuating CAR (Martin et al., 1993). This property is also shared by many neuroleptic agents such as haloperidol (Table 1). However, the degree of catalepsy produced by adenosine agonists was not sufficient to produce a nonspecific behavioral disruption of escape responding in the CAR procedure (Martin et al., 1993). Thus, the lack of a clear dose-dependence in the ability of CGS 21680 to induce catalepsy illustrates the possibility that an adenosine-A_2-selective agonist can demonstrate favorable antipsychotic activity without untoward motor impairment (Martin et al., 1993). The separation between therapeutic efficacy and adverse side effects remains the biggest challenge in the discovery and development of novel adenosine-based medicines (Jacobson et al., 1992; Bhagwat and Williams, 1995).

CONCLUSION

Caffeine is the psychomotor stimulant most widely consumed by man (Nehlig et al., 1992; Jacobson et al., 1991). The methylxanthines produce stimulus and hedonic properties that are similar to those of the classical psychomotor stimulants that function as dopaminergic agonists (Holtzman, 1990). The similarity in the psychomotor properties of these agents is not suprising, given the recent advances in our understanding of purine–dopamine interactions in the basal ganglia (Ferre et al., 1992), and suggests the possibility that comparable purinergic neuromodulation of dopaminergic function may occur in other brain regions, including the mesolimbic dopaminergic pathway. Currently there is interest in purinergics as therapeutic interventions in dopamine-mediated motor disfunction. In addition, purinergic modulation may be therapeutically beneficial in the treatment of depression and substance abuse based on similar neuromodulatory interactions. Much of the neuropsychopharmacology of adenosine has been devoted to the delineation of the mechanistic role of this purine in neurotransmission. Investigations at the level of the behaving organism have contributed greatly to our understanding of purinergic neuromodulation, and as noted above, helped to focus on the issue of pharmacological specificity in terms of the development of novel CNS therapeutic agents. There continues to be a need for molecularly diverse and receptor-selective ligands, as illustrated by the lack of agents with high affinity and/or selectivity for the A_{2b} receptor (Bhagwat and Williams, 1995). Additionally, other strategies are being explored for therapeutic utility, including the development of partial agonists and of indirect agonists (Williams, 1995).

ACKNOWLEDGMENT

The author would like to thank Ms. Helen Park for her comments on an earlier version of this manuscript.

REFERENCES

Ahiljanian M, Takemori AE (1986): Changes in adenosine receptor sensitivity in morphine-tolerant and -dependent mice. J Pharmacol Exp Ther 236:615–620.

Arnt J (1982): Pharmacological specificity of conditioned avoidance response inhibition in rats: Inhibition by neuroleptics and correlation to dopamine receptor blockade. Acta Pharmacol Toxicol 51:321–329.

Barraco RA, Aggarawai AK, Phillis JW, Moran MA, Wu PH (1986): Dissociation of locomotor and hypertensive effects of adenosine analogs in rat. Neurosci Lett 48:139–144.

Barraco RA, Swanson TH, Phillis JW, Berman RF (1984): Anticonvulsant effects of adenosine analogs on amygdaloid-kindled seizures in the rat. Neurosci Lett 46:317–322.

Barraco RA, Coffin VL, Altman HJ, Phillis JW (1983): Central effects of adenosine analogs on locomotor activity in mice and antagonism of caffeine. Brain Res 272:392–395.

Bhagwat SS, Williams M (1995): Recent progress in modulators of purinergic activity. Curr Opin Ther Patent 5:547–558.

Bitler CM, Howard BD (1986): Dopamine metabolism in hypoxanthine-guanine phosphoribosyltransferase-deficient variants of PC12 cells. J Neurochem 47:107–112.

Boulenger JP, Patel J, Post RM, Parma AM, Marangos PJ (1983): Chronic caffeine consumption increases the number of brain adenosine receptors. Life Sci 32:1135–1142.

Boyd EM, Dolman M, Knight LM, Sheppard EP (1965): The chronic oral toxicity of caffeine. Can J Pharmacol 43:995–1007.

Brown SJ, Gill R, Evenden JL, Iverson SD, Richardson PJ (1991): Striatal A_2 receptor regulates apomorphine-induced turning in rat with unilateral dopamine denervation. Psychopharmacology 103:78–82.

Buckholtz NS, Middaugh LD (1987): Effects of caffeine and L-phenylisopropyladenosine on locomotor activity of mice. Pharmacol Biochem Behav 28:179–185.

Caporali MG, de Carolis AS, Popoli P (1992): N-ethyl-carboxamide adenosine inhibits perioral dyskinesias induced by sulpiride + SKF 38393 in rabbits. Eur J Pharmacol 223Z:15–18.

Carney JM (1982): Effects of caffeine, theophylline and theobromide on schedule controlled responding in rats. Br J Pharmacol 75:451–454.

Carney JM, Christensen HD (1980): Discriminative stimulus properties of caffeine: Studies using pure and natural products. Pharmacol Biochem Behav 13:313–318.

Carney JM, Holloway FA, Modrow HE (1985a): Discriminative stimulus properties of methylxanthines and their metabolites in rats. Life Sci 36:913–920.

Carney JM, Holloway FA, Williams HL, Seale TW (1985b): Behavioral pharmacology of caffeine in experimental subjects. In Balster R, Seiden L (eds): "Behavioral Pharmacology: The Current Status." New York: Alan R Liss, pp 281–293.

Casas M, Ferre S, Cadafaleh J, Grau JM, Jane F (1989a): Rotational behavior induced by theophylline in 6-OHDA nigrostriatal denervated rats is dependent on the supersensitivity of striatal dopaminergic receptors. Pharmacol Biochem Behav 29:609–613.

Casas M, Ferre S, Cobos A, Grau JM, Jane F (1989b): Relationship between rotational behavior induced by apomorphine and caffeine in rats with unilateral lesions of the nigrostriatal pathway. Neuropharmacology 28:407–409.

Choi OH, Shamim MT, Padgett WL, Daly JW (1988): Caffeine and theophylline analogues: Correlation of behavioral effects with activity as adenosine receptor antagonists and as phosphodiesterase inhibitors. Life Sci 43:387–398.

Chou DT, Kan S, Forde J, Hirsh KR (1985): Caffeine tolerance: Behavioral electrophysicological and neurochemical evidence. Life Sci 36:2347–2358.

Coffin VL, Spealman RD (1989): Psychomotor-stimulant effects of 3-isobutyl-1-methylxanthine: Comparison with caffeine and 7-(2-chloroethyl)theophylline. Eur J Pharmacol 170:35–40.

Coffin VL, Spealman RD (1987): Behavioral and cardiovascular effects of analogs of adenosine in cynomolgus monkeys. J Pharmacol Exp Ther 241:76–83.

Coffin VL, Taylor JA, Phillis JW, Altman HJ, Barraco RA (1984): Behavioral interaction of adenosine and methylxanthines on central purinergic systems. Neurosci Lett 47:91–98.

Concannon JT, Braughler M, Schechter, MD (1983): Pre- and postnatal effects of caffeine on brain biogenic amines, cyclic nucleotides and behavior in developing rats. J Pharmacol Exp Ther 226:673–679.

Criswell H, Mueller RA, Breese GR (1988): Assessment of purine-dopamine interactions in 6-hydroxydopamine-lesioned rats: Evidence for pre- and postsynaptic influences by adenosine. J Pharmacol Exp Ther 244:493–500.

Durcan MJ, Morgan PF (1989): Evidence for adenosine A_2 receptor involvement in the hypomobility effects of adenosine analogues in mice. Eur J Pharmacol 168:285–290.

Erinoff L, Snodgrass SR (1986): Effects of adult or neonatal treatment with 6-hydroxydopamine or 5,7-dihydroxytryptamine on locomotor activity, monoamine levels, and response to caffeine. Pharmacol Biochem Behav 24:1039–1045.

Evans DB, Schenden JA, Bristol, JA (1982): Adenosine receptors mediating cardiac depression. Life Sci 31:2425–2432.

Ferre S, Fuxe K (1992): Dopamine denervation leads to an increase in the intramembrane interaction between adenosine A_2 and dopamine D_2 receptors in the neostriatum. Brain Res 594:124–130.

Ferre S, Popoli P, Gimenez-Llort L, Finnman UB, Martinez E, de Carolis AS, Fuxe K (1994): Postsynaptic antagonistic interaction between adenosine A_1 and dopamine D_1 receptors. Neuroreport 6:73–76.

Ferre S, Snaprud P, Fuxe K (1993): Opposing actions of adenosine A_2 receptor agonist and a GTP analogue on the regulation of dopamine D_2 receptors in rat neostriatal membranes. Eur J Pharmacol Mol Pharmacol 244:311–315.

Ferre S, Fuxe K, von Euler G, Johansson B, Fredholm BB (1992): Adenosine–dopamine interactions in the brain. Neuroscience 3:501–512.

Ferre S, von Euler G, Johansson B, Fredholm BB, Fuxe K (1991a): Stimulation of high affinity adenosine A_2 receptors decreases the affinity of dopamine D_2 receptors in rat striatal membranes. Proc Natl Acad Sci USA 88:7238–7241.

Ferre S, Herrera-Marschitz, Graboska-Anden M, Ungerstedt U, Casas M, Anden NE (1991b): Postsynaptic dopamine/adenosine interaction, I. Adenosine analogues inhibit dopamine D_2-mediated behaviour in short-term reserpinized mice. Eur J Pharmacol 192:25–30.

Ferre S, Herrera-Marschitz, Graboska-Anden M, Ungerstedt U, Casas M, Anden NE (1991c): Postsynaptic dopamine/adenosine interaction, II. Postsynaptic dopamine agonism and adenosine antagonism of methylxanthines in short-term reserpinized mice. Eur J Pharmacol 192:31–37.

Ferrer I, Costell M, Grisolia S (1982): Lesch-Nyhan syndrome like behavior in rats from caffeine ingestion: Changes in HGPRTase activity, urea and some nitrogen metabolism enzymes. FEBS Lett 141:275–278.

File SA, Baldwin HA, Johnston AL, and Wilks, LJ (1988): Behavioral effects of acute and chronic administration of caffeine in the rat. Pharmacol Biochem Behav 30:809–815.

Finn IB, Holtzman SG (1987): Pharmacologic specificity of tolerance to caffeine-induced stimulation of locomotor activity. Psychopharmacology 93:428–434.

Finn IB, Holtzman SG (1988): Tolerance and cross-tolerance to theophylline-induced stimulation of locomotor activity in rats. Life Sci 42:2475–2482.

Fisher CE, Hughes RN (1996): Effects of diazepam and cyclohexyladenosine on open-field behavior in rats perinatally exposed to caffeine. Life Sci 58:701–709.

Fredholm BB, Fuxe K, Agnati, L (1976): Effects of some phosphodiesterase inhibitors on central dopamine mechanisms. Eur J Pharmacol 38:31–38.

Fredholm BB, Herrara-Marschitz A, Jonzon B, Lindstrom K, Ungerstedt U (1983): On the mechanism by which methylxanthines enhance apomorphine induced rotational behavior in the rat. Pharmacol Biochem Behav 19:535–541.

Fuxe K, Ungerstedt U (1974): Action of caffeine and theophylline on supersensitive dopamine receptors: Considerable enhancement of receptor responses to treatment with dopa and dopamine. Med Biol 52:48–54.

Garrett BE, Holtzman SG (1996): Comparison of the effects of prototypical behavioral stimulants on locomotor activity and rotational behavior in rats. Pharamcol Biochem Behav 54:469–477.

Garrett BE, Holtzman SG (1994): Caffeine cross-tolerance to selective dopamine D_1 and D_2 receptor agonists but not to their synergistic interaction. Eur J Pharmacol 262:65–75.

Glowa JR, Spealman RD (1984): Behavioral effects of caffeine, N^6-(L-phenylisopropyl)adenosine and their combination in the squirrel monkey. J Pharmacol Exp Ther 231:665–670.

Glowa JR, Sobel E, Malaspina S, Dews PB (1985): Behavioral effects of caffeine, (−)N-((*R*)-1-methyl-2-phenylethyl)-adenosine (PIA) and their combination in the mouse. Psychopharmacology 87:421–424.

Goldberg SR, Prada JA, Katz, JL (1985): Stereoselective behavioral effects of N^6-phenylisopropyl-adenosine and antagonism by caffeine. Psychopharmacology 87:272–277.

Goldstein M, Kuga S, Kusano N, Meller E, Dancis J, Schwarcz R (1986): Dopamine agonist induced self-mutilative biting behavior in monkeys with unilateral ventromedial tegmental lesions of the brainstem: Possible pharmacological model for Lesch-Nyhan syndrome. Brain Res 367:114–120.

Green RM, Stiles GL (1986): Chronic caffeine ingestion sensitizes the A_1 adenosine receptor-adenylate cyclase system in rat cerebral cortex. J Clin Invest 77:222–227.

Green R, Proudfit HK, Yeung SH (1982): Modulation of striatal dopaminergic function by local injection of 5′-N-ethylcarboxamide adenosine. Science 218:58–61.

Griebel G, Soffroy-Spittler M, Misslin R, Remmy D, Vogel E, Bourguignon JJ (1991): Comparison of the behavioral effects of an adenosine A_1/A_2 antagonist CGS 15943, and an A_1-selective antagonist, DPCPX. Psychopharmacology 103:541–544.

Hamilton HW, Taylor MD, Steffen RP, Haleen SJ, Bruns RF (1987): Correlation of adenosine receptor affinities and cardiovascular activity. Life Sci 41:2295–2302.

Heffner TG, Wiley JN, Williams AE, Bruns RF, Coughenour LL, Downs, DA (1989): Comparison of the behavioral effects of adenosine agonists and dopamine antagonists in mice. Psychopharmacology 98:31–37.

Herrera-Marschitz M, Casas M, Ungerstedt U (1988). Caffeine produces contralateral rotation in rats with unilateral dopamine denervation: Comparisons with apomorphine-induced responses. Psychopharmacology 94:38–45.

Holloway FA, Michaelis RC, Huerta PL (1985a): Caffeine-phenylethylamine combinations mimic the amphetamine discriminative cue. Life Sci 36:723–730.

Holloway FA, Modrow HE, Michaelis RC (1985b): Methylxanthine discrimination in the rat: Possible benzodiazepine and adenosine mechanisms. Pharmacol Biochem Behav 22:815–824.

Holtzman SG (1990): Caffeine as a drug of abuse model. Trends Pharmacol Sci 11:355–356.

Holtzman SG (1987): Discriminative stimulus effects of caffeine: tolerance and cross-tolerance with methylphenidate. Life Sci 40:381–389.

Holtzman SG (1986): Discriminative stimulus properties of caffeine in the rat: Noradrenergic mediation. J Pharmacol Exp Ther 239:706–714.

Holtzman SG, Mante S, Minneman KP (1991): Role of adenosine receptors in caffeine tolerance. J Pharmacol Exp Ther 256:62–68.

Howell LL (1993): Comparative effects of caffeine and selective phosphodiesterase inhibitors on respiration and behavior in rhesus monkeys. J Pharmacol Exp Ther 266:894–903.

Howell LL, Byrd LD (1993): Effects of CGS 15943, a nonxanthine adenosine antagonist, on behavior in the squirrel monkey. J Pharmacol Exp Ther 267:432–439.

Hutchison AJ, Williams M, deJesus R, Oei HH, Ghai GR, Webb RL, Zoganas HC, Stone GA, Jarvis MF (1990): 2-Arylalkylamino-adenosine 5′-uronamides: A new class of highly selective adenosine A_2 agonists. J Med Chem 33:1683–1689.

Jacobson KA, Kim HO, Siddiq SM, Olah ME, Stiles GL, von Lubitz DKJE (1995): A_3-adenosine receptors: Design of selective ligands and therapeutic prospects. Drugs of the Future 20:689–699.

Jacobson KA, Nikodijevic O, Shi D, Gallo-Rodriguez C, Olah ME, Stiles GL, Daly JW (1993): A role for central A_3 adenosine receptors: Mediation of behavioral depressant effects. FEBS Lett 336:57–60.

Jacobson KA, Trivedi B, Churchill P, Williams M (1991): Novel therapeutics acting via purine receptors. Biochem Pharmacol 41:1399–1410.

Jacobson KA, van Galen P, Williams M (1992): Adenosine receptors: Pharmacology, structure activity relationships and therapeutic potential. J Med Chem 35:407–422.

Jacobson KA, von Lubitz DKJE, Daly JW, Fredholm BB (1996): Adenosine receptor ligands: Differences with acute and chronic treatment. Trends Pharmacol Sci 17:108–113.

Jarvis MF (1988): Autoradiographic localization and characterization of brain adenosine receptor subtypes. In Leslie F, Altar CA (eds): "Receptor Localization: Ligand Autoradiography." New York: Alan R Liss, pp 95–113.

Jarvis MF, Schulz R, Hutchinson A, Do U, Sills M, Williams M (1989): [^{3}H]CGS 21680, a selective A_2 adenosine receptor agonist directly labels A_2 receptors in rat brain. J Pharmacol Exp Ther 251:888–893.

Jarvis MF, Williams M (1987): Adenosine and dopamine function in the CNS. Trends Pharmacol Sci 8:330–332.

Jarvis MF, Williams M (1989): Direct autoradiographic localization of adenosine A_2 receptors in the rat brain using the A_2-selective agonist [^{3}H]CGS 21680. Eur J Pharmacol 168:243–246.

Jarvis MF, Williams M (1990): Adenosine in central nervous system function. In Williams M (ed): "Adenosine and Adenosine Receptors." Clifton NJ: Humana Press, pp 423–474.

Jarvis MF, Williams M, Do UH, Sills MA (1991): Characterization of the binding of a novel non-xanthine adenosine radioligand, [^{3}H]CGS 15943, to multiple affinity states of the adenosine A_1 receptor in rat cortex. Mol Pharmacol 39:49–54.

Ji X-d, von Lubitz D, Olah ME, Stiles GL, Jacobson KA (1994): Species differences in ligand binding affinity at central A_3 receptors. Drug Dev Res 33:51–59.

Jiang H, Jackson-Lewis V, Muthane U, Dollison A, Ferreira M, Espinosa A, Parsons B, Przedborski S (1993): Adenosine receptor antagonists potentiate dopamine receptor agonist-induced rotational behavior in 6-hydroxydopamine-lesioned rats. Brain Res 613:347–351.

Johansson B, Ahlberg S, van der Ploeg I, Brene S, Lindefors N, Persson H, Fredholm BB (1993): Effect of long term caffeine treatment on A_1 and A_2 adenosine receptor binding and on mRNA levels in rat brain. Naunyn Schmiedebergs Arch Pharmacol 347:407–414.

Josselyn SA, Beninger RJ (1991): Behavioral effects of intrastriatal caffeine mediated by adenosinergic modulation of dopamine. Biochem Pharmacol Behav 39:97–103.

Katims JJ, Annau Z, Snyder SH (1983): Interactions in the behavioral effects of methylxanthines and adenosine derivatives. J Pharmacol Exp Ther 227:167–173.

Klawans HL, Moses H, Beaulieu DM (1974): The influence of caffeine on *d*-amphetamine and apomorphine-induced stereotyped behavior. Life Sci 14:1493–1500.

Kowaluk EA, Kohlhaas KL, Daanen J, Lee L (1996): Anticonvulsant effects of adenosine kinase inhibitors. Drug Dev Res 37:190.

Kleven MS, Sparber SB (1987): Attenuation of isobutylmethylxanthine-induced suppression of operant behavior by pretreatment of rats with clonidine. Pharmacol Biochem Behav 28:235–241.

Kopin IJ (1981): Neurotransmitters in Lesch-Nyhan syndrome. N Eng J Med 305:1148–1150.

Kuribara H, Tadokoro S (1981): Correlation between antiavoidance activities and antipyschotic drugs in rats and daily clinical dose. Pharmacol Biochem Behav 14:181–192.

Lloyd KG, Hornykiewicz O, Davidson L, Shannak K, Farley I, Coldstein M, Shibuya M, Kelley WN, Fox IH (1981): Biochemical evidence of dysfunction of brain neurotransmitters in Lesch-Nyhan syndrome. N Eng J Med 305:1106–1111.

Lupica CR, Berman RF, Jarvis MF (1991): Chronic theophylline treatment increases adenosine A_1 but not A_2 receptor binding in the rat brain: An autoradiographic study. Synapse 9:95–102.

Martin GE, Rossi, DJ, Jarvis MF (1993): Adenosine agonists reduce conditioned avoidance responding in the rat. Pharmacol Biochem Behav 45:951–958.

Martin GE, Elgin RJ, Kesslick JM, Karkanias CJ, Mathiasen JR, Baldy WJ, Shank RP, DiStefano DL, Scott, MK (1989): Activity of serotonergic aromatic substituted phenylpiperazines lacking affinity for dopamine binding sites in a preclinical test of antipsychotic potency. J Med Chem 32:1052–1056.

Martinez-Mir MI, Probs A, Palacios JM (1991): Adenosine A_2 receptors: Selective localization in the human basal ganglia and alterations with disease. Neuroscience 3:697–706.

Modrow HE, Holloway FA, Carney JM (1981): Caffeine discrimination in the rat. Pharmacol Biochem Behav 14:683–688.

Morelli M, Fenu S, Pinna A, Di Chiara G (1994): Adenosine A_2 receptors interact negatively with dopamine D_1 and D_2 receptors in unilaterally 6-hydroxydopamine-lesioned rats. Eur J Pharmacol 251:21–25.

Muller K, Nyhan W (1982): Pharmacologic control of pemoline self-injurious behavior in rats. Pharmacol Biochem Behav 16:957–963.

Murphy KM, Snyder SH (1982): Heterogenity of adenosine A_1 receptor binding in brain tissue. Mol Pharmacol 22:260–267.

Murray TF (1982): Upregulation of rat cortical adenosine receptors following chronic administration of theophylline. Eur J Pharmacol 82:113–114.

Nehlig A, Daval JL, Derby G (1992): Caffeine and the central nervous system: Mechanisms of action, biochemical, metabolic and psychostimulant effects. Brain Res Rev 17:139–170.

Nikodijevic O, Jacobson KA, Daly JW (1990): Characterization of the locomotor depression produced by an A_2-selective adenosine agonist. FEBS Lett 261:67–70.

Nikodijevic O, Jacobson KA, Daly JW (1993): Locomotor activity in mice during chronic treatment with caffeine and withdrawal. Pharmacol Biochem Behav 44:199–216

Nikodijevic O, Sarges R, Daly JW, Jacobson KA (1991): Behavioral effects of A_1- and A_2-selective adenosine agonists and antagonists: Evidence for synergism and antagonism. J Pharmacol Exp Ther 259:286–294.

Nyhan WL (1973): The Lesch-Nyhan syndrome. Ann Rev Med 24:41–61.

Oei HH, Ghai GR, Zoganas HC, Stone GA, Field FP, Williams M (1989): Correlation between binding affinities for brain A_1 and A_2 receptors of adenosine agonists and antagonists and their effects on heart rate and coronary vascular tone. J Pharmacol Exp Ther 247:882–888.

Parkinson FE, Fredholm BB (1990): Autoradiographic evidence for G-protein coupled A_2-receptors in rat neostriatum using [^{3}H]CGS 21680 as a ligand. Naunyn Schmiedbergs Arch Pharmacol 342:85–89.

Parsons B, Togasaki DM, Kassir S, Predborski S (1995): Neuroleptics upregulate adenosine A_{2a} receptors in rat striatum: Implications for the mechanism and treatment of tardive dyskinesia. J Neurochem 65:2057.

Phillis JW, Barraco RA, DeLong RE, Washington DO (1986): Behavioral characteristics of centrally administered adenosine analogs. Pharmacol Biochem Behav 24:263–270.

Popoli P, Pezzola A, Reggio R, Caporali MG, de Carolis AS (1994): CGS 21680 antagonizes motor hyperactivity in a rat model of huntington's disease. Eur J Pharmacol 257:R5–R6.

Ricaurte GA, Guillery RW, Seiden LS, Schuster CR, Moore RY (1982): Dopamine nerve terminal degeneration produced by high doses of methylamphetamine in the rat brain. Brain Res 235:93–103.

Sanders RC, Murray TF (1988): Chronic theophylline exposure increases agonist and antagonist binding to A_1 adenosine receptors in rat brain. Neuropharmacology 27:757–760.

Schechter MD (1977): Caffeine potentiation of amphetamine: Implications for hyperkinetic therapy. Pharmacol Biochem Behav 6:359–361.

Seale TW, Abla KA, Shamin MT, Carney JM, Daly JW (1988): 3,7-dimethyl-1-propargylxanthine: A potent and selective in vivo antagonist of adenosine analogs. Life Sci 43:1671–1684.

Seale TW, Roderick TH, Johnson P, Logan L, Rennert OM, Carney JM (1986): Complex genetic determinants of susceptibility to methylxanthine-induced locomotor activity changes. Pharmacol Biochem Behav 24:1333–1341.

Shimada J, Suzuki F, Nonaka H, Ishii A, Ichikawa S (1992): (E)-1,3,dialkyl-7-methyl-8-(3,4,5-trimethoxystyryl) xanthines: Potent and selective adenosine antagonists. J Med Chem 35:2342–2345.

Sivam SP (1996): Dopamine, serotonin, and tachykinin in self-injurious behavior. Life Sci 58:2367–2375.

Snyder SH, Katims JJ, Annau Z, Bruns RF, Daly JW (1981): Adenosine receptors and behavioral actions of methylxanthines. Proc Natl Acad Sci USA 78:3260–3264.

Spealman RD (1988): Psychomotor stimulant effects of methylxanthines in squirrel monkeys: Relation to adenosine antagonism. Psychopharmacology 95:19–24.

Spealman RD, Coffin VL (1988): Discriminative-stimulus effects of adenosine analogs: Mediation by adenosine A_2 receptors. J Pharmacol Exp Ther 246:610–617.

Szot P, Sanders RC, Murray TF (1987): Theophylline-induced upregulation of A_1-adenosine receptors associated with reduced sensitivity to convulsants. Neuropharmacology 26:1173–1180.

Thithapandha A, Maling HM, Gillette GR (1972): Effects of caffeine and theophylline on activity of rats in relation to brain xanthine concentrations. Proc Soc Exp Biol Med 139:582–586.

Ungerstedt U (1971): Postsynaptic supersensitivity after 6-hydroxydopamine induced degeneration of the nigrostriatal dopamine system in the rat brain. Acta Physiol Scand 367:69–93.

Waldeck B (1975): Effect of caffeine on locomotor activity and central catecholamine mechanisms: A study with special reference to drug interaction. Acta Pharmacol Toxicol 36(suppl 4):1–23.

Watanabe H, Ikeda M, Watanabe K (1981): Properties of rotational behavior produced by methylxanthine derivatives in mice with unilateral striatal 6-hydroxydopamine-induced lesions. J Pharm Dyn 4:301–307.

White BC, Keller GE (1984): Caffeine pretreatment: Enhancement and attenuation of *d*-amphetamine-induced activity. Pharmacol Biochem Behav 20:383–386.

White BC, Simpson CC, Adams JE, Harkins D (1978): Monoamine synthesis and caffeine-induced locomotor activity. Neuropharmacology 17:511–513.

Williams M (1995): Purinoceptors in central nervous system function. In Bloom FE, Kupfer DJ (eds): "Psychopharmacology: The Fourth Generation of Progress." New York: Raven Press, pp 643–655.

Williams M, Francis J, Ghai G, Braundwalder A, Psychoyos S, Stone G (1987): Biochemical characterization of the triazolo-quinazoline, CGS 15943, a novel non-xanthine adenosine antagonist. J Pharmacol Exp Ther 214:415–420.

Wood PL, Kim HS, Boyar WC, Hutchison A (1989): Inhibition of nigrostriatal release of dopamine in the rat by adenosine receptor agonists: A_1 receptor mediation. Neuropharmacology 28:21–25.

Wotring LL, Townsend LB (1979): Study of the cytotoxicity and metabolism of 4-amino-3-carboxamido-1-(β-D-ribofuranosyl)pyrazolo[3,4-d]pyrimidine using inhibitors of adenosine kinase and adenosine deaminase. Cancer Res 39:3018–3023.

Zetterstrom T, Vernet T, Ungerstedt U, Tossman U, Jonzon B, Fredholm BB (1982): Purine levels in the intact rat brain. Studies with an implanted hollow fibre. Neurosci Lett 29:111–115.

CHAPTER 22

Adenosine and ATP in Epilepsy

LARS J. S. KNUTSEN and THOMAS F. MURRAY

MedChem Research I, Health Care Discovery, Novo Nordisk A/S, Novo Nordisk Park, DK 2760 Måløv, Denmark (L.J.S.K); College of Pharmacy, Oregon State University, Corvallis, OR 97331 (T.F.M.)

INTRODUCTION

Adenosine and Seizures

Early evidence of adenosine's involvement in protection against seizure states was obtained over 20 years ago when a protective effect, following intraperitoneal (i.p.) injection of adenosine **1,** was observed in mice sensitive to audiogenic seizures (Maitre et al., 1974). These initial studies have since been followed up at an increasing pace, so that researchers are now gaining a much clearer understanding of the anticonvulsant effect of endogenous adenosine and the potential for modulation of purinergic pathways in treatment of epilepsy (Dragunow, 1988; Daval et al., 1991; Knutsen et al., 1995).

Since adenosine was first recognized as a neuromodulator in the central nervous system (CNS) (Williams, 1984), there have been numerous investigations of possible therapeutic applications of this ubiquitous nucleoside in brain disease (Jarvis and Williams, 1990). Some of the other early studies of the seizure-protective potential of purinergic agents (Dunwiddie and Worth, 1982; Murray et al., 1985) took place around the time when this neuromodulatory effect of adenosine was first being established, and a subsequent review further outlined adenosine's neuromodulatory role applied to epilepsy (Chin, 1989). The overall therapeutic potential of agonists at adenosine receptors has been reviewed (Jacobson et al., 1992), including their potential for treatment of neurological diseases (Marangos, 1991), and a very recent review focuses on the role of adenosine receptors in the CNS (Fredholm, 1995). The opportunities for drug discovery within purinergic agents have been summarized (Williams, 1993); and recent progress in developing purinergic agents has been outlined, with a focus on pharmaceutical industry patenting (Bhagwat and Williams, 1995).

In contrast to adenosine 5′-triphosphate ATP **2,** adenosine modulates synaptic transmission in the CNS without itself acting as a neurotransmitter. This is due, at

Purinergic Approaches in Experimental Therapeutics, Edited by Kenneth A. Jacobson and Michael F. Jarvis
ISBN 0-471-14071-6 © 1997 Wiley-Liss, Inc.

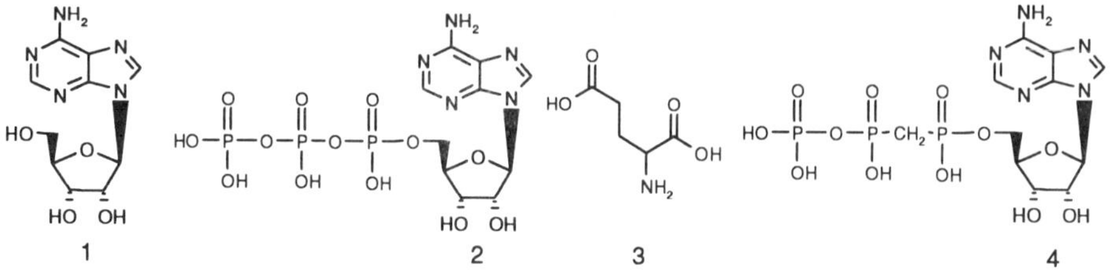

FIGURE 1.

least in part, to the ability of adenosine A_1 agonists to inhibit neurotransmitter release, especially the release of excitatory amino acids (EAAs) such as glutamate **3** (Dolphin and Prestwich, 1985; Fastbom and Fredholm, 1985; Fredholm et al., 1989; Barnie and Nicholls, 1993; Abbracchio et al., 1995; Simpson et al., 1992; Cantor et al., 1995).

ATP as an Excitatory Neurotransmitter in the CNS

The actions of ATP in the CNS were originally explored through an assessment of the behavioral effects of intracerebroventricular (icv) administration of ATP. The icv administration of high doses (1000 and 1500 μg) of ATP in rats evoked severe clonic-tonic convulsions, whereas lower doses of ATP or adenosine elicited an akinetic state with weakness (Buday et al., 1961). Subsequent electrophysiological experiments documented excitatory responses to ATP in a variety of neural preparations (Phyllis and Wu, 1981). Although calcium chelation often contributed to the depolarizing actions of ATP and adenine nucleotides, the demonstration that the dicalcium salt of ATP had a depolarizing action equivalent to that of the sodium salt suggested a specific excitatory effect of ATP on neurons (Phillis and Kirkpatrick, 1978).

More recent studies employing patch-clamp electrophysiological recordings of both whole cell and single-channel currents have characterized excitatory ATP receptors as ligand-gated ion channels (Dubyak and El-Moatassim, 1993). Ion channels activated by ATP comprise the P_{2X} family of P_2 purinergic receptors. These ATP-gated ion channels are expressed in a variety of muscle and nerve cells (Dubyak and El-Moatassim, 1993). ATP-gated channels are equally permeable to small cations such as Na^+, K^+ and Ca^{2+}, and respond to the application of ATP within milliseconds, indicating that the cation channels are directly activated by binding of ATP molecules (Bean, 1992; Zimmerman, 1994). The kinetics of ATP-gated ion channels therefore resemble those of other excitatory neurotransmitter-gated channels such as nicotinic cholinergic and glutamate receptor subtypes in the CNS.

There is now compelling evidence that ATP functions as a fast excitatory neurotransmitter in both the central and peripheral nervous systems. In the peripheral nervous systems, ATP was shown to mediate excitatory neurotransmission in cultured guinea pig celiac ganglia cells (Silinksy et al., 1992; Evans et al., 1992). These authors demonstrated that synaptic currents between these cells in culture were blocked by the P_2 receptor antagonist suramin and mimicked by ATP and α,β-methylene-ATP **4.** Similarly, in the CNS, miniature and elicited synaptic currents in the rat medial habenula were blocked by suramin and P_{2X} receptor desensitizing agonist α,β-methylene-ATP (Edwards et al., 1992). These results suggest that ATP functions as an excitatory transmitter between neurons. Indeed, the release of ATP by synaptosomes was first demonstrated by White (1978), and it is now appreciated that noncytolytic release of cytoplasmic ATP from neurons may occur via exocytosis with other neurotransmitters or alternatively by plasma membrane transporters (Dubyak and El-Moatassim, 1993).

P_{2X} Receptor Subtypes

Seven P_{2X} receptor cDNAs have recently been cloned. These receptors comprise a new class of ligand-gated ion channels. The individual P_{2X} receptor subtypes form a homologous group of proteins that differ functionally with respect to the actions of α,β-methylene-ATP, the rate of desensitization, and the sensitivity to antagonists such as suramin (Collo et al., 1996). The initial isolation of cDNAs that encode P_{2X} receptors employed the rat vas deferens (Valera et al., 1994) and NGF-differentiated PC12 cells (Brake et al., 1994). The vas deferens receptor (P_{2X1}) is sensitive to α,β-methylene-ATP and desensitizes when agonist exposure is maintained for a few seconds, whereas the PC12 receptor (P_{2X2}) was found to be insensitive to α,β-methylene-ATP and did not undergo desensitization. Northern blot analysis showed that the P_{2X2}, but not the P_{2X1}, receptor is expressed in rat brain. More recent studies have identified cDNAs for the P_{2X3}, P_{2X4}, P_{2X5}, P_{2X6}, and P_{2X7} receptors (Chen et al., 1995; Lewis et al., 1995; Bo et al., 1995; Beull et al., 1996; Seguela et al., 1996; Collo et al., 1996; Surprenant et al., 1996). Of these recently identified P_{2X} receptor subtypes, the P_{2X4} and P_{2X6} receptors were expressed at high levels in rat brain (Seguela et al., 1996; Buell et al., 1996; Surprenant et al., 1996). Although the properties of heterologously expressed homomeric P_{2X4} and P_{2X6} receptors do not adequately account for many of the functional responses to ATP observed in rat brain, the role of heteropolymerization of P_{2X} receptor subunits in the physiologic actions of ATP in the CNS remains to be elucidated.

STUDIES OF ADENOSINE IN HUMANS AND IN POSTMORTEM BRAIN TISSUE

There is a diverse and complex series of underlying disease states that result in the group of symptoms collectively known as epilepsy. Up to 5% of the population have at some time experienced seizure episodes and 30%–50% of epilepsy patients continue to suffer seizures when treated with current therapies. Many sufferers therefore have to endure a poor quality of life, and sedation due to medication is a further problem, so that there is a strong market pull for new, effective seizure treatments. There have been seven new drug launches since 1989, but prior to that the last antiepileptic drug to be launched was valproate **5** in 1976. Since epileptic seizures appear to be caused by an imbalance between inhibitory and excitatory transmission in the CNS, some of the newer drugs for epilepsy such as vigabatrin **6**, gabapentin **7**, and tiagabine **8** enhance inhibitory neurotransmission, often by reinforcing the effect of γ-aminobutyric acid (GABA) **9** (Dichter, 1994).

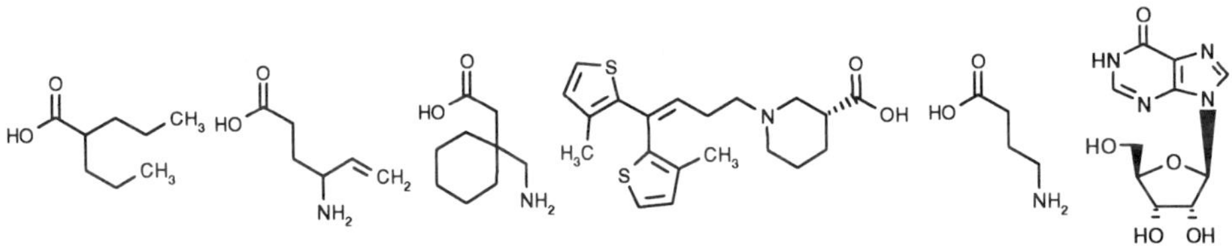

5 6 7 8 9 10

FIGURE 2.

A pivotal clinical study (During and Spencer, 1992) was carried out in four patients suffering from complex partial seizures using depth electrodes modified to include microdialysis probes and implanted in hippocampi for 10–16 days. The concentration of adenosine in extracellular fluid was raised from the basal level of approximately 2 μM to concentrations as high as 65 μM during seizures and this elevation was sustained postictally. In a separate investigation, concentrations of adenosine metabolites such as inosine **10** and hypoxanthine **11,** but not adenosine, were found to be increased in human cerebrospinal fluid following status epilepticus (Chin et al., 1995), findings which are consistent with adenosine's rapid metabolism *in vivo*. This and other evidence suggests that adenosine is the endogenous agent that causes seizure arrest in man; moreover, the evidence is consistent with previous studies indicating that status epilepticus may be caused by loss of adenosine anticonvulsant mechanisms (Young and Dragunow, 1994). This latter concept has been reinforced in a more recent study (Glass et al., 1996) using quantitative receptor autoradiographic methods to examine A_1 adenosine receptors, adenosine uptake sites, benzodiazepine receptors, and N-methyl-D-aspartate (NMDA) 12, ($\pm$)-α-amino-3-hydroxy-5-methylisoxazole-4-propionic acid (AMPA), and kainic acid receptors, in temporal lobes removed from patients suffering from complex partial seizures and in normal control postmortem temporal cortex. Binding to A_1 adenosine receptors and NMDA receptors was reduced in epileptic temporal cortex, while the other neurochemical parameters were unchanged. This A_1 receptor loss may be a consequence of rather than an initial cause of seizures. The authors conclude that loss of anticonvulsant A_1 receptors may contribute to the human epileptic condition.

In contrast, a quantitative autoradiographic study of the human neocortex (Angelatou et al., 1993) revealed a 48% increase of adenosine receptor binding in patients suffering from temporal lobe epilepsy. These adaptive changes may indicate an elevation of the depressant response of the brain to endogenous adenosine. In vitro studies of human epileptogenic cerebral cortex have shown that endogenous adenosine can inhibit epileptiform discharges (Kostopoulos et al., 1989), including the epileptiform activity mainly caused by the activation of NMDA receptors.

The marketed epilepsy drug, carbamazepine **13,** has effects on the purinergic system in rodents (see also Receptor Density Changes, below), acting potentially as an adenosine receptor antagonist (Marangos et al., 1987a) and causing upregulation of P_1 receptors (Marangos et al., 1985). Using synaptic membranes from human temporal lobe cerebral cortex from seven different autopsy cases, it was found that carbamazepine could displace the binding of both of the adenosine receptor agonists [^{3}H]-*R*-

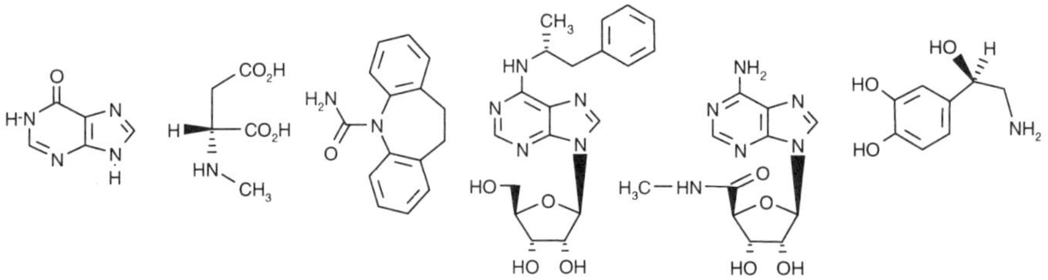

11 12 13 14 15 16

FIGURE 3.

phenylisopropyladenosine (R-PIA) **14** and [^{3}H]-N-ethylcarboxamidoadenosine (NECA) **15** from their binding sites (Dodd et al., 1986).

Administration of purinergic agents that result in activation of adenosine receptors, thus causing inhibition of neurotransmission, can be considered as an extension of current approaches to epilepsy treatment (Dichter, 1994).

MECHANISTIC STUDIES OF THE EFFECT OF ADENOSINE IN SEIZURES

Inhibition of Neurotransmitter Release

There is strong evidence, as noted in the introduction, for the presynaptic inhibition of EAA release by adenosine acting at A_1 receptors, and this topic has been reviewed recently (Higgins et al., 1994). There is apparently a stronger effect upon the release of excitatory than on inhibitory amino acids such as noradrenaline **16** and γ-aminobutyric acid (GABA) (Hollins and Stone, 1980; Jonzon and Fredholm, 1984), which helps to explain the overall inhibitory effect of adenosine.

Postsynaptically, adenosine induces hyperpolarization by increasing potassium conductance. Postsynaptic sensitivity of epileptiform activity to adenosine in hippocampal pyramidal cells is dependent on the extracellular concentrations of calcium and magnesium; removal of calcium results in desensitization (Hosseinzadeh and Stone, 1994). These observations were supported by studies of rat hippocampal slices that show that adenosine inhibits synaptic transmission and epileptiform activity by at least two mechanisms, a postsynaptic barium-sensitive increase in potassium conductance (g_k) and a presynaptic effect independent of this adenosine-evoked outward potassium conductance (Birnstiel et al., 1992).

Removal of the protective action of endogenous adenosine leads to an exaggeration of nerve cell activity. An extracellular accumulation of adenosine at or near the synaptic cleft in response to repetitive fiber activation may not only strengthen the postsynaptic effects of adenosine but also the depression of presynaptic transmitter release, for example that of the EAA glutamate, an essential pathogenic factor in the generation of epilepsy (Schubert, 1991). This investigator also provides evidence for the involvement of protein kinase C (PKC) in some of the observed actions of adenosine, such as the depression of burst discharges and neuronal calcium influx.

Receptor Density Changes

Autoradiography has proved to be a powerful technique for the investigation of functional changes following seizures. After seizures induced by bicuculline methiodide (BMI) **17,** a specific $GABA_A$ receptor antagonist, in developing rats, a widespread increase in binding sites for both [^{3}H]cyclohexyladenosine (CHA) **18** (an A_1 receptor agonist) and [^{3}H]-flunitrazepam **19** was observed. There were regional variations depending on which brain areas were investigated. Those mediating seizure activity had a particularly marked enhancement (Daval and Werck, 1991; Daval et al., 1992). Chronic twice-daily administration of theophylline in rats was found to elicit a significant increase in binding sites for [^{3}H]CHA, indicating an upregulation of A_1 receptors in cerebral cortex. There was a concomitant elevation of the threshold to BMI-induced seizures (Sanders and Murray, 1989).

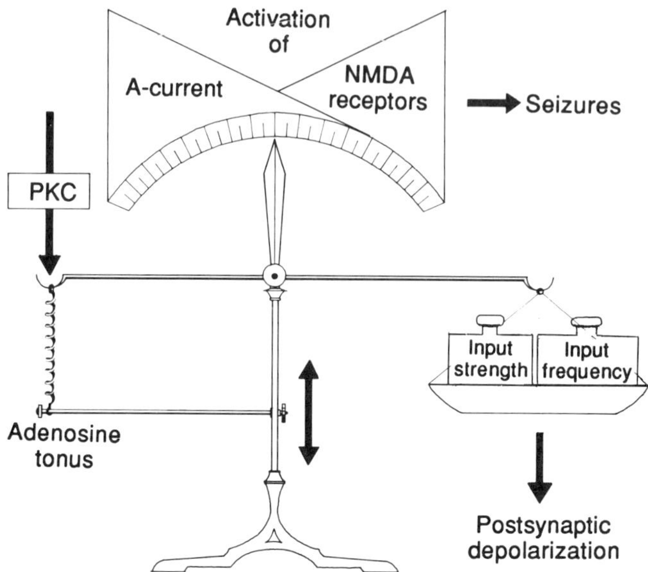

FIGURE 4. Presumed adenosine effects resulting from a nonsynaptic tuning of the postsynaptic adenosine potential. The action of adenosine is reflected by the tension of the spring, which counteracts the downward deflection of the balance. If the tension of the spring is reduced, a given input load will lead to a more pronounced deflection (membrane depolarization), and this in turn tends to block voltage-sensitive K^+ currents and facilitates the activation of NMDA receptors, thus favoring neuronal hyperactivity (Schubert, 1992; used with permission).

Pentylenetetrazole (PTZ) **20** is another prominent chemoconvulsant that appears to act as a selective GABA antagonist by interaction with the picrotoxin site on the GABA-benzodiazepine receptor complex. The effect of PTZ on A_1 receptor density in many brain regions following a period of twice-daily seizures in mice have been assessed by *in vitro* quantitative autoradiography, using [^{3}H]CHA as radioligand. The observed increases are in brain regions that mediate seizure activity, such as hippocampus, mamillary bodies, septum, substantia nigra, thalamic nuclei, and cerebral cortices. There was a significant downregulation of A_1 receptors in basal ganglia (Pagonopoulou et al., 1993). This increase in hippocampal [^{3}H]CHA binding sites was confirmed in a

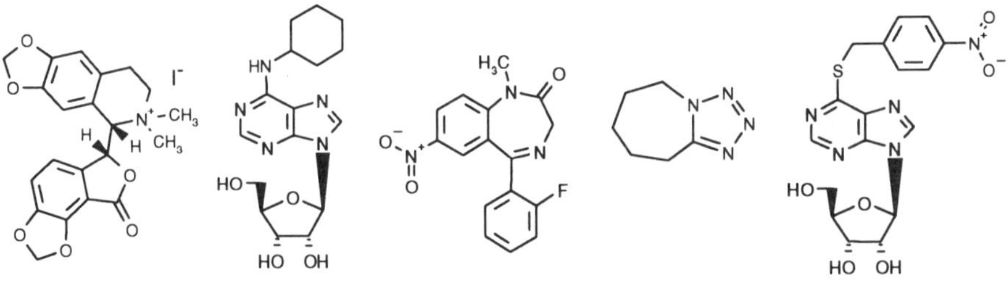

17 18 19 20 21

FIGURE 5.

subsequent study (Psarropolou et al., 1994). This study also revealed that the density of $GABA_A$ receptors was lowered in hippocampus (by 19%) and cortex (by 14%) of mice subjected to PTZ-induced seizures 24 hours and 48 hours before decapitation, when compared to untreated animals.

Following chemically induced seizures the adaptive changes in upregulation of A_1 receptors can clearly play a role in enhancing the depressive physiological response to adenosine and provide a greater anticonvulsant impact of endogenous adenosine.

Binding studies in rat cerebral cortex 30 minutes after a single electroconvulsive shock (ECS) indicated that no changes were observed in adenosine A_1 and A_2 receptors, but showed significantly higher [^{3}H]nitrobenzylthioinosine (NBI) **21** binding to an adenosine uptake site in striatum. However, after chronic ECS (10-fold) a 20% upregulation of A_1 receptors was observed; this was possibly indicative of an adaptive phenomenon (Gleiter et al., 1989). Likewise, carbamazepine treatment of rats for 2 weeks significantly elevated A_1 receptor number in cerebral cortex, hippocampus, and cerebellum (Marangos et al., 1987b). This elevation was found to be significant for 8 weeks after cessation of drug treatment.

Kindled seizures in rats are reportedly an animal model of temporal lobe epilepsy. It has been shown that several adenosine analogs have anticonvulsant effects in this model (Barraco et al., 1984; Berman et al., 1990). The brains of kindled rats have been investigated, and the affinity of [^{3}H]CHA for adenosine receptors is slightly increased in the hippocampus, but not in the cerebral cortex. CHA is potent in terms of inhibiting the outflow of D-aspartate **22** from hippocampal synaptosomes from kindled rats (Simonato et al., 1994). These data suggest that increased affinity of adenosine for A_1 receptors may contribute to the anticonvulsant effect of adenosine A_1 agonists in the kindling model.

In a separate study, CHA was injected bilaterally into the substantia nigra pars reticulata shortly before each of the first five of a series of daily kindling stimuli delivered to the rat amygdala. This did not affect the acquisition of kindled afterdischarges or the rate at which seizures developed over subsequent kindling sessions, but convulsions occurring 48–72 hours after treatment were significantly shortened (Herberg et al., 1993), emphasizing the role of the substantia nigra in controlling the generalization of seizures.

Anticonvulsant Profiles of non-A_1 Adenosine Receptor Agonists

Most of the studies of adenosine in epilepsy have focused on A_1 agonists, owing to their demonstrated involvement in the inhibition of neurotransmitter release (see Inhibition of Neurotransmitter Release, p. 427). The adenosine receptor agonist N[(2-

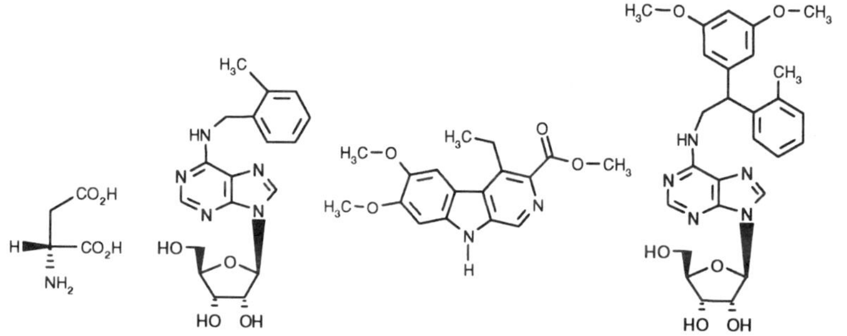

22 23 24 25

FIGURE 6.

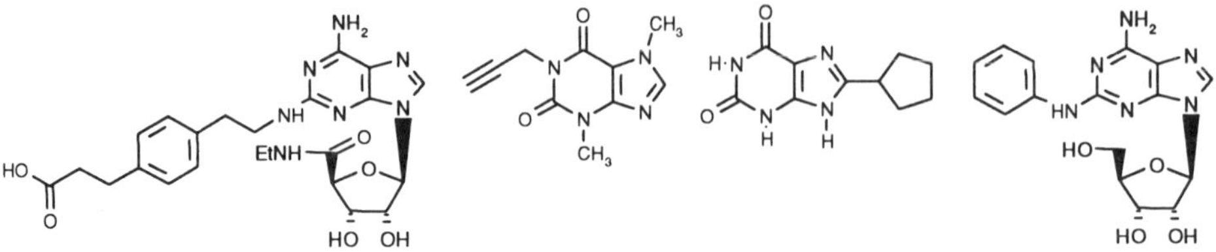

FIGURE 7.

methylphenyl)methyl]adenosine (metrifudil) **23,** which has a 4 : 1 selectivity for A_{2A} over A_1 receptors in binding studies, has been shown to be a potent anticonvulsant (ED_{50} = 0.5 mg/kg) against seizures induced by the inverse benzodiazepine agonist methyl 6,7-dimethoxy-4-ethyl-β-carboline-3-carboxylate (DMCM) **24** in mice (Klitgaard et al., 1993; Knutsen et al., 1995). Metrifudil has also been examined using both the amygdala kindling and O-Mg^{2+}-induced bursting hippocampal slice seizure models. Fully amygdala-kindled rats were injected with 1, 10, or 25 mg/kg intraperitoneal (ip) metrifudil or vehicle, 20 minutes prior to receiving the kindling stimulus. Metrifudil significantly reduced afterdischarge duration in a dose-related fashion compared to vehicle injections. Afterdischarge threshold and behavioral seizure stage were not significantly affected. In the *in vitro* hippocampal slice preparation, stimulation of Schaffer collateral axons under O-Mg^{2+} conditions resulted in spontaneous and evoked epileptiform burst activity in the CA1 region. The addition of 1, 10, or 100 mM metrifudil to the O-Mg^{2+} media abolished this epileptiform activity in a concentration-related fashion (Krahl et al., 1995).

Two highly selective A_2 agonists have also been studied, N-[2-(3,5-dimethoxyphenyl)-2-(2-methylphenyl)ethyl]adenosine (DPMA) **25** and the A_{2A} receptor–specific 2-*p*-(2-carboxyethyl)phenethylamino-5′-N-ethylcarboxamidoadenosine (CGS 21680) **26.** DPMA was found not to possess anticonvulsant effects in several models (Klitgaard et al., 1993) and potentiated seizures induced by the adenosine A_2 agonist 3,7-dimethyl-1-propargylxanthine (DMPX) **27.** Other investigators showed that intracaudate injections of CGS 21680 did not affect caudate seizures. However, seizure threshold was increased in the presence of CGS 21680 after blockade of the A_1 receptor with the antagonist 8-cyclopentylxanthine (CPX) **28** or following activation of the A_1 receptor with R-PIC or NECA (Janusz and Berman, 1993a). In a separate study, the anticonvulsant effect of CGS 21680 following focal injection into rat prepiriform cortex was studied in motor seizures induced by BMI. The conclusion was that the anticonvulsant effect was elicited by activation of adenosine A_1 receptors (Zhang et al., 1994). The A_2 agonist 2-phenylaminoadenosine (CV1808) **29** was also shown to lack seizure suppression activity (Franklin et al., 1989) in the BMI model.

Effects of the adenosine A_3 agonist N^6-(3-iodobenzyl)adenosine-5′-methylcarboxamide (IB-MECA) **30** in seizures (Von Lubitz et al., 1995a) are outlined in Excitatory Amino Acid-Induced Seizures, below. Some further mechanistic studies are summarized in the section on xanthines.

THE ROLE OF ATP IN THE PATHOGENESIS OF EPILEPSY

The participation of ATP-gated ion channels in excitatory neurotransmission in the CNS renders ATP and P_{2X} receptors of relevance to epileptogenic disorders. A reduced

Ca^{2+} ATPase activity has been suggested as a possible mechanism underlying sensitivity to audiogenic seizures in mice (Palayoor and Seyfried, 1984; Palayoor et al., 1986). The brain Ca^{2+}- or Mg^{2+}-stimulated ATPase appears to be localized to the outer surface of neural membranes and therefore may contribute to the regulation of extracellular ATP levels in the CNS, and reduced activity of this enzyme could be associated with elevated synaptic levels of ATP. A more recent genetic analysis, however, failed to document an association between the expression of Ca^{2+}-stimulated ecto-ATPase and susceptibility to audiogenic seizures in DBA/2J mice (Allen and Seyfried, 1994). As a result, the potential involvement of ecto-ATPase in vulnerability to audiogenic seizures remains to be established. Alterations in ecto-ATPase activity have also been examined in brain specimens derived from humans with focal epilepsy. Significant decreases in ecto-ATPase activity were documented in the temporal cortex of epileptic patients as compared to controls, while a significant increase in activity was observed in the posterior region of hippocampus from epileptic brain samples (Nagy et al., 1990). Whether the observed alterations in ecto-ATPase activity represent a causal or compensatory neurochemical factor is presently unknown. Additional studies are therefore required to assess the involvement of nucleotide degradative enzymes such as ecto-ATPase, ecto-ADPase, and ecto-5′-nucleotidase in epileptogenic events.

Another potential mechanism for ATP involvement in epileptogenesis derives from the ability of Zn^{2+} to potentiate the excitatory actions of ATP on mammalian neurons. ATP-gated inward currents are dramatically augmented by micromolar concentrations of Zn^{2+} (Li et al., 1993; Cloues et al., 1993). A release of Zn^{2+} from mossy fibers in the hippocampus and neurons in other limbic structures has been demonstrated to occur during seizure activity (Assaf and Chung, 1984; Charton et al., 1985). It is therefore conceivable that Zn^{2+} released during bursts of neuronal discharges could augment ATP signaling through P_{2X} receptors and as a result contribute to epileptogenesis.

A paucity of data also exists concerning ATP receptor mechanisms in epileptic disorders. In addition to the original report of Buday et al. (1961) documenting the ability of i.c.v. administration of ATP to evoke convulsions, high intracortical doses of the ATP analog β,γ-methylene-ATP **31,** produced massive epileptiform discharges in anesthetized rats pretreated with desacetyl lanatocid (Kesim et al., 1994). These observations have recently been extended through an assessment of the effects of focally administered ATP and ATP analogs in the rat prepiriform cortex. The prepiriform cortex represents the largest division of primary olfactory cortex and has been

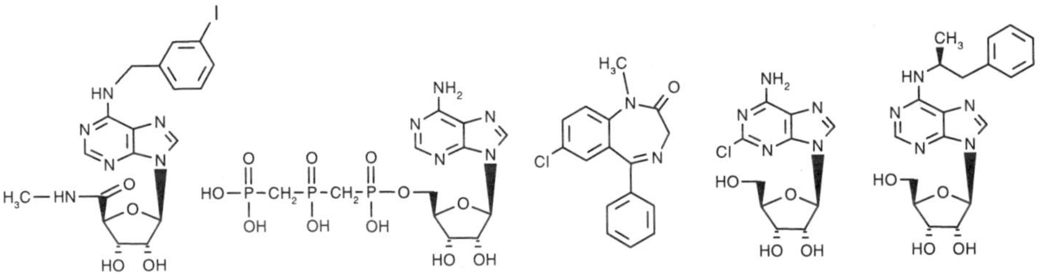

30 31 32 33 34

FIGURE 8.

shown to be an exquisitely sensitive site for the initiation of bilateral motor seizures following focal administration of excitatory amino acids, cholinergic agonists, or $GABA_A$ receptor antagonists (Piredda and Gale, 1985). Microinjection of the P_{2X} receptor agonists β,γ-methyleneATP and α,β-methyleneATP at a dose of 24 nmol induced generalized motor seizures in 84% and 54%, respectively, of rats implanted with cannulae in the prepiriform cortex (Figure 9). The behavioral characteristics of these ATP-analog-induced seizures were qualitatively identical to those elicited by NMDA or bicuculline methiodide in this brain area. These behavioral seizures are indistinguishable from the limbic seizures originally described by Racine (1972) in kindled rats, and consist of forelimb clonus; clonus of jaw and vibrissae, with or without head nodding; and rearing that progresses to a loss of balance after an incident of rearing. The behavioral seizures elicited by microinjection of ATP analogs in the prepiriform cortex are not a consequence of Ca^{2+} chelation inasmuch as a $CaCl_2$ vehicle was employed in these studies.

When ATP (100 nmol), the endogenous P_{2X} receptor agonist, is injected alone in the prepiriform cortex, a statistically insignificant seizure score was observed. This low order of potency for ATP is likely a consequence of the rapid degradation of ATP by ecto-ATPase in the CNS. The sequential degradation of nucleotides by ecto-ATPase,

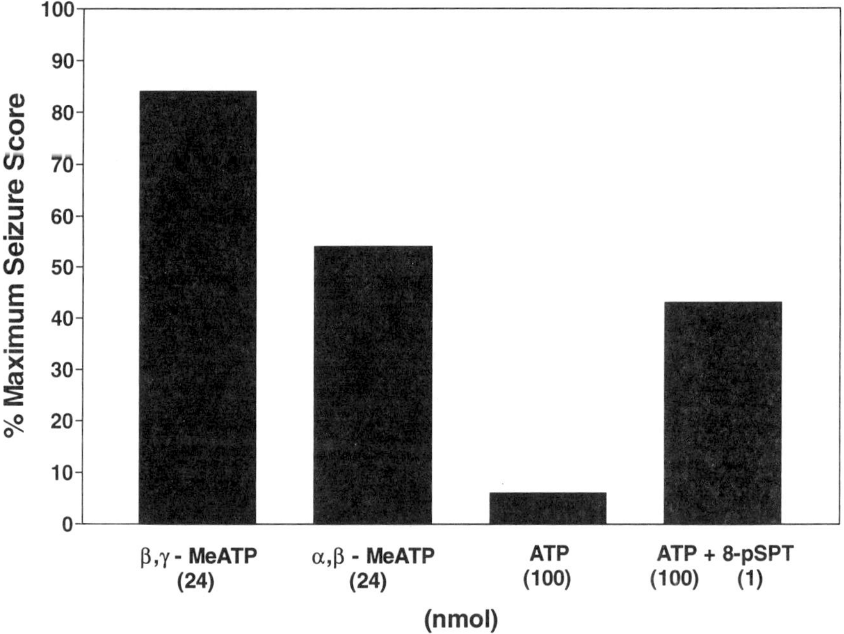

FIGURE 9. Convulsant effects of β,γ-methyleneATP (β,γ-MeATP), α,β-methyleneATP (α,β-MeATP), and ATP focally injected in rat prepiriform cortex. Intracerebral microinjections were performed using the method described previously (Franklin et al., 1989). ATP and analogs were administered in $CaCl_2$ vehicle (equimolar to nucleotide concentration) and were injected at a rate of 2 nl/sec in a volume of 500 nl. The adenosine receptor antagonist SPT **40** had no effect on behavior when administered alone. Values shown are means of four animals for each treatment group.

ecto-ADPase, and ecto-5′-nucleotidase results in the production of adenosine (James and Richardson, 1993). This production of adenosine derived from ATP metabolism would tend to mitigate the excitatory ATP-mediated currents through adenosine A_1 receptor–induced hyperpolarization of neurons (Edwards and Gibb, 1993; Edwards, 1995). To test this hypothesis we coinjected ATP and the adenosine A_1 receptor antagonist SPT **40** focally in the rat prepiriform cortex. As predicted, SPT enhanced the ability of microinjected ATP to evoke seizures from the prepiriform cortex (Figure 9). While having no effect on seizures when administered alone, SPT (1 nmol) increased the mean seizure score for ATP from 6% to 43% of the maximum response.

The relationship between the neuroactive purines, ATP and adenosine, in the CNS is therefore unique in that the release of ATP may result in a rapid yet transient excitation, followed by a delayed inhibitory response to its metabolite adenosine (Edward and Gibb, 1993; Edwards, 1995). The time course of ATP degradation to adenosine outside nerve terminals is somewhat complex, in that the production of adenosine displays a delay followed by a sustained production. The delay in adenosine production at cholinergic synapses has been shown to be proportional to the initial ATP concentration, and a consequence of feedforward inhibition of ADPase and 5′-nucleotidase (James and Richardson, 1993). Thus high-frequency release rate of ATP may not result in adneosine production until the end of a period of repetitive discharges. This characteristic of ATP metabolism could uncouple ATP-mediated excitatory currents associated with epileptic discharges from the normal inhibitory tone exerted by adenosine.

THE GABA/BENZODIAZEPINE COMPLEX AND ADENOSINE

Numerous investigators have shown that adenosine receptor agonists are uniquely potent against seizures induced by agents that act at the GABA/benzodiazepine receptor complex. A complex interaction seems to exist between adenosine- and benzodiazepine-receptor ligands and their recognition sites. This was recently demonstrated in a study where agonist R-PIA selectively antagonized seizures induced by DMCM when compared to seizures involving chemoconvulsants affecting other sites within the GABA/benzodiazepine receptor complex (Petersen, 1991). This observation was confirmed and extended in a study that confirmed the selective action of both adenosine A_1 receptor agonists against DMCM-induced seizures (Klitgaard et al., 1993).

Even though purinergic compounds have been implicated as possible endogenous ligands for the benzodiazepine receptor (Skolnick et al., 1980; Davies, 1985), *in vitro* binding studies showed that adenosine or adenosine receptor agonists are devoid of benzodiazepine receptor affinity (Williams and Risley, 1981). However, these findings do not exclude possible subtle interactions between adenosine and benzodiazepine recognition sites within the GABA/benzodiazepine receptor complex. Several lines of evidence suggest such a link (Phillis and O'Regan, 1988). Many benzodiazepines inhibit adenosine uptake (Phillis and Wu, 1983), and chronic diazepam administration downregulates the number of adenosine receptors in the rat brain (Hawkins et al., 1988), whereas chronic caffeine treatment upregulates the number of benzodiazepine receptors in the rat brain (Wu and Coffin, 1984), and chronic caffeine or theophylline treatment reduces the ability of GABA to facilitate the binding of flunitrazepam in chicken brain neurons (Roca et al., 1988, 1990). In the latter experiments, the theophylline-induced uncoupling of the GABA and benzodiazepine recognition sites could be counteracted by an adenosine receptor agonist, a finding that suggests an interaction between adenosine receptors and the GABA/benzodiazepine receptors.

Several competitive adenosine uptake inhibitors also reduce the binding of [^{3}H]diazepam **32** to brain receptors, suggesting an overlap of the recognition sites for adenosine uptake inhibitors and benzodiazepine ligands (Phillis and Wu, 1981). Thirty minutes after seizures in developing rats induced by the $GABA_A$ antagonist BMI, the specific binding of [^{3}H]CHA and [^{3}H]flunitrazepam, as assessed by autoradiography, is significantly augmented in numerous brain areas, especially in the substantia nigra pars reticula (Daval et al., 1992). It is therefore no coincidence that adenosine agonists protect against seizures induced by the chemoconvulsants studied in the next section, many of which modulate the GABA/benzodiazepine receptor complex.

PROTECTION AGAINST CHEMOCONVULSANT-INDUCED SEIZURES BY ADENOSINE DERIVATIVES

Pentylenetetrazole

Some of the archetypal observations in this field were made when documenting the effect of the adenosine agonists R-PIA, CHA, and 2-chloroadenosine **33** on PTZ-induced loss of equilibrium and other seizure types (Dunwiddie and Worth, 1982). These authors utilized PTZ dosed i.p. at 100 mg/kg in mice, with drugs dosed 30 minutes before the chemoconvulsant. At doses of 2 mg/kg, they observed that seizure latency was 7 times as long with R-PIA, 3 times as long for CHA, and 1.6 times as long for 2-chloroadenosine. Lower doses of these agonists produced marked sedation and hypothermia.

The effects of 2-chloroadenosine, CHA, R-PIA, and S-PIA **34** on the threshold for seizures in rats were also observed (Murray et al., 1985) by measuring the dose of PTZ, infused through a tail vein, that was required to elicit a myoclonic jerk. Using 1.5 mg/ml of PTZ dosed at a rate of 1.5 ml/min, the potencies observed (ED_{50} [median effective dose] values) were 0.033 mg/kg (R-PIA), 0.126 mg/kg (2-chloroadenosine), 1.44 mg/kg (CHA) and 2.62 mg/kg (S-PIA). This rank order of potency provided evidence for the involvement of the A_1 receptor. Antagonism of the elevation of the 2-chloroadenosine seizure threshold by the adenosine A_1 antagonist, theophylline, was noted.

The activity against PTZ-induced seizures of the adenosing A_3 agonist IB-MECA administered both acutely and chronically has been demonstrated (Von Lubitz et al., 1995a). Another group have used the modulation of PTZ-induced seizures in rats as a measure of the proconvulsive potential of new xanthine derivatives, a subject that will be discussed further in the section on xanthines below.

Methyl 6,7-dimethoxy-4-ethyl-β-carboline-3-carboxylate

An investigation of the effect of 2-chloroadenosine, CHA, and R-PIA was carried out (Petersen, 1991) in experiments where the infused inverse benzodiazepine agonist DMCM was the chemoconvulsant. When mice were dosed i.p. all three agonists exhibited anticonvulsant activity against clonic seizures, but R-PIA was found to be around 30 times more potent than the other two agents, with an ED_{50} of 0.30 mg/kg. R-PIA was found to be only very weakly active against seizures elicited by PTZ and BMI.

These key observations were confirmed and extended shortly afterwards by a study of the A_1 agonists R-PIA and N-cyclopentyladenosine (CPA) **35,** as well as of the A_2 receptor agonists DPMA and metrifudil (Klitgaard et al., 1993). The study confirmed the potent effect of the two A_1 agonists versus DMCM-induced seizures following ip dosing, as well as the potent effect of CPA, and the lack of seizure protection by R-PIA against a range of other chemoconvulsants acting at the GABA/benzodiazepine receptor complex. Results for the A_2 agonists showed that, unlike metrifudil, DPMA exhibited no protection against DMCM-induced seizures (Klitgaard et al., 1993), but this may be explained in part by DPMA's lower A_1 affinity. A recent study (Adami et al., 1995) indicates that both A_1 and A_{2A} receptor agonism appear to be involved in protection from PTZ-induced seizures in rats (see section on pentylenetetrazole, p. 434), and when both A_1 and A_{2A} receptor agonists are dosed repeatedly, the A_1-mediated effects are subject to tolerance in this model, whereas the effects depending on A_{2A} receptor activation are maintained. Adenosine A_2 receptor–mediated excitatory actions on the nervous system have been reviewed (Sebastiao and Ribeiro, 1996).

A systematic study of 17 examples from a series of structurally novel A_1 agonists and 7 reference compounds was recently published (Knutsen et al., 1995). All of these compounds were evaluated in DMCM-induced seizures in mice, using a paradigm whereby the agonists were dosed ip 30 minutes before clonic seizure induction by DMCM (18 mg/kg, i.p.). Several of the novel adenosine A_1 agonists gave ED_{50} values below 1 mg/kg. Both metrifudil and 2-chloro-N-(1-phenoxy-2-propyl)adenosine (NNC 21-0041) **37,** a new prototypical A_1 agonist with an optimized cardiovascular profile, appeared promising when the ratio between TD_{50} values for mouse rotarod (also termed neurotoxicity) and ED_{50} for DMCM clonic seizures were compared.

Seizures induced by PTZ, DMCM, and BMI could also be successfully inhibited by the A_1 agonist 2-chloro-N-cyclopentyladenosine (CCPA) **36** in mice, although it was shown that this agonist reduces the function of the GABA-coupled chloride channel function (Concas et al., 1993). 2-Chloroadenosine has been shown to be anticonvulsant when seizures were induced by β-carboline-3-carboxylic acid methyl amide (FG 7142) **38,** a ligand related to DMCM (Stevens and Weidmann, 1989).

Excitatory Amino Acid–Induced Seizures

The effects of the reference A_1 agonist CPA and the antagonist CPX were examined against seizures evoked by i.p. injection of the chemoconvulsant NMDA in mice (Von

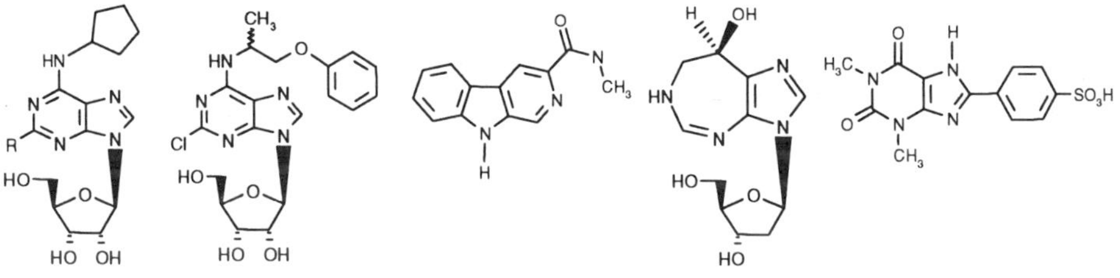

R = H, 35
R = Cl, 36

37

38

39

40

FIGURE 10.

Lubitz et al., 1993). When NMDA was administered at 60 mg/kg, a brief period of hyperactivity ensued, followed by clonic and/or tonic seizures, with some mortality being observed. Prior administration of 0.5 mg/kg CPA resulted either in a delay in seizure onset and unchanged mortality or, at higher doses, an elimination of tonic seizures and a significant reduction in postictal mortality. CPX pretreatment increased the mortality compared to that observed with NMDA alone.

There is a very close functional interaction between adenosine A_1 receptors and NMDA receptors; acute stimulation of NMDA receptors results in the release of adenosine. In an experiment where chronic dosing of mice for 8 weeks with NMDA (20 mg/kg ip daily) was followed by CPA (0.01 mg/kg) just before a convulsant dose of NMDA (Von Lubitz et al., 1995b), mortality was reduced to 10%. If CPA was replaced by 1.0 mg/kg of the adenosine deaminase inhibitor 2′-deoxycofomycin **39,** mortality was eliminated. Density of adenosine A_1 receptors in the cortex and hippocampus was increased in animals chronically treated with NMDA.

The effect of both adenosine A_1 and A_2 agonists on seizures caused by i.c.v. infusion of glutamate in mice have been investigated (Klitgaard et al., 1993). Following i.p. administration, the ED_{50} values for CPA and R-PIA in seizure protection were 0.4 and 1.1 mg/kg respectively; however this was markedly above the TD_{50} values (0.1 mg/kg) for impairment of rotarod performance. The A_2 receptor agonists metrifudil and DPMA exhibited no anticonvulsant effect in this seizure model.

The A_3 agonist IB-MECA administered acutely at 10 mg/kg prior to a 60 mg/kg ip dose of the proconvulsant NMDA in mice had no effect on either the incidence of seizures or the degree of neurological impairment, but a delay in the onset of convulsions was observed. At 100 mg/kg, IB-EMCA was protective in all the measures studied, including mortality (Von Lubitz et al., 1995a). Chronic dosing of IB-MECA (100 mg/kg) resulted in a reduction of neurological impairment and a delay in the onset of tonic seizures, as well as reduced mortality.

Bicuculline-Induced Seizures

Seizures induced by the $GABA_A$ antagonist BMI have been studied extensively, especially by the Murray group. In rats, a potent anticonvulsant effect was seen against seizures induced by BMI in the prepiriform cortex, a brain area which may play a significant role in the pathobiology of epilepsy. Animals were pretreated with 2-chloroadenosine injected into the same brain region. The anticonvulsant effect was completely reversed by the A_1 antagonist 8-(*p*-sulfophenyl)theophylline (SPT) **40** (Franklin et al., 1988). Subsequently this work was extended to a whole series of mainly A_1 adenosine agonists—NECA, CHA, CPA, 2-chloroadenosine—as well as the R- and S-PIA isomers, which were also microinjected into the rat prepiriform cortex (Franklin et al., 1989). These compounds all protected animals against BMI-induced seizures, with NECA effective at doses ≥6.8 pmol. In contrast, "heroic" doses of the A_2-selective CV1808 (>335 pmol) did not afford seizure protection, and rank orders of potency were indicative of an A_1 receptor–mediated effect.

A study of the effect of peripherally ip dosed agonists following subcutaneous administration of BMI has also been carried out (Klitgaard et al., 1993). The conclusion was that CPA and metrifudil showed potent protection against clonic convulsions, with ED_{50} values of 2.4 and 8.9 mg/kg respectively; however, R-PIA and DPMA were not effective. The A_1 agonist CCPA36 has also been shown to inhibit BMI-induced seizures (Concas et al., 1993), in accordance with previous studies of A_1 agonists. In addition, it has been demonstrated that BMI-induced seizures led to a widespread increase in both [^{3}H]CHA and [^{3}H]flunitrazepam binding sites in the brain, as assessed by quantitative autoradiography (Daval et al., 1992).

Metal Ion–Induced Seizures

An injection of cobalt chloride solution into the unilateral sensorimotor cortex of rats induces electrographic epileptic activity, followed by a peripheral motor disturbance. Adenosine receptor–mediated generation of cAMP was found to be altered in the primary region of the cobalt-induced epileptic cortex of these rats. The increase in cAMP accumulation was observed regardless of whether an adenosine uptake inhibitor, phosphodiesterase inhibitor, or adenosine deaminase was present. Increased cAMP accumulation was observed as early as 8 days after the injection, peaked at 17–19 days after injection, and returned to control levels 40–50 days after injection (Hattori et al., 1993). The authors concluded that the adenosine-sensitive cAMP-generating systems may play a role in central mechanisms underlying experimental models of epilepsy.

NON-AGONIST MODULATORS OF ADENOSINE

Xanthines

The trend expounded earlier in this chapter is that adenosine receptor agonists always exhibit anticonvulsant properties, but this supposition does not hold in all cases. The nature of the chemoconvulsive agent used in the particular seizure model has a major influence. For example, when seizures are induced by 8-[4-[[[[(2-aminoethyl)amino]carbonyl]methyl]oxy]phenyl]-1,3-dipropyl xanthine (xanthine amine congener [XAC] **41,** an A_1 receptor antagonist), adenosine A_1 agonists such as R-PIA and CPA are proconvulsant. Likewise, adenosine A_2 agonists such as DPMA and metrifudil are proconvulsant when seizures are induced by DMPX, an A_2 receptor antagonist (Klitgaard et al., 1993). Similar observations have been made in the case of NECA, which is proconvulsant when seizures are induced by caffeine (Morgan and Durcan, 1990). The general observation is that xanthine-based adenosine antagonists derived from caffeine and theophylline **42,** are proconvulsant, but there are some exceptions.

In humans, overdoses with theophylline frequently results in seizures that are particularly refractory to treatment with clinically useful anticonvulsants (Chu, 1981). Theophylline was found to be proconvulsant in rats at doses of 15–60 mg/kg in PTZ-induced myoclonic jerks (Murray et al., 1985), a conclusion confirmed by other investigators (Cotrufo et al., 1992). Other authors have noted the proconvulsant properties of theophylline and caffeine in the hippocampus (Ault et al., 1987). However, it should be noted that recent research implies that adenosine receptors may not be involved in theophylline-induced seizures in mice (Hornfeldt and Larson, 1994). This was also implied by authors who first noted the seizure-suppressant effects of A_1 agonists. They observed intriguingly that the anticonvulsant actions of R-PIA were not reversed by theophylline (Dunwiddie and Worth, 1982).

In mice, long-term treatment (14 days of 60–70 mg/kg/day) with caffeine via drinking water has been shown to lead to a decreased susceptibility to NMDA-induced clonic seizures. Control mice and mice during and following the caffeine administration were dosed with NMDA (150 ng/kg i.p.). Clonic seizures, wet dog shakes and mortality

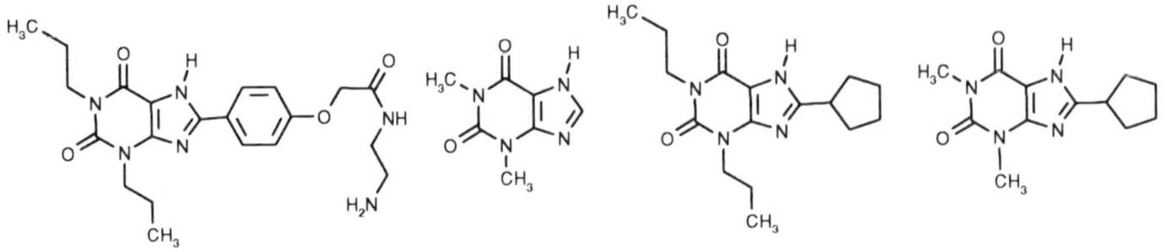

FIGURE 11.

were significantly reduced at the end of the long-term administration, but returned towards control levels at 1 and 2 days after caffeine withdrawal. Changes could not be ascribed to changes in adenosine A_1 receptor density since receptor density in gyrus dentatus of the hippocampus, assessed by quantitative receptor autoradiography with [^{3}H]CHA, was not significantly altered following the caffeine treatment (Georgiev et al., 1993).

In another study, the effects of chronic theophylline treatment on brain A_1 receptors and amygdala-kindled seizures were examined in rats. Repeated i.p. administration of theophylline (75 mg/kg/daily) for 14 days resulted in a significant increase in [^{3}H]CHA binding to adenosine A_1 receptors in thalamus, hippocampus, and cerebellum. The chronic theophylline treatment was shown to significantly prolong the immediate post-seizure refractory period (defined as sensitivity to a subsequent kindling stimulation), while both the severity and duration of an initial kindled seizure were unaffected, indicating that the postseizure refractory period can be influenced by the brain adenosine system (Kleinsorge et al., 1993). Similar conclusions were reached when rats were chronically dosed with theophylline (2 × 75 mg/kg daily). Under these conditions, there was a significant increase in [^{3}H]CHA binding sites and a corresponding increase in BMI seizure threshold (Sanders and Murray, 1989). Studies of 2-chloroadenosine and the antagonist aminophylline gave similar results (Handforth and Treiman, 1994).

Studies of the A_1 antagonist 1,3-dipropyl-8-cyclopentylxanthine (DPCPX) **43** in hippocampal CA3 neurons have demonstrated that endogenously released adenosine exerts a vigorous control on their excitability at both the pre- and postsynaptic sites (Alzheimer et al., 1993). The effects of the A_1 antagonist 8-cyclopentyl-1,3-dimethylxanthine (cyclopentyltheophylline, CPT) **44** have been tested on electrically induced seizures in rats. CPT prolonged secondary seizures into generalized motor seizures, but did not affect primary seizure duration (Dragunow and Robertson, 1987). In this context, an unusual result was obtained when it was shown that CPT had anticonvulsant properties when seizures were introduced by DMCM (ED_{50} 23 mg/kg, i.p.) or PTZ (ED_{50} 62 mg/kg, i.p.) in mice (Klitgaard and Knutsen, 1992; Klitgaard et al., 1993).

In contrast to some of the above studies, investigations on hippocampal slices imply that the epileptogenic concentrations of the xanthine derivatives 3-isobutyl-1-methylxanthine (IBMX) **45,** DPCPX, and 8-phenyltheophylline (8-PT) **46** strongly correlated with their affinities for the A_1 adenosine receptor. The authors suggest that the epileptogenic potency of xanthines is primarily due to the blockade of the A_1 receptors (Moraidis and Bingmann, 1994).

The Effects of Agents Augmenting Endogenous Adenosine

New adenosine-regulating agents have been targeted at CNS disorders (Foster et al., 1995a,b). Several abstracts from meetings of the Society for Neuroscience in-

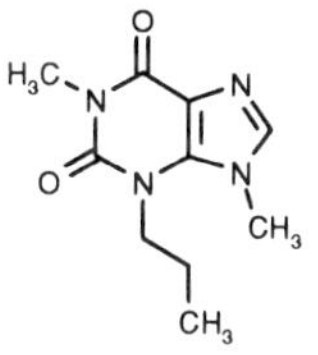

FIGURE 12.

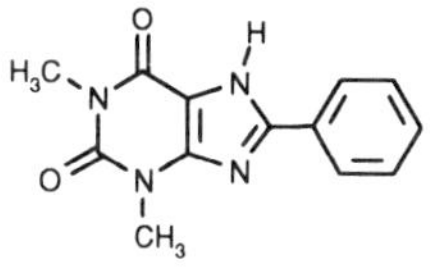

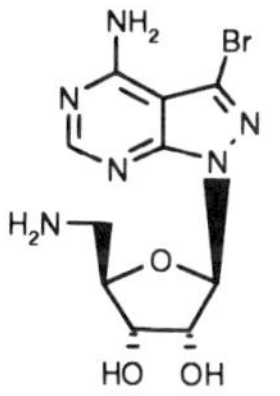

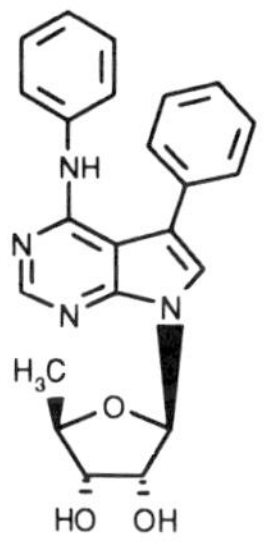

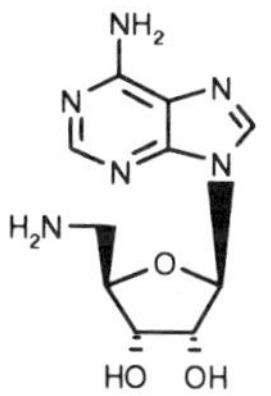

dicate the potential for the treatment of epilepsy by new adenosine kinase (AK) inhibitors such as 4-amino-1-(5′-amino-5′-deoxy-1-β-D-ribofuranosylpyrazolo[3,4-*d*]-pyrimidine (GP 515) **47** (Zimring et al., 1995) or 4-(N-phenylamino)-5-phenyl-7-(5′-deoxyribofuranosyl)pyrrolo[2,3-*d*]-pyrimidine (GP 683) **48** (Wiesner et al., 1995).

When dosed systemically in rats, GP 515 gave ED_{50} values of 15.9 and 23.4 mg/kg against maximal electroshock seizures (MES) and PTZ-induced seizures respectively. The corresponding rat PTZ seizure figures for GP 683 were ED_{50} values of 1.1 mg/kg and > 40 mg/kg; the reference AK inhibitor 5′-deoxy-5′-aminoadenosine **49** gave values of 153 and >300 mg/kg, presumably reflecting poor penetration of the blood-brain barrier by this latter agent.

More potent effects were recorded when GP 515 was dosed i.p. in mouse PTZ seizures, with ED_{50} values of 11.9 and 7.5 mg/kg respectively. 5′-Deoxy-5′-aminoadenosine and another reference AK inhibitor, iodotubericidin **50,** gave values of 277 and 2.1 mg/kg in MES-induced seizures, and 39.3 and 1.5 mg/kg in PTZ-induced seizures, respectively, when injected i.p. in mice. However, substantial hypothermia was observed when GP 683 was administered to mice (Weisner et al., 1996). It should also be noted that iodotubericidin gave an ED_{50} value of 2.0 mg/kg when injected IP 30 minutes before DMCM-induced seizures in mice, and an ED_{50} of 2.8 mg/kg was recorded in the mouse rotarod test (Knutsen et al., 1994).

The adenosine uptake inhibitor papaverine **51** has been shown to inhibit electrically induced seizures in rats (Dragunow, 1987). Rats with implanted dorsal hippocampal electrodes were injected with papaverine (35 mg/kg) and tested for seizure responses at 5, 20, and 60 minutes after injection, and compared to saline. Papaverine-treated rats had greatly elevated afterdischarge thresholds; effects were most pronounced 5

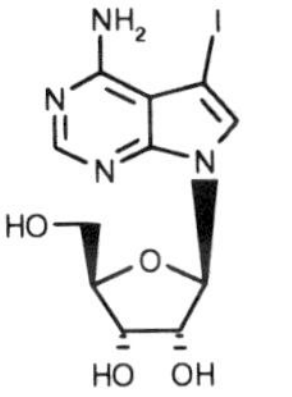

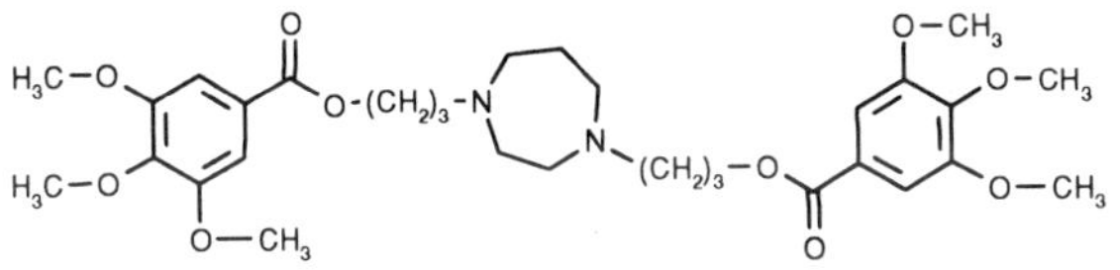

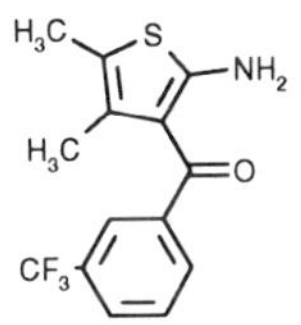

FIGURE 13.

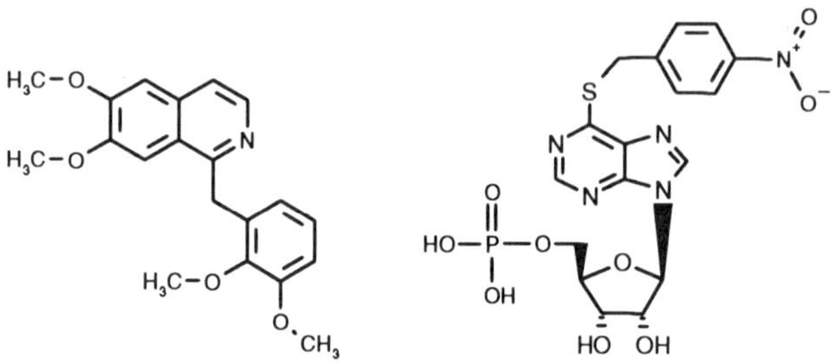

FIGURE 14.

minutes after injection and had largely dissipated after 60 minutes. Even though this drug has multiple actions, evidence pointed towards an adenosinergic mechanism.

The effects of adenosine and the adenosine-binding enhancer PD 81,723 **52** on low Mg^{2+}-induced bursting have been examined in the *in vitro* hippocampal slice. Extracellular recordings were obtained from the CA3 pyramidal cell layer while electrically stimulating in the stratum radiatum under low Mg^{2+} perfusion. Adenosine (6–100 μM) reduced the duration of epileptiform bursting in a dose-related manner, and the effect was reversible upon washout of adenosine. Application of PD 81,723 (50–100 μM) also resulted in a dose-dependent reduction in the duration of the triggered burst, which was irreversible. These results demonstrate the anticonvulsant activity of adenosine and this binding enhancer (Janusz and Berman, 1993b).

An important study has been carried out on the manipulation of BMI-induced seizures by the AK inhibitors 5′-deoxy-5′-aminoadenosine and iodotubercidin, the adenosine uptake inhibitors diazep **53** or NBI-5-monophosphate **54,** as well as the deaminase inhibitor 2′-deoxycoformycin (Zhang et al., 1993). Relatively large doses of focally injected exogenous adenosine (ED_{50} 48.1 $\pm$ 8.4 nmol) were required to produce seizure protection, and the effects were potentiated by the above agents, with 5′-deoxy-5′-aminoadenosine affording the strongest potentiation. The first three agents listed above, when administered alone, gave ED_{50} values of 2.6 $\pm$ 0.8, 4.0 $\pm$ 2.7, 5.6 $\pm$ 1.6 nmol respectively. This and other evidence further support the role of endogenous adenosine as a tonically active antiepileptogenic substance.

SUMMARY AND PROSPECTS FOR FUTURE DRUG DEVELOPMENT

In the relatively short period of 20 years since the first observations linking adenosine to epilepsy, this has now become a firmly established area of seizure pharmacology. Much evidence, especially *in vivo* microdialysis in humans (During and Spencer, 1992), now implicates adenosine as an endogenous agent bringing about seizure cessation in man.

One of the important early articles on the seizure-suppressive effects of A_1 agonists (Dunwiddie and Worth, 1982) concluded that it is unlikely that the adenosine derivatives then available would be clinically useful anticonvulsants because they elicit profound sedation at concentrations that have relatively weak anticonvulsant effects.

The separation of the desirable effects of adenosine from the unwanted has always been a problem, given the plethora of pharmacological activities ascribed to this nucleoside. The challenge ahead lies in separating the unwanted sedative and hypothermic activities from the highly desirable anticonvulsant features of adenosine discussed above, especially when designing drugs for chronic dosing. One future approach to drug discovery will be the utilization of modern medicinal chemistry to design more selective purinergic agents. Progress has already been made with novel agonists that have a substantially improved cardiovascular and hypothermic profile (Knutsen et al., 1995).

The agonist approach may be dependent for its future success upon the future availability of A_1 partial agonists or the identification of new adenosine receptor subtypes that can lead to more selective pharmacological manipulation. There is also potential for augmenting endogenous adenosine by a variety of agents, as shown by some of the newer adenosine regulating agents (Foster et al., 1995a,b), especially those agents acting via inhibition of adenosine kinase (Zhang et al., 1994).

The role of ATP as an excitatory neurotransmitter in the CNS is now established. Moreover the ubiquitous distribution of P_{2X} receptors in mammalian brain indicates a plethora of functional roles for these ligand-gated ion channels. Although P_{2X2}, P_{2X4}, and P_{2X6} receptors are expressed in the piriform cortex (Kidd et al., 1995; Seguela et al., 1996; Collo et al., 1996), which is the neuroanatomic substrate for the ATP analog–evoked seizures reported herein, the identity of homomeric or heteromeric P_{2X} receptors involved in pathologic ATP-mediated excitatory transmission remains to be delineated. The possible relationship between ATP-gated ion channels and epileptogenic events therefore renders the development of P_{2X} receptor antagonists a promising strategy for the introduction of novel antiepileptic drugs.

The combined endeavors of scientists engaged in these aspects of adenosine and ATP are aimed at ensuring that it does not take another 20 years of research for scientists to provide a CNS-effective drug with a purinergic mechanism of action.

ACKNOWLEDGMENTS

The authors wish to thank Nils Ole Dalby and Judith L. Knutsen for helpful comments on this chapter, and Hanne R. Petersen for assistance with its production.

REFERENCES

Abbracchio MP, Brambilla R, Camisa M, Rovati GE, Ferrari R, Canevari L, Dagain F, Cataben1 F (1995): Adenosine A_1 receptors in rat brain synaptosomes: Transductional mechanisms, effects on glutamate release, and preservation after metabolic inhibition. Drug Dev Res 35:119–129.

Adami M, Bertolli R, Ferri N, Foddi C, Ongini E (1995): Effects of repeated administration of selected adenosine A_1 and A_{2a} receptor aongists on pentylenetetrazole-induced convulsions in the rat. Eur J Pharmacol 294:383–389.

Allen KM, Seyfried TN (1994): Genetic analysis of nucleotide triphosphatase activity in the mouse brain. Genetics 137:257–265.

Alzheimer C, Sutor B, Ten Bruggengate G (1993): Disinhibition of hippocampal CA3 neurons induced by suppression of an adenosine A_1 receptor-mediated inhibitory tonus: Pre- and postsynaptic components. Neuroscience 57:565–575.

Angelatou F, Pagonopoulou O, Maraziotis T, Olivier A, Villemeure JG, Avoli M, Kostopoulos G (1993): Upregulation of A_1 receptors in human temporal lobe epilepsy: A quantitative autoradiographic study. Neurosci Lett 163:11–14.

Assaf SY, Chung SH (1984): Release of endogenous Zn^{2+} from brain tissue during activity. Nature 308:734–736.

Ault B, Olney MA, Joyner JL, Boyer CE, Notrica MA, Soroko FE, Wang M (1987): Proconvulsant actions of theophylline and caffeine in the hippocampus: Implications for the management of temporal lobe epilepsy. Brain Res 546:69–78.

Barraco RA, Swanson TK, Phillis JW, Berman RF (1984): Anticonvulsant effects of adenosine analogues on amygdaloid-kindled seizures in rats. Neurosci Lett 46:317–322.

Barrie AP, Nicholls DG (1993): Adenosine A_1 inhibition of glutamate exocytosis and protein kinase C-mediated decoupling. J Neurochem 60:1081–1086.

Bean B (1992): Pharmacology and electrophysiology of ATP-activated ion channels. Trends Pharmacol Sci 13:87–90.

Berman RF, Jarvis MF, Lupica CR (1990): Adenosine involvement in kindled seizures. In Wada J (ed): "Kindling 4, Advances in Behavioral Biology, Vol 37." New York: Plenum Press, pp 423–435.

Bhagwat SS, Williams M (1995): Recent progress in modulators of purinergic activity. Expert Opinion in Therapeutic Patients 5:547–558.

Birnstiel S, Gerber U, Greene RW (1992): Adenosine-mediated synaptic inhibition: Partial blockade by barium does not prevent anti-epileptiform activity. Synapse 11:191–196.

Bo X, Zhang Y, Massar M, Burnstock G, Schoepfer R (1995): A P_{2X} purinoceptor cDNA conferring a novel pharmacological profile. FEBS Lett 375:129–133.

Brake AJ, Wagenbach MH, Julius D (1994): New structural motif for ligand-gated ion channels defined by ionotropic ATP receptor. Nature 371:519–523.

Buday PV, Carr CJ, Myga TS (1961): A pharmacologic study of some nucleosides and nucleotides. J Pharm Pharmacol 13:290–299.

Buell G, Lewis C, Collo G, North RA, Surprenant A (1996): An antagonist-insensitive P_{2X} receptor expressed in epithelia and brain. EMBO J 15:55–62.

Cantor SL, Zornow MH, Miller LP, Yaksh TL (1995): The effect of cyclohexyladenosine on the periischemic increases of hippocampal glutamate and glycine in the rabbit. J Neurochem 59:1884–1892.

Charton G, Rovira C, Ben-Ari Y, Leviel V (1985): Spontaneous and evoked release of endogenous zinc in the hippocampal mossy fibres zone of the rat in situ. Exp Brain Res 58:202–205.

Chen CC, Akoplan AN, Sivilotti L, Colquhoun D, Burnstock G, Wood JN (1995): A P_{2X} purinoceptor expressed by a subset of sensory neurons. Nature 377:428–431.

Chin JH (1989): Adenosine receptors in brain: Neuromodulation and role in epilepsy. Ann Neurol 26:695–698.

Chin JH, Wiesner JB, Fujitaki J (1995): Increase in adenosine metabolites in human cerebrospinal fluid after status epilepticus. J Neurol Neurosurg Psychiat 58:513–514.

Chu NS (1981): Caffeine and aminophylline induced seizures. Epilepsia 22:85–94.

Cloues R, Jones S, Brown DA (1993): Zn^{2+} potentiates ATP-activated currents in rat sympathetic neurons. Pflugers Arch 424:152–158.

Collo G, North RA, Kawashima E, Merlo-Pich E, Neidhart S, Surprenant A, Buell G (1996): Cloning of P_{2X5} and P_{2X6} receptors and the distribution and properties for an extended family of ATP-gated ion channels. J Neurosci 16:2495–2507.

Concas A, Santoro G, Mascia MP, Maciocco E, Dazzi L, Ongini E and Biggio G (1993): Anticonvulsant doses of 2-chloro-N^6-cyclopentyladenosine, an adenosine A_1 receptor agonist, reduce GABAergic transmission in different areas of the mouse brain. J Pharm Exp Ther 267:844–851.

Cotrufo C, Bortot L, Giachetti A, Manzini S (1992): Differential effects of various xanthines on pentylenetetrazole-induced seizures in rats: An EEG and behavioural study. Eur J Pharmacol 222:1–6.

Daval J-L, Werck MC (1991): Regional changes in rat brain adenosine A1 receptors following seizures. Nucleosides and Nucleotides 10:1187–1188.

Daval J-L, Nehlig A, Nicolas F (1991): Physiological and pharmacological properties of adenosine: Therapeutic implications. Life Sci 49:1435–1453.

Daval J-L, Pereira de Vasconcelos A, El Hamdi G, Werck MC, Nehlig A (1992): Quantitative autoradiographic measurements of functional changes induced by generalized seizures in the developing rat brain: Central adenosine and benzodiazepine receptors and local cerebral glucose utilization. Epilepsy Res Suppl 9:83–92.

Davies LP (1985): Pharmacological studies on adenosine analogous isolated from marine organisms. Trends Pharmacol Sci 6:143.

Dichter MA (1994): Emerging insights into mechanisms of epilepsy: Implications for new antiepileptic drug development. Epilepsia 35:S51–S57.

Dodd PR, Watson WEJ, Johnston GAR (1986): Adenosine receptors in post-mortem human cerebral cortex and the effect of carbamazepine. Clin Exp Pharmacol Physiol 13:711–722.

Dolphin AC, Prestwich SA (1985): Pertussis toxin reverses adenosine inhibition of neural glutamate release. Nature 316:148–150.

Dragunow M (1987): Time-dependent effects of papaverine on electrically induced seizures in rats. Neurosci Lett 74:320–324.

Dragunow M (1988): Purinergic mechanisms in epilepsy. Prog Neurobiol 31:85–108.

Dragunow M, Robertson HA (1987): 8-cyclopentyl-1,3-dimethylxanthine prolongs epileptic seizures in rats. Brain Res 417:377–379.

Dubyak GR, E-Moatassim C (1993): Signal transductive via P_2-purinergic receptors for extracellular ATP and other nucleotides. Am J Physiol 34:C577–C606.

Dunwiddie TV, Worth T (1982): Sedative and anticonvulsant effects of adenosine analogs in mouse and rat. J Pharm Exp Ther 220:70–76.

During MJ, Spencer DD (1992): Adenosine: A potential mediator of seizure arrest and postictal refractoriness. Ann Neurol 32:618–624.

Edwards FA, Gibb AJ (1993): ATP—a fast neurotransmitter. FEBS Lett 325:86–89.

Edwards FA (1995): ATP receptors. Curr Opin Neurobiol 4:347–352.

Edwards FA, Gibb AJ, Colquhoun D (1992): ATP receptor-mediated synaptic currents in the central nervous system. Nature 359:144–147.

Evans RJ, Derkach V, Surprenant A (1992): ATP mediates fast synaptic transmission in mammalian neurons. Nature 357:503–505.

Fastbom J, Fredholm BB (1985): Inhibition of [^{3}H]-glutamate release from rat hippocampal slices by L-PIA. Acta Physiol Scand 125:121–123.

Foster AC, Miller LP, Weisner JB (1995a): Regulation of endogenous adenosine levels in the CNS: Potential for therapy in stroke, epilepsy and pain. In Sahota A, Taylor M (eds): "Purine and Pyrimidine Metabolism in Man, VIII." New York: Plenum Press, pp 427–430.

Foster AC, Zimring ST, Jelovich LA, Babb M, Weisner JB (1995b): Antiepileptic activity of adenosine regulating agents. Soc Neurosci Abs 21:151.2.

Franklin PH, Tripp ED, Zhang G, Gale K, Murray TF (1988): Adenosine receptor activation blocks seizures induced by bicuculline methiodide in the rat prepiriform cortex. Eur J Pharmacol 150:207–209.

Franklin PH, Zhang G, Tripp ED, Murray TF (1989): Adenosine A_1 receptor activation mediates suppression of (−)-bicuculline methiodide-induced seizures in rat prepiriform cortex. J Pharm Exp Ther 251:1229–1236.

Fredholm BB (1995): Adenosine receptors in the central nervous system. News in Physiological Sciences 10:122–128.

Fredholm BB, Proctor W, Van der Ploeg I, Dunwiddie TV (1989): In vivo pertussis toxin treatment attenuates some, but not all, adenosine A_1 effects in slices of the rat hippocampus. Eur J Pharmacol 172:249–262.

Georgiev V, Johanson B, Fredholm BB (1993): Long term caffeine treatment leads to a decreased susceptibility to NMDA-induced clonic seizures in mice without changes in adenosine receptor number. Brain Res 612:271–277.

Glass M, Faull RLM, Bullock JY, Jansen K, Mee EW, Walker EB, Synek BJL, Dragunow M (1996): Loss of A_1 adenosine receptors in human temporal lobe epilepsy. Brain Res 710:56–68.

Gleiter CH, Deckert J, Nutt DJ, Marangos PJ (1989): Electroconvulsive shock (ECS) and the adenosine neuromodulatory system: Effect of single and repeated ECS on the adenosine A_1 and A_2 receptors, adenylate cyclase and the adenosine uptake site. J Neurochem 52:641–646.

Handforth A, Treiman DM (1994): Effect of an adenosine antagonist and an adenosine agonist on status entry and severity in a model of limbic status epilepticus. Epilepsy Res 18:29–42.

Hattori Y, Moriwaki A, Hayashi Y, Hori Y (1993): Involvement of adenosine-sensitive cyclic AMP-generating systems in cobalt-induced epileptic activity in the rat. J Neurochem 61:2169–2174.

Hawkins M, Pravica M, Radulovacki M (1988): Chronic administration of diazepam downregulates adenosine receptors in the rat brain. Pharmacol Biochem Behav 30:303.

Herberg LJ, Rose IC and Mintz M (1993): Effect of an adenosine A_1 agonist injected into substantial nigra on kindling of epileptic seizures and convulsion duration. Pharmacol Biochem Behav 44:113–117.

Higgins MJ, Hosseinzadeh H, MacGregor DG, Ogilvy H, Stone TW (1994): Release and actions of adenosine in the central nervous system. Purine and Pyrimidine Metabolism 16:62–68.

Hollins C, Stone TW (1980): Adenosine inhibition of GABA release from slices of rat cerebral cortex. Br J Pharmacol 69:107–112.

Hornfeldt CS, Larson AA (1994): Adenosine receptors are not involved in theophylline induced seizures. Clin Toxicol 32:257–265.

Hosseinzadeh H, Stone TW (1994): The effect of calcium removal on the suppression by adenosine of epileptiform activity in the hippocampus: Demonstration of desensitization. Br J Pharmacol 112:316–322.

Jacobson KA, van Galen PJM, Williams M (1992): Adenosine receptors: Pharmacology, structure activity relationships, and therapeutic potential. J Med Chem 35:407–422.

James S, Richardson PJ (1993): Production of adenosine from extracellular ATP at the striatal cholinergic synapse. J Neurochem 60:219–277.

Janusz CA, Berman RF (1993a): Adenosinergic modulation of the EEG and locomotor effects of the A_2 agonist, CGS 21680. Pharmacol Biochem Behav 45:913–919.

Janusz CA, Berman RF (1993b): The adenosine binding enhancer, PD-81,723, inhibits epileptiform bursting in the hippocampal brain slice. Brain Res 619:131–136.

Jarvis MF, Williams M (1990): Adenosine in central nervous system function. In Williams M (ed): "Adenosine and Adenosine Receptors." Clifton, NJ: Humana Press, pp 423–474.

Jonzon B, Fredholm BB (1984): Adenosine receptor mediated inhibition of NA release from slices of the rat hippocampus. Life Sci 35:1971–1979.

Kesim Y, Marangoz C, Aggildiz M, Tasci N, Agar E, Sahinoglu H (1994): The effects of the purinergic system on digitalis-induced epileptiform activity. J Basic Clin Physiol Pharmacol 5:167–178.

Kidd EJ, Grahames BA, Simon J, Michel AD, Barnard EA, Humphrey PPA (1995): Localization of P_{2X} purinoceptor transcripts in the rat nervous system. Mol Pharmacol 45:569–573.

Kleinsorge RJ, Bowers LM, Jarvis MF, Berman RF (1993): Chronic theophylline prolongs the refractory period in amygdala-kindled rats. Drug Dev Res 29:287–291.

Klitgaard H, Knutsen LJS (1992): 8-cyclopentyl-1,3-dimethylxanthine (CPT), an adenosine A_1 receptor antagonist, with anticonvulsant properties. Int J Purine Pyrimidine Res 3:45.

Klitgaard H, Knutsen LJS, Thomsen C (1993): Contrasting effects of adenosine A_1 and A_2 receptor ligands in different chemoconvulsive models. Eur J Pharmacol 224:221–228.

Knutsen LJS, Lau J, Eskesen K, Sheardown MJ, Thomsen C, Weis JU, Judge ME, Klitgaard H (1995): Anticonvulsant actions of novel and reference adenosine agonists. In Belardinelli L, Pelleg A (eds): "Adenosine and Adenine Nucleotides: From Molecular Biology to Integrative Physiology." Boston: Kluwer, pp 479–487. See also: NNC-21-0041. Annual Drug Data Report, 1995, 17, 981.

Knutsen LJS, Lau J, Judge ME, Sheardown MJ, Thomsen C, Weis JU, Klitgaard H (1994): Anticonvulsant actions of adenosine agonists and antagonists. Drug Dev Res 31:286. (Abstract of lecture which included the effect of iodotubericidin in DMCM-induced seizures)

Kostopoulos GK, Drapeau C, Avoli M, Olivier M, Villemeure JG (1989): Endogenous adenosine can reduce epileptiform discharges in human epileptogenic cortex maintained in vitro. Neurosci Lett 106:119–124.

Krahl SE, Treas LM, Castle JD, Berman RF (1995): Attenuation of in vivo and in vitro seizure activity using the adenosine agonist, metrifudil. Drug Dev Res 34:30–34.

Lewis C, Neidhart S, Holy C, North RA, Buell G, Surprenant A (1995): Coexpression of P_{2X2} and P_{2X3} receptor subunits can account for ATP-gated currents in sensory neurons. Nature 377:432–435.

Li C, Peoples RW, Li Z, Weight FF (1993): Zn^{2+} potentiates excitatory action of ATP on mammalian neurons. Proc Natl Acad Sci USA 90:8264–8267.

Maitre M, Ciesielski L, Lehmann A, Kempe E, Mandel P (1974): Protective effect of adenosine and nicotinamide against audiogenic seizure. Biochem Pharmacol 23:2807–2816.

Marangos PJ (1991): Potential therapeutic roles for adenosine in neurologic disease. In Stone, TW (ed): "Adenosine in the nervous system." London: Academic Press, pp 217–226.

Marangos PJ, Montgomery P, Weiss SRB, Patel J, Post RM (1987b): Persistent upregulation of brain receptors in response to chronic carbamazepine treatment. Clin Neuropharmacol 10:443–448.

Marangos PJ, Patel J, Smith KD, Post RM (1987a): Adenosine antagonist properties of carbamazepine. Epilepsia 28:387–394.

Marangos PJ, Weiss SRB, Montgomery P, Patel J, Narang PK, Cappabianca AM, Post RM (1985): Chronic carbamazepine treatment increases brain adenosine receptors. Epilepsia 26:493–498.

Moraidis I, Bingmann D (1994): Epileptogenic actions of xanthines in relation to their affinities for adenosine A_1 receptors in CA3 neurons of hippocampal slices (guinea pig). Brain Res 640:140–145.

Morgan PF, Durcan MJ (1990): Caffeine-induced seizures: Apparent proconvulsant action of N-ethyl carboxamidoadenosine (NECA). Life Sci 47:1.

Murray TF, Franklin PH, Zhang G, Tripp E (1992): A_1 adenosine receptors express seizure-suppressant activity in the rat prepiriform cortex. Epilepsy Res Suppl 8:255–261.

Murray TF, Sylvester D, Schultz CS, Szot P (1985): Purinergic modulation of the seizure threshold for pentylenetetrazol in the rat. Neuropharmacology 24:761–766.

Nagy AK, House CR, Delgado-Escueta AV (1990): Synaptosomal ATPase activities in temporal cortex and hippocampal formation of humans with focal epilepsy. Brain Res 529:192–206.

Pagonopoulou O, Angelatou F, Kostopoulos G (1993): Effect of pentylenetetrazol-induced seizures on A_1 adenosine receptor regional density in the mouse brain: A quantitative autoradiographic study. Neuroscience 56:711–716.

Palayoor ST, Seyfried TN (1984): Genetic association between Ca^{2+}-ATPase activity and audiogenic seizures in mice. J Neurochem 42:1771–1774.

Palayoor ST, Seyfried TN, Bernard DJ (1986): Calcium ATPase activities in synaptic membranes of seizure-prone mice. J Neurochem 46:1370–1375.

Petersen EN (1991): Selective protection by adenosine receptor agonists against DMCM-induced seizures. Eur J Pharmacol 195:261–265.

Phillis JW, Kirkpatrick JR (1978): The actions of adenosine and various nucleosides and nucleotides on the isolated toad spinal cord. Gen Pharmacol 9:239–247.

Phillis JW, Wu PH (1981): The role of adenosine and its nucleotides in central synaptic transmission. Prog Neurobiol 16:187–239.

Phillis JW, Wu PH (1983): Roles of adenosine and adenine nucleotides in the CNS. In Daly JW, Kuroda Y, Phillis JW, Shimizu H, Ui M (eds): "Physiology and Pharmacology of Adenosine Derivatives." New York: Raven Press, p 219.

Phillis JW, O'Regan MH (1988): Benzodiazepine interaction with adenosine systems explains some anomalies in GABA hypothesis. Trends Pharmacol Sci 9:153.

Piredda S, Gale K (1985): A critical epileptogenic site in the deep prepiriform cortex. Nature 317:623–625.

Psarropoulou C, Matsokis N, Angelatou F, Kostopoulos G (1994): Pentylenetetrazol-induced seizures decrease γ-aminobutyric acid-mediated recurrent inhibition and enhance adenosine-mediated depression. Epilepsia 35:12–19.

Racine RJ (1972): Modification of seizure activity by electrical stimulation, II. Motor seizure. Electroenceph Clin Neurophysiol 32:281–293.

Roca DJ, Schiller GD, Farb DH (1988): Chronic caffeine or theophylline exposure reduces gamma aminobutyric acid/benzodiazepine receptor site interactions. Mol Pharmacol 30:481.

Roca DJ, Schiller GD, Friedman L, Rosenberg I, Gibbs TT, Farb DH (1990): Gamma aminobutyric acid A receptor regulation in culture: Altered allosteric interactions following prolonged exposure to benzodiazepines, barbiturates, and methylxanthines. Mol Pharmacol 37:710.

Sanders RC, Murray TF (1989): Temporal relationship between A_1 adenosine receptor upregulation and alterations in bicuculline seizure susceptibility in rats. Neurosci Lett 101:325–330.

Schubert P (1992a): Neurotransmitters in epilepsy. Epilepsy Res Suppl 8: 243–253.

Schubert P (1992b): Depression of burst discharges and of neuronal calcium influx by adenosine. In Avanzini G, Engel J, Fariello R, Heinemann U (eds): "Neurotransmitters in Epilepsy." (Epilepsy Res. Suppl 8), pp 243–253.

Sebastiao AM, Ribeiro JA (1996): Adenosine A_2 receptor-mediated excitatory actions on the nervous system. Prog Neurobiol 48:167–189.

Seguela P, Haghighi A, Soghomonian JJ, Cooper E (1996): A novel neuron P_{2X} ATP receptor ion channel with widespread distribution in the brain. J Neurosci 16:448–455.

Silinsky EM, Gerzanich V, Vanner SM (1992): ATP mediates excitatory synaptic transmission in mammalian neurons. Br J Pharmacol 106:762–763.

Simonato M, Varani K, Muzzolini A, Bianchi C, Beani L, Borea PA (1994): Adenosine A_1 receptors in the rat brain in the kindling model of epilepsy. Eur J Pharmacol 265:121–124.

Simpson RE, O'Regan MH, Perkins LM, Phillis JW (1992): Excitatory transmitter release from the ischemic rat cerebral cortex: Effects of adenosine receptor agonists and antagonists. J Neurochem 58:1683–1690.

Skolnick P, Paul SM, Marangos PJ (1980): Purines as endogenous ligands of the benzodiazepine receptor. Fed Proc 39:3050.

Stevens DN, Weidmann R (1989): Blockade of FG 7142 kindling by anticonvulsants acting at sites distant from the benzodiazepine receptor. Brain Res 492:89–98.

Surprenant A, Rassendren F, Jawashima E, North RA, Buell G (1996): The cytolytic P_{2X} receptor for extracellular ATP identified as a P_{2X} receptor (P_{2X7}). Science 272:735–738.

Valera S, Hussy N, Evans RJ, Adami N, North RA, Surprenant A, Buell G (1994): A new class of ligand-gated ion channel defined by P_{2X} receptor for extracellular ATP. Nature 371:516–519.

von Lubitz DKJE, Carter MF, Deutsch SI, Lin RC-S, Mastropaolo J, Meshulam Y, Jacobson KA (1995a): The effects of adenosine A_3 receptor stimulation on seizures in mice. Eur J Pharmacol 275:23–29.

von Lubitz DKJE, Kim J, Beenhakker MF, Carter MF, Lin RC-S, Meshulam Y, Daly JW, Shi D, Zhou L-M, Jacobson KA (1995b): Chronic NMDA receptor stimulation: Therapeutic implications of its effects on adenosine receptors. Eur J Pharmacol 283:185–192.

von Lubitz DKJE, Paul I, Carter MF, Jacobson KA (1993): Effects of N^6-cyclopentyl adenosine and 8-cyclopentyl-1,3-dipropylxanthine on N-methyl-D-aspartate induced seizures in mice. Eur J Pharmacol 249:265–270.

White TD (1978): Direct detection of depolarization-induced release of ATP from a synaptosomal preparation. Nature 267:67–68.

Wiesner JB, Ugarkar B, Castellino A, DaRe J, Ramirez-Weinhouse M, Zimring S, Barankiewicz J, Bullough D, Bayat S, Dumas D, Erion MD, Gruber HE, Foster AC (1995): Anticonvulsant activity of novel adenosine kinase inhibitors. Soc Neurosci Abs 21:314.10.

Wiesner JB, Zimring ST, Castellino AJ, Ugarkar BG, Shea M, Erion MD, Foster AC (1995): Anticonvulsant profile of a novel adenosine kinase inhibitor, GP683, in rodents. Epilepsia 36(4):F3.

Williams M (1984): Adenosine—a selective neuromodulator in the mammalian CNS? Trends Neurosci 7:164–168.

Williams M (1993): Purinergic drugs: Opportunities in the 1990s. Drug Dev Res 28:438–444.

Williams M, Risley EA (1981): Interaction of putative anxiolytic agents with central adenosine receptors. Can J Physiol Pharmacol 59:897.

Wu PH, Coffin VL (1984): Upregulation of brain [^{3}H]diazepam binding sites in chronic caffeine treated rats. Brain Res 294:186.

Young D, Dragunow M (1994): Status epilepticus may be caused by loss of adenosine anticonvulsant mechanisms. Neuroscience 58:245–261.

Zhang G, Franklin PH, Murray TF (1994): Activation of adenosine A_1 receptors underlies anticonvulsant effect of CGS21680. Eur J Pharmacol 255:239–243.

Zhang G, Franklin PH, Murray TF (1993): Manipulation of endogenous adenosine in the rat preperiform cortex modulates seizure susceptibility. J Pharm Exp Ther 264:1415–1424.

Zimmermann H (1994): Signalling via ATP in the nervous system. Trends Neurosci 17:420–426.

Zimring ST, Miller LP, Babb MJ, Hill GM, DaRe J, Ramirez-Weinhouse MM, Ugarkar HE, Gruber HE, Foster AC, Weisner JB (1995): Adenosine kinase inhibitors exert anticonvulsant activity in rats and mice. Soc Neurosci Abs 21:151.1.

CHAPTER 23

Adenosine and Acute Treatment of Cerebral Ischemia and Stroke—"Put Out More Flags"

DAG K. J. E. VON LUBITZ

Laboratory of Bioorganic Chemistry, Molecular Recognition Section, NIH/NIDDK/LBC, Bethesda, MD 20892

INTRODUCTION

The title of Evelyn Waugh's famous novel, so unashamedly borrowed for the present chapter, may well apply to the current status of adenosine-based concepts of therapies against brain damage caused by ischemia and stroke. Together with cancer and heart disease, stroke has the dubious distinction of being one of the three leading causes of death in the United States. The latest available data indicate that with over 500,000 incidents each year, bona fide stroke accounts for half of all patients hospitalized for acute neurological disease (National Foundation for Brain Research, 1992). The number of cases becomes alarmingly high when all instances of cerebral damage caused by cardiac arrest, drowning, and other instances of severe anoxia/hypoxia are incorporated. It is then quite paradoxical that, despite a great progress in understanding pathophysiology of the ischemic brain damage, the effective means of its clinical prevention or amelioration are as elusive as ever. Moreover, the late inclusion of adenosine among the therapies contemplated in the context of cerebral ischemia and stroke offers a paradox of its own. Well known for its inhibitory properties among cardiologists, adenosine (under the name Adenocard) is considered one of the best clinical means of treating supraventricular tachycardias (Jacobson et al., 1992). Yet, although neuroinhibitory and neuromodulatory properties of cerebral adenosine are known equally well (Phillis and Wu, 1981), the potential usefulness of adenosine in the treatment of cerebral ischemia and stroke is but scantily recognized by the world of clinical neurology.

Purinergic Approaches in Experimental Therapeutics, Edited by Kenneth A. Jacobson and Michael F. Jarvis
ISBN 0-471-14071-6 © 1997 Wiley-Liss, Inc.

The reasons for such late acceptance of the feasibility of adenosine-based therapies directed against ischemic neurodegeneration are multiple. While the understanding of its fundamental role in control of neuronal (or even cerebral) activity paralleled that of the understanding of pathophysiology of stroke, the limited spectrum of sufficiently selective adenosine receptor ligands posed one of the most important stumbling blocks in the development of viable drugs (Williams, 1993). Furthermore, the side effects of experimentally tested adenosine A_1 receptor agonists endowed with sufficient selectivity (e.g., receptor desensitization, hypotension, bradycardia) appeared to be insurmountable (Williams, 1993). Recent, dramatic progress in the medicinal chemistry of drugs acting at adenosine receptors (Van Galen et al., 1992) stimulated renewed interest on the part of the pharmacological industry in developing adenosine-based therapeutics. Thus, despite the initial difficulties, the clinical introduction of such drugs may be very near.

ADENOSINE AND BRAIN

Endogenous Adenosine and Its Receptor Types

Together with γ-aminobutyric acid (GABA), adenosine functions as the brain's principal inhibitory neuromodulator (Kostopoulos and Phillis, 1977) whose actions are mediated via the interaction with both pre- and postsynaptic receptors. The determination of the exact extracellular concentration of adenosine in the normal brain is, however, difficult (von Lubitz and Marangos, 1990; Phillis, 1990; Rudolphi et al., 1992), due to its short half-life, and fluctuations caused by the trauma produced by many of the methods used (von Lubitz and Marangos, 1990; Phillis, 1990; Rudolphi et al., 1992). For these reasons, the published levels of extracellular adenosine measured by microdialysis in awake, freely moving animals vary between 50 and 300 nM (Rudolphi et al., 1992).

Endogenously released adenosine acts at three major receptor subtypes (A_1, A_2, and A_3), whose cloning (Ji et al., 1994) showed that, despite some differences in their molecular sequence, all are members of the same G protein–coupled receptor family (Jacobson et al., 1992). Moreover, the functional properties of A_2 receptors necessitated their further subdivision into high-affinity A_{2a} and low-affinity A_{2b} receptors.

The main physiological distinction between A_1 and A_2 receptors is the effect on the level of intracellular cyclic AMP (cAMP; van Calker et al., 1979) induced by their activation. Thus, stimulation of the A_1 receptors decreases, and stimulation of the A_2 receptors increases, production of cAMP. A_3 receptors differ from the two preceding subtypes, and two distinct effector mechanisms participate in their operation, i.e., enhancement of phosphoinositide metabolism (via stimulation of phospholipase C) or, like the A_1 subclass, depression of adenylate cyclase (Ramkumar et al. 1993).

Cerebral Distribution of Adenosine Receptors

Although all three adenosine receptors are present in the brain, their distribution is uneven. A_1 receptors are abundant in the hippocampus, IV–VI cortical laminae, striatum, amygdala, and superior colliculus (Jarvis and Williams, 1989). Particularly dense populations of high-affinity A_{2a} receptors are found within the dopamine-rich regions of the brain, e.g., striatum (Jarvis and Williams, 1989), while low-affinity A_{2b} have been detected on the glia (Rudolphi et al., 1992). There is a possibility that neurons may be characterized by the presence of A_{2b} receptors as well (Daly et al., 1980). Finally, A_2 receptors are located on both the smooth muscle and endothelial cells of the cerebral blood vessels (Kalaria and Harik, 1986).

The still elusive A_3 receptors are widely distributed throughout the entire brain, albeit at densities that are significantly lower than those of either A_1 or A_2 subtypes (Ji et al., 1994). The exact localization of A_3 is unknown, but there are indications that they may be found both on neurons and within cerebral vasculature (Linden, 1994).

Neuronal Effects of Adenosine Receptor Stimulation

A_1 Receptors The inhibitory actions of adenosine are mediated via A_1 receptors whose stimulation enhances K^+ (Trussel and Jackson, 1985) and voltage-dependent, non-GABAergic Cl^- conductances (Mager et al., 1990), resulting in the depression of the membrane potential. Since A_1 receptor–evoked hyperpolarization of neuronal membrane attenuates the function of presynaptic voltage-regulated calcium channels, Ca^{2+} uptake (Wu et al., 1982; Schubert et al., 1986; Madison et al., 1987; Fredholm and Dunwiddie, 1988; Schubert and Kreutzberg, 1990) is also inhibited. The net effect of A_1 receptor–mediated diminution of the presynaptic Ca^{2+} influx is the depression in the release of several neurotransmitters, e.g., glutamate, acetylcholine, dopamine, noradrenaline, and serotonin (Fredholm and Dunwiddie, 1988). Thus, the overall functional consequence of A_1 receptor activation is the reduction of neuronal excitability and firing rate (Dunwiddie, 1980, 1985), accompanied by the reduction of neuronal metabolism (von Lubitz and Marangos, 1990).

A_2 Receptors Adenosine A_2 receptors have been traditionally considered as responsible for the regulation of cerebral blood flow (Phillis, 1989) and accumulation of cAMP in the brain (Rudolphi et al., 1992). However, recent studies indicate that excitatory A_2 receptors found in the hippocampus (Sebastião and Ribeiro, 1992) appear to potentiate calcium-dependent glutamate and acetylcholine release (O'Regan et al., 1992; Fredholm and Dunwiddie, this volume), while in the globus pallidus they may be involved in the modulation of GABA liberation (Mayfield et al., 1993). Moreover, stimulation of A_2 receptors found on neutrophils suppresses generation of superoxides when these cells become activated (Dianzani et al., 1994), and in platelets suppresses their aggregation (Sandoli et al., 1994).

A_3 Receptors Although information on the involvement of A_3 receptors in neuronal activity is still limited, the available data indicate that they may perform a number of important functions, particularly in the environment that has been pathologically modified. Sei et al. (unpublished) have shown that acute stimulation of cultured cerebellar granule cells with the A_3 receptor agonist 2 chloro-N^6-(3-iodobenzyl)adenosine-5′-N-me-thyluronamide (Cl-IB-MECA) results in their necrosis. The same authors have also shown that subthreshold doses of Cl-IB-MECA significantly increase toxicity of glutamate administered at subneurotoxic doses. Interestingly, exposure of promyelocytic human HL-60 cells or CHO cells transfected with human A_3 receptors to Cl-IB-MECA or N^6-(3-iodobenzyl)adenosine-5′-N-methyluronamide (IB-MECA) induces apoptosis rather than necrosis (Kohno et al., 1996; Brambilla et al., 1996), and the presence of both necrosis and apoptosis has been detected in the hippocampus of gerbils injected with IB-MECA immediately prior to brief forebrain ischemia (von Lubitz, Sei, Lin, and Jacobson, unpublished). Finally, while acute treatment with IB-MECA has no effect, chronic exposure to this drug results in a significant depression of cortical and hippocampal nitric oxide synthase (von Lubitz et al., 1996a). As indicated by Ceruti et al., (1996), astrocytes appear to be modulated by A_3 receptors as well. It must be mentioned that the operation of the entire brain might be affected by the hypotensive effect of acute A_3 receptor stimulation (Fozard and Carruthers, 1993). A_3-mediated hypotension appears to be the result of histamine release induced by degranulation of mast cells (Fozard et al., 1996).

Recently a number of specific intracellular effects of A_3 receptor stimulation have been described. Processes such as stimulation of extracellular Ca^{2+} influx and mobilization of internal Ca^{2+} stores (Kohno et al., 1996), stimulation of phospholipase C and phosphoinositol turnover (Abbracchio et al., 1995a), or inhibition of TNF-α (Sajjadi et al., 1996) are of particular significance in the context of stroke and global ischemia. However, the degree to which A_3 receptor–mediated effects contribute to the generation of ischemic damage is yet to be determined.

Interaction Between Different Adenosine Receptor Types

Direct evidence of specific interaction between adenosine receptors of different types does not exist. However, the puzzling increase of intracellular Ca^{2+} caused by the stimulation of astrocytic A_1 receptors in the presence of a subthreshold concentration of the metabotropic glutamate receptor ((*t*)-1-Aminocyclopentane-trans-1,3-dicarboxylic acid (t-ACPD) (Ogata et al., 1994) combined with the available evidence from nonneural systems led to a highly intriguing hypothesis of cooperative action of the normally antagonistic A_1 and A_2 receptors (Schubert et al., 1994; Rudolphi and Schubert, 1996) that may be of a significant interest in development of adenosine-based therapies against ischemic damage. The hypothesis is based on the fact that, while increased intracellular concentration of Ca^{2+} results in the activation of phosphokinase C, the enhanced level of the latter stimulates, in turn, the generation of cAMP (for details, see Schubert et al., 1994). Increased levels of cAMP result in a negative feedback control of a number of Ca^{2+} controlled processes (Nishizuka, 1986), many of which are particularly destructive in the context of cerebral ischemia (Siesjö, 1988). Hence, the A_1/A_2 receptor–mediated shift toward increased production of cAMP may have a highly beneficial effect and may help in explaining prolonged efficacy of brain-protective effects resulting from administration of A_1 receptor agonists as late as 18 hours postischemia (see below). Recent findings that A_3 receptor stimulation results in the liberation of intracellular Ca^{2+} stores (Kohno et al., 1996), and the fact that stimulation with very low doses of A_3 receptor agonists results in activation of astrocytes (Ceruti et al., 1996), strongly suggest that the cooperative effects of adenosine receptors may include the A_3 type as well.

Interaction Between Adenosine and Other Receptor Systems Involved in Neural Transmission

In several regions of the brain the distribution of adenosine A_1 closely follows that of N-methyl-D-aspartate (NMDA) receptors (Cotman et al., 1987; Daval et al., 1989). Recent studies (Craig and White, 1992; Manzoni et al., 1994; Pazzagli et al., 1985) have demonstrated that activation of NMDA receptors by glutamate results in the release of adenosine. The effect appears to be reciprocal (Poli et al., 1991). In the hippocampus, and possibly also in the cortex, adenosine liberated through NMDA receptor stimulation derives (at least in part) from interneurons, and it is likely that adenosine generated in this manner may inhibit further release of glutamate at more distant excitatory synapses (Craig and White, 1992; Manzoni et al., 1994). Activation of muscarinic receptors activates adenosine efflux as well (Pazzagli et al., 1984).

One of the curious aspects of the functional relation between NMDA and A_1 receptors is the finding that, while stimulation of NMDA receptors leads to the release of adenosine, activation of postsynaptic A_1 receptors has a powerful effect on the function of NMDA complexes. It is a well-known fact that the Mg^{2+} block of the NMDA receptor–gated ion channel is removed only when the membrane is depolarized beyond a critical level (Novak et al., 1984). The extent of this depolarization is determined by the input frequency of the stimulation (Dunwiddie and Lynch, 1978). However, as discussed above, activation of A_1 receptor–governed conductances results in depression of the postsynaptic membrane potential, and a higher input frequency is necessary to depolarize the membrane to the level of its critical instability (Schubert and Kreutzberg, 1990; Schubert and Mager 1991). For this reason, stimulation of postsynaptic A_1 receptors results in the depression (Schubert and Mager, 1991) or even inhibition (Mendonça et al., 1995) of NMDA receptor excitability (Schubert and Mager, 1991).

Functional links of A_2 receptors to other receptor systems have been also demonstrated. Thus, the agonist affinity of striatal dopamine D_2 receptors is attenuated by colocalized (an coexpressed) high-affinity A_2 receptors (Schiffman et al., 1993; Ferré et al., 1993). Since stimulation of metabotropic glutamate (probably mGluR2) receptors diminishes both forskolin-evoked cAMP production and adenosine release (Genazzani et al., 1993; Cassabona et al., 1994), the existence of functional interactions between A_2 and metabotropic glutamate receptors has been also postulated (Cassabona et al., 1994). However, even if such a link exists, its nature appears to be much more tenuous than in the preceding cases and requires further studies.

ADENOSINE AND THE ISCHEMIC BRAIN

Adenosine Release

Metabolic stress induced by seizures, hypoxia or ischemia results in a rapid increase of extracellular adenosine (Rudolphi et al., 1992). The level of presynaptic activity appears to determine the intensity of adenosine release from axonal terminals (Schubert et al., 1976) and, in ischemia, the increase may be as high as 50-fold (Hagberg, et al., 1987). Although several possibilities exist, the specific mechanism involved in the intraischemic adenosine liberation is unknown. The arrest of cerebral circulation is accompanied by a very rapid breakdown of ATP (Whittingham, 1990). Hence, adenosine—an intermediary product of this process—may diffuse directly across cellular membranes (Schubert and Dux, 1990). Excessive stimulation of glutamate receptors (Andiné et al., 1991) appears to be another possibility, as indicated by the attenuation of intraischemic adenosine release following pretreatment with the NMDA receptor antagonist AP5 (D-amino-5-phosphonovalerate; Hagberg et al., 1986). Involvement of glutamate receptors in the intraischemic adenosine release is supported by the results of *in vitro* studies demonstrating that electrical field stimulation of NMDA and

AMPA receptors enhances adenosine liberation (Manzoni et al., 1994; Hoehn and White, 1990a,b). Since both ATP and neurotransmitters may be contained within the same synaptic vesicle (Burnstock, 1986), there is also a possibility that extracellular adenosine derives from ATP released together with other neurotransmitters. Once released, ATP is then broken down to adenosine by the enzymes located on the outer surface of cells (von Lubitz and Marangos, 1990; Schubert and Dux, 1990).

ADENOSINE-MEDIATED EFFECTS AND PATHOPHYSIOLOGY OF ISCHEMIA

Adenosine and Ionic Conductances

Several studies have shown that neurons hyperpolarize during the early stages of ischemia (the "functional loss stage" (Hansen et al., 1982; Fujiware et al., 1987; Leblond and Krnjević, 1989)). Since stimulation of A_1 receptors enhances K^+ and voltage-dependent non-GABA-ergic Cl^- conductances (Greene and Haas, 1991; Mager et al., 1990; Schubert et al., 1990) at both pre- and postsynaptic terminals, there is a likelihood that intraischemically released adenosine is involved in that process.

As mentioned previously, A_1 receptor–mediated elevation of K+ conductance is an important factor in lowering membrane potential to the level at which both reduction of Ca^{2+} influx and subsequent depression of the neurotransmitter release occur (Fredholm and Dunwiddie, 1988). Moreover, elevation of postsynaptic K^+ conductance may be also responsible for attenuation of the activity-evoked membrane depolarization necessary for lifting the voltage-dependent Mg^{2+} block of NMDA receptor (Rudolphi et al., 1992). Finally, since excessive intracellular accumulation of Cl^- impairs this form of inhibition (Hugenard and Alger, 1986), the enhancement of Cl^- conductance associated with intraischemic activation of A_1 receptors may be an important factor in sustaining inhibitory efficiency of GABA (Schubert and Kreutzburg, 1990). In summary, the wide range of hyperpolarizing ionic shifts elicited by intraischemic stimulation of A_1 receptors may constitute an important mechanism whose activation delays the onset of the "hypoxic depolarization stage" (Lee and Lowenkopf, 1993), i.e., the stage at which most of the destructive intraischemic phenomena begin to take place (Obrenovitch et al., 1990; Leão, 1947).

Adenosine and Intraischemic Neurotransmitter Release

Intracerebral elevation of both adenosine and glutamate are directly related to the cerebral blood flow rate. Importantly, extracellular accumulation of adenosine precedes that of glutamate and occurs at a flow of approximately 25 ml $\cdot$ $100g^{-1}$ $\cdot$ min^{-1}, while further reduction (20 ml $\cdot$ $100g^{-1}$ $\cdot$ min^{-1}) is necessary to induce a measurable increase of glutamate (Matsumoto et al., 1992). It seems, therefore, that endogenous adenosine interacts with presynaptic A_1 receptors and counteracts the release of excitatory neurotransmitters already at the very early stages of ischemia. The plausibility of such sequence is corroborated by the experimental evidence showing that pretreatment with adenosine uptake inhibitors (or A_1 receptor agonists) results in a significant reduction of intraischemic release of excitatory amino acids (Andiné et al., 1991; Hagberg et al., 1990; Phillis et al., 1991; Simpson et al., 1992). Conversely, preocclusive antagonism of A_1 receptors with theophylline increases the amount of liberated glutamate and aspartate (Sciotti et al., 1992).

Adenosine and Intraischemic Glutamate Uptake

There is no direct evidence that stimulation of adenosine receptors affects the uptake of either glutamate or other neurotransmitters. However, Anderson et al. (1993) showed that even a brief ischemia results in a prolonged upregulation of high-affinity excitatory amino acid uptake sites, while Schmidt et al. (1993) demonstrated that a similar increase is also induced when normoxic hippocampus slices are briefly exposed to adenosine. There is thus a probability that intraischemic stimulation of A_1 receptors results in the depression of neurotransmitter release, while the concomitant activation of astrocytic A_2 receptors (and possibly A_1—see Schubert et al., 1994) causes increased removal of excitatory amino acids from the extracellular space (von Lubitz et al., 1995a). A close spatial interrelationship of relevant structural elements, i.e., neurotransmitter-releasing presynaptic terminals and astrocytic processes that surround synaptic complexes (Schubert and Kreutzberg, 1990; von Lubitz and Diemer, 1982) and contain glutamate uptake sites (Anderson et al., 1993), makes this possibility even more plausible.

Adenosine and Intraischemic Hyperactivation of the NMDA Receptor

The intensity of the excitatory synaptic input depends on the amount of NMDA receptor–mediated Ca^{2+} influx (Herron et al., 1986). However, it has been shown that even at physiological concentrations of extracellular adenosine, the evoked Ca^{2+} influx is closely controlled by postsynaptic A_1 receptors (Schubert et al., 1993). Hence, at least during the initial stages of ischemia, A_1 receptors may constitute one of the principal mechanisms whose activation delays opening of NMDA channels (Schubert and Mager, 1991). Moreover, A_1 receptor–mediated attenuation of glycine release (Cantor et al., 1992) may slow intraischemic activation of NMDA receptors even further. Thus, the cumulative effect of intraischemic stimulation of pre- and postsynaptic A_1 receptors is highly adverse to the optimal activation of NMDA receptors, for which simultaneous availability of adequate concentration of the agonist and coagonist (glutamate and glycine), and a sufficient degree of membrane depolarization to lift the Mg^{2+} block of the ion channel, are required.

One of the less direct but equally interesting consequences of intraischemic adenosine A_1–NMDA receptor interaction may be its effect on voltage-sensitive K^+ currents. The NMDA receptor–mediated depolarization seen in cerebral ischemia interrupts these currents and leads to enhanced neuronal excitability and firing rate (Segal et al., 1984). It is conceivable that A_1 receptor–mediated attenuation of NMDA receptor may force the membrane potential below the voltage at which depolarization-dependent inhibition of potassium currents is likely to occur. Hence, the ensuing decrease in axonal activity may represent another aspect of the endogenous complex of neuroprotective effects induced by intraischemic stimulation of A_1 receptors.

Adenosine—Postischemia

Intraischemically increased level of extracellular adenosine normalizes within 30 to 60 minutes after the onset of reperfusion (Hagberg et al., 1990; Phillis et al., 1987) and coincides with the normalization of glutamate concentration (Hagberg et al., 1990; Globus et al., 1990). Preischemic blockade of adenosine A_1 receptors with the antagonist theophylline significantly delays reversal in the postischemic levels of glutamate and aspartate (Sciotti et al., 1992). Hence, it is quite likely that, during the initial postocclusive period, the sustained activation of adenosine A_1 receptors may play a substantial role in reducing neurotoxic phenomena evoked by the continuously elevated concentration of excitatory amino acids. The intensity of neuronal protection afforded through postischemic interaction of endogenous adenosine with A_1 receptors is enhanced even further by the concomitant reduction of metabolic perturbations (Tominaga et al., 1992) and neuronal hyperactivity (Suzuki et al., 1983).

Adenosine and its A_2 receptors are well known as the regulators of cerebral blood flow in the normal brain (Phillis, 1989). Following ischemia, however, less direct cerebrovascular effects may be equally important in restoration of the postocclusive blood flow (Morii et al., 1987). Thus, stimulation of A_2 receptors results in the amelioration of endothelial damage induced by free radicals and cytotoxins released by the activated neutrophils (Engler, 1987; Cronstein et al., 1986). Moreover, even modestly elevated concentrations of endogenous adenosine reduce the platelet aggregation (Sollevi, 1986) that constitutes a significant factor in generation of postischemic damage (Fujimoto et al., 1985).

Even if brief, the continuous postischemic stimulation of its receptors by endogenously released adenosine results in a number of interlinked damage-limiting processes that affect both neurons and cerebral vasculature. Ultimately, the combination of all of these effects may be a major factor contributing to the absence of both neuropathological and neurological changes following very brief periods of cerebrocirculatory arrest.

Adenosine Receptors—Postischemia

Both focal and global ischemia cause a rapid decrease in the density, but not the affinity, of the adenosine A_1 receptor population (Daval et al., 1989; Onodera et al., 1989; Nagasawa et al., 1994; von Lubitz and Marangos, 1990). However, postischemic downregulation of A_1 receptors is inhomogeneous, and there are striking differences among individual brain regions. In the hippocampus, A_1 receptor loss precedes physical disappearance of neurons. Already by 1–2 hours after ischemia, the number of hippocampal A_1 receptors is decreased by as much as 10%–20% (Onodera and Kogure, 1985, 1990; Lee et al., 1986). Several months later, their density is reduced even more (Araki et al., 1992). In the cortex and striatum, on the other hand, the decrease observed 2 hours postischemia is marginal (Lee et al., 1986).

Long-term downregulation in the hippocampus is undoubtedly the consequence of the loss of pyramidal cells (Onodera nad Kogure, 1986). However, despite the substantial loss of cortical and striatal neurons after ischemia of moderate severity (von Lubitz et al., 1989, 1994a), the number of adenosine A_1 receptors in these regions returns to the preischemic level within approximately 24 hours after the insult (Lee et al., 1986), and no subsequent fluctuations take place (Daval et al., 1989; Araki et al., 1992).

No explanations exist for the nonuniform postischemic response of adenosine A_1 receptors. One may speculate, however, that the fate of neurons in the selectively vulnerable brain regions may depend, at least to a degree, on the divergent pattern of those density shifts. It is well known, for example, that enkephalin- and substance P–containing neurons of the dorsolateral striatum are particularly vulnerable and are among the first neurons to disappear even after a brief period of ischemia (Chesselet et al., 1990). Glutamatergic corticostriatal fibers provide afferent input to these neurons (Cepeda et al., 1993), which are also characterized by the presence of colocalized dopamine D_2 and adenosine A_2 receptors (Schiffman et al., 1993). Ischemia causes a long-lasting increase in the concentration of intrastriatal dopamine and cAMP (Prado et al., 1992). Since cAMP increase is most likely the consequence of A_2 receptor activation, the elevated level of cAMP and dopamine indicates that both D_2 and A_2 receptors are intensely stimulated both during and after the ischemic episode. However, while activation of dopamine D_2 receptor attenuates the efficacy of glutamatergic stimulation (Cepeda et al., 1993), activation of A_2 receptors diminishes the affinity of D_2 sites to dopamine (Ferré et al., 1992) and may amplify glutamatergic excitation. Loss of glutamate release modulation resulting from the lowered density of A_1 receptors (Lee et al., 1983) on the presynaptic terminals of the corticostriatal fibers may exacerbate this effect.

Adenosine and Postischemic Response of Glia

Activation of astrocytes and microglia constitutes a very significant aspect of postischemic pathology. Ischemia-induced hypertrophy of astrocytes has been known for more than a decade (Petito, 1976; Petito and Babiak, 1982). The response of astrocytes indicated by the elevation of the glial fibrillary acidic protein (GFAP) and morphological changes (Petito et al., 1990; Petito and Halaby, 1993) is now recognized as the consequence of the modified neuronal environment rather than the result of direct astrocyte activation induced by ischemic stress (Petito and Halaby, 1993). While astrocytosis might be considered an indicator of the postocclusive repair processes (Abbracchio et al., 1995b), microglial activation has been shown to contribute to a series of highly injurious events, e.g., release of nitric oxide, glutamate, and free radicals (Banati et al., 1993).

Although the presence of A_1 and low-affinity A_{2B} receptors had been detected several years ago (van Calker et al., 1979), the interest in the role of adenosine in both normal and pathological functions of astrocytes is relatively new. *In vitro* studies of cultured astrocytes reveal that stimulation of their A_1 receptors in conjunction with simultaneous activation of metabotropic receptors enhances mobilization of intracellular Ca^{2+} (Ogata et al., 1994). Moreover, interdependence between A_1/A_2 receptor–mediated effects on intracellular concentration of cAMP and activation of phosphokinase C (PKC) has been also postulated (Schubert et al., 1995, see above). Since cAMP regulates astrocytic uptake of glutamate and glycogen metabolism (Hansson and Rönnbäck, 1991; Sorg and Magistretti, 1992), stimulation of A_1 and/or A_2 receptors, and its impact upon the intracellular concentration of both Ca^{2+} and cAMP, provides astrocytes with a range of discrete responses suitable to the prevailing conditions in their immediate environment. A_3 receptors appear to be involved in these processes as well, since stimulation of cultured astrocytes with the A_3 receptor agonist IB-MECA prevents or induces their apoptotic death, depending on whether concentration of the agonist is low or high (Ceruti et al., 1996).

Adenosine receptors may be also involved in control of microglia, as evidenced by Si et al. (1996), who have shown that introduction of propentofylline to the medium significantly inhibited microglial proliferation in culture. Since propentofylline, apart from being a weak A_1 receptor inhibitor, acts also as a blocker of adenosine membrane transporter and phosphodiesterases (Abbracchio et al., 1995b), it is quite likely that adenosine receptor involvement in microglial control is highly complex and may extend beyond the level of simple activation of either A_1 or A_2 receptors. Such a conclusion is supported by that fact that, while the inhibitory effects of the selective A_1 receptor agonist CCPA are similar to those of propentofylline, the selective antagonist CPX cannot reverse them, while the A_2 agonist CGS 21680 is ineffective (Si et al., 1996).

The complexity of purinergic involvement in control of glial cells is underlined by pronounced effects elicited by agonists and antagonists of ATP receptors. The latter have been extensively reviewed in a recent paper (Abbracchio et al., 1995b).

ADENOSINE AND THE EXPERIMENTAL TREATMENT OF GLOBAL AND FOCAL ISCHEMIA

Acute therapies based on the concept of interaction between adenosine and adenosine receptors may be instituted in several ways, i.e., through administration of adenosine itself, through manipulation of the level of endogenous adenosine (e.g., either enhancement of its release or attenuation of its breakdown), or by direct stimulation of adenosine receptors by exogenously administered drugs.

Adenosine

Adenosine (Adenocard™) is used in treatment of supraventricular tachycardias. However, its implementation in the context of stroke and/or cerebral ischemia appears impractical for several reasons. A bolus injection of adenosine will have either no or minimal effect due to its very short biological half-life. Although employed clinically either as a cardioprotective or cerebroprotective agent in cardiopulmonary bypass and cerebral aneurysm surgery (Sollevi, 1986), a continuous infusion of adenosine may produce cardiovascular side effects such as bradycardia or hypotension (Sollevi, 1986; Powers, 1993) that would adversely affect the outcome of cerebral ischemia/stroke.

Enhanced Release of Endogenous Adenosine

As shown in the studies of myocardial ischemia, agents that enhance the release of adenosine also augment its extracellular concentration at the injury site in both an event-specific and site-specific manner (Gruber et al., 1989). Improvement of postischemic cardiac recovery (Hori et al., 1989) following treatment with acadesine (AICA-riboside [5-amino-1-β-D-ribofuranosyl-imidazole-4-carboxamide]) indicated potential usefulness of this drug in cerebral ischemia. Subsequently, reduction of brain damage by acadesine has been shown in experimental models of global ischemia and stroke (Clough-Helfman and Phillis, 1990; Dietrich et al., 1995). However, despite initially encouraging clinical trials (Leung et al., 1992), the ultimate results were inconclusive.

Blockade of Adenosine Uptake or Breakdown

Reports on the effect of adenosine transport inhibition as a means of reducing ischemic brain damage are not uniformly positive. In a single study where two uptake inhibitors NBI (6-(4-nitrobenzyl)-thioinosine) and NBG (6-(2-hydroxy-benzyl)-thioguanosine) were administered 15 minutes prior to moderate ischemia in gerbils, no beneficial effect could be detected (DeLeo et al., 1987). Administration of propentofylline, on the other hand, resulted in a consistent improvement of the outcome in focal, global, and spinal cord ischemia (Rudolphi and Schubert, 1996; Danielisova et al., 1994). As mentioned previously, the drug has been shown to inhibit microglial proliferation in culture (Si et al., 1996), which may have a significant bearing upon its effectiveness when given several hours postischemia (von Lubitz et al., unpublished). While it is only a moderately potent adenosine transport inhibitor (Fredholm and Lindstrom, 1986), propentofylline is also an inhibitor of both cAMP and cGMP phosphodiesterases (Rudolphi and Schubert, 1996). Hence, its actions may extend beyond the level of adenosine receptors. An intriguing review of the neuronal ramifications of propentofylline treatment has been recently published by Rudolphi and Schubert (1996).

In vitro exposure of cortical neurons subjected to N_2 hypoxia and glucose deprivation to the clinically available adenosine transport inhibitor dipyridamole (Persantine^{TM}) resulted in neuronal death and increased glutamate release (Lobner and Choi, 1994). Moreover, although *in vivo* administration of this drug has not been attempted, preischemic treatment may be ineffective since dipyridamole penetrates poorly through the blood-brain barrier (BBB) (Sollevi, 1986). Furthermore, since *in vitro* studies indicate that dipyridamole is not a very effective uptake inhibitor (Rudolphi et al., 1992), these results may explain the lack of the protective effect in hypoxic neuronal culture, and may also indicate that postischemic *in vivo* treatment with this drug, i.e., at a time when BBB opens, may be equally disappointing.

Interestingly, neuroprotective actions of L-type calcium channel blockers nimodipine and flunarizine may involve an adenosine-related component since, at clinically active concentrations, both drugs prove to be very efficient inhibitors of adenosine transport (Deckert and Gleiter, 1994).

Studies in which inhibition of the enzymatic breakdown of extracellular adenosine has been attempted by means of deoxycoformycin—an adenosine deaminase inhibitor—produced inconclusive results (Phillis and Clough-Helfman, 1990; Phillis et al., 1990). The effect of "allosteric enhancers" of the activity of endogenous adenosine on the outcome of cerebral ischemia is yet to be determined.

A_1 Receptor Agonists

Neurochemical and morphological correlates of protection against hypoxic/ischemic injury by adenosine A_1 receptors have been described using *in vitro* models (Goldberg et al., 1988). Several recent reviews provide details of experimental *in vivo* therapies (von Lubitz and Marangos, 1990; Phillis, 1990; Rudolphi et al., 1992; Miller and Hsu, 1992; von Lubitz et al., 1995a; Rudolphi and Schubert, 1996). Both focal and global ischemia have been studied, with the duration of insults varying between 5 and 30 minutes. Drugs were administered either prior to, or shortly after, cerebrocirculatory arrest, and several measures were employed in determination of the outcome, e.g., mortality, neuronal loss in selectively vulnerable regions, or the extent of neurological dysfunction. Overwhelmingly, the reports indicated a significant amelioration of all studied parameters (Rudolphi et al., 1992; Miller and Hsu, 1992; Dudolphi and Schubert, 1996). Very recent results (von Lubitz et al., 1996b,c) indicate that either pre- or postischemic administration of very low doses of the selective A_1 receptor agonist ADAC (adenosine amine congener) results in prevention of postischemic mortality, a very high degree of neuroprotection, and a virtually complete elimination of postischemic memory and learning deficits. While ADAC reduces glutamate release from gerbil cortical slices in a dose-dependent manner (Boyd et al., 1996), other mechanisms are probably involved as well, as evidenced by the fact that postischemic treatment with ADAC protected against neuronal damage and memory loss, even when instituted as late as 12 hours postischemia (von Lubitz et al., 1996b). Furthermore, a significant reduction in degeneration of microtubule associated protein 2 (MAP-2) was evident when ADAC was given up to 18 hours following 10-minute bilateral carotid occlusion in gerbils. The finding that ADAC is neuroprotective when the treatment is instituted at such late stages after ischemia conflicts with the notion that the "time window" for A_1 receptor therapy is rather brief (Rudolphi and Schubert, 1996; and above). It does, however, support the emerging notion that the interplay between the effects elicited by stimulation of A_1 receptors located on different cell types (i.e., neurons, glia) may be exceedingly complex and far from fully elucidated (von Lubitz et al., 1995a; Rudolphi and Schubert, 1996). The finding that A_1 receptor agonist treatment also causes an impressive reduction of neuronal injury in the hyperglycemic global ischemia model in rats (Hsu et al., 1995) emphasizes the powerful nature of this form of therapy.

The effectiveness of acute adenosine A_1 receptor stimulation in preventing neuronal damage has been stressed by several experiments in which both selective (von Lubitz et al., 1994a) and nonselective (Rudolphi et al., 1987) A_1 receptor antagonists have been used. Invariably, treatment with these agents resulted in a significant worsening of postischemic necrosis (Rudolphi et al., 1987) and mortality (von Lubitz et al., 1994a). The effects seen with chronic stimulation of A_1 receptors are highly inconsistent and the results may depend on several factors, e.g., the applied drug, duration of treatment, or both. The problems related to chronic treatment with adenosine receptor–acting drugs are the subject of a recent exhaustive review (Jacobson et al., 1996).

Stimulation of Adenosine A_2 and A_3 Receptors

Very few reports describe the effect of adenosine A_2 receptor manipulation on the consequences of cerebral ischemia (Gao and Phillis, 1994; Phillis, 1995; von Lubitz et al., 1995b). Surprisingly, although stimulation of A_2 receptors improves postischemic cerebral blood perfusion (von Lubitz et al., 1995b), treatment with A_2 receptor agonists has no effect on either survival or neuronal preservation (Gao and Phillis, 1994; von Lubitz et al., 1995b). Significant improvement of postischemic morphology is, however, obtained following administration of A_2 receptors antagonists (Gao and Phillis, 1994; Phillis, 1995; von Lubitz et al., 1995c). The effect is most likely due to the inhibitory effect of these drugs on excitatory amino acid release.

Acute stimulation of A_3 receptors results in a very significant worsening of the outcome of 10-minute ischemia in gerbils (von Lubitz et al., 1994b), while chronic treatment with A_3 receptor agonist reduces both hippocampal damage and postischemic mortality (von Lubitz et al., 1994b). Moreover, a preliminary report indicates postischemic protection of cognitive functions following chronic treatment with the same A_3 receptor agonist (von Lubitz et al., 1996b). Since acute stimulation of A_3 receptors leads to a pronounced delay in normalization of the postischemic blood flow, it is unknown whether the effect of both acute and chronic preischemic activation is purely neuronal or involves cerebrovascular events as well (von Lubitz et al., 1994b, 1995b).

Side Effects of Adenosine-Based Therapies

Hypothermia, hypotension, and bradycardia are cited most frequently as the side effects of greatest concern (von Lubitz and Marangos, 1990; Rudolphi et al., 1992; Miller and Hsu, 1992).

It has been postulated that protection obtained through adenosine A_1 receptor stimulation is chiefly due to severe hypothermia induced by A_1 receptor agonists. However, similar protection has been also obtained in *in vitro* experiments (Goldberg et al., 1988), and in the experiments where normothermic environment has been carefully maintained (von Lubitz et al., 1994). Irrespective of the "purist" view on the nature of brain sparing by A_1 agonists, it must be reemphasized that, since mild hypothermia has been shown to be independently protective (Busto et al., 1989), the depression of body and brain temperature induced by A_1 agonists is most likely the consequence of tissue metabolism reduction induced by A_1 receptor stimulation. Thus, the "adverse effect" may, as a matter of fact, constitute but one of several aspects of the neuroprotective spectrum evoked by the activated A_1 receptors.

Hypertension is a frequent companion of recovery after stroke. It is, however, a matter of debate whether lowering blood pressure may lead to a clinically beneficial outcome (Powers, 1993; Donnan et al., 1994), and since postischemic reduction of blood pressure can lead to further progression of injury (Powers, 1993), the hypotensive effect of adenosine A_1 agonists may constitute a significant drawback.

Recent studies have demonstrated that both bradycardiac and hypotensive actions of A_1 receptor agonists can be reduced. A Danish company, Novo Nordisk, introduced a series of novel selective A_1 agonists that possess diminished cardiovascular side effects but retain their neuroprotective and anticonvulsant potency (Knutsen et al., 1995; Sheardown et al., 1994). ADAC, when given at the therapeutic doses of 20–100 μg/kg, is equally free of adverse effects of either cardiac rate or blood pressure although, at higher amounts, both hypotension and bradycardia are manifested (von Lubitz et al., 1996b, and unpublished results). Due to the very rapid progress in the medicinal chemistry of A_1 receptor–active drugs, its can be expected that a number of similar agents will become available in the near future.

Therapeutic Window

In the hippocampus, the strength of adenosine modulation of neural functions appears to be directly related to the density of the adenosine A_1 receptor population (Lee et al., 1983). It is likely that a similar relationship exists in other brain regions as well. Experimental results indicate that the neuroprotective effects of A_1 receptor agonists disappear approximately 18–24 hours after ischemia (Rudolphi and Schubert, 1996; von Lubitz et al., 1996b). However, in view of the rapid postischemic deterioration of A_1 receptor density, the feasibility of a clinically efficient treatment with A_1 agonists appears to diminish at times later than 3–6 hours after cerebrocirculatory arrest (von Lubitz et al., 1996b). With the treatment administered at later times, some degree of protection may be expected, although its scale will be significantly reduced (von Lubitz et al., 1996b).

CONCLUSION

With the exception of propentophylline treatment the effects of therapies based on the implementation of endogenous adenosine, either through the enhancement of its release or depression of its breakdown rate are inconsistent. Moreover, since phase III clinical trials of the adenosine-releasing agent aicariboside (Protara™) resulted in a statistically insignificant reduction of myocardial infarctions, it is possible that the clinical application of this class of drugs may also be equally ineffective in the treatment of cerebral ischemia/stroke.

The effects of stimulation or antagonism of A_2 and A_3 receptors require further experimental studies. However, both pre- and postischemic treatment with highly selective A_1 receptor agonists results in a consistent enhancement of recovery following even protracted episodes of experimental global ischemia and stroke. Moreover, while hypotensive and bradycardiac side effects of several hitherto tested A_1 receptor agonists represent a potential source of concern, new drugs with a much lower profile of cardiovascular activity offer solutions to these problems (Knutsen et al., 1995; von Lubitz et al., 1996b). Ultimately, it must not be forgotten that the medicinal chemistry of agents acting with a very high selectivity at any of the three adenosine receptors subtypes is in its most vigorous phase ever (Sidiqqi et al., 1995; Jacobson et al., 1995). Hence, the concentrated effort of basic and clinical neurology and of the pharmaceutical industry may open a rational path leading to a viable clinical treatment of cerebral ischemia and stroke based upon drugs acting as adenosine receptors.

REFERENCES

Abbracchio MP, Brambilla R, Ceruti S, Kim HO, von Lubitz DKJE, Jacobson KA, Cattabeni F (1995a): G protein-dependent activation of phospholipases C by adenosine A3 receptors in rat brain. Molec Pharmacol 48:1038–1045.

Abbracchio MP, Ceruti, Burnstock G, Cattabeni F (1995b): Purinoceptors on glial cells of the central nervous system: Functional and pathologic implications. In Belardinelli L, Pelleg A (eds): "Adenosine and Adenine Nucleotides: From Molecular Biology to Integrative Physiology." Boston: Kluwer Academic, 271–280.

Anderson KJ, Nellgård B, Wieloch T (1993): Ischemia-induced upregulation of excitatory amino acid transport sites. Brain Res 622:93–98.

Andiné P, Orwar O, Jacobson I, Sandberg M, Hagberg H (1991): Changes in extracellular amino acids and spontaneous neuronal activity during ischemia and extended reflow in the CA1 of the rat hippocampus. J Neurochem 57:222–229.

Araki T, Kato H, Kogure K, Kanai Y (1992): Long term changes in gerbil brain neurotransmitter receptors following transient cerebral ischemia. Br J Pharmacol 107:437–442.

Banati RB, Gehrmann J, Schubert P, Kreutzberg GW (1993): Cytotoxicity of microglia. Glia 7:111–118.

Brambilla R, Abbracchio MP, Ceruti S, Franceschi C, Malorni W, von Lubitz DKJE, Jacobson KA, Lohse M, Klotz KN, Cattabeni F (1996): Morphological effects induced by adenosine A3 receptor agonists on CHO cells transfected with the human A3 receptor. Abstract, Purines 96, International Symposium, Milan. Drug Dev Res 3:181.

Boyd M, Meshulam Y, Jacobson KA, von Lubitz DKJE (1996): Effect of A1 adenosine receptor agonist and antagonist on glutamic acid release from cortical tissue in vitro. FASEB J 10:A159.

Burnstock G (1986): Purines as cotransmitters in adrenergic and cholinergic neurones. Prog Brain Res 68:193–203.

Busto R, Dietrich WD, Globus MYT, Ginsberg MD (1989): The importance of brain temperature in cerebral ischemic injury. Stroke 20:1113–1114.

Cantor SL, Zornow MH, Miller LP, Yaksh TL (1992): The effect of cyclohexyladenosine on the periischemic increases of hippocampal glutamate and glycine in the rabbit. J Neurochem 59:1884–1892.

Cassabona G, L'Episcopo MR, Di Iorio P, Shinozaki H, Nicoletti F, Caciagli F (1994): Interaction between metabotropic receptors and purinergic transmission in rat hippocampal slices. Brain Res 645:13–18.

Cepeda N, Buchwald A, Levine MS (1993): Neuromodulatory actions of dopamine in the neostriatum are dependent upon the excitatory amino acid receptor subtypes activated. Proc Natl Acad Sci USA 90:9576–9580.

Ceruti S, Barbieri D, Franceschi C, Giammarioli AM, Malorni W, Kim HO, von Lutbitz DKJE, Jacobson KA, Cattabeni F, Abbracchio MP (1996): Effect of adenosine A3 receptor agonists on astrocytes: induction of cell protection at low and cell death at high concentrations. Abstract Purines 96, Milan. Drug Dev Res 3:177.

Chesselet MF, Gonzales C, Lin C-S, Polsky K, Jin B-K (1990): Ischemic damage in the striatum of adult gerbils: Relative sparing of somatostatinergic and cholinergic inter neurons contrasts with loss of efferent neurons. Exp Neurol 110:209–218.

Clough-Helfman C, Phillis JW (1990): 5-aminoimidazole-4-carboxamide riboside (AICAr) administration reduces cerebral ischemia damage in Mongolian gerbil. Brain Res Bull 25:203–206.

Cotman CW, Monaghan DT, Ottersen OP, Storm-Mathiesen J (1987): Anatomical organization of excitatory amino acid receptors and their pathways. Trends Neurosci 10:273–280.

Craig CG, White TD (1992): Low-level N-methyl-D-aspartate receptor activation provides a purinergic inhibitory threshold against further n-methyl-D-aspartate mediated neurotransmission. J Pharmacol Exp Ther 260:1278–1284.

Cronstein BN, Levin RI, Belanoff J, Weissmann G, Hirschhorn R (1986): Adenosine: An endogenous inhibitor of neurtrophil-mediated injury to endothelial cells. J Clin Invest 78:760–770.

Daly JW, McNeal E, Partington C, Neuwirth M, Creveling CR (1980): Accumulations of cyclic AMP in adenine-labeled cell-free preparations from Guinea pig cerebral cortex: Role of α-adrenergic and H_1-histaminergic receptors. J Neurosci 35:326–337.

Danielisova V, Chavko M, Schubert PH (1994): Effect of propentofylline (HWA 285) on metabolic and functional recovery in the spinal cord after ischemia. Neuropharmacology 2:199–204.

Daval J-L, von Lubitz DKJE, Deckert J, Redmond DJ, Marangos PJ (1989): Protective effect of cyclohexyladenosine on adenosine A_1 receptors, guanine nucleotide and forskolim binding sites following transient brain ischemia: A quantitative autoradiographic study. Brain Res 491:212–226.

Deckert J, Gleiter CH (1994): Adenosine—An endogenous neuroprotective metabolite and neuromodulator. J Neural Transm Suppl 43:23–31.

DeLeo J, Toth L, Schubert P, Rudolphi K, Kreutzberg GW (1987): Ischemia-induced neuronal cell death, calcium accumulation, and glial response in the hippocampus of the Mongolian gerbil and protection by propentofylline (HWA 285). J Cereb Blood Flow Metab 7:745–752.

DeLeo J, Schubert P, Kreutzberg CW (1988): Propentofylline (HWA 285) protects hippocampal neurons of Mongolian gerbils against ischemic damage in the presence of an adenosine antagonist. Neurosci Lett 84:307–311.

Dianzani C, Brunelleschi S, Viano I, Fantozzi R (1994): Adenosine modulation of primed human neutrophils. Eur J Pharmacol 263:223–226.

Diemer NH, von Lubitz DKJE (1982): Cerebral ischemia in rat: Increased permeability of postsynaptic membranes to horseradish peroxidase in the early postischemic period. Neuropathol Appl Neurobiol 8:197–215.

Dietrich WD, Miller LP, Prado R et al., (1995): Acadesine reduces indium-labeled platelet desposition after photothrombosis of the common carotid artery in rats. Stroke 26:111–116.

Donnan GA, Thrift A, You RX, McNeil JJ (1994): Hypertension and stroke. J Hypertension 12:865–869.

Dunwiddie TV (1980): Endogenously released adenosine regulates excitability in the in vitro hippocampus. Epilepsia 21:541–548.

Dunwiddie TV (1985): The physiological role of adenosine in the central nervous system. Int Rev Neurobiol 27:64–139.

Dunwiddie TV, Lynch G (1978): Long term potentiation and depression of synaptic responses in the hippocampus. Localization and frequency dependency. J Physiol 276:353–367.

Engler R (1987): Consequences of activation and adenosine-mediated inhibition of granulocytes during myocardial ischemia. Fed Proc 46:2407–2412.

Ferré S, Fuxe K, von Euler G, Johansson B, Fredholm BB (1992): Adenosine-dopamine interactions in the brain. Neuroscience 3:501–512.

Ferré S, Snaprud P, Fuxe K (1993): Opposing actions of an adenosine A_2 and a GTP analogue on the regulation of dopamine D_2 receptors in rat neostriatal membranes. Eur J Pharmacol (Mol Pharm Sect) 244:311–315.

Fozard JR, Carruthers AM (1993): Adenosine A3 receptors mediate hypotension in the angiotensin II-supported circulation of the pithed rat. Br J Pharmacol 109:3.

Fozard JR, Pfannkuche HJ, Schuurman HJ (1996): Mast cell degranulation following adenosine A3 receptor activation in rats. Eur J Pharmacol 298:293–297.

Fredholm BB, Dunwiddie TV (1988): How does adenosine inhibit transmitter release? Trends Pharmacol Sci 9:130–134.

Fredholm BB, Lindstrom K (1986): The xanthine derivative 1-(5′-oxohexyl)-3-methyl-7-propylxanthine (HWA 285) enhances the action of adenosine. Acta Pharmacol Toxicol 58:187–192.

Fujimoto T, Suzuki H, Tanoue K, Fukushima Y, Yamazaki H (1985): Cerebrovascular injuries and brain edema following activation of platelets. In Inaba Y, Klatzo I, Spatz M (eds): "Brain Edema." Tokyo: Springer Verlag, pp 310–316.

Fujiwara N, Higashi H, Shimoji K, Yoshimura M (1987): Effects of hypoxia on rat hippocampal neurones in vitro. J Physiol (London) 384:131–151.

Gao Y, Phillis JW (1994): CGS 15943, an adenosine A_2 receptor antagonist, reduces cerebral ischemic injury in the Mongolian gerbil. Life Sci 55:61–65.

Genazzani AA, Casabona G, L'Episcope MR, Condorelli DF, Dell'Albani P, Shinozaki H, Nicoletti F (1993): Characterization of metabotropic glutamate receptors negatively linked to adenylyl cyclase in brain slices. Brain Res 622:132–138.

Globus MY-T, Martines E, Valdes I, Busto R (1990): The comparison between ischemia induced release of neurotransmitters in vulnerable and non-vulnerable brain regions. In Krieglstein J, Oberpichler H (eds): "Pharmacology of Cerebral Ischemia." Stuttgart: Wissenschaftliche Verlagsgesselschaft, pp 197–203.

Goldberg MP, Monyer H, Weiss JH, Choi DW (1988): Adenosine reduces cortical neuronal injury induced by oxygen and glucose deprivation in vitro. Neurosci Lett 89:323–327.

Greene RW, Haas HL (1991): The electrophysiology of adenosine in the mammalian central nervous system. Progr Neurobiol 36:329–341.

Gruber HE, Hoffer ME, McAllister DR, Laikind PK, Lane TA, Schmid-Schoenbein GW, Englrer R (1989): Increased adenosine concentration in blood from ischemic myocardium by AICA riboside. Circulation 80:1400–1411.

Hagberg H, Andersson P, Lazarewicz J, Jacobson I, Butcher S, Sandberg M (1987): Extracellular adenosine, inosine, hypoxanthine, and xanthine in relation tissue nucleotides and purines in rat striatum during transient ischemia. J Neurochem 49:227–231.

Hagberg H, Andersoon P, Ostwald C et al. (1986): Ischemia evoked release of neuroactive compounds and acute effect of N-methyl-D-aspartate receptor blockade. In Krieglstein J (ed): "Pharmacology of Cerebral Ischemia." Amsterdam: Elsevier Science, pp 298–303.

Hagberg H, Andiné P, Fredholm BB, Rudolphi (1990): Effect of the adenosine uptake inhibitor propentophylline on extracellular adenosine and glutamate and evaluation of its neuroprotective efficacy after ischemia in neonatal and adult rats. In Krieglstein J, Oberpichler H (eds):

"Pharmacology of Cerebral Ischemia." Stuttgart: Wissenschaftliche Verlagsgesselschaft, pp 427–437.

Hansen AJ, Hounsgaard J, Jahnsen H (1982): Anoxia increases potassium conductance in hippocampal nerve cells. Acta Physiol Scand 115:301–310.

Hansson E, Rönnbäck L (1991): Receptor regulation of the glutamate, GABA and taurine high-affinity uptake into astrocytes in primary culture. Brain Res 548:215–221.

Herron CE, Lester RJ, Coan EJ, Collingridge GL (1986): Frequency-dependent involvement of NMDA receptors in the hippocampus: A novel synaptic mechanism. Nature 322:265–268.

Hoehn K, White TD (1990a): N-menthyl-D-aspartate, kainate and quisqualate release endogenous adenosine from rat cortical slices. Neuroscience 39:441–450.

Hoehn K, White TD (1990b): Role of excitatory amino acid receptors in K^+ and glutamate-evoked release of endogenous adenosine from rat cortical slices. J Neurochem 54:256–265.

Hori M, Tamai J, Kitakaze M, et al. (1989): Adenosine-induced hyperemia attenuates myocardial ischemia in coronary microembolization in dogs. Am J Physiol 257:H244–251.

Hsu SS-F, Meno J, Zhou J-G, Gordon R, Winn HR (1995): Hyperglycemic ischemia and reperfusion: Effects on adenosine and adenine nucleotides. In Belardinelli L, Pelleg A (eds): "Adenosine and Adenine Nucleotides: From Molecular Biology to Integrative Physiology." Boston: Kluwer Academic Press, pp 399–411.

Hugenard JR, Alger BE (1986): Whole-cell voltage clamp of the fading of GABA activated currents in acutely dissociated hippocampal neurons. J Neurophysiol 56:1–18.

Jacobson KA, van Galen PJM, Williams M (1992): Perspective, adenosine receptors: Pharmacology, structure activity relationships, and therapeutic potential. J Med Chem 35:407–422.

Jacobson KA, Kim HO, Sidiqqi SM, Olah ME, Stiles GL, von Lubitz DKJE (1995): The A3-adenosine receptor: design of selective ligands and therapeutic prospects. Drugs of the Future 20:689–699.

Jacobson KA, von Lubitz DKJE, Daly JW, Fredholm BB (1996): Adenosine receptor ligands: Differences with acute versus chronic treatment. Trends Physiol Sci 17:108–113.

Jarvis MF, Williams M (1989): Adenosine in central nervous system function. In Williams M (ed): "Adenosine and Adenosine Receptors." Clifton: Humana Press, pp 423–474.

Ji X-D, von Lubitz D, Olah ME, Stiles GL, Jacobson KA (1994): Species differences in ligand affinity at central A3 adenosine receptors. Drug Dev Res 33:51–59.

Kalaria RN, Harik SI (1986): Adenosine receptors of cerebral microvessels and choroid plexus. J Cereb Blood Flow Metab 6:463–470.

Knutsen LJS, Lau J, Sherardown M, Eskesen K, Thomsen C, Weis JU, Judge ME, Klitgaard H (1995): Anticonvulsant actions of novel and reference adenosine agonists. In Belardinelli L, Pelleg A (eds): "Adenosine and Adenine Nucleotides." Philadelphia: Kluwer Academic, pp 479–487.

Kohno Y, Sei Y, Koshiba M, Kim HO, Jacobson KA (1996): Induction of apoptosis in HL-60 human promyelocytic leukemia cells by adenosine A3 receptor agonists. Biochem Biophys Res Commun 219:904–910.

Kostopoulos GK, Phillis JW (1977): Purinergic depression of neurons in different areas of the brain. Exp Neurol 55:719–724.

Leão AAP (1947): Further observations on the spreading depression of activity in the cerebral cortex. J Neurophysiol 10:409.

Leblond J, Krnjević K (1989): Hypoxic changes in hippocampal neurons. J Neurophysiol 62:1–14.

Lee KS, Lowenkopf T (1993): Endogenous adenosine delays the onset of hypoxic depolarization in the rat hippocampus in vitro via an action at A_1 receptors. Brain Res 609:313–315.

Lee KS, Reddington M, Schubert P, Kreutzberg GW (1983): Regulation of the strength of adenosine modulation in the hippocampus by a differential distribution of the density of A1 receptors. Brain Res 260:156–159.

Lee KS, Tetzlaffand W, Kreutzberg GW (1986): Rapid downregulation of hippocampal adenosine receptors following brief anoxia. Brain Res 380:155–158.

Leung J, Stanley T, Matthew J, et al. (1992): Effects of acadesine on perioperative cardiac morbidity in a placebo controlled blind study. J Am Coll Cardiol 19:112A.

Linden J (1994): Cloned adenosine A_3 receptors: Pharmacological properties, species differences and receptor functions. Trends Pharmacol Sci 15:298–306.

Lobner D, Choi DW (1994): Dipyridamole increases oxygen-glucose deprivation-induced injury in cortical cell culture. Stroke 25:2085.

Madison DV, Fox AP, Tsien RW (1987): Adenosine reduces an inactivating component of calcium current in hippocampal CA3. Proc Biophys J 51:30A. Abstract.

Mager R, Ferroni S, Schubert P (1990): Adenosine modulates a voltage-dependent chloride conductance in cultured hippocampal neurons. Brain Res 532:58–62.

Manzoni OJ, Manabe T, Nicoll RA (1994): Release of adenosine by activation of NMDA receptors in the hippocampus. Science 265:2098–2101.

Matsumoto K, Graf R, Rosner G, Schimada N, Weiss WD (1992): Flow thresholds for extracellular purine catabolite elevation in rat focal ischemia. Brain Res 579:309–314.

Mayfield RD, Suzuki F, Zahniser N (1993): Adenosine A_{2a} receptor modulation of electrically evoked endogenous GABA release from slices of rat globus pallidus. J Neurochem 60:2334–2337.

Mendonça, A, Sebastião AM, Ribeiro AJ (1995): Inhibition of NMDA receptor-mediated currents in isolated rat hippocampal neurones by adenosine A1 receptor activation. Neuroreport 6:1097–1100.

Miller LP, Hsu Ch (1992): Therapeutic potential for adenosine receptor activation in ischemic brain injury. J Neurotrauma 9(suppl 2):S563–577.

Morri S, Ngai AC, Ko KR, Winn HR (1987): Role of adenosine in regulation of cerebral blood flow: Effect of theophylline during normoxia and hypoxia. Am J Physiol 253:H165–175.

Nagasawa H, Araki T, Kogure K (1994): Alteration of adenosine A1 binding in the postischemic brain. Neuroreport 5:1453–1456.

National Foundation for Brain Research (1992): "The Cost of Disorders of the Brain." Washington, DC.

Nishizuka Y (1986): Studies and perspectives of protein kinase C. Science 233:305–312.

Novak L, Bregestovski P, Asher P, Herbert A, Prociants A (1984): Magnesium gates glutamate activated channels in mouse central neurones. Nature 307:462–465.

Obrenovitch TP, Gurcharan SS, Symon L (1990): Ionic homeostasis and neurotransmitter changes. In Krieglstein J, Oberpichler H (eds): "Pharmacology of Cerebral Ischemia." Stuttgart: Wiesenschaftliche Verlagsgesselschaft, pp 97–112.

Ogata T, Nakamura Y, Tsuji K, Kataoka K, Schubert P (1994): Adenosine enhances intracellular Ca^{2+} mobilization in conjunction with metabotropic glutamate receptor activation by t-ACPD in cultured hippocampal astrocytes. Neurosci Lett 170:5–8.

Onodera H, Kogure K (1985): Autoradiographic visualization of adenosine A1 receptors in the gerbil brain hippocampus: changes in the receptor density after transient ischemia. Brain Res 345:406–408.

Onodera H, Kogure K (1990): Differential localization of adenosine A_1 receptors in the rat hippocampus: Quantitative autoradiographic study. Brain Res 458:212–217.

Onodera H, Sato G, Kogure K (1989): Quantitative autoradiographic analysis of muscarinic, cholinergic, and adenosine A1 binding sites after transient forebrain ischemia in the gerbil. Brain Res 415:309–322.

O'Regan MH, Simpson RE, Perkins LM, Phillis JW (1992): The selective adenosine A_2 receptor agonist CGS 21680 enhances excitatory transmitter amino acid release from the ischemic rat cerebral cortex. Neurosci Lett 138:169–172.

Pazzagli M, Corsi C, Latini S, Pedata F, Pepeu G (1984): In vivo regulation of extracellular adenosine levels in the cerebral cortex by NMDA and muscarinic receptors. Eur J Pharmacol 254:277–282.

Petito CK (1976): Transformation of postischemic perineuronal glial cells. In Raichle ME, Powers WJ (eds): "Cerebrovascular Diseases." New York: Raven Press, pp 103–106.

Petito CK, Babiak T (1982): Early proliferative changes in astrocytes in postischemic noninfarcted rat brain. Ann Neurol 11:510–518.

Petito CK, Halaby IA (1993): Relationship between ischemia and ischemic neuronal necrosis to astrocyte expression of glial fibrillary acidic protein. In J Dev Neurosci 2:239–247.

Petito CK, Morgello S, Felix JC, Lesser ML (1990): The two patterns of reactive astrocytosis in postischemic rat brain. J Cereb Blood Flow Metab 10:850–859.

Phillis JW (1989): Adenosine in the control of cerebral circulation. Cerebrovasc Brain Metab Rev 1:26–54.

Phillis JW (1990): Adenosine, inosine, and oxypurines in cerebral ischemia. In Schurr A, Rigor BM (eds): "Cerebral Ischemia and Resuscitation." Boca Raton, FL: CRC Press, pp 189–204.

Phillis JW (1995): The effects of selective A1 and A2a adenosine receptor antagonists on cerebral ischemic injury in the gerbil. Brain Res 705:79–84.

Phillis JW, Clough Helfman C (1990): Oxypurinol, but not deoxycoformycin, administered post-ischemia, protects against CA1 hippocampal damage in the gerbil. Int J Purine Pyrimidine Res 1:31–35.

Phillis JW, Wu PH (1981): The role of adenosine and its nucleotides in central synaptic transmission. Prog Neurobiol 16:187–239.

Phillis JW, Simpson RE, Walter GA (1990): Brain adenosine and transmitter amino acid release from the ischemic rat cerebral cortex: Effects of adenosine deaminase inhibitor deoxycoformycin. Brain Res 524:336–338.

Phillis JW, Walter GA, O'Regan MH, Stair R (1987): Increases in cerebral cortical perfusate adenosine and inosine during hypoxia and ischemia. J Cereb Blood Flow Metab 7:679–686.

Phillis JW, Walter GA, Simpson RE (1991): Brain adenosine and transmitter amino acid release from the ischemic rat cerebral cortex: Effects of the adenosine deaminase inhibitor deoxycoformycin. J Neurochem 56:644–650.

Poli A, Lucchi R, Vibio M, Barnabei M (1991): Adenosine and glutamate modulate each other's release from rat hippocampal synaptosomes. J Neurochem 57:298–306.

Powers WJ (1993): Acute hypertension after stroke: The scientific basis for treatment decisions. Neurology 43:461–467.

Prado R, Busto R, Globus MYT (1992): Ischemia-induced changes in extracellular levels of striatal cyclic AMP: Role of dopamine neurotransmission. J Neurochem 59:1581–1584.

Ramkumar V, Stiles GL, Beaven MA, Ali H (1993): The A_3 adenosine receptor is the unique adenosine receptor which facilitates release of allergic mediators in mast cells. J Biol Chem 268:16887–16890.

Rudolphi K, Schubert P (1996): Purinergic interventions in traumatic and ischemic injury. In Peterson PL, Phillis JW (eds): "Novel Therapies for CNS Injuries: Rationales and Results." Boca Raton, FL: CRC Press, pp 327–337.

Rudolphi KA, Keil M, Fastbom J, Fredholm BB (1989): Ischaemic damage in gerbil hippocampus is reduced following upregulation of adenosine A_1 receptors by caffeine treatment. Neurosci Lett 103:275–280.

Rudolphi KA, Keil M, Hinze HJ (1987): Effect of theophylline on ischemically induced hippocampal damage in Mongolian gerbils: A behavioural and histopathological study. J Cereb Blood Flow Metab 7:74–81.

Rudolphi KA, Schubert P, Parkinson FE, Fredholm BB (1992): Adenosine and brain ischemia. Cerebrovasc Brain Metab Rev 4:346–369.

Sajjadi FG, Takabayashi K, Foster AC, Domingo AC, Firestein GS (1996): Inhibition of TNF-alpha expression by adenosine. J Immunol 9:3435–3442.

Sandoli D, Chiu PJS, Chintala M, Dionisotti S, Ongini E (1994): In vivo and ex vivo effects of adenosine A1 and A2 receptor agonists on platelet aggregation in the rabbit. Eur J Pharmacol 259:43–49.

Schiffmann SN, Halleux P, Menu R, Vanderhaeghen J-J (1993): Adenosine A_{2a} receptor expression in striatal neurons: Implications for basal ganglia pathophysiology. Drug Dev Res 28:381–385.

Schmidt W, Wolf G, Grüngreiff K, Linke K (1993): Adenosine influences the high affinity uptake of transmitter glutamate and aspartate under conditions of hepatic encephalopathy. Metab Brain Dis 8:73–80.

Schubert P, Dux E (1990): Selective neuronal death in cerebral ischemia and protective mechanisms. In Schurr A, Rigor BM (eds): "Cerebral Ischemia and Resuscitation." Boca Raton, FL: CRC Press, pp 259–269.

Schubert P, Kreutzberg GW (1990): Neuroprotective mechanisms of endogenous adenosine action and pharmacological implications. In Krieglstein J, Oberpichler H (eds): "Pharmacology of Cerebral Ischemia." Stuttgart: Wissenschaftliche Verlagsgesselschaft, pp 417–426.

Schubert P, Mager R (1991): The critical input frequency for NMDA-mediated neuronal Ca^{2+} frequency depends on endogenous adenosine. Int J Purine Pyrimidine Res 2:11–16.

Schubert P, Ferroni S, Mager R (1990): Pharmacological blockade of Cl^- pumps or Cl^- channels reduces the adenosine mediated depression of stimulation train-evoked Ca^{2+} fluxes in rat hippocampal slices. Neurosci Lett 124:174–177.

Schubert P, Heinemann U, Kolb R (1986): Differential effects of adenosine on pre- and postsynaptic calcium fluxes. Brain Res 376:382–386.

Schubert P, Keller F, Rudolphi KA (1993): Depression of synaptic transmission and evoked NMDA Ca^{2+}influx in hippocampal neurons by adenosine and its blockade by LTP or ischemia. Drug Dev Res 28:399–405.

Schubert P, Lee K, West M, Deadwhyler S, Lynch G (1976): Stimulation-dependent release of ^{3}H-adenosine derivatives from central axon terminals to target neurones. Nature 260:541–542.

Schubert P, Pintor J, Miras-Portugal MT (1995): Inhibitory action of adenosine and adenine dinucleotides on synaptic transmission in the central nervous system. In Belardinelli L, Pelleg A (eds): "Adenosine and Adenine Nucleotides: From Molecular Biology to Integrative Physiology." Boston: Kluwer Scientific, pp 281–288.

Schubert P, Rudolphi K, Fredholm F, Nakamura Y (1994): Modulation of nerve and glial cell function by adenosine—role in development of ischemic brain damage. Int J Biochem 26:1227–1236.

Sciotti VM, Roche FM, Grabb MC, van Wylen DGL (1992): Adenosine receptor blockade arguments interstitial fluid levels of excitatory amino acids during cerebral ischemia. J Cereb Blood Flow Metab 12:646–655.

Sebastião AM, Ribeiro JA (1992): Evidence for the presence of A_2 adenosine receptors in the rat hippocampus. Neurosci Lett 138: 41–44.

Segal M, Rogawski MA, Barker JL (1984): A transient potassium conductance regulates the excitability of cultured hippocampal and spinal neurons. Brain Res 4;604–609.

Sheardown M, Hansen AJ, Thomsen C, Judge ME, Knutsen LJS (1994): Novel adenosine agonists: A strategy for stroke therapy. In Grotta J, Miller LP, Buchan AM, Sussman C (eds): "Ischemic Stroke: Recent Advances in Understanding and Therapy." Southborough, MA: pp 187–214.

Si QS, Nakamura Y, Schubert P, Rudolphi K, Kataoka K (1996): Adenosine and propentofylline inhibit proliferation of cultured microglial cells. Exp Neurobiol 137:345–349.

Sidiqqi SM, Jacobson KA, Esker JL, et al. (1995): Search for new purine- and ribose-modified adenosine analogues as selective agonists and antagonists at adenosine receptors. J Med Chem 1174–1188.

Siesjo BK (1988): Historical overview: Calcium, ischemia, and death of brain cells, Ann NY Acad Sci 522:638–661.

Simpson RE, O'Reagan MH, Perkins LM, Phillis JW (1992): Excitatory transmitter amino acid release from the ischemic rat cerebral cortex. Effects of adenosine agonists and antagonists. J Neurochem 58:1683–1690.

Sollevi A (1986): Cardiovascular effects of adenosine in man: Possible clinical implications. Prog Neurobiol 26:319–349.

Sorg O, Magistretti PJ (1992): Vasoactive intestinal peptide and noradrenaline exert long-term control on glycogen levels in astrocytes: Blockade by protein synthesis inhibition. J Neurosci 12:4923–4931.

Suzuki R, Yamaguchi T, Choh-luh L, Klatzo I (1983): The effects of 5 minute ischemia in Mongolian gerbils, II. Changes of spontaneous neuronal activity in cerebral cortex and CA1 sector of hippocampus. Acta Neuropathol (Berl) 60:217–222.

Tominaga K, Shibata S, Watanabe S (1992): A neuroprotective effect of adenosine A1-receptor agonists on ischemia-induced decrease in 2-deoxyglucose uptake in rat hippocampal slices. Neurosci Lett 145:67–70.

Trussel LO, Jackson BJ (1985): Adenosine-activated potassium conductance in cultured striatal neurons. Proc Natl Acad Sci USA 82:4857–4861.

van Calker D, Müller M, Hamprecht G (1979): Adenosine regulates via two different types of receptors the accumulation of cyclic AMP in cultured brain cells. J Neurochem 33:999–1005.

Van Galen PJM, Stiles GL, Michaels G, Jacobson KA (1992): Adenosine A1 and A2 receptors: Structure-function relationship. Med Res Rev 5:423–471.

von Lubitz DKJE, Diemer NH (1982): Ultrastructural morphometry of hippocampal CA1 region in rats with cerebral ischemia. Polish Neuropathol suppl 1.

von Lubitz DKJE, Marangos PJ (1990): Self-defense of the brain: Adenosinergic strategies. In Marangos PJ, Lal H, (eds): "Emerging Strategies in Neuroprotection." Boston: Birkhauser, pp 151–186.

von Lubitz DKJE, Beenhakker M, Lin RCS, Carter MF, Paul IA, Bischofberger N, Jacobson KA (1996c): Reduction of postischemic brain damage and memory deficits following treatment with the selective A1 receptor agonist. Eur J Pharmacol (in press).

von Lubitz DKJE, Carter MF, Beenhakker M, Lin RC-S, Jacobson KA (1995a): Adenosine: A prototherapeutic concept in neurodegeneration. In Trembly B, Slikker B Jr, (eds): "Neuroprotective Agents. Proceedings of II International Conference." New York: Ann NY Acad Sci 765:163–178.

von Lubitz DKJE, Dambrosia JM, Kempski O (1986): Postischemic application of cyclohexyladenosine (CHA): Improvement of survival and of preservation of selectively vulnerable areas in gerbil. Stockholm X Int Congr Neuropath 108 Abstract.

von Lubitz DKJE, Deutsch Si, Carter MF, Lin RC-S, Mastropaolo J, Jacobson KA (1995b): The effects of adenosine A_3 receptor stimulation on seizures in mice. Eur J Pharmacol 275:23–29.

von Lubitz DKJE, Lin RC-S, Jacobson KA (1995c): Cerebral ischemia in gerbils: Effects of acute and chronic treatment with adenosine A2a receptor agonist and antagonist. Eur J Pharmacol 287:295–302.

von Lubitz DKJE, Lin RC-S, Melman N, Ji X-d, Carter MF, Jacobson KA (1994a): Chronic administration of adenosine A1 receptor agonist or antagonist in cerebral ischemia. Eur J Pharmacol 256:161–167.

von Lubitz DKJE, Lin RC-S, Paul LA, Beenhakker M, Boyd M, Bischofberger N, Jacobson KA (1996b): Postischemic administration of adenosine amine conger (ADAC): Analysis of recovery in gerbils. Eur J Pharmacol (in press).

von Lubitz DKJE, Lin RC-S, Popika P, Carter MF, Jacobson KA (1994b): Adenosine A_3 receptor stimulation and cerebral ischemia. Eur J Pharmacol 263:59–67.

von Lubitz DKJE, Lin RC-S, Sei Y, Boyd M, Abbracchio MP, Bischofberger N, Jacobson KA (1996a): Adenosine A3 receptors and ischemic brain injury: A hope or a disaster? Abstract Purines 96, Milan. Drug Dev Res (in press).

Whittingham TS (1990): Aspects of brain energy metabolism and cerebral ischemia. In Schurr A, Rigor BM (eds): "Cerebral Ischemia and Resuscitation." Boca Raton, FL: CRC Press, pp 101–121.

Williams M (1993): Purinergic drugs: Opportunities in the 1990s. Drug Dev Res 28:438–444.

Wu Ph, Phillis JW, Thierry DL (1982): Adenosine receptor agonists inhibit K^+ evoked Ca^{2+} uptake by rat brain cortical synaptosomes. J Neurochem 39:700–708.

CHAPTER 24

Purines in Anesthesia

ATSUO F. FUKUNAGA

Department of Anesthesiology, Harbor-UCLA Medical Center, Torrance, CA 90509

INTRODUCTION

It is almost seven decades since Drury and Szent-Györgyi (1929) described the cardiovascular effects of extracellular adenosine and related compounds. Twenty years later, Green and Stoner (1950) reported that the hypotension induced with ATP closely resembled traumatic and hypovolemic shock. Research on the physiology and pharmacology of the purines was renewed in 1963 when Berne (Berne, 1963) proposed that adenosine formation in the heart was a homeostatic mechanism for matching local tissue blood flow to local metabolic demand. A decade later, Burnstock (Burnstock, 1972) identified ATP as the putative transmitter for a purinergic division of the autonomic nervous system and proposed the "non-cholinergic, non-adrenergic" neurotransmission theory. In 1978, a classification of purinoceptors as P_1 and P_2 was proposed, based on the extracellular actions of purine nucleosides (adenosine) and nucleotides (ATP) on a wide variety of tissues (Burnstock, 1978). The discovery of adenosine receptors provided an effective means of studying adenosine-related effects in a variety of mammalian tissues (Van Calker et al., 1979; Londos et al., 1980), and the number of studies devoted to adenosine receptors rose dramatically. During this time, a vast amount of research was directed to elucidating the molecular mechanisms of specific receptor-mediated signal transduction, by which extracellular purines may act to elicit their array of biological responses. Purines, particularly adenosine and ATP, have recently received considerable attention because of the multiple protective effects that these compounds exert in a variety of physiopathological conditions in several organs, including the brain (Rudolphi et al., 1992; Miller and Hsu, 1992) and the heart (Ely and Berne, 1992; Gatell et al., 1993). Hence, the potential therapeutic uses of these compounds have been recognized and explored.

The principal objective of anesthesia is to protect the patient from the pain and stress of surgical trauma and to preserve homeostasis during the perioperative period (Nunn et al., 1989). In view of the many interesting physiological and pharmacological

Purinergic Approaches in Experimental Therapeutics, Edited by Kenneth A. Jacobson and Michael F. Jarvis

ISBN 0-471-14071-6 © 1997 Wiley-Liss, Inc.

properties of the purines, particularly their neuromodulatory and protective role, we can outline only the most general points regarding how we can apply the vast amount of available knowledge to identify and develop therapeutic strategies for the use of these compounds in perioperative care.

It was our objective to assess the activity by which these purines can inhibit stress, modify pain perception and stress responses to surgical stimulation since (i) pharmacotherapy requires specific goals of treatment, (ii) the objective endpoint should be related as closely as possible to the clinical goals of therapy, and (iii) the practice of anesthesiology usually involves control of pain and stress or stress responses inflicted by surgical intervention. Thus, the analgesic/anesthetic effects of the purines were tested in acute animal experimental models simulating clinical anesthesia. We have included a summary of our laboratory studies demonstrating that the behavioral sedative/antinociceptive properties of adenosine and ATP were not secondary to decrease in blood pressure, and that significant analgesia could be achieved without causing hypotension. Further, we characterized the analgesic properties of the purines to develop therapeutic strategies for their use in clinical anesthesia, taking advantage of their rapid acting antiadrenergic and sympatholytic analgesic qualities. This chapter provides a brief overview of certain properties of adenosine and ATP that are desirable in clinical anesthesia.

BACKGROUND

Adenosine and ATP play particularly conspicuous role in the cardiovascular and central nervous systems (CNS). In the CNS, adenosine and its nucleotides inhibit neuronal cell firing and release of several neurotransmitters (Phillis and Wu, 1981). It has been recognized that the purines function as neuromodulators or neurotransmitters (Fredholm and Hedqvist, 1980; Dunwiddie, 1985; Williams, 1989). Moreover, there has been evidence that the sedative activity of some of the drugs used in anesthesia, such as the anxiolytic and sedative/hypnotic agents, may be mediated in part via central purine-linked mechanisms (Barraco et al., 1991). Indeed, the relationship between adenosine and benzodiazepines has been well documented (Phillis et al., 1980; Barraco et al., 1984; Phillis and O'Regan, 1988), and the adenosine receptor antagonists caffeine and theophylline were effective in reversing the benzodiazepine-induced anesthesia in animals and humans (Phillis et al., 1979; Roache and Griffiths, 1987; Gurel et al., 1987). In addition, there is considerable evidence to implicate the purines in some of the mechanisms modulating pain perception and/or analgesia. (Sawynok and Sweeney, 1989). There seems to be an interaction of endogenous adenosine with morphine (Perkins and Stone, 1980; Wu et al. 1982), and it has been suggested that some of the spinal analgesic activity of morphine is medicated in part by adenosine mechanism (Sawynok et al., 1989).

Since the antinociceptive effects of adenosine analogs were described in 1975 (Vapaatalo et al., 1975) there has been great interest on the part of a number of researchers in the possibility of developing adenosine compounds as analgesics. Nonetheless, due to the multitude of adenosine effects, it has been suggested that such a goal might be frustrating (Sawynok and Sweeney, 1989). Indeed, the peripheral cardiovascular actions (e.g., hypotension, bradycardia) are so overwhelmingly potent that it was difficult to dissociate hypotension from centrally mediated antinociception. In addition, with regard to pain perception, adenosine seems to play a paradoxical role, and has been reported to have both algogenic (Bleehen and Keele, 1977; Pappagallo et al., 1993) and antinociceptive effects. However, adenosine's direct stimulating actions on the peripheral sensory systems (Biaggioni, 1992), including carotid chemoreceptors (oppressive tachydyspnea) and cardiac sensory fibers (angina-like chest pain) (Sylven et al., 1986), are transient and mediated by different mechanisms, whereas our studies show the central antinociceptive (analgesic) effects to be long-lasting. We have determined the analgesic/anesthetic effects of the purines in several paradigms, which are summarized in the following sections.

CHARACTERIZATION OF THE ANTINOCICEPTIVE EFFECTS OF ADENOSINE AND ATP

To estimate the antinociceptive activity of the purines, the standard method for measuring anesthetic potency (MAC, minimum alveolar concentration of an inhalational anesthetic that produces immobility in 50% of animals exposed to tail clamping stimulus; Eger et al., 1965; Quasha et al., 1980) was first used in a series of experiments with rabbits (Fukunaga et al., 1989). In this study, inhibitory responses were noted only with very high doses of adenosine or ATP; similar data have been reported in dogs (Seitz et al., 1990). At such dosages and modes of administration, severe hypotension occurred. Notwithstanding, our laboratory observations following administration of the purines indicated that the inhibitory effects appeared to be more selective to pain perception. Motor behavior and response to touch stimulation remained unimpaired, although the animals were unresponsive to painful stimulation (Ginsburg et al., 1990). Therefore, we questioned whether assessment was done with the appropriate pain stimulus. Tail clamping using a surgical hemostat entails a stimulus that involves multiple and confounded stimuli. Since the effects of the purines did not appear to modify tactile sensation, a more discriminative and quantifiable stimulus was required to assess properly the antinociceptive activity of the purines. Electrical stimulation, although is not a selective stimulation, was useful to mimic surgical stimulation.

In the study, the sedative-hypnotic/analgesic dose-responses of an inhalational anesthetic, isoflurane, and intravenous (IV) adenosine and ATP were assessed in the rabbit model using electrical tail stimulation (ETS) with a graded degree of electrical intensities. A pair of subcutaneous platinum needle electrodes were placed at the base of the shaved tail, and variable electrical current with square pulse wave (5 Hz, 1 msec, 1–100 V) was delivered from a nerve stimulator. Two distinct behavioral responses to ETS were observed and their threshold response curves were determined in sedated animals: as the stimulation intensity increases, the sedated animal responded initially with an arousal response, in which the animals usually opened the eyes and lifted the head (head lift, HL; sedative-hypnotic index); as the intensity was further increased, an aversive response, generally more vigorous than HL, was observed, (escape movement EM, analgesic index). The stimulus-response relationship is sensitive, consistent, and reproducible; and this type of electrical stimulus has the advantage of not causing tissue damage, even after prolonged experiments.

Which of the two distinct behavioral responses is observed depends on whether the predominant drug action is sedative-hypnotic or analgesic. For example isoflurane, a hypnotic/anesthetic, almost equally and concomitantly elevated both the HL and

EM threshold response curves in a dose-related manner (Figure 1), whereas with opioids, the EM response curve elevated more than the HL curve, which characterizes the analgesic property (Fukunaga et al., 1991). Likewise, with adenosine and ATP, the EM and HL response curves diverged even more widely than with the opioids, in a dose-dependent manner (Fukunaga et al., 1992a) findings which suggest that the purines have a predominantly analgesic/anesthetic profile as compared to the hypnotic/anesthetic profile of volatile inhalational agents (see Figure 1). Most importantly, such antinociceptive effects were not the result of a fall in blood pressure (BP) or decrease in body temperature (hypothermia). Since the analgesic activities of the purines could be attained without causing side effects or hypotension, it follows that clinical application could be considered; purines could enhance the anesthetic effectiveness of inhalational agents or intravenous sedative-hypnotic drugs (e.g., benzodiazepines, barbiturates, propofol), much like the opioids and other analgesics presently used during anesthesia.

REDUCTION OF THE INHALATIONAL ANESTHETIC REQUIREMENTS BY ATP

ATP combined with nitrous oxide (N_2O) produced anesthetic effects comparable to those of N_2O combined with enflurane. To assess the antinociceptive activity of ATP, abolition of movement in response to tail clamp stimulation was used as the endpoint.[1] In addition, antinociception was tested with electrical tail stimulations (ETS) as a graded and quantitative measurement. The behavioral and cardiovascular responses were assessed simultaneously in the tracheotomized and intravascularly cannulated rabbit placed in a sling that allowed the head and legs to move freely; this model provided a reliable and easy way to assess multiple responses simultaneously in a relatively unrestrained animal. After complete recovery from the surgical preparation and anesthesia, N_2O (60% in oxygen) was administered via endotracheal tube as the basal anesthetic throughout the study, and the baseline control measurements were made at stage I with N_2O alone. Subsequently, enflurane was added stepwise. The animal was equilibrated with each test dose of the administered drugs for at least 30 minutes. Two antinociceptive responses were tested consecutively, first tail clamp and then ETS threshold responses. Sufficient recovery time was allowed between the two stimulations. As shown in Figure 2, addition of enflurane to N_2O indeed increased the antinociceptive responses in a dose-related manner, but BP decreased concomitantly in a dose-related fashion. Complete abolition of movement in response to tail clamp in all animals was achieved with 2.5% enflurane and N_2O (stage II).

To determine ATP's antinociceptive effects, doses of enflurane were decreased stepwise and, after positive movement response to tail clamp and decrease of ETS threshold were first confirmed, infusion doses of ATP were gradually increased until abolition of movement in response to tail clamp was achieved. This procedure was repeated at each step with enflurane decrements until no enflurane (0%) was given. Thus, ATP could completely replace enflurane without resulting in a movement response to tail clamp (stage III in Figure 2); the effect was comparable to the anesthetic effect achieved with 2.5% enflurane. Surprisingly, however, the BP was restored to near control, normal value with this dose of ATP combined with N_2O (see Figure 2).

[1] This stimulation has been used as the standard method to determine anesthetic potency in experimental animals.

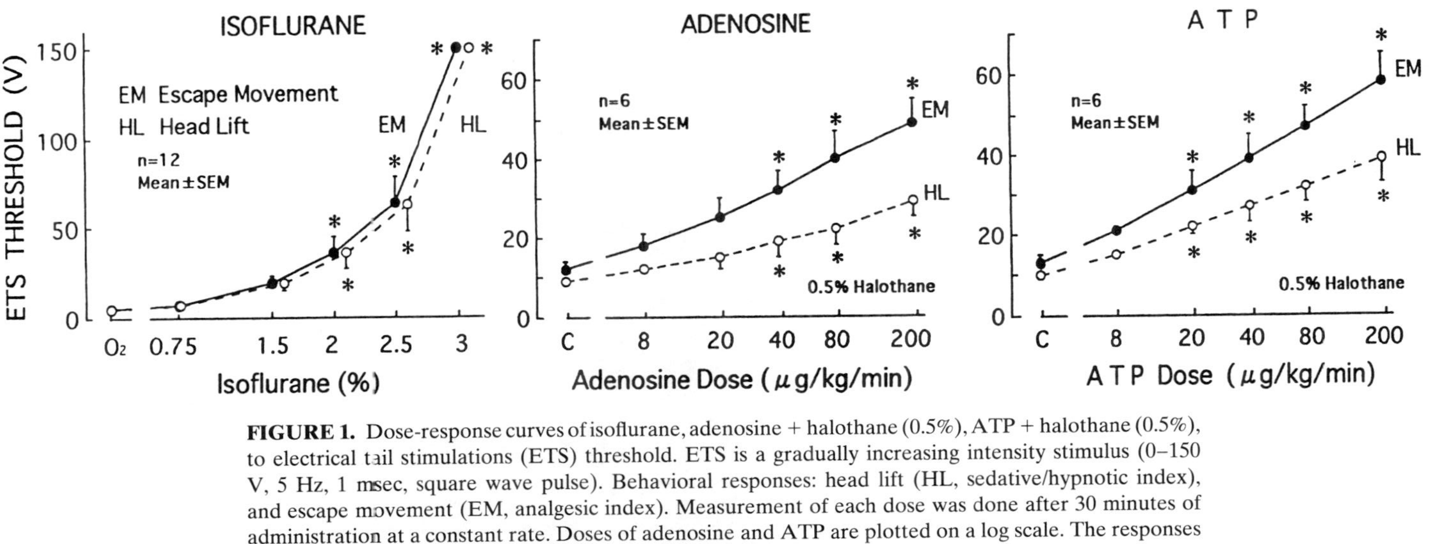

FIGURE 1. Dose-response curves of isoflurane, adenosine + halothane (0.5%), ATP + halothane (0.5%), to electrical tail stimulations (ETS) threshold. ETS is a gradually increasing intensity stimulus (0–150 V, 5 Hz, 1 msec, square wave pulse). Behavioral responses: head lift (HL, sedative/hypnotic index), and escape movement (EM, analgesic index). Measurement of each dose was done after 30 minutes of administration at a constant rate. Doses of adenosine and ATP are plotted on a log scale. The responses for each individual rabbit were the average of responses tested in triplicate. Statistical analysis was done with ANOVA followed by Dunnett's test; * $p < 0.05$ was considered significant.

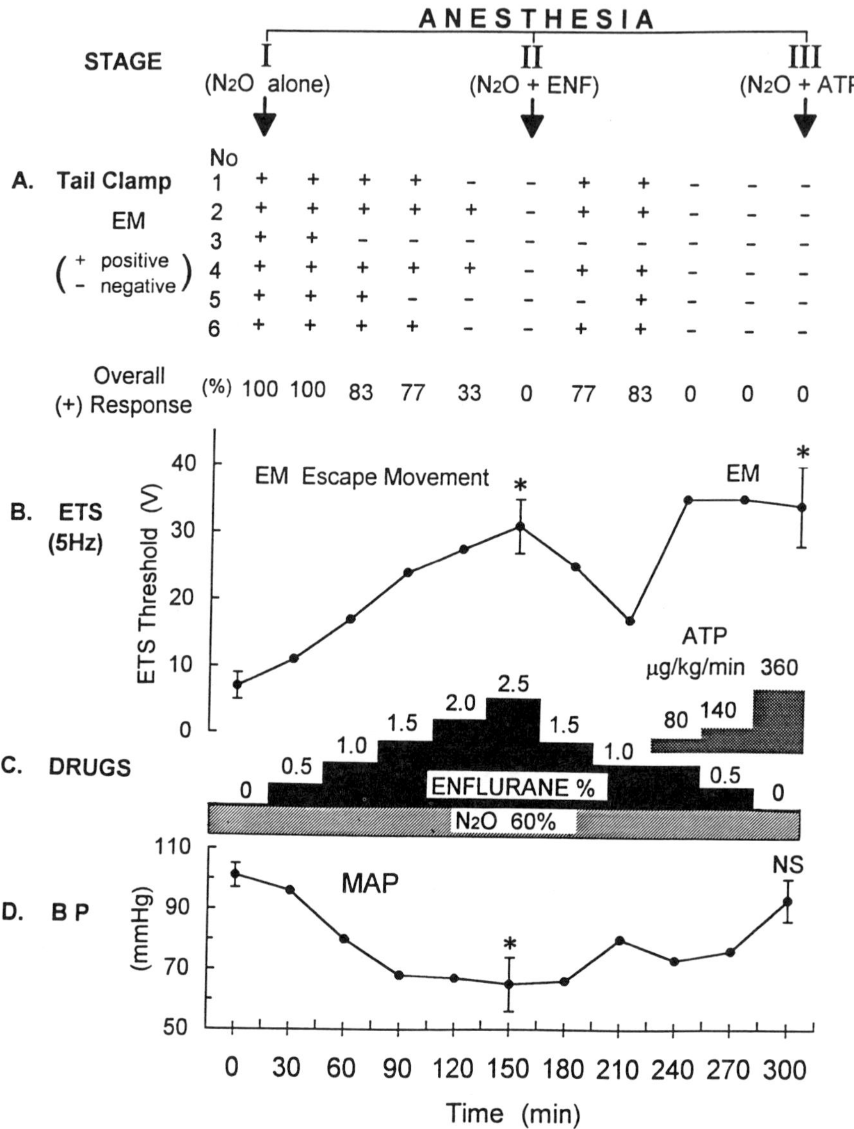

FIGURE 2. Behavior (escape movement, EM) and blood pressure (BP) responses determined at three stages (I) N_2O, (II) N_2O + enflurane (ENF), and (III) N_2O + ATP; with two anesthetic tests, tail clamp and electrical tail stimulation (ETS). Notice that ATP completely replaced enflurane; ATP infusion doses ($\mu g \cdot kg^{-1} \cdot min^{-1}$): 83 ± 67, 142 ± 117, and 363 ± 165, at enflurane 1%, 0.5% and 0% respectively. The hypotensive effect observed with enflurane (2.5%, stage II) is in contrast to the normal BP with ATP at stage III. The mean value ± SEM of the results obtained from 6 rabbits and data were compared versus stage I. $p < 0.05$ was considered significant.* The criterion for statistical significance using ANOVA followed by Dunnett's test; NS = not significant.

This was in contrast to the significantly low BP produced by the combination of N_2O and enflurane (2.5%). Moreover, as shown in Table 1, other circulatory and respiratory variables remained unchanged during ATP infusion.

INHIBITION OF THE CARDIOVASCULAR RESPONSES TO NOXIOUS STIMULATION BY ADENOSINE OR ATP: AN INDICATION OF SYMPATHOLYTIC AND ANALGESIC EFFECT

In clinical anesthesia, maintenance of hemodynamic stability during surgical stimulation is considered an indication of anesthetic effectiveness. Measurements of BP and heart rate (HR) have been used as indirect indices of the level of sympathetic activity to assess depth of anesthesia. We have demonstrated that continuous infusion of adenosine (170–190 $\mu g \cdot kg^{-1} \cdot min^{-1}$) titrated according to BP responses during continuous electrical tail stimulation could effectively inhibit the behavioral movement, as well as the cardiovascular (BP and HR) responses to noxious stimulation in the rabbit. In the study (adenosine, n = 5; ATP, n = 5) baseline sedation was maintained with halothane (0.5% in O_2) in spontaneously breathing rabbits throughout the study period. Electrical tail stimulation (5 Hz, 1 msec, 10–20 V) was applied continuously through a nerve stimulator. Stimulation was gradually increased until behavioral and hemodynamic responses were evidenced, at which point infusion of either adenosine or ATP was started, while maintaining constant the intensity of electrical stimulation. During the purine infusion, the animals appeared sedated and movement was effectively inhibited. Furthermore, as shown in Figure 3, remarkable hemodynamic stability could be maintained throughout the period of purine infusion (120 minutes) despite continuous noxious stimulation. After IV or IT administration of aminophylline, the animals responded with vigorous BP and movement responses (Figure 3 and Tables 2 and 3). Such cardiovascular and behavioral responses to stimulation were consistent with increased sympathetic activity. Adenosine or ATP could effectively inhibit both sympathetic and somatic responses caused by noxious stimulation while maintaining circula-

TABLE 1 Circulatory, Respiratory and Blood Gas Data During Nitrous Oxide, Enflurane/Nitrous Oxide, and ATP/Nitrous Oxide Anesthesia in Spontaneously Breathing Rabbits

Drugs	Stage I (N_2O Only)	Stage II (Enflurane + N_2O)	Stage III (ATP + N_2O)
BP (mmHg)			
Systolic	120 ± 15	69 ± 22*	112 ± 20
Diastolic	91 ± 11	49 ± 18*	83 ± 16
HR (beats/min)	267 ± 15	255 ± 33*	278 ± 26
Blood gas			
pH	7.43 ± 0.02	7.42 ± 0.04*	7.45 ± 0.08
$Paco_2$ (mmHg)	22 ± 2	25 ± 1*	22 ± 3
Pao_2 (mmHg)	147 ± 11	148 ± 16	154 ± 19
BE (mEq/l)	−7 ± 1	−7 ± 2	−7 ± 4
Respiratory rate (breath/min)	85 ± 5	72 ± 19*	81 ± 14
Rectal temperature (°C)	38.7 ± 0.3	38.8 ± 0.7	38.6 ± 0.4

Nitrous oxide (N_2O), 60% in O_2; enflurane, 2.5 vol %; ATP, 363 ± 165 ($\mu g \cdot kg^{-1} \cdot min^{-1}$ IV); mean ± SD, (n = 6); comparison versus Stage I; *$p < 0.05$ was considered significant.

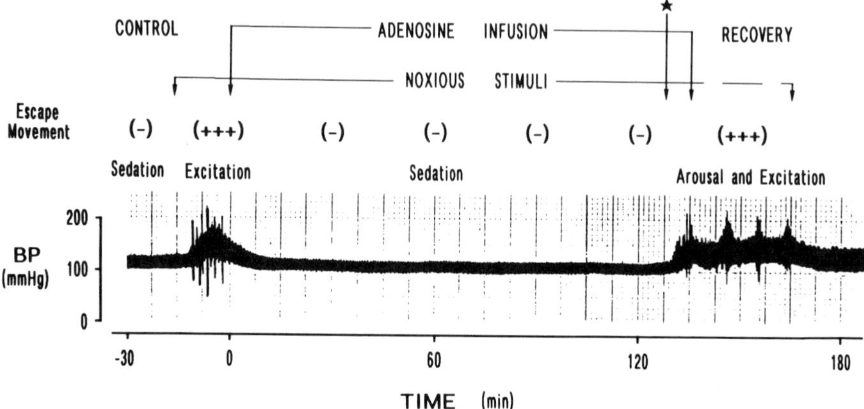

FIGURE 3. Blood pressure (BP) responses during continuous noxious stimulation and adenosine infusion in a spontaneously breathing rabbit sedated with halothane (0.5%). Notice the remarkably stable blood pressure (sympatholytic action) during adenosine infusion (170–190 $\mu g \cdot kg^{-1} \cdot min^{-1}$). In contrast, sympathetic excitation caused by noxious stimulation is evidenced before adenosine and after aminophylline* (1 mg, injected intrathecally at high spinal cervical level). Noxious stimulation was applied by continuous electrical tail stimulation (15 V, 50 Hz, 1 msec, square pulse wave).

tory and respiratory functions within normal levels (Tables 2 and 3). We concluded that this is indicative of the sympatholytic and analgesic property of the purines.

SUSTAINED ANALGESIA PRODUCED BY INTRAVENOUS ADENOSINE

Because adenosine disappears from the circulation very rapidly, and because of the notion that adenosine does not easily penetrate the blood-brain barrier, it was surprising that intravenously administered adenosine could produce profound and sustained analgesia, even after discontinuing the infusion of adenosine (Fukunaga et al., 1993). In the experiments, adenosine was continuously infused in eight rabbits, with baseline sedation maintained with propofol infusion (200 $\mu g \cdot kg^{-1} \cdot min^{-1}$ IV). Analgesic effects were assessed by two tests: (i) tail clamp and (ii) electrical tail stimulation (ETS). In previous experiments, in order to have a comparable index of anesthetic responses to ETS, commonly used inhalational anesthetics were tested with N_2O (60% in oxygen, n = 6), and 1%, 2%, or 3% isoflurane (n = 17) respectively. Furthermore, we determined that propofol, at doses given in the study, produced only sedation, without increasing the analgesic thresholds throughout the study period as indicated by the control (propofol only) value.

As can be appreciated from Figure 4, a total dose of approximately 100 mg/kg adenosine administered by IV infusion at a rate of 2.6 ± 3.2 $mg \cdot kg^{-1} \cdot min^{-1}$ progressively elevated both HL (sedative-hypnotic) and EM (analgesic) thresholds. This profound analgesia, comparable to anesthesia produced by 3% isoflurane, was sustained well over 60 minutes after adenosine infusion was discontinued, and it was achieved without deleterious signs in cardiorespiratory variables, except for the hypotension during the adenosine infusion period (Table 4), which in most cases was treated with simultaneous

TABLE 2 Continuous Adenosine Infusion in Rabbits Spontaneously Breathing Halothane (0.5%): Circulatory, Respiratory, and Blood Gas Data During Continuous Electrical Tail Stimulation

			Adenosine Infusion				
	Control	ETS Onset	30 min	60 min	90 min	120 min	Aminophylline
Infusion Dose ($\mu g \cdot kg^{-1} \cdot min^{-1}$)		0	192 ± 66	176 ± 73	172 ± 76	176 ± 74	176 ± 74
BP (mmHg)							
Systolic	93 ± 13	101 ± 12	82 ± 8	87 ± 10	88 ± 7	88 ± 5	96 ± 24
Diastolic	68 ± 6	72 ± 4	53 ± 12	56 ± 15	54 ± 10	55 ± 13	68 ± 17
Mean	78 ± 9	83 ± 7	65 ± 14	66 ± 13	66 ± 8	66 ± 10	77 ± 19
Heart rate (beats/min)	262 ± 19	282 ± 34	294 ± 22	295 ± 21	295 ± 23	294 ± 20	292 ± 48
Respiratory Rate (breath/min)	97 ± 25	109 ± 23	90 ± 14	87 ± 12	89 ± 15	89 ± 16	119 ± 44
Blood Gas							
pH	7.51 ± 0.06	7.51 ± 0.07	7.51 ± 0.09	7.51 ± 0.08	7.52 ± 0.10	7.50 ± 0.10	7.45 ± 0.08
$Paco_2$ (mmHg)	29 ± 3	28 ± 6	26 ± 6	26 ± 4	24 ± 6	24 ± 5	27 ± 6
Pao_2 (mmHg)	499 ± 32	472 ± 35	479 ± 34	481 ± 41	485 ± 57	492 ± 77	516 ± 37
HCO_3 (mEq/L)	24 ± 4	22 ± 3	21 ± 2	20 ± 1	19 ± 3	19 ± 2	19 ± 2*
BE (mEq/L)	−2 ± 4	−1 ± 4	−1 ± 2	−1 ± 3	−2 ± 3	−2 ± 3	−4 ± 2
Rectal temperature (°C)	37.6 ± 0.9	38.0 ± 1.4	38.3 ± 1.0	38.5 ± 0.6	38.8 ± 0.7	39.1 ± 0.5	39.3 ± 0.7

ETS = electrical tail stimulation (square pulse wave, 50 Hz, 1 msec, 10–20 V); aminophylline (10 mg/kg IV); comparison versus control data; Mean ± SD (n = 5); *$p < 0.05$ was considered significant.

TABLE 3 Continuous ATP Infusion in Rabbits Spontaneously Breathing Halothane (0.5%): Circulatory, Respiratory, and Blood Gas Data During Continuous Electrical Tail Stimulation

	Control	ETS Onset	ATP Infusion				
			30 min	60 min	90 min	120 min	Aminophylline
Infusion Dose ($\mu g \cdot kg^{-1} \cdot min^{-1}$)		0	190 ± 90	150 ± 68	140 ± 69	130 ± 76	130 ± 76
BP (mmHg)							
Systolic	95 ± 22	104 ± 21	87 ± 10	81 ± 9	82 ± 11	82 ± 9	97 ± 10
Diastolic	72 ± 11	74 ± 12	60 ± 7	56 ± 4	58 ± 3	55 ± 6	72 ± 9
Mean	81 ± 15	86 ± 17	71 ± 7	66 ± 5	67 ± 6	67 ± 7	80 ± 9
Heart rate (beat/min)	259 ± 19	270 ± 53	273 ± 41	265 ± 16	262 ± 23	264 ± 16	315 ± 35
Respiratory rate (breath/min)	109 ± 36	122 ± 41	111 ± 36	105 ± 35	106 ± 33	99 ± 29	96 ± 20
Blood gas							
pH	7.51 ± 0.03	7.51 ± 0.05	7.54 ± 0.05	7.54 ± 0.03	7.55 ± 0.03	7.56 ± 0.03	7.52 ± 0.05
$Paco_2$ (mmHg)	22 ± 4	21 ± 3	19 ± 4	19 ± 4	18 ± 2	18 ± 2	17 ± 2
Pao_2 (mmHg)	512 ± 39	517 ± 23	502 ± 30	529 ± 29	536 ± 39	555 ± 9	533 ± 17
HCO_3 (mEq/L)	18 ± 3	17 ± 4	16 ± 2	16 ± 3	16 ± 2	16 ± 3	14 ± 3
BE (mEq/L)	−3 ± 3	−4 ± 4	−4 ± 1	−4 ± 3	−4 ± 2	−4 ± 3	−6 ± 4
Rectal temperature (°C)	39.2 ± 0.6	39.6 ± 0.9	39.9 ± 0.9	39.8 ± 0.8	39.8 ± 0.7	39.8 ± 0.4	39.6 ± 0.2

ETS = electrical tail stimulation (square pulse wave, 50 Hz, 1 msec, 10–20 V); aminophylline (10 mg/kg IV); comparison versus control data; Mean ± SD (n = 5); $^*p < 0.05$ was considered significant.

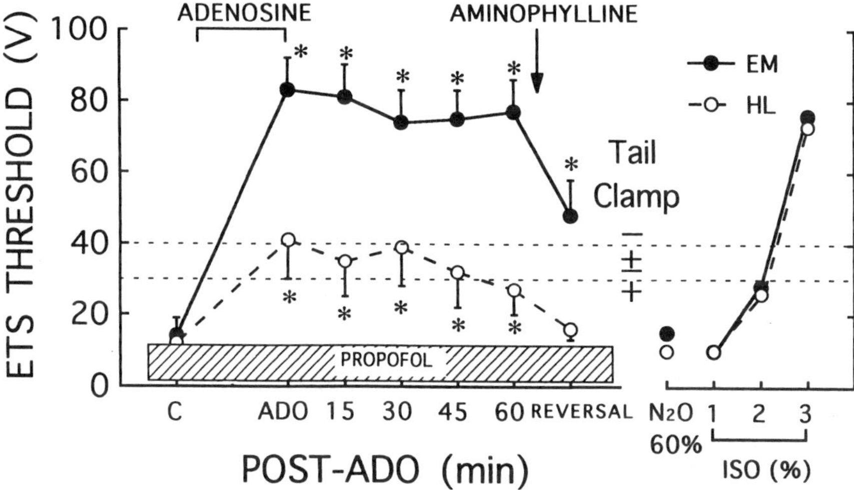

FIGURE 4. Time course of the antinociceptive effect produced after intravenous adenosine of approximately 100 mg/kg. Adenosine was continuously infused at a rate of 2.6 ± 3.2 mg·kg^{-1}·min^{-1} over a period of 83 ± 67 minutes in spontaneously breathing rabbits (n = 8) sedated with propofol (200 μg·kg^{-1}·min^{-1}). Nociception was tested by tail clamp and electrical tail stimulation (ETS: 0–100 V, 5 Hz, 1 msec, square wave pulse), EM (escape movement, analgesic index), and HL (head lift, sedative/hypnotic index). Aminophylline (34 ± 8.9 mg/kg IV) was given to antagonize the sedative/analgesic effects. For a potency comparison, the antinociceptive responses of 60% nitrous oxide (N_20, n = 6) and 1%, 2%, or 3% isoflurane (ISO, n = 17) are on the right side; n = number of animals. Each point on the curve represents a mean value ± SEM. Data are compared versus control (C, propofol only); The criterion for statistical significance, $^*p < 0.05$, was considered significant, using the Wilcoxon signed rank sum test.

TABLE 4 Circulatory, Respiratory and Blood Gas Data During and After Adenosine in Propofol-Sedated Rabbits Breathing Mixed Air and Oxygen Spontaneously

	Control	Adenosine (ADO)	Post-ADO (60 min)	Aminophylline
BP (mmHg)				
Systolic	119 ± 14	89 ± 28*	114 ± 11	112 ± 22
Diastolic	83 ± 8	43 ± 10*	81 ± 11	78 ± 14
HR (beats/min)	237 ± 33	194 ± 23*	231 ± 21	273 ± 38
Blood Gas				
pH	7.37 ± 0.10	7.32 ± 0.11	7.27 ± 0.07*	7.28 ± 0.12*
$Paco_2$ (mmHg)	34 ± 12	30 ± 15	31 ± 14	28 ± 16
Pao_2 (mmHg)	128 ± 24	140 ± 35	140 ± 28	114 ± 31
BE (mEq/L)	−5 ± 4	−10 ± 4*	−13 ± 5*	−14 ± 3*
Respiratory rate (breath/min)	74 ± 29	64 ± 26	70 ± 21	64 ± 23
Rectal temperature (°C)	38.6 ± 0.9	37.9 ± 1.0	37.8 ± 0.9	38.0 ± 0.3

ADO, 2.6 ± 3.2 mg/kg per minute IV; aminophylline, 34 ± 9 mg/kg IV; comparison versus control data mean ± SD; (n = 8); $^*p < 0.05$ was considered significant.

infusion of variable doses of phenylephrine to maintain the BP. Significant and sustained analgesia could be achieved with continuous and prolonged infusion time (over 83 ± 67 minute period) of high doses of adenosine. Naloxone or flumazenil could not reverse the analgesic effects, but aminophylline (34 ± 8.9 mg/kg IV) could antagonize almost all the behavioral responses, thus indicating the direct involvement of the adenosine receptor mechanism.

POTENCY AND DURATION OF THE ANALGESIC EFFECTS OF INTRAVENOUS ADENOSINE AND ATP: DISSOCIATION OF ANALGESIA FROM HYPOTENSION

The magnitude and duration of the antinociceptive effect of adenosine and ATP in nonmedicated rabbits were assessed. The behavioral and cardiovascular (BP, HR) responses were simultaneously monitored. By supporting the BP pharmacologically, we could prove that the analgesic effect was not secondary to the hypotensive action. For a comparison of analgesic potency, the well-known drug morphine was tested, using the same electrical tail stimulation as in previous experiments.

In 24 unmedicated rabbits, 100 mg/kg of either adenosine (n = 12) or ATP (n = 12) consistently produced analgesia comparable to that produced by morphine (10 mg/kg IV; see Figure 5). As expected, during infusion of adenosine or ATP alone, at a rate of 1600 $\mu g \cdot kg^{-1} \cdot min^{-1}$ for over a 1-hour period produced significant hypotension. However, HR did not change significantly, and as soon as the infusion was stopped, BP returned to near control levels without rebound hypertension. In half of the experiments adenosine (n = 6) or ATP (n = 6) was coadministered with norepinephrine, using the method described elsewhere (Fukunaga, 1995). This resulted in no significant change in BP during infusion (Figure 5). Furthermore, all other variables including cardiovascular and blood gas data were maintained without significant changes in either group, regardless of whether purines were administered alone or coadministered with norepineophrine. The profound sedative/analgesic effect after stopping the infusion was sustained for over 5 hours with stable BP and HR, and without evidence of respiratory depression, hypothermia, or metabolic acidosis. By contrast, and in agreement with its well-recognized effects, morphine caused severe respiratory depression and bradycardia (see Figure 5). Most of the analgesic effects of the purines were antagonized by IV 8-phenyltheophylline, a specific adenosine receptor antagonist. In addition, yohimbine, an alpha-2 adrenoceptor antagonist was used to completely reverse the sedative and analgesic effects in the group receiving purines and norepinephrine. Accordingly, the experimental data demonstrated that intravenous adenosine or ATP could produce profound, and sustained analgesia that was not secondary to a decrease in blood pressure. The analgesic properties of adenosine or ATP are most likely mediated by adenosine receptors, since such effects were almost completely antagonized by 8-phenyltheophylline.

CLINICAL IMPLICATIONS

Laboratory results demonstrating the sedative and analgesic properties in a variety of species and tests could be supported in humans as well. Intravenous ATP administered in a peripheral vein of the forearm in volunteers could potentiate the sedative

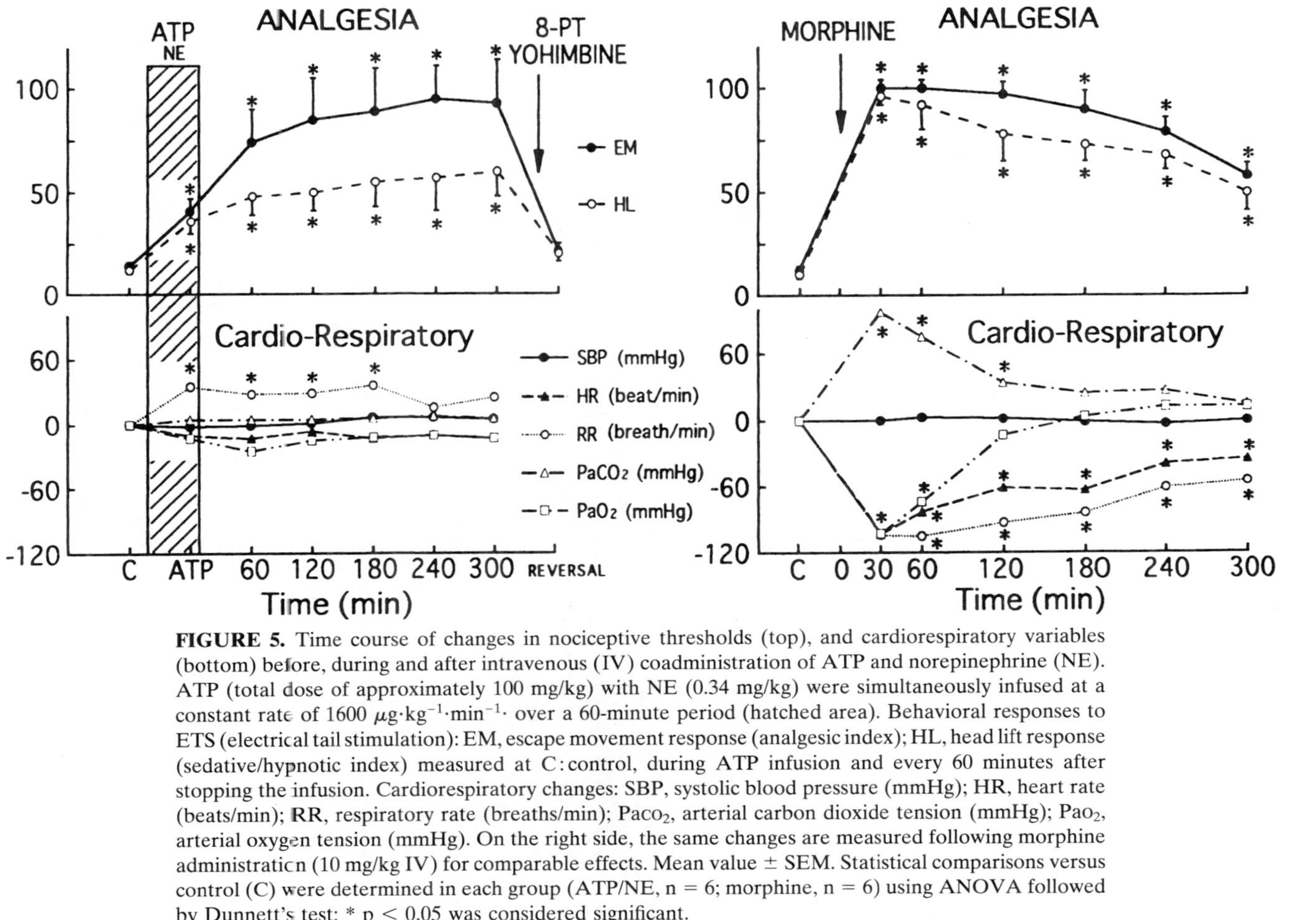

FIGURE 5. Time course of changes in nociceptive thresholds (top), and cardiorespiratory variables (bottom) before, during and after intravenous (IV) coadministration of ATP and norepinephrine (NE). ATP (total dose of approximately 100 mg/kg) with NE (0.34 mg/kg) were simultaneously infused at a constant rate of 1600 $\mu g \cdot kg^{-1} \cdot min^{-1}$. over a 60-minute period (hatched area). Behavioral responses to ETS (electrical tail stimulation): EM, escape movement response (analgesic index); HL, head lift response (sedative/hypnotic index) measured at C: control, during ATP infusion and every 60 minutes after stopping the infusion. Cardiorespiratory changes: SBP, systolic blood pressure (mmHg); HR, heart rate (beats/min); RR, respiratory rate (breaths/min); $Paco_2$, arterial carbon dioxide tension (mmHg); Pao_2, arterial oxygen tension (mmHg). On the right side, the same changes are measured following morphine administration (10 mg/kg IV) for comparable effects. Mean value ± SEM. Statistical comparisons versus control (C) were determined in each group (ATP/NE, n = 6; morphine, n = 6) using ANOVA followed by Dunnett's test; * $p < 0.05$ was considered significant.

and hypnogenic effects of midazolam (benzodiazepine) in a double-blind placebo-controlled study (Kaneko et al., 1992). The degree of sedation, amnesia, anxiety level, sleepiness, and psychomotor function (Romberg test) were assessed. It was found that coadministration of ATP and midazolam significantly elevated the sedation scores in both subjective and objective evaluations as compared to midazolam alone. In agreement with laboratory results using adenosine (Fukunaga et al., 1992b), ATP potentiated the sedative effects of midazolam without producing side effects like hypotension, respiratory depression, nausea, vomiting, headaches, or an intolerable feeling of chest oppression.

Tooth pulp electrical stimulation has been used to assess pain by several researchers (Gracely, 1989). Therefore, this stimulation was used to assess the antinociceptive actions of ATP in volunteers in a double-blind placebo-controlled study. Infusion of ATP at a rate of 100 $\mu g \cdot kg^{-1} \cdot min^{-1}$ significantly elevated the pain thresholds of tooth pulp electrical stimulation. The subjects tolerated this dosage of ATP well, and did not complain of side effects (headaches or respiratory depression) during or after ATP administration. The results of the study showed that nonhypotensive doses of ATP could produce analgesia as assessed by the tooth pulp electrical stimulation test (Fukunaga et al., 1994).

It has been reported that in an experimentally induced ischemic pain model, adenosine was found to reduce ischemic muscle pain in a randomized, single-blind, and placebo-controlled study in humans. The authors found that adenosine (70 $\mu g \cdot kg^{-1} \cdot min^{-1}$) could produce the same effect as morphine (0.1 mg/kg) and ketamine (0.1 mg/kg) in reducing pain, without producing significant changes in BP (Segerdahl et al., 1994). Likewise, in volunteer studies, IV adenosine reduced cutaneous thermal pain perception (Ekblom et al., 1995) and decreased touch-evoked allodynia induced by mustard oil (Segerdahl et al., 1995a). Based on the experimental data, the authors suggested that adenosine decreased activation of nociceptive neurons may be linked to adenosine receptors in the CNS. Furthermore, intrathecal administration of phenylisopropyladenosine (R-PIA) was reported to have strikingly potent analgesic effects in a patient with well-characterized intractable neurogenic pain. Although the report was on a single case, it gives indications that R-PIA was effective in abolishing pain and the allodynia evoked by touch and vibration in a patient afflicted by severe neurogenic pain for more than 10 years (Karlsten and Gordh, 1995). Similar analgesic effects have been reported with intravenous infusion of adenosine in patients suffering peripheral neuropathic pain (Belfrage et al., 1995).

ADENOSINE AND ATP AS ANESTHETIC ADJUVANTS IN CLINICAL ANESTHESIA

In most anesthetic practices, combinations of drugs are used to fulfill the requirements of general anesthesia. Because the agents used in anesthesia are extremely potent and must be administered in a very short period of time, small doses of various drugs are usually used in combination, to supplement each other while avoiding the adverse effects of using a large dose of any single drug (Warner and Dubbink, 1992). Indeed, the low therapeutic ratio (toxic dose/therapeutic dose) and the inherent dangers of the commonly used anesthetics are well recognized and make the continuous monitoring of respiratory and circulatory variables during anesthesia a standard practice. Accordingly, a drug with potent sympatholytic analgesic properties that is devoid of respiratory depression, capable of reducing the requirement of inhalational anesthetics or intravenous hypnotic drugs, and which is easily titratable appears to have potential benefit as an anesthetic adjuvant.

The significance of the animal studies was confirmed by demonstration of the analgesic action in clinical anesthesia. Indeed, intravenous infusion of ATP could effectively reduce the requirement of inhaled anesthetics, including enflurane and nitrous oxide, in patients undergoing major surgical procedures such as maxillofacial orthognathic surgery, and surgery of the abdomen and extremities (Fukunaga et al., 1995). In agreement with the purines' sympatholytic analgesic activity evidenced in our laboratory studies, ATP could attenuate the autonomic responses to stressful surgical stimulation. In operations involving painful intra-abdominal manipulations, such as resection and retraction of visceral organs, where acute hypertensive responses accompanied by tachyarrhythmias are common clinical problems, ATP, in combination with nitrous oxide and midazolam (Kikuta et al., 1992), could maintain hemodynamic stability intraoperatively as shown in Figure 6. Notice in the figure that whereas the BP fluctuates before ATP administration, by titrating the infusion rate, ATP could effectively attenuate the BP fluctuations.

Sternotomy is considered to be one of the most painful surgical stimulations and may cause severe stress responses associated with sympathetic activation and concomitant

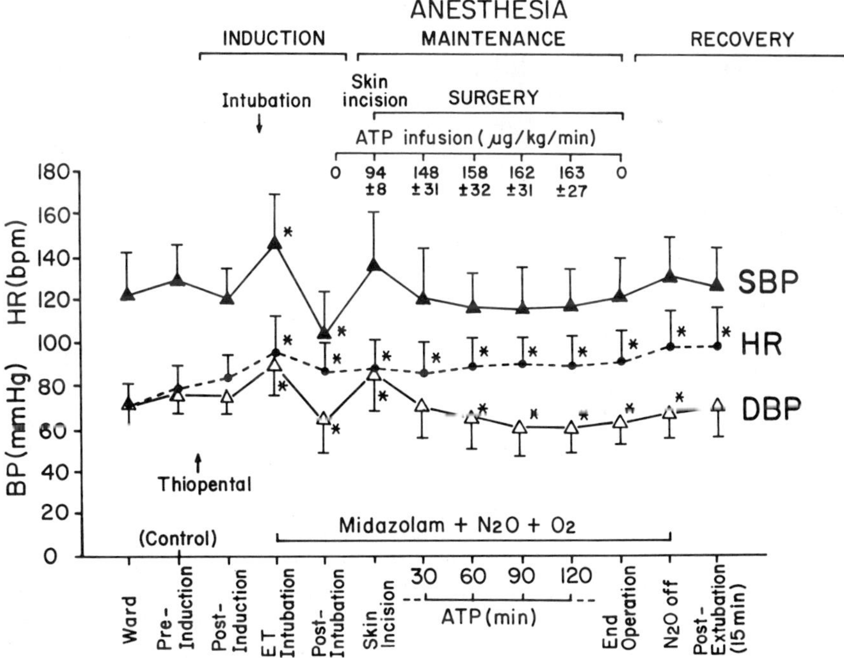

FIGURE 6. Cardiovascular changes during midazolam–nitrous oxide–ATP anesthesia. BP, blood pressure; SBP, systolic blood pressure (mmHg); HR, heart rate (beats/min); DBP, diastolic blood pressure (mmHg); N_2O, nitrous oxide (60% in oxygen). Mean ± SD. Statistical comparison versus control (pre-induction values), used ANOVA followed by Dunnett's test; * $p < 0.05$ was considered significant.

increased levels of catecholamines (Sonntag et al., 1989). BP and HR may acutely increase, imposing stress on the cardiovascular system, an effect particularly undesirable in patients prone to coronary ischemia. It was shown that ATP in combination with fentanyl effectively inhibited the cardiovascular responses to sternotomy and sternal spread, and improved myocardial oxygen balance in patients undergoing coronary artery bypass surgery (Taniguchi et al., 1993). We theorize that a variety of ATP effects act in concert to maintain the integrity of circulatory responses to severe surgical stress and that these may have accounted for such efficient activity. In particular, the potent vasodilating, antiadrenergic (counteracting circulating catecholamine effects), sympatholytic (presynaptic inhibition of noradrenaline release), and analgesic properties of ATP appeared to have been the primary mechanisms for the cardiovascular inhibitory responses to intense surgical stimulation. It is noteworthy that intraoperative infusion of ATP provided postoperative analgesia in a significant number of patients. This finding was consistent with the long-lasting analgesic effect demonstrated in our experimental animals. Moreover, hypertension, a common occurrence during the early postoperative period, was attenuated, even after ATP and nitrous oxide were discontinued (see Figure 6), an effect that may also be an indication of ATP's sympatholytic analgesia.

Adenosine infusion could replace perioperative opioid administration during isoflurane-nitrous oxide anesthesia (Sollevi, 1992). Further, adenosine was found to reduce the requirements for isofiurane and postoperative analgesics during the first 24 hours after surgery. In this randomized, double-blind study, the authors (Segerdahl, 1995b) reported a number of beneficial effects of adenosine, including reduction of the inhalation agent requirement, a long-lasting analgesia that was observed postoperatively, and reduction of blood loss perioperatively, despite of the fact that blood pressure was not significantly decreased.

ADENOSINE AND ATP AS VASODILATOR AGENTS DURING ANESTHESIA FOR INDUCTION OF DELIBERATE HYPOTENSION

Induced arterial hypotension or deliberate hypotension is a technique used to lower BP electively in order to decrease blood loss during surgery and to provide a dry surgical field in certain kinds of surgery such as a major orthopedic, orthognathic, and neurosurgical procedures. The most frequently used agents for inducing deliberate hypotension include sodium nitroprusside, nitroglycerin, trimetaphan, and hydralazine—sodium nitroprusside being the most commonly used peripheral vasodilator. However, these agents suffer from disadvantages such as cyanide toxicity, rebound hypertension, tachyphylaxis, and tachycardia.

Although ATP was reported as being used clinically to induce hypotension during anesthesia in Japan (Maruyama et al., 1978; Numajiri et al., 1979), it was not until the beginning of the 1980s that the pharmacological characterization of adenosine and ATP as a vasodilator for deliberate hypotension was clearly established (Fukunaga et al., 1982a). An important factor was the technological availability of the infusion pump. The hypotensive actions of adenosine and ATP are rapid in onset and dissipate almost as rapidly after discontinuation. The marked and well-maintained hypotension without tachycardia, tachyphylaxis, or rebound hypertension that are beneficial for controlled hypotension have been supported by clinical studies (Fukunaga et al., 1982b; Ito et al., 1994). Intravenous administration of adenosine or ATP decreases systemic arterial pressure in a dose-dependent manner, while cardiac output is increased or unaffected. During blood pressure decrease to mean arterial pressure (MAP): 50 mmHg, adenosine and ATP have been shown to decrease peripheral vascular resistance, increase cardiac output, coronary blood flow, reduce myocardial and whole body oxygen consumption, and continued aerobic metabolism as indicated by low plasma lactate levels (Bloor et al., 1985; Sollevi, 1986). Hence, adenosine has been advocated in Sweden for inducing hypotension in cerebral aneurysm surgical patients as an alternative to the above-mentioned vasodilating drugs (Sollevi et al., 1984; Owall et al., 1988).

ADENOSINE AND ATP FOR PREVENTING AND/OR ATTENUATING HYPERTENSIVE CRISIS IN THE PERIOPERATIVE THERAPY

Laryngoscopy and Tracheal Intubation

It has been well recognized that laryngoscopy and tracheal intubation causes hemodynamic responses such as hypertension, tachycardia and arrhythmias, which may be of concern, particularly in patients with compromised cerebrovascular and cardiovascular disease. Reports of increases in BP of 115 mmHg systolic and 54 mmHg diastolic, paralleled by increases in plasma noradrenaline, underline the sympathetic involvement (Russell, 1981). This problem has traditionally been treated mainly by drugs that act peripherally to reduce the cardiovascular responses, including α- or β-adrenergic blockers, and vasodilators such as nitroprusside. The alternative use of purines would reduce the sympathetic outflow centrally, counteract the effects of circulating catecholamines, and dilate peripheral vessels, to result in a more salutary reduction of BP and HR. Indeed, ATP has been shown to effectively attenuate the hypertensive response to laryngoscopy and tracheal intubation (Mikawa et al., 1991).

Tumor Removal of Pheochromocytoma and Neuroblastoma

The prevention of hypertensive crisis is important for an adequate anesthetic management in patients during surgery for pheochromocytoma. Severe hypertension and cardiac arrhythmias occur as a result of massive release of catecholamine during resection of the tumor. Administration of α-adrenergic and β-adrenoceptor blocking agents such as phentolamine, phenoxybenzamine, and propranolol have been used. However, their effects are far from ideal. Adenosine or ATP have been found effective in controlling BP, preventing hypertensive crisis, and avoiding ventricular arrhythmias during surgical resection of pheochromocytoma because of their potent antiadrenergic and vasodilating effects with rapid onset and offset of action (Murata et al., 1987; Gröndal et al., 1988; Mora et al., 1994). Moreover, during dissection and manipulation of a sympathetic nervous system tumor, the blood pressure was successfully controlled by IV adenosine in a 1-month-old infant with a neuroblastoma in the pelvic fossa who suffered severe hypertension and elevated plasma norepinephrine (Sellden et al., 1995).

WOUND HEALING

Postoperative immunologic and inflammatory responses are part of a progression of complex, interrelated events that occur in response to stress and tissue injury induced by surgical trauma. Potential inflammatory mediators such as cytokines can be released by activated monocytes/macrophages and may lead to considerable cell damage in tissues (Weissman, 1990). It has been reported that purines are involved in modulation of inflammation (Cronstein and Hirschhorn, 1990), and that adenosine release may be a cellular response to tissue stimulation or injury. Adenosine exhibits a variety of anti-inflammatory activities, including reduced cytokine production and inhibition of activated neutrophil-mediated generation of oxygen free radicals (Cronstein, 1994). This suggests that adenosine could prevent the injurious effects of inflammatory cells on the surrounding, unaffected tissues. Furthermore, adenosine and ATP have been reported to prevent joint tissue injury in the arthritic rat model (Green et al., 1991), and adenosine receptor agonists have shown anti-inflammatory effects in the acute model of carrageenan-induced pleural inflammation (Schrier et al., 1990).

In the process of wound healing, progression of neovascularity is evident. Capillaries normally proliferate during inflammation and tissue repair. Adenosine has been suggested to play a role in angiogenesis (Meninger and Granger, 1991). In addition to the increased blood flow implicated as a physical factor involving capillary growth, direct chemical stimulation from locally increased concentrations of adenosine is believed to partly influence the neovascularization process, due to its ability to stimulate endothelial cell proliferation and migration. Long-term administration of adenosine has been reported to produce increase in capillary growth in the heart and skeletal muscle (Ziada et al., 1984).

Delivery of blood and oxygen to tissue is probably the most important determinant for wound healing and resistance to infection. It is known that impaired oxygen supply is a contributory cause of infection and defective repair in surgical patients. Increasing blood supply permits more rapid accumulation of collagen, enhances the angiogenic response, and accelerates wound healing. Since systemic administration of purines is known to decrease peripheral vascular resistance and to increase systemic and regional blood flow (Sollevi, 1991), it might be logical to assume that adenosine can increase oxygen tension in the extracellular wound environment and conceivably promote wound healing

PURINES IN POSTOPERATIVE CARE

Postoperative thromboembolism is a major cause of morbidity and mortality that occurs in early postoperative surgical patients. Anesthetic drugs, surgical trauma, and subsequent postoperative pain and stress accompanied by increased sympathetic activity and high plasma catecholamines can affect platelet functions (i.e., increase platelet aggregation, decrease platelet survival) and promote thrombus formation (Vlachakis et al., 1981), thus predisposing postoperative patients to vasospasm and thromboembolization. In addition to having sympatholytic, antiadrenergic, and direct peripheral vasodilating effects, purines are reported to regulate human platelet function. Adenosine- and ATP-mediated inhibition of platelet aggregation appear to be antithrombogenic, and these processes have been implicated in reducing the risk of thrombosis (Cusack and Hourani, 1991).

There is a need for pain control following surgery and after regaining consciousness from general anesthesia. Presently, narcotic analgesics are an important component of modern pain management. However, besides side effects such as pruritus, urinary retention, nausea, and vomiting, the major concern is respiratory arrest, which may occur shortly after injection or several hours later. Delayed respiratory depression has been reported anywhere from 3 to 24 hours after injection (Knill et al., 1981; Rauck, 1992). Hence, the sufficient use of narcotic analgesics is usually restricted to intensive care settings where constant vigilance can diminish the potential for serious sequelae (Gilbert, 1992).

Intraoperative analgesia rendered by the purines permits extension of analgesia into the postoperative period. Our laboratory experiments and clinical observations have shown that the analgesic effects are long-lasting (unpublished results): adenosine analgesia has been reported well beyond the point at which infusion has been stopped and inhalational anesthetics have been discontinued in patients (Segerdahl et al., 1995b). Continuing the infusion of purines into the postoperative period may provide additional benefits. Postoperative patients often experience hypertension and dysrhythmias that appear to be due to sympathetic activation and catecholamine release (Halter, 1977; Derbyshire and Smith, 1984). By titrating the purine infusion rate, hypertension and dysrhythmias may be prevented or attenuated. Besides analgesia, the anti-inflammatory and antithrombic activity of purines, as well as associated increase in local tissue blood flow and oxygenation, may improve wound healing, thereby resulting in a faster postoperative recovery.

FUTURE PERSPECTIVE

Unlike therapy for a specific disease, perioperative care must deal acutely with multiple variables. Surgery inflicts continuous stress caused by fear, anxiety, and pain of varying degrees and character, before, during, and for a variable period after an operation. The purines' wide array of actions interact and influence each other during surgery; their retaliatory and neuromodulatory effects contribute to reestablishing homeostasis. Surgical trauma and the anesthetics themselves affect numerous physiological activities at once; which causes sometimes rather severe physiological disorders. The body responds to such insult with a complex series of neuroendocrine and metabolic changes. It is conceivable that endogenous purines might play a homeostatic, neuromodulatory role under the chaotic physiopathological conditions caused by surgery. However, their concentrations may not be sufficient to successfully counteract all the excessive stimuli induced by the massive aggression of surgical intervention. We hypothesized that exogenously administered adenosine or ATP may enhance the defensive action of the endogenous purines. If adenosine's ubiquitous potential is activated in concert, at all levels, the most effective and comprehensive protective and self-preserving functions may be promoted to help accomplish the ultimate goal of anesthesia: that of protecting the patient from the pain and stress of surgical intervention, and maintaining global homeostasis during the perioperative periods. The analgesic properties of adenosine and ATP have been shown to be effective when used as anesthetic adjuvants, and these purines may deserve a place among the wide spectrum of drugs used in anesthesia.

REFERENCES

Barraco RA (1991): Behavioral actions of adenosine and related substances. In Phillis JW (ed): "Adenosine and Adenine Nucleotides as Regulators of Cellular Function." Boca Raton, FL: CRC Press, pp 339–366.

Barraco RA, Phillis JW, Delong RE (1984): Behavioral interaction of adenosine and diazepam in mice. Brain Res 323:159–163

Belfrage M, Sollevi A, Segerdahl M, Sjolund KF, Hansson P (1995): Systemic adenosine infusion alleviates spontaneous and stimulus evoked pain in patients with peripheral neuropathic pain. Anesth Analg 81:713–717.

Berne RM (1963): Cardiac nucleotides in hypoxia: Possible role in regulation of coronary blood flow. Am J Physiol 204:317–322.

Biaggioni I (1992): Contrasting excitatory and inhibitory effects of adenosine in blood pressure regulation. Hypertension 20:457–465.

Bleehen T, Keele CA (1977): Observations on the algogenic actions of adenosine compounds on the human blister base preparation. Pain 3:367–377.

Bloor BC, Fukunaga AF, Ma CC, Flacke WE, Ritter J, Van Etten A, Olewine S (1985): Myocardial hemodynamics during induced hypotension: A comparison between sodium nitroprusside and adenosine triphosphate. Anesthesiology 63:517–525.

Burnstock G (1972): Purinergic nerves. Pharmacol Rev 24:509–581.

Burnstock G (1978): A basis for distinguishing two types of purinergic receptor. In Straub RW, Bolis L (eds): "Cell Membrane Receptors for Drugs and Hormones: A Multidisciplinary Approach." New York: Raven Press, pp 107–118.

Cronstein BN (1994): Adenosine, an endogenous anti-inflammatory agent. J Appl Physiol 76:5–13.

Cronstein BN, Hirschhorn R (1990): Adenosine and host defense. In Williams M (ed): "Adenosine and Adenosine Receptors." Clifton, NJ: Humana Press, pp 475–500.

Cusack NJ, Hourani MO (1991): Adenosine, adenine nucleotides, and platelet function. In Phillis JW (ed): "Adenosine and Adenine Nucleotides as Regulators of Cellular Function." Boca Raton, FL: CRC Press, pp 121–131.

Derbyshire DR, Smith G (1984): Sympathoadrenal responses to anaesthesia and surgery. Br J Anaesth 56:725–739.

Drury AN, Szent-Györgyi A (1929): The physiological activity of adenine compounds with special reference to their action upon the mammalian heart. J Physiol (London) 68:213–237.

Dunwiddie TV (1985): The physiological role of adenosine in the central nervous system. Intl Rev Neurobiol 27:63–139.

Eger EI II, Saidman LJ, Brandstater B (1965): Minimum alveolar anesthetic concentration: A standard of anesthetic potency. Anesthesiology 26:756–763.

Ekblom A, Segerdahl M, Sollevi A (1995): Adenosine increases the cutaneous heat pain threshold in healthy volunteers. Acta Anaesthesiol Scand 39:717–722.

Ely SW, Berne RM (1992): Protective effects of adenosine in myocardial ischemia. Circulation 85:893–904.

Fredholm BB, Hedqvist P (1980): Modulation of neurotransmission by purine nucleotides and nucleosides. Biochem Pharmacol 29:1635–1643.

Fukunaga AF (1995): Intravenous administration of large dosages of adenosine or adenosine triphosphate with minimal blood pressure fluctuation. Life Sci 56:PL209–PL218.

Fukunaga A, Aida H, Fukunaga B, Ichinohe T, Mamiya H, Kaneko Y (1994): Intravenous infusion of adenosine triphosphate elevates the pain threshold of electrical tooth pulp stimulation. Drug Dev Res 31:272.

Fukunaga AF, Flacke WE, Bloor BC (1982a): Hypotensive effects of adenosine and adenosine triphosphate compared with sodium nitroprusside. Anesth Analg 61:273–278.

Fukunaga AF, Fukunaga B, Ichinohe T (1993): Intravenous adenosine produces profound and sustained analgesia in spontaneously breathing rabbits. Anesthesiology 79:A423.

Fukunaga AF, Fukunaga BM, Kikuta Y (1991): Assessing and characterizing the hypnotic and analgesic effects of anesthetic drugs. Anesthesiology 75:A45.

Fukunaga AF, Fukunaga BM, Kikuta Y (1992a): Assessment and characterization of the anesthetic effects of intravenous adenosine in the rabbit. Anesth Analg 74:S103.

Fukunaga A, Fukunaga B, Kikuta Y (1992b): Adenosine potentiates the sedative effect of midazolam without respiratory depression in rabbits. Abstracts. 10th World Congress of Anaesthesiologists, The Hague, The Netherlands, June 12–19, 1992, Abstract 223.

Fukunaga AF, Ikeda K, Matsuda I (1982b): ATP-induced hypotensive anesthesia during surgery. Anesthesiology 57:A65

Fukunaga AF, Miyamoto TA, Kikuta Y, Kaneko Y, Ichinohe T (1995): Role of adenosine and adenosine triphosphate as anesthetic adjuvants. In Belardinelli L, Pelleg A (eds): "Adenosine and Adenine Nucleotides: From Molecular Biology to Integrative Physiology." Boston: Kluwer Academic, pp 511–523.

Fukunaga AF, Taniguchi Y, Kikuta Y (1989): Effects of intravenously administered adenosine and ATP on halothane MAC and its reversal by aminophylline in rabbits. Anesthesiology 71:A260.

Gatell JA, Barner HB, Shevde K (1993): Adenosine and myocardial protection. J Cardiothorac Vasc Anesth 7:466–480.

Gilbert HC (1992): Postoperative pain management. In Vender JS, Spiess BD (eds): "Post Anesthesia Care." Philadelphia: WB Saunders, pp 292–314.

Ginsburg R, Fukunaga B, Sasaki S, Kikuta Y, Fukunaga AF (1990): Analgesic activity of intravenous adenosine: a comparison of potency with morphine sulphate. Anesthesiology 73:A362.

Gracely RH (1989): Methods of testing pain mechanisms in normal man. In Wall PD, Melzack, R (eds): "Textbook of Pain." New York: Churchill Livingstone, pp 257–268.

Green HN, Stoner HB (1950): "Biological Actions of the Adenine Nucleotides." London: Lewis.

Green PG, Basbaum AI, Helms C, Levine JD (1991): Purinergic regulation of bradykinin-induced plasma extravasation and adjuvant-induced arthritis in the rat. Proc Natl Acad Sci USA 88:4162–4165.

Gröndal S, Bindslev L, Sollevi A, Hamberger B (1988): Adenosine: A new antihypertensive agent during pheochromocytoma removal. World J Surg 12:581–585.

Gurel A, Elevli M, Hamulu A (1987): Aminophylline reversal of flunitrazepam sedation. Anesth Analg 66:333–336.

Halter JB, Pflug AE, Porte D Jr (1977): Mechanism of plasma catecholamine increases during surgical stress in man. J Clin Endocrinol Metab 45:936–944.

Ito H, Fukunaga A, Jinno S, Hirose Y, Kubota Y (1994): Adenosine triphosphate for prolonged deliberate hypotension during oral and maxillofacial surgery. Drug Dev Res 31:280.

Kaneko Y, Fukunaga AF, Suzuki M, Sakurai S, Sakurai M (1992): Intravenous ATP potentiates the sedative and hypnogenic effects of midazolam in man. Anesthesiology 77:A9.

Karlsten R, Gordh T Jr (1995): An A_1-selective adenosine agonist abolishes allodynia elicited by vibration and touch after intrathecal injection. Anesth Analg 80:844–847.

Kikuta Y, Okada K, Fukunaga AF (1992): Hemodynamic stability during ATP and midazolam-N_2O balanced anesthesia in surgical patients. Abstracts. 10th World Congress of Anaesthesiologists. The Hague, The Netherlands, June 12–19, 1992, Abstract 397.

Knill RL, Clement JL, Thompson WR (1981): Epidural morphine causes delayed and prolonged ventilatory depression. Can Anaesth Soc J 28:537–543.

Londos C, Cooper DMF, Wolff J (1980): Subclasses of external adenosine receptors. Proc Natl Acad Sci USA 77:2551–2554.

Maruyama M, Saito K, Shimoji K (1987): Hypotensive anesthesia using ATP. Masui 27:1533–1540.

Meninger CJ, Granger HJ (1991): Role of adenosine in angiogenesis. In Phillis JW (ed): "Adenosine and Adenine Nucleotides as Regulators of Cellular Function." Boca Raton, FL: CRC Press, pp 241–246.

Mikawa K, Maekawa N, Kaetsu H, Goto R, Yaku H, Obara H (1991): Effects of adenosine triphosphate on the cardiovascular response to tracheal intubation. Br J Anaesth 67:410–415.

Miller LP, Hsu C (1992): Therapeutic potential for adenosine receptor activation in ischemic brain injury. J Neurotrauma 9:S563–S577.

Mora A, Cortes C, Lopez G, Ballve M, Cabarrocas E (1994): Adenosine triphosphate in the management of hypertensive crises and abnormal heart rhythm during surgery to remove pheochromocytoma. Rev Esp Anestesiol Reanim 41:262–267.

Murata K, Sodeyama O, Ikeda K, Fukunaga AF (1987): Prevention of hypertensive crisis with ATP during anesthesia for pheochromocytoma. J Anesthesia 1:162–167.

Numajiri Y, Mokuji T, Kagami K, Niki H, Yoshida K, Hosoyamada A (1979): ATP induced hypotensive anesthesia for total hip replacement surgery. J Clin Anesth (Japan) 3:279–284.

Nunn JF, Utting JE, Brown BR Jr (1989): Introduction. In Nunn JF, Utting JE, Brown BR Jr (eds): "General Anaesthesia," 5th ed. London: Butterworths, pp 1–6.

Owall A, Lagerkranser M, Sollevi A (1988): Effects of adenosine-induced hypotension on myocardial hemodynamics and metabolism during cerebral aneurysm surgery. Anesth Analg 67:228–232.

Pappagallo M, Gaspardone A, Tomai F, Iamele M, Crea F, Gioffre PA (1993): Analgesic effect of bamiphylline on pain induced by intradermal injection of adenosine. Pain 53:199–204.

Perkins MN, Stone TW (1980): Blockade of striatal neurone responses to morphine by aminophylline: Evidence for adenosine mediation of opiate action. Br J Pharmacol 69:131–137.

Phillis JW, O'Regan MH (1988): The role of adenosine in the central actions of the benzodiazepines. Prog Neuropsychopharmacol Biol Psychiatry 12:389–404.

Phillis JW, Wu PH (1981): The role of adenosine and its nucleotides in central synaptic transmission. Prog Neurobiol 16:187–239.

Phillis JW, Edstrom JP, Ellis SW, Kirkpatrick JR (1979): Theophylline antagonizes flurazepam-induced depression of cerebral cortical neurons. Can J Physiol Pharmacol 57:917–920.

Phillis JW, Siemens RK, Wu PH (1980): Effects of diazepam on adenosine and acetylcholine release from rat cerebral cortex: Further evidence for a purinergic mechanism in action of diazepam. Br J Pharmacol 70:341–348.

Quasha AL, Eger EI II, Tinker JH (1980): Determination and applications of MAC. Anesthesiology 53:315–334.

Rauck RL (1992): Postoperative analgesia. In Brown DL (ed): "Risk and Outcome in Anesthesia," 2nd ed. Philadelphia: JB Lippincott, pp 450–487.

Roache JD, Griffiths RR (1987): Interactions of diazepam and caffeine: Behavioral and subjective dose effects in humans. Pharmacol Biochem Behav 26:801–812.

Rudolphi KA, Schubert P, Parkinson FE, Fredholm BB (1992): Neuroprotective role of adenosine in cerebral ischemia. Trends Pharmacol Sci 13:439–445.

Russell WJ, Morris RG, Frewin DB, Drew SE (1981): Changes in plasma catecholamine concentrations during endotracheal intubation. Br J Anaesth 53:837–839.

Sawynok J, Sweeney MI (1989): The role of purines in nociception. Neuroscience 32:557–569.

Sawynok J, Sweeney MI, White TD (1989): Adenosine release may mediate spinal analgesia by morphine. Trends Pharmacol Sci 10:186–189.

Schrier DJ, Lesch ME, Wright CD, Gilbertsen RB (1990): The antiinflammatory effects of adenosine receptor agonists on the carrageenan-induced pleural inflammatory response in rats. J Immunol 145:1874–1879.

Segerdahl M, Ekblom A, Sollevi A (1994): The influence of adenosine, ketamine, and morphine on experimentally induced ischemic pain in healthy volunteers. Anesth Analg 79:787–791.

Segerdahl M, Ekblom A, Sjolund KF, Belfrage M, Forsberg C, Sollevi A (1995a): Systemic adenosine attenuates touch evoked allodynia induced by mustard oil in humans. Neuroreport 6:753–756.

Segerdahl M, Ekblom A, Sandelin K, Wickman M, Sollevi A (1995b): Preoperative adenosine infusion reduces the requirements for isoflurane and postoperative analgesics. Anesth Analg 80:1145–1149.

Seitz PA, ter Riet M, Rush W, Merrell WJ (1990): Adenosine decreases the minimum alveolar concentration of halothane in dogs. Anesthesiology 73:990–994.

Sellden H, Kogner P, Sollevi A (1995): Adenosine for per-operative blood pressure control in an infant with neuroblastoma. Acta Anaesthesiol Scand 39:705–708.

Sollevi A (1986): Cardiovascular effects of adenosine in man; possible clinical implications. Prog Neurobiol 27:319–349.

Sollevi A (1991): Clinical studies on the effect of adenosine. In Imai S, Nakazawa M (eds): "Role of Adenosine and Adenine Nucleotides in the Biological System." Amsterdam: Elsevier, pp 525–537.

Sollevi A (1992): Adenosine infusion during isoflurane-nitrous oxide anaesthesia: Indications of perioperative analgesic effect. Acta Anaesthesiol Scand 36:595–599.

Sollevi A, Lagerkranser M, Irestedt L, Gordon E, Lindquist C (1984): Controlled hypotension with adenosine in cerebral aneurysm surgery. Anesthesiology 61:400–405.

Sonntag H, Stephan H, Lange H, Rieke H, Kettler D, Martschausky N (1989): Sufentanil does not block sympathetic responses to surgical stimuli in patients having coronary artery revascularization surgery. Anesth Analg 68:584–592.

Sylvén C, Beermann B, Jonzon B, Brandt R (1986): Angina pectoris-like pain provoked by intravenous adenosine in healthy volunteers. Br Med J 293:227–230.

Taniguchi Y, Fukunaga AF, Harano K, Totoki T, Nakamura Y (1993): ATP attenuates the cardiovascular responses to sternotomy during fentanyl anesthesia in CABG surgery. Anesthesiology 79:A147.

Van Calker D, Müller M, Hamprecht B (1979): Adenosine regulates via two different types of receptors, the accumulation of cyclic AMP in cultured brain cells. J Neurochem 33:999–1005.

Vapaatalo H, Onken D, Neuvonen PJ, Westermann E (1975): Stereospecificity in some central and circulatory effects of Phenylisopropyl-adenosine (PIA). Arzneimittel forschung (Drug Res) 25:407—410.

Vlachakis ND, Pratilas V, Pratila (M) (1981): Raised plasma catecholamines. Possible consequences following major surgery. N Y State J Med 81:27–35.

Warner MA, Dubbink DA (1992): Parenteral anesthetic agents. In Liu PL (ed): "Principles and Procedures in Anesthesiology." Philadelphia: JB Lippincott, pp 175–190.

Weissman C (1990): The metabolic response to stress: An overview and update. Anesthesiology 73:308–327.

Williams M (1989): Adenosine: The prototypic neuromodulator. Neurochem Int 14:249–264.

Wu PH, Phillis JW, Yuen H (1982): Morphine enhances the release of ^{3}H-purines from rat brain cerebral cortical prisms. Pharmacol Biochem Behav 17:749–755.

Ziada AMAR, Hudlicka O, Tyler KR, Wright AJA (1984). The effect of long term vasodilatation on capillary growth and performance in rabbit heart and skeletal muscle. Cardiovasc Res 18:724–732.

CHAPTER 25

Purines and Nociception

JANA SAWYNOK

Department of Pharmacology, Dalhousie University, Halifax, Nova Scotia B3H 4H7, Canada

INTRODUCTION

Adenosine and adenosine 5′-triphosphate (ATP) play complex yet potentially very significant roles in the modulation of nociception. Complexity arises partly from the ability of these purines to act at multiple sites within the pain transmission system. Thus, both can act at the periphery to produce pain-initiating or pain-enhancing properties, as well as in the spinal cord, where adenosine produces antinociception via activation of P_1 receptors while ATP can produce pronociception via activation of P_2 receptors or antinociception via breakdown to adenosine. Additional complexity is conferred by activation of selective receptor subtypes (which may produce opposing effects) and interactions with endogenous mediators. Both purines may play significant endogenous roles under certain conditions. Thus, adenosine may be an endogenous anti-inflammatory autacoid that contributes to the pain signal under conditions of inflammation, while ATP may be released in the spinal cord following activation of large diameter sensory afferents and contribute to the suppression of pain signaling under such conditions. The possible recruitment of endogenous systems by indirectly acting agents (e.g., adenosine kinase inhibitors) merits consideration. Caffeine, a nonselective adenosine receptor antagonist, is widely used as an analgesic adjuvant in combination with nonsteroidal anti-inflammatory agents and/or acetaminophen, and understanding its actions requires an appreciation of the various roles of endogenous adenosine in the pain process. A final issue that merits close attention is raised in some case reports where adenosine and its analogs have unique efficacy in neuropathic pain conditions, which are typically refractory to many other analgesic regimens. An understanding of the multiple facets of the role of purines in nociception is essential for understanding the outcome of exogenous administration of adenosine and ATP, and their analogs, antagonists, and breakdown inhibitors, as well as for the recruitment of the purine system in pursuit of the therapeutic goal of pain relief.

Purinergic Approaches in Experimental Therapeutics, Edited by Kenneth A. Jacobson and Michael F. Jarvis

ISBN 0-471-14071-6 © 1997 Wiley-Liss, Inc.

PERIPHERAL PRONOCICEPTIVE EFFECTS OF ADENOSINE ON PAIN TRANSMISSION

Algogenic and Hyperalgesic Effects of Adenosine

Administered peripherally, adenosine can produce algogenic or pain-initiating effects. The earliest demonstration of this was in a human blister base model where the direct application of adenosine to the exposed blister base induced a pain response (Bleehen and Keele, 1977). Intravenous and intracoronary administration of adenosine provoke angina-like pain responses both in normal subjects (Sylvén et al., 1986; Lagerqvist et al., 1990) and in patients with stable angina (Sylvén et al., 1988; Lagerqvist et al., 1990, 1992; Crea et al., 1990, 1992; Sylvén, 1993). In the latter instance, the quality of pain is described as indistinguishable from typical anginal pain (Sylvén et al., 1988; Crea et al., 1990). Intravenous (IV) administration of adenosine also can provoke pain in the head, jaw, neck, back, chest, and abdomen (Conradsson et al., 1987; Watt et al., 1987). Muscular ischemic pain can be provoked by injection of adenosine into the brachial (Sylvén et al., 1988) or iliac arteries (Gaspardone et al., 1994). Pain-inducing effects of adenosine are blocked by methylxanthines such as theophylline or aminophylline; the ability of these agents to reduce ischemic pain implicates endogenously released adenosine as a factor in the generation of ischemic pain (Jonzon et al., 1989; Crea et al., 1990).

Pain facilitatory or hyperalgesic effects of adenosine also have been observed directly in animal models of pain. The intradermal application of adenosine into the hind paw produces a decrease in mechanical nociceptive threshold in rats (Taiwo and Levine, 1990). In an inflammatory test, coadministration of adenosine with formalin augments both the phase 1 (Karlsten et al., 1992) and phase 2 (Doak and Sawynok, 1995) behavioral response to formalin. The phase 1 response has been attributed to a direct activation of sensory afferents by formalin, while the phase 2 response is due to an inflammatory response (Tjølsen et al., 1992).

Adenosine Receptor Subtypes and Pain

The generation of pain following intradermal injection of adenosine (Pappagallo et al., 1993) and of cardiac and muscle pain following intracoronary or intrailiac infusion, respectively, in humans (Gaspardone et al., 1995) has been attributed to adenosine A_1 receptor activation, as it is reduced by bamiphylline, which has an appreciable selectivity for adenosine A_1 compared to A_2 receptors. Cardiac pain induced by adenosine is understood to result from an activation of cardiac afferents by adenosine, as it is not observed following afferent denervation (Bertolet et al., 1993). Adenosine has been observed to cause excitation of a variety of autonomic afferent nerves (see Mechanisms of Pain Activation, below), and activation of cardiac sympathetic afferents has been attributed to activation of adenosine A_1 receptors (Dibner-Dunlap et al., 1993). In contrast to these human studies, the enhancement of the pain signal produced by the peripheral administration of adenosine in animal models has been attributed to activation of adenosine A_2 receptors (Tawio and Levine, 1990; Karlsten et al., 1992; Doak and Sawynok, 1995). Thus, mechanical hyperalgesia is mimicked by 5′-N-ethyl-carboxamidoadenosine (NECA) and 2-phenylaminoadenosine (CV 1808), but not by N^6-cyclohexyladenosine (CPA) (Taiwo and Levine, 1990), while in the formalin test, behavioral responses are enhanced by 2-[2-aminoethylamino)-carbonylethylphenyl ethylamino]-5-*N*-ethylcarboxamidoadenosine (APEC), NECA, 2-[*p*-(2-carboxyethyl)phenethylamino]5′-N-ethylcarboxamidoadenosine (CGS 21680), but not by R-N^6-(2-phenylisopropyl)-adenosine (R-PIA) or N^6-cyclohexyladenosine (CHA) (Karlsten et al., 1992; Doak and Sawynok, 1995). There is additional evidence for a peripheral A_1 receptor–mediated suppression of the pain response in a number of these studies (Taiwo and Levine, 1990; Karlsten et al., 1992; Doak and Sawynok, 1995; Khasar et al., 1995). This raises the possibility of there being opposing influences of A_1 and A_2 receptors on the peripheral generation of the pain signal.

The discrepancy between adenosine A_1 receptors promoting pain in human studies and adenosine A_2 receptors promoting pain in rodent studies could reflect a species difference. A differential location of receptors (peripheral versus cardiac afferents) does not suffice as an explanation, since adenosine A_1 receptors also have been implicated in pain generated by intradermal adenosine (Pappagallo et al., 1993), as well as in cardiac pain (Gaspardone et al., 1994). Similarly, the absence or presence of inflammation is not a significant factor, as adenosine A_2 receptors are implicated in a pressure hyperalgesia model in which there is no apparent inflammation (Taiwo and Levine, 1990), as well as in a model of inflammation (Karlsten et al., 1992; Doak and Sawynok, 1995). The issue of adenosine receptor subtypes involved in pronociception will require further clarification.

Mechanisms of Pain Activation

Both direct actions of adenosine on sensory nerve terminals and indirect effects via release of, and interactions with, endogenously released mediators may be involved in the pain enhancing properties of adenosine (Figure 1).

Direct Actions A direct visualization of adenosine receptors on sensory nerve terminals has not been made. However, adenosine receptors are present on dorsal root ganglion neuron cell bodies (Dolphin et al., 1986; MacDonald et al., 1986) and small-diameter afferent nerve terminals in the spinal cord (Santiacioli et al., 1992), making it likely that adenosine receptors are also transported to the peripheral nerve terminal of sensory neurons. The time course of action of adenosine and CV 1808 is considered to be consistent with a directly mediated effect (Taiwo and Levine, 1990). Adenosine has been shown to increase nerve traffic in the carotid sinus, pulmonary vagal, and cardiac sympathetic afferents (McQueen and Ribeiro, 1981, 1986; Runold et al., 1987; Biaggioni et al., 1991; Dibner-Dunlap et al., 1993), providing support for the concept of a direct stimulation of sensory nerve terminals. Both A_1 (Dibner-Dunlap et al., 1993) and A_2 receptors (Ribeiro, 1991) have variously been implicated in afferent stimulation by adenosine, providing support for the implication of both receptor subtypes in pain activation in cardiac and peripheral models.

The direct activation of sensory nerve terminals by adenosine A_2 agonists appears to be mediated by an increase in cyclic AMP production within the nerve terminal itself (Taiwo and Levine, 1990; Khasar et al., 1995). Interestingly, the peripheral analgesic action of adenosine A_1 agonists may be mediated by a pertussis toxin–sensitive G protein, which interacts with the stimulation of adenylate cyclase produced by hyperalgesic agents (Khasar et al., 1995).

Indirect Actions Some of the pronociceptive effects of adenosine may be mediated by indirect effects due to interactions with other cell types, the subsequent release of endogenous mediators, and interactions with these mediators.

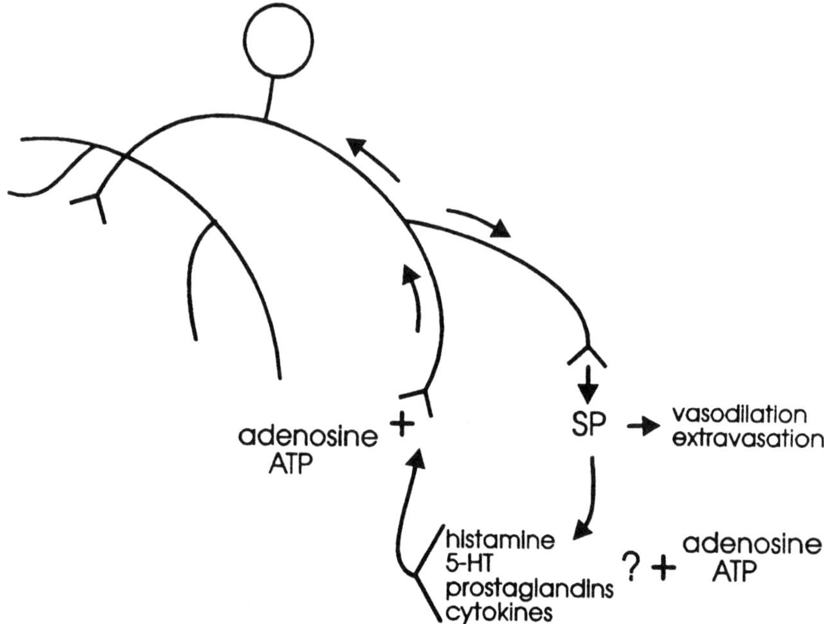

FIGURE 1. Schematic summary of effects of adenosine and ATP on peripheral sensory neurotransmission.

Substance P Substance P can be released from peripheral nerve terminals by sensory nerve activation (Dimitriadou et al., 1991) and may subsequently promote the release of inflammatory mediators such as histamine, arachidonic acid products, interleukins, and tumor necrosis factor (Lowman et al., 1988; Lotz et al., 1988; Mousli et al., 1992). These can then potentially further stimulate the nerve terminals, forming a positive stimulatory circuit (Rang et al., 1991). Substance P has recently been shown to augment the cardiac and muscle pain induced by adenosine (Gaspardone et al., 1994).

Histamine Histamine can promote itch and pain by actions at sensory nerve terminals (Keele and Armstrong, 1964). While effects of adenosine on mast cells can be complex, adenosine, via activation of adenosine A_2 (Marquardt et al., 1994) and A_3 receptors (Linden, 1994), can release histamine from mast cells. Interactions between histamine and adenosine on sensory signaling have not been reported, but adenosine has been reported to produce a selective modulation of histamine responses in central neural tissue (Paoletti et al., 1992).

5-Hydroxytryptamine 5-HT, which is released from platelets and can produce peripheral pain-enhancing effects (Richardson et al., 1985), also augments the pain signal induced by adenosine (G. Doak and J. Sawynok, unpublished observations). The pain of inflammation is considered to result from a number of interactions between pain mediators (Handwerker, 1991), and it may be that the effect of adenosine is dependent on the copresence of other endogenous mediators (cf. Euchner-Wamser et al., 1994).

Adenosine produces a spectrum of anti-inflammatory actions, particularly on neutrophil function, and adenosine has been proposed as an endogenous anti-inflammatory autacoid (Cronstein, 1994). Part of the pain signal generated by the inflammatory process appears to result from adenosine receptor activation, as behavioral responses to formalin are inhibited by the peripheral coadministration of the adenosine A_2 receptor antagonist 3,7-dimethyl-1-propargylxanthine (DMPX) (Doak and Sawynok, 1995). The contribution of adenosine to the pain signal under different conditions of inflammation remains to be assessed.

CENTRAL ANTINOCICEPTIVE/ANALGESIC EFFECTS OF ADENOSINE

Spinal Activity in Nociceptive and Neuropathic Pain Tests

Nociceptive Tests The direct spinal administration of adenosine and adenosine analogs via acute lumbar puncture or via an implanted intrathecal (IT) cannula produces antinociception or analgesia in a range of nociceptive test systems. These include the thermal threshold tail flick, tail immersion, and hot plate tests (Post, 1984; Sawynok et al., 1986; DeLander and Hopkins, 1986, 1987; Sosnowski et al., 1989; Karlsten et al., 1990); the chemically mediated acetic acid writhing test (Sosnowski et al., 1989); and the intrathecal substance P, excitatory amino acid and capsaicin behavioral tests (Hunskaar et al., 1986; Doi et al., 1987; DeLander and Wahl, 1988). More recently, the spinal administration of adenosine analogs has been shown to produce antinociception in the formalin test, a tonic pain test with inflammatory and neuroplastic components (Malmberg and Yaksh, 1993; Poon and Sawynok, 1995), and in the carrageenan thermal hyperalgesia inflammatory test (Poon and Sawynok, 1996).

The spinal administration of adenosine analogs produces additive interactions in combination with μ opioids (DeLander and Keil, 1994), but synergistic interactions in combination with α_2-adrenoceptor agonists (Aran and Proudfit, 1990a,b) and δ and κ opioid agonists (DeLander and Keil, 1994) in thermal threshold tests. In the formalin test, there is an additive interaction of R-PIA with a nonsteroidal anti-inflammatory agent (Malmberg and Yaksh, 1993).

The adenosine analogs examined in nociceptive tests have been primarily CHA, R-PIA and NECA, and on the basis of their actions, activation of both A_1 and A_2 receptors was originally implicated in spinal antinociception (reviewed Sawynok and Sweeney, 1989). However, more recent studies have noted that more selective adenosine A_2 agonists such as CV 1808 (DeLander and Wahl, 1988; Karlsten et al., 1991) and CGS 21680 (Poon and Sawynok, 1995) lack analgesic activity, and that the activity of a more extensive range of agents correlates with affinity at adenosine A_1 receptors (Karlsten et al., 1991). Spinal adenosine A_1 receptors may thus be more significant than A_2 receptors in inhibiting nociceptive input in the dorsal spinal cord.

Many of the studies mentioned above also observed that IT administration of adenosine analogs produces motor impairment in addition to antinociception, but only a limited number assessed motor impairment directly. These two actions can be dissociated in terms of time course (onset and offset of motor effects is delayed) and dose (motor effects require higher doses with some agents) (DeLander and Hopkins, 1987; Karlsten et al., 1990). Adenosine receptors are located in both the dorsal and ventral spinal cord (see Adenosine Receptors in the Spinal Cord, below); antinociception and motor effect potentially could result from actions on nociceptive processing in the substantia gelatinosa and on motor neurons in the ventral horn (Phillis and Kirkpatrick, 1978), respectively.

Neuropathic Pain In addition to the above mentioned test systems, which generally are models of nociceptive pain in which the stimulus is transmitted by small-diameter afferent fibres, adenosine analogs also reduce allodynia (touch-evoked agitation) in which the stimulus is transmitted by large-diameter afferent fibres. Thus, adenosine analogs suppress allodynia resulting from the IT administration of strychnine (Sosnowski and Yaksh, 1989; Sosnowski et al., 1989) and prostaglandin $F_{2\alpha}$ (Minami et al., 1992). Interestingly, IT R-PIA was recently shown to abolish allodynia induced by vibration and touch following a postsurgical lesion in a case report (Karlsten and Gordh, 1995). This agent has undergone some toxicological testing and appears to produce no overt changes that would preclude its use (Karlsten et al., 1992, 1993). It will be interesting to see if adenosine analogs exhibit efficacy in a larger patient population and against various forms of neuropathic pain, as this condition is relatively refractory to treatment with many commonly used analgesic agents (Arnér and Meyerson, 1988; Portenoy, 1991). The mechanisms by which purines relieve allodynia may differ from those involved in nociceptive pain (see Spinal Mechanisms of Antinociception, below), and may involve inhibition of the actions of excitatory amino acids (cf. DeLander and Wahl, 1988) or inhibition of the release of aspartate or glutamate (Corradetti et al., 1984; Burke and Nadler, 1988) with a consequent reduction in spinal sensitization phenomena (Reeve and Dickenson, 1995). Excitatory amino acids have been clearly implicated in mechanisms of plasticity and neuropathic pain within the spinal cord (Coderre et al., 1993).

Adenosine Receptors in the Spinal Cord

Adenosine A_1 and A_2 receptors have been localized in the spinal cord using both binding and autoradiographic techniques (Goodman and Snyder, 1982; Geiger et al., 1984; Choca et al., 1987, 1988a). Adenosine A_1 receptor levels are somewhat higher in the dorsal than the ventral spinal cord, but there is less differentiation for A_2 receptor levels (Geiger et al., 1984; Choca et al., 1987). The highest levels of both subtypes are found in the substantia gelatinosa (Goodman and Snyder, 1982; Geiger et al., 1984; Choca et al., 1988a). Receptors are located predominantly on interneurons (Geiger et al., 1984) and on projection neurons (Choca et al., 1988b) within the dorsal horn.

Spinal Mechanisms of Antinociception

There is evidence for actions of adenosine on primary afferent nerve terminals, on interneurons, and on projection neurons in the dorsal spinal cord, each of which could contribute to spinal antinociception as expressed in nociceptive tests (Figure 2).

Presynaptic Actions While there is no direct evidence of adenosine receptors on afferent nerve terminals in binding studies (Geiger et al., 1984; Choca et al., 1988a), adenosine analogs have been shown to inhibit the electrically evoked (but not K^+-evoked) release of peptides contained in sensory afferents (substance P, calcitonin gene–related peptide) (Vasko and Ono, 1990; Santicioli et al., 1992). This inhibitory action is mediated by adenosine A_1 receptors (Santicioli et al., 1993). The ability of adenosine analogs to inhibit behavioral effects produced by IT capsaicin also has been

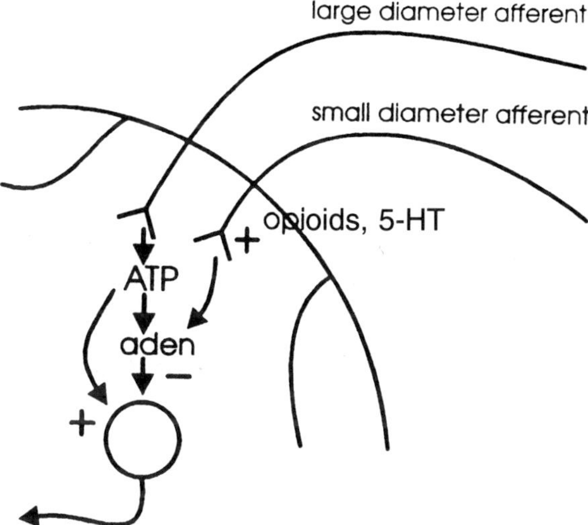

FIGURE 2. Schematic summary of effects of adenosine (aden) and ATP on sensory transmission within the dorsal horn of the spinal cord.

interpreted as reflecting a presynaptic action (Hunskaar et al., 1986). Earlier studies had demonstrated that adenosine A_1 analogs inhibited Ca^{2+} conductances in cultured sensory neurons (Dolphin et al., 1986; MacDonald et al., 1986) and provided a potential mechanistic basis for these actions.

Interneurons Adenosine's inhibition of Ca^{2+} currents in cultured spinal cord interneurons (Sah, 1990) provides some evidence for an inhibitory action at this site. There is also some electrophysiological evidence in support of an inhibitory action on interneuronal communication (Li and Perl, 1994).

Projection Neurons When adenosine and its analogs are coadministered with substance P or excitatory amino acids, there is a suppression of the behavioral effects elicited by these agents (biting, licking, scratching) (Doi et al., 1987; DeLander and Wahl, 1988). As these effects have been interpreted as reflecting a direct postsynaptic activation of sensory pathways, such observations are consistent with a direct action on transmission neurons. R-PIA has been shown to displace substance P binding from spinal cord preparations (Stiller et al., 1991), and the interaction with substance P potentially could represent a receptor-based interaction. Electrophysiological studies have provided additional evidence to support a direct postsynaptic suppression of sensory transmission (Salter and Henry, 1985; Salter et al., 1993; Li and Perl, 1994). The electrophysiological inhibition results from hyperpolarization and involves activation of an ATP- sensitive K^+ channel (Salter et al., 1992).

Supraspinal Antinociception by Adenosine

The supraspinal administration of adenosine and a number of its analogs via intracisternal (Yarbrough and McGuffin-Clineschmidt, 1981) or intracerebroventricular (ICV)

(Holmgren et al., 1986; Herrick-Davis et al., 1989; Contreras et al., 1990) injection has been shown to produce antinociception in nociceptive tests. Adenosine A_1 receptors are widely distributed in brain, while adenosine A_2 receptors exhibit a more restricted distribution (Jarvis and Williams, 1990). Adenosine A_1 receptors are present in a number of central nervous system regions known to be involved in the transmission and integration of pain (Reppert et al., 1991). In the ventrobasal complex, an area that receives input from spinothalamic tract projection neurons, these receptors are located on intrinsic neurons (Choca et al., 1989).

There is additional but less direct information on the role of supraspinal adenosine influences on nociception. Caffeine, a nonselective adenosine receptor antagonist, produces adjuvant analgesia in combination with nonsteroidal anti-inflammatory agents in humans (reviewed Sawynok and Yaksh, 1993) and produces intrinsic antinociception in threshold and inflammatory tests in animals (Sawynok et al., 1995a, and references therein). While caffeine potentially could produce antinociception by blocking peripheral pronociceptive effects of adenosine (see Peripheral Pronociceptive Effects of Adenosine on Pain Transmission, above) coadministration of caffeine with formalin does not produce antinociception (Sawynok et al., 1995a; Doak and Sawynok, 1995). The analgesic action of caffeine has been attributed to supraspinal action of caffeine, as spinal sites cannot account for such activity (Sawynok et al., 1995a). Central noradrenergic and serotonergic systems are implicated in antinociception by caffeine (Sawynok and Reid, 1996a), and it may be that blockade of adenosine receptors in areas such as the locus coeruleus (Pan et al., 1995) or dorsal and median raphe nuclei (Geyer et al., 1975) contribute to the central regulation of pain transmission. The implication of a supraspinal site of action in analgesia by caffeine presents a paradox in view of observations that supraspinal adenosine agonists produce antinociception. This issue will need to be addressed directly, using direct microinjection techniques for selective adenosine agonists and antagonists.

SYSTEMIC EFFECTS OF ADENOSINE AND ADENOSINE ANALOGS

In view of the peripheral pronociceptive effects (see Peripheral Pronociceptive Effects of Adenosine on Pain Transmission, above) and central antinociceptive effects (see Central Antinociceptive Analgesic Effects of Adenosine, above) of adenosine and its analogs, the systemic administration of these agents has the potential to be somewhat complex. In general however, systemic delivery by intraperitoneal (IP) or subcutaneous (SC) routes produces antinociception in nociceptive threshold and chemical tests in rodents (Holmgren et al., 1986; Ahlijanian and Takemori, 1985; Herrick-Davis et al., 1989), probably reflecting a spinal site of action (Holmgren et al., 1986). Such administration also enhances antinociception by morphine (Ahlijanian and Takemori, 1985; Contreras et al., 1990) (but see Mantegazza et al., 1984).

While the intravenous (IV) administration (bolus, infusions) of adenosine to humans generally provokes pain (see Algogenic and Hyperalgesic Effects of Adenosine, above), IV infusion of adenosine to humans has been reported to produce analgesia in an experimental paradigm in awake subjects (Segerdahl et al., 1994) and to decrease perioperative analgesic requirements in anesthetized subjects (Sollevi, 1992; Segerdahl et al., 1995). Preoperative adenosine infusion also reduces anesthetic requirements (Segerdahl et al., 1995). Interestingly, IV infusion of low doses of adenosine relieves spontaneous pain and tactile allodynia in neuropathic pain conditions (Sollevi et al.,

1995) (see Spinal Activity in Nociceptive and Neuropathic Pain Tests, above). While the action of IV adenosine may reflect spinal actions of adenosine on pain transmission, it should be noted that adenosine has a very short half life in blood (about 10 sec) and it is not clear to what extent adenosine crosses the blood-brain barrier (see Segerdahl et al., 1994). In addition, the duration of its effects on neuropathic pain are surprisingly long compared to the duration of infusions (Sollevi et al., 1995), raising the possibility that adenosine produces some kind of significant interruption of mechanisms involved in neuroplasticity.

RECRUITMENT OF ENDOGENOUS ADENOSINE SYSTEMS

While adenosine is ubiquitously distributed in cells, it may play a specific role in small-diameter afferent neurons in the dorsal horn of the spinal cord. Thus, adenosine deaminase is highly localized in the substantia gelatinosa and is present in capsaicin sensory afferents (Nagy and Daddona, 1985), as are nitrobenzylthioinosine-binding sites that label adenosine uptake sites (Geiger and Nagy, 1985). Adenosine can be released from dorsal spinal cord preparations by depolarization with K^+ or capsaicin (Sweeney et al., 1987, 1989), and this release is sensitive to a pretreatment with the sensory neurotoxin capsaicin. There is some evidence for tonic release of adenosine in the spinal cord, which then modulates nociceptive thresholds, since IT administration of methylxanthines produces hyperalgesia, providing that the stimulus intensity is mild (Jurna, 1984; Sawynok et al., 1986). Endogenously released adenosine may be recruited pharmacologically to produce spinal analgesia, as the IT administration of the adenosine kinase inhibitor 5′-amino-deoxyadenosine (but not the adenosine deaminase inhibitor deoxycoformycin) produces antinociception in the tail flick test (Keil and DeLander, 1992) and the formalin test (Poon and Sawynok, 1995). Both 5′-amino-deoxyadenosine and deoxycoformycin augment the release of adenosine from the spinal cord, with the degree of effect depending on the paradigm (Golembiowska et al., 1995; 1996). In addition to intrinsic actions, 5′-amino-deoxyadenosine also augments spinal antinociception by opioids (Keil and DeLander, 1994).

ENDOGENOUS ADENOSINE AS A MEDIATOR OF ANTINOCICEPTION BY OPIOIDS AND 5-HYDROXYTRYPTAMINE

The spinal release of adenosine contributes to antinociception by morphine and other μ opioids, as methylxanthines reduce spinal analgesia by morphine (Jurna, 1984; DeLander and Hopkins, 1986; Sweeney et al., 1987) and other μ opioids (DeLander et al., 1992; Cahill et al., 1995). Morphine and μ opioids release adenosine from spinal cord preparations (Sweeney et al., 1987, 1989; Cahill et al., 1995). This release occurs at nanomolar concentrations, is Ca^{2+}-dependent, requires Ca^{2+} entry via an N-type channel and arises from capsaicin-sensitive afferents (Sweeney et al., 1989; Cahill et al., 1993, 1995). There is some discrepancy in the effects of methylxanthines on antinociception by δ agonists, as both inhibition (DeLander et al., 1992) and augmentation (Cahill et al., 1995) of the action of these agents has been reported. δ Receptor agonists do not induce release of adenosine in a neurochemical paradigm (Cahill et al., 1995). Antinociception by κ agonists is not altered by methylxanthines (DeLander and Keil, 1994). The adenosine released by opioids subsequently activates spinal adenosine A_1 receptors, since it is blocked by a selective adenosine A_1 receptor antagonist (cyclopentyltheophylline) but not by a selective adenosine A_2 receptor antagonist (DMPX) (C. Cahill and J. Sawynok, unpublished observations). This provides further evidence for the relative importance of spinal adenosine A_1 receptors in the spinal regulation of nociception (see Spinal Activity in Nociceptive and Neuropathic Pain Tests, above).

Adenosine release also contributes to spinal antinociception by 5-HT, as methylxanthines inhibit spinal antinociception by 5-HT (DeLander and Hopkins, 1987; Sawynok and Reid, 1991), and 5-HT releases adenosine (which originates as nucleotide, probably cyclic AMP) from *in vitro* and *in vivo* spinal cord preparations (Sweeney et al., 1988, 1990). This effect may be due to activation of a 5-HT_1-like receptor, as the action of certain 5-HT_1 (but not 5-HT_2 or 5-HT_3 receptor ligands is also reduced by methylxanthines; a 5-HT_4 receptor also could be involved due to the positive implication of cyclic AMP) (Sawynok and Reid, 1996b).

Supraspinal administration of opioids also can induce a spinal release of adenosine (Sweeney et al., 1991), and antinociception by supraspinally administered opioids can be reduced by IT administration of methylxanthines (DeLander and Hopkins, 1986; DeLander et al., 1992; Sweeney et al., 1991; Sawynok et al., 1991). This spinal release of adenosine arises secondary to activation of spinally projecting 5-HT pathways (Sweeney et al., 1991; Sawynok et al., 1991).

PERIPHERAL EFFECTS OF ATP ON PAIN TRANSMISSION

Adenosine triphosphate is a ubiquitous component of cells, and tissue damage is likely to result in accumulation of ATP at sensory nerve endings. Local application of ATP to the human blister base produces a pain response (Bleehen and Keele, 1977), while coadministration of ATP with low concentrations of formalin enhances the phase 2 behavioral response to formalin (Sawynok et al., 1995b). These actions may be due to direct actions of ATP on the sensory nerve terminal, since ATP produces a direct depolarization of sensory neurons, which results from an increase in cation permeability (Krishtal et al., 1983, 1988; Bean, 1990). The receptor subtype mediating this response has recently been definitively characterized as a P_{2x} heteromer using electrophysiological approaches (Chen et al., 1995; Lewis et al., 1995).

Some of the effects of ATP on initiation of the pain signal also may be due to interactions with other mediators. Thus, ATP receptors have been identified on neutrophils, macrophages, and monocytes, which could potentially result in the release of various inflammatory mediators such as cytokines and prostanoids, or prime the effector cell for an enhanced response to other mediators (Dubyak and El-Moatassim, 1993). The possible involvement of such indirect actions of ATP with cellular elements, and of interactions with various inflammatory mediators, in the initiation of the pain signal has not been addressed directly, but is likely to occur [cf. delay in onset in augmentation of response to formalin (Sawynok et al., 1995b)]. Interactions with 5-HT (Bleehen and Keele, 1977), bradykinin (Green et al., 1991), and substance P (Patra and Westfall, 1996) occur in related systems, and these are likely co-agonists.

CENTRAL AND SYSTEMIC EFFECTS OF ATP ON NOCICEPTION

There has been interest in the role of ATP in sensory transmission since ATP was reported to be released from sensory nerve terminals by the antidromic activation of sensory nerves (Holton, 1959). In cultured spinal cord preparations, ATP excites a population of dorsal horn neurons (Jahr and Jessell, 1983). When applied *in vivo* to functionally identified neurons, ATP excites some nociceptive (Salt and Hill, 1983; Salter and Henry, 1985), but predominantly non-nociceptive, neurons (Salt and Hill, 1983; Fyffe and Perl, 1984; Salter and Henry, 1985). Some of the excitatory effects of ATP may be facilitatory with other agents (e.g., glutamate; Li and Perl, 1995). Applied to wide dynamic range neurons, ATP produces a mix of excitatory (via P_2 receptors) and inhibitory (via P_1 receptors) actions (Salter and Henry, 1985). ATP is released from spinal cord preparations (dorsal > ventral) by depolarization (Yoshioka and Jessell, 1984; White et al., 1985), and there is both electrophysiological (Fyffe and Perl, 1984; Salter and Henry, 1987) and biochemical evidence (Sawynok et al., 1993) in support of the concept of release of ATP from large-diameter afferent nerve terminals. ATP released from large-diameter sensory afferents with a subsequent activation of P_1 receptors has been implicated in vibration-induced analgesia in animals (Salter and Henry, 1987) and in transcutaneous electrical nerve stimulation in humans (Marchand et al., 1995). ATP clearly has the potential to influence spinal nociceptive processing both directly, by activation of P_2, and indirectly, by activation of P_1 receptors (Figure 2).

In behavioral experiments, the IT application of analogs of ATP (α,β-MeATP, 2-MeSATP) produces hyperalgesia in the tail flick test (Driessen et al., 1994). Conversely, P_2 receptor antagonists (suramin, Evans blue, reactive blue, trypan blue) produce antinociception (Driessen et al., 1994), which is suggestive of a facilitatory effect of spinal ATP on nociceptive transmission. Although the order of potency of the ATP analogs (α,β-MeATP > 2-MeSATP) was suggestive of P_{2X} receptor involvement, the lack of effect of a selective P_{2X} antagonist in this spinal action of ATP, suggests caution in interpretation (Driessen et al., 1994).

The systemic (SC) administration of the P_2 antagonist suramin produces antinociception in a thermal threshold and chemical test (Ho et al., 1992), as well as in the formalin test (Sawynok et al., 1995b). This activity may be due to actions at a spinal site, because the IT administration of suramin produces a reduction in responses to formalin as well as tail flick antinociception (Driessen et al., 1994). While such observations are suggestive of a spinal role of ATP in pain transmission, it is important to appreciate that suramin has a number of pharmacological actions in addition to blocking P_2 receptors (Dalziel and Westfall, 1994), and some caution in interpretation is required.

Both the IT and IV application of ATP also can produce antinociception due to breakdown to adenosine. Thus, IT ATP (also ADP and AMP) mimics the action of adenosine in reducing behavioral responses to IT substance P, and these actions are completely blocked by theophylline (Doi et al., 1987). With IV administration, ATP produces both an intrinsic analgesia (Gomaa, 1987; Kikuta et al., 1990) and an anesthetic sparing action (Fukunaga et al., 1989). Both actions are blocked by aminophylline (Kikuta et al., 1990; Fukunaga et al., 1989). A similar anesthetic sparing effect has been reported for IV ATP in humans (Fukunaga et al., 1990). While the effects of IV ATP are attributed to central actions via P_1 receptors, the half-life of these purines in blood is short, and it is not clear to what extent these agents cross the blood-brain barrier.

CONCLUSIONS

Adenosine agents (adenosine and its analogs or breakdown inhibitors) show considerable promise for development as analgesics. In general, the utility of any class of agent will depend upon the route of administration (systemic vs. spinal vs. peripheral) and the pain condition (nociceptive vs. inflammatory vs. neuropathic). More specifically, the utility of an exogenous agonist will rely on the juxtaposition of adenosine receptors with critical elements of the condition (e.g., following neuroplastic changes induced by neuropathic pain), while that of indirectly acting agents will rely upon the amount of endogenously expressed adenosine in a particular condition. It is important to note that adenosine and its analogs have already been applied to humans with some unique benefit in neuropathic pain conditions. Adenosine itself is an endogenous agent with a short half-life and these factors contribute to its safety. Some toxicology has been performed with R-PIA following spinal administration, and the evidence has been encouraging in that no effect which would preclude its use has been observed. As mechanisms involved in the spinal regulation of pain emerge, adenosine A_1 receptors are being increasingly implicated in the spinal suppression of pain. It will therefore be important to evaluate the toxicology of agents with greater adenosine A_1 receptor selectivity such as CHA or CPA. A limitation to the use of adenosine analogs may be the appearance of motor effects. While, this is less of an issue for A_1 agonists than for A_2 agonists, it still requires direct attention. Interestingly, indirectly acting adenosine agents appear devoid of motor effects at doses with marked effects on nociception, such that this class of agents shows some promise. With respect to the use of adenosine antagonists as analgesic agents, the issue of adenosine A_1 versus A_2 receptors being involved in pronociception at the sensory nerve terminal will need to be adequately resolved. If an adenosine A_2 receptor is involved, the use of a selective A_2 receptor antagonist may not be useful in ischemic cardiac pain, as this would additionally block the beneficial vasodilatory effects induced by adenosine A_2 receptors in the vasculature. It is important to note that caffeine, a nonselective adenosine receptor antagonist, is widely used as an analgesic adjuvant, and its efficacy in this regard is established. A consideration of the role of purines in nociception clearly merits close attention.

ACKNOWLEDGMENTS

Work conducted in the author's laboratory and quoted herein was supported by the Medical Research Council of Canada.

REFERENCES

Ahlijanian MK, Takemori AE (1985): Effects of (−)-N^6-(R-phenylisopropyl)-adenosine (PIA) and caffeine on nociception and morphine-induced analgesia, tolerance and dependence in mice. Eur J Pharmacol 112:171–179.

Aran S, Proudfit HK (1990a): Antinociceptive interactions between intrathecally administered α noradrenergic agonists and 5′-*N*-ethylcarboxamide adenosine. Brain Res 519:287–293.

Aran S, Proudfit HK (1990b): Antinociception produced by interactions between intrathecally administered adenosine agonists and norepinephrine. Brain Res 513:255–263.

Arnér S, Meyerson BA (1988): Lack of analgesic effect of opioids on neuropathic and idiopathic forms of pain. Pain 33:11–23.

Bean BP (1990): ATP-activated channels in rat and bullfrog sensory neurons: Concentration dependence and kinetics. J Neurosci 10:1–10.

Bertolet B, Belardinelli L, Hill JA (1993): Absence of adenosine-induced chest pain after total cardiac afferent denervation. Am J Cardiol 72:483–484.

Biaggioni I, Killian TJ, Mosqueda-Garcia R, Robertson RM, Robertson D (1991): Adenosine increases sympathetic nerve traffic in humans. Circulation 83:1668–1675.

Bleehen T, Keele CA (1977): Observations on the algogenic actions of adenosine compounds on the human blister base preparation. Pain 3:367–377.

Burke SP, Nadler JV (1988): Regulation of glutamate and aspartate release from slices of the hippocampal CA1 area: Effects of adenosine and baclofen. J Neurochem 51:1541–1551.

Cahill CM, White TD, Sawynok J (1993): Morphine activates ω-conotoxin-sensitive Ca^{2+} channels to release adenosine from spinal cord synaptosomes. J Neurochem 60:894–901.

Cahill CM, White TD, Sawynok J (1995): Spinal opioid receptors and adenosine release: Neurochemical and behavioural characterization of opioid subtypes. J Pharmacol Exp Ther 275:84–93.

Chen CC, Akoplan AN, Sivilotti L, Colquhoun D, Burnstock G, Wood JN (1995): A P_{2x} purinoreceptor expressed by a subset of sensory neurons. Nature 377:428–431.

Choca JI, Green RD, Proudfit HK (1988a): Adenosine A_1 and A_2 receptors of the substantia gelatinosa are located predominantly on intrinsic neurons: An autoradiographic study. J Pharmacol Exp Ther 247:757–764.

Choca JI, Green RD, Proudfit HK (1988b): Autoradiographic evidence that spinothalamic tract neurons possess adenosine receptors. Soc Neurosci Abstr 14:776.

Choca JI, Green RD, Proudfit HK (1989): Adenosine A_1 receptors are located on the intrinsic neurons of the ventrobasal complex of the thalamus. Soc Neurosci Abstr 15:232.

Choca JI, Proudfit HK, Green RD (1987): Identification of A_1 and A_2 adenosine receptors in the rat spinal cord. J Pharmacol Exp Ther 242:905–910.

Coderre TJ, Katz J, Vaccarino AL, Melzack R (1993): Contribution of central neuroplasticity to pathological pain: Review of clinical and experimental evidence. Pain 52:259–285.

Conradsson TB, Dixon CMS, Clark B, Barnes PJ (1987): Cardiovascular effects of infused adenosine in man: Potentiation by dipyridamole. Acta Physiol Scand 129:387–391.

Contreras E, Germany A, Villar M (1990): Effects of some adenosine analogs on morphine-induced analgesia and tolerance. Gen Pharmacol 21:763–767.

Corradetti R, Lo Conte G, Moroni F, Passani MB, Pepeu G (1984): Adenosine decreases aspartate and glutamate release from rat hippocampal slices. Eur J Pharmacol 104:19–26.

Crea F, Gaspardone A, Kaski JC, Davies G, Maseri A (1992): Relation between stimulation site of cardiac afferent nerves by adenosine and distribution of cardiac pain: Results of a study in patients with stable angina. J Am Coll Cardiol 20:1498–1502.

Crea F, Pupita G, Galassi AR, El-Tamimi H, Kaski JC, Davies G, Maseri A (1990): Role of adenosine in pathogenesis of anginal pain. Circulation 81:164–172.

Cronstein BN (1994): Adenosine, an endogenous anti-inflammatory agent. J Appl Physiol 76:5–13.

Dalziel HH, Westfall DP (1994): Receptors for adenine nucleotides and nucleosides: Subclassification, distribution, and molecular characterization. Pharmacol Rev 46:449–466.

DeLander GE, Hopkins CJ (1986): Spinal adenosine modulates descending antinociceptive pathways stimulated by morphine. J Pharmacol Exp Ther 239:88–93.

DeLander GE, Hopkins CJ (1987): Involvement of A_2 adenosine receptors in spinal mechanisms of antinociception. Eur J Pharmacol 139:215–223.

Delander GE, Keil GJ (1994): Antinociception induced by intrathecal coadministration of selective adenosine receptor and selective opioid receptor agonists in mice. J Pharmacol Exp Ther 268:943–951.

DeLander GE, Wahl JJ (1988): Behavior induced by putative nociceptive neurotransmitters is inhibited by adenosine or adenosine analogs coadministered intrathecally. J Pharmacol Exp Ther. 246:565–570.

DeLander GE, Mosberg HI, Porreca F (1992): Involvement of adenosine in antinociception produced by spinal or supraspinal receptor-selective opioid agonists: Dissociation from gastrointestinal effects in mice. J Pharmacol Exp Ther 263:1097–1104.

Dibner-Dunlap ME, Kinugawa T, Thames MD (1993): Activation of cardiac sympathetic afferents: Effects of exogenous adenosine and adenosine analogues Am J Physiol 265:H395–H400.

Dimitriadou V, Buzzi MG, Moskowitz MA, Theoharides TC (1991): Trigeminal sensory fiber stimulation induces morphological changes reflecting secretion in rat dura mater mast cells. Neuroscience 44:97–112.

Doak GJ, Sawynok J (1995): Complex role of peripheral adenosine in the genesis of the response to subcutaneous formalin in the rat. Eur J Pharmacol (in press).

Doi T, Kuzuna S, Maki Y (1987): Spinal antinociceptive effects of adenosine compounds in mice. Eur J Pharmacol 137:227–231.

Dolphin AC, Forda SR, Scott RH (1986): Calcium-dependent currents in cultured rat dorsal root ganglion neurons are inhibited by an adenosine analogue. J Physiol (London) 373:47–61.

Driessen B, Reimann W, Selve N, Friderichs E, Bültmann R (1994): Antinociceptive effect of intrathecally administered P_2-purinoceptor antagonists in rats. Brain Res 666:182–188.

Dubyak GR, El-Moatassim C (1993): Signal transduction via P_2-purinergic receptors for extracellular ATP and other nucleotides. Am J Physiol 265:C577–C606.

Euchner-Wamser I, Meller ST, Gebhart GF (1994): A model of cardiac nociception in chronically instrumented rats: behavioural and electrophysiological effects of pericardial administration of algogenic substances. Pain 58:117–128.

Fukunaga AF, Kaneko Y, Ichinohe T, Igarashi O, Nakakuki T (1990): Intravenous ATP attenuates surgical stress responses and reduces inhalation anesthetic requirements in humans. Anesthesiology 73:A400.

Fukunaga AF, Taniguchi Y, Kikuta Y (1989): Effects of intravenously administered adenosine and ATP on halothane MAC and its reversal by aminophylline in rabbits. Anesthesiology 71:A259.

Fyffe REW, Perl ER (1984): Is ATP a central synaptic mediator for certain primary afferent fibres from mammalian skin? Proc Natl Acad Sci USA 81:6890–6893.

Gaspardone A, Crea F, Tomai F, Iamele M, Crossman DC, Pappagallo M, Versaci F, Chiariello L, Gioffrè PA (1994): Substance P potentiates the algogenic effects of intraarterial infusion of adenosine. J Am Coll Cardiol 24:477–482.

Gaspardone A, Crea F, Tomai F, Versaci F, Iamele M, Gioffrè G, Chiariello L, Gioffrè PA (1995): Muscular and cardiac adenosine-induced pain is mediated by A_1 receptors. J Am Coll Cardiol 25:251–257.

Geiger JD, Nagy JI (1985): Localization of [^{3}H]nitrobenzylthioinosine binding sites in rat spinal cord and primary afferent neurons. Brain Res 347:321–327.

Geiger JD, LaBella FS, Nagy JI (1984): Characterization and localization of adenosine receptors in rat spinal cord. J Neurosci 4:2303–2310.

Geyer MA, Dawsey WJ, Mandell AJ (1975): Differential effects of caffeine, D-amphetamine and methylphenidate on individual raphe cell fluorescence; a microspectrofluorimetric demonstration. Brain Res 85:135–139.

Golembiowska K, White TD, Sawynok J (1995): Modulation of adenosine release from rat spinal cord by adenosine deaminase and adenosine kinase inhibitors. Brain Res 699:315–320.

Golembiowska K, White TD, Sawynok J (1996): Adenosine kinase inhibitors augment release of adenosine from spinal cord slices. Eur J Pharmacol 307:157–162.

Gomaa AA (1987): Characteristics of analgesia induced by adenosine triphosphate. Pharmacol Toxicol 61:199–202.

Goodman RR, Snyder SH (1982): Autoradiographic localization of adenosine receptors in rat brain using [^{3}H]cyclohexyladenosine. J Neurosci 2:1230–1241.

Green PG, Basbaum AI, Helms G, Levine JD (1991): Purinergic regulation of bradykinin-induced plasma extravasation and adjuvant-induced arthritis in the rat. Proc Natl Acad Sci USA 88:4162–4165.

Handwerker HO (1991): What peripheral mechanisms contribute to nociceptive transmission and hyperalgesia? In Basbaum AI, Besson J-M (eds): "Towards a New Pharmacotherapy of Pain." New York: John Wiley & Sons, pp 5–19.

Herrick-Davis K, Chippari S, Luttinger D, Ward SJ (1989): Evaluation of adenosine agonists as potential analgesics. Eur J Pharmacol 162:365–369.

Ho BT, Huo YY, Lu JG, Newman RA, Levin VA (1992): Analgesia activity of anticancer agent suramin. Anti-Cancer Drugs 3:91–94.

Holmgren M, Hedner J, Mellstrand T, Nordberg G, Hedner T (1986): Characterization of the antinociceptive effects of some adenosine analogues in the rat. Naunyn Schmiedebergs Arch Pharmacol 334:290–293.

Holton P (1959): The liberation of adenosine triphosphate on antidromic stimulation of sensory nerves. J Physiol 145:494–504.

Hunskaar S, Post C, Fasmer OB, Arwestrom E (1986): Intrathecal injection of capsaicin can be used as a behavioural nociceptive test in mice. Neuropharmacology 25:1149–1153.

Jahr CE, Jessell TM (1983): ATP excites a subpopulation of rat dorsal horn neurones. Nature 304:730–733.

Jarvis MF, Williams M (1990): Adenosine in central nervous system function. In Williams M (ed): "Adenosine and Adenosine Receptors." Clifton, NJ: Humana Press, pp 423–474.

Jonzon B, Sylvén C, Kaijser L (1989): Theophylline decreases pain in the ischaemic forearm test. Cardiovasc Res 9:807–809.

Jurna I (1984): Cyclic nucleotides and aminophylline produce different effects on nociceptive motor and sensory responses in the rat spinal cord. Naunyn Schmiedebergs Arch Pharmacol 327:23–30.

Karlsten R, Gordh T (1995): An A_1-selective adenosine agonist abolishes allodynia elicited by vibration and touch after intrathecal injection. Anesth Analg 50:844–847.

Karlsten R, Gordh T, Hartvig P, Post C (1990): Effects of intrathecal injection of the adenosine receptor agonists R-phenylisopropyl-adenosine and *N*-ethylcarboxamide-adenosine on nociception and motor function in the rat. Anesth Analg 71:60–64.

Karlsten R, Gordh T, Svensson BA (1993): A neurotoxicological evaluation of the spinal cord after chronic intrathecal injection of R-phenylisopropyl adenosine (R-PIA) in the rat. Anesth Analg 77:731–736.

Karlsten R, Kristensen JD, Gordh T (1992): R-phenylisopropyl-adenosine increases spinal cord blood flow after intrathecal injection in the rat. Anesth Analg 75:972–976.

Karlsten R, Post C, Hide I, Daly JW (1991): The antinociceptive effect of intrathecally administered adenosine analogs in mice correlates with the affinity for the A_1-adenosine receptor. Neurosci Lett 121:267–270.

Keele CA, Armstrong D (1964): "Substances Producing Pain and Itch." Baltimore: Williams & Wilkins, pp 124–151.

Keil GJ II, DeLander GE (1992): Spinally-mediated antinociception is induced in mice by an adenosine kinase-, but not by an adenosine deaminase-, inhibitor. Life Sci 51:171–176.

Keil GJ, DeLander GE (1994): Adenosine kinase and adenosine deaminase inhibition modulate spinal adenosine- and opioid agonist-induced antinociception in mice. Eur J Pharmacol 271:37–46.

Khasar SG, Wang J-F, Taiwo YO, Heller PH, Green PG, Levine JD (1995): Mu-opioid agonist enhancement of prostaglandin-induced hyperalgesia in the rat: A G-protein $\beta\gamma$ subunit-mediated effect? Neuroscience 67:189–195.

Kikuta Y, Fukunaga AF, Ginsburg R, Fukunaga B, Okada K (1990): Effect of intravenous ATP on enflurane-N_2O MAC in spontaneously breathing rabbits: Assessment of cardio-respiratory effects. Anesthesiology 73:A401.

Krishtal OA, Marchenka SM, Obukhov AG (1988): Cationic channels activated by extracellular ATP in rat sensory neurons. Neuroscience 27:995–1000.

Krishtal OA, Marchenko SM, Pidoplichko VI (1983): Receptor for ATP in the membrane of mammalian sensory neurones. Neurosci Lett 35:41–45.

Lagerqvist B, Sylvén C, Beermann B, Helmius G, Waldenström A (1990): Intracoronary adenosine causes angina pectoris like pain—an inquiry into the nature of visceral pain. Cardiovasc Res 24:609–613.

Lagerqvist B, Sylvén C, Theodorsen E, Kaijser L, Helmius G, Waldenström A (1992): Adenosine induced chest pain—a comparison between intracoronary bolus injection and steady state infusion. Cardiovasc Res 26:810–814.

Lewis C, Neldhart S, Holy C, North RA, Buell G, Suprenant A (1995): Coexpression of P_{2x_2} and P_{2x_3} receptor subunits can account for ATP-gated currents in sensory neurons. Nature 377:432–435.

Li J, Perl ER (1994): Adenosine inhibition of synaptic transmission in the substantia gelatinosa. J Neurophysiol 72:1611–1621.

Li J, Perl ER (1995): ATP modulation of synaptic transmission in the spinal substantia gelatinosa. J Neurosci 15:3357–3365.

Linden J (1994): Cloned adenosine A_3 receptors; pharmacological properties, species differences and receptor functions. Trends Pharmacol Sci 15:298–306.

Lotz M, Vaughan JH, Carson DA (1988): Effect of neuropeptides on production of inflammatory cytokines by human monocytes. Science 241:1218–1221.

Lowman MA, Benyon RC, Church MK (1988): Characterization of neuropeptide-induced histamine release from human dispersed skin mast cell. Br J Pharmacol 95:121–130.

MacDonald RL, Skerritt JH, Werz MA (1986): Adenosine agonists reduce voltage-dependent calcium conductance of mouse sensory neurons in cell culture. J Physiol (London) 370:75–90.

Malmberg AB, Yaksh TL (1993): Pharmacology of the spinal action of ketorolac, morphine, ST-91, U50488H, and L-PIA on the formalin test and an isobolographic analysis of the NSAID interaction. Anesthesiology 79:270–281.

Mantegazza P, Tammiso R, Zambotti F, Zecca L, Zonta N (1984): Purine involvement in morphine antinociception. Br J Pharmacol 83:883–888.

Marchand S, Li J, Charest J (1995): Effects of caffeine on analgesia from transcutaneous electrical nerve stimulation. N Engl J Med 333:325–326.

Marquardt DL, Walker LL, Heinemann S (1994): Cloning of two adenosine receptor subtypes from mouse bone marrow-derived mast cells. J Immunol 152:4508–4515.

McQueen DS, Ribeiro JA (1981): Effect of adenosine on carotid chemoreceptor activity in the cat. Br J Pharmacol 74:129–136.

McQueen DS, Ribeiro JA (1986): Pharmacological characterization of the receptor involved in chemoexcitation induced by adenosine. Br J Pharmacol 88:615–620.

Minami T, Uda R, Horiguchi S, Ito S, Hyodo M, Hayashi O (1992): Allodynia evoked by intrathecal administration of prostaglandin $F_{2\alpha}$ to conscious mice. Pain 50:223–229.

Mousli M, Fischer T, Landry Y (1992): Role of phospholipase A_2 and pertussis toxin-sensitive G proteins in histamine secretion induced by substance P. Agents Actions 35:C305–C307.

Nagy JI, Daddona PE (1985): Anatomical and cytochemical relationships of adenosine deaminase-containing primary afferent neurons in the rat. Neuroscience 15:799–813.

Pan WJ, Osmanović SS, Shefner SA (1995): Characterization of the adenosine A_1 receptor-activated potassium current in rat locus coeruleus neurons. J Pharmacol Exp Ther 273:537–544.

Paoletti AM, Balduini W, Cattabeni F, Abbracchio MP (1992): Adenosine modulation of histamine-stimulated phosphoinositides metabolism in rat cortical slices. Neurosci Res Commun 11:19–26.

Patra PB, Westfall DP (1996): Potentiation by bradykinin and substance P of purinergic neurotransmission in urinary bladder. J Urol 156:532–535.

Pappagallo M, Gaspardone A, Tomai F, Iamele M, Crea F, Gioffré PA (1993): Analgesic effect of bamiphylline on pain induced by intradermal injection of adenosine. Pain 53:199–204.

Phillis JW, Kirkpatrick JR (1978): The actions of adenosine and various nucleosides and nucleotides on the isolated toad spinal cord. Gen Pharmacol 9:239–247.

Poon A, Sawynok J (1995): Antinociception by adenosine analogs and an adenosine kinase inhibitor: dependence on formalin concentration. Eur J Pharmacol 286:177–184.

Poon A, Sawynok J (1996): Spinal analgesia by adenosine agonists and inhibitors of adenosine metabolism, with synergy between an adenosine kinase and an adenosine deaminase inhibitor. Br J Pharmacol 119(Suppl.) 204P

Portenoy RK (1991): Issues in the management of neuropathic pain. In Bausbaum AI, Besson J-M (eds): "Towards a New Pharmacotherapy of pain." New York: John Wiley & Sons, pp 393–414.

Post C (1984): Antinociceptive effects in mice after intrathecal injection of 5′-N-ethylcarboxamide adenosine. Neurosci Lett 51:325–330.

Rang HP, Bevan S, Dray A (1991): Chemical activation of nociceptive peripheral neurones. Br Med Bull 47:534–548.

Reeve AJ, Dickenson AH (1995): The roles of spinal adenosine receptors in the control of acute and more persistent nociceptive responses of dorsal horn neurons in the anaesthetized rat. Br J Pharmacol 116:2221–2228.

Reppert SM, Weaver DR, Stehle JH, Rivkees SA (1991): Molecular cloning and characterization of a rat A_1-adenosine receptor that is widely expressed in brain and spinal cord. Mol Endocrinol 5:1037–1048.

Ribiero JA (1991): Adenosine and the central nervous system control of autonomic function. In: "Adenosine in the Nervous System." London: Academic Press pp 229–246.

Richardson BP, Engel G, Donatsch P, Stadler PA (1985): Identification of serotonin M-receptor subtypes and their specific blockade by a new class of drugs. Nature 316:126–131.

Runold M, Prabhakar NR, Mitra J, Cherniack NS (1987): Adenosine stimulates respiration by acting on vagal receptors. Fed Proc 46:825.

Sah DWY (1990): Neurotransmitter modulation of calcium current in rat spinal cord neurons. J Neurosci 10:136–141.

Salt TE, Hill RG (1983): Excitation of single sensory neurones in the rat caudal trigeminal nucleus by iontophoretically applied adenosine 5′-triphosphate. Neurosci Lett 35:53–57.

Salter MW, Henry JL (1985): Effects of adenosine 5′-monophosphate and adenosine 5′-triphosphate on functionally identified units in the cat spinal dorsal horn. Evidence for a differential effect of adenosine 5′-triphosphate on nociceptive vs non-nociceptive units. Neuroscience 3:815–825.

Salter MW, Henry JL (1987): Evidence that adenosine mediates the depression of spinal dorsal horn neurons induced by peripheral vibration in the cat. Neuroscience 22:631–650.

Salter MW, De Koninck Y, Henry JL (1992): ATP-sensitive K^+ channels mediate in IPSP in dorsal horn neurones elicited by sensory stimulation. Synapse 11:214–220.

Salter MW, De Koninck Y, Henry JL (1993): Physiological roles for adenosine and ATP in synaptic transmission in the spinal dorsal horn. Prog Neurobiol 41:125–156.

Santicioli P, Del Bianco E, Maggi CA (1993): Adenosine A_1 receptors mediate the presynaptic inhibition of calcitonin gene-related peptide release by adenosine in the rat spinal cord. Eur J Pharmacol 231:139–142.

Santicioli P, Del Bianco E, Tramontana M, Maggi CA (1992): Adenosine inhibits action potential-dependent release of calcitonin gene-related peptide- and substance P-like immunoreactivities from primary afferents in rat spinal cord. Neurosci Lett 144:211–214.

Sawynok J, Reid A (1991): Noradrenergic and purinergic involvement in spinal antinociception by 5-hydroxytryptamine and 2-methyl-5-hydroxytryptamine. Eur J Pharmacol 204:301–309.

Sawynok J, Reid A (1996a): Neurotoxin-induced lesions to central serotonergic, noradrenergic and dopaminergic systems modify caffeine-induced antinociception in the formalin test and locomotor stimulation in rats. J Pharmacol Exp Ther 277:646–653.

Sawynok J, Reid A (1996b): Interactions of descending serotonergic systems with other neurotransmitters in the modulation of nociception. Behav Brain Res 73:63–68.

Sawynok J, Sweeney MI (1989): The role of purines in nociception. Neuroscience 32:557–569.

Sawynok J, Yaksh TL (1993): Caffeine as an analgesic adjuvant: A review of pharmacology and mechanisms of action. Pharmacol Rev 45:43–85.

Sawynok J, Downie JW, Reid AR, Cahill CM, White TD (1993): ATP release from dorsal spinal cord synaptosomes: characterization and neuronal origin. Brain Res 610:32–38.

Sawynok J, Espey MJ, Reid A (1991): 8-Phenyltheophylline reverses the antinociceptive action of morphine microinjected into the periaquaductal gray region of the rat. Neuropharmacology 30:871–877.

Sawynok J, Reid AR, Doak GJ (1995a): Caffeine antinociception in the rat hot plate and formalin tests and locomotor stimulation: Involvement of noradrenergic mechanisms. Pain 61:203–213.

Sawynok J, Reid AR, Doak GJ (1995b): ATP and α,β-MeATP potentiate while suramin inhibits nociception in the formalin test. Soc Neurosci Abstr 21:387.

Sawynok J, Sweeney MI, White TD (1986): Classification of adenosine receptors mediating antinociception in the rat spinal cord. Br J Pharmacol 88:923–930.

Segerdahl M, Ekblom A, Sandelin K, Wickman M, Sollevi A (1995): Peroperative adenosine infusion reduces the requirements for isoflurane and postoperative analgesics. Anesth Analg 80:1145–1149.

Segerdahl M, Ekblom A, Sollevi A (1994): The influence of adenosine, ketamine and morphine on experimentally induced ischemic pain in healthy volunteers. Anesth Analg 79:787–791.

Sollevi A (1992): Adenosine infusion during isoflurane-nitrous oxide anaesthesia: Indications of perioperative analgesic effect. Acta Anaesthesiol Scand 36:595–599.

Sollevi A, Belfrage M, Lundeberg T, Sergerdahl M, Hansson P (1995): Systemic adenosine infusion: A new treatment modality to alleviate neuropathic pain. Pain 61:155–158.

Sosnowski M, Yaksh TL (1989): Role of spinal adenosine receptors in modulating the hyperesthesia produced by spinal glycine receptor antagonism. Anesth Analg 69:587–592.

Sosnowski M, Stevens CW, Yaksh TL (1989): Assessment of the role of A_1/A_2 adenosine receptors mediating the purine antinociception, motor and autonomic function in the rat spinal cord. J Pharmacol Exp Ther 250:915–922.

Stiller CO, Fastbom J, Fredholm BB, Brodin E (1991): The adenosine analogue R-PIA interacts with substance P binding in rat brain and spinal cord. Acta Physiol Scand 141:573–574.

Sweeney MI, White TD, Sawynok J (1987): Involvement of adenosine in the spinal antinociceptive effects of morphine and noradrenaline. J Pharmacol Exp Ther 243:657–665.

Sweeney MI, White TD, Sawynok J (1988): 5-Hydroxytryptamine releases adenosine from primary afferent terminals in the spinal cord. Brain Res 462:346–349.

Sweeney MI, White TD, Sawynok J (1989): Morphine, capsaicin and K^+ release purines from capsaicin-sensitive primary afferent nerve terminals in the spinal cord. J Pharmacol Exp Ther 248:447–454.

Sweeney MI, White TD, Sawynok J (1990): 5-Hydroxytryptamine releases adenosine and cyclic AMP from primary afferent nerve terminals in the spinal cord in vivo. Brain Res 528:55–61.

Sweeney MI, White TD, Sawynok J (1991): Intracerebroventricular morphine releases cyclic AMP and adenosine from the spinal cord via a serotonergic mechanism. J Pharmacol Exp Ther 259:1013–1018.

Sylvén C (1993): Mechanisms of pain in angina pectoris—a critical review of the adenosine hypothesis. Cardiovasc Drugs Ther 7:745–759.

Sylvén C, Beermann B, Edlund A, Lewander R, Jonzon B, Mogensen L (1988): Provocation of chest pain in patients with coronary insufficiency using the vasodilator adenosine. Eur Heart J 9(suppl N):6–10.

Sylvén C, Beermann B, Jonzon B, Brandt R (1986): Angina pectoris-like pain provoked by intravenous adenosine in healthy volunteers. Br Med J 293:227–230.

Taiwo YO, Levine JD (1990): Direct cutaneous hyperalgesia induced by adenosine. Neuroscience 38:757–762.

Tjølsen A, Berge OG, Hunskaar S, Rosland JH, Hole K (1992): The formalin test: An evaluation of the method. Pain 51:5–17.

Vasko MR, Ono H (1990): Adenosine analogs do not inhibit the potassium-stimulated release of substance P from rat spinal cord slices. Naunyn Schmiedebergs Arch Pharmacol 342:441–446.

Watt AH, Lewis DJM, Horne JJ, Smith PM (1987): Reproduction of epigastric pain of duodenal ulceration by adenosine. Br Med J 294:10–12.

White TD, Downie JW, Leslie RA (1985): Characteristics of K^+- and veratridine-induced release of ATP from synaptosomes prepared from dorsal and ventral spinal cord. Brain Res 334:372–374.

Yarbrough GG, McGuffin-Clineschmidt JC (1981): In vivo behavioural assessment of central nervous system purinergic receptors. Eur J Pharmacol 76:137–144.

Yoshioka K, Jessell TM (1984): ATP release from the dorsal horn of rat spinal cord. Soc Neurosci Abstr 10:993.

CHAPTER 26

Adenosine Effects in Sleep Apnea

DAVID CARLEY and MIODRAG RADULOVACKI

University of Illinois College of Medicine at Chicago, Chicago, IL 60612

ADENOSINE CENTRAL FUNCTIONS AND RECEPTORS

Our initial interest in adenosine was in its possible hypnotic role, since behavioral stimulant effects of methylxanthines have been shown to involve a blockade of central adenosine receptors (Snyder et al., 1981). In addition, experiments with iontophoretic application of adenosine had depressant effects on the responses of neurons in several brain regions (Phillis and Wu, 1981), and general neurophysiologic effects of adenosine were shown to be inhibitory (Snyder et al., 1981; Stone, 1981). Thus, it was conceivable that stimulation of adenosine receptors may produce sedation or sleep. Moreover, preliminary data in dogs suggested a possible hypnotic role for adenosine, whose effects were thought to be mediated via serotonin (Haulica et al., 1973). Administration of relatively low doses of adenosine analogs to mice and rats produced marked sedation and hypothermia (Snyder et al., 1981; Dunwiddie and Worth, 1982).

Our own experiments in rats (Radulovacki et al., 1982, 1984; Yanik et al., 1987) have confirmed the hypothesis that stimulation of adenosine receptors may produce sleep. The hypnotic action of adenosine may be related to its inhibition of the release of the excitatory amino acid glutamate (Dolphin et al., 1985), as well as acetylcholine (Sawynok and Jhamandas, 1976; Rainnie et al., 1994), dopamine (Michaelis et al., 1979), norepinephrine (Hedquist and Fredholm, 1976), and serotonin (Harms et al., 1979) into the synaptic cleft. Recent findings of Haas and Greene (1988; Greene and Haas, 1985) showed that endogenous adenosine reduces neuronal excitability. There is no direct evidence of whether endogenous adenosine is also a behavioral depressant, but our work with deoxycoformycin (Radulovacki et al., 1983), a potent inhibitor of adenosine deaminase, and caffeine (Radulovacki et al., 1982), an adenosine receptor antagonist (Snyder et al., 1981), leads us to believe that it may be.

At least two distinct types of extracellular adenosine receptors have been identified in the mammalian CNS. A_1 (or R_1) receptors exhibit the adenosine agonist potency series of N^6-cyclohexyladenosine (CHA) > R-N^6-phenylisopropyladenosine (R-

Purinergic Approaches in Experimental Therapeutics, Edited by Kenneth A. Jacobson and Michael F. Jarvis
ISBN 0-471-14071-6 © 1997 Wiley-Liss, Inc.

PIA) > 5′-N-ethylcarboxamidoadenosine (NECA) > N^6(L-2-phenylisopropyl) adenosine (L-PIA) and inhibitory coupling to adenylate cyclase, resulting in decreased 3′,5′-cyclic adenosine monophosphate (cAMP) concentrations. A_2 (or R_a) adenosine receptors exhibit the adenosine agonist potency series of NECA > R-PIA > L-PIA > CHA and stimulatory coupling to adenylate cyclase, resulting in increased cAMP concentrations (van Calker et al., 1979; Londos et al., 1980; Bruns et al., 1986, 1987a; Stiles, 1986).

Recently, adenosine receptor ligands with greater *in vivo* selectivity between A_1 and A_2 receptors than previously available compounds have been developed and/or characterized. To date, the ligands for A_1 adenosine receptors, both agonists and antagonists, display substantially greater selectivity than do A_2 receptor ligands. For example, the most selective A_1 adenosine receptor agonists are N^6-cyclopentyladenosine (CPA) and CHA, which exhibit 780- and 390-fold selectivity, respectively; and the most selective A_1 receptor antagonists are 8-cyclopentyl-1,3-dipropylxanthine (CPX) and 8-cyclopentyltheophylline (CPT) which exhibit 740- and 130-fold selectivity, respectively. In contrast, the most selective known A_2 adenosine receptor agonists 2-(phenylamino) adenosine (CV-1808) and 2-(4-methoxyphenyl) adenosine (CV-1674) were only 4.8 and 2.2 times respectively, more potent at A_2 than at A_1 adenosine receptors (Bruns et al., 1986, 1987a,b). Antagonist ligands for A_2 adenosine receptors also exhibit considerably less selectivity than do A_1 antagonists. Among the most selective A_2 antagonist ligands *in vivo* are 1-propargyl-3,7-dimethylxanthine (PDX), the triasoloquinazoline CGS1594A, and alloxazine (ALX), which exhibit 11.0-, 6.2, and 1.98-fold selectivity, respectively, for A^2 receptors (Bruns et al., 1986, 1987a; Ukena et al., 1986; Williams et al., 1987).

A series of N^6-(*p*-sulfophenyl)alkyl and N^6-sulfoalkyl derivatives was synthesized, revealing that N^6-(*p*-sulfophenyl)adenosine (SPA) is a moderately potent (K_i vs ([^{3}H)]PIA in rat cortical membranes was 74 nM) and A_1-selective (120-fold) adenosine agonist, of exceptional aqueous solubility of 1.5 g/ml (3M). At a dose of 0.1 mg/kg intraperitoneally in rats, the drug inhibited lipolysis (a peripheral A_1 effect) by 85% after 1 hour. This *in vivo* effect was reversed using the peripherally selective A_1 antagonist 1,3-dipropyl-8-[*p*-(carboxyethynyl)phenyl]xanthine (BW1433)(Jacobson et al., 1992). These data suggest that the newly developed water-soluble adenosine A_1 receptor agonist and antagonist show poor CNS penetration.

ADENOSINE AND RESPIRATION

Adenosine production appears to be related to the balance between oxygen supply and demand within a tissue. Newby (1984) summarized evidence indicating that adenosine may exert feedback effects tending to restore balance in such supply and demand. Watt and Routledge (1985) proposed that adenosine release, possibly in the carotid body, may play a role in the ventilatory response to hypoxia, since hypoxia was shown to release considerable amounts of adenosine in various tissues (Berne, 1974). In their study on the effect of carotid body denervation on arousal response to hypoxia in sleeping dogs, Bowes et al. (1981) found that, following carotid body resection, all dogs exhibited hypoventilation during wakefulness, slow-wave sleep, and REM sleep. The authors concluded that the arousal response to hypoxia was primarily dependent on carotid chemoreceptor function. It has also been demonstrated that hypoxic stimulation of the carotid chemoreceptors produces an excitation of the reticular activating system (Hugelin et al., 1959), and that the arousal threshold to hypoxia is greatly depressed in sleep, particularly during REM sleep (Phillipson et al., 1978). In order to test the responsiveness of carotid body chemoreceptors to stimulation by adenosine or hypoxia, McQueen and Ribeiro (1986) infused anesthetized cats with adenosine receptor antagonists, theophylline and 8-phenyltheophylline, and reduced the chemoexcitation induced by hypoxia or adenosine. These results can be taken as indirect evidence for the involvement of adenosine, released in the carotid body during hypoxia, in chemosensory excitation.

Most authors agree that adenosine administered peripherally has stimulant effects on respiration, while its central administration depresses breathing. Indeed, infusion of adenosine to humans (Watt and Routledge, 1985; Maxwell et al., 1986; Biaggioni et al., 1987; Jonzon et al., 1989; Gleeson and Zwillich, 1992) has been shown to stimulate ventilation. Since adenosine has a short plasma half-life (Klabunde, 1983) and crosses the blood-brain barrier slowly (Berne et al., 1974), it is likely that peripheral mechanisms involving arterial chemoreceptors or intrapulmonary receptors are involved. In agreement with adenosine's inhibitory effect on central neurons, intracerebroventricular administration of adenosine-related compounds to anesthetized rats and cats depressed ventilation (Hedner et al., 1982; Eldridge et al., 1984; Wessberg et al., 1984). Ventilation was also depressed in vagotomized and glomectomized cats when a long-acting adenosine analog (L-PIA) was given either systemically or in the third cerebral ventricle (Eldridge et al., 1985). When another adenosine agonist, NECA, was injected into the nucleus tractus solitarius of anesthetized rats, it depressed respiration (Barraco et al., 1990). Thus, the central action of adenosine on respiration follows the pattern of adenosine inhibitory action on the CNS neurons and results in depression of respiration.

In an attempt to pharmacologically characterize the adenosine receptor subtypes in the carotid bodies in anesthetized cats, McQueen and Ribeiro (1986) have infused adenosine A_1 agonist L-PIA, adenosine A_1/A_2 agonist NECA, and adenosine antagonist 8-phenyltheophylline, and recorded chemoreceptor activity from the peripheral end of a sectioned carotid sinus nerve. They concluded that the adenosine receptors in the cat carotid bodies were xanthine-sensitive and appeared to be of the A_2 receptor subtype.

ADENOSINE AND SLEEP APNEA

Spontaneous sleep-related respiratory pauses have been reported in several rat strains (Mendelson et al., 1988; Sato et al., 1990; Monti et al., 1995a,b; Thomas et al., 1992, 1995). In view of the associated loss of phasic diaphragmatic excitation (Sato et al., 1990), these pauses have been interpreted as central sleep apneas. Although the exact mechanisms underlying central sleep apneas in man and rat remain undetermined (Önal, 1988), the peripheral chemoreceptors have been clearly implicated in the regulation of respiratory pattern during sleep (Phillipson et al., 1978; Phillipson, 1978; Bowes et al., 1980, 1981; Lahiri and Data, 1992). Both experimental hypoxia (Lahiri et al., 1983, 1984; Carley and Shannon, 1988b) and mathematical modeling (Grodins et al., 1967; Khoo et al., 1982; Carley and Shannon 1988a) have demonstrated that carotid chemoreceptor stimulation can induce central apneas. The potential of peripheral chemoreceptors to modulate apnea expression, coupled with the likely role of adenosine in peripheral chemoreceptor function, has prompted recent investigations into the role of adenosinergic receptors in apnea genesis in the rat (Monti et al., 1995a,b).

Monti et al. (1995a) examined the effects of intraperitoneal (IP) bolus injection of the A_1 receptor agonist L-PIA or A_2 receptor agonist 2-*p*-(2-carboxyethyl)-phenethylamino-5′-N-ethylcarboxamido-adenosine hydrochloride (CGS 21680) on apnea expression in 26 adult male Sprague-Dawley rats surgically implanted with recording electrodes for cortical electroencephalogram (EEG) and nuchal muscle electromyogram (EMG). Each animal was recorded for 6 hours (1000–1600), beginning 15 minutes after the injection of vehicle or active compound. Respiration was recorded by single chamber plethysmography and apnea was defined as respiratory cessation for at least 2.5 seconds. Level of consciousness was staged as wake, non–rapid eye movement (NREM) sleep, or rapid eye movement (REM) sleep by visual analysis of 60-second epochs.

Figure 1 illustrates a typical apnea recorded during NREM sleep. The apnea corresponds to approximately four "missed" breaths. Figure 2 reflects the dose-dependent effects of L-PIA on the apnea index (apneas per hour of each sleep stage). This dose–response characteristic appears to be non-monotonic, as significant suppression of apnea occurs only at the intermediate concentration of the A1 agonist (1.0 mg/kg). Statistically significant apnea suppression occurs only during NREM sleep. The lack of effect during REM may, however, be artifactual in that few animals exhibited REM sleep at the two higher L-PIA doses (Monti et al., 1995a). Figure 2 also illustrates that the baseline rate of apnea expression is higher ($p < .01$) during REM with respect to NREM sleep and that this sleep state dependence is unaffected by L-PIA ($p > .2$ for interaction term in two-way ANOVA). The increased apnea index during REM sleep has also been observed in other studies (Mendelson et al., 1988; Sato et al., 1990; Monti et al., 1995b).

Figure 3 characterizes the dose response of the A_2 agonist CGS 21680 on apnea index. As for L-PIA, the effect is limited to NREM sleep, but in this case the NREM

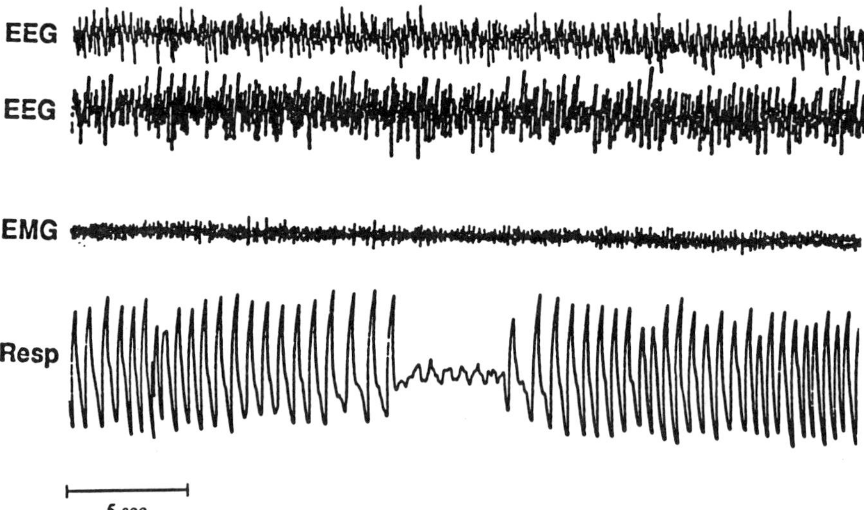

FIGURE 1. Typical spontaneous central apnea recorded during NREM sleep in an adult male Sprague-Dawley rat. The loss of tidal respiration, or apnea, corresponds to approximately four "missed" breaths.

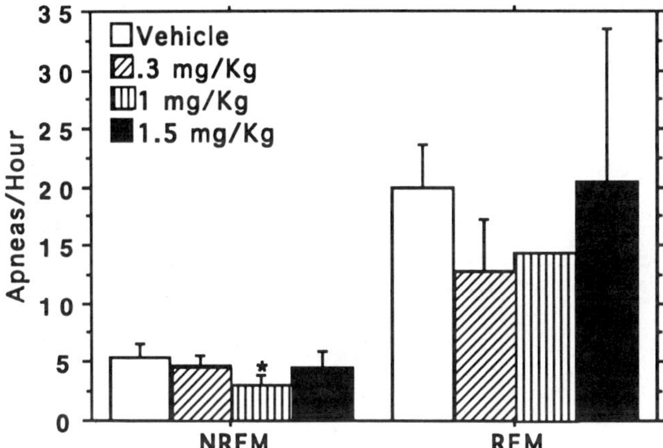

FIGURE 2. Dose dependent suppression of apnea index (apneas per hour) associated with L-PIA administration. For the doses examined, L-PIA had no effect on apnea index during REM sleep ($p > .2$ for all doses) and caused significant apnea suppression during NREM sleep only at the intermediate dose (*$p < .02$). For all doses, the apnea index was higher during REM than during NREM sleep ($p < .01$).

apnea index is decreased ($p < .03$) at each of the two highest doses (150 μg/g and 300 μg/kg).

Because L-PIA and CGS 21680 may cross the blood-brain barrier and may therefore have activity at both central and peripheral receptors, Monti et al. (1995b) also examined the effects of the recently characterized peripherally selective A_1 agonist SPA on apnea expression. These effects are demonstrated in Figure 4. SPA causes a mono-

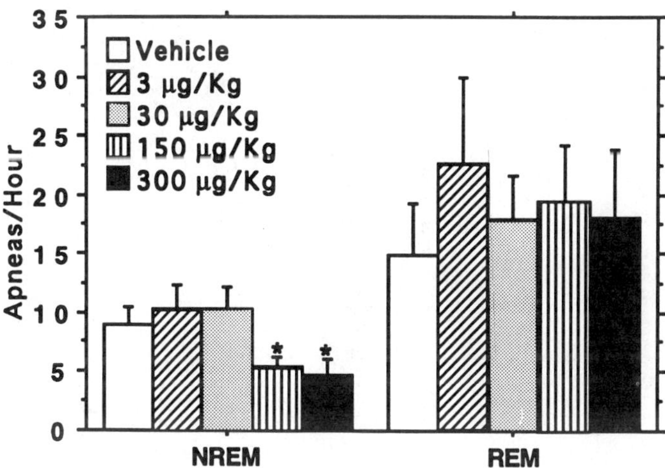

FIGURE 3. Suppression of NREM apnea by A_2 receptor agonist CGS 21680. Significant effect are associated with doses of 150 μg/kg and 300 μg/kg (*$p < .03$). CGS 21680 had no effect on apnea index during REM sleep.

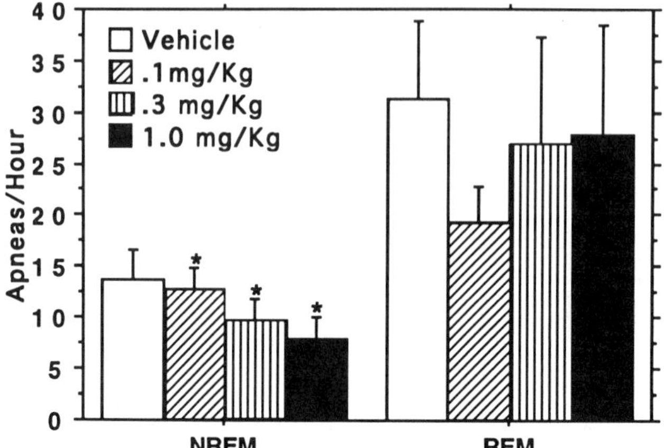

FIGURE 4. Reduction of NREM apnea associated with all doses of peripherally selective A_1 receptor agonist SPA (*$p < .05$). Apnea index during REM sleep was unaffected by any tested dose of SPA.

tonic, dose-dependent suppression of NREM apnea index. No effect of SPA on REM apneas occurs (Monti et al., 1995b).

Each of these adenosine compounds also demonstrated small, but statistically significant, effects on sleep architecture as demonstrated in Figures 5 through 7. Despite the fact that central actions of adenosine promote sleep (Radulovacki et al., 1982, 1984), each of these three compounds reduces sleep volumes when administered by IP injection. Only SPA yielded significant suppression of apnea index at a concentration (.01 mg/kg) that had no effect on sleep stage architecture (Figures 4 and 7). This CNS

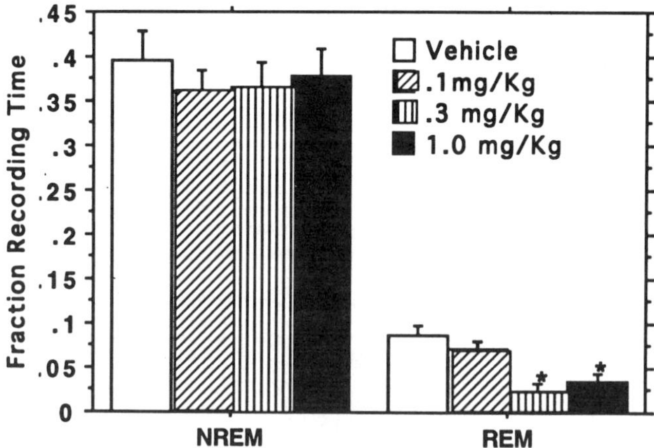

FIGURE 5. Effects of L-PIA on sleep stage architecture. Fraction of recording time spent in NREM sleep was unaffected by L-PIA. 1.0 mg/kg and 1.5 mg/kg doses were associated with reduced REM sleep time ($p < .0001$).

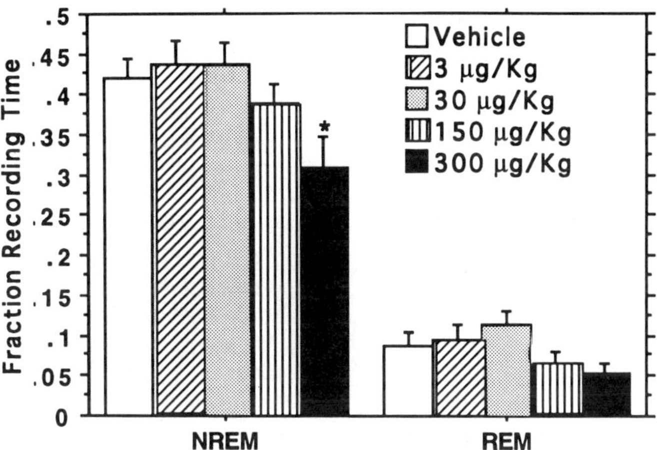

FIGURE 6. Effects of CGS 21680 on sleep architecture. REM sleep time was unaffected by CGS 21680. 300 μg/kg was associated with reduced time in NREM sleep (*p < .009).

activation may have resulted from peripheral chemoreceptor stimulation, which is well known to elicit arousal (Bowes et al., 1980, 1981; Phillipson et al., 1978).

Despite the clear and consistent modulation of apnea index by adenosine A_1 and A_2 receptor agonists demonstrated by Monti et al., (1995a,b), the precise mechanisms by which this effect occurs remain uncertain. The most probable mechanism that would account for the observed apnea suppression is direct stimulation of peripheral chemoreceptors by adenosine receptor agonists. This interpretation is also supported by our observation that mild hypoxia was associated with respiratory stimulation and apnea suppression during sleep.

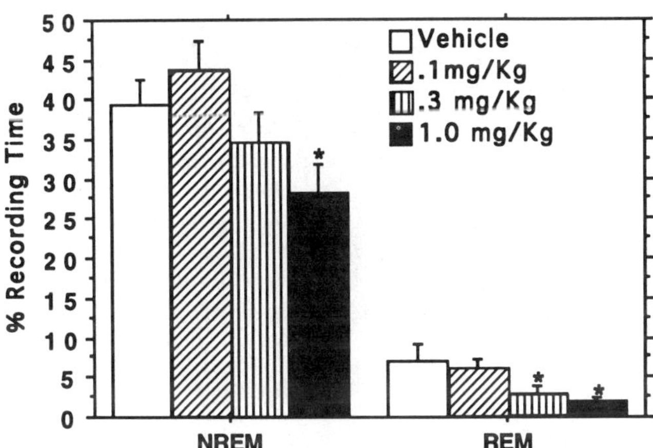

FIGURE 7. Effects of SPA on sleep architecture. NREM sleep time was reduced by 1.0 mg/kg and REM sleep time was reduced by .3 mg/kg and 1.0 mg/kg SPA (*p < .05).

Functionally significant carotid body and abdominal periaortic chemoreceptors have been demonstrated in the rat (Hollinshead, 1946; Andrews et al., 1972; Howe et al., 1981; Martin-Body et al., 1985, 1986; Montiero and Ribeiro, 1987; Sapru and Kreiger, 1977). McQueen and Ribeiro (1986) presented pharmacological evidence that the adenosine-related stimulation of carotid chemoreceptors in the cat is mediated primarily by A_2 receptors. This could potentially explain the effects of CGS 21680 on apnea expression via chemoreceptor stimulation, but does not explain the effects of the A_1 receptor agonists L-PIA and SPA. It is possible that species effects are important; peripheral chemoreceptors may be affected by both A_1 and A_2 receptors in the rat. It is also possible that nonspecific peripheral or, in the cases of L-PIA and CGS 21680, central actions of adenosine agonists may contribute to the observed apnea suppression.

Central activity of L-PIA at a dose of 1.5 mg/kg might account, at least in part, for the loss of effect on apnea index at this dose. If the peripheral effect of L-PIA is to suppress apnea via chemoreceptor stimulation, the central depressant effects on respiration (Hedner et al., 1982; Eldridge et al., 1984, 1985; Wessberg et al., 1984) might offset or even reverse these peripheral actions at sufficient doses. Central activity of adenosine agonists is also known to produce hypothermia (Wager-Srdar et al., 1983), which would be expected to have a depressant effect on respiration. Despite these potential central effects of L-PIA and CGS 21680, the overall stimilarity of NREM apnea suppression evoked by L-PIA, CGS 21680, and SPA suggests that the apnea suppression of all three compounds is mediated primarily by peripheral activity.

Systemic hypotension is another effect known to be associated with adenosine agonists at the doses employed (Hutchison et al., 1989). Hypotension may have contributed to, or been primarily responsible for, the observed apnea suppression. Ohtake and Jennings (1992) demonstrated that even a 10% reduction in mean arterial pressure was associated with significant respiratory stimulation in awake dogs. If similar respiratory stimulation attends hypotension in rats, the hypotensive actions of adenosine agonists may indirectly oppose apnea expression by stimulating respiration. The findings of Ohtake and Jennings suggest that in future studies designed to investigate the mechanisms of apnea in the rat or in man, it will be important to take into account even small changes in arterial pressure.

In summary, the endogenous nucleoside adenosine has been implicated in respiratory control at multiple levels. Pharmacological evidence links adenosine receptors to peripheral chemoreceptor function and to the reflex actions responsible for carbon dioxide homeostasis. These actions appear to be receptor-mediated, as they can be blocked by specific antagonists. Peripheral actions of adenosine agonists have also been demonstrated to significantly modulate apnea expression in a recently described model of spontaneous central apnea in the rat. The target tissues and mechanisms underlying this adenosinergic apnea suppression remain to be determined. Elucidating these mechanisms in the rat may provide insight regarding mechanisms of and therapeutic strategies to oppose central apnea in man.

REFERENCES

Andrews WHH, Deane DM, Howe A, Orback J (1972): Abdominal chemoreceptors in the rat. J Physiol 222:84–85P.

Barraco RA, Janusz CA, Schoener EP, Simpson LL (1990): Cardiorespiratory function is altered by picomole injections of 5′-N-ethylcarboxamido-adenosine into the nucleus solitarius of rats. Brain Res 507:234–246.

Berne RM, Rubio R, Curnish RR (1974): Release of adenosine from ischemic brain. Effect of cerebral vascular resistance and incorporation into cerebral adenine nucleotides. Circ Res 35:262–271.

Biaggioni I, Olafson B, Robertson R, Hallister AS, Robertson D (1987): Cardiovascular and respiratory effects of adenosine in conscious man: Evidence of thermoreceptor activation. Circ Res 6:779–788.

Bowes G, Townsend ER, Kozar LF, Bromley SM, Phillipson EA (1981): Effect of carotid body denervation on arousal response in sleeping dogs. J Appl Physiol 51:40–45.

Bowes G, Woolf GM, Sullivan CE, Phillipson EA (1980): Effect of sleep fragmentation on ventilatory and arousal responses of sleeping dogs to respiratory stimuli. Am Rev Respir Dis 122:899–908.

Bruns RF, Lu GH, Pugsley TA (1987a): Adenosine receptor subtypes: Binding studies. In Gerlach E, Becker BF (eds): "Topics and Perspective in Adenosine Research." Berlin Springer-Verlag, pp 59–73.

Bruns RF, Fergus JH, Badger EW, Bristol JA, Santay LA, Hartman JD, Hays SJ, Huang CC (1987b): Binding of the A_1-selective adenosine antagonist 8-cyclopentyl-1, 3-dipropylxanthine to rat brain membranes. Naunyn Schmiedebergs Arch Pharmacol 335:59–63.

Bruns RF, Lu GH, Pugsley TA (1986): Characterization of the A adenosine receptor labeled by [^{3}H]NECA in rat striatal membranes. J Pharmacol Exp Ther 29:331–346.

Carley DW, Shannon DC (1988a): A minimal mathematical model of human periodic breathing. J Appl Physiol 65:1400–1409.

Carley DW, Shannon DC (1988b): Relative stability of human respiration during progressive hypoxia. J Appl Physiol 65:1389–1399.

Daly JW, Butts-Lamb P, Padgett W (1983): Subclasses of adenosine receptors in the central nervous system: Interactions with caffeine and related methylxanthines. Cell Molec Neurobiol 3:69–80.

Dolphin AC, Prestwich SA, Forda SR (1985): Presynaptic modulation by adenosine analogues: relationship to adenylate cyclase. In Srefanovich V, Rudolphi V, Schubert P (eds): "Adenosine Receptors and Modulation of Cell Function." Oxford: IRL Press, p 107.

Dunwiddie TV, Worth T (1982): Sedative and anti-convulsant effects of adenosine analogs in mouse and rat. J Pharmacol Exp Ther 220:70–76.

Eldridge FL, Millhorn DE, Kiley JP (1984): Respiratory effects of a long-acting analog of adenosine. Brain Res 301:273–280.

Eldridge FL, Millhorn DE, Kiley JP (1985): Antagonism by the theophylline of respiratory inhibition induced by adenosine. J Appl Physiol 59:1428–1433.

Gleeson K, Zwillich CW (1992): Adenosine stimulation, ventilation, and arousal from sleep. Am Rev Respir Dis 145:453–457.

Greene RW, Haas HL (1985): Adenosine actions of CA1 pyramidal neurons in rat hippocampal slices. J Physiol 366:119–127.

Grodins FS, Buell J, Bart A (1967): A mathematical analysis and digital simulation of the respiratory control system. J Appl Physiol 22:260–276.

Haas HH, Greene RW (1988): Endogenous adenosine inhibits hippocampal CA1 neurons: Further evidence from extra- and intracellular recordings. Naunyn Schmiedebergs Arch Pharmacol 337:561–565.

Harms HH, Wardeh G, Mulder AH (1979): Effects of adenosine depolarization-induced release of various radiolabelled neurotransmitters from slices of rat corpus striatum. Neuropharmacology 18:577–580.

Haulica I, Ababeih, Branisteanu D, Topoliceanu F (1973): Preliminary data on the possible hypnogenic role of adenosine. J Neurochem 21:1019–1020.

Hedner T, Hedner J, Wessberg P, Jonason J (1982): Regulation of breathing in the rat: Indication for a role of central adenosine mechanisms. Neurosci Lett 33:147–151.

Hedquist P, Fredholm BB (1976): Effects of adenosine on adrenergic neurotransmission: Prejunctional inhibition and postjunctional enhancement. Naunyn Schniedebergs Arch Pharmacol 293:217–224.

Hollinshead WH (1946): The function of the abdominal chemoreceptors of the rat and mouse. Am J Physiol 147:654–660.

Howe A, Pack RJ, Wise JCM (1981): Arterial chemoreceptor-like activity in the abdominal vagus of the rat. J Physiol 320:309–318.

Hugelin A, Bonvallet M, Dell P (1959): Activation reticulaire et corticale d'origine chemoreceptive au cours de l'hypoxie. Electroencephalogr Clin Neurophysiol 11:325–340.

Hutchison AJ, Webb RL, Oei HH, Ghai GR, Zimmerman MB, Williams M (1989): CGS 21680C, an A2 selective adenosine receptor agonist with preferential hypotensive activity. J Pharmacol Exp Ther 251:47–55.

Jacobson KA, et al. (1992): Synthesis and biological activity of N6-(*p*-sulfophenyl)alkyl and N^6-sulfoalkyl derivatives of adenosine: Water soluble and peripherally selective adenosine agonists. J Med Chem 35:4144–4149.

Jonzon B, Sylven C, Beerman B, Brant R (1989): Adenosine receptor mediated stimulation of ventilation of man. Eur J Clin Invest 19:65–71.

Khoo MCK, Kronauer RE, Strohl KP, Slutsky AS (1982). Factors inducing periodic breathing in humans: A general model. J Appl Physiol 53:644–659.

Klabunde RE (1983): Dipyridamole inhibition of adenosine metabolism in human blood. Eur J Pharmacol 93:21–26.

Lahiri S, Data PG (1992): Chemosensitivity and regulation of ventilation during sleep at high altitudes. Int J Sports Med 13:S31–S33.

Lahiri S, Maret K, Sherpa MG (1983): Dependence of high altitude sleep apnea on ventilatory sensitivity to hypoxia. Respir Physiol 52:281–301.

Lahiri S, Maret KH, Sherpa MG, Peters RM (1984): Sleep and periodic breathing at high altitude: Sherpa natives versus sojourners. In West JB, Lahiri S (eds): "High Altitude and Man." Bethesda: American Physiological Society, pp 73–90.

Londos C, Cooper MF, Wolff J (1980): Subclasses of external receptors. Proc Natl Acad Sci USA 77:2551–2554.

Martin-Body RL, Robson GJ, Sinclair JD (1985): Respiratory effects of sectioning the carotid sinus glossopharyngeal and abdominal vagal nerves in the awake rat. J Physiol 361:35–45.

Martin-Body RL, Robson GJ, Sinclair JD (1986): Restoration of hypoxic respiratory responses in the awake rat after carotid body denervation by sinus nerve section. J Physiol 380:61–73.

Maxwell DL, Fuller RW, Nolop KB, Dixon CMS, Hughes JMB (1986): Effects of adenosine on ventilatory responses to hypoxia and hypercapnia in humans. J Appl Physiol 61:1762–1766.

McQueen DS, Ribeiro JA (1986): Pharmacological characterization of the receptor involved in chemoexcitation induced by adenosine. Br J Pharmacol 88:615–620.

Mendelson WB, Martin JV, Perlis JM, Giesen H, Wagner R, Rapoport SI (1988): Periodic cessation of respiratory effort during sleep in audlt rats. Physiol Behav 43:229–234.

Michaelis ML, Michaelis EK, Myers SL (1979): Adenosine modulation of synaptosomal dopamine release. Life Sci 24:2083–2092.

Monti D, Carley DW, Radulovacki M (1995a): Adenosine analogs modulate the incidence of sleep apneas in rats. Pharmacol Biochem Behav 51:125–131.

Monti D, Carley DW, Christon J, Radulovacki M (1995b). p-SPA, a peripheral adenosine A_1 analogue, reduces sleep apneas in rats. Pharmacol Biochem Behav 53:341–345.

Montiero EC, Ribeiro JA (1987): Ventilatory effects of adenosine mediated by carotid body chemoreceptors in the rat. Naunyn Schmiedebergs Arch Pharmacol 335:143–148.

Newby AC (1984): Adenosine and the concept of "retaliatory metabolites." Trends Biochem Sci 9:42–44.

Ohtake PJ, Jennings DB (1992): Ventilation is stimulated by small reductions in arterial pressure in the awake dog. J Appl Physiol 73:1549–1557.

Önal E (1988): Central sleep apnea. Seminars in Respiratory Medicine 9:547–553.

Phillipson EA, Sullivan CE, Read DJC, Murphy E, Kozar LF (1978): Ventilatory and waking responses to hypoxia in sleeping dogs. J Appl Physiol 44:512–520.

Phillipson EA (1978): Control of breathing during sleep. Am Rev Respir Dis 118:909–939.

Phillis JW, Wu PH (1981): The role of adenosine and its nucleotides in central synaptic transmission. Prog Neurobiol 16:187–239.

Radulovacki M, Miletich RS, Green RD (1982): N^6(L-phenylisopropyl) adenosine (L-PIA) increases slow-wave sleep (S_2) and decreases wakefulness in rats. Brain Res 246:178–180.

Radulovacki M, Virus RM, Djuricic-Nedelson M, Green RD (1983): Hypnotic effects of deoxycoformycin in rats. Brain Res 271:392–395.

Radulovacki M, Virus RM, Djuricic-Nedelson M, Green RD (1984): Adenosine analogs and sleep in rats. J Pharmacol Exp Ther 228:268–274.

Rainnie DG, Grunze HC, McCarley RW, Greene RW (1994): Adenosine inhibition of mesopontine cholinergic neurons: Implications for EEG arousal. Science 263:689–669.

Sapru HN, Kreiger AJ (1977): Carotid and aortic chemoreceptor function in the rat. J Appl Physiol 42:344–348.

Sato T, Saito H, Seto K, Takatsuji H (1990): Sleep apneas and cardiac-arrhythmias in freely moving rats. Am J Physiol 259:R282–R287.

Sawynok J, Jhamandas KH (1976): Inhibition of acetylcholine release from cholinergic nerves by adenosine, adenosine nucleotides, and morphine: Antagonism by theophylline. J Pharmacol Exp Ther 197:379–390.

Snyder SH, Sklar P (1984): Behavioral and molecular actions of caffeine: Focus on adenosine. J Psychiatr Res 18:91–106.

Synder SH, Katims JJ, Annau S, Bruns RF, Daly JW (1981): Adenosine receptors and behavioral actions of methylxanthines. Proc Natl Acad Sci USA 78:3260–3264.

Stiles GL (1986): Adenosine receptors: Structure, function and regulation. Trends Pharmacol Sci 7:486–490.

Stone TW (1981): Physiological roles of adenosine and adenosine 5′-triphosphate in the nervous system. Neuroscience 6:523–552.

Thomas AJ, Austin W, Friedman L, Strohl KP (1992): A model of ventilatory instability induced in the unrestrained rat. J Appl Phyiol 73:1530–1536.

Thomas AJ, Friedman L, MacKenzie CN, Strohl KP (1995): Modification of conditioned apneas in rats: Evidence for cortical involvement. J Appl Physiol 73:1530–1536.

Ukena D, Shamim MT, Padgett W, Daly JW (1986): Analogs of caffeine: Antagonists with selectivity for A_2 adenosine receptors. Life Sci 39:743–750.

van Calker D, Muller M, Hambrecht V (1979): Adenosine regulates, via two different types of receptors, the accumulation of cyclic AMP in cultured brain cells. J Neurochem 33:999–1005.

Wager-Srdar SA, Oken MM, Morley JE, Levine AS (1983): Thermoregulatory effects of purines and caffeine. Life Sci 33:2431–2438.

Watt AH, Routledge PA (1985): Adenosine stimulates respiration in man. Br J Clin Pharmacol 20:503–506.

Wessberg P, Hedner J, Hedner T, Persson B, Jonason J (1984): Adenosine mechanisms in the regulation of breathing in the rat. Eur J Pharmacol 106:59–67.

Williams M, Francis J, Ghai G, Braunwalder A, Psychoyos S, Stone GA, Cash WD (1987): Biochemical characterization of the triazoloquinazoline, CGS 15843A, a novel non-xanthine adenosine antagonist. J Pharmacol Exp Ther 241:415–420.

Yanik G, Glaum S, Radulovacki M (1987): The dose-response effects of caffeine on sleep in rats. Brain Res 403:177–180.

CHAPTER 27

The Influence of Adenosine on the Developing Fetus

SCOTT A. RIVKEES

Department of Pediatrics, Yale Medical School, New Haven, CT 06520

INTRODUCTION

Increasing evidence suggests a potential role for the adenosinergic system in fetal physiology. Because local adenosine levels are dynamically regulated by tissue oxygenation, adenosine is uniquely suited to mediate fetal responses to changes in oxygenation. This concept is supported by physiology studies demonstrating adenosine-mediated cardiovascular responses. The recent discovery of adenosine receptors in the fetal heart at early stages of gestation further supports the concept that the adenosine system influences the fetal cardiovascular system. The observation that adenosine receptors (ARs) are expressed in the fetal brain identifies the nervous system as a target of adenosine action at early stages of development as well.

REGULATION OF FETAL ADENOSINE LEVELS

The nucleoside adenosine is released by all cells under basal conditions [1]. Following increased tissue activity, decreased blood flow, or hypoxia, there is rapid breakdown of cellular ATP stores to adenosine. This results in large increases in local adenosine levels [1]. Thus, changes in local adenosine levels provide a barometer of tissue activity and oxygenation.

Currently, our understanding of the factors that regulate fetal adenosine levels are at an early stage. In sheep under normoxic conditions (Po_2, 24 mm Hg), fetal plasma adenosine levels are about 1 μM [2,3]. These values are nearly four times the adenosine levels present in the maternal circulation [4]. With mild hypoxia (Po_2, 15 mm Hg), fetal plasma adenosine levels more than double [2,3]. During simulated birth, fetal adenosine levels increase to a similar degree [4]. When adenosine levels are examined

Purinergic Approaches in Experimental Therapeutics, Edited by Kenneth A. Jacobson and Michael F. Jarvis
ISBN 0-471-14071-6 © 1997 Wiley-Liss, Inc.

in the brains of fetal sheep, basal levels are 35 nM [5]. With hypoxia, adenosine levels rise nearly three-fold in the brainstem [5]. However, because adenosine is rapidly broken down in the circulation [6], it is likely that changes in local adenosine levels are greatly in excess of circulating levels.

INFLUENCE OF ADENOSINE ON FETAL PHYSIOLOGY

Direct support for the concept that changes in local adenosine levels influence fetal physiology comes largely from studies in sheep. In fetal sheep, intra-arterial infusion of adenosine results in significant increases in fetal heart rate and decreases in blood pressure [7,8]. This observation is duplicated using the A2aAR agonist CGS-21680, suggesting a role for A2aARs in this process [7]. Adenosine infusion in fetal sheep has also been shown to increase myocardial blood flow [9].

Pulmonary blood flow is also influenced by adenosine in the fetus. Infusions of adenosine into the right atria result in pulmonary vasodilation and increased fetal pulmonary blood flow [10]. It has therefore been suggested that adenosine is one of several mediators of the postnatal increase in pulmonary blood flow seen immediately after birth.

Effects of adenosine on fetal respirations have also been observed. In fetal sheep, peripherally administered adenosine suppresses fetal respiration [3,8]. This effect is blocked by the antagonist theophylline [8,11], which also stimulates fetal breathing during hypoxia [11]. When the adenosine analog R-N^6phenylisopropyladenosine is infused into the fourth ventricle of fetal sheep, suppression of fetal respiration is observed [12]. Thus, it has been suggested that adenosine is an important mediator of hypoxia-induced decreases in fetal breathing activity.

Available evidence also suggests a role for adenosine in influencing the fetal neuroendocrine axis. During fetal hypoxia, large increases in arginine vasopressin are seen [13]. Infusions of adenosine to late gestation sheep mimic this response, and adenosine antagonists block hypoxia-induced arginine vasopressin secretion [13].

The fetal cerebrovascular system is also a target of adenosine action. During hypoxia, a prompt increase in cerebral blood flow is observed [14]. Dose–response studies suggest that these effects are mediated via A2aARs [14]. This premise is strongly supported by recent observations showing that A2aARs are present in small cerebral blood vessels in rats [15].

INFLUENCE OF ADENOSINE ON PLACENTAL FUNCTION

In addition to directly affecting the fetus, adenosine influences placental physiology [16,17]. As in the fetus, placental adenosine levels are regulated by tissue oxygenation. Studies using perfused human placental cotyledons have found that placental adenosine levels range between 3.5 and 52 nM [18]. With ischemia imposed by decreasing maternal placental perfusion, placental adenosine levels increase sixfold [18].

In sheep, the placental vasculature responds to adenosine in a biphasic manner [19]. Following adenosine administration to the fetus, there is transient vasoconstriction in fetal placental vasculature followed by vasodilatation [19]. Studies of isolated human placental lobules have revealed that adenosine induces placental vasodilation in a dose-dependent manner [20].

It is likely that these events are receptor mediated. In humans, A1ARs are present in the placenta [21]. It has also been suggested that A2aARs may be present in human placenta, although these data are less convincing [21].

INFLUENCE OF CAFFEINE ON THE DEVELOPING FETUS

Caffeine acts as a nonselective adenosine antagonist [22]. Thus, studies of prenatal caffeine exposure may provide insights into the role of adenosine during development. Because caffeine is widely consumed during pregnancy, the influence of caffeine on the developing fetus has received considerable attention. It is important to note that whereas caffeine can act as an adenosine antagonist, caffeine also acts via non–adenosine receptor mechanisms at high doses [23]. Thus, non–adenosine receptor–mediated events may mediate methylxanthine action in the fetus.

Caffeine readily crosses the placenta to reach the fetus [24,25]. During pregnancy, caffeine clearance is reduced, as the half-life increases from 5 to 18 hours [26–28]. Because fetuses and newborns do not metabolize caffeine well [26–28], maternal caffeine ingestion therefore results in prolonged fetal caffeine exposure.

Several experimental studies have examined the influence of prenatal caffeine exposure on the developing fetus. Studies of rodent fetuses exposed to very high doses of caffeine (100 mg $\cdot$ kg^{-1} $\cdot$ day^{-1}; equivalent to 50 cups of coffee daily) have revealed a variety of complications ranging from intrauterine growth retardation to malformations [29–31]. In rats, the administration of large doses of caffeine to pregnant dams (100 mg $\cdot$ kg^{-1} $\cdot$ day^{-1}) from gestation days 6 to 20, results in reduced maternal weight and decreases in fetal weight and length [32]. In some fetuses of dams treated with 100 mg $\cdot$ kg^{-1} $\cdot$ day^{-1}, ectodactyly has been observed [33]. In mice, exposure to (175 mg $\cdot$ kg^{-1} $\cdot$ day^{-1} of caffeine) at gestation days 11 and 12, has been associated with paw malformations [34]. When pregnant rats are given caffeine in drinking water (0.04%) before and during pregnancy, reduced placental weight is observed, along with reduced fetal body and brain weight [31]. Neural tube defects and skeletal defects have also been observed in pregnant rats given 25 mg $\cdot$ kg^{-1} $\cdot$ day^{-1} of caffeine [35].

In contrast to rodent studies, human studies have not found a clear association between prenatal caffeine exposure and birth defects. Independent studies by Rosenberg and coworkers [36], Kurppa et al. [38] and Linn et al. [37] involving more than 14,000 infants have not detected an increased incidence of birth detects associated with maternal caffeine consumption. Heavy maternal caffeine consumption may lead to decreased birth weight [39–43]. Maternal caffeine exposure also is associated with increased fetal loss in a dose-dependent manner [44,45].

Physiological effects of caffeine on human fetuses have been examined. When mothers consume high amounts of caffeine (>500 mg/day; 1 cup of coffee contains 50–150 mg of caffeine), fetuses spend less time in active sleep and more time in arousal than when maternal caffeine intake is <200 mg/day [46]. Fetal caffeine exposure has also been shown to increase fetal cardiac work [47,48].

The long-term effects of fetal caffeine exposure in neural function are currently unknown. It was suggested that high prenatal caffeine exposure is associated with neurobehavioral problems [44]. Other reports have not found adverse effects on child neurodevelopment [49]. However, as discussed by James and Pauli [50], methodological problems related to assessment of caffeine intake, associated nicotine use, and the lack of adequate control groups, prevent firm conclusions from being drawn about the role of caffeine on the human fetus. Based on the results of animal studies, the FDA has taken a cautionary stand in regard to recommendations for maternal caffeine exposure during pregnancy [51]. It is recommended that maternal caffeine exposure be minimized during pregnancy [51].

DEVELOPMENTAL EXPRESSION OF ADENOSINE RECEPTORS

Until recently, it was not known if adenosine receptors were expressed in fetuses. Recent studies of A1 and A2aARs, however, have provided new insights into fetal sites of adenosine action [15,52]. At present, we are unaware of reports describing prenatal sites of A2b and A3AR receptor expression. Whereas studies of prenatal adenosine action have focused on sheep, developmental expression studies of A1ARs have involved rats.

A2aAR EXPRESSION IN THE DEVELOPING BRAIN

Studies of adenosine receptor expression during development first involved investigations of A2aARs [15,53]. These studies have revealed prominent A2aAR expression in the developing striatum and transient expression of A2aARs in fetal cerebral blood vessels.

In adult animals, the highest levels of A2aARs are found in the brain [15,53], with the striatum containing the greatest concentration of A2aARs [53]. When the ontogeny of A2aAR gene expression is examined, A2aAR mRNA is detected in the developing striatum as soon as the striatum can be identified, on gestational day (GD) 14 [15] (Figure 1). This corresponds with when the first set of striatal neurons complete neurogenesis [15].

In the striatum of adult rats, A2aARs are expressed in medium spiny neurons that express the D_2 dopamine receptor and proenkephalin genes [53]. During development, prominent A2aAR expression in these developing cells is also seen. This results in the appearance of patches interspersed among different populations of striatal cells making up the striatal matrix [15]. However, after birth the patch compartments become more evenly distributed, and the patches become less prominent [15].

When A2aAR expression is examined using receptor autoradiography, A2aAR binding sites are detectable in rat striatum at GD 16 [54]. However, because a tritiated

FIGURE 1. A_{2a}AR gene expression in rat striatum on GD 14, GD 18, and PD 0. Left panels: Brightfield photomicrographs. Right panels: Darkfield photomicrographs from emulsion autoradiograms; hydribization signal appears white. Arrowheads indicate the surface of the striatal ventricular zone (VZ) on GD 18 and PD 0. Note the presence of hybridization in the subventricular zone (SVZ) on GD 18 and the patchiness of hybridization in the caudate-putamen (CP) on PD 0. GD, gestational day; Acb, accumbens nucleus; OT, olfactory tubercle. Modified from Weaver [15], with permission.

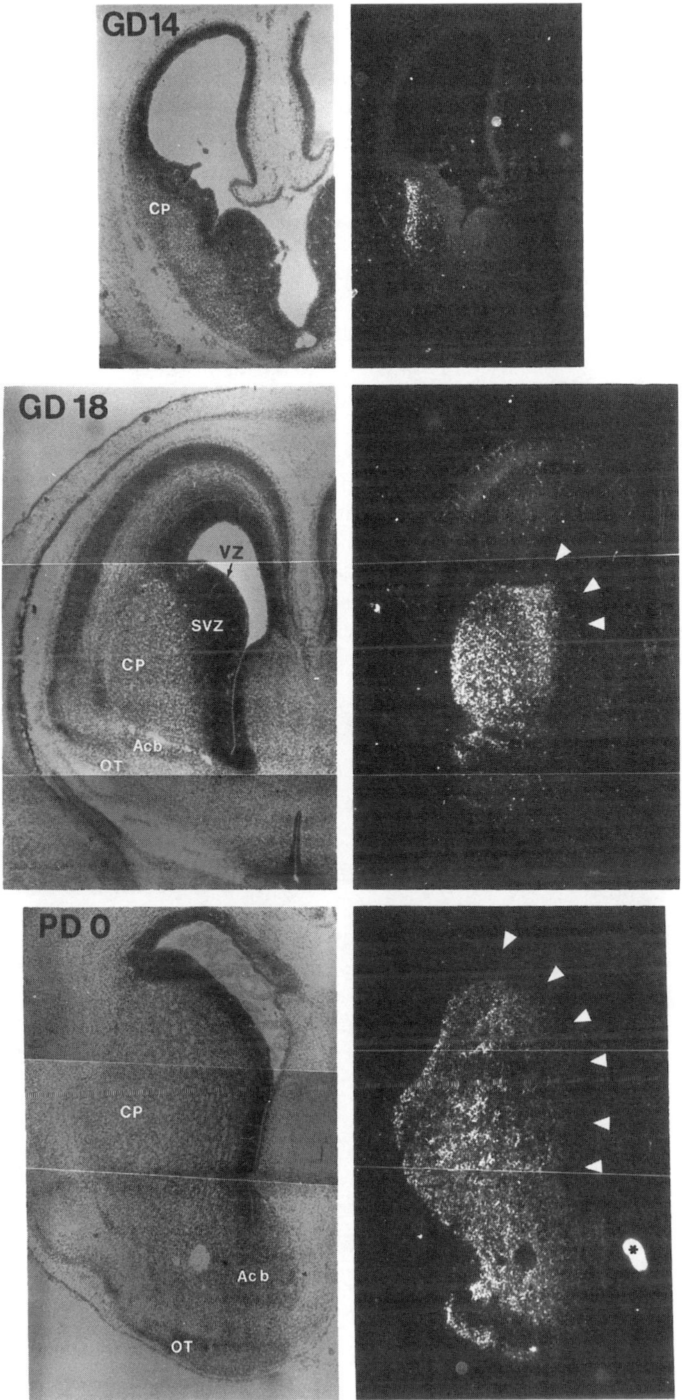
GD14
CP
GD 18
VZ
SVZ
CP
Acb
OT
PD 0
CP
Acb
OT

ligand was used with relatively short exposure periods in those studies [53], it is possible that A2aARs are expressed at even earlier ages. As observed in mRNA expression studies, striatal labeling is present in patches [53].

Although striatal A2aAR expression persists into adulthood [15], A2aAR expression in other brain regions is limited to the fetus. From GD 16 to 20, A2aAR mRNA expression is seen in the cerebral cortex, subiculum, parafascicularis nucleus of the thalamus, locus caeruleus, area postrema, motor nucleus of the trigeminal, and in the dorsolateral facial nucleus [15]. At older ages, A2aAR expression is much less, or not detectable, in these sites [15].

Another interesting site of A2aAR mRNA expression at early stages of development is the cerebral vasculature. From GD 14 to postnatal day (PD) 0, A2aAR mRNA expression is detectable over the pial surface containing small vessels (Table 1, Figure 2). Beginning at GD 16, labeling is present over radial vessels extending from the pial surface into cerebral cortex [15]. This pattern of expression persists until PD 7 [15]. Labeling is not observed over larger blood vessels. Studies in sheep have shown that adenosine influences cerebral blood flow in a manner most consistent with A2aAR-mediated events [14]. Thus, the observations of Weaver [15] strongly support the notion that adenosine influences blood flow in the fetal brain.

Heavy A2aAR expression also is observed in the fetal carotid body [15]. Intracarotid administration of adenosine stimulates respiration via the carotid body through a receptor with A2a-like specificity [55]. Carotid body A2aARs may therefore be peripheral sensors of circulating adenosine levels in the fetus. A2aAR mRNA expression also is diffusely present throughout the pituitary gland at early stages of development. By GD 21, pituitary A2aAR expression becomes restricted to the neurointermediate lobe.

ONTOGENESIS OF A1AR EXPRESSION

Whereas studies of A2aAR expression have focused on receptor expression in the brain from GD 12 to birth, studies of A1AR expression have involved studies of whole rat fetuses from GD 8 onwards [52]. This level of anlaysis has revealed new insights into adenosine action in the fetus [52], showing early A1AR expression in the brain and identifying A1ARs as one of the earliest expressed G protein–coupled receptors in the fetal heart (Figure 3).

TABLE 1 Ontogeny of A2aAR mRNA Expression in Cerebral Vasculature

Site	GD 14	GD 16	GD 18	GD 20	PD 0	PD 7	Adult
Pial surface	1	2	2	2-3	1	0	0
Radial vessels	0	1	2	2	1	1	0
Choroid plexus	0	0	1	1	1	0	0

Semi-quantitative evaluation of the intensity of hybridization in various structures. Values are based on examination of emulsion autoradiograms using the following scale: 0, no specific hybridization; 1, faint specific hybridization detected; 2, moderate specific hybridization detected; 3, intense specific hybridization detected.
Modified from Weaver [15] with permission.
GD, gestational day. PD, postnatal day.

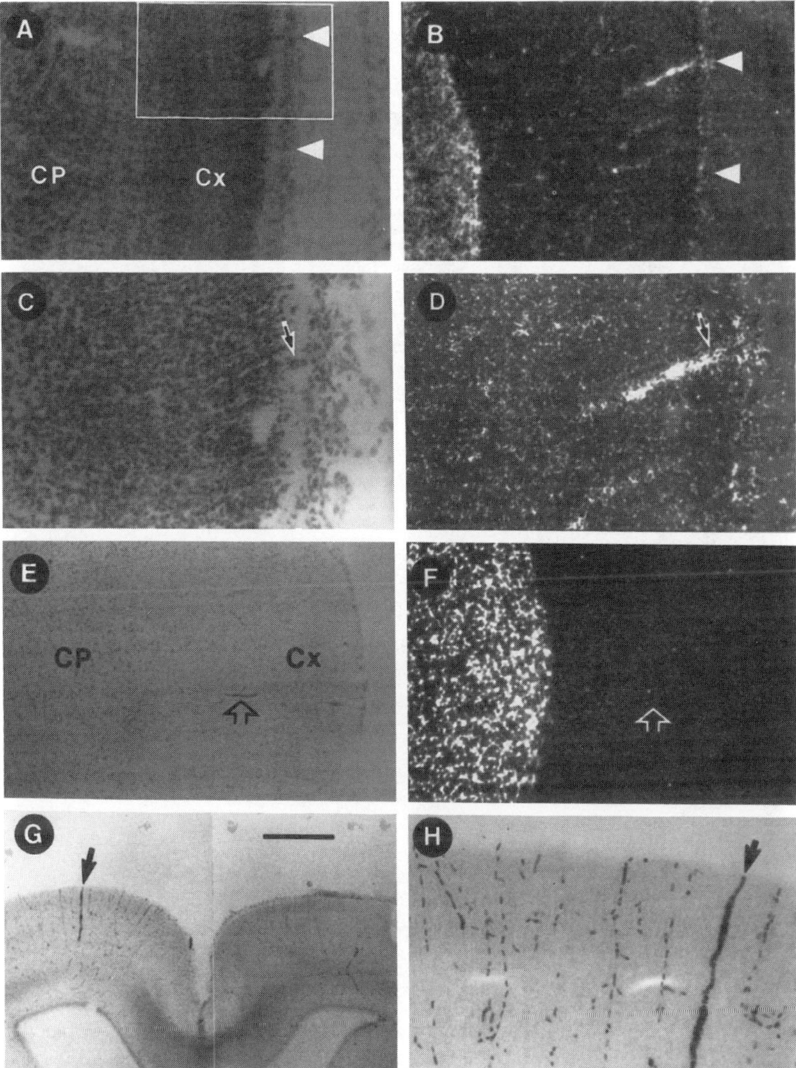

FIGURE 2. $A_{2a}AR$ gene expression at GD 20 in cerebral blood vessels. Left panels: Brightfield photomicrographs. Right panels: Darkfield photomicrographs from emulsion autoradiograms; hybridization signal appears white. (**A**) and (**B**) are low-power images showing specific labeling of A2aAR. mRNA is shown over radial vessels in the cerebral cortex. (**C**) and (**D**) are high power images of (A) and (B). (**E**) and (**F**) are images generated from an adult animal. Arrow in (F), indicates the absence of A2aAR mRNA labeling in the adult animal. (**G**) and (**H**) are low- and high-power images of diaminobenzadine stained GD 20 fetus showing the location of radial blood vessels in the cerebral cortex. GD, gestational day. Scale bar, 0.1 m. Figure courtesy of Dr. David R. Weaver.

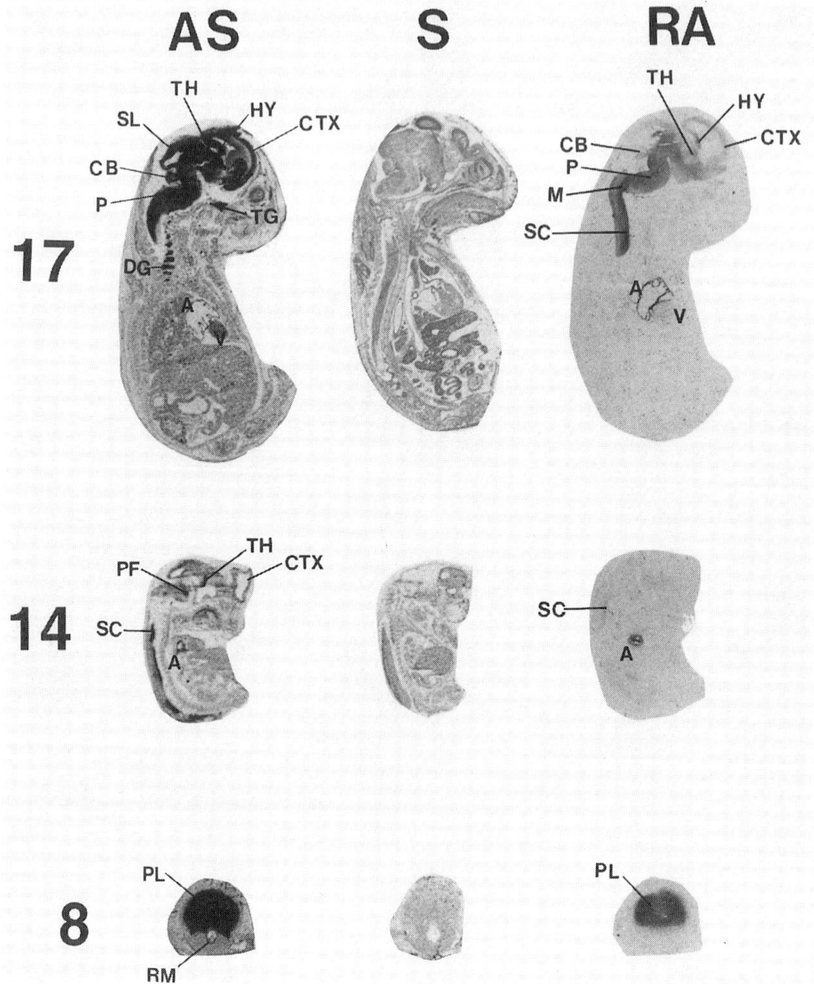

FIGURE 3. Film autoradiographs generated from A1AR in situ hybridization and receptor labeling autoradiography (RA) studies of fetuses at GD 8, GD 14, and GD 17 in side-by-side experiments. Antisense (AS) and sense (S; nonspecific labeling) images generated from in situ hybridization studies are shown, along with images of labeling generated using [^{3}H]DPCPX (4 nm). Areas of specific mRNA labeling appear as dark images in AS images. Nonspecific labeling using [^{3}H]DPCPX is equal to the film background and is thus not shown. Sagittal images are shown at GD 14 and 17; transverse sections are shown at GD 8. In situ hybridization exposures at 28 days. [^{3}H]DPCPX exposures are at 120 days. GD, gestational day; PL, placenta; RM, Reichert's membrane; SC, spinal cord; A, atria; V, ventricles; P, pons; PF, pontine flexure; TH, thalamus; CTX, cerebral cortex; SL, superior colliculus; TG, trigeminal ganglia; DG, dorsal root ganglia; HT, hypothalamus; TH, thalamus; HY, hippocampus; BG, basal ganglia; SC, ventral horn of spinal cord. Modified from Rivkees [52].

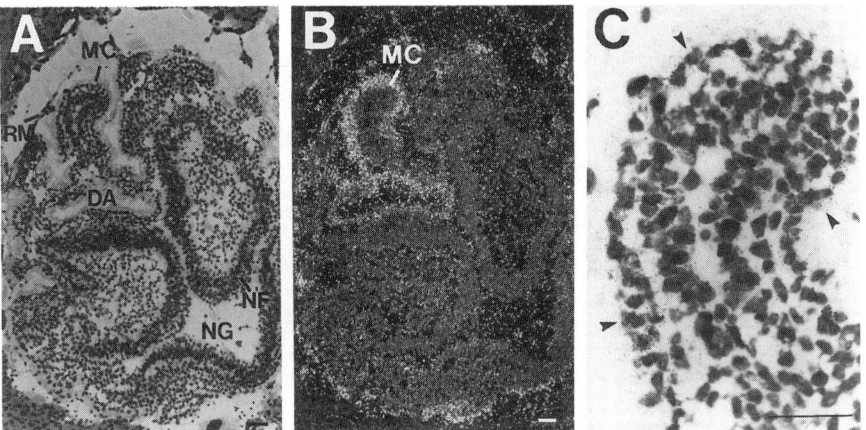

FIGURE 4. In situ hybridization studies showing labeling of A1AR mRNA of fetal heart at GD 8 in transverse embryo sections (**A**) Brightfield image of toluidine blue stained tissue sections used to generate emulsion autoradiograph. (**B**) Darkfield image of emulsion autoradiograph. Areas of specific labeling appear as white grains. Specific labeling is present over the myocardium and Reichert's membrane. (**C**) High-power, brightfield image of the myocardium. Arrows depict labeling of the outer myocardial region. MC, myocardium; DA, dorsal aorta; NF, neural fold; NG neural grove; RM, Reichert's membrane. Scale bars = 25 μm. Modified from Rivkees [52].

At GD 8, when the primitive cardiac cylinder appears, a very prominent A1AR mRNA hybridization signal is seen over the developing myocardium by *in situ* hybridization (Figure 4). No labeling of other embryonic structures is discerned at this age. At GD 11, when the atria and ventricle are distinct, a clear hybridization signal for A1AR mRNA is seen over the myocardium. Receptor autoradiography studies reveal a visible signal over the developing heart at this age as well.

When fetuses are studied at GD 14, *in situ* hybridization studies reveal A1AR mRNA expression in the heart, spinal cord, and kidney. Of these structures, receptor autoradiography shows heaviest A1AR expression over the heart. At GD 17, the somatic pattern of A1AR expression is similar to that seen at GD 14, with A1AR

TABLE 2 [^{3}H]DPCPX Binding in Fetal and Adult Tissues from Rats

	Specific [^{3}H]DPCPX Bound (fmol/mg of protein)		
Age	Heart	Brain	Liver
GD 12	n.d.	n.d.	n.d.
GD 14	1.73 ± 0.33	0.75 ± 0.21	n.d.
GD 18	1.95 ± 0.26	3.40 ± 0.28	0.25 ± 0.11
GD 21	2.30 ± 0.70	17.4 ± 2.0	n.d.
Adult	a1.22 ± 0.25 v0.28 ± 0.33	32.3 ± 2.7	0.27 ± 0.15

Values are means ± S.E.M. n.d., no specific binding detectable; a, atria; v, ventricles. [^{3}H]DPCPX concentration = 4 nM. Binding to crude tissue preparations is shown. Modified from Rivkees [52].
GD, gestational day.

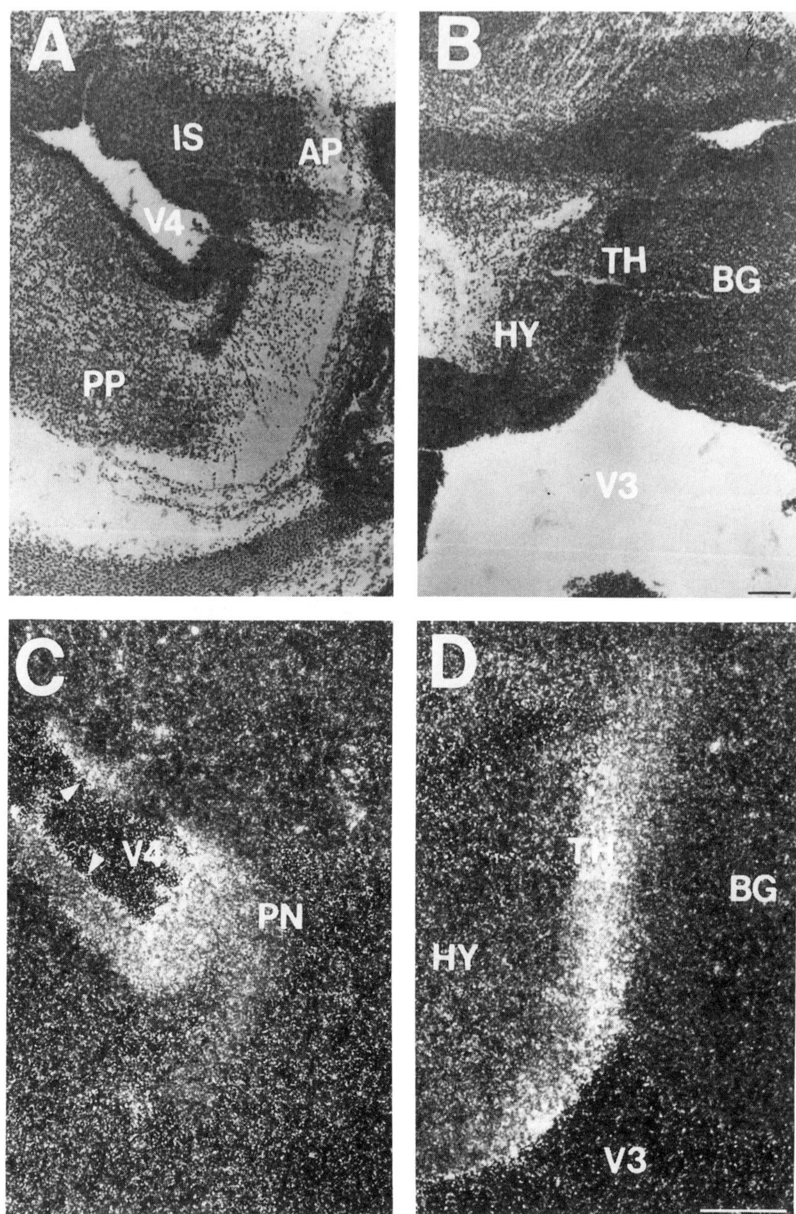

FIGURE 5. In situ hybridization studies showing labeling of A1AR MRNA of fetal brain structures at GD 14. (**A**) Brightfield image of toluidine blue stained tissue sections at level of the pons used to generate emulsion autoradiograph. (**C**) 2× Darkfield image of emulsion autoradiograph showing heavy labeling of the pontine neuroepithelium. Arrows depict heavily labeling of neuroepithelial cells lining the 4th ventricle. Areas of specific labeling appear as white gains. (**B**) Brightfield image of toluidine blue stained tissue section at level of the mid-brain used to generate emulsion autoradiograph. (**D**) 2× Darkfield image of emulsion autoradiograph showing heavy labeling of the thalamus. Sagittal images are shown. IS, pontine isthmus; AP, anterior pons; PP, posterior pons; V4, fourth ventricle; PN, pontine neuroepithelium; TH thalamus; HY, hypothalamus; BG, basal ganglia; V3, third ventricle. Scale bars = 50 μm. Modified from Rivkees [52].

expression being greatest in the heart. At both GD 14 and 17, A1AR binding is more prominent over the atria than over the ventricles (Table 2).

Although A1AR mRNA expression is observed in the heart as early as GD 8, A1AR mRNA expression is not observed over neural tissue until GD 14 (Figure 5). At GD 17, A1AR mRNA expression in neural tissue is much more intense and widespread than seen at GD 14. However, despite prominent A1AR mRNA expression in the brains of fetuses, prenatal A1AR concentrations are considerably less than postnatal levels. Recently, A1ARs have been discovered predominantly on axons in several brain regions [56]. Thus, differences in A1AR mRNA and binding site expression may represent temporal differences among cell body and axonal development.

Observations of prenatal A1AR expression in the brain at early stages of fetal development raise the possibility that adenosine affects the developing nervous system during critical periods of neurogenesis and neruonal migration. In rats, neuronal formation and migration is very active over the second half of gestation [57,58], when A1AR expression is detected in brains. In several brain regions, A1AR expression is seen in neuroepithelium regions, which contain differentiating neuronal precursors and migrating postmitotic neurons. Knowing the temporal and spatial patterns of A1AR expression in the fetal brain, it will be important to determine if adenosine and methylxanthine affect brain development during early periods of receptor expression.

Although A1AR binding sites in hearts and brains can be detected in mammalian fetuses, we do not know the effector systems that are prenatally regulated by A1ARs. In adult animals, A1ARs have been shown to inhibit adenylate cyclase activity, influence ion channels, and regulate phosphoinositol metabolism [59]. During development, G_i expression has been detected in rodent fetuses as early as GD 6.5 [60]. Thus, prenatal A1AR–G protein coupling is possible. However, because second-messenger system regulation may be less efficient before birth [61,62] studies are needed to examine fetal A1AR-effector coupling.

Recently, we have directly shown that A1ARs are potent regulators of fetal heart activity in mammals. Studying splanchnopleural cultures of mice, which differentiate into contractile cardiac tissue *in vitro,* we find that adenosine agonists slow and can even arrest fetal cardiac contractions. Pharmacological studies suggest that this influence is mediated via A1ARs. A1ARs therefore appear to exert potent inhibitory affects on fetal cardiac cells. Studying cultured chick atrial myocytes, Liang and co-workers have also identified functional A1ARs that slow contractility rates [63].

CONCLUSION

The observation that fetal adenosine levels are dynamically regulated, coupled with observations that adenosine receptors are expressed in fetuses, points to the presence of a functional adenosinergic system in the fetus. As shown in fetal physiology studies, adenosine plays a prominent role in regulating cardiovascular and pulmonary physiology. The identification of A1ARs in the mammalian heart at very early stages of development also identifies the heart as a direct site of prenatal adenosine action. The findings of A1 and A2aAR receptor expression in the fetus during periods of neurogenesis and neuronal migration raises the possibility that adenosine influences the nervous system at very early stages. Currently, it is not known if activation of adenosine receptors at critical stages of development influences morphogenesis. Future studies are needed both to delineate the role of adenosine on fetal physiology and to determine if activation of fetal adenosine receptors influences mammalian structural development.

ACKNOWLEDGMENTS

S.A.R. is supported by a Grant-in-Aid from the American Heart Association, the James Whitcomb Riley Memorial Association, and N.I.H. grant RO1NS326224. The author is indebted to Dr. David R. Weaver for his assistance.

REFERENCES

1. Dunwiddie TV (1990): Adenosine release. In "Adenosine and Adenosine Receptors." Williams, M (ed): Clifton, NJ: Humana, pp 173–224.
2. Yoneyama Y, Power GG (1992): Plasma adenosine and cardiovascular responses to dipyridamole in fetal sheep. J Dev Physiol 18:203–209.
3. Koos BJ, Doany W (1991): Role of plasma adenosine in breathing responses to hypoxia in fetal sheep. J Dev Physiol 16:81–85.
4. Sawa R, Asakura H, Power GG (1991): Changes in plasma adenosine during simulated birth of fetal sheep. J Appl Physiol 70:1524–1528.
5. Koos BJ, Mason BA, Punla O, Adinolfi AM (1994): Hypoxic inhibition of breathing in fetal sheep: Relationship to brain adenosine concentrations. J Appl Physiol 77:2734–2739.
6. Linden J (1989): Adenosine deaminase for removing adenosine: how much is. Trends Pharmacol Sci 10:260–262.
7. Koos BJ, Mason BA, Ducsay CA (1993): Cardiovascular responses to adenosine in fetal sheep: Autonomic blockade. Am J Physiol 264:H526–H532.
8. Koos BJ, Matsuda K (1990): Fetal breathing, sleep state, and cardiovascular responses to adenosine in sheep. J Appl Physiol 68:489–495.
9. Reller MD, Morton MJ, Giraud GD, Wu DE, Thornburg KL, (1992): Severe right ventricular pressure loading in fetal sheep augments global myocardial blood flow to submaximal levels. Circulation 86:581–588.
10. Konduri GG, Theodorou AA, Mukhopadhyay A, Deshmukh DR (1992): Adenosine triphosphate and adenosine increase the pulmonary blood flow to postnatal levels in fetal lambs. Pediatr Res 31:451–457.
11. Bissonnette JM, Hohimer AR, Chao CR, Knopp SJ Notoroberto NF (1990): Theophylline stimulates fetal breathing movements during hypoxia. Pediatr Res 28:83–86.
12. Bissonnette JM, Hohimer AR, Knopp SJ (1991): The effect of centrally administered adenosine on fetal breathing movements. Respir Physiol 84:273–285.
13. Koos BJ, Mason BA, Ervin MG (1994): Adenosine mediates hypoxic release of arginine vasopressin in fetal sheep. Am J Physiol 266:R215–R220.
14. Kurth CD, Wagerle LC (1992): Cerebrovascular reactivity to adenosine analogues in 0.6–0.7 gestation and near-term fetal sheep. Am J Physiol 262:H1338–H1342.
15. Weaver DR (1993): A2a adenosine receptor gene expression in developing rat brain. Brain Res Mol Brain Res 20:313–327.
16. Carney EW (1994): An integrated perspective on the developmental toxicity of ethylene glycol. Reproductive Toxicology 8:99–113.

17. Leal M, Barletta M, Carson S. (1990): Maternal-fetal electrocardiographic effects and pharmacokinetics after an acute i.v. administration of caffeine to the pregnant rat. Reproductive Toxicology 4:105–112.
18. Slegel P, Kitagawa H, Maguire MH (1988): Determination of adenosine in fetal perfusates of human placental cotyledons using fluorescence derivatization and reversed-phase high-performance liquid chromatography. Analyt Biochem 171:124–134.
19. Reid DL, Davidson SR, Phernetton TM, Rankin JH (1990): Adenosine causes a biphasic response in the ovine fetal placental vasculature. J Dev Physiol 13:237–240.
20. Read MA, Boura AL, Walters WA (1993): Vascular actions of purines in the foetal circulation of the human placenta. Br J Pharmacol 110:454–460.
21. Work C, Hutchison K, Prasad M, Bruns RF, Fox IH (1989): Characteristics of an adenosine A1 binding site in human placental membranes. Arch Biochem Biophys 268:191–202.
22. Bruns RF (1990): Adenosine receptors. Roles and pharmacology. Ann NY Acad Sci 603:211–25.
23. Stiles GL (1992): Adenosine receptors. J Biol Chem 267:6451–6454.
24. Jiritano L, Bortolotti A, Gaspari F, Bonati M (1985): Caffeine disposition after oral administration to pregnant rats. Xenobiotica 15:1045–1051.
25. Kimmel CA, Kimmel GL, White CG, Grafton TF, Young JF, Nelson CJ (1984): Blood flow changes and conceptal development in pregnant rats in response to caffeine. Fundam Appl Toxicol 4:240–247.
26. Aldridge A, Neims AH (1980): Relationship between the clearance of caffeine and its 7-N-demethylation in developing beagle puppies. Biochem Pharmacol 29:1909–1914.
27. Aranda JA, Louridas AT, Vitullo BB, Thom P, Aldridge A, Haber R (1979): Metabolism of theophylline to caffeine in human fetal liver. Science 206:1319–1321.
28. Aldridge A, Aranda JV, Neims AH (1979): Caffeine metabolism in the newborn. Clin Pharmacol Therap 25:447–453.
29. Yazdani M, Joseph F Jr, Grant S, Hartman AD, Nakamoto T (1990): Various levels of maternal caffeine ingestion during gestation affects biochemical parameters of fetal rat brain differently. Dev Pharmacol Therap 14:52–51.
30. Tanaka H, Nakazawa K, Arima M (1987): Effects of maternal caffeine ingestion on the perinatal cerebrum. Biol Neonate 51:332–339.
31. Tanaka H, Nakazawa K, Arima M, Iwasaki S (1984): Caffeine and its dimethylxanthines and fetal cerebral development in rat. Brain Dev 6:355–361.
32. Smith SE, McElhatton PR, Sullivan FM (1987): Effects of administering caffeine to pregnant rats either as a single daily dose or as divided doses four times a day. Food Chem Toxicol 25:125–133.
33. Beaulac-Baillargeon L, Desrosiers C (1987): Caffeine-cigarette interaction on fetal growth. Am J Obstet Gynecol 157;1236–1240.
34. York RG, O'Flaherty EJ, Scott WJ Jr, Shukla R (1987): Alteration of effective exposure of dam and embryo to caffeine and its metabolites by treatment of mice with beta-naphthoflavone. Toxicol Appl Pharmacol 88:282–293.
35. Ross CP, Persaud TV (1989): Neural tube defects in early rat embryos following maternal treatment with ethanol and caffeine. Anatomischer Anzeiger 169:247–252.
36. Rosenberg L, Mitchell AA, Shapiro S, Slone D (1982): Selected birth defects in relation to caffeine-containing beverages. JAMA 247:1429–1432.
37. Linn S, Schoenbaum SC, Monson RR, Rosner B, Stubblefield PG, Ryan KJ (1982): No association between coffee consumption and adverse outcomes of pregnancy. N Engl J Med 306:141–145.
38. Kurppa K, Holmberg PC, Kuosma E, Saxen L (1983): Coffee consumption during pregnancy and selected congenital malformations: A nationwide case-control study. Am J Public Health 73:1397–1399.

39. Caan BJ, Goldhaber MK (1989): Caffeinated beverages and low birthweight: A case-control study. Am J Public Health 79:1299–1300.

40. Streissguth AP, Sampson PD, Barr HM (1989): Neurobehavioral dose-response effects of prenatal alcohol exposure in humans from infancy to adulthood. Ann NY Acad Sci 562:145–158.

41. Martin TR, Bracken MB (1987): The association between low birth weight and caffeine consumption during pregnancy. Am J Epidemiol 126:813–821.

42. Furuhashi N, Sato S, Suzuki M, Hiruta M, Tanaka M, Takahashi T (1985): Effects of caffeine ingestion during pregnancy. Gynecol Obstet Invest 19:187–191.

43. Barr HM, Streissguth AP, Martin DC, Herman CS (1984): Infant size at 8 months of age: Relationship to maternal use of alcohol, nicotine, and caffeine during pregnancy. Pediatrics 74:336–341.

44. Eskenazi B (1993): Caffeine during pregnancy: Grounds for concern? JAMA 270:2973–2974.

45. Infante-Rvard C, Fernandez A, Gautheir R, David M, Rivard GE (1993): Fetal loss associated with caffeine intake before and during pregnancy. JAMA 270:2940–2943.

46. Devoe LD, Murray C, Youssif A, Arnaud M (1993): Maternal caffeine consumption and fetal behavior in normal third-trimester pregnancy. Am J Obstet Gynecol 168:1105–1111. disc.

47. Temples TE, Geoffray DJ, Nakamoto T, Hartman AD, Miller HI (1987): Effect of chronic caffeine intake on myocardial function during early growth. Pediatr Res 21:391–395.

48. Temples TE, Geoffray DJ, Nakamoto T, Hartman AD, Miller HI (1985): Effects of chronic caffeine ingestion on growth and myocardial function. Proc Soc Exp Biol Med 179: 388–395.

49. Barr HM, Streissguth AP (1991): Caffeine use during pregnancy and child outcome: A 7-year prospective study. Neurotoxicol Teratol 13:441–448.

50. James JE, Paull I (1985): Caffeine and human reproduction. Rev Environ Health 5:151–167.

51. Goyan JE (1980): Food and Drug Administration, News Release No. 80-36, Washington DC.

52. Rivkees SA (1995): The ontogeny of cardiac and neural A1 adenosine receptor expression in rats. Dev Brain Res. 89:202–213.

53. Fink JS, Weaver DR, Rivkees SA, Peterfreund RA, Pollack AE, Adler EM, Reppert SM (1992): Molecular cloning of the rat A2 adenosine receptor: Selective co-expression with D2 dopamine receptors in rat striatum. Brain Res Mol Brain Res 14:186–195.

54. Schiffmann SN, Vanderhaeghen JJ (1992): Ontogeny of gene expression of adenosine A2 receptor in the striatum: early localization in the patch compartment. J Comp Neurol 317:117–128.

55. Martinez-Mir MI, Probst A, Palacios JM (1986): Pharmacological characterization of the receptor involved in chemoexcitiation induced by adenosine. Br J Pharmacol 88:615–620.

56. Swanson TH, Drazba JA, Rivkees SA (1995): Adenosine A1 receptors are located predominantly on axons in the rat hippocampal formation. J Comp Neurol 363:517–531.

57. Altman, J (1992): The early stages of nervous system development: Neurogenesis and neuronal migration. In Bjorklund T, Hokfelt T, Tohyama M (eds.): "Handbook of Chemical Neuroanatomy, Volume 10. Ontogeny of Transmitters and Peptides in the CNS." Amsterdam: Elsevier pp 1–31.

58. Altman J, Bayer SA (1995): "Atlas of Prenatal Rat Brain Development" Boca Raton, FL: CRC Press.

60. Jones J, Logan CY, Schultz RM (1991): Changes in temporal and spatial patterns of Gi protein expression in postimplantation mouse embyros. Dev Biol 145:128–138.

61. Lin W, Seidler FJ, McCook EG, Slotkin TA (1992): Overexpression of alpha 2-adrenergic receptors in fetal rat heart: Receptors in search of a function. J Dev Physiol 17:183–187.

62. Slotkin TA, Lau C, Seidler J (1994): Beta-adrenergic receptor overexpression in the fetal rat: Distribution, receptor subtypes, and coupling to adenylate cyclase activity via G-proteins. Toxicol Appl Pharmacol 129:223–234.

63. Liang, BT (1989): Characterization of the adenosine receptor in cultured embryonic chick atrial myocytes: Coupling to modulation of contractility and adenylate cyclase activity and identification by direct radioligand binding. J Pharm Exp Ther 249:775–784.

5. Cancer

CHAPTER 28

ATP in the Treatment of Cancer

ELIEZER RAPAPORT

Worcester Foundation for Biomedical Research, Shrewsbury, MA 01545

INTRODUCTION

The intravenous administration of adenosine 5′-triphosphate (ATP) is now under clinical development by Medco Research for the treatment of advanced cancer and aspects of cancer cachexia associated with the advanced disease. Phase I clinical trial indicated that blood levels of ATP (red blood cells ATP pools) can easily be expanded after a few hours of intravenous infusions of ATP at levels of 50 $\mu g \cdot kg^{-1} \cdot min^{-1}$ in patients with advanced cancer (Haskell et al., 1994). The phase I studies have suggested that a 96-hour infusion schedule every 28 days is safe and was associated with inhibition of weight loss, stabilization of Karnofsky performance status, and improvements in overall survival of patients with advanced (stage IIIB or IV) non–small cell lung cancer. A multicenter phase II clinical study of ATP infusions in the treatment of patients with advanced non–small cell lung cancer was recently completed by Medco Research and the data are currently being analyzed. These recent clinical trials demonstrated few partial tumor responses along with improved or stabilized quality of life parameters (by EORTC instruments) in patients with extensive disease as well as confirming the phase I results.

This review will focus on the mechanisms of action of expanded ATP pools and their utilization for the treatment of advanced cancer, as established by *in vitro* and *in vivo* preclinical studies. Extensive preclinical data indicate that administration of ATP to a host produces a broad spectrum of anticancer activities which will be reviewed here. These activities include cytostatic and cytotoxic effects on the tumor, anticachexia effects and improvements in organ function, modulation of blood flow, antianemia effects, analgesic activities, improvements in oxygen delivery to peripheral sites, enhancement of superoxide anion production by phagocytic cells, and antithrombotic and profibrinolytic activities.

Purinergic Approaches in Experimental Therapeutics, Edited by Kenneth A. Jacobson and Michael F. Jarvis
ISBN 0-471-14071-6 © 1997 Wiley-Liss, Inc.

EXPANSIONS OF ORGAN, RED BLOOD CELL, AND BLOOD PLASMA (EXTRACELLULAR) ATP POOLS AFTER ADMINISTRATION OF ADENINE NUCLEOTIDES

Administration of adenine nucleotides (ATP, AMP, or any other adenine nucleotide) into the systemic circulation or extravascular sites results in the rapid degradation of the nucleotides to adenosine and inorganic phosphate due to the strong ectoenzymatic and soluble enzymatic activities. These catabolic activities are followed by the incorporation of adenosine and inorganic phosphate into liver ATP pools yielding significant expansions of the liver ATP pools (Rapaport and Fontaine, 1989a,b). Detailed radioactive precursor labeling experiments in mice demonstrated that the turnover of the elevated liver ATP pools supply the increased levels of precursors in the hepatic sinusoids for the salvage synthesis of ATP in red blood cells. Red blood cells utilize only salvage (mostly adenosine) precursors for ATP synthesis by glycolytic pathways and do not engage in either *de novo* synthesis of ATP or in oxidative phosphorylation. The elevated liver ATP pools thus yield elevated red blood cell ATP pools, which subsequently produce a slow release of micromolar levels of ATP from the red blood cells into the blood plasma compartment in a nonhemolytic process. The continuous release of micromolar amounts of ATP from red blood cells results in steady-state levels of extracellular ATP that are elevated in spite of the catabolic enzymatic activities present intravascularly, and these in turn lead to the production of increased extracellular levels of adenosine. The mechanisms of expansion of organ ATP levels, other than liver, after administration of exogenous ATP, proceed by both the increased supply of the major purine precursor for salvage ATP synthesis in cells (adenosine) and the interaction of extracellular ATP with membrane P_2 purine receptors, which signal an enhanced intracellular ATP synthesis via a signal tarnsduction pathway. Most of the expansions of total blood (red blood cell) ATP pools occur due to increased supply of purines to the mature erythrocyte in the hepatic sinusoids, where these purine precursors (mostly adenosine) arise from the turnover of increased hepatic ATP pools. A significant increase in red blood cell ATP pools of the magnitude observed *in vivo* after ATP administration cannot be obtained *in vitro* (Rapaport and Fontaine, 1989a). Similar elevations of mouse red blood cell ATP pools and blood plasma ATP levels after intraperitoneal as well as intravenous administration of ATP to tumor-bearing mice were recently reported (Estrella et al., 1995).

Expansions of organ ATP pools have been linked to improved function in liver (Rapaport, 1990; Soni and Mehendale, 1994), kidney (Kanwar et al., 1992), splenic immune functions (Meldrum et al., 1991), gut absorptive capacity (Singh et al., 1993) and pancreatic β cell insulin secretion (Rapaport, 1995). More importantly whereas depletion in ATP levels produced by several pathophysiological conditions led to significant declines in organ and cellular functions, infusions of ATP that restored the original organ and cellular ATP pools also yielded significant improvements in normal differentiated functions. Two recent studies demonstrated that in pigs and mice, expansion of organ ATP levels in non-tumor-bearing normal animals yielded significant protection from the cytotoxic effects and organ damage caused by high doses of radiation (Senagore et al., 1992; Szeinfeld with DeVilliers, 1992).

About 20 years ago, we demonstrated that *in vivo* administration of adenosine at levels below those that produce cardiovascular side effects can expand liver ATP pools (Rapaport and Zamecnik, 1976). Adenosine later became the key component in the solutions used for solid-organ preservation by cold storage, after graft harvesting and exposure to ischemic damage, and prior to graft transplantation (Beltzer and Southard, 1988). The successful clinical transplantation of livers and kidneys (Beltzer and Southard, 1988) and the experimental success in pancreas transplantations (Fujino et al., 1993) are dependent on the ability of adenosine to expand the damaged graft ATP pools, which, in turn, are directly responsible for the initiation of metabolic processes vital to the maintenance of cellular viability and differentiated functions as well as the repair of damaged cells. Successful transplantation of ischemically damaged pancreas graft is directly related to the ATP pools of the pancreas graft, and adenosine—rather than adenine, hypoxanthine, or inosine—is the only precursor that can expand the pancreas graft ATP pools (Fujino et al., 1993), as was shown for the expansions of liver ATP pools (Rapaport and Zamecnik, 1976).

PRECLINICAL ANTICANCER ACTIVITIES OF ATP

Initial experiments have demonstrated that low levels of extracellular ATP inhibit the growth of a variety of human tumor cells and subsequently yield substantial cell killing in *in vitro* systems (Rapaport, 1983; Rapaport et al., 1983). The mechanism of tumor cell killing is attributed to the permeabilization of tumor cell membrane by extracellular ATP (Rapaport, 1983), as well as to other non-receptor-mediated Na^{2+} channel opening (Wiley et al., 1990), P_2 purinergic receptor–mediated opening of plasma membrane Ca^{2+} channels, and release of Ca^{2+} from internal stores (Fang and Wu, 1993). Several reports have now identified the ATP receptor responsible for the cell killing by exposure of cells to low levels of ATP as P_{2Z} purine receptors (Murgia et al., 1993; Spranzi et al., 1993). These receptors produce a nonselective pore that is permeable to ions and metabolites via a gated channel mechanism (Spranzi et al., 1993), although other recent reports have suggested the involvement of a second-messenger pathway in the P_2 receptor–mediated cell killing (Fang and Wu, 1993; Fang et al., 1992). The increases in intracellular calcium concentrations $[Ca^{2+}]_i$ after ATP-induced recruitment from internal and external Ca^{2+} sources were demonstrated to initiate cell killing through apoptosis in hormonally sensitive breast tumor cells (Vandewalle et al., 1994).

The administration of ATP or AMP to murine hosts carrying cachectic tumors resulted in inhibition of tumor growth and host weight loss. Although tumor size is directly related to the rate of weight loss in tumor-bearing animals, a detailed analysis of both parameters led to the conclusion that the inhibition of tumor growth and host weight loss did not exhibit a cause-and-effect relationship (Rapaport and Fontaine, 1989a,b; Rapaport, 1990, 1993). Fluid retention or alterations in food consumption by the treated animals were eliminated as possible causes of the inhibition of weight loss in the ATP- or AMP-treated tumor-bearing hosts. Weight loss is a frequent and poorly understood aspect of advanced cancer. The prognostic effect of weight loss on the response rate to chemotherapy in patients with a variety of tumor types has been established. Chemotherapy response rates were lower in patients with weight loss and reversal of poor nutritional state was shown to affect the outcome in cases where the anticancer therapy was inherently effective (Van Eys, 1982). Intravenous infusions of nutritional substrate for support of tumor-bearing hosts (total parenteral nutritional support) were shown to be ineffective. Tumor growth is generally increased to a greater extent than host tissue growth (Popp et al., 1983). Although cancer cachexia in man, as well as in experimental animals, produces not only host weight loss but also anorexia, hormonal aberrations, and depletion and redistribution of host components, all of which lead to a progressive decline in vital host functions, much attention has been focused on the alterations in energy metabolism in cancer patients (Nelson et al., 1994). Gluconeogenesis from lactate and amino acids was proposed to account for lactate recycling and the large expenditures of energy that occur in the liver and kidney cortex during cancer cachexia (Stein, 1978). Whereas gluconeogenesis is costly in terms of the number of ATP molecules required for glucose synthesis from lactate in the liver and kidney cortex, anaerobic glycolysis of the newly synthesized glucose at another site, namely the tumor, occurs at a rapid rate and produces fewer molecules of ATP. This type of host-tumor interplay was suggested as the main reason for the progressive weight loss associated with cancer cachexia. Aberrations in glucose metabolism showing marked increases in Cori cycle activity (the cyclic metabolic pathway in which glucose is coverted by the tumor to lactic acid, which is in turn utilized for glucose resynthesis in the liver via gluconeogenesis) have been demonstrated using radioactively labeled precursors in human cancer patients suffering from progressive weight loss (Holyroyde, 1975). Furthermore, the significant depletion of host visceral energy stores by a progressively growing tumor was recently demonstrated in experimental animals (Scheenberger et al., 1989). Improvements in liver function were demonstrated in mice carrying cachectic tumors after administration of AMP or ATP, along with the significant inhibition of host weight losses (Rapaport, 1990).

Several groups now propose that the cytolytic activity of extracellular ATP against tumor cells accounts for at least part of the activity of certain cytolytic T lymphocytes (Correale et al., 1995). These cytolytic T lymphocytes release ATP that is stored in their cellular granules, in response to the target cell interaction with a T-cell receptor. The extracellular ATP released in the immediate vicinity of the target tumor cell is proposed to deliver the lethal hit. The target cell receptor that is required for lymphocyte-mediated cytotoxicity towards tumor cells was recently identified. The purified membrane protein, shown to be the cytolytic lymphocyte ligand by several criteria, was identical with the beta subunit of mitochondrial H^+ transporting ATP synthase, thus suggesting again the involvement of extracellular ATP in the lytic processes (Das et al., 1994). Several other physiological processes that are related to the presence of P_2 purinoceptors on a variety of tissues, as well as on vascular endothelial cells, are induced by the presence of elevated blood plasma ATP pools. Two of these activities that are related to cancer, the enhancement of intratumoral blood flow in malignant gliomas after intra-arterial administration of ATP (Natori et al., 1992) and the remarkable antipain activities of intravenously administered ATP (Fukunaga et al., 1992), were recently demonstrated in humans. The analgesic and sedative effects of ATP in humans and experimental animals (Gomma, 1987; Gomma et al., 1989; Fukunaga et al., 1990) after intravenous administration of ATP persisted after the injections or infusions had stopped and were shown to be intrinsic to the interactions of ATP with P_2 purine receptors. Increases in regional blood flow after administration of ATP can be exploited for enhanced delivery of anticancer agents to a tumor without additional effects on extratumoral sites that are susceptible to the drug's toxicity, as is suggested in the case of malignant gliomas (Natori et al., 1992). Since the intravenous infusions of low levels of ATP in humans (below 0.1 mg $\cdot$ kg^{-1} $\cdot$ min^{-1} are known to produce an increase in cardiac output or pulmonary artery flow without inducing systemic hypotension or increasing heart rate (Gaba and Prefaut, 1990), combinations of ATP and mainstay anticancer regiments could constitute a treatment modality for lung cancers in which a variety of activities of ATP will be utlized. Equally important are the circulatory effects induced by ATP administration in humans that result in increases in cardiac output and stroke volume without affecting heart rate. The reduction of systemic vascular resistance after ATP administration would be expected to account for improvements in quality of life parameters due to enhancement of oxygen delivery to peripheral sites.

Animal studies underestimate the efficacies of the activities of extracellular ATP since it is much easier to achieve elevated blood plasma levels of ATP in humans than in experimental animals. The reason lies in the much higher catabolic soluble blood plasma and ectoenzymatic activities that catalyze the degradation of extracellular ATP in animals as compared to the same activities in humans. These enzymatic activities are related to the higher basal metabolic rates in animals and their need for an increased supply of salvage precursors for the biosynthesis of nucleotides. Therefore, much lower levels of ATP are required for intravenous or intraperitoneal administrations in humans as compared to animals for sustaining elevated blood plasma ATP pools without systemic toxicity (Rapaport and Fontaine, 1989a; Ho and Frei, 1970; LePage et al., 1972).

Antianemia effects of AMP were demonstrated both in humans, where intramuscular injections of AMP were effective in increasing the viability of red blood cells in a patient with a hemolytic disease (Teitel et al., 1964), and in rabbits that were rendered anemic by lead poisoning (Gajdos et al., 1963). Low levels of adenine nucleotides have been known to stimulate regional cerebral blood flow (rCBF) in experimental animals (Forrester et al., 1979). Striking improvements in the motor functions of Parkinson's disease patients were recently obtained by Birkmayer after infusions of low levels of reduced nicotinamide adenine dinucleotide (NADH) (Birkmayer et al., 1993). The CNS activity of low levels of NADH in Parkinson's patients and similar types of CNS activities reported for AMP administered to multiple sclerosis patients (Lowry et al., 1950) are the result of the stimulation of rCBF by elevated (extracellular) blood plasma ATP levels produced after the rapid degradation of the nucleotides to adenosine and inorganic phosphate. Stimulation of rCBF is expected to increase regional metabolism, and since metabolism has been closely linked to neuronal function in the brain, the increases in rCBF would tend to lead to short-term improvements in neuronal function. Elevated red blood cell ATP pools are known to produce significant increases in red blood cell 2,3-diphospholyglycerate (2,3-DPG) (Brewer and Eaton, 1971; Salari et al., 1991), and these increases in turn cause a shift in the hemoglobin oxygen affinity curves to the right, leading to enhanced oxygen delivery under normal nonhypoxic conditions (Brewer and Eaton, 1971). This improvement in oxygen delivery to peripheral sites after the expansion of red blood cell ATP pools is expected to lead to improvements in the performance status of patients suffering from advanced cancers. Recently, significant declines in superoxide anion (O_2^-) production in patients with advanced cancers were documented and it was proposed that these comprise part of the host deficiencies in defending against tumors (Hara et al., 1992). There are several studies demonstrating that extracellular ATP stimulates superoxide anion production by phagocytic cells and that extracellular ATP can achieve this physiological function *in vivo* (Ward et al., 1990).

There is considerable evidence that platelet aggregation and fibrin deposition are important factors supporting the growth and metastasis of tumors. Thus the use of anticoagulants and antiplatelet drugs has been proposed for the treatment of cancer (Rickles et al., 1988). The antithrombotic effects of ATP were recently demonstrated in *in vivo* rat and rabbit models after short-term infusions of ADP (Aursnes and Stenberg-Nilsen, 1992), which by itself is pro-aggregatory, but when administered *in vivo* yields elevated blood and blood plasma ATP levels by mechanisms discussed earlier. The anti-aggregatory and antithrombotic effects of elevated ATP *in vivo* are, in all likelihood, the results of a direct inhibition of platelet aggregation by ATP (Soslau et al., 1995), as well as the result of increased prostacyclin (PGI_2) synthesis induced by increased extracellular ATP levels (Salari et al., 1991). Administration of ATP to a rat arthritis model produced significant anti-inflammatory activities via a mechanism that was proposed to involve enhanced plasma extravasation (Green et al., 1991) and is thus expected to augment the anti-inflammatory activities of prostacyclin.

In summary, a broad spectrum of anticancer activities is expected to be induced in patients by the administration of ATP, over and above the direct cytotoxic effects against the tumor cells. The elevated hepatic red blood cell and blood plasma ATP pools inhibit host weight loss and significantly improve parameters related to cancer cachexia and quality of life. The vast improvements in blood flow, due to the lowering of systemic vascular resistance by ATP at levels that do not affect mean arterial blood pressure, as well as the improvements in the delivery of oxygen to peripheral sites and the stimulation of regional cerebral blood flow are expected to produce significant beneficial effects on the asthenia or general weakness and malaise that is associated with the advanced disease. The only approved use of an adenine nucleotide in the United States is that of intramuscular injections of AMP for the symptomatic relief of varicose vein complications associated with stasis dermatitis. The injections result in improved blood flow due to the lowering of systemic vascular resistance by the increased extracellular (blood plasma) ATP pools.

The treatment of advanced cancers with continuous intravenous infusions (civ) of ATP has now been through a phase I and a multicenter phase II clinical trials with encouraging results. Administration of ATP in humans results in a rapid increase in red blood cell ATP pools, an increase that in experimental animals followed increases in liver ATP pools. Recent advances have demonstrated the release of micromolar levels of ATP into the extracellular environment through specific channels that secrete ATP and exist in a wide variety of cell membranes (Abraham et al., 1993; Al-Awqati, 1995). This type of continuous release produces, after the expansion of intracellular organ and red blood cell ATP pools, the steady-state levels of extracellular ATP that are required for pharmacological effects in tumor bearing hosts. The expansion of intracellular hepatic ATP pools after the administration of adenine nucleotides significantly improves hepatic functions in the cancer cachectic host, functions which due to the central role of the liver in host physiology and metabolism yield those clinical benefits that are becoming apparent in advanced cancer patients.

REFERENCES

Abraham EH, Prat AG, Gerweck L, Seneveratne T, Arceci RJ, Kramer R, Guidotti G, Cantiello HF (1993): The multidrug resistance (mdr1) gene product functions as an ATP channel. Proc Natl Acad Sci USA 90:312–316.

Al-Awqati Q (1995): Regulation of ion channels by ABC transporters that secrete ATP. Science 269:805–806.

Aursnes I, Stenberg-Nilsen H (1992): Low dose infusion of adenosine diphosphate prolongs bleeding time in rats and rabbits. Thrombosis Res 68:67–74.

Belzer FO, Southard JH (1988): Principles of solid-organ preservation by cold storage. Transplantation 45:673–676.

Birkmayer JGD, Vrecko C, Volc D, Birkmayer W (1993): Nicotinamide adenine dinucleotide (NADH)—a new therapeutic approach to Parkinson's disease. Comparison of oral and parenteral applications. Acta Neurol Scand 87(suppl 146):32–35.

Brewer GJ, Eaton JW (1971): Erthrocyte metabolism interaction with oxygen transport. Science 171:1205–1211.

Correale P, Giuliano M, Taglinferri P, Guarrasi R, Caraglia M, Marinetti MR, Iezzi T, Bianco AF, Procopio A (1995): Role of adenosine 5′-triphosphate in lymptokine activated (LAK) killing of human tumor cells. Res Commun Chem Pathol Pharmacol 87:67–69.

Das B, Mondragon MOH, Sadeghian M, Hatcher VB, Norin AJ (1994): A novel ligand in lymphocyte-mediated cytotoxicity: Expression of the β subunit of H^+ transporting ATP synthese on the surface of tumor cell lines. J Exp Med 180:273–281.

Estrela JM, Obrador E, Navarro J, Lasso De La Vega MC, Pellicer JA (1995): Elimination of Ehrlich tumors by ATP-induced growth inhibition, glutathione depletion and X-rays. Nature Medicine 1:84–88.

Fang WG, Pimia F, Bang YJ, Myers CE, Trepel JB (1992): P_2-purinergic receptor agonists inhibit the growth of androgen-independent prostate carcinoma cells. J Clin Invest 89:191–196.

Fang W, Wu B (1993): Differential growth regulation of a metastatic human lung carcinoma cell line through activation of phosphatidyl inositol turnover signal transduction pathway. Clin Exp Metastasis 11:330–336.

Forrester T, Harper AM, Mackenzie ET, Thomson EM (1979): Effect of adenosine triphosphate and some derivatives on cerebral blood flow and metabolism. J Physiol 296:343–355.

Fujino Y, Juroda Y, Morita A, Tanioka Y, Ku Y, Saitoh Y (1993): The effect of fasting and exogenous adenosine on ATP tissue concentration and viability of canine pancreas grafts during preservation by the two layer method. Transplantation 56:1083–1086.

Fukunaga A, Kaneko Y, Harano K, Kiuta Y, Fukunaga B, Okutsu Y (1992): Anesthetic effects of adenosine and ATP in animals and in humans. Int J Purine and Pyrimidine Res 3:50.

Fukunaga AF, Kaneko Y, Ichinohe T, Igarashi O, Nakakuki T (1990): Intravenous ATP attenuates surgical stress responses and reduces inhalation anesthetic requirements in humans. Anesthesiology 73:A400.

Gaba SJM, Prefaut C (1990): Comparison of pulmonary and systemic effects of adenosine triphosphate in chronic obstructive pulmonary disease-ATP: A pulmonary controlled vasoregulator? Eur Respir J 3:450–455.

Gajdos A, Dantchev D, Bnard H (1963): Action de l'acid adenosine 5′-monophosphorique sur la survie des globules rouges chez le lapin inoxique par le plomb ou la phenylhydrazine. Rev Franc Etud Clin Biol 8:62–66.

Gomaa AA (1987): Characteristics of analgesia induced by adenosine triphosphate. Pharmacol Toxicol 61:199–202.

Gomaa AA, Moustafa SA, Farghali AA (1989): Adenosine triphosphate blocks opiate withdrawal symptoms in rats and mice. Pharmacol Toxicol 64:111–115.

Green PG, Basbaum AI, Helms C, Levine JD (1991): Purinergic regulation of bradykinin-induced plasma extravasation and adjuvant-induced arthritis in the rat. Proc Natl Acad Sci USA 88:4162–4165.

Hara J, Ichinose Y, Asoh H, Yano T, Kawasaki M, Ohta M (1992): Superoxide anion generating activity of polymorphonuclear leukocytes and monocytes in patients with lung cancer. Cancer 69:1682–1687.

Haskell CM, Wong M, Lee LY, Williams A, Blahd W (1994): A phase I study of ATP in the treatment of advanced cancer. Drug Dev Res 31:276A.

Ho DWH, Frei E III (1970): Pharmacological studies of the antitumor agent 6-methylthiopurine ribonucleoside. Cancer Res 30:2852–2857.

Holroyde CP, Baguzda RC, Putnam PP, Reichard GA (1975): Altered glucose metabolism in metastatic carcinoma. Cancer Res 35:3710–3714.

Kanwar Y, Yoshinaga Y, Liu Z, Wallner E, Carone F (1992): Biosynthetic regulation of proteoglycans by aldohexoses and ATP. Proc Natl Acad Sci USA 89:8621–8625.

LePage GA, Lin TT, Orth RE, Gottlieb JA (1972): 5′-Nucleotides as potential formulations for administering nucleoside analogs in man. Cancer Res 32:2441–2444.

Lowry MI, Moore RW, Cailliet R (1953): Adenosine-5′-monophosphate in the treatment of multiple sclerosis. Am J Med Sci 226:73–83.

Meldrum DR, Ayala A, Wang P, Ertel W, Chaudry JH (1991): Association between decreased splenic ATP levels and immunodepression; amelioration with ATP-$MgCl_2$. Am J Physiol 261:R351–R357.

Murgia M, Hanau S, Pizzo P, Rippa M, Divirgilio F (1993): Oxidized ATP is an irreversible inhibitor of the macrophage purinergic P_{2Z} receptor. J Biol Chem 268:8199–8203.

Natori U, Moriguchi M, Fugiwara S, Takeshita I, Fukui M, Iwaki I, Kanaaide H (1992): Effects of L-NMMA and L-NNA on the selective ATP-induced enhancement of intratumoral blood flow. J Cereb Blood Flow Metab 12:120–127.

Nelson KA, Walsh D, Sheehan FA (1994): The cancer anorexia-cachexia syndrome. J Clin Oncol 12:213–225.

Popp MB, Wagner SC, Brito OJ (1983): Host and tumor responses to increasing levels of intravenous nutritional support. Surgery 94:300–308.

Rapaport E (1995): Involvement of elevated intracellular and extracellular ATP in the regulation of insulin secretion: Therapeutic targets in non-insulin-dependent diabetes mellitus. Amer J Therap 2:283–289.

Rapaport E (1993): Anticancer activities of adenine nucleotides in tumor bearing hosts. Drug Dev Res 28:428–431.

Rapaport E (1990): Mechanisms of anticancer activities of adenine nucleotides in tumor bearing hosts. Ann NY Acad Sci 603:142–150.

Rapaport E (1983): Treatment of human tumor cells with ADP or ATP yields arrest of growth in the S phase of the cell cycle. J Cell Physiol 114:279–283.

Rapaport E, Fontaine J (1989a): Anticancer activities of adenine nucleotides in mice are mediated through expansion of erythrocyte ATP pools. Proc Natl Acad Sci USA 86:1662–1666.

Rapaport E, Fontaine J (1989b): Generation of extracellular ATP in blood and its mediated inhibition of host weight loss in tumor-bearing mice. Biochem Pharmacol 38:4261–4266.

Rapaport E, Fishman RF, Gercel C (1983): Growth inhibition of human tumor cells in soft-agar cultures by treatment with low levels of adenosine 5′-triphosphate. Cancer Res 43:4402–4406.

Rapaport E, Zamecnik PC (1976): Incorporation of adenosine into ATP: Formation of compartmentazlied ATP. Proc Natl Sci USA 73:3122–3125.

Rickles FR, Hancock WW, Edwards RL, Zacharski LR (1988): Antimetastatic agents, I. Role of cellular procagulants in the pathogenesis of fibrin deposition in cancer and the use of anticoagulants and/or antiplatelet drugs in cancer treatment. Semin Thromb Hemost 14:88–94.

Salari PC, Marni A, Parise M, Calvani AB, Spaggiari PG, Gaja G, Ferrero ME (1991): Are increased ATP-blood levels responsible for PGI_2 release from endothelium? Int J Tiss React 13:219–223.

Scheenberger AL, Thompson RT, Driedger AA, Finley RJ, Inculet RI (1989): Effects of cancer on the in vivo energy state of rat liver and skeletal muscle. Cancer Res 49:1160–1164.

Senagore AJ, Milson JW, Walshaw RK, Mostoskey V, Dunstan R, Chaudry IH (1992): Adenosine triphosphate-magnesium chloride in radiation injury. Surgery 112:933–939.

Singh G, Chaudry KI, Chaudry IH (1993): ATP-$MgCl_2$ restores gut absorptive capacity early after trauma-hermorrhagic shock. Am J Physiol 264:R977–R983.

Soni MG, Mehendale HM (1994): Adenosine triphosphate protection of chlordecone-amplified CCl_4 hepatotoxicity and lethality. J Hepatol 20:267–274.

Soslau G, McKenzie RJ, Brodsky I, Devlin TM (1995): Extracellular ATP inhibits agonist-induced mobilization of internal calcium in human platelets. Biochim Biophys Acta 1268:73–80.

Spranzi E, Djeu JY, Hoffman SL, Epling-Burnette PK, Blanchard DK (1993): Lysis of human monocytic leukemia cells by extracellular adenosine triphosphate: Mechanism and characterization of the adenosine triphosphate receptor. Blood 82:1578–1585.

Stein TP (1978): Cachexia, gluconeogenesis and progressive weight loss in cancer patients. J Theor Biol 73:51–59.

Szeinfeld D, Devilliers N (1992): Radioprotective properties of ATP and modification of acid phosphatase response after a lethal dose of whole body p(66MeV)/Be neutron radiation of BALB/C mice. Cancer Biochem Biophys 13:123–132.

Tietel P, Bratu V, Nenakis A, Butoianu E (1964): Favorable therapeutic effect of adenosine 5′-monophosphate in a case of chronic compensation hemolytic disease with an impaired erythrocyte energy metabolism. Proc 10th Congr Int Soc Blood Transf Stockholm, pp 557–560.

Van Eys J (1982): Effect of nutritional status on response to therapy. Cancer Res 42(suppl):747S–753S.

Vandewalle B, Hornez L, Revillion F, Lefebvre J (1994): Effect of extracellular ATP on breast tumor cell growth, implication of intracellular calcium. Cancer Lett 85:47–54.

Ward PA, Walker BAM, Hogenlocker BE (1990): Functional consequences of interactions between human neutrophils and ATP, ATP-S and adenosine. Ann NY Acad Sci 603:108–119.

Wiley JS, Jamieson GP, Mayger W, Cragoe EJ Jr, Jopson M (1990): Extracellular ATP stimulates an amiloride-sensitive sodium influx in human lymphocytes. Arch Biochem Biophys 280:263–268.

INDEX

3336